Reference Series in Biomedical Engineering

Tissue Engineering and Regeneration

Series Editor

Heinz Redl
Ludwig Boltzmann Institute for Experimental
and Clinical Traumatology, AUVA Research
Center Austrian Cluster for Tissue Regeneration
Wien, Austria

This series Tissue Engineering and Regeneration consists of comprehensive reference texts encompassing the biological basis of tissue regeneration, basic principles of tissue engineering and the current state-of-the-art in tissue engineering of specific tissues and organs. Each volume combines established fundamentals and the latest developments, thus forming an invaluable collection for both experienced researchers as well as practitioners from other areas of expertise. The spectrum of topics ranges from the use of cells for tissue regeneration and tissue engineering, growth factors and biological molecules affecting tissue development and regeneration, to the specific roles of biophysical factors in tissue development and regeneration.

Tissue engineering lies at the crossroads of medicine, life sciences and engineering. The field has developed extensively over the last two decades, addressing the requirements of tissue and organ replacement as well as regeneration in a variety of congenital, traumatic, disease and aging-related conditions, including some of the most critical unmet challenges in modern medicine. Both our increased understanding of the biological basis of tissue engineering as well as significant technological advances mean that engineering design principles can now be used for the de novo construction of functional tissue replacements that meet the requirements of research and clinical applications.

More information about this series at http://www.springer.com/series/13441

Daniel Eberli • Sang Jin Lee
Andreas Traweger
Editors

Organ Tissue Engineering

With 61 Figures and 28 Tables

Springer

Editors
Daniel Eberli
Universitäts Spital Zürich
Zürich, Switzerland

Sang Jin Lee
Institute for Regenerative Medicine
Wake Forest University
Winston Salem, NC, USA

Andreas Traweger
Institute of Tendon and Bone Regeneration
Paracelsus Medical University
Salzburg, Austria

ISBN 978-3-030-44210-1 ISBN 978-3-030-44211-8 (eBook)
ISBN 978-3-030-44212-5 (print and electronic bundle)
https://doi.org/10.1007/978-3-030-44211-8

This Springer imprint is published by the registered company Springer Nature Switzerland AG.
The registered company address is: Gewerbestrasse 11, 6330 Cham, Switzerland

Preface

Tissue engineering may offer new treatment alternatives for organ replacement or repair of deteriorated organs. Among the many clinical applications of tissue engineering are the production of artificial skin for burn patients, tissue-engineered trachea, cartilage for knee-replacement procedures, urinary bladder replacement, urethra substitutes, and cellular therapies for the treatment of urinary incontinence. The tissue engineering approach has major advantages over traditional organ transplantation and circumvents the problem of organ shortage. Tissues reconstructed from readily available biopsy material induce only minimal or no immunogenicity when reimplanted in the patient.

This Springer Reference e-book on organ tissue engineering is aimed at anyone interested in the application of tissue engineering in different organ systems. It offers insights into a wide variety of state-of-the-art strategies applying the principles of tissue engineering to tissue and organ regeneration and is thus intended to be both an introductory and deepening textbook for research and teaching. As this reference book is continuously updated, it will reflect the latest developments and findings of both basic and clinical research and will discuss current trends.

Contents

About the Editors

Daniel Eberli
Universitäts Spital Zürich
Zürich, Switzerland

M.D., Ph.D., is a scientific physician working in the translational field of urologic tissue engineering. He has a medical degree from the Medical School in Zurich, Switzerland, and a Ph.D. in molecular medicine from Wake Forest University, Winston Salem, NC. He has a faculty position in the Department of Urology at the University Hospital Zurich, where he devotes half of his time to patient care. Together with his research team, he is working on novel biomaterials for bladder reconstruction, improving autonomic innervation, cellular treatment of incontinence, and tracking of stem cells.

Sang Jin Lee
Institute for Regenerative Medicine
Wake Forest University
Winston Salem, NC, USA

Sang Jin Lee, Ph.D., is currently a tenured associate professor at Wake Forest Institute for Regenerative Medicine (WFIRM), Wake Forest School of Medicine. Dr. Lee received his Ph.D. in chemical engineering from Hanyang University, Seoul, Korea, in 2003 and took a postdoctoral fellowship in the Laboratories for Tissue Engineering and Cellular Therapeutics at Harvard Medical School and Children's Hospital Boston and the WFIRM, where he is currently a faculty member. He is also cross-appointed to the Virginia Tech–WFU Biomedical Engineering and Science. Dr. Lee has authored more than 140 scientific publications and reviews, has edited 2 textbooks, and has written 34 chapters in several books. Dr. Lee has an extensive knowledge and experience in biomaterials science, especially, biodegradable polymers and tunable hydrogels, with specific training and expertise in key research areas for tissue engineering and regenerative medicine. His research team has developed various biomaterial systems that improve cellular interactions by providing appropriate environmental cues. These biomaterial systems consist of drug/protein delivery system, nano/micro-scaled topographical feature, and hybrid materials that can actively participate in functional tissue regeneration. Recently, his team is utilizing automated 3D bioprinting technology to manufacture complex, multicellular living tissue constructs that mimic the structure of native tissues. This can be accomplished by optimizing the formulation of biomaterials to serve as bioinks for 3D bioprinting, and by providing the biological microenvironment needed for the successful delivery of cells and biomaterials to discrete locations within the 3D structure.

Andreas Traweger
Institute of Tendon and Bone Regeneration
Paracelsus Medical University
Salzburg, Austria

Andreas Traweger currently holds a research professorship for regenerative biology at the Paracelsus Medical University in Salzburg, Austria. He received his Ph.D. in genetics from the University of Salzburg and completed his postdoctoral training at the Samuel Lunenfeld Research Institute in Toronto, Canada. He was then R&D manager at Baxter (Vienna, Austria) for 4 years before joining Paracelsus Medical University.

His research is interdisciplinary in nature, focusing on both high-quality fundamental science and translation for human health. His interests lie in promoting the understanding of tendon biology in general and to devise novel strategies to improve tendon and ligament healing. He has a considerable track record in the research area of tendinopathy, focusing on the role of neoangiogenesis and inflammatory processes in tendon health and disease. Further, he is co-founder of Celericon Therapeutics GmbH (2018), a biotech startup focusing on the use of MSC-derived extracellular vesicles to improve tissue regeneration.

Contributors

Christabella Adine Department of Biomedical Engineering, Faculty of Engineering, National University of Singapore, Singapore, Singapore

Emilio I. Alarcon BioEngineering and Therapeutic Solutions (BEaTS), Division of Cardiac Surgery, University of Ottawa Heart Institute, Ottawa, ON, Canada

Department of Biochemistry, Microbiology, and Immunology, Faculty of Medicine, University of Ottawa, Ottawa, ON, Canada

Helena Almeida 3B's Research Group, I3Bs – Research Institute on Biomaterials, Biodegradables and Biomimetics, University of Minho, Headquarters of the European Institute of Excellence on Tissue Engineering and Regenerative Medicine, Barco/Guimarães, Portugal

ICVS/3B's–PT Government Associate Laboratory, Braga/Guimarães, Portugal

Peter Angele Department of Trauma Surgery, University Hospital of Regensburg, Regensburg, Germany

Sporthopaedicum Regensburg, Regensburg, Germany

Amnie Ashour Department of Surgery, Harvard Medical School, Beth Israel Deaconess Medical Center, Boston, MA, USA

Renaissance School of Medicine, Stony Brook University, Stony Brook, NY, USA

Anthony Atala Wake Forest Institute for Regenerative Medicine, Wake Forest School of Medicine, Medical Center Boulevard, Winston-Salem, NC, USA

Pedro M. Baptista Instituto de Investigación Sanitaria Aragón (IIS Aragón), Zaragoza, Spain

Centro de Investigación Biomédica en Red de Enfermedades Hepáticas y Digestivas (CIBERehd), Zaragoza, Spain

Instituto de Investigación Sanitaria de la Fundación Jiménez Díaz, Madrid, Spain

Departamento de Ingeniería Biomédica y Aeroespacial, Universidad Carlos III de Madrid, Madrid, Spain

Fundación ARAID, Zaragoza, Spain

Ori Bar-Nur Laboratory of Regenerative and Movement Biology, Institute of Human Movement Sciences and Sport, Department of Health Sciences and Technology, Swiss Federal Institute of Technology (ETH) Zurich, Schwerzenbach, Switzerland

Paola Bonfanti Epithelial Stem Cell Biology and Regenerative Medicine Laboratory, The Francis Crick Institute, London, UK

Matthew Burton International Centre for Eye Health, London School of Hygiene and Tropical Medicine, London, UK

Yilin Cao Shanghai Tissue Engineering Research Key Laboratory, Shanghai Ninth People's Hospital, Shanghai Jiao Tong University School of Medicine, Shanghai, People's Republic of China

Elliot L. Chaikof Program in Health Sciences and Technology, Massachusetts Institute of Technology, Cambridge, MA, USA

Department of Surgery, Harvard Medical School, Beth Israel Deaconess Medical Center, Boston, MA, USA

Wyss Institute for Biologically Inspired Engineering, Harvard University, Cambridge, MA, USA

Paolo De Coppi Stem Cell and Regenerative Medicine Section, Great Ormond Street Institute of Child Health, University College London, London, UK

Specialist Neonatal and Paediatric Surgery, Great Ormond Street Hospital, London, UK

Seraina A. Domenig Laboratory of Regenerative and Movement Biology, Institute of Human Movement Sciences and Sport, Department of Health Sciences and Technology, Swiss Federal Institute of Technology (ETH) Zurich, Schwerzenbach, Switzerland

Rui M. A. Domingues 3B's Research Group, I3Bs – Research Institute on Biomaterials, Biodegradables and Biomimetics, University of Minho, Headquarters of the European Institute of Excellence on Tissue Engineering and Regenerative Medicine, Barco/Guimarães, Portugal

ICVS/3B's–PT Government Associate Laboratory, Braga/Guimarães, Portugal

Natalie Durkin Stem Cell and Regenerative Medicine Section, Great Ormond Street Institute of Child Health, University College London, London, UK

Maximilian Y. Emmert Institute for Regenerative Medicine (IREM), University of Zurich, Zurich, Switzerland

Wyss Translational Center Zurich, University and ETH Zurich, Zurich, Switzerland

Department of Cardiovascular Surgery, Charité – Universitätsmedizin Berlin, Berlin, Germany

Department of Cardiothoracic and Vascular Surgery, German Heart Center Berlin, Berlin, Germany

João Ferreira Exocrine Gland Biology and Regeneration Research Group, Faculty of Dentistry, Chulalongkorn University, Bangkok, Thailand

Faculty of Dentistry, National University of Singapore, Singapore, Singapore

Emanuela S. Fioretta Institute for Regenerative Medicine (IREM), University of Zurich, Zurich, Switzerland

Eric K. N. Gähwiler Institute for Regenerative Medicine (IREM), University of Zurich, Zurich, Switzerland

Matías Garrido Epithelial Stem Cell Biology and Regenerative Medicine Laboratory, The Francis Crick Institute, London, UK

Manuela E. Gomes 3B's Research Group, I3Bs – Research Institute on Biomaterials, Biodegradables and Biomimetics, University of Minho, Headquarters of the European Institute of Excellence on Tissue Engineering and Regenerative Medicine, Barco/Guimarães, Portugal

ICVS/3B's–PT Government Associate Laboratory, Braga/Guimarães, Portugal

Manuel Gómez-Florit 3B's Research Group, I3Bs – Research Institute on Biomaterials, Biodegradables and Biomimetics, University of Minho, Headquarters of the European Institute of Excellence on Tissue Engineering and Regenerative Medicine, Barco/Guimarães, Portugal

ICVS/3B's–PT Government Associate Laboratory, Braga/Guimarães, Portugal

Ana I. Gonçalves 3B's Research Group, I3Bs – Research Institute on Biomaterials, Biodegradables and Biomimetics, University of Minho, Headquarters of the European Institute of Excellence on Tissue Engineering and Regenerative Medicine, Barco/Guimarães, Portugal

ICVS/3B's–PT Government Associate Laboratory, Braga/Guimarães, Portugal

Miguel Gonzalez-Andrades Maimonides Biomedical Research Institute of Cordoba (IMIBIC), Department of Ophthalmology, Reina Sofia University Hospital and University of Cordoba, Cordoba, Spain

May Griffith Département d'ophtalmologie, Université de Montréal and Centre de Recherche de l'Hopital Maisonneuve-Rosemont, Montréal, QC, Canada

Institut de génie biomédical, Université de Montréal and Centre de recherche du Centre hospitalier de l'Université de Montréal, Montréal, QC, Canada

Jan Willem Hoekstra Department of Biomaterials, Radboud University Medical Center, Nijmegen, The Netherlands

Simon P. Hoerstrup Institute for Regenerative Medicine (IREM), University of Zurich, Zurich, Switzerland

Wyss Translational Center Zurich, University and ETH Zurich, Zurich, Switzerland

Jun Homma Institute of Advanced Biomedical Engineering and Science, Tokyo Women's Medical University, Tokyo, Japan

Victor H. Hu International Centre for Eye Health, London School of Hygiene and Tropical Medicine, London, UK

John D. Jackson Wake Forest Institute for Regenerative Medicine, Wake Forest School of Medicine, Medical Center Boulevard, Winston Salem, NC, USA

John Jansen Department of Biomaterials, Radboud University Medical Center, Nijmegen, The Netherlands

Pushpinder Kanda Department of Ophthalmology, University of Ottawa Eye Institute, The Ottawa Hospital, Ottawa, ON, Canada

Gauri Kulkarni Wake Forest Institute for Regenerative Medicine, Wake Forest School of Medicine, Medical Center Boulevard, Winston Salem, NC, USA

Monica M. Laronda Stanley Manne Children's Research Institute, Ann & Robert H. Lurie Children's Hospital of Chicago, Chicago, IL, USA

Department of Pediatrics, Feinberg School of Medicine, Northwestern University, Chicago, IL, USA

Sander Leeuwenburgh Department of Biomaterials, Radboud University Medical Center, Nijmegen, The Netherlands

Zihan Li Tufts University, Boston, MA, USA

Hongbing Liao Department of Prosthodontics, College of Stomatology, Guangxi Medical University, Nanning, China

Department of Biomaterials, Radboud University Medical Center, Nijmegen, The Netherlands

Diana Lim Wake Forest Institute for Regenerative Medicine, Wake Forest School of Medicine, Medical Center Boulevard, Winston-Salem, NC, USA

Yu Liu National Tissue Engineering Center of China, Shanghai, People's Republic of China

Kamal Malhotra Département d'ophtalmologie, Université de Montréal and Centre de Recherche de l'Hopital Maisonneuve-Rosemont, Montréal, QC, Canada

Ana M. Matos 3B's Research Group, I3Bs – Research Institute on Biomaterials, Biodegradables and Biomimetics, University of Minho, Headquarters of the European Institute of Excellence on Tissue Engineering and Regenerative Medicine, Barco/Guimarães, Portugal

ICVS/3B's–PT Government Associate Laboratory, Braga/Guimarães, Portugal

David Miranda-Nieves Program in Health Sciences and Technology, Massachusetts Institute of Technology, Cambridge, MA, USA

Department of Surgery, Harvard Medical School, Beth Israel Deaconess Medical Center, Boston, MA, USA

Wyss Institute for Biologically Inspired Engineering, Harvard University, Cambridge, MA, USA

Sara Morini Instituto de Investigación Sanitaria Aragón (IIS Aragón), Zaragoza, Spain

Department of Bioengineering and iBB-Institute for Bioengineering and Biosciences, Instituto Superior Técnico, Universidade de Lisboa, Lisbon, Portugal

Sarah E. Motta Institute for Regenerative Medicine (IREM), University of Zurich, Zurich, Switzerland

Wyss Translational Center Zurich, University and ETH Zurich, Zurich, Switzerland

Thomas Nau Ludwig Boltzmann Institute for Experimental and Clinical Traumatology, AUVA Research Center, Vienna, Austria

The Austrian Cluster for Tissue Regeneration, Vienna, Austria

Mohammed Bin Rashid University and Health Sciences – MBRU, Dubai, United Arab Emirates

Andrew S. Palmer Laboratory of Regenerative and Movement Biology, Institute of Human Movement Sciences and Sport, Department of Health Sciences and Technology, Swiss Federal Institute of Technology (ETH) Zurich, Schwerzenbach, Switzerland

Heung Jae Park Department of Urology, Kangbuk Samsung Hospital, Sungkyunkwan University School of Medicine, Seoul, Republic of Korea

Iris Pla-Palacín Instituto de Investigación Sanitaria Aragón (IIS Aragón), Zaragoza, Spain

Nikolaos Poulis Institute for Regenerative Medicine (IREM), University of Zurich, Zurich, Switzerland

Márcia T. Rodrigues 3B's Research Group, I3Bs – Research Institute on Biomaterials, Biodegradables and Biomimetics, University of Minho, Headquarters of the European Institute of Excellence on Tissue Engineering and Regenerative Medicine, Barco/Guimarães, Portugal

ICVS/3B's–PT Government Associate Laboratory, Braga/Guimarães, Portugal

Daniela Rubio-Soto Instituto de Investigación Sanitaria Aragón (IIS Aragón), Zaragoza, Spain

Natalia Sánchez-Romero Instituto de Investigación Sanitaria Aragón (IIS Aragón), Zaragoza, Spain

Cytes Biotechnologies, Barcelona, Spain

Hidekazu Sekine Institute of Advanced Biomedical Engineering and Science, Tokyo Women's Medical University, Tokyo, Japan

Soichi Shibuya Stem Cell and Regenerative Medicine Section, Great Ormond Street Institute of Child Health, University College London, London, UK

Tatsuya Shimizu Institute of Advanced Biomedical Engineering and Science, Tokyo Women's Medical University, Tokyo, Japan

Andreas Teuschl The Austrian Cluster for Tissue Regeneration, Vienna, Austria

Department Life Science Engineering, University of Applied Sciences Technikum Wien, Vienna, Austria

Joop Wolke Department of Biomaterials, Radboud University Medical Center, Nijmegen, The Netherlands

Pamela C. Yelick Tufts University, Boston, MA, USA

Xian Lin Yi Wake Forest Institute for Regenerative Medicine, Wake Forest School of Medicine, Winston Salem, NC, USA

Department of Urology, Tumor Hospital of GuangXi Medical University, Nanning, China

James J. Yoo Wake Forest Institute for Regenerative Medicine, Wake Forest School of Medicine, Medical Center Boulevard, Winston-Salem, NC, USA

Johannes Zellner Department of Trauma Surgery, Caritas Hospital St. Josef, Regensburg, Germany

Department of Trauma Surgery, University Hospital of Regensburg, Regensburg, Germany

Weibo Zhang Tufts University, Boston, MA, USA

Part I

Circulatory and Respiratory Systems

Bioinspired Vascular Grafts

David Miranda-Nieves, Amnie Ashour, and Elliot L. Chaikof

Contents

D. Miranda-Nieves · E. L. Chaikof (✉)
Program in Health Sciences and Technology, Massachusetts Institute of Technology, Cambridge, MA, USA

Department of Surgery, Harvard Medical School, Beth Israel Deaconess Medical Center, Boston, MA, USA

Wyss Institute for Biologically Inspired Engineering, Harvard University, Cambridge, MA, USA
e-mail: dmiranda@mit.edu; echaikof@bidmc.harvard.edu

A. Ashour
Department of Surgery, Harvard Medical School, Beth Israel Deaconess Medical Center, Boston, MA, USA

Renaissance School of Medicine, Stony Brook University, Stony Brook, NY, USA
e-mail: amnie.ashour@stonybrookmedicine.edu

D. Eberli et al. (eds.), *Organ Tissue Engineering*, Reference Series in Biomedical Engineering, https://doi.org/10.1007/978-3-030-44211-8_15

Abstract

A durable, synthetic small-caliber bypass graft has not been identified for revascularization of vessels less than 6 mm in diameter, and there exists limited availability of autologous conduits suitable for transplant. Consequently, alternative approaches have focused on designing arterial prostheses through the mimicry of some or all of the characteristics of the arterial wall. Notwithstanding early reports of promising results, important limitations remain associated with tissue engineering strategies, and the design of a living arterial substitute remains elusive. This chapter aims to describe the structural, mechanical, and biological properties of blood vessels and discuss the design considerations that must be implemented to realize the promise of bioinspired vascular grafts.

1 Introduction

Cardiovascular disease (CVD) affects over 80 million adults in the United States and represents a leading cause of death globally (Benjamin et al. 2019). CVD is most often due to progressive atherosclerosis, which may lead to plaque rupture and arterial occlusion with attendant clinical consequences of myocardial infarction, stroke, or amputation. While mild to moderate CVD can often be treated with modification in diet and lifestyle, as well as medications that limit the progression of atherosclerosis or risk of thrombosis, surgical and catheter-based interventions remain a mainstay of treatment, particularly for symptomatic disease. Catheter-based procedures, such as angioplasty, stenting, and atherectomy are commonly used to treat stenotic vessels or obstructive lesions (DeRubertis et al. 2007; Mwipatayi et al. 2008; McKinsey et al. 2008). Nonetheless, over 500,000 surgical bypass procedures, using a synthetic or autologous conduit, are performed annually in the United States (Go et al. 2013). Despite advances in minimally invasive, catheter-based interventions, including drug-coated stents and balloons, bypass surgery continues to be required for many patients and represents an optimal choice for durable long-term revascularization (Weintraub et al. 2012).

Despite a role for both synthetic and autologous conduits, their clinical performance remains suboptimal. The reconstruction of a large diameter vessel (>6 mm), such as the aorta, can be successfully performed using a synthetic prosthesis (Brewster 1997; Qu and Chaikof 2010). However, synthetic polymeric grafts display limited long-term patency when used in the femoral-popliteal position and exhibit very poor patency for tibial revascularization (Pereira et al. 2006; Van Der Slegt et al. 2014). The 1-year patency of synthetic polymeric conduits in the femoral-popliteal position is approximately 70% (Johnson and Lee 2000; Piffaretti et al. 2018). In general, a durable small-caliber synthetic bypass graft has not been identified for revascularization of vessels that are less than 6 mm in diameter, such as the coronary arteries.

Autologous grafts display higher patency rates than synthetic conduits but also suffer from a number of limitations (Klinkert et al. 2004; Goldman et al. 2004).

When used for coronary artery bypass graft (CABG) surgery, internal mammary artery (IMA) grafts exhibit patency as high as 93% at 10 years, (Loop et al. 1986; Berger et al. 2004; Taggart et al. 2019), and radial artery (RA) grafts have likewise been associated with low rates of stenosis and graft failure (Gaudino et al. 2018). However, the availability and length of IMA and RA conduits are limited. For this reason, saphenous vein grafts continue to be widely used, especially for multi-vessel revascularization. However, vein graft patency is less durable than one might expect, with failure of at least one vein graft commonly occurring within 12 to 18 months after CABG surgery (Mehta et al. 2011; Hess et al. 2014) and the occurrence of a hemodynamically significant stenosis or graft occlusion in 40% of vein grafts 1 year after lower extremity bypass surgery (Conte et al. 2006). Moreover, a suitable venous conduit is often unavailable in many older adults that have had prior surgery or who present with additional comorbid conditions (Kumar et al. 2011).

Vein graft failure has typically been attributed to intimal hyperplasia, which, when hemodynamically significant, leads to acute graft thrombosis (Sottiurai et al. 1983; Motwani and Topol 1998). Factors at the time of initial harvest of a vein graft that contribute to the development of intimal hyperplasia include ischemia-reperfusion injury of the vein wall that may cause endothelial and smooth muscle cell damage with release of pro-inflammatory factors, uncontrolled smooth muscle cell (SMC) proliferation, and extracellular matrix (ECM) production, particularly at the site of venous valves (Clowes 1993; Lemson et al. 2000; De Vries et al. 2016). A number of reports have also demonstrated that a mismatch in mechanical compliance between synthetic polymeric vascular conduits (0.2–1.9%/100 mmHg), autologous vein grafts (0.5–3%/100 mmHg), and native artery (5–15%/100 mmHg) can also initiate maladaptive biological responses that contribute to anastomotic intimal hyperplasia and subsequent graft failure (Abbott et al. 1987; Ballyk et al. 1997). Additional limitations of polymeric grafts include their susceptibility to bacterial colonization and infection and an inability of the synthetic conduit to grow in pediatric patients necessitating subsequent surgical intervention. Thus, there remains a critical need to address these shortcomings.

A current perspective held in tissue engineering is that an ideal vascular conduit would reproduce both the structure of a native artery and its related biological and mechanical characteristics (Fig. 1). In this chapter, we summarize past and current

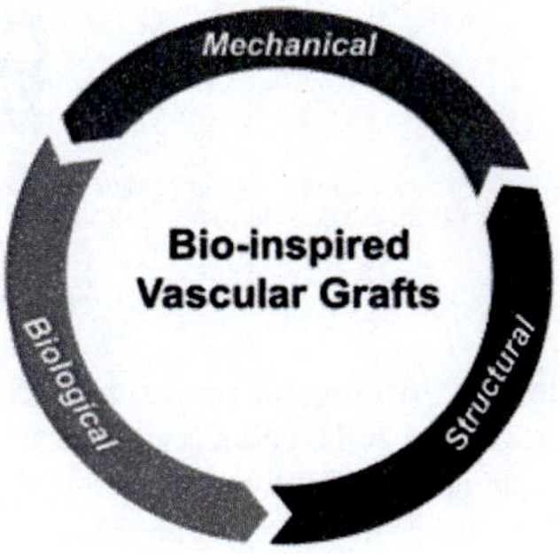

Fig. 1 Bioinspired vascular grafts should aim to recapitulate the structural, mechanical, and biological characteristics of native arteries

approaches to engineer a living blood vessel; the ability of each of these strategies to recapitulate the structural, mechanical, and biological characteristics of a native artery; and the existing shortcomings of these schemes.

2 Structural Considerations

The arterial wall, like many tissues in the body, can be represented as a reinforced composite of structural proteins that protect and orient living cells. It is characterized by a well-defined muscular layer that is responsible for providing strength, controlling vascular tone, and determining overall biomechanical responses. For this reason, the design of a bioinspired vascular conduit must be informed by (1) identifying the key cellular and acellular components responsible for the functional properties of a blood vessel (Fig. 2) and (2) an understanding of the architecture and organization of each component within the vessel wall (Table 1).

2.1 Structural Proteins: Collagen and Elastin

Collagen represents 20% to 50% of the dry weight of the arterial wall, present in the basement membrane and the interstitial matrix, and plays a crucial role in cell

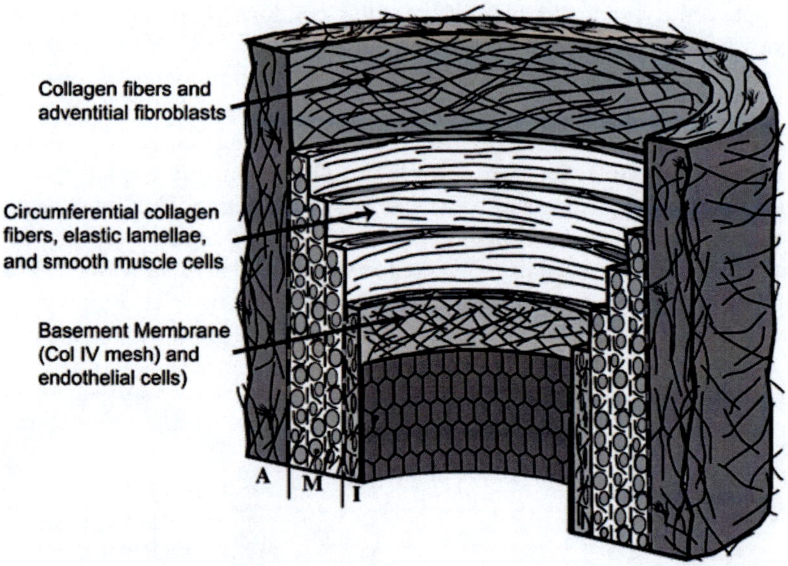

Fig. 2 *Histological representation of a native artery.* The Intima (I) consists of a monolayer of endothelial cells, and a basement membrane, a mesh-like substrate of type IV collagen. The Media is composed of networks of elastin (elastic lamellae) and circumferentially aligned smooth muscle cells and crimped collagen fibers (Col I and III). The Adventitia (A) consists of randomly aligned collagen fibers (Col I and III) and fibroblasts. (Modified with permission from Gasser et al. 2006)

Table 1 Structural components of the arterial wall

	Source	Location	Function	Organization
Collagen (I, III, IV)	Fibroblasts and SMCs	Tunica intima, media, adventitia	Provides mechanical support and strength and influences cell function through mechanotransduction pathways	Col I, III: Circumferentially aligned, crimped fibers Col IV: Mesh-like structure
Elastin	SMCs	Tunica media (elastic lamellae)	Component of arterial ECM that provides elasticity/recoil and allows interlamellar communication	Concentric 3 μm thick lamellae
Endothelial cells	Mesoderm	Luminal side of tunica intima	Regulates thrombotic and inflammatory responses	Monolayer of polygonal shaped cells
Smooth muscle cells	Ectoderm, Mesoderm	Tunica media	ECM production, vaso-responsiveness, and regulates inflammatory responses	Circumferential orientation
Fibroblasts	Mesoderm	Tunica adventitia	ECM production and vessel wall regulation	Isotropic alignment

behavior and vessel wall biomechanics (Linsenmayer 1991; Shoulders and Raines 2009). Closest to the luminal side, the subendothelial basement membrane, comprised mainly of type IV collagen, serves as a mesh-like physical barrier that protects and regulates normal endothelial function (Sand et al. 2016). The interstitial matrix is composed of type I and III collagens in the medial and adventitial layers (Shekhonin et al. 1985). As opposed to type IV, collagens I and III are structured into circumferentially aligned bundles of collagen fibers, with a characteristic 10 to 200 μm crimped or undulating morphology, which is an important determinant of passive and active biomechanical responses (Rezakhaniha et al. 2012; Robertson and Watton 2013). Genetic defects of either fibril-forming collagen can affect vascular wall strength and increases the likelihood of aneurysm formation, as in the case of Ehlers-Danlos syndrome (Sasaki et al. 1987).

Elastin is the second most common structural protein in the arterial wall and is secreted by vascular smooth muscle cells as tropoelastin, which undergoes post-translational modifications to form cross-linked, mature fibers (Parks et al. 1993). Elastin fibers are 1000 times more flexible than collagen and are found in high abundance in the aorta, where it comprises approximately 30% of the dry weight of the vessel wall (Debelle and Tamburro 1999). Elastin forms unique structures known as elastic lamellae, which consists of fibers arranged in 3 μm thick concentric fenestrated lamellae that confer elastic recoil and resilience to the arterial wall, and permits transmural delivery of nutrients and electrolytes (Mithieux and Weiss 2005).

2.2 Cellular Components: Endothelial Cells, Smooth Muscle Cells, and Adventitial Fibroblasts

Endothelial cells (ECs) populate the innermost layer of the arterial wall, where they are in direct contact with blood. This mesoderm-derived specialized epithelium is organized into a semipermeable monolayer that adheres to the basal lamina. Within the native artery, endothelial cell morphology is defined by polygonal-shaped cells, elongated in the direction of blood flow, measuring 12 to 25 μm in length (Garipcan et al. 2011). ECs are a unique cellular subset, due to the presence of tight intercellular and adherens junctions, which serve to regulate cell permeability and membrane polarity, as well as modulate endothelial cell growth through contact inhibition (Bazzoni and Dejana 2004; Lampugnani 2012). As the blood-contacting surface in the arterial wall, endothelial cells have been found to express various molecules that regulate blood homeostasis (Sumpio et al. 2002; Esmon 2005). For example, thrombomodulin, found in the membrane of ECs, inhibits blood coagulation by catalyzing thrombin-induced activation of the protein C pathway (Esmon 1989; Stearns-Kurosawa et al. 1996). Similarly, heparan sulfate is a surface proteoglycan on the luminal aspect of the endothelium, which contains a unique pentasaccharide motif that is recognized by antithrombin III (ATIII), binds to factor IIa (thrombin) and factor Xa, and inhibits clot formation (Bernfield et al. 1999; Rabenstein 2002). Heparan sulfate also facilitates leukocyte adhesion and diapedesis (Parish 2006).

Vascular smooth muscle cells (SMCs) are contractile, bi- or multinucleated cells, with a spindle-shaped morphology that populate the arterial tunica media. Most SMCs have been reported to be mesoderm derived; however, those that populate the aorta and pulmonary arteries are derived from neural crest cells (Le Lièvre and Le Douarin 1975). Within the arterial wall, SMCs are tightly packed in a circumferentially aligned manner, and, unlike endothelial cells, each cell is surrounded by a 40 to 80 nm thick basal lamina suspended in a collagen fibril matrix with alternating rings of elastic lamellae (Rhodin 1979; Clark and Glagov 1985). In healthy adults, SMCs are non-proliferative, are metabolically quiescent, and are not actively migrating or proliferating (Bacakova et al. 2018). As the primary cellular component of the arterial media, SMCs regulate arterial tone and local tissue oxygen delivery through vasomotor control. SMCs are vaso-responsive to Ca^{2+}, myogenic stretch, as well as endothelin, nitric oxide, and prostacyclin secreted by endothelial cells (Wilson 2011).

Adventitial fibroblasts populate the outer most layer of the arterial wall, where they regulate production and organization of undulated collagen fibers that serve to limit vessel overdistension (Stenmark et al. 2013). For years, the adventitia was commonly considered a supporting tissue. However, recent studies have identified adventitial fibroblasts as key regulators of the vessel wall response to hormonal, inflammatory, and environmental stresses, such as hypoxia, ischemia, or hypertension (Stenmark et al. 2011). Activated adventitial fibroblasts have been found to proliferate and upregulate the release of chemokines leading to adventitial remodeling and neointimal hyperplasia (Shi et al. 1996; Sartore et al. 2001). Overall, adventitial fibroblasts are capable of regulating vascular structure and function through the secretion of growth factors, cytokines, and chemokines that serve to communicate with adventitial neural cells, circulating inflammatory cells, and neighboring SMCs and ECs (Sorrell and Caplan 2009).

2.3 Structural Considerations in Vascular Grafts

In 1986, Crispin Weinberg and Eugene Bell published the first report of “a blood vessel model” for the replacement of small-caliber vessels. They encapsulated fibroblasts and smooth muscle cells in casted collagen tubes, supported with a Dacron mesh, lined with endothelial cells (Weinberg and Bell 1986). Although the end result was a weak construct that “grossly resembled a muscular artery,” this work motivated further investigations aimed at engineered vessels that mimicked the structure of native arteries.

Buijtenhuijs et al. subsequently developed a semi-aligned, porous scaffold consisting of collagen and elastin fibers and seeded it with vascular smooth muscle cells (Buijtenhuijs et al. 2004). They were among the first to explore strategies for controlling structural protein morphology and orientation within a vessel wall by tuning the freeze-drying of a suspension of insoluble type I collagen and elastin. Since then, electrospinning and wet-spinning have also been used to control the organization of collagen and elastin fibers (Buttafoco et al. 2006; McClure et al. 2010; Huang et al. 2013; Ahn et al. 2015). These modalities consist of the continuous formation of polymer filaments by either mechanical extrusion into coagulation baths or electrostatic repulsion between polymer solutions and charged surfaces (Miranda-Nieves and Chaikof 2017). As an example, Caves et al. developed a continuous wet-spinning system for the extrusion of synthetic collagen fibers into a 10 wt% polyethylene glycol bath and, after embedding within a recombinant elastin-like protein matrix, generated a vascular graft that resembled the reinforced composite structure of the arterial wall and the circumferential alignment of collagen fibers within the tunica media (Caves et al. 2010a, b).

Approximating the crimped morphology of native collagen fibrils has been possible through molding and chemical treatment (Caves et al. 2010c; Liu et al. 2015). For example, Naik et al. developed a MEMS-based micromolding approach capable of producing in-plane crimped microfibers with a crimp-periodicity of about 100 μm (Naik et al. 2014), and Kumar et al. used excimer-laser technology to ablate collagen lamellae without protein denaturing (Kumar et al. 2014). In both cases, sheets of crimped collagen fibrils were generated, which exhibited orthotropic tensile properties.

On another hand, various groups have focused on recapitulating the structure of native vessels through an approach driven by cellular engineering. For example, vascular conduits have been produced by rolling sheets of fibroblasts or smooth muscle cells produced over a 3- to 7-week period (L’Heureux et al. 1998, 2006). Cell-sheet engineered grafts have shown to recapitulate the lamellar structure of arteries and the position of vascular cells. Complementary strategies, such as patterned polydimethylsiloxane (PDMS) substrates and dynamic mechanical stimulation, have been successfully employed to guide cell sheet alignment during growth or after maturation, in order to better approximate the arterial microstructure (Xing et al. 2017; Rim et al. 2018). For instance, Isenberg et al. cultured smooth muscle cells on gelatin-coated, micropatterned PDMS and produced cell sheets with alignment in the same direction as the pattern (Isenberg et al. 2012). Similarly, Gauvin et al. reported that mechanical stimulation of 10% strain at a frequency of 1 Hz for 3 days significantly enhanced fibroblasts cell sheet alignment (Gauvin et al. 2011).

Other approaches have leveraged advances in polymer science to generate synthetic, biodegradable scaffolds for direct implantation or ex-vivo cell seeding and mechanical preconditioning (Roh et al. 2009; Dahl et al. 2011; Syedain et al. 2016). The goal of these approaches is to rely upon cell-mediated synthesis of ECM, with a number of strategies applied to control ECM composition, organization, and architecture. Successful strategies have included the use of perfusion bioreactors with control over pulse rate and cyclic distension to increase production of structural proteins (Solan et al. 2003; Syedain et al. 2008); culture medium supplementation with organic and inorganic compounds that enhance matrix remodeling and crosslinking (Neidert et al. 2002; Dahl et al. 2005); and controlled release of growth factors and recombinant chemokines in order to modulate inflammation and promote cellular migration (Wu et al. 2012; Yu et al. 2012). As an example, Huang et al. discovered that biaxial preconditioning of SMC-seeded, polyglycolic acid (PGA)-based scaffolds enhances the formation of mature elastin fiber and undulated collagen fibrils (Huang et al. 2015, 2016). Similarly, Roh et al. reported that the release of monocyte chemoattractant protein-1 (MCP-1) modulated monocyte recruitment, promoted migration and proliferation of adjacent vascular wall cells, and overall resulted in the in situ remodeling of poly(L-lactide-co-caprolactone) [PLCL] scaffolds (Roh et al. 2010).

Notwithstanding reports of promising results (McAllister et al. 2009; Syedain et al. 2017; Kirkton et al. 2019), cell-based strategies remain limited by the absence of protocols for the rapid and scalable production of patient-specific vascular wall cells and need for decellularization before implantation (Wystrychowski et al. 2014; Lawson et al. 2016). More specifically, studies have revealed feasibility challenges when using cells isolated from the intended patient population (>65 years old), or allogeneic sources, including reduced proliferation and ECM production due to telomere shortening (Poh et al. 2005), and immune rejection associated with HLA mismatching. Recent reports have sought to address these limitations by deriving SMCs and ECs from patient-specific human induced pluripotent stem cells (hiPSCs), which have unlimited proliferation capacity (Wang et al. 2014; Patsch et al. 2015; Gui et al. 2016; Luo et al. 2017). Similarly, multiplex genome editing has been employed to selectively ablate HLA class I and II molecules and introduce immunoregulatory factors in order to evade both the adaptive and innate immune mechanisms of immune rejection (Deuse et al. 2019; Xu et al. 2019; Han et al. 2019). However, the scalability and efficacy of these approaches remains uncertain.

3 Mechanical Considerations

The biomechanical characterization of engineered arterial substitutes has been largely limited to evaluating burst pressure and suture retention, with the goal of matching these discrete values to those reported for arteries or saphenous vein. Although important metrics, these parameters provide relatively limited insight into the biomechanical properties of the engineered conduit, which influence cellular responses, tissue remodeling, and long-term conduit durability. In this section, we will discuss (1) the mechanical behavior of native arteries and (2) unique mechanical considerations in the design of an engineered living blood vessel that influence both early and late performance characteristics of the conduit.

3.1 Hyperelasticity and Compliance

Arteries are constantly subject to cyclic mechanical stress and can compensate for alterations in intravascular blood volume with minimal changes in pressure through modulating vascular tone. Collagen and elastin, and their unique structural organization, are responsible for the nonlinear responses of arteries to loading forces (Wagenseil and Mecham 2009). At low radial stretch, less than 10% of collagen fibers are engaged and aligned with unfolding of elastic lamellae dictating mechanical behavior. Large radial changes result in marginal increases in pressure. Beyond the range of physiological pressure (80 to 120 mmHg), collagen bears the load with the recruitment, circumferential alignment, and straightening of undulated fibers, leading to marginal radius changes with increasing pressure (Fig. 3). Recent studies have shown that elastic stretching and architecture, as well as collagen fibril recruitment and straightening varies within the arterial wall, in order to compensate the

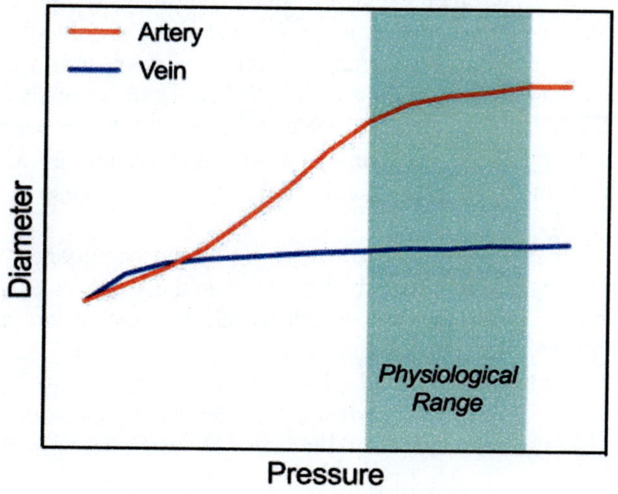

Fig. 3 *Pressure-diameter curves comparing murine carotid artery and murine inferior vena cava.* Values were recorded using pressure myograph

circumferential stretch experienced during radial distension (Zeinali-Davarani et al. 2015; Yu et al. 2018a, b).

When arteries are pressurized, they are subjected to distention in all directions. Many constitutive models have been used to calculate the stress-strain response of the arterial wall from pressure-diameter curves (Başar and Weichert 2000). The most commonly used formulation is the hyperelasticity model in which the vessel is treated as an orthotropic, cylindrical body in which all net strains are oriented along the circumferential, longitudinal, and radial directions (Dorbin 1978; Gasser et al. 2006). Longitudinal and circumferential stresses *(σ)* are calculated as:

$$\sigma = \frac{P\lambda}{ALt} \tag{1}$$

where P is the inflation pressure, λ is the stretch, L is the initial length, t is the thickness of the tissue, and A is the cross-sectional area of the cylinder. Similarly, strain *(E)* can be defined by:

$$E = \frac{1}{2}\left(\lambda^2 - 1\right) \tag{2}$$

Compliance mismatch is an important failure mode, which can be characterized through the arterial pressure-diameter relationship. Compliance *(c)* represents the percent change in diameter over a physiologic range of pressure and is calculated as:

$$\mathrm{c} = \frac{\mathrm{d}_{120} - \mathrm{d}_{80}}{\mathrm{d}_{80}\left(\mathrm{P}_{120} - \mathrm{P}_{80}\right)} \times 10^4 \tag{3}$$

where d is diameter and P is inflation pressure (Robertson and Watton 2013). Arterial compliance (%/100 mmHg) varies with vessel type and location, ranging from 8.0 to 17.0, 6.5 to 12.0, and 6.0 to 14.1%/100 mmHg, for coronary, internal thoracic, and the femoral arteries, respectively (Kumar et al. 2011). Precise methodologies, such as laser micrometry and pressure myography, should be employed to accurately measure outer diameter and pressure and successfully quantify compliance in the 80 to 120 mmHg pressure range. Laser micrometry uses a laser to scan a field, and by detecting the time during which the laser path is obstructed, the dimensions of any sample can be calculated (Syedain et al. 2011). Pressure myography relies upon recording changes in diameter using a high-resolution camera placed over a conduit that is mounted onto small cannulae during the course of pressurization (Schjørring et al. 2015).

The pressure-diameter, stress-strain, and compliance characteristics of a saphenous vein are significantly different than those responses measured for arterial blood vessels (Li 2018). Due to reduced elastin content and a lower number of elastic lamellae, stiffening occurs at lower pressures (Fig. 3), with the saphenous vein exhibiting a significantly lower compliance (0.7–2.6%/100 mmHg) (Lee et al. 2013). For this reason, when used as an arterial substitute, elevated arterial pressure induces increased stress in the vein wall, promotes smooth muscle cell proliferation and matrix production, and, as a consequence, increases the risk of intimal hyperplasia (Li 2018).

3.2 Residual Stress

In the 1960s, Bergel performed an experiment in which he prepared cross-sectional rings of excised, intact, unloaded arteries, and noted that when cut radially "an artery will unroll itself" (Bergel 1960). This was the first report of stress in an artery even when there is no distending pressure. The cut ring "opens" to minimize stored strain-energy, as the inner wall is in compression and the outer wall in tension (Humphrey 2002). This residual stress is the result of differences in the waviness of the elastic lamellae between the inner and outer wall (Yu et al. 2018a). Direct quantification of residual stress (Λ) is not a simple task. Thus, surrogate measures of residual strain have been employed, such as measurements of the opening angle after a radial cut (Matsumoto et al. 2015).

$$\Lambda = \frac{\pi\left(R_a^2 - R_i^2\right)}{\boldsymbol{\Theta}\left(r_a^2 - r_i^2\right)} \tag{4}$$

where R is the adventitial or intimal radius prior to cutting, $\boldsymbol{\Theta}$ is the observed opening angle, and r is the adventitial or intimal radius after cutting. Residual stress provides an indirect measurement of arterial wall stress and the mechanical microenvironment within the vessel wall.

3.3 Mechanical Considerations in Vascular Grafts

Reports of engineered arterial substitutes have focused almost entirely on optimizing burst pressure and suture retention strength while lacking attention to many critical biomechanical parameters discussed in this chapter. Nonetheless, some reports have identified these limitations and proposed modified design strategies (Dahl et al. 2007). For instance, failure to match the hyperelastic behavior of native tissues due to inferior collagen and elastin organization has been addressed by tuning initial polymer concentration (Cummings et al. 2004; Lai et al. 2012) and establishing fabrication protocols with increased control over ECM composition, organization, and architecture (Hall et al. 2016; Xing et al. 2017; Yokoyama et al. 2017). In particular, construct fabricated with pre-stretch, electrospun PLCL fibers at various orientations recapitulated the nonlinear stress-strain behavior of native arteries (Niu et al. 2019).

Variations in polymer composition, organization, and architecture, as well as cyclic pre-conditioning, have proven tunable strategies for designing constructs with control over compliance and residual stress (Niklason et al. 2001; Huang et al. 2016; Niu et al. 2019). For example, McClure et al. reported that altering combinations of collagen, elastin, and synthetic polymers in electrospun grafts allowed precise control over the compliance of the conduits (McClure et al. 2010). Caves et al. employed a fabrication scheme in which a range of collagen fiber orientation and volume fractions were investigated, and reported compliance from 2.8 to 8.4%/100 mmHg, matching values reported for major arteries of interest (Caves et al. 2010a). Huang et al. fabricated constructs through biaxial preconditioning of cell-seeded PGA scaffolds with increased conduit compliance due to the formation of mature elastin fiber and crimped collagen fibrils (Huang et al. 2015, 2016).

Noteworthy, the capacity for biomechanical properties to vary post-implantation has also been reported. In a clinical study, 25 patients were enrolled in an arteriovenous (A-V) shunt safety trial, and the compliance 6-month post-implantation increased approximately threefold (L'Heureux et al. 2007; Konig et al. 2009), suggesting that consideration should also be given to the role of in vivo remodeling in the long-term performance of vascular grafts.

4 Biological Considerations

The biological characterization of engineered living arterial grafts has almost exclusively been focused on evaluating cell infiltration and stem cell differentiation. Limited insight has been obtained surrounding the phenotypic variations of cellular components within vascular grafts. In this section, we will discuss (1) the key phenotypic biomarkers of vascular smooth muscle cells, endothelial cells, and adventitial fibroblasts and (2) biological considerations in the design of living blood vessels.

4.1 Smooth Muscle Cell Phenotype

Smooth muscle cells are a highly specialized and differentiated cell. Under normal conditions, SMCs are elongated, spindle-shaped, non-proliferative, metabolically quiesced, and functionally contractile (Rensen et al. 2007). They are associated with elevated expression of contractile apparatus proteins, such as α-smooth muscle actin (α-SMA), calponin, smoothelin, and smooth muscle myosin heavy chain (SM-MHC) (Owens 1995). However, SMCs are remarkably plastic and can shift phenotype in order to adapt to fluctuating environmental cues, physical stressors, and biochemical alterations. In response to vessel injury, SMCs have been documented to proliferate, migrate, and over-secrete EMC molecules, in particular fibronectin (Gomez and Owens 2012). This de-differentiated synthetic state is associated with loss of contractile proteins, a rhomboid morphology, and expression of proteolytic enzymes and inflammatory cytokines, such as MCP-1 (Bennett et al. 2016). Overall, human arteries may contain a heterogeneous mixture of contractile and synthetic functions (Hao et al. 2003).

4.2 Endothelial Cell Phenotype

Mechanical forces, soluble growth factors, and cytokines, as well as contact with tissue-based cells and ECM protein, regulate endothelial cell phenotype. Under normal conditions, ECs are quiescent, anticoagulant, anti-inflammatory, anti-oxidative, and non-angiogenic with a low replicative capacity (Cines et al. 1998). Quiescent ECs exhibit cobblestone morphology and express markers such as VE-cadherin, PECAM-1, CD34, and CD36 (Lin et al. 2000). However, in response to physical and biological stressors, endothelial cells can de-differentiate into an activated, pro-inflammatory state (Liao 2013). Activated ECs are associated with loss of vascular integrity, causing efflux of fluid into the extravascular space; expression of leucocyte adhesion molecules, such as selectins, ICAM-1, and VCAM-1; prothrombogenicity, due to loss of surface expressed thrombomodulin; and increased cytokine production, such as IL-6, IL-8, and MCP-1 (Hunt and Jurd 1998).

4.3 Adventitial Fibroblast Phenotype

The principal function of adventitial fibroblasts is the deposition of isotropic collagen bundles that provide support and prevent overdistension. However, in response to hormonal, inflammatory, and environmental stresses such as hypoxia, ischemia, or hypertension, adventitial fibroblasts undergo phenotypic switching into myofibroblasts, characterized by the expression of contractile proteins, in particular α-SMA; secretion of pro-inflammatory cytokines, such as TGF- β; increased proliferation and synthetic activity; as well as enhanced migration and functional contractility (Strauss and Rabinovitch 2000; Maiellaro and Taylor 2007). Adventitial fibroblast activation significantly alters the vessel wall microstructure, with the over-accumulation of ECM proteins, such as collagen, elastin, and fibronectin, influencing vessel elasticity and flow dynamics (Desmoulière et al. 2005), and increased cell migration associated with neointimal hyperplasia (Kalra et al. 2000; Misra et al. 2010).

4.4 Biological Considerations in Vascular Grafts

For most reports of tissue engineered vascular grafts, the extent of biological characterization has been limited to identifying post-implant cell infiltration or assessing stem cells differentiation in pre-seeded scaffolds (Koobatian et al. 2016; Syedain et al. 2017; Kirkton et al. 2019). Although an indication of tissue remodeling, such information provides relatively limited insight into the late performance characteristics of vascular conduits. A recent viewpoint in the tissue engineering of living blood vessels is that cellular phenotype should be a key consideration when selecting design strategies. More specifically, fabrication approaches should aim to generate constructs that induce contractile SMCs, quiescent ECs, and inactivated fibroblasts. As an example, Yokoyama et al. fabricated arterial grafts by varying hydrostatic pressure and evaluated expression levels of SMC markers, such as fibronectin, collagen, and elastin, to identify the ideal conditions that resulted in a contractile phenotype (Yokoyama et al. 2017). Similarly, iPSC-derived SMCs were examined for an array of biomarkers, including α-SMA, calponin, smoothelin, SM-MHC, collagen, and elastin, in order to determine serum and growth factor concentrations optimal for cell maturation (Wanjare et al. 2013, 2014).

5 Conclusions

Surgical revascularization of small diameter vessels remains a clinical challenge given limited availability of suitable arterial conduits. Synthetic and autologous grafts exhibit poor 1-year patency rates and are often constrained by availability and length. These challenges have motivated the development of tissue-engineering strategies. Preclinical and early clinical reports describe the use of vascular grafts produced by cellular engineering or from a variety of natural and synthetic biodegradable scaffolds with the potential for ex vivo preconditioning or direct in vivo implantation. Likewise, polymer- and cellular-based approaches have been explored to better recapitulate properties of the arterial wall. However, the design of a durable synthetic small-caliber bypass graft remains elusive.

A current perspective in blood vessel engineering suggests that an ideal living arterial conduit should reproduce both the structure of a native artery and its related biological and mechanical characteristics. In this chapter, we have summarized the characteristics of native arteries and discussed approaches to engineer living blood vessels that replicate these features. It has been noted that many published fabrication schemes lack a holistic view of the native artery as a unique structure that dictates overall biomechanical properties and influences cellular responses, tissue remodeling, and long-term conduit durability. For this reason, we conclude that future design strategies should be informed by the structural, mechanical, and biological considerations discussed herein.

References

Abbott WM, Megerman J, Hasson JE et al (1987) Effect of compliance mismatch on vascular graft patency. J Vasc Surg 5:376–382

Ahn H, Ju YM, Takahashi H et al (2015) Engineered small diameter vascular grafts by combining cell sheet engineering and electrospinning technology. Acta Biomater 16:14–22

Bacakova L, Travnickova M, Filova E et al (2018) The role of vascular smooth muscle cells in the physiology and pathophysiology of blood vessels. In: Muscle cell and tissue – current status of research field. InTech, London, UK

Ballyk PD, Walsh C, Butany J, Ojha M (1997) Compliance mismatch may promote graft–artery intimal hyperplasia by altering suture-line stresses. J Biomech 31:229–237

Başar Y, Weichert D (2000) Nonlinear continuum mechanics of solids. Springer, Berlin/Heidelberg

Bazzoni G, Dejana E (2004) Endothelial cell-to-cell junctions: molecular organization and role in vascular homeostasis. Physiol Rev 84:869–901
Benjamin EJ, Paul Muntner C, Chair Alvaro Alonso V et al (2019) Heart disease and stroke statistics – 2019 update. Circulation 139:56–528
Bennett MR, Sinha S, Owens GK (2016) Vascular smooth muscle cells in atherosclerosis. Circ Res 118:692–702
Bergel DH (1960) The visco-elastic properties of the arterial wall. University of London, London, UK
Berger A, MacCarthy PA, Siebert U et al (2004) Long-term patency of internal mammary artery bypass grafts: relationship with preoperative severity of the native coronary artery stenosis. Circulation 110:II36–II40
Bernfield M, Götte M, Park PW et al (1999) Functions of cell surface Heparan Sulfate proteoglycans. Annu Rev Biochem 68:729–777
Brewster DC (1997) Current controversies in the management of aortoiliac occlusive disease. J Vasc Surg 25:365–379
Buijtenhuijs P, Buttafoco L, Poot AA et al (2004) Tissue engineering of blood vessels: characterization of smooth-muscle cells for culturing on collagen-and-elastin-based scaffolds. Biotechnol Appl Biochem 39:141–149
Buttafoco L, Kolkman NG, Engbers-Buijtenhuijs P et al (2006) Electrospinning of collagen and elastin for tissue engineering applications. Biomaterials 27:724–734
Caves JM, Kumar VA, Martinez AW et al (2010a) The use of microfiber composites of elastin-like protein matrix reinforced with synthetic collagen in the design of vascular grafts. Biomaterials 31:7175–7182
Caves JM, Kumar VA, Wen J et al (2010b) Fibrillogenesis in continuously spun synthetic collagen fiber. J Biomed Mater Res – Part B Appl Biomater 93:24–38
Caves JM, Kumar VA, Xu W et al (2010c) Microcrimped collagen fiber-elastin composites. Adv Mater 22:2041–2044
Cines DB, Pollak ES, Buck CA et al (1998) Endothelial cells in physiology and in the pathophysiology of vascular disorders. J Am Soc Hematol 91:3527–3561
Clark JM, Glagov S (1985) Transmural organization of the arterial media. The lamellar unit revisited. Arteriosclerosis 5:19–34
Clowes AW (1993) Intimal hyperplasia and graft failure. In: Cardiovascular pathology. Elsevier, Amsterdam, Netherlands, pp 179–186
Conte MS, Bandyk DF, Clowes AW et al (2006) Results of PREVENT III: a multicenter, randomized trial of edifoligide for the prevention of vein graft failure in lower extremity bypass surgery. J Vasc Surg 43:742–751
Cummings CL, Gawlitta D, Nerem RM, Stegemann JP (2004) Properties of engineered vascular constructs made from collagen, fibrin, and collagen-fibrin mixtures. Biomaterials 25:3699–3706
Dahl SLM, Rucker RB, Niklason LE (2005) Effects of copper and cross-linking on the extracellular matrix of tissue-engineered arteries. Cell Transplant 14:861–868
Dahl SLM, Rhim C, Song YC, Niklason LE (2007) Mechanical properties and compositions of tissue engineered and native arteries. Ann Biomed Eng 35:348–355
Dahl SLM, Kypson AP, Lawson JH et al (2011) Readily available tissue-engineered vascular grafts. Sci Transl Med 3:68ra9
De Vries MR, Simons KH, Jukema JW et al (2016) Vein graft failure: from pathophysiology to clinical outcomes. Nat Rev Cardiol 13:451–470
Debelle L, Tamburro AM (1999) Elastin: molecular description and function. Int J Biochem Cell Biol 31:261–272
DeRubertis BG, Faries PL, McKinsey JF et al (2007) Shifting paradigms in the treatment of lower extremity vascular disease. Ann Surg 246:415–424
Desmoulière A, Chaponnier C, Gabbiani G (2005) Tissue repair, contraction, and the myofibroblast. Wound Repair Regen 13:7–12
Deuse T, Hu X, Gravina A et al (2019) Hypoimmunogenic derivatives of induced pluripotent stem cells evade immune rejection in fully immunocompetent allogeneic recipients. Nat Biotechnol 37:252–258

Dorbin PB (1978) Mechanical properties of arteries. Physiol Rev 58:397–460
Esmon CT (1989) The roles of protein C and thrombomodulin in the regulation of blood coagulation. J Biol Chem 264:4743–4746
Esmon CT (2005) The interactions between inflammation and coagulation. Br J Haematol 131:417–430
Garipcan B, Maenz S, Pham T et al (2011) Image analysis of endothelial microstructure and endothelial cell dimensions of human arteries - a preliminary study. Adv Eng Mater 13: B54–B57
Gasser TC, Ogden RW, Holzapfel GA (2006) Hyperelastic modelling of arterial layers with distributed collagen fibre orientations. J R Soc Interface 3:15–35
Gaudino M, Benedetto U, Fremes S et al (2018) Radial-artery or saphenous-vein grafts in coronary-artery bypass surgery. N Engl J Med 378:2069–2077
Gauvin R, Parenteau-Bareil R, Larouche D et al (2011) Dynamic mechanical stimulations induce anisotropy and improve the tensile properties of engineered tissues produced without exogenous scaffolding. Acta Biomater 7:3294–3301
Go AS, Mozaffarian D, Roger VL et al (2013) Heart disease and stroke statistics – 2013 update. Circulation 127:e6–e245
Goldman S, Zadina K, Moritz T et al (2004) Long-term patency of saphenous vein and left internal mammary artery grafts after coronary artery bypass surgery: results from a Department of Veterans Affairs Cooperative Study. J Am Coll Cardiol 44:2149–2156
Gomez D, Owens GK (2012) Smooth muscle cell phenotypic switching in atherosclerosis. Cardiovasc Res 95:156–164
Gui L, Dash BC, Luo J et al (2016) Implantable tissue-engineered blood vessels from human induced pluripotent stem cells. Biomaterials 102:120–129
Hall MS, Alisafaei F, Ban E et al (2016) Fibrous nonlinear elasticity enables positive mechanical feedback between cells and ECMs. Proc Natl Acad Sci U S A 113:14043–14048
Han X, Wang M, Duan S et al (2019) Generation of hypoimmunogenic human pluripotent stem cells. Proc Natl Acad Sci 116:10441–10446
Hao H, Gabbiani G, Bochaton-Piallat M-L (2003) Arterial smooth muscle cell heterogeneity. Arterioscler Thromb Vasc Biol 23:1510–1520
Hess CN, Lopes RD, Gibson CM et al (2014) Saphenous vein graft failure after coronary artery bypass surgery. Circulation 130:1445–1451
Huang NF, Okogbaa J, Lee JC et al (2013) The modulation of endothelial cell morphology, function, and survival using anisotropic nanofibrillar collagen scaffolds. Biomaterials 34:4038–4047
Huang AH, Lee Y-U, Calle EA et al (2015) Design and use of a novel bioreactor for regeneration of Biaxially stretched tissue-engineered vessels. Tissue Eng Part C Methods 21:841–851
Huang AH, Balestrini JL, Udelsman BV et al (2016) Biaxial stretch improves elastic Fiber maturation, collagen arrangement, and mechanical properties in engineered arteries. Tissue Eng Part C Methods 22:524–533
Humphrey JD (2002) The normal arterial wall. In: Cardiovascular solid mechanics. Springer Science and Business Media, New York, pp 249–364
Hunt BJ, Jurd KM (1998) Endothelial cell activation. A central pathophysiological process. BMJ 316:1328–1329
Isenberg BC, Backman DE, Kinahan ME et al (2012) Micropatterned cell sheets with defined cell and extracellular matrix orientation exhibit anisotropic mechanical properties. J Biomech 45:756–761
Johnson WC, Lee KK (2000) A comparative evaluation of polytetrafluoroethylene, umbilical vein, and saphenous vein bypass grafts for femoral-popliteal above-knee revascularization: a prospective randomized Department of Veterans Affairs cooperative study. J Vasc Surg 32:268–277
Kalra M, Miller VM, Miller VM (2000) Early remodeling of saphenous vein grafts: proliferation, migration and apoptosis of adventitial and medial cells occur simultaneously with changes in graft diameter and blood flow. J Vasc Res 37:576–584

Kirkton RD, Santiago-Maysonet M, Lawson JH et al (2019) Bioengineered human acellular vessels recellularize and evolve into living blood vessels after human implantation. Sci Transl Med 11:1–12

Klinkert P, Post P, Breslau P, van Bockel J (2004) Saphenous vein versus PTFE for above-knee Femoropopliteal bypass. A review of the literature. Eur J Vasc Endovasc Surg 27:357–362

Konig G, McAllister TN, Dusserre N et al (2009) Mechanical properties of completely autologous human tissue engineered blood vessels compared to human saphenous vein and mammary artery. Biomaterials 30:1542–1550

Koobatian MT, Row S, Smith RJ et al (2016) Successful endothelialization and remodeling of a cell-free small-diameter arterial graft in a large animal model. Biomaterials 76:344–358

Kumar VA, Brewster LP, Caves JM, Chaikof EL (2011) Tissue engineering of blood vessels: functional requirements, progress, and future challenges. Cardiovasc Eng Technol 2:137–148

Kumar VA, Martinez AW, Caves JM et al (2014) Microablation of collagen-based substrates for soft tissue engineering. Biomed Mater 9:011002

L'Heureux N, Pâquet S, Labbé R et al (1998) A completely biological tissue-engineered human blood vessel. FASEB J 12:47–56

L'Heureux N, Dusserre N, Konig G et al (2006) Human tissue engineered blood vessel for adult arterial revascularization. Nat Med 12:361–365

L'Heureux N, McAllister TN, de la Fuente LM (2007) Tissue-engineered blood vessel for adult arterial revascularization. N Engl J Med 357:1451–1453

Lai VK, Frey CR, Kerandi AM et al (2012) Microstructural and mechanical differences between digested collagen-fibrin co-gels and pure collagen and fibrin gels. Acta Biomater 8:4031–4042

Lampugnani MG (2012) Endothelial cell-to-cell junctions: adhesion and signaling in physiology and pathology. Cold Spring Harb Perspect Med 2:a006528

Lawson JH, Glickman MH, Ilzecki M et al (2016) Bioengineered human acellular vessels for dialysis access in patients with end-stage renal disease: two phase 2 single-arm trials. Lancet 387:2026–2034

Le Lièvre CS, Le Douarin NM (1975) Mesenchymal derivatives of the neural crest: analysis of chimaeric quail and chick embryos. Development 34:125–154

Lee YU, Naito Y, Kurobe H et al (2013) Biaxial mechanical properties of the inferior vena cava in C57BL/6 and CB-17 SCID/bg mice. J Biomech 46:2277–2282

Lemson MS, Tordoir JHM, Daemen MJAP, Kitslaar PJEHM (2000) Intimal hyperplasia in vascular grafts. Eur J Vasc Endovasc Surg 19:336–350

Li W (2018) Biomechanical property and modelling of venous wall. Prog Biophys Mol Biol 133:56–75

Liao JK (2013) Linking endothelial dysfunction with endothelial cell activation. J Clin Invest 123:540–541

Lin Y, Weisdorf DJ, Solovey A, Hebbel RP (2000) Origins of circulating endothelial cells and endothelial outgrowth from blood. J Clin Invest 105:71–77

Linsenmayer TF (1991) Collagen. In: Cell biology of extracellular matrix. Springer, Boston, pp 7–44

Liu W, Lipner J, Moran CH et al (2015) Generation of electrospun nanofibers with controllable degrees of crimping through a simple, plasticizer-based treatment. Adv Mater 27:2583–2588

Loop FD, Lytle BW, Cosgrove DM et al (1986) Influence of the internal-mammary-artery graft on 10-year survival and other cardiac events. N Engl J Med 314:1–6

Luo J, Qin L, Kural MH et al (2017) Vascular smooth muscle cells derived from inbred swine induced pluripotent stem cells for vascular tissue engineering. Biomaterials 147:116–132

Maiellaro K, Taylor WR (2007) The role of the adventitia in vascular inflammation. Cardiovasc Res 75:640–648

Matsumoto T, Sugita S, Yaguchi T (2015) Biomechanics of blood vessels: structure, mechanics, and adaptation. In: Advances in metallic biomaterials. Springer, Berlin, Germany, pp 71–98

McAllister TN, Maruszewski M, Garrido SA et al (2009) Effectiveness of haemodialysis access with an autologous tissue-engineered vascular graft: a multicentre cohort study. Lancet 373:1440–1446

McClure MJ, Sell SA, Simpson DG et al (2010) A three-layered electrospun matrix to mimic native arterial architecture using polycaprolactone, elastin, and collagen: a preliminary study. Acta Biomater 6:2422–2433

McKinsey JF, Goldstein L, Khan HU et al (2008) Novel treatment of patients with lower extremity ischemia: use of percutaneous Atherectomy in 579 lesions. Trans Meet Am Surg Assoc 126:160–169

Mehta RH, Ferguson TB, Lopes RD et al (2011) Saphenous vein grafts with multiple versus single distal targets in patients undergoing coronary artery bypass surgery: one-year graft failure and five-year outcomes from the project of ex-vivo vein graft engineering via transfection (PREVENT) IV trial. Circulation 124:280–288

Miranda-Nieves D, Chaikof EL (2017) Collagen and elastin biomaterials for the fabrication of engineered living tissues. ACS Biomater Sci Eng 3:694–711

Misra S, Fu AA, Misra KD et al (2010) Hypoxia-induced phenotypic switch of fibroblasts to Myofibroblasts through a matrix metalloproteinase 2/tissue inhibitor of metalloproteinase-mediated pathway: implications for venous neointimal hyperplasia in hemodialysis access. J Vasc Interv Radiol 21:896–902

Mithieux SM, Weiss AS (2005) Elastin. Adv Protein Chem 70:437–461

Motwani JG, Topol EJ (1998) Aortocoronary saphenous vein graft disease: pathogenesis, predisposition, and prevention. Circulation 97:916–931

Mwipatayi BP, Hockings A, Hofmann M et al (2008) Balloon angioplasty compared with stenting for treatment of femoropopliteal occlusive disease: a meta-analysis. J Vasc Surg 47:461–469

Naik N, Caves J, Chaikof EL, Allen MG (2014) Generation of spatially aligned collagen fiber networks through microtransfer molding. Adv Healthc Mater 3:367–374

Neidert MR, Lee ES, Oegema TR, Tranquillo RT (2002) Enhanced fibrin remodeling in vitro with TGF-B1, insulin and plasmin for improved tissue-equivalents. Biomaterials 23:3717–3731

Niklason LE, Abbott W, Gao J et al (2001) Morphologic and mechanical characteristics of engineered bovine arteries. J Vasc Surg 33:628–638

Niu Z, Wang X, Meng X et al (2019) Controllable fiber orientation and nonlinear elasticity of electrospun nanofibrous small diameter tubular scaffolds for vascular tissue engineering. Biomed Mater 14:035006

Owens GK (1995) Regulation of differentiation of vascular smooth muscle cells. Physiol Rev 75:487–517

Parish CR (2006) The role of heparan sulphate in inflammation. Nat Rev Immunol 6:633–643

Parks WC, Richard AP, Katherine AL, Mecham RP (1993) Elastin. Adv Mol Cell Biol 6:133–181

Patsch C, Challet-Meylan L, Thoma EC et al (2015) Generation of vascular endothelial and smooth muscle cells from human pluripotent stem cells. Nat Cell Biol 17:994–1003

Pereira CE, Albers M, Romiti M et al (2006) Meta-analysis of femoropopliteal bypass grafts for lower extremity arterial insufficiency. J Vasc Surg 44:510–517.e3

Piffaretti G, Dorigo W, Castelli P et al (2018) Results from a multicenter registry of heparin-bonded expanded polytetrafluoroethylene graft for above-the-knee femoropopliteal bypass. J Vasc Surg 67:1463–1471

Poh M, Boyer M, Dahl SLM et al (2005) Blood vessels engineered from human cells. Lancet 365:2122–2124

Qu Z, Chaikof EL (2010) Prosthetic grafts. In: Rutherford's vascular surgery. Elsevier, USA

Rabenstein DL (2002) Heparin and heparan sulfate: structure and function. Nat Prod Rep 19:312–331

Rensen SSM, Doevendans PAFM, van Eys GJJM (2007) Regulation and characteristics of vascular smooth muscle cell phenotypic diversity. Neth Hear J 15:100–108

Rezakhaniha R, Agianniotis A, Schrauwen JTC et al (2012) Experimental investigation of collagen waviness and orientation in the arterial adventitia using confocal laser scanning microscopy. Biomech Model Mechanobiol 11:461–473

Rhodin J (1979) Architecture of the vessel wall. In: Berne RM (ed) Handbook of physiology, Section 2, vol 2. American Physiological Society, Maryland, USA

Rim NG, Yih A, Hsi P et al (2018) Micropatterned cell sheets as structural building blocks for biomimetic vascular patches. Biomaterials 181:126–139
Robertson AM, Watton PN (2013) Mechanobiology of the arterial wall. Transp Biol Media 1: 275–347
Roh JD, Nelson GN, Brennan MP et al (2009) Small-diameter biodegradable scaffolds for functional vascular tissue engineering in the mouse model. Biomaterials 29:1454–1463
Roh JD, Sawh-Martinez R, Brennan MP et al (2010) Tissue-engineered vascular grafts transform into mature blood vessels via an inflammation-mediated process of vascular remodeling. Proc Natl Acad Sci 107:4669–4674
Sand JMB, Genovese F, Karsdal MA (2016) Type IV collagen. Biochem Collagens Laminins Elastin 2:31–41
Sartore S, Chiavegato A, Faggin E et al (2001) Contribution of adventitial fibroblasts to neointima formation and vascular remodeling from innocent bystander to active participant. Circ Res 89:1111–1121
Sasaki T, Arai K, Ono M et al (1987) Ehlers-Danlos syndrome: a variant characterized by the deficiency of Proa2 chain of type I Procollagen. Arch Dermatol 123:76
Schjørring OL, Carlsson R, Simonsen U (2015) Pressure Myography to study the function and structure of isolated small arteries. Humana Press, New York, pp 277–295
Shekhonin BV, Domogatsky SP, Muzykantov VR et al (1985) Distribution of type I, III, IV and V collagen in normal and atherosclerotic human arterial wall: immunomorphological characteristics. Coll Relat Res 5:355–368
Shi Y, O'Brien JE, Fard A et al (1996) Adventitial Myofibroblasts contribute to neointimal formation in injured porcine coronary arteries. Circulation 94:1655–1664
Shoulders MD, Raines RT (2009) Collagen structure and stability. Annu Rev Biochem 78:929–958
Solan A, Mitchell S, Moses M, Niklason L (2003) Effect of pulse rate on collagen deposition in the tissue-engineered blood vessel. Tissue Eng 9:579–586
Sorrell JM, Caplan AI (2009) Fibroblasts – a diverse population at the center of it all, Chapter 4. In: International review of cell and molecular biology. Elsevier, Amsterdam, Netherlands, pp 161–214
Sottiurai VS, Yao JS, Flinn WR, Batson RC (1983) Intimal hyperplasia and neointima: an ultrastructural analysis of thrombosed grafts in humans. Surgery 93:809–817
Stearns-Kurosawa DJ, Kurosawa S, Mollica JS et al (1996) The endothelial cell protein C receptor augments protein C activation by the thrombin-thrombomodulin complex. Proc Natl Acad Sci U S A 93:10212–10216
Stenmark KR, Nozik-Grayck E, Gerasimovskaya E et al (2011) The adventitia: essential role in pulmonary vascular remodeling. Compr Physiol 1:141–161
Stenmark KR, Yeager ME, El Kasmi KC et al (2013) The adventitia: essential regulator of vascular wall structure and function. Annu Rev Physiol 75:23–47
Strauss BH, Rabinovitch M (2000) Adventitial fibroblasts defining a role in vessel wall remodeling. Perspect Am J Respir Cell Mol Biol 22:1–3
Sumpio B, Riley J, Dardik A (2002) Cells in focus: endothelial cell. Int J Biochem Cell Biol 34:1508–1512
Syedain ZH, Weinberg JS, Tranquillo RT (2008) Cyclic distension of fibrin-based tissue constructs: evidence of adaptation during growth of engineered connective tissue. Proc Natl Acad Sci 105:6537–6542
Syedain ZH, Meier LA, Bjork JW et al (2011) Implantable arterial grafts from human fibroblasts and fibrin using a multi-graft pulsed flow-stretch bioreactor with noninvasive strength monitoring. Biomaterials 32:714–722
Syedain Z, Reimer J, Lahti M et al (2016) Tissue engineering of acellular vascular grafts capable of somatic growth in young lambs. Nat Commun 7:1–9
Syedain ZH, Graham ML, Dunn TB et al (2017) A completely biological "off-the-shelf" arteriovenous graft that recellularizes in baboons. Sci Transl Med 9:eaan4209
Taggart DP, Benedetto U, Gerry S et al (2019) Bilateral versus single internal-thoracic-artery grafts at 10 years. N Engl J Med 380:437–446

Van Der Slegt J, Steunenberg SL, Donker JMW et al (2014) The current position of precuffed expanded polytetrafluoroethylene bypass grafts in peripheral vascular surgery. J Vasc Surg 60:120–128
Wagenseil JE, Mecham RP (2009) Vascular extracellular matrix and arterial mechanics. Physiol Rev 89:957–989
Wang Y, Hu J, Jiao J et al (2014) Engineering vascular tissue with functional smooth muscle cells derived from human iPS cells and nanofibrous scaffolds. Biomaterials 35:8960–8969
Wanjare M, Kuo F, Gerecht S (2013) Derivation and maturation of synthetic and contractile vascular smooth muscle cells from human pluripotent stem cells. Cardiovasc Res 97:321–330
Wanjare M, Kusuma S, Gerecht S (2014) Defining differences among perivascular cells derived from human pluripotent stem cells. Stem Cell Rep 2:1–15
Weinberg C, Bell E (1986) A blood vessel model constructed from collagen and cultured vascular cells. Science. (80–) 231:397–400
Weintraub WS, Grau-Sepulveda MV, Weiss JM et al (2012) Comparative effectiveness of revascularization strategies. N Engl J Med 366:1467–1476
Wilson DP (2011) Vascular smooth muscle structure and function. University of Adelaide Press, Adelaide, Australia
Wu W, Allen RA, Wang Y (2012) Fast-degrading elastomer enables rapid remodeling of a cell-free synthetic graft into a neoartery. Nat Med 18:1148–1153
Wystrychowski W, McAllister TN, Zagalski K et al (2014) First human use of an allogeneic tissue-engineered vascular graft for hemodialysis access. J Vasc Surg 60:1353–1357
Xing Q, Qian Z, Tahtinen M et al (2017) Aligned Nanofibrous cell-derived extracellular matrix for anisotropic vascular graft construction. Adv Healthc Mater 6:1–9
Xu H, Wang B, Ono M et al (2019) Targeted disruption of HLA genes via CRISPR-Cas9 generates iPSCs with enhanced immune compatibility. Cell Stem Cell 24:566–578.e7
Yokoyama U, Tonooka Y, Koretake R et al (2017) Arterial graft with elastic layer structure grown from cells. Sci Rep 7:1–16
Yu J, Wang A, Tang Z et al (2012) The effect of stromal cell-derived factor-1α/heparin coating of biodegradable vascular grafts on the recruitment of both endothelial and smooth muscle progenitor cells for accelerated regeneration. Biomaterials 33:8062–8074
Yu X, Turcotte R, Seta F, Zhang Y (2018a) Micromechanics of elastic lamellae: unravelling the role of structural inhomogeneity in multi-scale arterial mechanics. J R Soc Interface 15:20180492
Yu X, Wang Y, Zhang Y (2018b) Transmural variation in elastin fiber orientation distribution in the arterial wall. J Mech Behav Biomed Mater 77:745–753
Zeinali-Davarani S, Wang Y, Chow M-J et al (2015) Contribution of collagen Fiber undulation to regional biomechanical properties along porcine thoracic aorta. J Biomech Eng 137:051001

Heart Valve Bioengineering

Emanuela S. Fioretta, Sarah E. Motta, Eric K. N. Gähwiler, Nikolaos Poulis, Maximilian Y. Emmert, and Simon P. Hoerstrup

Contents

E.S. Fioretta and S.E. Motta contributed equally to this work and are shared first authors

E. S. Fioretta · E. K. N. Gähwiler · N. Poulis
Institute for Regenerative Medicine (IREM), University of Zurich, Zurich, Switzerland
e-mail: Emanuela.Fioretta@irem.uzh.ch; eric.gaehwiler@irem.uzh.ch; nikolaos.poulis@irem.uzh.ch

S. E. Motta · S. P. Hoerstrup (✉)
Institute for Regenerative Medicine (IREM), University of Zurich, Zurich, Switzerland

Wyss Translational Center Zurich, University and ETH Zurich, Zurich, Switzerland
e-mail: sarah.motta@uzh.ch; simon.hoerstrup@irem.uzh.ch

M. Y. Emmert
Institute for Regenerative Medicine (IREM), University of Zurich, Zurich, Switzerland

Wyss Translational Center Zurich, University and ETH Zurich, Zurich, Switzerland

Department of Cardiovascular Surgery, Charité – Universitätsmedizin Berlin, Berlin, Germany

Department of Cardiothoracic and Vascular Surgery, German Heart Center Berlin, Berlin, Germany
e-mail: maximilian.emmert@uzh.ch

D. Eberli et al. (eds.), *Organ Tissue Engineering*, Reference Series in Biomedical Engineering, https://doi.org/10.1007/978-3-030-44211-8_4

Abstract

Valve degeneration and dysfunction affect an increasing number of aging patients in the developed countries, while congenital and rheumatic diseases affect younger patients worldwide. When possible, valvular disease is treated by valve repair; however, severe pathologies require a surgical or transcatheter valve replacement procedure. Nowadays, clinical-grade valvular prostheses consist of mechanical or bioprosthetic valves, which still present serious disadvantages, such as thrombogenicity and limited durability.

In this chapter, we describe how tissue-engineered heart valves (TEHVs) can solve the shortcomings of current clinically adopted valvular replacements by providing a prosthesis with potential regenerative and remodeling capabilities and, hence, a lifelong durability. Different tissue engineering (TE) approaches, such as in vitro, in vivo, and in situ are described, with particular focus on preclinical and clinical validation of TEHVs. In this context, we will also discuss in silico, in vitro, and in vivo models that can be used to manufacture TEHVs, test their functionality, and predict their remodeling before moving to preclinical and clinical settings. Finally, we address the scientific, regulatory, and clinical challenges that pace the safe clinical translation of TEHVs.

1 Introduction

The human heart functions as a pump, providing continuous blood flow through the body, and has evolved from an initial tube-like structure to an organ consisting of four chambers (right atrium, right ventricle (RV), left atrium, left ventricle (LV) (Fig. 1a)). The right and left atriums act as blood reservoirs, while the RV and LV act as pumps, responsible for the circulation of the blood in the pulmonary and systemic circulation, respectively (Rehman and Rehman 2018) (Fig. 1b). Each

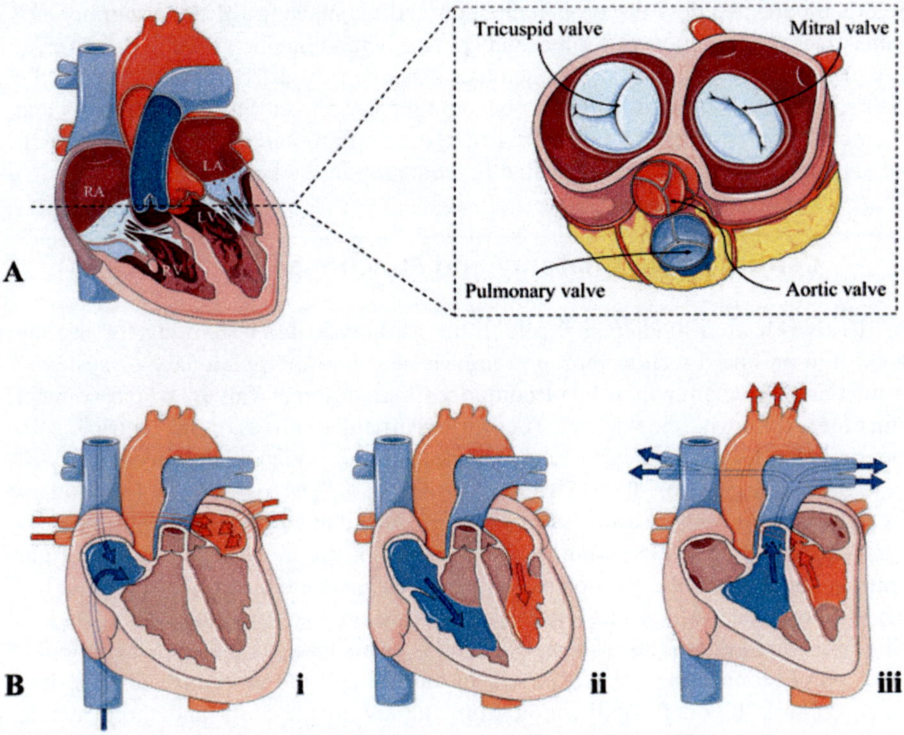

Fig. 1 Schematic representation of the heart anatomy and physiology. (**a**) The right atrium (RA) and right ventricle (RV) are connected via the tricuspid valve. The RV is connected to the pulmonary circulation via the pulmonary valve. The right heart works at low-pressure conditions (0–25 mmHg). The left atrium (LA) and left ventricle (LV) are connected via the mitral valve. The LV is connected to the systemic circulation via the aortic valve. The left heart operates at high-pressure conditions (80–120 mmHg). (**b**) The blood cycle through the heart starts in the RA (i) where deoxygenated blood from the systemic circulation is collected. (ii) The deoxygenated blood flows into the RV, where the blood is then pumped into the pulmonary artery and the lung circulation to get oxygenated. Then, the blood flows back to the heart in the LA in order to be transferred into the LV. (iii) In the LV, the oxygenated blood is pumped in the systemic circulation passing through the aortic valve and into the aorta to supply the organs with oxygen-rich blood. (Images adapted from Servier Medical Art under a creative common attribution 3.0 unported license)

chamber is delimited by specific cardiac valves: the atrioventricular valves (tricuspid and mitral) located between atria and ventricles and the semilunar valves (pulmonary and aortic) (Fig. 1b). The four heart valves act as important inlet and outlet checkpoints during every cardiac cycle, opening and closing in response to different pressure gradients, therefore allowing the unidirectional flow of blood.

An average adult human heart beats between 60 and 70 times per minute, which leads the valves to open and close up to three billion times during a lifetime (Tao et al. 2012). These high performances are enabled by the continuous remodeling and turnover of heart valve cells and the extracellular matrix (ECM) surrounding them.

Both components allow the proper function and maintenance of the organ homeostatic state in a variation of biomechanical and hemodynamic stimuli. However, it should not be surprising that degenerative diseases may affect these organs in the elderly, causing the valve leaflets to become thicker and stiffer due to fibrosis and calcifications (Fishbein and Fishbein 2019). In addition, inflammatory and congenital disease may impact on heart valve functionality in the young.

2 Cardiac Valve Anatomy and Functionality

Heart valves have a fundamental role in the cardiovascular system, by forcing the blood flow in one direction through the heart and into the pulmonary or systemic circulation. The mammalian heart comprises four different valves, which connect either the atrium with the ventricle (i.e., atrioventricular valves) or the ventricle with the artery that leaves the heart (i.e., semilunar valves). Function and morphology of these valves depend on their location in the heart. The right heart (pulmonary circulation) is a low-pressure system (0–25 mmHg) that comprises the atrioventricular tricuspid valve, which controls the direction of the blood flow from the right atrium to the RV, and the pulmonary valve, which leads the blood flow from the RV into the pulmonary artery. The left heart (systemic circulation), on the other hand, is a high-pressure system (80–120 mmHg) that is characterized by the atrioventricular mitral valve, connecting the left atrium to the LV, and the semilunar aortic valve, which controls blood flow direction from the LV to the aorta and the systemic circulation. During the cardiac cycle, the pulmonary circulation delivers blood to the lungs for an oxygen recharge and carbon dioxide discharge. The systemic circulation delivers oxygenated nutrient-rich blood from the heart to the brain and peripheral tissues.

Heart valve leaflets present a unique tri-layered structure that allows them to withstand the specific mechanical environment caused by the differential blood pressure and flow on each side such as pressure, shear stress, and stretch (Merryman et al. 2006).

Valvular heart dysfunction is a disease status for which heart valves are insufficient or stenotic. Nowadays, these pathologies are considered a major societal burden, with a higher incidence of structural diseases in western countries. To understand such pathologies and find a competitive treatment strategy, it is therefore important to comprehend valve anatomy, structure, and function.

2.1 Valve Anatomy and Functionality

2.1.1 Atrioventricular Valves

Both the mitral and the tricuspid valves enable the unidirectional blood flow from the atria to the ventricles and prevent the retrograde blood flow from the ventricles back to the atria. The atrioventricular valves are characterized and supported by a complex network of collagen and elastic fibers attached to the leaflets, known as chordae tendineae. These structures connect the valve leaflets with the papillary muscles, a series of muscles located within the ventricle that, when contract, prevent leaflet prolapse during systole (Anderson et al. 2000; Schoen 2008).

The combination of papillary muscles, chordae tendineae, and leaflets forms a functional unit that controls and ensures the opening and closure of the atrioventricular valves during the cardiac cycle (Millington-Sanders et al. 1998; McCarthy et al. 2010). Despite the similar function, the anatomy of these valves is slightly different.

The mitral valve is composed of four complex structures known as annulus, leaflets, chordae tendineae, and papillary muscles. The annulus designates the opening part of the mitral valve and enables horizontal and vertical geometry changes in the orifice area during cardiac cycle. The mitral valve possesses only two leaflets, the anterior and the posterior, which come together at the two commissures known as the anterolateral and posteromedial commissure (Misfeld and Sievers 2007; Dal-Bianco and Levine 2013).

The tricuspid valve has three thin and translucent leaflets, known as anterior, ventral, and posterior. The leaflets form a complex unit with the annulus, whose structure is subject to dynamic changes during the cardiac cycle (Yacoub and Cohn 2004). Leaflet morphology varies among the three leaflets, with the anterior leaflet being the largest, followed by the medial leaflet, and, finally, posterior leaflet which is the smallest (Buzzatti et al. 2018).

2.1.2 Semilunar Valves

The semilunar valves (i.e., aortic and pulmonary) have the function of dividing the ventricles from the main arteries and preventing retrograde flow into the ventricles during diastole. Semilunar valves comprise three leaflets, which are attached in a semilunar pattern to the annulus, with no external support from papillary muscles and/or chordae tendineae. The basal point of attachment of every leaflet is characterized by a fundamental anatomical feature, the sinuses of Valsalva, which represents the main load-bearing component of the cusps during diastole. The sinuses collect the regurgitant blood during early diastolic phase, influencing the leaflet closure dynamics, ensuring the blood washout, and, for the aortic valve, allowing blood supply to the coronary arteries (Toninato et al. 2016). This specific geometry allows the valves to be functional and to withstand hemodynamic forces throughout a lifetime.

The aortic valve connects the LV with the aorta and the systemic circulation. Its leaflets are anchored to the aortic root by a fibrous annulus, which supports the aortic valve and has a key role in the proper functioning of the heart, by ensuring the appropriate coronary perfusion and by maintaining the laminar flow in the vascular system (Anderson 2000). The pulmonary valve connects the heart to the pulmonary root thus to the pulmonary circulation. Similar to the aortic valve, the pulmonary valve is composed of the annulus, the sinuses, and the leaflets (Bateman et al. 2013).

The commissures are the highest connection points of the leaflets to the arterial wall and are located above the interleaflet triangles, where they define the sinotubular junctions. The interleaflet triangles are of high importance for the opening and closing motions of the valves.

All the structures composing the aortic valve such as the annulus, the sinuses of Valsalva, commissures, and the leaflets operate together to allow proper valve function.

The three aortic valve leaflets are characterized by three fundamental anatomical features: the sinuses of Valsalva, the lannula, and the noduli of Arantii. As previously mentioned, the Valsalva sinuses correspond to the attachment point of the leaflets to the annulus and transmit the stresses endured by the leaflets to the aortic wall. Importantly, when the leaflets are in closed configuration, they reveal the coronary artery ostia, the access point for perfusing the coronary arteries, which give the name to the three aortic leaflets as left coronary, right coronary, and non-coronary (Underwood et al. 2000). The noduli of Arantii is located in the middle of the free edge of the coapting surface. Next to each side of the nodulus, there is a thin semilunar-shaped structure known as the lannula, which is linked to the aortic root wall close to the commissures and which ensures valve coaptation.

2.2 Valve Cell and Tissue Composition

Heart valve leaflets are subject to three different stress states, bending, shear stress, and stretch caused by blood pressure (Merryman et al. 2006). In order to withstand those mechanical stresses and enable high flow rate with low resistance, the valve leaflets evolved to form a complex microscopic tissue structure. The semilunar valve leaflets have a highly organized three-layered ECM architecture (i.e., fibrosa, spongiosa, and ventricularis/atrialis) populated by valve interstitial cells (VICs) and valve endothelial cells (VECs). This specific tissue composition and structure allows the proper function of the valves, thus enabling unidirectional flow of the blood (Balachandran et al. 2011; Hinton and Yutzey 2011).

2.2.1 Tissue Structure

The three layers that compose the leaflets allow them to withstand perpetual changes of pressure during the cardiac cycle (Lincoln et al. 2006; Balachandran et al. 2011) (Fig. 2, Table 1). The fibrosa layer is mainly composed of collagen fibers (types I and III) and is restricted to the outflow surface (Schoen 2012; Schoen and Gotlieb 2016). In order for the leaflet to have the necessary tensile strength during the opening phase of the valve and to transmit the forces during the closing phase, the collagen fibers position themselves circumferentially (Lincoln et al. 2006; Balachandran et al. 2011). The spongiosa is the intermediate layer characterized by sparse collagen fibers and abundance of proteoglycans and glycosaminoglycans (GAGs). This layer composition acts as a shock absorbent when the valve leaflets are closed, and it confers the ability to compress when the valve is in closed configuration (Sacks et al. 2009). The third and final layer, the ventricularis in the case of the semilunar valves or atrialis in the case of the atrioventricular valves, is located at the inflow surface. The ECM of the ventricularis/atrialis is mainly composed of elastin fibers

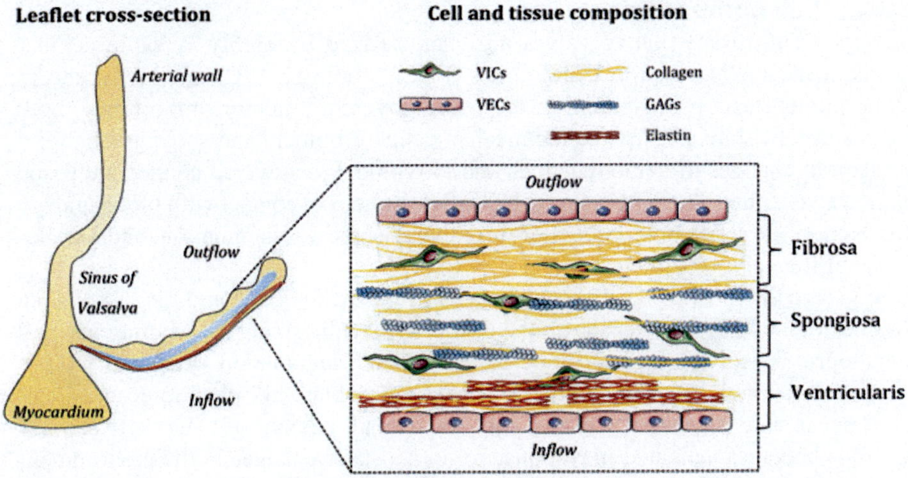

Fig. 2 Schematic representation of a semilunar heart valve leaflet structure and composition. A cross-section of the heart valve leaflet, with the proximal portion of the valve connecting to the myocardium and the distal portion connecting to the arterial wall. The leaflet connects at the hinge region with the annulus and forms a sinus of Valsalva. A semilunar heart valve leaflet is characterized by three layers: the fibrosa, mainly consisting in circumferentially aligned collagen; the spongiosa, rich in glycosaminoglycans (GAGs); and the ventricularis, with radially oriented elastin. Valve interstitial cells (VICs) are distributed throughout the leaflet, while valve endothelial cells (VECs) are covering the leaflet surfaces in contact with blood. (The image was inspired by the work of Rutkovskiy et al. 2017, and re-created using images adapted from Servier Medical Art under a creative common attribution 3.0 unported license)

Table 1 Valve leaflet composition. Localizations and function of the different cellular and extracellular matrix (ECM) components characterizing a heart valve

Extracellular matrix (ECM) components		
	Location	**Function**
Collagen	Main ECM component of the fibrosa layer	Strength during opening and closing motion
GAGs	Main ECM component of the spongiosa layer	Absorption of forces during diastole
Elastin	Main ECM component of the ventricularis/atrialis layer	Flexibility and recoiling of the leaflets
Cellular components		
	Location	**Function**
VICs	Resident tissue cells of the fibrosa, spongiosa, and ventricularis/atrialis	ECM synthesis and remodeling
VECs	Covering all the valve/leaflet surfaces in contact with blood	Regulation of inflammation, thromboresistance, maintenance of the structure integrity

(e.g., fibrillin and elastin) with radial orientation. The elastin fibers allow better movement of the leaflet by extending and recoiling during the opening and closing motions (Hinton and Yutzey 2011).

2.2.2 Cell Composition

Being a living tissue, native valves are characterized by highly versatile cellular components, such as the VICs and VECs (Fig. 2, Table 1). VICs and VECs ensure tissue homeostasis by maintaining the leaflet layered structure. Importantly, both VICs and VECs are sensitive to the mechanical environment and show great ability to adapt to changes of hemodynamics and to remodel the layered connective tissue of the valve leaflet. The VECs cover the leaflet surfaces in contact with blood and are involved in the regulation process of thromboresistance, inflammation, and maintenance of the valvular structure.

VICs are the most abundant resident cells of the leaflet ECM and are responsible for the remodeling and repair of the leaflet by synthesizing ECM proteins and metalloproteinases to regulate tissue formation and degradation (Liu et al. 2007). In an adult heart, the majority of the VICs have a fibroblast-like phenotype and are in a quiescent state (quiescent VICs). However, a small amount of cells (between 2% and 5%) becomes activated in response to homeostatic changes in the environment (Schoen 2008), and four more phenotypes have been described: (1) embryonic progenitor VICs that act as an internal source, (2) activated VICs that are mostly present during active ECM remodeling processes and tissue repair, (3) osteoblastic VICs that are responsible for valve calcification (Liu et al. 2007), and (4) mesenchymal VICs.

It is this balanced interaction between VECs and VICs that allows to maintain valve tissue integrity throughout life (Butcher and Nerem 2006).

2.3 Cellular Mechanisms of Valvular Disease

Valvular heart diseases are usually progressive diseases, actively regulated by VECs and VICs cellular interactions, that cause damages to the leaflet, disruption of the hemodynamics, and/or inflammation (Gould et al. 2013; Mathieu et al. 2015). The typical resulting inflammation process triggers the production of inflammatory cytokines and chemokines, attracting cells from the immune system into the leaflet. These events determine the continuous activation of VECs and VICs and can result in chronic inflammation which leads to fibrotic thickening, followed by an extensive calcification of the valve (Rutkovskiy et al. 2017; Schoen and Gotlieb 2016).

Leaflet calcification can occur in normal and in congenitally malformed valves, such as bicuspid aortic valves (Roberts and Ko 2005; Rajamannan et al. 2011). Briefly, due to progressive degeneration and/or fibrosis, changes in leaflet stiffness progressively increase the risk of developing calcification. Calcium deposits onto the leaflets reduce the effective orifice area of the valve (i.e., stenosis) and cause an outflow obstruction by limiting leaflet mobility. The molecular and cellular processes regulating calcific valve stenosis are not yet fully characterized. However, VICs and VECs are responsible for the maintenance of valve homeostasis and leaflet integrity by constantly renewing the leaflet ECM and by preserving the characteristic layered composition. Calcific deposit on the valve leaflets are mediated by the resident VICs that, upon pathological cues, can differentiate toward an activated myofibroblast-like and osteoblast-like phenotype (Demer and Tintut 2008; Yip et al. 2009). In addition, as a result of endothelial injury, the immune system has been shown to be involved in VIC activation, causing ECM remodeling, and osteoblastic differentiation, inducing progressive valve mineralization (Coté et al. 2013). Moreover, increased endothelium permeability allows the diffusion of different molecules (i.e., cytokines, growth hormones, low-density lipoproteins) into the valve leaflets that cause the recruitment of immune cells, generating a chronic inflammation and subsequently initiating a phenotype switch of quiescent VICs toward an osteogenic phenotype (Mohty et al. 2008; Mathieu et al. 2015; Schoen and Gotlieb 2016).

On the other hand, myxomatous degeneration is a noninflammatory progressive disease, which is characterized by a disrupted leaflet structure caused by a defect in the synthesis and/or remodeling of collagen. This results in heart valves showing collagen fiber disorganization in the fibrosa layer, resulting in thickened leaflets, and demonstrating thickening of the spongiosa layer due to excessive proteoglycan deposition. These events, driven by improper ECM turnover from the VICs, significantly alter the structural integrity of the leaflet, leading to the mechanical weakening of the valve (Grande-Allen et al. 2003; Schoen and Gotlieb 2016) and causing stretching and elongation (Vesely et al. 1988).

The activation of VICs and VECs on a cellular level results, therefore, in morphological changes of the valve (e.g., leaflet thickening and stiffening) that have a strong impact on valve functionality, causing pathologies such as valve stenosis and valve insufficiency, as detailed in Sect. 3.

3 Heart Valve Pathology

Valvular heart disease constitutes an important clinical problem, causing approximately 25,000 deaths annually in the United States only (Schoen 2018) and contributing to an increased mortality rate (Benjamin et al. 2017), with aortic valve disease accounting for more than half of the deaths (Coffey et al. 2016).

In industrialized countries, aging is the principal cause of the continuous rising of valve disease incidence in patients over 65 years old (Schoen 2018). During an average person lifetime, the valves will open and close up to 100,000 times each day. Hence, it should not be surprising that degenerative diseases may affect these organs in the elderly, causing the valve leaflets to become thicker and stiffer due to fibrosis and calcifications (Fishbein and Fishbein 2019). In developing countries, on the other hand, rheumatic fever (RF) is a leading cause of valvular pathologies and remains a major cause of death and disability in children and young adults (Leal et al. 2019). Congenital heart disease (CHD), affecting up to 2% of the newborn, may also lead to poor valve functionality either early in life (i.e., as in the case of unicuspid aortic valve) or once adulthood is achieved (i.e., bicuspid aortic valve) (Fishbein and Fishbein 2019). Current advancements in pediatric cardiac surgery have resulted in an increasing number of adults with CHD that are followed up at later stages (Marelli et al. 2007). Heart valve disease can affect all the heart valves and can be classified into two categories by their impact on valve functionality: *valve stenosis* and *valve*

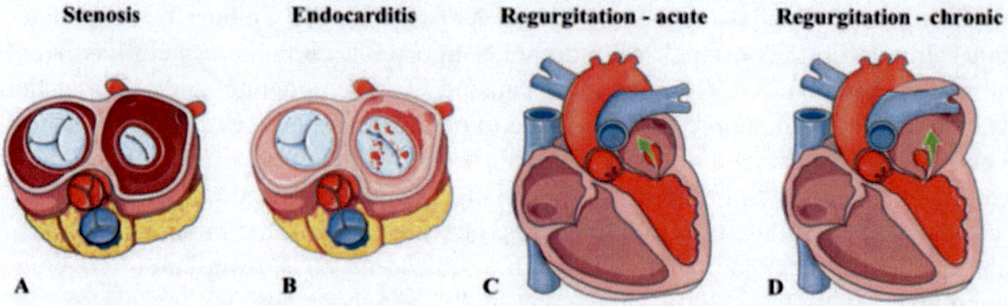

Fig. 3 Schematic representations of the most common valvular diseases. (**a**) Valve stenosis, (**b**) inflammatory disease such as endocarditis, (**c**) acute, and (**d**) chronic valve insufficiency (or regurgitation). For representative purpose, the affected valve in the schematics is the mitral valve; however, all the heart valves may be subjected to these diseases. (Images adapted from Servier Medical Art under a creative common attribution 3.0 unported license)

insufficiency (Fig. 3). It is important to consider that most of the time, valve diseases are a consequence of changes in the hemodynamics that impact valvular cell phenotype and functionality.

3.1 Valve Stenosis

Valve stenosis is a condition characterized by obstruction of the blood flow due to limited opening of the valve (Fig. 3a). The etiologies of valve stenosis can be subdivided into age-related calcific and degenerative disease, CHD, and inflammatory disease.

Age-related valve stenosis mostly affects the aortic valve, and it is caused by stiffening and thickening of the valve cusps due to calcification and fibrosis. Aortic valve stenosis is, therefore, becoming a prevalent pathology in developed countries, due to aging of the population (Chong et al. 2019). Repeated clinical follow-ups are fundamental to assess the progress of aortic stenosis in elderly patients and timely plan a suitable valve replacement.

Congenital malformations are the most common cause of valvular stenosis in pediatric and young adult patients. Aortic bicuspid or unicuspid valves (i.e., the aortic valve is composed of only two cusps with similar size or of one single leaflet, respectively) are prone to progressive degeneration and stenosis about 10–15 years earlier than normal aortic valves. Hence, aortic bicuspid or unicuspid valves are becoming the most common cause of aortic stenosis in patients age 50–70 years old (Fishbein and Fishbein 2019) with an average rate reduction of the valve orifice area that has been estimated to be $\sim$0.12 cm^2/year. Pulmonary stenosis, on the other hand, is less prevalent but can be caused by rare malformations that induce fusion of adjacent leaflets due to CHD.

Inflammatory disease, such as RF and endocarditis (Fig. 2b), can affect multiple valves and is the most prevalent cause of stenosis to affect the atrioventricular mitral and tricuspid valves. Similar to degenerative stenosis, post-inflammatory stenosis is also characterized by cusp thickening, calcification, and fusion of adjacent leaflet at the commissures (Fishbein and Fishbein 2019) that determine limited blood flow through the valve orifice.

Since the progression of valve stenosis can vary significantly between patients, clinical follow-ups are fundamental to assess the hemodynamic parameters and grade the severity of the stenosis (Table 2) via spectral Doppler echocardiography. The recommended primary values for grading aortic stenosis are reported in Table 2.

3.2 Valve Insufficiency

Valve insufficiency (also known as regurgitation or incompetence) refers to the incomplete closure of the valve leaflets, thereby inducing reverse flow in the atrium (when atrioventricular valves are affected) or in the ventricle (when semilunar valves are regurgitant) (Fig. 3).

Compared to valvular stenosis, valve regurgitation is caused by a more extended list of diseases (Fishbein and Fishbein 2019) that differs between acute and chronic regurgitations (Table 3 and Fig. 3). Chronic regurgitation refers to valve damage that determines a progressive increase in regurgitating blood volume over time, causing dilatation of the atrium (when atrioventricular valves are affected) or of the ventricle (when semilunar valves are affected). In contrast, acute aortic regurgitation may lead to sudden elevation of left ventricular filling pressure, reduction in cardiac output, and/or sudden death.

Table 2 Hemodynamic parameters used to evaluate the severity of aortic valve stenosis (Chong et al. 2019) and insufficiency (Maurer 2006)

Aortic valve stenosis			
Parameter/grade	Mild	Moderate	Severe
Peak velocity (m/sec)	2.6–2.9	3.0–4.0	>4.0
Mean gradient (mmHg)	<20	20–40	>40
Valve orifice area (cm^2)	>1.5	1.0–1.5	<1
Aortic valve insufficiency			
Parameter/grade	Mild	Moderate	Severe
Central jet width compared to LVOT (%)	<25	25–65	>65
Regurgitation volume (ml/beat)	<30	30–60	>60
Regurgitation fraction (%)	<30	30–50	>50
Effective regurgitant orifice area (cm^2)	<0.10	0.10–0.30	>0.30

Table 3 Etiologies of acute and chronic aortic regurgitation

Aortic valve insufficiency	
Acute	Infective endocarditis, traumatic injuries, aortic dissection
Chronic	Aortic root dilation, congenital bicuspid valve, calcific degeneration, rheumatic disease, hypertension, Marfan syndrome, syphilitic aortitis, ventricular septal defect, cusp prolapse

Aortic valve regurgitation may be a consequence of valve leaflet or valve root abnormalities due to congenital and/or degenerative disease, such as dilation of the annulus, cusp prolapse, leaflet retraction, cusp perforation, or rupture. In addition, it may be caused by a variety of other diseases, either inflammatory (i.e., syphilis, endocarditis) or noninflammatory (connective tissue disease, i.e., Marfan syndrome), that usually determine a dilation of the valve root that prevents complete leaflet coaptation. The principal tool to grade aortic valve insufficiency is color Doppler echocardiography, a highly sensitive method to visualize the regurgitant jet (Maurer 2006).

Pulmonary insufficiency is often a consequence of pulmonary hypertension, rheumatic disease, and infective endocarditis. On the other hand, mitral regurgitation is one of the most common forms of valvular disease, and it is often caused by dysfunction of any of the mitral valve structures (i.e., annulus, papillary muscle, chordae tendinae). Specifically, mitral valve prolapse (i.e., the displacement of a mitral valve thickened leaflet into the left atrium during systole) is the most prevalent cause of severe mitral regurgitation in both developed and developing countries (Althunayyan et al. 2019).

4 Current Treatment Options for Valvular Disease

Currently, there is no lifelong medication for valvular disease, and the treatment options and clinical outcome significantly depend on the valve involved, the development of the disease, the overall heart functionality and compensatory mechanisms, as well as the age of the patient.

Whenever possible, valvular repair is the first choice of treatment for a diseased valve, because it comes with several conceptual advantages, such as reduced risk of infection, reduced need for anticoagulants, preservation of valve anatomy, and restoration of patient's valve functionality, without the need for a prosthesis (Girdauskas et al. 2018). However, the decision between valve repair and valve replacement is evaluated by taking into consideration the severity of the disease, the age of the patient, and whether there are other heart pathologies to correct or comorbidities to consider.

4.1 Valve Repair

In the beginning of the nineteenth century and till the Second World War, approximate techniques of valve repair, such as inserting a finger to free the fused leaflets of a stenotic valve (Dr. Theodore Tuffier, 1912) (Ellis et al. 1996), were used to treat valvular disease. In addition, during and after the Second World War and thanks to the establishment of hypothermia procedures, novel techniques for valve repair were developed as wounded soldiers were injured in the myocardium by explosions. With the introduction of cardiopulmonary bypass in the 1950s and the increased knowledge of cardiac surgeons, heart valve-sparing procedures for semilunar and atrioventricular valves had been constantly developing.

Pulmonary valve stenosis, caused by congenital anomalies that lead to thickening and fusion of the valve leaflets and/or to a bicuspid pulmonary valve, can be efficiently treated via valve-sparing procedures (Parikh et al. 2017). Many patients with pulmonary stenosis are operated with a transannular patch to relieve the stenotic outflow tract of the RV. Briefly, the arterial wall is incised, and a polytetrafluorethylene (PTFE) patch is sutured to enlarge the outflow track (Parikh et al. 2017). While this is an effective procedure to solve the stenosis, pulmonary insufficiency post-intervention, caused by fibrotic tissue formation along the suture, is still a common complication that causes regurgitation and leaflet prolapse (Said et al. 2016). To correct valve insufficiency, an artificial leaflet, made of PTFE or glutaraldehyde-fixed bovine pericardium, is transplanted to recreate a functional three-leaflet pulmonary valve and to preserve the annulus size and the pressure gradient of the outflow tract (Said et al. 2016).

To correct aortic insufficiency, valve repair focusing on ensuring full leaflet coaptation in diastole is used in about 25% of low-risk patients, and it showed good outcomes with low mortality rates even in elderly subjects (Bisleri 2016). As described in, several are the causes of aortic insufficiency (e.g., leaflet retraction, prolapse, calcifications, thickening, and valve annulus dilation). Leaflet retraction and bicuspid aortic valve regurgitation can be corrected by suturing the commissure points or by shortening the belly of the cusps with plicating sutures, also used to solve valve leaflet enlargements, cuspid elevations, and Valsalva sinus remodeling (Girdauskas et al. 2018). Finally, aortic root dilation, commonly observed in bicuspid aortic valves, was initially treated by reducing the perimeter of the aortic root. However, this procedure caused re-dilation of the aortic root over time. To solve this issue, current techniques focus on the placement of an external ring on the aortic root to prevent re-dilation events (Bisleri 2016).

Valve repair is the optimal solution to treat degenerative mitral regurgitation. Together with various surgical techniques, annuloplasty using PTFE rings are typically used to minimize annulus size and, thereby, limit insufficiency (Grasso et al. 2015). In cases of posterior mitral leaflet prolapse, instead, the mitral chordae tendinae are replaced by PTFE- or glutaraldehyde-fixed bovine pericardium-based implants (Shah and Jorde 2019). In elderly and otherwise inoperable patients, valve insufficiency can nowadays also be solved by an edge-to-edge repair of the mitral leaflets using the MitraClip or PASCAL transcatheter system (Mendirichaga et al. 2017). Recently, similar transcatheter strategies have been also developed for the treatment of tricuspid regurgitation. However, current transcatheter strategies for tricuspid regurgitation are still in their early stages (Tabata et al. 2019).

Overall, valve repair is commonly preferred for young patients, because it allows for somatic growth of developing hearts and do not require lifelong anticoagulation treatment. However, valve repair techniques are not always sufficient for the treatment of severe valvular dysfunction associated with degenerative or congenital diseases.

4.2 Valve Replacement

Heart valve repair options provide little improvement for severe conditions, and patients often need a replacement procedure shortly after repair intervention (Gasser et al. 2019). On the other hand, heart valve replacement methods, either surgical or transcatheter, can provide long-term results with low risk of reoperation, in particular for elderly patients (Baumgartner et al. 2017).

Currently, both surgical and transcatheter approaches are in routine clinical use in developed countries, and the selection of one method over the other depends on the pathology, patient age and specific anatomy, gender, and presence of comorbidities.

Surgical valve replacement (SVR) is the first treatment option to treat patients affected by severe valvular disease, as it provides good perioperative and long-term results (Baumgartner et al. 2017). SVR is based on an invasive open-heart surgery that involves the use of temporary cardiac arrest and of cardiopulmonary bypass to establish extracorporeal circulation and ventilation during the time required by the surgeon to replace the valve. Considering the consequences of these techniques on the recovery time, patients may be considered inoperable or at a high risk for surgery due to comorbidities. Minimally invasive transcatheter valve replacement (TVR) was developed in the early 2000s (Bonhoeffer et al. 2000; Cribier et al. 2002) (Fig. 4), and, since then, it has revolutionized the treatment options for patients affected by valvular disease. TVR was initially developed to treat patients considered inoperable or at high risk for SVR (Rodés-Cabau et al. 2012), because the minimally invasive implantation procedure significantly reduces the hospital stay and patient recovery and rehabilitation time (Chatterjee et al. 2019). Over the course of two decades, great progresses have been made to deliver valves in a minimally invasive fashion, with specific focus on the replacement of the aortic valve (Rodés-Cabau et al. 2012). Transfemoral aortic valve replacement allows for a fully percutaneous implantation via the retrograde insertion of a long catheter through the femoral artery. However, transfemoral TVR is not suitable to all patients because of poor femoral access and presence of vascular complications (Fioretta et al. 2017). Hence, other access routes have been investigated, such as the transapical (i.e., access to the valve via ventricular apex (Ye et al. 2006)) and the transaortic (i.e., access to the aortic valve from a mini-sternotomy and a puncturing of the aortic wall (Bauernschmitt et al. 2009)) approaches. As of today, TVR has been approved to intermediate- (Leon et al. 2016; Reardon et al. 2017) and low-risk (Mack et al. 2019; Popma et al. 2019) patient cohorts, becoming therefore an available treatment option for all patients affected by aortic stenosis.

4.3 Heart Valve Prostheses Options

4.3.1 Mechanical Valve Prostheses: A Durable Solution

After the first ball-in-cage mechanical valve replacement (Russo et al. 2017), major progresses in heart valve prostheses design and material selection have been made, until the development of the first bi-leaflet tilting disc mechanical valve (Head et al. 2017)

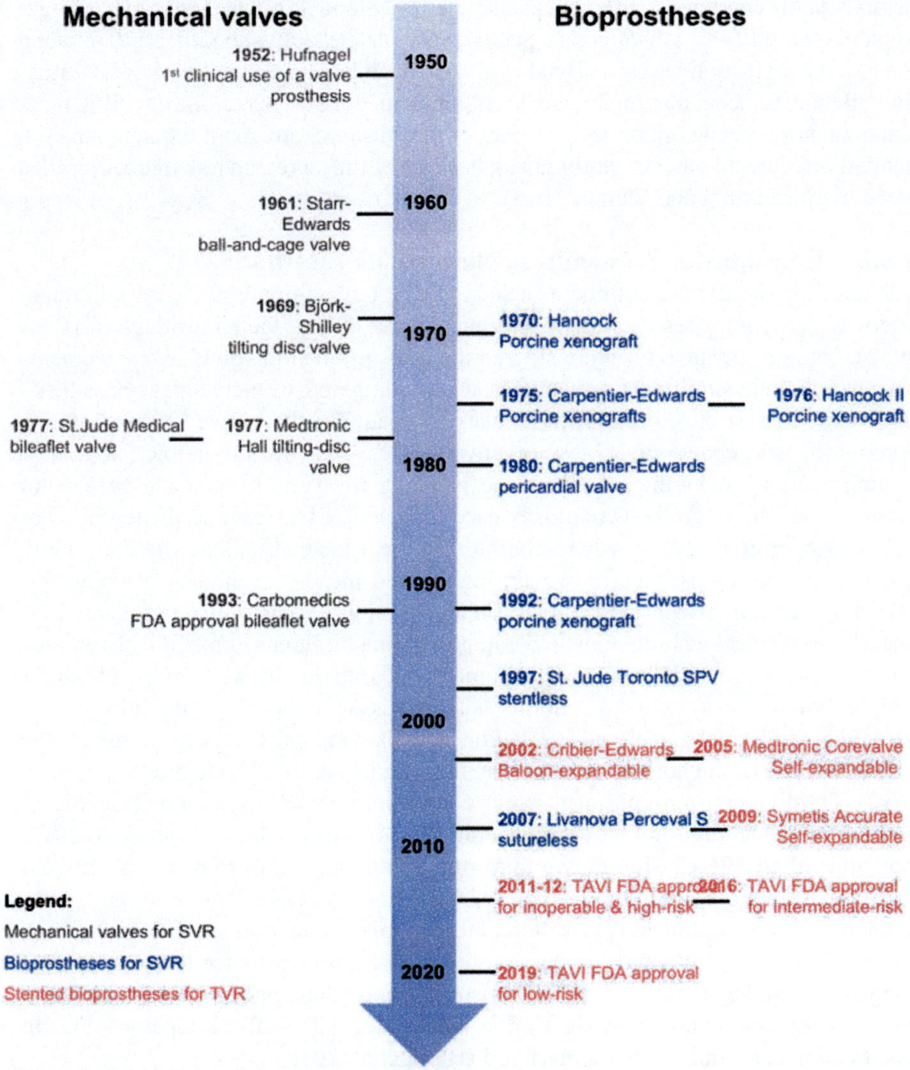

Fig. 4 Evolution of clinical heart valve prostheses and replacement technique options. Mechanical valves, only eligible for surgical valve replacement (SVR), are in a blue font color. Bioprostheses, on the other hand, were firstly developed as SVR option to overcome the limitations of mechanical valves and are here represented by the red color font. Starting in the 2000s, stented bioprostheses compatible with minimally invasive transcatheter valve replacement (TVR) techniques were developed and are here represented by the green color font

(Fig. 4). Mechanical valves, based on long-lasting metals such as titanium, are characterized by a mechanical durability that lasts more than 20 years, deeming them as the gold standard for the majority of the SVR procedures in younger patients. Despite the remarkable long-term performances, mechanical valves are responsible for altered hemodynamic conditions, turbulent and accelerated blood flow that determine damages to the red blood cells, which consequently cause platelet activation, thereby initiating the blood coagulation cascade (Head et al. 2017). Hence, to prevent thrombus formation, lifelong anticoagulation therapy is recommended. However, blood-thinning medication is not a viable option for a variety of patients (e.g., pregnant women, athletes, patients affected by other comorbidities), because of the increased risk of uncontrolled bleeding (Nishimura and Warnes 2015).

4.3.2 Homografts: A Promising Native-Like Alternative

Valvular homografts are defined as a section of a human donor aorta or pulmonary artery which comprises the intact semilunar valve. Since their introduction (Ross 1962), homografts have been considered the most promising valvular replacements because of their significant advantages when compared to mechanical prostheses (Delmo Walter et al. 2012). In particular, by retaining the native valve anatomy, homografts succeeded to achieve a physiological-like hemodynamics with low thrombogenicity over the course of the patient's lifespan (Lisy et al. 2017). For these reasons, homografts were mostly used for severe CHD, such as the tetralogy of Fallot (Meijer et al. 2019), where complex surgical reconstruction where required. Long-term follow-ups have, however, highlighted the disadvantages of using this particular valvular prosthesis for SVR. Firstly, homografts can only be implanted via open-heart surgical technique, which complexity may influence the clinical outcome (Delmo Walter et al. 2012). Secondly, homograft tissue quality was reported to differ due to not donor-to-donor variability and to tissue banking methodologies. In particular, differences in cryopreservation and thawing protocols may impact the ECM structure of the homografts, thereby limiting its durability (Delmo Walter et al. 2012). Thirdly, long-term performance is compromised by chronic immunogenicity, onset of calcifications, and valve degeneration over time (Delmo Walter et al. 2012; Bonetti et al. 2019), causing only a third of the implanted graft to reach the 20-year follow-up (Delmo Walter et al. 2012). Last, the limited availability of donor tissues hinders the broader clinical application of these replacement options. Taken together, these limitations significantly impact on the use of homografts for the treatment of congenital and degenerative diseases in pediatric and young patients, where stenosis, degeneration upon implantation, leaflet shortening, and leaflet thickening due to calcifications and fibrosis were observed (Blum et al. 2018).

4.3.3 Xenogeneic Bioprosthetic Valves: An Alternative to Homografts and Mechanical Valves

To overcome the limitations of mechanical and homograft prostheses, animal-derived materials (i.e., tissues derived from bovine or porcine sources treated with glutaraldehyde to limit their immunogenicity) were introduced to manufacture valves with improved hemodynamics and physiological-like tissue composition, based on collagenous ECM (Fig. 4, Table 4). Due to their native-like geometry, bioprostheses are more hemocompatible, and they allow for a reduced need of anticoagulants (Manji et al. 2015). Xenogenic tissues have been also used to manufacture stented bioprostheses compatible with minimally invasive TVR techniques, demonstrating

Table 4 Overview of the different clinically available bioprosthetic valve replacements

Bioprosthetic valve replacement options	
Stented porcine valves	A three-leaflet porcine valve replacement sutured onto artificial struts or stents to provide support and retain a specific geometry. When balloon-expandable or self-expandable stents are used, these valves are compatible with minimally invasive transcatheter implantation procedures
Stentless porcine valves	A replacement consisting of the porcine aorta portion comprising the native valve root (i.e., the tissue below (proximal) and above (distal) the valve). These prostheses are implanted via surgical valve replacement procedure
Bovine pericardial valves	A three-leaflet valve is manufactured by using bovine pericardium that is then sutured onto artificial struts or stents to provide support and retain a specific geometry. When balloon-expandable or self-expandable stents are used, these valves are compatible with minimally invasive transcatheter implantation procedures
Homografts	A replacement consisting of the human aorta portion comprising the native valve root. These prostheses are implanted via surgical valve replacement procedure
Autografts	A replacement consisting of valve and root excised from the patient and re-implanted in the same patient to substitute a different valve. An example of valve autograft is the Ross procedure

successful preoperational crimping and loading into the delivery catheter (Wiegerinck et al. 2016; Arsalan and Walther 2016). After the promising outcome of recent clinical trials, TVR indication has been recently extended to low-risk patient cohorts (Mack et al. 2019; Popma et al. 2019). Hence, while previously designed to treat the elderly, TVR compatible bioprostheses will be now used also in younger patients (Manji et al. 2015). Nevertheless, bioprostheses are prone to progressive degeneration and, thus, require reoperation after 10–20 years (David 2010; Arsalan and Walther 2016). The major causes of bioprosthesis failure are leaflet degeneration, stiffening, calcification, and immunogenic reactions (Human and Zilla 2017; Fishbein and Fishbein 2019). Importantly, these degenerative phenomena are more prone to occur in pediatric patients, where the immune reaction to the xenogenic material causes severe inflammation (Rabkin-Aikawa et al. 2005).

4.3.4 Non-resorbable Polymeric Valves: A Cost-Effective Solution

Non-resorbable polymers, such as silicon, polytetrafluoroethylene, and polyurethanes, were introduced from the late 1960s as an alternative material to achieve valve prostheses with a physiological design and improved durability (Roe 1969; Mackay et al. 1996). While some polymeric prostheses were successfully used as pulmonary replacement for the treatment of congenital disease (Ando and Takahashi 2009), non-resorbable polymeric valves have shown major flaws, including limited durability, impaired functionality, leaflet stiffening, and thrombogenicity (Hilbert et al. 1987; Nistal et al. 1990; Daebritz et al. 2004). In addition, discrepancies in the manufacturing methods and variability in the polymer batches have impeded their broad clinical translation (Kheradvar et al. 2015). More recently, a siloxane poly-urethane urea polymeric valve has been shown to fully comply with the International Organization for Standardization (ISO)-norm requirements for in vitro testing and demonstrated good functionality with no evidence of calcification or other degenerative phenomena in a preclinical animal model (Bezuidenhout et al. 2015). Next, another novel polymeric aortic valve has been granted permission for clinical trials (ClinicalTrials.gov Identifier: NCT03851068). Ultimately, polymeric valves could be the grounding foundations of cost-effective off-the-self solutions for heart valve replacements using TVR approaches (Bezuidenhout et al. 2015; Scherman et al. 2018). However, it remains a challenge to couple durability of these valves with biocompatibility and proceed further to the clinical evaluation.

4.4 Clinical Impact and Burden on Society

In developed countries, the prevalence of heart valve disease is highest in patients older than 75 years, reaching up to 13.3% (Huygens et al. 2018), and the number of valve replacement procedure is expected to increase to 800,000 annually worldwide by 2050 (Yacoub and Takkenberg 2005). The growing aging population affected by degenerative aortic stenosis that requires a valve replacement procedure will considerably impact on society, with healthcare costs reaching above 1 billion euro just in European countries (Huygens et al. 2018).

TAVR has drastically changed the field of valve replacement over the past 20 years, with the introduction of minimally invasive techniques. Every year, approximately 180,000 patients undergo TAVR in the European Union and Northern America alone. This number is expected to rise up to 270,000 annually (Durko et al. 2018) as recent clinical trials proved non-inferiority of TAVR over SAVR also in intermediate- (Leon et al. 2016; Reardon et al. 2017) and low-risk patients (Mack et al. 2019; Popma et al. 2019). Following these promising results, TAVR valves Sapien 3 and CoreValve Evolut have been approved also for low-risk patients by the FDA (August 2019). This approval makes TAVR available for all patients with severe, symptomatic aortic stenosis, significantly increasing the estimated number of TAVR procedure in European and North American countries of almost 54% (Table 5) (Durko et al. 2018).

Despite this tremendous technical evolution in the field of transcatheter techniques, only little progress has been made on the bioprosthetic materials used for TAVR. Currently available heart valve prostheses are still based on non-regenerative materials (namely, glutaraldehyde-fixed xenogenic or allogenic tissues, titanium, or carbon) that can cause several prostheses-associated problems upon implantation: thrombogenicity, progressive degeneration, limited durability, and, most importantly, the inability to remodel and grow with the patient (Henaine et al. 2012; Head et al. 2017). It should not be of surprise that the quality of life and life expectancy of patients with a valvular replacement is significantly impacted and healthcare costs are considerably higher in the first 3 years post-surgery compared to age-matched healthy individuals (Table 6) (Huygens et al. 2018). To these early costs, further expenses should be taken into account in particular for pediatric and

Table 5 Estimated annual number of transcatheter aortic valve replacement (TAVR) procedures in European and North American countries and the forecasted increase after the FDA approval of TAVR for low-risk patients in 2018, as reported by Durko et al. (2018)

Country	Estimated TAVR procedures as per 2017 guidelines	Estimated TAVR procedure as per 2018 guidelines (including low-risk patients)	Increase in %
United States	51,998	80,076	54.00
Canada	6826	10,516	54.06
Austria	1892	2919	54.28
Belgium	2431	3740	53.85
Bulgaria	1726	2642	53.07
Croatia	956	1468	53.56
Czech Republic	2259	3465	53.39
Denmark	1261	1938	53.69
Estonia	293	456	55.63
Finland	1314	2006	52.66
France	14,632	22,607	54.50
Germany	20,466	31,596	54.38
Greece	2718	4190	54.16
Hungary	2108	3262	54.74
Ireland	717	1106	54.25
Italy	15,784	24,368	54.38
Latvia	460	709	54.13
Lithuania	655	1009	54.05
Netherlands	3614	5524	52.85
Norway	998	1537	54.01
Poland	7008	10,797	54.07
Portugal	2516	3892	54.69
Romania	4035	6224	54.25
Slovakia	906	1391	53.53
Slovenia	444	681	53.38
Spain	10,274	15,783	53.62
Sweden	2279	3526	54.72
Switzerland	1756	2696	53.53
United Kingdom	13,787	21,133	53.28

young adult patients, that they will be more likely to undergo multiple reoperation to substitute an outgrown or degenerated valve prosthesis.

Therefore, there is a clear need for new durable valve replacements compatible with transcatheter procedures that are based on novel materials capable to regenerate, remodel, repair, and adjust to the functional and somatic growth of the patient. This led to the development of multidisciplinary approaches, combining cell

Table 6 Annual healthcare costs as reported by Huygens et al. (2018) during the first three post-surgery (PS) years of surgical valve replacement. Costs are compared to age-matched healthy subjects. The average fold increase of healthcare costs is calculated over 3 years

Age category	Costs for patients with a heart valve replacement (€)			Costs for control healthy subjects age-matched (€)			Average fold increase over 3 years
	1 year PS	*2 years PS*	*3 years PS*	*1 year PS*	*2 years PS*	*3 years PS*	
Children	11,766	5495	5015	796	802	786	9.34
Young adults	15,060	7596	6633	2944	2972	2955	3.31
Middle-aged	16,104	7832	7764	4612	4679	4681	2.27
Elderly	18,255	10,478	10,701	9236	9273	9200	1.42

biology, engineering, and medicine to create tissue-engineered heart valves (TEHVs), a novel concept of valvular prostheses that could remodel and even grow with the patient.

5 Heart Valve Tissue Engineering

"The loss or failure of an organ or tissue is one of the most frequent, devastating, and costly problems in human health care. A new field, tissue engineering, applies the principles of biology and engineering to the development of functional substitutes for damaged tissue," with those exact words in 1993, Robert Langer and Joseph P. Vacanti opened up a new era for the fields of medicine, biology, and engineering (Langer and Vacanti 1993). The concept was easy and straightforward: generate a new living tissue starting from cells directly isolated from the patient (autologous) and placed on or within biodegradable scaffold matrices. The original in vitro tissue engineering (TE) paradigm embraces the use of three components: (1) a 3D scaffolds; (2) autologous cells, seeded onto the scaffold; and (3) an in vitro bioreactor system to induce ECM formation by simulating the physiological tissue conditions (Langer and Vacanti 1993; Mayer et al. 1997). Once the new extracellular tissue is formed, the living patient-specific organ substitute can be implanted in the patient enabling the direct function of the prosthesis with further in vivo tissue growth and remodeling. Few years before, an American surgeon pioneer in heart valve surgery highlighted the essential features of the ideal heart valve substitute which included the capacity to grow, self-repair, remodel, adjust to functional changes, be durable over time, prevent thrombus formation, and resist to infections (Harken 1989). Nowadays, the ideal heart valve prostheses which comprise these unique and fundamental properties of native heart valves still do not exist. In this sense, TE offers the possibility to create biodegradable and biocompatible living valve replacements with regenerative capabilities which could potentially overcome the shortcomings of the current clinically used heart valve prostheses. Since the first application of the in vitro TE traditional dogma, various alternative approaches have been developed to produce off-the-shelf available TEHV prostheses by reducing the production costs and time. This chapter deals with the different approaches, requirements, and potential benefits that characterize TEHVs and the various experimental and technological challenges that are being faced to achieve a broad use of TE in clinics. Section 5 deals with the different technological concepts characterizing TEHVs nowadays. Section 6 describes the multitude of cell sources and scaffold types that are used for engineering heart valves in vitro, in vivo, and in situ. To conclude, Sects. 7 and 8 provide an outlook on the current testing platforms for TEHVs and on the next-in-line steps towards clinical translation of such prostheses.

Given the extensive body of literature on the multiple approaches existing to generate TEHVs, this chapter gives the reader a general overview of the main concepts involved in heart valve TE field. Conceptually, we can distinguish three main categories: (1) in vitro TE (the classical paradigm), (2) in vivo TE, and (3) in situ TE.

5.1 In Vitro Heart Valve Tissue Engineering

The classic in vitro heart valve TE concept (Fig. 5) starts with the isolation and expansion of autologous cells from the patient using in vitro cell culture techniques. When in sufficient amount, cells are seeded onto a biodegradable, biocompatible porous scaffold, either of synthetic or biological origin, which provides a temporary structural support for the cells to produce their own ECM. This cell-scaffold construct is then transferred and cultured into a bioreactor system over a predetermined period of time. Here, cells are exposed to different biomechanical and/or biochemical cues to induce cell proliferation first and in a second time tissue formation. The purpose of the bioreactor is to mimic the physiological environment of native heart valves, which tissue organization is heavily influenced by the hemodynamic loading of the cardiac cycle (Barron et al. 2003). Hence, mechanical properties, pressure, and shear stresses are culturing conditions which have been optimized in the years to enable the engineering of lab-grown valves that mimic their native counterparts as best as possible.

Cells of various origin, such as adipose-derived stem cells, endothelial cells, myofibroblasts, dermal fibroblasts, valve interstitial cells, bone marrow-derived stem cells, blood progenitor cells, umbilical cord vascular cells, and/or amniotic fluid cells, have been used for seeding TEHVs over the course of the years (Jana et al. 2016). Vascular-derived myofibroblasts isolated from saphenous veins or forearm vein have been extensively investigated as a relevant cell source for autologous cardiovascular TE applications (Jana et al. 2016). These cells have, indeed, good proliferation potential and can produce collagenous ECM in vitro.

First attempt to generate TEHVs were first introduced in 1995 by Shinoka et al. (1995), where polyglycolic acid (PGA) sheets were seeded with autologous myofibroblasts and endothelial cells and used to reconstruct and replace the right

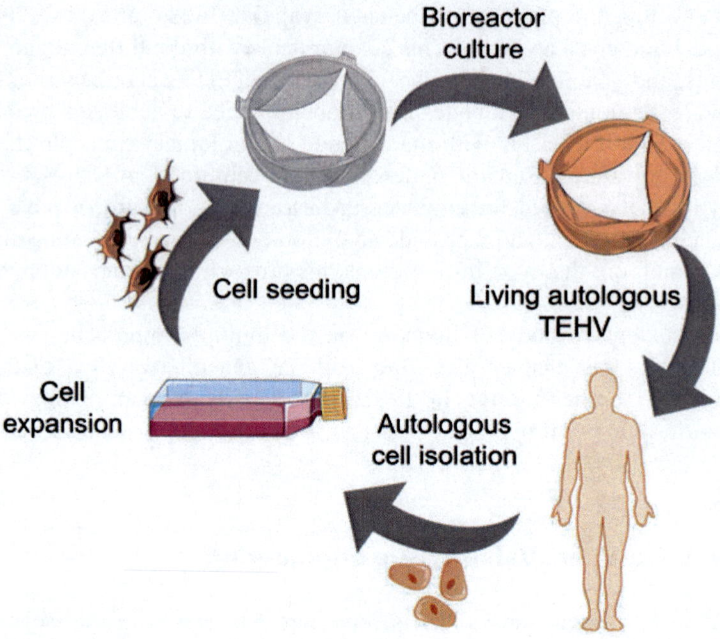

Fig. 5 Schematic representation of in vitro heart valve tissue engineering. This approach aims at the development of an autologous TEHV by isolating cells from the patient. After expansion, the cells are seeded onto a bioresorbable scaffold, cultured in a bioreactor system to ensure nutrient exchange and chemical and mechanical stimulation to favor ECM deposition. After a predetermined culture time, the autologous TEHV is ready for implantation into the patient. (Images adapted from Servier Medical Art under a creative common attribution 3.0 unported license)

posterior pulmonary heart valve leaflet in lambs. These first reports were then followed by proof-of-concept studies showing the complete valve replacement via in vitro engineered pulmonary heart valves in lambs (Dijkman et al. 2012a). Since then, many approaches for the production of TEHVs were reported in literature and reviewed elsewhere (Mol et al. 2009; Fioretta et al. 2017), and significant progresses have been made in the development of different scaffold materials and cell sources for the development of in vitro TE. However, the long-term proof of such concepts when applied to TEHVs is still pending. Indeed, latest studies investigating TEHV performance and tissue remodeling into preclinical animal models have been characterized by gradual loss of functionality within few months due to adverse remodeling, which was caused by the development of leaflet retraction and valvular incompetence (Flanagan et al. 2009; Gottlieb et al. 2010; Schmidt et al. 2010; Weber et al. 2013; Driessen-Mol et al. 2014; Syedain et al. 2015; Reimer et al. 2017; Motta et al. 2018). Finally, from a translational perspective, in vitro TE is recognized to be logistically complex, costly, and time-consuming, all aspects that hinder the scalability and translatability into clinical trials and commercialization.

5.2 In-Body Heart Valve Tissue Engineering

In-body TE, also known as in vivo TE (Fioretta et al. 2017), is an additional research line developed to exploit the foreign body reaction of the host upon subcutaneous implantation of a non-degradable mold (Hayashida et al. 2007, 2008). The aim is to generate a fibrotic collagen-rich tissue which is able to encapsulate and take the shape of the implanted mold. The final TEHV product, which researcher in the field has named *biovalve*, can be harvested and implanted as autologous non-immunogenic replacement in the same host (Fig. 6). Several prototypes and different generations of biovalves have been already tested in vitro and in vivo under pulmonary and aortic conditions (Takewa et al. 2013; Funayama et al. 2015a, b; Sumikura et al. 2015), even in combination to minimally invasive implantation techniques (Nakayama et al. 2015; Funayama et al. 2015b; Sumikura et al. 2015). In addition, tools to noninvasively monitor tissue formation have been developed to ensure complete tissue formation around the mold over time (Funayama et al. 2015c).

Despite the early positive results demonstrated in both in vitro and in vivo studies, even in vascular graft applications (Terazawa et al. 2019), the in-body manufacturing of such matrices in humans is questionable: (1) the approach is highly invasive; (2) the technique is expected to require at least 4 months to obtain a biovalve in human; (3) foreign body response and fibrotic capsule formation are uncontrolled phenomena that may lead to unpredictable tissue thickness over time; (4) tissue durability may be limited by the lack of elastin in the construct (Hayashida et al. 2007, 2008); and (5) the regenerative potential of fibrotic tissues in human is limited.

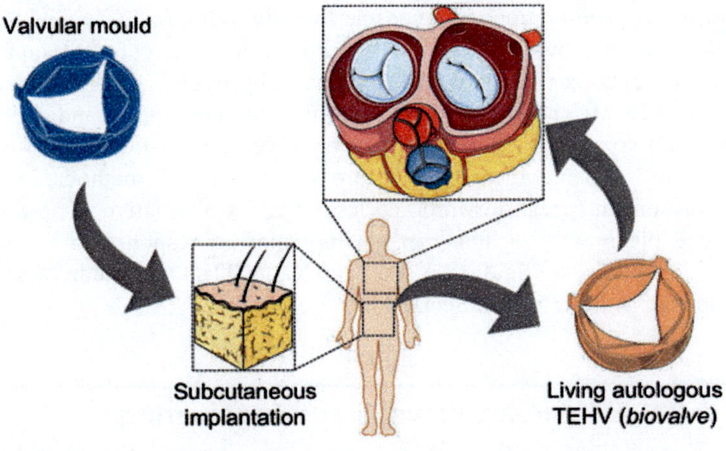

Fig. 6 Schematic representation of in-body heart valve tissue engineering. This approach creates an autologous TEHV by implanting subcutaneously a valvular mold into the patient. Due to the host response, the mold is covered by fibrotic collagenous tissue and, after a predetermined amount of time, is removed to obtain a TEHV. The valve is then implanted back into the patient heart as valvular substitute. (Images adapted from Servier Medical Art under a creative common attribution 3.0 unported license)

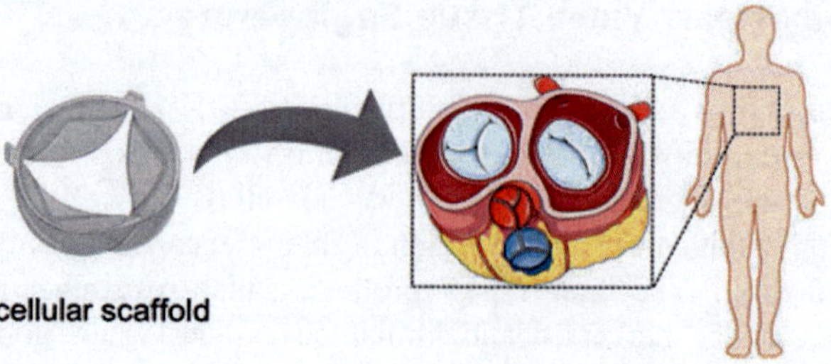

Fig. 7 Schematic representation of in situ heart valve tissue engineering. This approach simplifies the logistical complexity and reduces the costs of the in vitro TE method. In situ TE aims at the implantation of an off-the-shelf available, cell-free valve replacement directly into the patient. Following the natural inflammatory response to the implanted material, the remodeling cascade will be initiated, with host cell adhesion and differentiation and de novo ECM deposition, while scaffold degradation will occur. TEHVs used for this approach can be based on different scaffold materials: (a) a biodegradable polymers, (b) decellularized xenogenic tissue, (c) decellularized allogenic tissue, or (d) in vitro manufactured tissue-engineered matrix. (Images adapted from Servier Medical Art under a creative common attribution 3.0 unported license)

5.3 In Situ Heart Valve Tissue Engineering

Also known as "the remodeling within the host," in situ TE attempts to regenerate implanted tissues by harnessing the natural regenerative potential of the human body. This approach exploits the active or passive recruitment of endogenous cells to the acellular implant to induce remodeling and regeneration of the new prosthesis into a native-like tissue (Fioretta et al. 2017; Wissing et al. 2017) (Fig. 7). In this regard, biomaterials might be designed to enhance cell infiltration and adhesion into the scaffold, favor proliferation and overtime formation of new ECM, and reabsorb and degrade gradually over time. After its initial role as mechanical support for endogenous cells, the scaffold is slowly degraded by hydrolysis and replaced by newly deposited ECM. These implants are designed to favor, guide, and control cell recruitment and tissue remodeling to achieve a native-like functional substitute that can last a lifetime. When compared to in vitro and in-body TE methods, in situ TE approach represents a straightforward and less complex alternative to produce off-the-shelf available prostheses that can be implanted in response to the specific patient's needs. Within the field of in situ heart valve TE, a multitude of scaffolds have been tested, as described in Sect. 6.

6 Scaffolds for Heart Valve Tissue Engineering

The field of heart valve TE can count on a multitude of scaffold materials that are suitable for either in vitro and in situ applications: (1) native tissue-derived scaffolds, which are based on existing native tissues, either from human (allogenic) or animal (xenogenic) origin; (2) tissue-engineered matrices (TEM), obtained by in vitro culture of cells onto a scaffold and subsequent off-the-shelf treatments, such as decellularization; and (3) natural, synthetic, or hybrid polymers. In the following paragraphs, we will focus on the scaffold materials used for in situ TEHV concepts, referring the reader to other publications where more broad information about scaffolds for cardiovascular TE can be found (Fioretta et al. 2012; Generali et al. 2014; Jana et al. 2014; Dijkman et al. 2016).

For in situ TEHV applications, the scaffold of choice plays an important role as it needs to provide sufficient strength to sustain in vivo valve functionality immediately upon implantation, by providing mechanical stability and durability over time. At the same time, the scaffold should favor host cell infiltration and promote new tissue formation and ECM remodeling.

6.1 Native Tissue-Derived Scaffolds

Native tissue-derived TEHVs, based on human or animal matrix depleted of cells, aim to mimic the characteristics and function of native tissues by retaining the physiological ECM composition and eliminating the immunogenic cellular component via decellularization. Decellularization is a technique used to preserve the complex structure and protein composition of the ECM by physical, chemical, and/or enzymatic removal of the cellular and nuclear components (Crapo et al. 2011). When compared to cryopreservation or glutaraldehyde fixation techniques, decellularization of allogenic and xenogenic matrices results in a more promising concept where off-the-shelf tissues with remodeling potential are achieved and immunogenicity and disease transmission risks are reduced.

6.1.1 Decellularized Homografts

Pioneering studies investigating the employment of decellularized homografts (or allografts) date back to nearly two decades ago. Human heart valve homografts still represent the most valid alternative as heart valve substitute due to characteristics such as maintained anatomy and physiological-like hemodynamics, limited immunological reactions, low infection risk, and low thromboembolic risks compared to clinically available prostheses. For preclinical and clinical evaluation (Table 7), homografts are firstly decellularized to reduce potential immunological responses and then cryopreserved until further use. In this context, results have showed reduced immune response, repopulation by endogenous cells, positive remodeling, promising midterm functionality, and no signs of degeneration (Miller et al. 2006; Dohmen et al. 2007). Additionally, midterm results of decellularized homografts in young patients as pulmonary and aortic valve replacement demonstrated good functionality and, in comparison to standard xenogeneic prostheses, reduced reoperation rates (Cebotari et al. 2011; Sarikouch et al. 2016). However, a recent 10-year study involving decellularized aortic homografts reported extensive fibrosis, calcification, minimal recellularization, and degeneration, questioning the safety of such prostheses on the longer term (Helder et al. 2016). Notwithstanding, their potential, decellularized homografts present several disadvantages. First, there is limited availability of human tissue donors that, combined with the increasing

Table 7 Overview of the (A) clinical and (B) preclinical studies evaluating functionality and remodeling potential of decellularized homograft-based TEHVs for in situ regeneration

(A) Clinical studies		
Decellularized pulmonary valve homografts (DPVH) for surgical PVR		
Ref.	Control valve	Main findings
2019 (Boethig et al. 2019)	–	DPH functionality was excellent and regurgitation was trivial, with improved freedom from reoperation. DPH proved to be safe and effective
2016 (Sarikouch et al. 2016)	Cryopreserved homograft Bovine jugular vein	DPH showed good functionality and 100% freedom from reoperation and endocarditis after 10 years
2011 (Cebotari et al. 2011)	Bovine jugular vein	After 5 years, DPH showed good functionality, with low-pressure gradients, no dilation, no leaflet thickening, and 100% freedom from reoperation. Importantly, DPH also showed signs of adaptive growth
2011 (Brown et al. 2011)	Cryopreserved homograft	DPH demonstrated comparable outcome to cryopreserved homografts, with good functionality, and no need for reoperation
2010 (Burch et al. 2010)	Cryopreserved homograft	DPH showed comparable outcome in terms of valve functionality and a slightly improved freedom from reoperation to cryopreserved homografts
2003 (Hawkins et al. 2003)	Cryopreserved homograft	After 1 year, DPH demonstrated comparable functionality and a significantly lower panel-reactive antibody level for both class I and class II antibodies compared to cryopreserved homografts
2019 (Bobylev et al. 2019)	–	DPH proved to be a good alternative option for young patients, with superior midterm results in children and young adults for surgical PVR
Decellularized aortic valve homografts (DAVH) for surgical AVR		
Ref.	Control valve	Main findings
2019 (Bobylev et al. 2019)	–	Early functionality data on using DAH for surgical AVR are promising, making DAH a new surgical option for young patients
2016 (Tudorache et al. 2016)	–	DAH can withstand the systemic circulation, with good functionality (trivial regurgitation, no dilation) and no calcification or degeneration
2016 (Helder et al. 2016)	Cryopreserved homograft	DAH showed promising early functionality. However, after 10 years, freedom from reoperation is significantly better for cryopreserved homografts compared to DAH
2010 (Da Costa et al. 2010)	–	DAH showed good functionality with only trivial to mild regurgitation. Histological evaluation of explanted DAH demonstrated low cellularization but intact matrix structure. Low degree of calcification was detected

(continued)

Table 7 (continued)

(A) Clinical studies			
2005 (Zehr et al. 2005)	–	After 1 year, DAH demonstrated comparable functionality with low transvalvular gradients and negative panel-reactive antibody level in 95% of patients	
(B) Preclinical studies			
Decellularized pulmonary valve homografts (DPVH) for surgical PVR			
Ref.	Control valve	Animal model	Main findings
2018 (Goecke et al. 2018)	Decellularized porcine pulmonary valve	Sheep	DPVH processed via freeze-drying showed promising results in terms of early hemodynamic performance and cell infiltration. On the other hand, decellularized porcine pulmonary valves showed evidence of immunological reaction
2013 (Hopkins et al. 2013)	Decellularized human and porcine pulmonary valves	Baboon	Decellularized human and baboon pulmonary valves showed native-like hemodynamics with limited antibody response when compared to porcine cryopreserved pulmonary valves

PVR pulmonary valve replacement, *AVR* aortic valve replacement, *DAH* decellularized aortic homograft, *DPH* decellularized pulmonary homograft, *ECM* extracellular matrix

number of valve replacement procedures, makes this approach a nonviable solution for broad clinical adoption. Second, growth potential of this prosthesis is still debated, with only one published study where signs of growth were reported (Cebotari et al. 2011). Third, decellularized homografts can be only implanted via high-risk open-heart SVR, and, because of the limited remodeling and growth in human, multiple reoperations may be required during the course of the life, in particular for younger patients, thereby reducing their life quality and increasing risk of comorbidities.

6.1.2 Decellularized Xenografts

Decellularized heart valves based on xenogeneic-derived materials (e.g., bovine pericardium, pig heart valves) have been investigated as a potential alternative to glutaraldehyde-fixed bioprostheses in both preclinical and clinical studies (Table 8). Similar to human homografts, decellularized xenografts preserve the native valve ECM structure, providing a physiological-like template for the host cells and good hemodynamic performance. Despite the first promising preclinical results, which demonstrated complete cell infiltration, good hemodynamics, and absence of calcifications (Table 8B), clinical experiences resulted in stenosis, inflammation, pseudo-aneurysm, and dilation (Table 8A). Most importantly, the implantation of xenograft-based TEHVs in pediatric patients resulted in severe immune reactions, driven by an incomplete decellularization of the implant, which unfortunately led to early fatal failure of the implanted prosthesis in three patients (Simon et al. 2003). These results suggested that the use of cell-free xenografts is not recommended in human applications because of the severe immune and inflammatory response observed in multiple clinical studies, which lead to early valve failure for stenosis and/or insufficiency, and a high incidence of reoperation (Simon et al. 2003; Woo et al. 2016). However, improved decellularization protocols capable of eradicating any residual antigens of the xenogeneic valves may give rise to improved xenogenic TEHVs with remodeling potential (Helder et al. 2017).

6.2 In Vitro-Derived TEM-Based Scaffolds

Tissue-engineered matrices (TEM) obtained via classic in vitro TE concept by using non-autologous cell sources can be successfully decellularized to achieve an off-the-shelf and ready-to-use constructs for in situ applications (Fig. 8).

Common cell sources for this approach are easily accessible cells, such as umbilical cord vascular cells or dermal fibroblasts. Both these fibroblastic cells have shown potential to generate dense and organized collagenous matrices for vascular (Niklason et al. 1999; Syedain et al. 2014, 2016; Lawson et al. 2016; Kirkton et al. 2019) and valvular applications (Syedain et al. 2015; Reimer et al. 2017; Motta et al. 2018, 2019; Lintas et al. 2018).

Decellularization of TEM-based TEHVs dates back to 2012, when it was firstly introduced to simplify the manufacturing procedures of in vitro engineered valves obtained with autologous cells. In addition, decellularization was able to prevent the leaflet retraction observed immediately upon leaflet separation in in vitro cultured TEHVs, with significant impact on in vivo valve functionality (Dijkman et al. 2012b; Driessen-Mol et al. 2014). Since then, a multitude of studies have shown the potential of using decellularized TEM-based TEHVs in preclinical animal models, showing good functionality, host cell repopulation, and remodeling over time in both pulmonary and aortic positions (Table 9). Remarkably, TEM-based TEHVs proved to be competent and functional as aortic valve replacement in sheep, with good functionality and almost complete cellular repopulation 6 months after implantation (Syedain et al. 2015). Remarkably, TEM-based TEHVs were proved to even be compatible with transcatheter implantation techniques in several preclinical studies as pulmonary (Weber et al. 2013; Driessen-Mol et al. 2014; Emmert et al. 2018; Motta et al. 2018, 2019) and aortic (Lintas et al. 2018) replacements.

Till 2018, the first cause of failure for all TEM-based TEHV was the development of progressive insufficiency in vivo, most probably ascribable to leaflet thickening and/or shortening and retraction (Flanagan et al. 2009; Gottlieb et al. 2010; Schmidt et al. 2010; Weber et al. 2013; Driessen-Mol et al. 2014; Syedain et al. 2015; Reimer et al. 2017; Motta et al. 2018).

Table 8 Overview of the (A) clinical and (B) preclinical studies evaluating functionality and remodeling potential of decellularized xenograft-based TEHVs for in situ regeneration

(A) Clinical studies			
Porcine decellularized pulmonary valve xenografts (DPVX) for PVR			
Ref.	Commercial name	Procedure	Main findings
2014 (Backhoff et al. 2014)	Matrix P	Surgical	Porcine DPX showed limited valve performance with severe stenosis, moderate regurgitation, hypertrophic right ventricle, and aneurysms formation. Inflammatory or foreign body reactions were not observed
2014 (Breitenbach et al. 2014)	Matrix P Plus	Surgical	Porcine DPX use is not recommended in adults because of the massive inflammation, tissue necrosis, and conduit stenosis observed upon implantation
2013 (Voges et al. 2013)	Matrix P and Matrix P Plus	Surgical and transcatheter	More than half the patients needed a redo surgery because of stenosis and moderate insufficiency. Histological evaluation showed severe foreign body reaction, inflammation, wall thickening, and lack of endothelialization
2012 (Perri et al. 2012)	Matrix P and Matrix P Plus	Surgical	35.5% of the implanted porcine DPX failed, with stenosis, pseudo-aneurysm, or dilatation. Histological evaluation showed lack of cell infiltration and presence of inflammatory giant cells
2011 (Konertz et al. 2011)	Matrix P and Matrix P Plus	Surgical	Porcine DPX used in patients with congenital heart disease showed unremarkable functionality and normal morphology with lack of calcification
2010 (Rüffer et al. 2010)	Matrix P Plus	Surgical	10 months after implantation, 38% of the implanted porcine DPX were stenotic because of severe inflammation and calcification
2005 (Konertz et al. 2005)	Matrix P	Surgical	In 50 patients, the porcine DPX showed physiological-like hemodynamic and low-pressure gradients
2003 (Simon et al. 2003)	SynerGraft 500 and 700	Surgical	Porcine DPX used in pediatric settings showed good postoperative functionality. However, structural valve failure and degeneration caused the sudden death of three patients. Histological evaluation showed a severe inflammatory response and calcifications. Incomplete decellularization of the original porcine DPX was demonstrated

(continued)

Table 8 (continued)

(A) Clinical studies			
(B) Preclinical studies			
Decellularized pulmonary valve xenografts (DPVX)			
Ref.	Xenograft and control valves	Procedure and animal model	Main findings
2018 (Goecke et al. 2018)	*Porcine DPX* *Ovine DPH*	Surgical PVR in sheep	Porcine DPX processed with freeze-drying showed immunological reaction that did, however, not affect early functionality or cell infiltration
2013 (Hopkins et al. 2013)	*Human DPX* *Baboon DPH*	Surgical PVR in baboon	Human DPX and baboon DPH showed native-like hemodynamic with limited inflammatory response
2005 (Dohmen et al. 2005)	*Porcine DPX*	Surgical PVR in sheep	Evaluation of explanted porcine DPX showed good remodeling potential in sheep, with smooth and pliable leaflets covered by endothelial cells and populated by fibroblasts. De novo ECM was observed. No thrombosis and no calcification were detected
2003 (Leyh et al. 2003)	*Porcine DPX pre-seeded with aVEC and aMFB*	Surgical PVR in sheep	Porcine DPX contained endogenous retrovirus DNA that was, however, not transmitted to the host animal blood samples. Remodeling potential was confirmed after 6 months, as indicated by the presence of host endothelial and interstitial cells
2003 (Dohmen et al. 2003)	*Porcine DPX pre-seeded with aVECs*	Surgical PVR in sheep	The explanted porcine DPX showed a confluent endothelial cell monolayer and an increasing number of fibroblasts. No signs of calcification were observed
Decellularized aortic valve xenografts (DAVX)			
Ref.	Xenograft and control valves	Procedure and animal model	Main findings
2017 (Hennessy et al. 2017)	Porcine DAX	Surgical PVR in sheep	Porcine DAX showed sufficient functionality as PVR in sheep. Histological evaluation showed host cell infiltration with myofibroblasts and endothelial cells
2012 (Dohmen and da Costa 2012)	Porcine DAX	Surgical AVR in sheep	Porcine DAX showed good early functionality as AVR in sheep, with no regurgitation. At explantation, the valve was presented with smooth and pliable leaflets. Histological evaluation showed leaflet cellularization with host interstitial cells

(continued)

Table 8 (continued)

(A) Clinical studies			
2007 (Iwai et al. 2007)	Porcine DAX	Surgical PVR in dog	Porcine DAX were efficiently decellularized and caused minimal immune response and calcification. Stable valve functionality was observed, with no regurgitation. Interstitial cell repopulation and endothelialization occurred within 2 months
2005 (Ota et al. 2005)	Porcine DAX with/ without Fn-HGF	Surgical PVR in dog	Porcine DAX were functionalized with Fn-HGF to favor endothelialization and resulted in complete endothelial coverage within 1 month. Host cell infiltration was greater in the Fn-HGF functionalized valve compared to the control
1999 (O'Brien et al. 1999)	Porcine DAX	Surgical PVR in sheep	Porcine DAX showed good functionality as PVR in sheep, with no signs of calcifications or leaflet degeneration. Over time, DAX were repopulated by host fibroblasts

PVR pulmonary valve replacement, *AVR* aortic valve replacement, *aVEC*, autologous vascular endothelial cells, *aMFB* autologous myofibroblasts, *Fn-HGF* fibronectin-hepatocyte growth factor, *ECM* extracellular matrix, *DPVH* decellularized pulmonary valve homografts

To solve this long-standing problem, Emmert et al. introduced the use of computational modeling tools to optimize the design of a TEM-based TEHV (Sanders et al. 2016; Emmert et al. 2018) to control tissue remodeling and to ensure long-term functionality upon implantation in vivo. As more thoroughly explained in Sect. 6.1.2, the simplified TEM-based valve geometry, lacking a belly region, used in these studies lead to radial leaflet compression under physiological hemodynamics, with consequent leaflet shortening (Loerakker et al. 2016). To prevent leaflet retraction, computational modeling suggested a new valve geometry that was achieved by using a constraining bioreactor insert during TEHV culture (Sanders et al. 2016). TEHVs manufactured with the novel geometry were then implanted in the pulmonary position in the ovine model for up to 1 year, reporting excellent performance, host cell repopulation, native-like remodeling, and sustained durability, without leaflet thickening nor shortening (Emmert et al. 2018). In this regard, valve design, hemodynamic loading, and cell contractile forces have been predicted to be crucial in influencing the remodeling outcome under physiological conditions.

6.3 Bioresorbable Polymeric Scaffolds

Bioresorbable, natural, synthetic, or hybrid polymeric scaffolds are widely used in the medical field due to their high versatility and associated low costs. The main

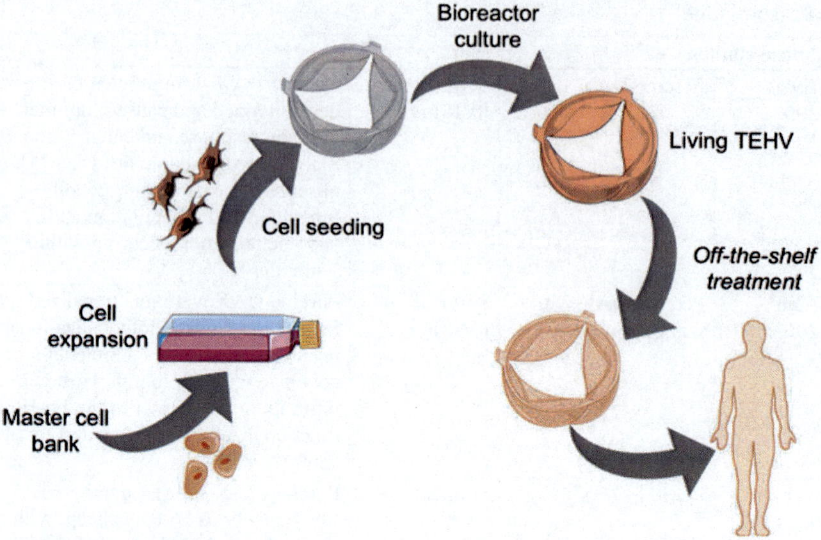

Fig. 8 Schematic representation of in situ heart valve tissue engineering using a tissue-engineered matrix (TEM). This method uses classic TE methodologies to obtain a scaffolds consisting of in vitro grown ECM depleted of cells (i.e., tissue-engineered matrix (TEM)). Compared to the autologous in vitro TE (Fig. 5), this approach uses allogenic cells to create the TEHV. After culture, off-the-shelf availability and immunocompatibility are granted by the decellularization process. The TEHV can be efficiently stored until implantation. (Images adapted from Servier Medical Art under a creative common attribution 3.0 unported license)

advantage of using bioresorbable polymers for in situ TE applications is the possibility to obtain a cell-free TEHV with tunable scaffold architecture, mechanical properties, stability, and degradation rate, hence assuring reproducibility, unlimited supply, and scalability in the production process. The polymeric valve has inherent off-the-shelf availability and can be directly implanted into the patient when needed, without the use of further expensive and complicated in vitro techniques (Fig. 8). Natural polymers are mostly represented in the form of hydrogels and comprise components of the native tissue composition (e.g., collagen, fibrin, chitosan, silk, and keratin) as starting material (Jana et al. 2014). Such protein-based scaffolds are fast degrading and nontoxic materials, with low immunogenicity and tunable architecture, degradation rate, and porosity. However, they are limited by the inherently low mechanical properties of hydrogels, which make them unsuitable as starting matrix for in situ TEHV applications.

6.3.1 Bioresorbable Synthetic Polymeric Scaffolds

Biodegradable synthetic polymers, such as polyglycolic acid, polylactic acid, polycaprolactone, and bisurea-modified polycarbonate, have been extensively used as starting matrices for both in vitro and in situ TE applications because they can be degraded and absorbed by the human body. Such polymers possess tunable mechanical, chemical, and structural properties that can be tailored for the development of

Table 9 Overview of the preclinical studies evaluating functionality and remodeling potential of decellularized TEM-based TEHVs for in situ regeneration

Preclinical studies			
Decellularized human cell-derived tissue-engineered matrix (hTEM)			
Ref.	TEM	Procedure and animal model	Main findings
2019 (Motta et al. 2019)	*PGA/P4HB-based hTEM*	Transcatheter PVR in sheep	TEHVs with integrated Valsalva sinuses were manufactured using hTEM. Acute feasibility testing was performed in sheep, showing excellent valve functionality as PVR
2018 (Lintas et al. 2018)	*PGA/P4HB-based hTEM*	Transcatheter AVR in sheep	Principal feasibility of performing a TAVR using an hTEM-based TEHV. Positioning was good (i.e. no coronary obstruction) and acute valve performance was promising (no stenosis, no central nor paravalvular regurgitation)
2013 (Weber et al. 2013)	*PGA/P4HB-based hTEM*	Transcatheter PVR in baboon	Good early post-procedural performance, but mild to moderate regurgitation was observed at 8 weeks post implantation. Explant analyses showed thin and pliable leaflets that, however, shortened overtime. Histological evaluation confirmed the remodeling potential of the hTEM with fast cellular repopulation
Decellularized animal cell-derived tissue-engineered matrix (TEM)			
Ref.	TEM	Procedure and animal model	Main findings
2018 (Emmert et al. 2018)	*PGA/P4HB-based ovine TEM*	Transcatheter PVR in sheep	An optimized TEM-based TEHV design was achieved using computational modeling tools. The resulting valve showed excellent in vivo performance (no regurgitation, no stenosis, retained coaptation) for up to 1 year. Histological evaluation showed profound tissue remodeling with complete endothelialization and de novo collagen and elastin synthesis towards native-like remodeling characteristics, while maladaptive remodeling phenomena were absent
2018 (Motta et al. 2018)	*PGA/P4HB-based ovine TEM*	Transcatheter PVR in sheep	This TEM-based TEHV was integrated on a customized nitinol stent to introduce Valsalva sinuses. Acute valve performance was promising, but leaflet shortening and regurgitation were observed after 16 weeks. Histological evaluation showed host cell infiltration, endothelialization, and ECM remodeling

(continued)

Table 9 (continued)

Preclinical studies			
2017 (Reimer et al. 2017)	*Fibrin-based ovine TEM*	Surgical PVR in sheep	While early valve functionality was promising, regurgitation increased over time due to leaflet shortening. Histological evaluation indicated promising remodeling with host cell infiltration and collagen and elastin deposition
2015 (Syedain et al. 2015)	*Fibrin-based ovine TEM*	Surgical AVR in sheep	First evaluation of a TEM-based TEHV as AVR in sheep for up to 6 months. Valve functionality was good, with no stenosis. Histological evaluation showed ECM remodeling, collagen and elastin synthesis, and host endothelial and interstitial cells
2014 (Driessen-Mol et al. 2014)	*PGA/P4HB-based ovine TEM*	Transcatheter PVR in sheep	While early valve functionality was promising, mild regurgitation started at 8 weeks and progressed to moderate at 24 weeks due to a compromised leaflet coaptation and shortening. Remodeling potential of the TEM was confirmed by quick cellular repopulation, new ECM synthesis, and absence of calcifications

PVR pulmonary valve replacement, *AVR* aortic valve replacement, *PGA* polyglycolic acid, *P4HB*, poly(4-hydroxybutyrate), *ECM* extracellular matrix

TEHVs with long-term durability, gradual degradation rates, flexible and elastic leaflets, and potential for further functionalization (Fioretta et al. 2012; Wissing et al. 2017). Synthetic materials are attractive for their fast production, unlimited availability, and lack of disease transmission. In light of these properties, numerous preclinical large animal studies investigating such synthetic materials in the context of TEHVs have been performed, even in combination to transcatheter implantation techniques (Table 10). In this context, TEHVs based on bioresorbable supramolecular elastomeric polymers (i.e., bisurea-modified poly(carbonate) and 2-ureido-4[1H]-pyrimidinone-modified polycaprolactone) were manufactured and implanted with both surgical and transcatheter approaches, demonstrating reasonable performance for up to 1 and 2 years as pulmonary valve replacement. In these studies, ECM deposition, partial degradation, and endogenous cellularization were observed. Following these results supramolecular polymer-based valve replacements have advanced into clinical trials, with however heterogenous outcomes. Hence, further chronic animal studies are required to further assess the long-term durability and remodeling of such prostheses. Indeed, a recent study showed important differences in tissue remodeling between valves but also in between leaflets of the very same valve (Fioretta et al. 2019a). These latter results suggest that a further investigation of polymer degradation mechanisms and tissue remodeling pathways is needed to ensure safe clinical translation of this technology into patients.

6.3.2 Autologous Cell Pre-seeding onto Bioresorbable Polymeric TEHVs

To favor early remodeling responses, researchers have also investigated the use of autologous cell pre-seeding onto biodegradable scaffold materials. These on-the-fly concepts proved to be attractive especially when combined with bone marrow-derived mononuclear cells (BMMNCs) because they could be performed in a one-step procedure prior implantation (Table 10) and where previously proved to favor vascular graft remodeling in small animal models. In light of these results, BMMNC were used to pre-seed synthetic scaffolds for both pulmonary and aortic TVR procedures (Table 10). While most of these studies are proof of concept with an acute or short-term follow-up (up to 4 weeks) (Weber et al. 2011; Emmert et al. 2011, 2012, 2014), a recent publication showed detrimental effects of BMMNC pre-seeding onto bioresorbable supramolecular TEHVs, with loss of functionality over time due to regurgitation, leaflet thickening, and calcifications (Fioretta et al. 2019a) within 16 weeks from implantation.

6.3.3 Hybrid Polymeric Scaffolds

Natural proteins (e.g., gelatin, alginate, hyaluronic acid) can be efficiently combined to synthetic polymeric scaffolds to form hybrids, mainly manufactured via rotary jet spinning, electrospinning, or 3D printing techniques. The advantages of applying synthetic materials to create hybrid scaffolds are multiple and allow for tunable mechanical properties, controlled biodegradation, off-the-shelf availability, scalability, and reproducibility. Additionally, to further improve polymer biocompatibility and reduce thrombogenicity, scaffolds can be functionalized using natural proteins. In this regard, synthetic polymers are linked to natural proteins via physical absorption, chemical binding, or co-spinning (Rossi and Van Griensven 2014). Hence, hybrid material organization should ideally mimic the physical properties and structure of native tissues, thereby improving the cellular adhesion, distribution, and proliferation. As an example, cell-free biomimetic TEHVs were recently manufactured starting from a blend of synthetic polymer (i.e., poly (4-hydroxybutyrate) and natural proteins (i.e., gelatin) processed via rotary jet spinning (Capulli et al. 2017). These valves proved to be compatible with transcatheter pulmonary valve implantation techniques proving feasibility and functionality in an acute proof-of-concept study in sheep with early cellular adhesion and infiltration (Capulli et al. 2017).

Beyond mechanical stability and scaffold architecture properties, hybrid and synthetic scaffolds offer the possibility to include bioactive molecules and serve as vehicles for the delivery of components such as peptides, growth factors, antibodies, and drugs (Fioretta et al. 2012). This functionalization aims at improving scaffold performance by favoring the recruitment of host cells, promoting tissue formation, ensuring hemocompatibility, and controlling the early steps of the inflammatory cascade (Roh et al. 2010). Additionally, the inflammatory response triggered by those treatments might be modulated toward a positive and native-like tissue remodeling (Muylaert et al. 2016).

Table 10 Overview of the preclinical studies evaluating functionality and remodeling potential of bioresorbable polymer-based TEHV and their combination with a one-step pre-seeding procedure using autologous bone marrow mononuclear cells (BMMNCs)

Preclinical studies			
Bioresorbable polymeric TEHVs			
Ref.	Polymer	Procedure and animal model	Main findings
2019 (Coyan et al. 2019)	*Polycarbonate urethane urea + AZ31 magnesium alloy stent*	Surgical PVR in pig	Acute feasibility test of a novel bioresorbable valve combined with a biodegradable stent. Normal leaflet functionality was observed with no signs of thrombosis, regurgitation, or degradation
2019 (Fioretta et al. 2019a)	*Bisurea-polycarbonate pre-seeded with BMMNC*	Transcatheter PVR in sheep	BMMNC pre-seeding caused severe regurgitation and calcification. Valve remodeling was observed with cellular repopulation, ECM synthesis, and scaffold degradation. Independently of pre-seeding, differential valve and leaflet remodeling were observed
2018 (Bennink et al. 2018)	*UPy-polyester-urethanes*	Surgical PVR in sheep	The remodeling of this bioresorbable polymeric valve was described as an ongoing process characterized by neointima formation and scaffold degradation over the 12 months post-implantation
2017 (Miyazaki et al. 2017)	*UPy-polyester-urethanes (XPV)*	Transcatheter AVR in sheep	When used as transcatheter AVR, the valves showed good early functionality, comparable to commercially available valves in acute settings
2017 (Soliman et al. 2017)	*UPy-polyester-urethanes (XPV)*	Surgical PVR in sheep	Valve performance as PVR in sheep is promising, with no stenosis and no severe regurgitation observed
2017 (Kluin et al. 2017)	*Bisurea-polycarbonate*	Surgical PVR in sheep	Proof-of-concept study where a bioresorbable polymeric TEHV was implanted in sheep for 12 months. The valve showed sustained performance, polymer reabsorption and de novo tissue deposition (collagens, elastin) and cellular repopulation
2017 (Capulli et al. 2017)	*P4HB-gelatin*	Transcatheter PVR in sheep	Acute feasibility study where a hybrid TEHV based on gelatin and P4HB processed via jet spinning was implanted as PVR in sheep. The valve demonstrated appropriate acute performance with retained leaflet coaptation

(continued)

Table 10 (continued)

Preclinical studies			
2014 (Emmert et al. 2014)	*PGA/P4HB pre-seeded with BMMNCs*	Transcatheter AVR in sheep	In this feasibility study, the transcatheter AVR procedure was combined with BMMNC pre-seeding. Good valve performance was demonstrated in the acute settings, with correct positioning and free coronary flow
2012 (Emmert et al. 2012)	*PGA/P4HB pre-seeded with BMMNCs*	Transcatheter AVR in sheep	Short-term feasibility study (up tp 2 weeks) demonstrating adequate leaflet mobility and valve performance of the BMMNC pre-seeded TEHV implanted as transcatheter AVR. Explant analysis showed intact leaflets, cell infiltration and early signs of tissue remodeling
2011 (Emmert et al. 2011)	*PGA/P4HB pre-seeded with BMMNCs*	Transcatheter AVR in sheep	Proof-of-concept study to assess the possibility of combining minimally invasive aortic replacement techniques with tissue engineering methods
2011 (Weber et al. 2011)	*PGA/P4HB pre-seeded with BMMNCs*	Transcatheter PVR in baboon	BMMNC pre-seeded TEHV showed intact leaflets and sufficient valve performance after 5 months in vivo. Substantial cell infiltration and implant remodeling was observed

PVR pulmonary valve replacement, *AVR* aortic valve replacement, *PGA* polyglycolic acid, *P4HB* poly(4-hydroxybutyrate), *UPy* ureido-pyrimidinone, *ECM* extracellular matrix, *BMMNCs* bone marrow mononuclear cells

7 Testing of Tissue-Engineered Heart Valves

Physiologically, heart valves open once every second, and, for each cardiac cycle, the valve leaflets are exposed to complex deformation and hemodynamic forces. The hemodynamics is a combination of mechanical forces that control blood flow and blood pressure in the body. Non-physiological values of blood flow and/or pressure are associated with changes in valvular structure and, therefore, play a significant role in the etiology of valvular disease (Chandran 2010). However, there is a new general consensus that these forces are also responsible for adaptive or maladaptive remodeling of TEHVs (Emmert et al. 2018). Hence, it is important to validate a novel TEHV by using an array of testing platform consisting of in silico models, in vitro models, and, finally, in vivo preclinical animal models.

7.1 In Silico Models to Optimize Valve Design

In silico models are capable of simulating the complex hemodynamic environment of a heart valve. Computational simulations are, therefore, becoming an important tool to improve our understanding of heart valve physiology and pathology, but also to assess valve prosthesis functionality, and to improve surgical planning using precision medicine tools.

The hemodynamics plays a key role in controlling the physiological or pathological remodeling of the native leaflet. By using computational modeling, the hemodynamic forces on the valve can be simulated to understand structural and biological changes in response to (non-)physiological stresses; as an example, higher values of blood pressure can lead to increased leaflet deformation that, on a microscopic scale, induces changes in the valvular interstitial and endothelial cells (Gould et al. 2013). As explained in Sect. 2, this may result in VIC activation toward a fibrotic or osteoblastic phenotype, setting the basis for fibrosis and calcification. Hence, understanding the distribution of stress and strain in the valve leaflets is fundamental to predict the remodeling potential of TEHVs.

7.1.1 Design Optimization for Artificial Heart Valve Replacements

In silico models have significantly advanced in the past decades, and they are now a valid tool to better understand the complexity of the valve hemodynamics: cusp opening and closure pattern, flow pattern, and stress and strain distribution in the leaflets (Chandran 2010). More importantly, computational simulations have been used to assess the efficacy of novel heart valve prosthesis designs, significantly reducing the number of prototype manufacturing, bench testing, and in vivo testing (Morris et al. 2016).

Finite element analysis (FEA) is a modeling technique that focus on the assessment of stress and strain distribution in a region of interest. FEA is used to optimize valve design by implementing valve geometry and dimension into the computational tool. Initially, FEA was using simplified symmetric geometries and valves in closed configuration (Cataloglu et al. 1977; Ghista and Reul 1977; Sabbah et al. 1985). However, with the increased computational power of modern computers, the dynamic opening and closure configuration of a heart valve can be also modeled and combined to patient-specific geometries based on a detailed 3D reconstruction of the valve via cardiac imaging techniques (Gnyaneshwar et al. 2002; Sripathi et al. 2004).

More recently, a proof-of-concept study developed machine learning techniques to estimate stress and strain forces on transcatheter valves, starting with a set of TAVI-compatible leaflet designs (Liang and Sun 2019). The results were promising, with the model being able to accurately estimate the deformed leaflet geometries and stress distribution within seconds, therefore considerably reducing the modeling and simulation complexity of standard FEA techniques.

Computational fluid dynamics (CFD) is another method to provide a quantitative description of flow characteristics. CFD is commonly used to calculate the values of wall shear stress in a noninvasive manner, providing information on the flow pattern characteristics (i.e., laminar or turbulent flow) (Jin et al. 2004). CFD is also efficiently used to understand the flow pattern through a valve prosthesis, providing crucial information to optimize valve design in order to limit thrombotic events (Kelly 2002; Yoganathan et al. 2005; Simon et al. 2010; Zakaria et al. 2017).

Finally, fluid-structure interaction (FSI) models focus on the interactions between valve structures (i.e., leaflets) and blood flow. Specifically, FSI has been used in combination to platelet activation models to assess the hemodynamics of valvular replacement and to understand the interaction between blood cells and valve structures (Borazjani 2015). FSI models can, for example, provide detailed information on platelet distribution and accumulation and on blood flow patterns within a specific valve design. Hence, FSI models provide important information on the thrombogenic risk profile of the tested valve design and suggest where improvement is needed to prevent thrombotic events (Piatti et al. 2015).

7.1.2 Design Optimization for Tissue-Engineered Heart Valves

Design optimization in silico is, therefore, a common application in artificial valves to predict the consequences of changes in artificial valve design on the overall outcome.

However, computational modeling has hardly been utilized to improve the performance of TEHVs, nor has it been validated in clinically relevant in vivo models (Soares et al. 2014a, b).

Recently, a multidisciplinary team of researchers showed the potential of using computational modeling tools to optimize the design of a TEHV (Sanders et al. 2016; Emmert et al. 2018) to manufacture a TEM-based TEHV capable of controlling tissue remodeling to ensure long-term functionality upon implantation in vivo. Indeed, most TEHVs based on TEM lose their functionality within a few months due to uncontrolled (adverse) tissue remodeling phenomena which translate into leaflet shortening, resulting in valve insufficiency (Flanagan et al. 2009; Schmidt et al. 2010; Weber et al. 2013; Driessen-Mol et al. 2014; Reimer et al. 2015, 2017; Schmitt et al. 2016; Motta et al. 2018).

Computational simulations showed that the simplified valve geometry used in most of these studies led to radial leaflet compression when exposed to physiological pulmonary pressure and in the presence of contractile cells (Loerakker et al. 2016). To counteract cusp tissue compaction, a new valve geometry was firstly computationally derived and then implemented in TEHVs by using a constraining bioreactor insert during culture (Sanders et al. 2016). The novel valve geometry showed limited radial tissue compression compared to previous designs, suggesting that these TEHVs would be less prone to host cell-mediated tissue retraction upon implantation.

To confirm this hypothesis, the researchers implanted such TEHVs minimally invasively as pulmonary valve replacements in sheep. The results of this study were outstanding, with preserved and good long-term in vivo performance for up to 1 year, as predicted by and consistent with our computational modeling. Additionally, tissue remodeling was profound, with collagen and elastin deposition, polymer reabsorption, and even the initial formation of sinuses of Valsalva in some of the explants. This study indicated the high relevance of an integrated in silico, in vitro, and in vivo bioengineering approach for developing functional TEHVs.

7.2 In Vitro Models to Test Valve Functionality

Similar to artificial valve prostheses, TEHVs should be tested in vitro to characterize valve performance by measuring parameters such as pressure difference, regurgitation, and durability using hydrodynamic studies. Two systems are frequently used in the characterization and in vitro validation of a valvular replacement: the pulse duplicator system and the durability system. More recently, cardiac biosimulator platform have been developed to assess valve performance inside a real heart. These testing platforms are also regulated by the ISO 5840 (cardiovascular implants: cardiac valve prostheses) that provides a series of requirements a valve prosthesis should fulfil before clinical translation.

7.2.1 Pulse Duplicator Systems

Pulse duplicator systems (e.g., HDT-500 Heart Valve Pulse Duplicator System by BDC Laboratories or the ViVitro Pulse Duplicator System by ViVitroLabs) (Fig. 9) can be used to simulate left or right heart functionality and to measure forward and backflow flow and pressure gradients along with an accurate visualization of the valve's opening and closing behavior from optical visualization ports. This system is usually characterized by two chambers connected by a fixture where the valve prosthesis is allocated for the testing. A pulsatile pump provides the desired pressure and flow conditions to mimic the left or right heart, which are monitored real time with pressure and flow sensors in the cardiac chambers. The system is finally connected to a computer where a software provides you with the acquired information on valve performance. The pulse duplicator systems can be used to simulate physiological and non-physiological cardiac conditions by varying the resistance and compliance of the system.

While initially developed to test mechanical and bioprosthetic valves, these systems have been recently extended also for the testing of TEHVs by using customized fixtures to allocate TEHVs with different stent designs (Sanders et al. 2016; Kluin et al. 2017; Emmert et al. 2018; Lintas et al. 2018; Motta et al. 2019) and even biological valved conduits (Buse et al. 2016).

However, it is important to notice that considerable differences were observed in the hydrodynamic performance (i.e., pressure gradient and backflow leakage) of a single valve tested in different ISO 5840 compliant pulse duplicators (Retta et al. 2017). These results suggest the importance of using a known reference valve in hydrodynamic performance testing to assess the individual measurement conditions in the duplicator.

7.2.2 Durability Systems

Valve durability systems (e.g., VDT-3600i Heart Valve Durability System by BDC Laboratories and DuraPulse Heart Valve Test Instrument, TA Instruments) (Fig. 9) perform accelerated durability tests to assess the fatigue behavior of the valve (Black et al. 1994; Sanders et al. 2016; Buse et al. 2016). The platform can consist of multiple independent testing stations where different valves can be allocated and tested independently at accelerated frequencies (up to 40 Hz) to simulate long-term

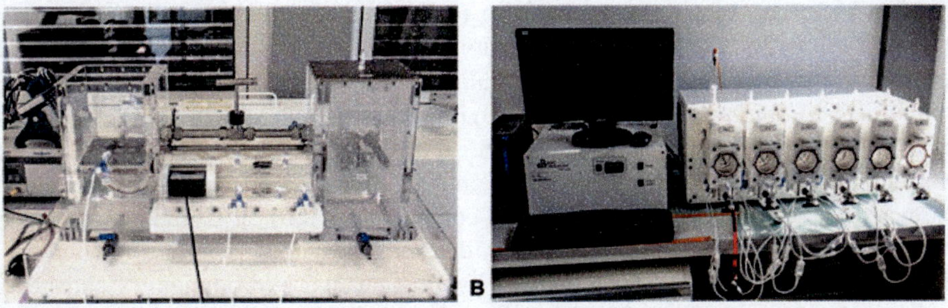

A B

Fig. 9 Representative images of in vitro valve testers. (**a**) A pulse duplicator system (HDT-500 Heart Valve Pulse Duplicator System by BDC Laboratories) and (**b**) a durability testing platform (VDT-3600i Heart Valve Durability System by BDC Laboratories). (Pictures courtesy of Nikolaos Poulis, Institute for Regenerative Medicine, University of Zurich, Switzerland)

in vivo performance. Despite the wide range of accelerated frequencies that can be used to test the valve, it is important to ensure that even at high frequency, the valve is subjected to a complete opening and closing cycle to prevent incorrect and/or insufficient loading of the valve cusps that would affect valve durability (Sanders et al. 2016).

The physiological environment can be further simulated by replicating blood viscosity via the addition of xanthan gum or polyacrylamide to the water solution (Pohl et al. 1996). This method will mimic the complex viscoelastic behavior of blood and will improve the accuracy of the recorded leakage flow and leakage volume.

ISO5840 provides guidelines to assess artificial valve prosthesis durability and performance, suggesting that bioprostheses should sustain at least 200 million cycles while mechanical valves 400 million cycles (corresponding to 5 and 10 years in vivo, respectively). However, these norms have been specifically designed for artificial non-regenerative valve prostheses and may not be applicable to TEHVs. Indeed, TEHV functionality and durability are closely related to the in vivo remodeling potential, and performing a durability test on TEHVs will most likely result in underestimation of valve durability due to the lack of remodeling potential in the in vitro settings. As an example, TEM-based TEHV durability in vitro was reported to be the equivalent of 16 weeks (Sanders et al. 2016). However, upon implantation in a preclinical sheep model, these valves showed good functionality for up to 1 year (Emmert et al. 2018), thereby proving that upon remodeling, durability was improved. Hence, there is a clear need to improve guidelines for the testing of regenerative heart valves, in order to take into account their regenerative potential.

7.2.3 Cardiac Biosimulator and Beating Heart Platforms

Pulse duplicator and durability systems have the advantage of providing automatic feedback control and independent test chambers to ensure that test conditions are maintained throughout the test. However, assessing valve functionality inside a real heart will provide even more insights on the efficacy and efficiency of the developed

TEHV. A cardiac biosimulator platform (e.g., Cardiac Biosimulator Platform by LifeTec Group) consists of a mock apparatus where a full (animal) heart is connected to a pump that actuates the heart to simulate the hemodynamic environment in an anatomical setting, thereby providing in vitro the most physiological-like environment to assess valve performance (Leopaldi et al. 2015). Modern imaging techniques can be connected to the platform to obtain precise information on valve hemodynamics and performance in the heart (Leopaldi et al. 2012). Compared to pulse duplicator and durability systems, the cardiac biosimulator platform allows to test the valve in a pressurized heart with physiological-like hydrodynamics and heart wall movements, which will efficiently replicate the in vivo animal model in assessing functionality and implantability of the valve, but with limited ethical concerns.

Another interesting setup aimed at testing new cardiovascular products in a relevant physiological environment is the beating heart platform (e.g., PhysioHeart by LifeTec Group). Compared to the cardiac biosimulator, the heart used in this platform is the actual pump that, by retained physiological contraction, provides realistic physiological cardiac dynamic conditions analogous to an animal model. However, it has the advantage of allowing endoscopy to study valve implantation, efficacy, and performance over time (de Weger et al. 2010).

Both the cardiac biosimulator and the beating heart apparatus have been validated with different mammalian hearts and have been successfully used to train cardiac surgeon in novel surgical and transcatheter interventions (Leopaldi et al. 2018).

7.3 In Vivo Animal Models to Assess Remodeling

Heart valve functionality depends on a complex combination of mechanical, biological, and physical processes that are only partly replicated by in vitro models. In vitro model systems are, therefore, not yet able of fully replicating the complexity of a living being, in particular regarding remodeling and growth potential, fundamental aspect of a tissue-engineered prosthesis (Fioretta et al. 2019b). Therefore, clinical translation of valvular prostheses still relays on in vivo experiments. Research aiming at understanding the basic mechanisms involved in heart organogenesis can be performed in non-mammalian models, such as the zebrafish, because of easy reproduction, external embryonic development, and transparency of the embryos (Cesarovic et al. 2020).

Small mammalians, such as mice and rats, are, on the other hand, extremely helpful in assessing the molecular mechanisms of heart physiology and pathophysiology, since they are genetically similar to humans, they have a four-chambered heart that better represent the human anatomy compared to non-mammalian animal models, and they can be genetically modified to express certain disease characteristics (Cesarovic et al. 2020). In addition, they have been also extensively used to assess functionality and host cell-mediated remodeling of tissue-engineered vascular grafts (Roh et al. 2010; Hibino et al. 2011; Lee et al. 2014). However, due to size and hemodynamic differences, they do not provide a suitable animal model for the assessment of heart valve replacements.

The ideal animal model to assess valve implantation and functionality should closely resemble the human heart in size, anatomy, and hemodynamics. In addition, similar coagulation cascade and immune response will be fundamental to better interpret the remodeling potential of a TEHV. While guidelines are provided for the in vivo preclinical validation of artificial valve prostheses (i.e., ISO5840) and to design and execute animal experiments to evaluate medical devices (ISO10993), there is not yet a standard protocol to test TEHV prostheses. TEHVs have been tested in different large mammals, such as sheep, pig, and nonhuman primates, which provide a comparable size, anatomy, functionality, and physiology to the human heart. These similarities allow for the use of clinical-grade imaging and diagnostic devices to determine feasibility of valve implantation procedure and to assess valve functionality in clinical-like settings.

7.3.1 Nonhuman Primate Model

Nonhuman primates, such as baboons, are the animal model that is most similar to human in terms of anatomy, physiology, and growth curve, making them an ideal candidate for long-term chronic studies and growth assessment.

Only a handful of studies have tested TEHVs in baboons. Decellularized human and baboon pulmonary valves have been implanted as pulmonary replacement in baboon and the results compared to clinical-grade porcine valved conduits (Hopkins et al. 2013). All animals showed excellent valve functionality upon implantation. While decellularized valves remained functional over time, a reduction of effective orifice area was detected for the porcine conduits, and it was associated with a more intense inflammatory response. In another study, bioresorbable valves were pre-seeded with autologous bone marrow mononuclear cells prior to implantation and delivered as pulmonary valve replacement in baboon (Weber et al. 2011). Valve functionality was adequate up to 4 weeks upon implantation, with substantial remodeling and cellular infiltration. Finally, TEM-based TEHVs were implanted as pulmonary replacement in baboons to assess functionality and remodeling (Weber et al. 2013) and the results compared to decellularized human valves. The results indicated better cellular infiltration for TEM-based TEHVs but leaflet shortening over time associated with moderate regurgitation.

Nonhuman primates are, however, a strictly regulated and costly animal model, which is also often prohibited for ethical and regulatory reasons (Rashid et al. 2004; Cesarovic et al. 2020).

7.3.2 Swine Model

Swine heart anatomy and physiology closely resemble the human heart, making pigs a good animal model for cardiovascular research (Suzuki et al. 2011), in particular for the investigation of coronary artery disease and myocardial infarction (Patterson and Kirk 1983). In addition, other biological processes, such as the immune systems, the coagulation cascade, and the endothelialization potential, are similar to humans (Gallo et al. 2017).

Despite these advantages, pigs have a fast-growing curve that may determine size mismatches with the implanted prosthesis during long-term follow-ups, thereby affecting the chronic study outcome (Rashid et al. 2004; Kheradvar et al. 2017). In the field of heart valve TE, pigs have been used to investigate feasibility of the implantation procedure of novel fully bioresorbable stented prostheses in acute studies (Coyan et al. 2019).

To limit size-related issues, the Vietnamese mini-pig has been used to investigate functionality and remodeling of TEHVs based on decellularized porcine aortic roots (Bottio et al. 2010; Gallo et al. 2016, 2017). This homologous replacement showed promising results over a 15-month follow-up, with retained functionality, host cell repopulation, tissue remodeling, and no calcification observed. Taken together, these results suggest that pigs can be used for assessing TEHV performance and may become a valid preclinical animal model in the future to evaluate regenerative replacements.

7.3.3 Ovine Model

The sheep is the FDA-approved and ISO-norm recommended animal model for the in vivo preclinical validation of heart valve replacements, with similar heart anatomy, heart rate, pressure values, and, most importantly, heart valve sizes to human's (Kheradvar et al. 2017). In addition, the high calcium level in the lamb serum (up to 12 months old), makes this animal model "the worst-case scenario" (Taramasso et al. 2015) to investigate durability and susceptibility to calcification of an implanted valve prosthesis (Barnhart et al. 1982). Finally, the growth curve of a sheep is moderate compared to the swine model, making it a valid choice for the assessment of regenerative tissue-engineered replacements with growth potential, as extensively demonstrated in literature (Iop et al. 2014; Zafar et al. 2015; Quinn et al. 2016; Reimer et al. 2017; Hennessy et al. 2017).

The sheep is, therefore, the gold standard animal model for assessing TEHV functionality and remodeling when implanted as pulmonary or aortic valve replacement, as well as for determining growth potential of TEHVs (Tables 6, 7, 8, and 9).

Although the heart anatomy is similar to the human's, animal models still present several limiting factors. Firstly, the orientation of the great arteries and of the heart differs considerably in quadrupeds compared to humans, making it difficult to test devices (e.g., transcatheter delivery systems) that are designed for the human anatomy. In addition, large animal models do not acquire the same valvular disease than human, and there is no animal model capable of replicating valve stenosis, insufficiency, or calcification. Moreover, due to financial and ethical constraints, long-term studies (>1 year) in large animal models are rare (Kheradvar et al. 2017; Cesarovic et al. 2020).

8 Challenges Toward Clinical Translation

By simplifications of manufacturing methods to grant off-the-shelf availability and scalability, TEHVs with in situ remodeling potential could potentially see a way to commercialization in the near future. However, despite the promising outcomes in both preclinical and clinical trials, routinely application of TEHVs into clinics is to date limited. From a scientific point of view, TEHVs need to provide proof of successful acute feasibility, short- and long-term functionality, over time durability, self-regeneration, and growth capabilities. However, several other aspects need to be equally considered.

8.1 Regulatory Challenges

Following FDA guidelines, cardiac valve replacements are devices classified as class III, according to their risk-based classification scheme. More specifically, such devices belong to the group of "support or sustainment of human life, which are of substantial importance in preventing impairment of human health or present a potential, unreasonable risk of illness or injury." However, the precise classification of such TEHVs might be difficult due to the variety of manufacturing approaches reported in literature and the heterogeneity in the international guidelines of the different regulatory agencies (Ram-Liebig et al. 2015). For instance, class III devices require a premarket approval (PMA), showing that for this particular device, there is no approved/existing equivalent device and demonstrating via research studies that the device is safe and effective. Preclinical studies for the translation of TEHVs need to withstand additional technical requirements in compliance to the guidelines ISO-5840 of the International Organization for Standardization (ISO). However, ISO-5840 document was developed for mechanical and bioprosthetic valves only, and therefore, current standards and regulations must be adapted to determine preclinical safety and efficacy of TEHVs.

8.2 Logistical Challenges

With the introduction in 1978 of the Good Laboratory Practice (GLP) concepts, standardized premarket evaluation rules were established. Hence, the translation of TEHV into clinics will only be effective when device production processes will also follow good manufacturing practices (GMP). Both regulation systems help the simplification and standardization of the clinical translation of TEHVs. Hence, researchers need to collaborate with GMP-compliant facility and perform GLP-compliant testing in order to further advance their prototypes into clinical trials. However, these facilities are not usually broadly available, and renting of special laboratories, such as clean rooms, is very expensive, thereby limiting the possibilities for academic researchers. In this regard, foundations to help accelerate the translation process are fundamental. These foundations may provide funding, facilities, and/or expertise to ensure that academic research and scientific discoveries are translated into clinical applications, even without the collaboration of industry and investors.

8.3 Clinical Requirements

The complete regeneration and repair of the implanted scaffold material is of utmost importance for any in situ approach. However, the regenerative capabilities of the host recipient play a pivotal role in the acceptance and remodeling of the implanted engineered construct, hence directly influencing the quality and success rate of the TEHVs. An in-depth clinical screening of the patient to determine his/her own regenerative potential is one of the requirements to anticipate when translating TEHVs into clinical practice and to ensure the best chance of remodeling upon implantation. Specifically, some patient-specific risk factors (e.g., diabetes, age, and smoke) are associated with impaired regenerative processes by the cells, hence possibly affecting TEHV remodeling and, therefore, functionality and durability on the long term. In addition, implant monitoring strategies using noninvasive imaging techniques, such as echocardiography, CT scan, and/or MRI, will be required to assess the status of the remodeling over time and ensure correct valve functionality. Finally, emergency treatments and strategies to adopt in case of failure should be considered and planned for safe clinical translation.

9 Conclusions

The prevalence of valvular diseases requiring a replacement is growing in light of our aging population and the ability to correct congenital heart defects. There are various treatment options for patients affected by valvular disease, such as valve repair and surgical or transcatheter valve replacement techniques. However, current clinically adopted valve replacements are a suboptimal solution, requiring the need for multiple reoperations either to accommodate for patient's growth or to substitute a degenerated valve. In addition, after the recent FDA approval of TAVI for low-risk case of aortic stenosis, the number of (younger) patients now eligible for TAVI has significantly increased. There is, therefore, a clear need for novel valve replacement solutions compatible with TVR techniques that could remodel and even grow upon implantation, ensuring a lifelong durability even for younger patient cohorts. Off-the-shelf available TEHVs with in situ regenerative and growth potential may overcome the limitations of currently used heart valve replacements, providing a novel durable option even compatible with transcatheter techniques. Clinical translation of decellularized homografts showed promising performance and recellularization potential (Table 7). In contrast, decellularized xenografts failed because of severe inflammation and stenosis (Table 8). Based on encouraging results from a first clinical trial using tissue-engineered vascular grafts based on supramolecular polymers (i.e., extracardiac total cavopulmonary connection to treat CHD) (Bockeria et al. 2017), clinical pilot trials using bioresorbable supramolecular polymer-based TEHVs are underway (trial number NCT02700100, NCT03022708, and NCT03405636). Furthermore, it is expected that TEM-based transcatheter TEHVs (Emmert et al. 2018; Lintas et al. 2018), will reach clinical translation in the near future.

By providing native-like remodeling, and long-term durability, TEHVs will be also able to reduce the healthcare costs.

Acknowledgments S.P. Hoerstrup is a shareholder at Xeltis BV and LifeMatrix AG. M.Y. Emmert is a shareholder at LifeMatrix AG. The other authors have nothing to disclose.

References

Althunayyan A, Petersen SE, Lloyd G, Bhattacharyya S (2019) Mitral valve prolapse. Expert Rev Cardiovasc Ther 17:43–51. https://doi.org/10.1080/14779072.2019.1553619

Anderson RH (2000) Clinical anatomy of the aortic root. Heart 84:670–673

Anderson RH, Ho SY, Becker AE (2000) Anatomy of the human atrioventricular junctions revisited. Anat Rec 260:81–91. https://doi.org/10.1002/1097-0185(20000901)260:1<81::AID-AR90>3.0.CO;2-3

Ando M, Takahashi Y (2009) Ten-year experience with handmade trileaflet polytetrafluoroethylene valved conduit used for pulmonary reconstruction. J Thorac Cardiovasc Surg 137:124–131. https://doi.org/10.1016/j.jtcvs.2008.08.060

Arsalan M, Walther T (2016) Durability of prostheses for transcatheter aortic valve implantation. Nat Rev Cardiol 13:360–367. https://doi.org/10.1038/nrcardio.2016.43

Backhoff D, Steinmetz M, Sigler M, Schneider H (2014) Formation of multiple conduit aneurysms following Matrix P® conduit implantation in a boy with tetralogy of Fallot and pulmonary atresia. Eur J Cardiothorac Surg 46:500–502. https://doi.org/10.1093/ejcts/ezt635

Balachandran K, Sucosky P, Yoganathan AP (2011) Hemodynamics and Mechanobiology of Aortic Valve Inflammation and Calcification. Int J Inflamm 2011:1–15. https://doi.org/10.4061/2011/263870

Barnhart GR, Jones M, Ishihara T et al (1982) Bioprosthetic valvular failure. Clinical and pathological observations in an experimental animal model. J Thorac Cardiovasc Surg 83:618–631

Barron V, Lyons E, Stenson-Cox C et al (2003) Bioreactors for cardiovascular cell and tissue growth: a review. Ann Biomed Eng 31:1017–1030

Bateman MG, Quill JL, Hill AJ, Iaizzo PA (2013) The anatomy and function of the semilunar valves. In: Heart valves: from design to clinical implantation. Springer US, Boston, pp 27–43

Bauernschmitt R, Schreiber C, Bleiziffer S et al (2009) Transcatheter aortic valve implantation through the ascending aorta: an alternative option for no-access patients. Heart Surg Forum 12. https://doi.org/10.1532/HSF98.20081112

Baumgartner H, Falk V, Bax JJ et al (2017) 2017 ESC/EACTS Guidelines for the management of valvular heart disease. Eur Heart J 38:2739–2791. https://doi.org/10.1093/eurheartj/ehx391

Benjamin EJ, Blaha MJ, Chiuve SE et al (2017) Heart disease and stroke statistics – 2017 update: a report from the American Heart Association. Circulation 135:e146–e603. https://doi.org/10.1161/CIR.0000000000000485

Bennink G, Torii S, Brugmans M et al (2018) A novel restorative pulmonary valved conduit in a chronic sheep model: mid-term hemodynamic function and histologic assessment. J Thorac Cardiovasc Surg 155:2591–2601.e3. https://doi.org/10.1016/j.jtcvs.2017.12.046

Bezuidenhout D, Williams DF, Zilla P (2015) Polymeric heart valves for surgical implantation, catheter-based technologies and heart assist devices. Biomaterials 36:6–25

Bisleri G (2016) Aortic valve repair. Curr Opin Cardiol 31:581–584. https://doi.org/10.1097/HCO.0000000000000334

Black MM, Hose DR, Lamb CJ et al (1994) In vitro heart valve testing: steady versus pulsatile flow. J Heart Valve Dis 3:212–215

Blum KM, Drews JD, Breuer CK (2018) Tissue-engineered heart valves: a call for mechanistic studies. Tissue Eng Part B Rev 24:240–253. https://doi.org/10.1089/ten.teb.2017.0425

Bobylev D, Sarikouch S, Tudorache I et al (2019) Double semilunar valve replacement in complex congenital heart disease using decellularized homografts. Interact Cardiovasc Thorac Surg 28:151–157. https://doi.org/10.1093/icvts/ivy212

Bockeria L, Svanidze O, Kim A et al (2017) Total cavopulmonary connection with a new bioabsorbable vascular graft: first clinical experience. J Thorac Cardiovasc Surg 153:1542–1550

Boethig D, Horke A, Hazekamp M et al (2019) A European study on decellularized homografts for pulmonary valve replacement: initial results from the prospective ESPOIR Trial and ESPOIR Registry data†. Eur J Cardiothorac Surg 56:503–509. https://doi.org/10.1093/ejcts/ezz054

Bonetti A, Marchini M, Ortolani F (2019) Ectopic mineralization in heart valves: new insights from in vivo and in vitro procalcific models and promising perspectives on noncalcifiable bioengineered valves. J Thorac Dis 11:2126–2143. https://doi.org/10.21037/jtd.2019.04.78

Bonhoeffer P, Boudjemline Y, Saliba Z et al (2000) Percutaneous replacement of pulmonary valve in a right-ventricle to pulmonary-artery prosthetic conduit with valve dysfunction. Lancet 356:1403–1405. https://doi.org/10.1016/S0140-6736(00)02844-0

Borazjani I (2015) A review of fluid-structure interaction simulations of prosthetic heart valves. J Long-Term Eff Med Implants 25:75–93

Bottio T, Tarzia V, Dal Lin C et al (2010) The changing hydrodynamic performance of the decellularized intact porcine aortic root: considerations on in-vitro testing. J Heart Valve Dis 19:485–491

Breitenbach I, El-Essawi A, Pahari D et al (2014) Early failure of decellularized xenogenous pulmonary valve conduit (Matrix-P-Plus) for reconstruction of the right ventricular outflow tract in the ross procedure. Thorac Cardiovasc Surg 62. https://doi.org/10.1055/s-0034-1367197

Brown JW, Ruzmetov M, Eltayeb O et al (2011) Performance of synergraft decellularized pulmonary homograft in patients undergoing a ross procedure. Ann Thorac Surg 91:416–423. https://doi.org/10.1016/j.athoracsur.2010.10.069

Burch PT, Kaza AK, Lambert LM et al (2010) Clinical performance of decellularized cryopreserved valved allografts compared with standard allografts in the right ventricular outflow tract. Ann Thorac Surg 90:1301–1306. https://doi.org/10.1016/j.athoracsur.2010.05.024

Buse EE, Hilbert SL, Hopkins RA, Converse GL (2016) Pulse duplicator hydrodynamic testing of bioengineered biological heart valves. Cardiovasc Eng Technol 7:352–362. https://doi.org/10.1007/s13239-016-0275-9

Butcher JT, Nerem RM (2006) Valvular endothelial cells regulate the phenotype of interstitial cells in co-culture: effects of steady shear stress. Tissue Eng 12:905–915. https://doi.org/10.1089/ten.2006.12.905

Buzzatti N, De Bonis M, Moat N (2018) Anatomy of the tricuspid valve, pathophysiology of functional tricuspid regurgitation, and implications for percutaneous therapies. Interv Cardiol Clin 7:1–11

Capulli AK, Emmert MY, Pasqualini FS et al (2017) JetValve: rapid manufacturing of biohybrid scaffolds for biomimetic heart valve replacement. Biomaterials 133:229–241. https://doi.org/10.1016/j.biomaterials.2017.04.033

Cataloglu A, Clark RE, Gould PL (1977) Stress analysis of aortic valve leaflets with smoothed geometrical data. J Biomech 10:153–158

Cebotari S, Tudorache I, Ciubotaru A et al (2011) Use of fresh decellularized allografts for pulmonary valve replacement may reduce the reoperation rate in children and young adults: early report. Circulation 124:115–124. https://doi.org/10.1161/CIRCULATIONAHA.110.012161

Cesarovic N, Lipiski M, Falk V, Emmert MY (2020) Animals in cardiovascular research. Eur Heart J 41:200–203. https://doi.org/10.1093/eurheartj/ehz933

Chandran KB (2010) Role of computational simulations in heart valve dynamics and design of valvular prostheses. Cardiovasc Eng Technol 1:18–38. https://doi.org/10.1007/s13239-010-0002-x

Chatterjee A, Bhatia N, Torres MG et al (2019) Comparison of transcatheter pulmonic valve implantation with surgical pulmonic valve replacement in adults (from the National Inpatient Survey Dataset). Am J Cardiol. https://doi.org/10.1016/j.amjcard.2019.09.031

Chong A, Senior R, Wahi S (2019) Contemporary imaging of aortic stenosis. Heart Lung Circ 28:1310–1319. https://doi.org/10.1016/j.hlc.2019.05.177

Coffey S, Cairns BJ, Iung B (2016) The modern epidemiology of heart valve disease. Heart 102:75–85. https://doi.org/10.1136/heartjnl-2014-307020

Coté N, Mahmut A, Bosse Y et al (2013) Inflammation is associated with the remodeling of calcific aortic valve disease. Inflammation 36:573–581. https://doi.org/10.1007/s10753-012-9579-6

Coyan GN, D'Amore A, Matsumura Y et al (2019) In vivo functional assessment of a novel degradable metal and elastomeric scaffold-based tissue engineered heart valve. J Thorac Cardiovasc Surg 157:1809–1816. https://doi.org/10.1016/j.jtcvs.2018.09.128

Crapo PM, Gilbert TW, Badylak SF (2011) An overview of tissue and whole organ decellularization processes. Biomaterials 32:3233–3243. https://doi.org/10.1016/j.biomaterials.2011.01.057

Cribier A, Eltchaninoff H, Bash A et al (2002) Percutaneous transcatheter implantation of an aortic valve prosthesis for calcific aortic stenosis: first human case description. Circulation 106:3006–3008

Da Costa FDA, Costa ACBA, Prestes R et al (2010) The early and midterm function of decellularized aortic valve allografts. Ann Thorac Surg 90:1854–1860. https://doi.org/10.1016/j.athoracsur.2010.08.022

Daebritz SH, Fausten B, Hermanns B et al (2004) New flexible polymeric heart valve prostheses for the mitral and aortic positions. Heart Surg Forum 7:E525–E532. https://doi.org/10.1532/HSF98.20041083

Dal-Bianco JP, Levine RA (2013) Anatomy of the mitral valve apparatus. Role of 2D and 3D echocardiography. Cardiol Clin 31:151–164

David TE (2010) Reoperations after the ross procedure. Circulation 122:1139–1140. https://doi.org/10.1161/CIRCULATIONAHA.110.977991

de Weger A, van Tuijl S, Stijnen M et al (2010) Direct endoscopic visual assessment of a transcatheter aortic valve implantation and performance in the physioheart, an isolated working heart platform. Circulation 121:e261–e262. https://doi.org/10.1161/CIR.0b013e3181d9b879

Delmo Walter EM, de By TMMH, Meyer R, Hetzer R (2012) The future of heart valve banking and of homografts: perspective from the Deutsches Herzzentrum Berlin. HSR Proc Intensive Care Cardiovasc Anesth 4:97–108

Demer LL, Tintut Y (2008) Vascular calcification: pathobiology of a multifaceted disease. Circulation 117:2938–2948. https://doi.org/10.1161/CIRCULATIONAHA.107.743161

Dijkman PE, Driessen-Mol A, de Heer LM et al (2012a) Trans-apical versus surgical implantation of autologous ovine tissue-engineered heart valves. J Heart Valve Dis 21:670–678

Dijkman PE, Driessen-Mol A, Frese L et al (2012b) Decellularized homologous tissue-engineered heart valves as off-the-shelf alternatives to xeno- and homografts. Biomaterials 33:4545–4554. https://doi.org/10.1016/j.biomaterials.2012.03.015

Dijkman PEPE, Fioretta ESES, Frese L et al (2016) Heart valve replacements with regenerative capacity. Transfus Med Hemother 43:282–290. https://doi.org/10.1159/000448181

Dohmen P, da Costa F (2012) An experimental study of decellularized xenografts implanted into the aortic position with 4 months of follow up. J Clin Exp Cardiol s4. https://doi.org/10.4172/2155-9880.s4-004

Dohmen PM, Ozaki S, Nitsch R et al (2003) A tissue engineered heart valve implanted in a juvenile sheep model. Med Sci Monit 9:BR97–BR104

Dohmen PM, da Costa F, Lopes SV et al (2005) Results of a decellularized porcine heart valve implanted into the juvenile sheep model. Heart Surg Forum 8:100. https://doi.org/10.1532/HSF98.20041140

Dohmen PM, Lembcke A, Holinski S et al (2007) Mid-term clinical results using a tissue-engineered pulmonary valve to reconstruct the right ventricular outflow tract during the ross procedure. Ann Thorac Surg 84:729–736. https://doi.org/10.1016/j.athoracsur.2007.04.072

Driessen-Mol A, Emmert MY, Dijkman PE et al (2014) Transcatheter implantation of homologous "off-the-shelf" tissue-engineered heart valves with self-repair capacity: long-term functionality and rapid in vivo remodeling in sheep. J Am Coll Cardiol 63:1320–1329. https://doi.org/10.1016/j.jacc.2013.09.082

Durko AP, Osnabrugge RL, Van Mieghem NM et al (2018) Annual number of candidates for transcatheter aortic valve implantation per country: current estimates and future projections. Eur Heart J 39:2635–2642. https://doi.org/10.1093/eurheartj/ehy107

Ellis JT, Healy TM, Fontaine AA et al (1996) Velocity measurements and flow patterns within the hinge region of a Medtronic Parallel bileaflet mechanical valve with clear housing. J Heart Valve Dis 5:591–599

Emmert MY, Weber B, Behr L et al (2011) Transapical aortic implantation of autologous marrow stromal cell-based tissue-engineered heart valves. JACC Cardiovasc Interv 4:822–823. https://doi.org/10.1016/j.jcin.2011.02.020

Emmert MY, Weber B, Wolint P et al (2012) Stem cell-based transcatheter aortic valve implantation: first experiences in a pre-clinical model. JACC Cardiovasc Interv 5:874–883. https://doi.org/10.1016/j.jcin.2012.04.010

Emmert MY, Weber B, Behr L et al (2014) Transcatheter aortic valve implantation using anatomically oriented, marrow stromal cell-based, stented, tissue-engineered heart valves: technical considerations and implications for translational cell-based heart valve concepts. Eur J Cardiothorac Surg 45:61–68. https://doi.org/10.1093/ejcts/ezt243

Emmert MY, Schmitt BA, Loerakker S, et al (2018) Computational modeling guides tissue-engineered heart valve design for long-term in vivo performance in a translational sheep modelSci Transl Med 10. https://doi.org/10.1126/scitranslmed.aan4587

Fioretta ESES, Fledderus JO, Burakowska-Meise EAEA et al (2012) Polymer-based scaffold designs for in situ vascular tissue engineering: controlling recruitment and differentiation behavior of endothelial colony forming cells. Macromol Biosci 12. https://doi.org/10.1002/mabi.201100315

Fioretta ESES, Dijkman PEPE, Emmert MYMY, Hoerstrup SPSP (2017) The future of heart valve replacement: recent developments and translational challenges for heart valve tissue engineering. J Tissue Eng Regen Med. https://doi.org/10.1002/term.2326

Fioretta ES, Lintas V, Mallone A et al (2019a) Differential leaflet remodeling of bone marrow cell pre-seeded versus nonseeded bioresorbable transcatheter pulmonary valve replacements. JACC Basic Transl Sci. https://doi.org/10.1016/j.jacbts.2019.09.008

Fioretta ES, von Boehmer L, Motta SE et al (2019b) Cardiovascular tissue engineering: from basic science to clinical application. Exp Gerontol 117:1–12. https://doi.org/10.1016/j.exger.2018.03.022

Fishbein GA, Fishbein MC (2019) Pathology of the aortic valve: aortic valve stenosis/aortic regurgitation. Curr Cardiol Rep 21:81. https://doi.org/10.1007/s11886-019-1162-4

Flanagan TC, Sachweh JS, Frese J et al (2009) In vivo remodeling and structural characterization of fibrin-based tissue-engineered heart valves in the adult sheep model. Tissue Eng Part A 15:2965–2976. https://doi.org/10.1089/ten.tea.2009.0018

Funayama M, Furukoshi M, Moriwaki T, Nakayama Y (2015a) Development of an in vivo tissue-engineered valved conduit (type S biovalve) using a slitted mold. J Artif Organs 18:382–386. https://doi.org/10.1007/s10047-015-0856-7

Funayama M, Matsui Y, Tajikawa T et al (2015b) Successful implantation of autologous valved conduits with self-expanding stent (stent-biovalve) within the pulmonary artery in beagle dogs. J Vet Cardiol 17:54–61. https://doi.org/10.1016/j.jvc.2014.12.003

Funayama M, Takewa Y, Oie T et al (2015c) In situ observation and enhancement of leaflet tissue formation in bioprosthetic "biovalve". J Artif Organs 18:40–47. https://doi.org/10.1007/s10047-014-0793-x

Gallo M, Bonetti A, Poser H et al (2016) Decellularized aortic conduits: could their cryopreservation affect post-implantation outcomes? A morpho-functional study on porcine homografts. Heart Vessel 31:1862–1873. https://doi.org/10.1007/s00380-016-0839-5

Gallo M, Poser H, Bottio T et al (2017) The Vietnamese pig as a translational animal model to evaluate tissue engineered heart valves: promising early experience. Int J Artif Organs 40:142–149. https://doi.org/10.5301/ijao.5000568

Gasser S, von Stumm M, Sinning C et al (2019) Can we predict failure of mitral valve repair? J Clin Med 8:526. https://doi.org/10.3390/jcm8040526

Generali M, Dijkman PE, Hoerstrup SP (2014) Bioresorbable scaffolds for cardiovascular tissue engineering. EMJ Interv Cardiol 1:91–99

Ghista DN, Reul H (1977) Optimal prosthetic aortic leaflet valve: design parametric and longevity analyses: development of the Avcothane-51 leaflet valve based on the optimum design analysis. J Biomech 10:313–324

Girdauskas E, Petersen J, Sachweh J et al (2018) Aortic valve repair in adult congenital heart disease. Cardiovasc Diagn The 8:811–813

Gnyaneshwar R, Kumar RK, Balakrishnan KR (2002) Dynamic analysis of the aortic valve using a finite element model. Ann Thorac Surg 73:1122–1129

Goecke T, Theodoridis K, Tudorache I et al (2018) In vivo performance of freeze-dried decellularized pulmonary heart valve allo- and xenografts orthotopically implanted into juvenile sheep. Acta Biomater 68:41–52. https://doi.org/10.1016/j.actbio.2017.11.041

Gottlieb D, Kunal T, Emani S et al (2010) In vivo monitoring of function of autologous engineered pulmonary valve. J Thorac Cardiovasc Surg 139:723–731. https://doi.org/10.1016/j.jtcvs.2009.11.006

Gould ST, Srigunapalan S, Simmons CA, Anseth KS (2013) Hemodynamic and cellular response feedback in calcific aortic valve disease. Circ Res 113:186–197. https://doi.org/10.1161/CIRCRESAHA.112.300154

Grande-Allen KJ, Griffin BP, Ratliff NB et al (2003) Glycosaminoglycan profiles of myxomatous mitral leaflets and chordae parallel the severity of mechanical alterations. J Am Coll Cardiol 42:271–277. https://doi.org/10.1016/S0735-1097(03)00626-0

Grasso C, Capodanno D, Tamburino C, Ohno Y (2015) Current status and clinical development of transcatheter approaches for severe mitral regurgitation. Circ J 79:1164–1171

Harken DE (1989) Heart valves: ten commandments and still counting. Ann Thorac Surg 48:S18–S19. https://doi.org/10.1016/0003-4975(89)90623-1

Hawkins JA, Hillman ND, Lambert LM et al (2003) Immunogenicity of decellularized cryopreserved allografts in pediatric cardiac surgery: comparison with standard cryopreserved allografts. J Thorac Cardiovasc Surg 126:247–253. https://doi.org/10.1016/S0022-5223(03)00116-8

Hayashida K, Kanda K, Yaku H et al (2007) Development of an in vivo tissue-engineered, autologous heart valve (the biovalve): preparation of a prototype model. J Thorac Cardiovasc Surg 134:152–159. https://doi.org/10.1016/j.jtcvs.2007.01.087

Hayashida K, Kanda K, Oie T et al (2008) "In vivo tissue-engineered" valved conduit with designed molds and laser processed scaffold. J Cardiovasc Nurs 23:61–64. https://doi.org/10.1097/01.JCN.0000305053.50506.97

Head SJ, Çelik M, Kappetein AP (2017) Mechanical versus bioprosthetic aortic valve replacement. Eur Heart J 38:2183–2191. https://doi.org/10.1093/eurheartj/ehx141

Helder MRKK, Kouchoukos NT, Zehr K et al (2016) Late durability of decellularized allografts for aortic valve replacement: a word of caution. J Thorac Cardiovasc Surg 152:1197–1199. https://doi.org/10.1016/j.jtcvs.2016.03.050

Helder MRK, Stoyles NJ, Tefft BJ et al (2017) Xenoantigenicity of porcine decellularized valves. J Cardiothorac Surg:12. https://doi.org/10.1186/s13019-017-0621-5

Henaine R, Roubertie F, Vergnat M, Ninet J (2012) Valve replacement in children: a challenge for a whole life. Arch Cardiovasc Dis 105:517–528. https://doi.org/10.1016/j.acvd.2012.02.013

Hennessy RSRR, Go JL, Hennessy RSRR et al (2017) Recellularization of a novel off-the-shelf valve following xenogenic implantation into the right ventricular outflow tract. PLoS One 12:e0181614. https://doi.org/10.1371/journal.pone.0181614

Hibino N, Yi T, Duncan DR et al (2011) A critical role for macrophages in neovessel formation and the development of stenosis in tissue-engineered vascular grafts. FASEB J 25:4253–4263. https://doi.org/10.1096/fj.11-186585

Hilbert SL, Ferrans VJ, Tomita Y et al (1987) Evaluation of explanted polyurethane trileaflet cardiac valve prostheses. J Thorac Cardiovasc Surg 94:419–429

Hinton RB, Yutzey KE (2011) Heart valve structure and function in development and disease. Annu Rev Physiol 73:29–46. https://doi.org/10.1146/annurev-physiol-012110-142145

Hopkins RA, Bert AA, Hilbert SL et al (2013) Bioengineered human and allogeneic pulmonary valve conduits chronically implanted orthotopically in baboons: hemodynamic performance and immunologic consequences. J Thorac Cardiovasc Surg 145:1098–1107.e3. https://doi.org/10.1016/j.jtcvs.2012.06.024

Human P, Zilla P (2017) The neglected villain of bioprosthetic degeneration: inflammatory and immune processes. J Long-Term Eff Med Implants 27:159–180. https://doi.org/10.1615/JLongTermEffMedImplants.v27.i2-4.50

Huygens SA, Goossens LMA, Van Erkelens JA et al (2018) How much does a heart valve implantation cost and what are the health care costs afterwards? Open Heart 5:e000672. https://doi.org/10.1136/openhrt-2017-000672

Iop L, Bonetti A, Naso F et al (2014) Decellularized allogeneic heart valves demonstrate self-regeneration potential after a long-term preclinical evaluation. PLoS One:9. https://doi.org/10.1371/journal.pone.0099593

Iwai S, Torikai K, Coppin CM, Sawa Y (2007) Minimally immunogenic decellularized porcine valve provides in situ recellularization as a stentless bioprosthetic valve. J Artif Organs 10:29–35. https://doi.org/10.1007/s10047-006-0360-1

Jana S, Tefft BJ, Spoon DB, Simari RD (2014) Scaffolds for tissue engineering of cardiac valves. Acta Biomater 10:2877–2893. https://doi.org/10.1016/j.actbio.2014.03.014

Jana S, Tranquillo RT, Lerman A (2016) Cells for tissue engineering of cardiac valves. J Tissue Eng Regen Med 10:804–824. https://doi.org/10.1002/term.2010

Jin S, Yang Y, Oshinski J et al (2004) Flow patterns and wall shear stress distributions at atherosclerotic-prone sites in a human left coronary artery – an exploration using combined methods of CT and computational fluid dynamics. Conf Proc IEEE Eng Med Biol Soc 5:3789–3791. https://doi.org/10.1109/IEMBS.2004.1404062

Kelly SGD (2002) Computational fluid dynamics insights in the design of mechanical heart valves. Artif Organs 26:608–613. https://doi.org/10.1046/j.1525-1594.2002.07083.x

Kheradvar A, Groves EM, Dasi LP et al (2015) Emerging trends in heart valve engineering: part I. Solutions for future. Ann Biomed Eng 43:833–843. https://doi.org/10.1007/s10439-014-1209-z

Kheradvar A, Zareian R, Kawauchi S et al (2017) Animal models for heart valve research and development. Drug Discov Today Dis Model 24:55–62. https://doi.org/10.1016/j.ddmod.2018.04.001

Kirkton RD, Santiago-Maysonet M, Lawson JH et al (2019) Bioengineered human acellular vessels recellularize and evolve into living blood vessels after human implantation. Sci Transl Med 11:eaau6934. https://doi.org/10.1126/scitranslmed.aau6934

Kluin J, Talacua H, Smits AIPMAIPM et al (2017) In situ heart valve tissue engineering using a bioresorbable elastomeric implant – from material design to 12 months follow-up in sheep. Biomaterials 125:101–117. https://doi.org/10.1016/j.biomaterials.2017.02.007

Konertz W, Dohmen PM, Liu J et al (2005) Hemodynamic characteristics of the Matrix P decellularized xenograft for pulmonary valve replacement during the Ross operation. J Heart Valve Dis 14:78–81

Konertz W, Angeli E, Tarusinov G et al (2011) Right ventricular outflow tract reconstruction with decellularized porcine xenografts in patients with congenital heart disease. J Heart Valve Dis 20:341–347

Langer R, Vacanti JP (1993) Tissue engineering. Science 260:920–926. https://doi.org/10.1126/science.8493529

Lawson JH, Glickman MH, Ilzecki M et al (2016) Bioengineered human acellular vessels for dialysis access in patients with end-stage renal disease: two phase 2 single-arm trials. Lancet 387:2026–2034. https://doi.org/10.1016/S0140-6736(16)00557-2

Leal MTBC, Passos LSA, Guarçoni FV et al (2019) Rheumatic heart disease in the modern era: recent developments and current challenges. Rev Soc Bras Med Trop 52:e20180041. https://doi.org/10.1590/0037-8682-0041-2019

Lee Y-U, Yi T, Tara S et al (2014) Implantation of inferior vena cava interposition graft in mouse model. J Vis Exp. https://doi.org/10.3791/51632

Leon MB, Smith CR, Mack MJ et al (2016) Transcatheter or surgical aortic-valve replacement in intermediate-risk patients. N Engl J Med 374:1609–1620. https://doi.org/10.1056/NEJMoa1514616

Leopaldi AM, Vismara R, Lemma M et al (2012) In vitro hemodynamics and valve imaging in passive beating hearts. J Biomech 45:1133–1139. https://doi.org/10.1016/j.jbiomech.2012.02.007

Leopaldi AM, Vismara R, van Tuijl S et al (2015) A novel passive left heart platform for device testing and research. Med Eng Phys 37:361–366. https://doi.org/10.1016/j.medengphy.2015.01.013

Leopaldi AM, Wrobel K, Speziali G et al (2018) The dynamic cardiac biosimulator: a method for training physicians in beating-heart mitral valve repair procedures. J Thorac Cardiovasc Surg 155:147–155. https://doi.org/10.1016/j.jtcvs.2017.09.011

Leyh RG, Wilhelmi M, Walles T et al (2003) Acellularized porcine heart valve scaffolds for heart valve tissue engineering and the risk of cross-species transmission of porcine endogenous retrovirus. J Thorac Cardiovasc Surg 126:1000–1004. https://doi.org/10.1016/s0022-5223(03)00353-2

Liang L, Sun B (2019) A proof of concept study of using machine-learning in artificial aortic valve design: from leaflet design to stress analysis. Bioengineering 6:104. https://doi.org/10.3390/bioengineering6040104

Lincoln J, Lange AW, Yutzey KE (2006) Hearts and bones: shared regulatory mechanisms in heart valve, cartilage, tendon, and bone development. Dev Biol 294:292–302

Lintas V, Fioretta ES, Motta SE et al (2018) Development of a novel human cell-derived tissue-engineered heart valve for transcatheter aortic valve replacement: an in vitro and in vivo feasibility study. J Cardiovasc Transl Res. https://doi.org/10.1007/s12265-018-9821-1

Lisy M, Kalender G, Schenke-Layland K et al (2017) Allograft heart valves: current aspects and future applications. Biopreserv Biobanking 15:148–157. https://doi.org/10.1089/bio.2016.0070

Liu AC, Joag VR, Gotlieb AI (2007) The emerging role of valve interstitial cell phenotypes in regulating heart valve pathobiology. Am J Pathol 171:1407–1418. https://doi.org/10.2353/ajpath.2007.070251

Loerakker S, Ristori T, Baaijens FPTT (2016) A computational analysis of cell-mediated compaction and collagen remodeling in tissue-engineered heart valves. J Mech Behav Biomed Mater 58:173–187. https://doi.org/10.1016/j.jmbbm.2015.10.001

Mack MJ, Leon MB, Thourani VH et al (2019) Transcatheter aortic-valve replacement with a balloon-expandable valve in low-risk patients. N Engl J Med 380:1695–1705. https://doi.org/10.1056/NEJMoa1814052

Mackay TG, Wheatley DJ, Bernacca GM et al (1996) New polyurethane heart valve prosthesis: design, manufacture and evaluation. Biomaterials 17:1857–1863. https://doi.org/10.1016/0142-9612(95)00242-1

Manji RA, Lee W, Cooper DKC (2015) Xenograft bioprosthetic heart valves: past, present and future. Int J Surg 23:280–284. https://doi.org/10.1016/j.ijsu.2015.07.009

Marelli AJ, Mackie AS, Ionescu-Ittu R et al (2007) Congenital heart disease in the general population: changing prevalence and age distribution. Circulation 115:163–172. https://doi.org/10.1161/CIRCULATIONAHA.106.627224

Mathieu P, Bouchareb R, Boulanger M-C (2015) Innate and adaptive immunity in calcific aortic valve disease. J Immunol Res 2015:851945. https://doi.org/10.1155/2015/851945

Maurer G (2006) Aortic regurgitation. Heart 92:994. https://doi.org/10.1136/HRT.2004.042614

Mayer JE, Shin'oka T, Shum-Tim D (1997) Tissue engineering of cardiovascular structures. Curr Opin Cardiol 12:528–532. https://doi.org/10.1097/00001573-199711000-00005

McCarthy KP, Ring L, Rana BS (2010) Anatomy of the mitral valve: understanding the mitral valve complex in mitral regurgitation. Eur J Echocardiogr 11:i3–i9

Meijer FMM, Kies P, Jongbloed MRM et al (2019) Excellent durability of homografts in pulmonary position analysed in a predefined adult group with tetralogy of Fallot†. Interact Cardiovasc Thorac Surg 28:279–283. https://doi.org/10.1093/icvts/ivy242

Mendirichaga R, Singh V, Blumer V et al (2017) Transcatheter mitral valve repair with mitraclip for symptomatic functional mitral valve regurgitation. Am J Cardiol 120:708–715

Merryman WD, Youn I, Lukoff HD et al (2006) Correlation between heart valve interstitial cell stiffness and transvalvular pressure: implications for collagen biosynthesis. Am J Physiol Heart Circ Physiol 290. https://doi.org/10.1152/ajpheart.00521.2005

Miller DV, Edwards WD, Zehr KJ (2006) Endothelial and smooth muscle cell populations in a decellularized cryopreserved aortic homograft (SynerGraft) 2 years after implantation. J Thorac Cardiovasc Surg 132:175–176. https://doi.org/10.1016/j.jtcvs.2006.02.038

Millington-Sanders C, Meir A, Lawrence L, Stolinski C (1998) Structure of chordae tendineae in the left ventricle of the human heart. J Anat 192:573–581. https://doi.org/10.1046/j.1469-7580.1998.19240573.x

Misfeld M, Sievers HH (2007) Heart valve macro- and microstructure. Philos Trans R Soc Lond B Biol Sci 362:1421–1436

Miyazaki Y, Soliman O, Abdelghani M et al (2017) Acute performance of a novel restorative transcatheter aortic valve: preclinical results. EuroIntervention 13:e1410–e1417. https://doi.org/10.4244/EIJ-D-17-00554

Mohty D, Pibarot P, Després J-P et al (2008) Association between plasma LDL particle size, valvular accumulation of oxidized LDL, and inflammation in patients with aortic stenosis. Arterioscler Thromb Vasc Biol 28:187–193. https://doi.org/10.1161/ATVBAHA.107.154989

Mol A, Smits AIPM, Bouten CVC, Baaijens FPT (2009) Tissue engineering of heart valves: advances and current challenges. Expert Rev Med Devices 6:259–275. https://doi.org/10.1586/erd.09.12

Morris PD, Narracott A, von Tengg-Kobligk H et al (2016) Computational fluid dynamics modelling in cardiovascular medicine. Heart 102:18–28. https://doi.org/10.1136/heartjnl-2015-308044

Motta SE, Fioretta ES, Dijkman PE et al (2018) Development of an off-the-shelf tissue-engineered sinus valve for transcatheter pulmonary valve replacement: a proof-of-concept study. J Cardiovasc Transl Res 11:182–191. https://doi.org/10.1007/s12265-018-9800-6

Motta SE, Lintas V, Fioretta ES et al (2019) Human cell-derived tissue-engineered heart valve with integrated Valsalva sinuses: towards native-like transcatheter pulmonary valve replacements. npj Regen Med 4:14. https://doi.org/10.1038/s41536-019-0077-4

Muylaert DEP, van Almen GC, Talacua H et al (2016) Early in-situ cellularization of a supramolecular vascular graft is modified by synthetic stromal cell-derived factor-1α derived peptides. Biomaterials 76:187–195. https://doi.org/10.1016/j.biomaterials.2015.10.052

Nakayama Y, Takewa Y, Sumikura H et al (2015) In-body tissue-engineered aortic valve (Biovalve type VII) architecture based on 3D printer molding. J Biomed Mater Res B Appl Biomater 103:1–11. https://doi.org/10.1002/jbm.b.33186

Niklason LE, Gao J, Abbott WM et al (1999) Functional arteries grown in vitro. Science 284:489–493

Nishimura RA, Warnes CA (2015) Anticoagulation during pregnancy in women with prosthetic valves: evidence, guidelines and unanswered questions: Table 1. Heart 101:430–435. https://doi.org/10.1136/heartjnl-2014-306500

Nistal F, Garcia-Martinez V, Arbe E et al (1990) In vivo experimental assessment of polytetrafluoroethylene trileaflet heart valve prosthesis. J Thorac Cardiovasc Surg 99(6):1074–1081

O'Brien MF, Goldstein S, Walsh S et al (1999) The SynerGraft valve: a new acellular (nonglutaraldehyde-fixed) tissue heart valve for autologous recellularization first experimental studies before clinical implantation. Semin Thorac Cardiovasc Surg 11:194–200

Ota T, Sawa Y, Iwai S et al (2005) Fibronectin-hepatocyte growth factor enhances reendothelialization in tissue-engineered heart valve. Ann Thorac Surg 80:1794–1801. https://doi.org/10.1016/j.athoracsur.2005.05.002

Parikh KN, Shah NC, Clark JB, Myers JL (2017) Pulmonary valve restitution following transannular patch repair of tetralogy of fallot. Interact Cardiovasc Thorac Surg 25:985–986. https://doi.org/10.1093/icvts/ivx215

Patterson RE, Kirk ES (1983) Analysis of coronary collateral structure, function, and ischemic border zones in pigs. Am J Phys 244:H23–H31. https://doi.org/10.1152/ajpheart.1983.244.1.H23

Perri G, Polito A, Esposito C et al (2012) Early and late failure of tissue-engineered pulmonary valve conduits used for right ventricular outflow tract reconstruction in patients with congenital heart disease. Eur J Cardiothorac Surg 41:1320–1325. https://doi.org/10.1093/ejcts/ezr221

Piatti F, Sturla F, Marom G et al (2015) Hemodynamic and thrombogenic analysis of a trileaflet polymeric valve using a fluid-structure interaction approach. J Biomech 48:3641–3649. https://doi.org/10.1016/j.jbiomech.2015.08.009

Pohl M, Wendt MO, Werner S et al (1996) In vitro testing of artificial heart valves: comparison between Newtonian and non-Newtonian fluids. Artif Organs 20:37–46

Popma JJ, Deeb GM, Yakubov SJ et al (2019) Transcatheter aortic-valve replacement with a self-expanding valve in low-risk patients. N Engl J Med. https://doi.org/10.1056/NEJMoa1816885

Quinn RW, Bert AA, Converse GL et al (2016) Performance of allogeneic bioengineered replacement pulmonary valves in rapidly growing young lambs. J Thorac Cardiovasc Surg 152:1156–1165.e4. https://doi.org/10.1016/j.jtcvs.2016.05.051

Rabkin-Aikawa E, Mayer JE, Schoen FJ (2005) Heart valve regeneration. Adv Biochem Eng Biotechnol 94:141–179

Rajamannan NM, Evans FJ, Aikawa E et al (2011) Calcific aortic valve disease: not simply a degenerative process: a review and agenda for research from the national heart and lung and blood institute aortic stenosis working group. Circulation 124:1783–1791. https://doi.org/10.1161/CIRCULATIONAHA.110.006767

Ram-Liebig G, Bednarz J, Stuerzebecher B et al (2015) Regulatory challenges for autologous tissue engineered products on their way from bench to bedside in Europe. Adv Drug Deliv Rev 82:181–191. https://doi.org/10.1016/j.addr.2014.11.009

Rashid ST, Salacinski HJ, Hamilton G, Seifalian AM (2004) The use of animal models in developing the discipline of cardiovascular tissue engineering: a review. Biomaterials 25:1627–1637

Reardon MJ, Van Mieghem NM, Popma JJ et al (2017) Surgical or transcatheter aortic-valve replacement in intermediate-risk patients. N Engl J Med 376:1321–1331. https://doi.org/10.1056/NEJMoa1700456

Rehman I, Rehman A (2018) Anatomy. Thorax, Heart

Reimer JM, Syedain ZH, Haynie BHT, Tranquillo RT (2015) Pediatric tubular pulmonary heart valve from decellularized engineered tissue tubes. Biomaterials 62:88–94. https://doi.org/10.1016/j.biomaterials.2015.05.009

Reimer J, Syedain Z, Haynie B et al (2017) Implantation of a tissue-engineered tubular heart valve in growing lambs. Ann Biomed Eng 45:439–451. https://doi.org/10.1007/s10439-016-1605-7

Retta SM, Kepner J, Marquez S et al (2017) In-vitro pulsatile flow measurement in prosthetic heart valves: an inter-laboratory comparison. J Heart Valve Dis 26:72–80

Roberts WC, Ko JM (2005) Frequency by decades of unicuspid, bicuspid, and tricuspid aortic valves in adults having isolated aortic valve replacement for aortic stenosis, with or without associated aortic regurgitation. Circulation 111:920–925. https://doi.org/10.1161/01.CIR.0000155623.48408.C5

Rodés-Cabau J, Webb JG, Cheung A et al (2012) Long-term outcomes after transcatheter aortic valve implantation. J Am Coll Cardiol 60:1864–1875. https://doi.org/10.1016/j.jacc.2012.08.960

Roe BB (1969) Late follow-up studies on flexible leaflet prosthetic valves. J Thorac Cardiovasc Surg 58:59–61

Roh JD, Sawh-Martinez R, Brennan MP et al (2010) Tissue-engineered vascular grafts transform into mature blood vessels via an inflammation-mediated process of vascular remodeling. Proc Natl Acad Sci U S A 107:4669–4674. https://doi.org/10.1073/pnas.0911465107

Ross DN (1962) Homograft replacement of the aortic valve. Lancet 280:487. https://doi.org/10.1016/S0140-6736(62)90345-8

Rossi F, Van Griensven M (2014) Polymer functionalization as a powerful tool to improve scaffold performances. Tissue Eng A 20:2043–2051

Rüffer A, Purbojo A, Cicha I et al (2010) Early failure of xenogenous de-cellularised pulmonary valve conduits – a word of caution! Eur J Cardiothorac Surg 38:78–85. https://doi.org/10.1016/j.ejcts.2010.01.044

Russo M, Taramasso M, Guidotti A et al (2017) The evolution of surgical valves. Cardiovasc Med 20:285–292. https://doi.org/10.4414/cvm.2017.00532

Rutkovskiy A, Malashicheva A, Sullivan G et al (2017) Valve interstitial cells: the key to understanding the pathophysiology of heart valve calcification. J Am Heart Assoc 6:e006339.

https://doi.org/10.1161/JAHA.117.006339 (available here: https://www.ncbi.nlm.nih.gov/pmc/articles/PMC5634284/)

Sabbah HN, Hamid MS, Stein PD (1985) Estimation of mechanical stresses on closed cusps of porcine bioprosthetic valves: effects of stiffening, focal calcium and focal thinning. Am J Cardiol 55:1091–1096

Sacks MS, David Merryman W, Schmidt DE (2009) On the biomechanics of heart valve function. J Biomech 42:1804–1824

Said SM, Mainwaring RD, Ma M et al (2016) Pulmonary valve repair for patients with acquired pulmonary valve insufficiency. Ann Thorac Surg 101:2294–2301. https://doi.org/10.1016/j.athoracsur.2016.01.035

Sanders B, Loerakker S, Fioretta ESES et al (2016) Improved geometry of decellularized tissue engineered heart valves to prevent leaflet retraction. Ann Biomed Eng 44:1061–1071. https://doi.org/10.1007/s10439-015-1386-4

Sarikouch S, Horke A, Tudorache I et al (2016) Decellularized fresh homografts for pulmonary valve replacement: a decade of clinical experience. Eur J Cardiothorac Surg 50:281–290. https://doi.org/10.1093/ejcts/ezw050

Scherman J, van Breda B, Appa H et al (2018) Transcatheter valve with a hollow balloon for aortic valve insufficiency. Multimed Man Cardiothorac Surg 2018. https://doi.org/10.1510/mmcts.2018.012

Schmidt D, Dijkman PE, Driessen-Mol A et al (2010) Minimally-invasive implantation of living tissue engineered heart valves. J Am Coll Cardiol 56:510–520. https://doi.org/10.1016/j.jacc.2010.04.024

Schmitt B, Spriestersbach H, H-Icí DO et al (2016) Percutaneous pulmonary valve replacement using completely tissue-engineered off-the-shelf heart valves: six-month in vivo functionality and matrix remodelling in sheep. EuroIntervention 12:62–70. https://doi.org/10.4244/EIJV12I1A12

Schoen FJ (2008) Evolving concepts of cardiac valve dynamics: the continuum of development, functional structure, pathobiology, and tissue engineering. Circulation 118:1864–1880

Schoen FJ (2012) Mechanisms of function and disease of natural and replacement heart valves. Annu Rev Pathol Mech Dis 7:161–183. https://doi.org/10.1146/annurev-pathol-011110-130257

Schoen FJ (2018) Morphology, clinicopathologic correlations, and mechanisms in heart valve health and disease. Cardiovasc Eng Technol 9:126–140. https://doi.org/10.1007/s13239-016-0277-7

Schoen FJ, Gotlieb AI (2016) Heart valve health, disease, replacement, and repair: a 25-year cardiovascular pathology perspective. Cardiovasc Pathol 25:341–352. https://doi.org/10.1016/j.carpath.2016.05.002

Shah M, Jorde UP (2019) Percutaneous mitral valve interventions (repair): current indications and future perspectives. Front Cardiovasc Med 6:88. https://doi.org/10.3389/fcvm.2019.00088

Shinoka T, Breuer CK, Tanel RE et al (1995) Tissue engineering heart valves: valve leaflet replacement study in a lamb model. Ann Thorac Surg 60:S513–S516. https://doi.org/10.1016/0003-4975(95)00733-4

Simon P, Kasimir MT, Seebacher G et al (2003) Early failure of the tissue engineered porcine heart valve SYNERGRAFT in pediatric patients. Eur J Cardiothorac Surg 23:1002–1006; discussion 1006

Simon HA, Ge L, Sotiropoulos F, Yoganathan AP (2010) Simulation of the three-dimensional hinge flow fields of a bileaflet mechanical heart valve under aortic conditions. Ann Biomed Eng 38:841–853. https://doi.org/10.1007/s10439-009-9857-0

Soares ALF, Oomens CWJ, Baaijens FPT (2014a) A computational model to describe the collagen orientation in statically cultured engineered tissues. Comput Methods Biomech Biomed Engin 17:251–262. https://doi.org/10.1080/10255842.2012.680192

Soares ALF, van Geemen D, van den Bogaerdt AJ et al (2014b) Mechanics of the pulmonary valve in the aortic position. J Mech Behav Biomed Mater 29:557–567. https://doi.org/10.1016/j.jmbbm.2013.07.009

Soliman OI, Miyazaki Y, Abdelghani M et al (2017) Midterm performance of a novel restorative pulmonary valved conduit: preclinical results. EuroIntervention 13:e1418–e1427. https://doi.org/10.4244/EIJ-D-17-00553

Sripathi VC, Kumar RK, Balakrishnan KR (2004) Further insights into normal aortic valve function: role of a compliant aortic root on leaflet opening and valve orifice area. Ann Thorac Surg 77:844–851. https://doi.org/10.1016/S0003-4975(03)01518-2

Sumikura H, Nakayama Y, Ohnuma K et al (2015) In vitro hydrodynamic evaluation of a biovalve with stent (tubular leaflet type) for transcatheter pulmonary valve implantation. J Artif Organs 18:307–314. https://doi.org/10.1007/s10047-015-0851-z

Suzuki Y, Yeung AC, Ikeno F (2011) The representative porcine model for human cardiovascular disease. J Biomed Biotechnol 2011:1–10. https://doi.org/10.1155/2011/195483

Syedain ZH, Meier LA, Lahti MT et al (2014) Implantation of completely biological engineered grafts following decellularization into the sheep femoral artery. Tissue Eng Part A 20:1726–1734. https://doi.org/10.1089/ten.tea.2013.0550

Syedain Z, Reimer J, Schmidt J et al (2015) 6-Month aortic valve implantation of an off-the-shelf tissue-engineered valve in sheep. Biomaterials 73:175–184. https://doi.org/10.1016/j.biomaterials.2015.09.016

Syedain Z, Reimer J, Lahti M et al (2016) Tissue engineering of acellular vascular grafts capable of somatic growth in young lambs. Nat Commun 7:1–9. https://doi.org/10.1038/ncomms12951

Tabata N, Sugiura A, Tsujita K et al (2019) Percutaneous interventions for mitral and tricuspid heart valve diseases. Cardiovasc Interv Ther. https://doi.org/10.1007/s12928-019-00610-z

Takewa Y, Yamanami M, Kishimoto Y et al (2013) In vivo evaluation of an in-body, tissue-engineered, completely autologous valved conduit (biovalve type VI) as an aortic valve in a goat model. J Artif Organs 16:176–184. https://doi.org/10.1007/s10047-012-0679-8

Tao G, Kotick JD, Lincoln J (2012) Heart valve development, maintenance, and disease. The role of endothelial cells. In: Current topics in developmental biology. Academic Press, London, pp 203–232

Taramasso M, Emmert MY, Reser D et al (2015) Pre-clinical in vitro and in vivo models for heart valve therapies. J Cardiovasc Transl Res 8:319–327. https://doi.org/10.1007/s12265-015-9631-7

Terazawa T, Furukoshi M, Nakayama Y (2019) One-year follow-up study of iBTA-induced allogenic biosheet for repair of abdominal wall defects in a beagle model: a pilot study. Hernia 23:149–155. https://doi.org/10.1007/s10029-018-1866-1

Toninato R, Salmon J, Susin FM et al (2016) Physiological vortices in the sinuses of Valsalva: an in vitro approach for bio-prosthetic valves. J Biomech 49:2635–2643. https://doi.org/10.1016/j.jbiomech.2016.05.027

Tudorache I, Horke A, Cebotari S et al (2016) Decellularized aortic homografts for aortic valve and aorta ascendens replacement. Eur J Cardiothorac Surg 50:89–97. https://doi.org/10.1093/ejcts/ezw013

Underwood MJ, El Khoury G, Deronck D et al (2000) The aortic root: structure, function, and surgical reconstruction. Heart 83:376–380

Vesely I, Boughner D, Song T (1988) Tissue buckling as a mechanism of bioprosthetic valve failure. Ann Thorac Surg 46:302–308. https://doi.org/10.1016/S0003-4975(10)65930-9

Voges I, Bräsen JH, Entenmann A et al (2013) Adverse results of a decellularized tissue-engineered pulmonary valve in humans assessed with magnetic resonance imaging. Eur J Cardiothorac Surg 44:272–279. https://doi.org/10.1093/ejcts/ezt328

Weber B, Scherman J, Emmert MY et al (2011) Injectable living marrow stromal cell-based autologous tissue engineered heart valves: first experiences with a one-step intervention in primates. Eur Heart J. https://doi.org/10.1093/eurheartj/ehr059

Weber B, Dijkman PE, Scherman J et al (2013) Off-the-shelf human decellularized tissue-engineered heart valves in a non-human primate model. Biomaterials 34:7269–7280. https://doi.org/10.1016/j.biomaterials.2013.04.059

Wiegerinck EMA, Van Kesteren F, Van Mourik MS et al (2016) An up-to-date overview of the most recent transcatheter implantable aortic valve prostheses. Expert Rev Med Devices 13:31–45. https://doi.org/10.1586/17434440.2016.1120665

Wissing TB, Bonito V, Bouten CVC, Smits AIPM (2017) Biomaterial-driven in situ cardiovascular tissue engineering – a multi-disciplinary perspective. npj Regen Med 2:18. https://doi.org/10.1038/s41536-017-0023-2

Woo JS, Fishbein MC, Reemtsen B (2016) Histologic examination of decellularized porcine intestinal submucosa extracellular matrix (CorMatrix) in pediatric congenital heart surgery. Cardiovasc Pathol 25:12–17. https://doi.org/10.1016/j.carpath.2015.08.007

Yacoub MH, Cohn LH (2004) Novel approaches to cardiac valve repair: from structure to function: part II. Circulation 109:1064–1072

Yacoub MH, Takkenberg JJM (2005) Will heart valve tissue engineering change the world? Nat Clin Pract Cardiovasc Med 2:60–61. https://doi.org/10.1038/ncpcardio0112

Ye J, Cheung A, Lichtenstein SV et al (2006) Transapical aortic valve implantation in humans. J Thorac Cardiovasc Surg 131:1194–1196. https://doi.org/10.1016/j.jtcvs.2006.01.026

Yip CYY, Chen JH, Zhao R, Simmons CA (2009) Calcification by valve interstitial cells is regulated by the stiffness of the extracellular matrix. Arterioscler Thromb Vasc Biol 29:936–942. https://doi.org/10.1161/ATVBAHA.108.182394

Yoganathan AP, Chandran KB, Sotiropoulos F (2005) Flow in prosthetic heart valves: state-of-the-art and future directions. Ann Biomed Eng 33:1689–1694. https://doi.org/10.1007/s10439-005-8759-z

Zafar F, Hinton RB, Moore RA et al (2015) Physiological growth, remodeling potential, and preserved function of a novel bioprosthetic tricuspid valve: tubular bioprosthesis made of small intestinal submucosa-derived extracellular matrix. J Am Coll Cardiol 66:877–888. https://doi.org/10.1016/j.jacc.2015.06.1091

Zakaria MS, Ismail F, Tamagawa M et al (2017) Review of numerical methods for simulation of mechanical heart valves and the potential for blood clotting. Med Biol Eng Comput 55:1519–1548. https://doi.org/10.1007/s11517-017-1688-9

Zehr KJ, Yagubyan M, Connolly HM et al (2005) Aortic root replacement with a novel decellularized cryopreserved aortic homograft: postoperative immunoreactivity and early results. J Thorac Cardiovasc Surg 130:1010–1015. https://doi.org/10.1016/j.jtcvs.2005.03.044

Cell Sheets for Cardiac Tissue Engineering

Hidekazu Sekine, Jun Homma, and Tatsuya Shimizu

Contents

Abstract

Recent studies have reported that the injection of isolated cells can improve cardiac function in models of myocardial infarction. However, the loss of transplanted cells from the target site due to local hypoxia and cell washout remains a major problem. To overcome these limitations, we have developed cell sheet-based tissue engineering that allows the generation of confluent cultured cells, stacked cell sheets, and three-dimensional (3D) cell-dense tissues. Cell sheet-based patches can improve the function of damaged hearts in animal models. Stacked cardiac cell sheets beat synchronously both in vitro and in vivo and have the characteristic structure of native heart tissue. Upscaling of this technology through multistep transplantation of triple-layered cell sheets allowed the construction of functional cardiac tissue about 1 mm thick. Furthermore, we

H. Sekine (✉) · J. Homma · T. Shimizu
Institute of Advanced Biomedical Engineering and Science, Tokyo Women's Medical University, Tokyo, Japan
e-mail: sekine.hidekazu@twmu.ac.jp; homma.jun@twmu.ac.jp; shimizu.tatsuya@twmu.ac.jp

D. Eberli et al. (eds.), *Organ Tissue Engineering*, Reference Series in Biomedical Engineering, https://doi.org/10.1007/978-3-030-44211-8_3

succeeded in bioengineering 3D cardiac tissue containing a vascular network by in vitro perfusion culture of cell sheets stacked sequentially on a vascular bed obtained from resected tissue. Since the vascular bed was excised with its artery and vein intact, the bioengineered tissue could be transplanted by anastomosis of its vessels with those of the host animal. Following the creation of cardiac patches for direct implantation onto a damaged heart, the next challenge will be to engineer organs with tubular or spherical structures that can function as pumps to provide circulatory support. The goal for the future is to develop this technology to create functional organ-like tissues with vascular networks that can be used in patients as an alternative to conventional organ transplantation.

1 Introduction

Regenerative medicine is receiving great attention as a potential new therapy for illnesses that cannot be completely cured by pharmacological or surgical methods. Regenerative medicine also is a promising technique for creating transplantable tissues that could substitute for conventional organ transplants in the future. To date, the clinical application of regenerative medicine has mainly involved the injection of autologous or allogeneic cell suspensions in order to regenerate defective tissue. However, ongoing research is further developing bioengineering techniques with the aim of creating transplantable tissues using biodegradable scaffolds made from synthetic or natural polymers. Tissue engineering technology has already been applied clinically to tissues with a low cell density and low vascular requirement such as bone, cartilage, and skin.

Conventional tissue engineering techniques are limited with regard to the thickness and function of the tissue that can be constructed because they rely on diffusion for the supply of oxygen and nutrients and the removal of waste products. The construction of complex and cell-dense tissues such as the heart, liver, and kidney will require innovative technologies to realize a functional vascular network within the engineered tissue.

In this chapter, we introduce three related technologies that have the potential to be developed into novel regenerative therapies: cardiac cell sheet-based tissue engineering for myocardial regeneration, creation of three-dimensional tissue with a functional vascular network, and fabrication of a bioartificial pump for circulatory support.

2 Myocardial Regenerative Therapy

Research on myocardial cell transplantation was initiated by Soonpaa et al. in the early 1990s, who demonstrated that mouse fetal cardiomyocytes successfully engrafted onto host myocardium after transplantation (Soonpaa et al. 1994). Subsequently it was reported that the transplantation of suspensions of various cell types, including cardiomyocytes, was able to help restore cardiac function (Laflamme and Murry 2005). Skeletal muscle myoblasts are considered relatively resistant to ischemia and have been used in place of cardiac cells for myocardial regeneration. The MAGIC II trial in 2003 reported that cardiac function was improved by the injection of myoblasts collected from the patient's own skeletal muscle in combination with coronary artery bypass surgery (Menasche et al. 2003, 2008). However, because there were some cases of arrhythmia and death, the combined use of antiarrhythmic drugs and implantable defibrillators was considered essential when this technique was utilized.

In recent years, embryonic stem (ES) cells and induced pluripotent stem (iPS) cells have been actively studied as a source of cells with a high ability to proliferate and differentiate into cardiomyocytes. A method for inducing the differentiation of human stem cells into cardiomyocytes has also been established. Differentiated cardiomyocytes have been shown to engraft onto a host heart and recover cardiac function in animal models of heart failure (Caspi et al. 2007; Nelson et al. 2009; Shiba et al. 2016). Since iPS cells circumvent the moral and ethical issues associated with the use of ES cells, including the destruction of human embryos, it is anticipated that allogeneic transplantation based on cell banks will be realized in the future (Wilmut et al. 2015).

3 Scaffold-Based Engineering of Cardiac Tissue

The concept of tissue engineering was first proposed by Prof. Robert Langer at the Massachusetts Institute of Technology and Prof. Joseph Vacanti at Harvard University in the late 1980s. Tissue engineering is an interdisciplinary study born from the fusion of medicine and engineering that aims to reproduce the tissue structures of the body in vivo or in vitro. The generation of tissue requires an extracellular matrix (ECM) that acts a scaffold for the cells and the growth factors that promote cell differentiation and proliferation. In this method, cells are seeded on a three-dimensional biodegradable scaffold, cultured and then transplanted into the body. The scaffold is typically a biodegradable material composed of polyglycolic acid (PGA) or poly-L-lactic acid (PLLA) and its copolymer. Since the scaffold is slowly degraded and absorbed in the body and replaced with ECM produced by the cells, tissue structures similar to native tissues can be regenerated (Langer and Vacanti 1993). A major advantage of tissue engineering techniques is the ability to overcome the loss of cells due to the washout and necrosis that inevitably occur with cell infusion-based therapy. In addition, tissue engineering potentially allows the treatment of defective structures, such as those associated with congenital heart disease, which cannot be corrected by cell infusion- or cytokine-based therapy (Zandonella 2003). Gelatin, alginic acid, and PGA porous sponges have all been used as a scaffold for cell seeding (Fig. 1a) (Bursac et al. 1999; Li et al. 1999; Leor et al. 2000). In addition, three-dimensional myocardial tissue has been constructed by mixing a collagen solution with cardiac cells and culturing in a mold (Fig. 1b) (Zimmermann et al. 2006). Moreover, the imposition of a stretching load in vitro was

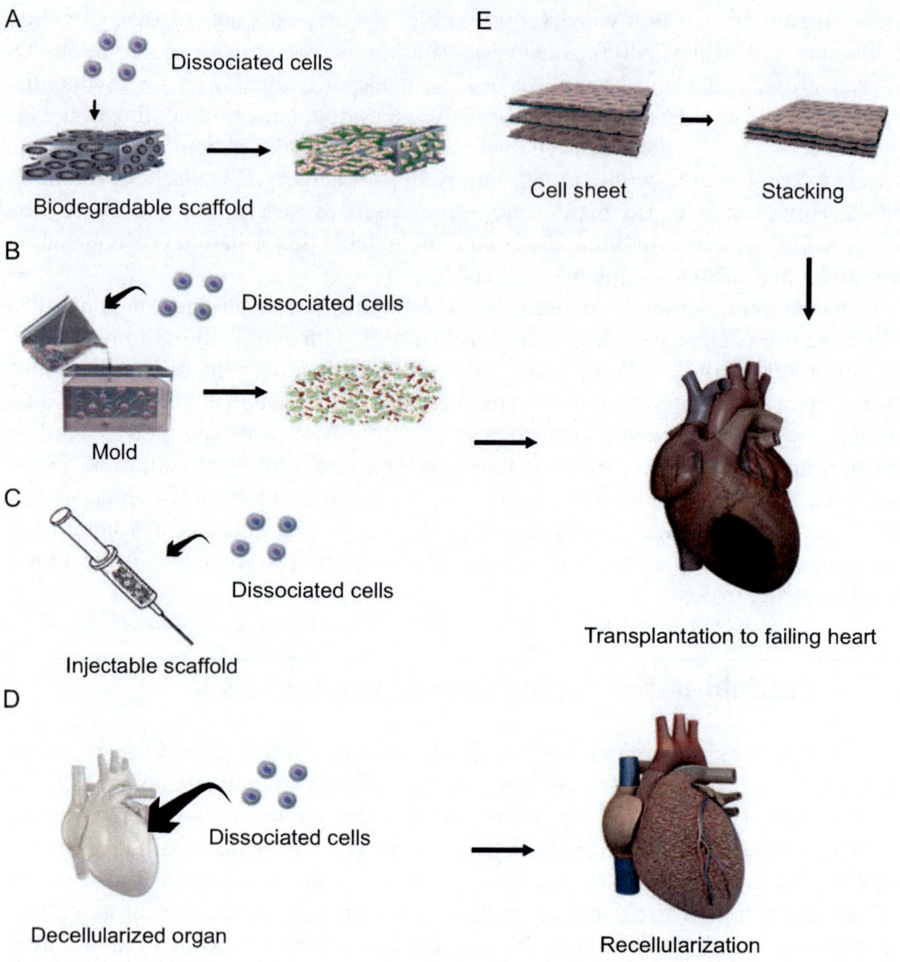

Fig. 1 Various type of tissue engineering approach for myocardial tissue reconstruction

shown to impart orientation to the myocardial tissue and enlarge the cardiomyocytes (Zimmermann et al. 2006). An intermediate approach between cell injection-based therapy and tissue engineering, which reduced the problem of cell loss, was also reported in which cells were mixed with fibrin glue or a collagen-containing solution and then injected into failing myocardium (Fig. 1c) (Christman and Lee 2006).

Various studies have investigated the transplantation of bioengineered cardiac tissue in animal models of heart failure. Li et al. created a graft by seeding rat fetal cardiac cells on a degradable gelatin mesh, and transplantation of this graft onto scar tissue in a cryoinjured rat heart resulted in an improvement in left ventricular systolic pressure when compared with the transplantation of gelatin mesh without cardiac cells (Li et al. 1999). Leor et al. seeded rat fetal cardiac cells onto a porous alginate scaffold and transplanted the resulting construct onto myocardial scar tissue in rats with experimental myocardial infarction. Although there was no improvement in left ventricular contractility, transplantation was found to suppress left ventricular dilatation due to myocardial remodeling (Leor et al. 2000). Zimmermann et al. generated three-dimensional myocardial tissue by mixing rat neonatal cardiomyocytes with collagen solution and culturing in a silicone mold. After transplantation into rats with experimentally induced myocardial infarction, the engineered heart tissue was observed to couple electrically with the host heart, improve left ventricular contractility, and inhibit left ventricular dilatation (Zimmermann et al. 2006).

4 Cell Sheet-Based Tissue Engineering for the Generation of Cardiac Tissue

One of the limitations of using a scaffold as a cellular foothold during tissue construction is that it can be difficult to seed a sufficient number of cells within the scaffold, resulting in a construct with a small cellular component and a large amount of connective tissue. Although scaffold-based techniques are suitable for the production of tissues that are sparsely populated by cells, such as heart valves and cartilage, alternative methods are needed to create cell-dense tissues with complex structures and functions such as the heart, kidney, and liver. Therefore, we developed an original technology known as cell sheet-based tissue engineering that permits the generation of tissue without the use of a scaffold. Cell sheet-based tissue engineering relies on the controlled use of temperature changes to regulate the attachment and detachment of cells from the surface of a culture dish. Poly(*N*-isopropylacrylamide) (PIPAAm) is a thermoresponsive polymer with a critical solution temperature of 32 °C in water. When this polymer is immobilized covalently by an electron beam, it becomes hydrophobic at 37 °C, which enables cells to adhere to it. However, the polymer surface becomes hydrophilic when the temperature is lowered to 32 °C, resulting in the detachment of cells from it (Yamada et al. 1990; Okano et al. 1993). Conventionally, a protease such as trypsin is used to harvest cultured cells, but this method degrades not only the proteins that adhere the cells to the surface of the culture dish but also other proteins on the cell membrane. Thus, a major advantage of using a temperature-responsive culture dish is that a simple lowering of the temperature allows cells to be recovered as a sheet without any disruption of cell-cell adhesion or the structure and function of the ECM (Fig. 2) (Kushida et al. 1999). Furthermore, three-dimensional tissues can be constructed by laminating multiple cell sheets (Fig. 1d). Since a tissue constructed by lamination consists only of cells and the small amount of ECM that they produce, this technique avoids the problems associated with the use of scaffolds (Yang et al. 2005).

When two neonatal rat cardiac cell sheets prepared using a temperature-responsive culture dish were stacked together in vitro, morphological and electrical connections were established between the pair of cardiac cell sheets within several tens of minutes. Furthermore, synchronously beating myocardial tissue was created successfully by the lamination of cardiac cell sheets (Shimizu et al. 2002; Haraguchi et al. 2006). When layered cardiac cell sheets were transplanted onto the dorsal

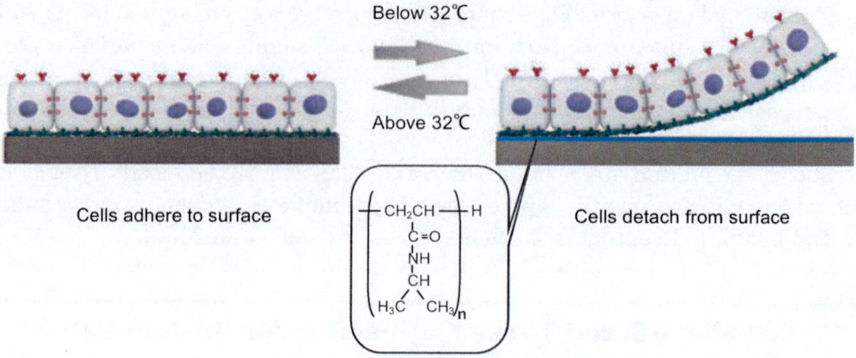

Fig. 2 Cell sheet technology using temperature-responsive culture dish

subcutaneous tissue of a rat, the transplanted myocardial graft exhibited pulsations visible to the naked eye and generated unique electrical potentials that were detected by electrography. Notably, the transplanted cardiac tissue was observed to contain newly formed capillary networks and structures characteristic of native myocardial tissue such as sarcomeres, gap junctions, and desmosomes (Shimizu et al. 2002). It was also shown that this transplanted myocardial tissue continued to engraft for a further 2 years while maintaining its spontaneous beating (Shimizu et al. 2006a).

5 Transplantation of Cardiac Patches onto Ischemic Hearts

When engineered myocardial tissue is transplanted onto an ischemic heart, it is essential that the graft connects electrically and synchronizes with the host heart to enable it to support cardiac function. To evaluate whether our bioengineered construct would possess this ability, we transplanted a stratified cardiac cell sheet onto the heart of an animal with ectopically produced myocardial infarction and evaluated whether electrical connections were formed between the host heart and graft. Morphological analysis demonstrated cell-cell junctions between the cardiomyocytes of the graft and those of the host non-infarcted tissue 1 week after graft transplantation. Immunohistochemistry experiments and transmission electron microscopy (TEM) confirmed the expression of connexin-43 and the presence of intervening plate between host cells and graft cells. In addition, it was shown that a low molecular weight fluorescent dye was able to migrate between connected cells. Immunohistochemistry also revealed the dedifferentiation of epicardial mesothelial cells that were sandwiched between the host and graft. Based on the above results, it was concluded that the transplantation of a multilayered cardiac cell sheet onto the ischemic region of a host heart resulted in the loss of mesothelial cell function and the formation of gap junctions between the cells of the graft and those of the non-infarcted myocardial tissue. This raised the possibility that the function of an ischemic heart could be improved by a graft beating synchronously with normal myocardium (Sekine et al. 2006a).

The implantation of various cell sheets has been shown to restore cardiac function in small animal models of severe heart failure. The main mechanisms underlying the improvement in cardiac function are thought to be the promotion of angiogenesis by various cytokines secreted from the transplanted cell sheets as well as the suppression of fibrosis, apoptosis, and left ventricular remodeling (Memon et al. 2005; Miyagawa et al. 2005, 2010; Hata et al. 2006; Kondoh et al. 2006; Miyahara et al. 2006). Nevertheless, echocardiography has demonstrated a reduction in left ventricular end systolic dimension and an increase in fractional shortening ratio after the transplantation of cardiac cell sheets, implying that the improvement in host heart function by cardiac cell sheets involves not only the effects of cytokines but also an enhancement of cardiomyocyte contractility (Miyagawa et al. 2005; Sekine et al. 2011). This latter effect is thought to be related to the formation of gap junctions between graft and host cells and the synchronization of contractile function between the transplanted cardiac cell sheet and host myocardium (Sekine et al. 2006a). When we evaluated our cardiac cell sheet in an animal model of cardiac injury, the survival of transplanted cells (evaluated by in vivo luminescence imaging and immunohistochemistry) and the improvement in host cardiac function (assessed by echocardiography) were significantly greater following the transplantation of cardiac cell sheets than after the injection of a cell suspension (Sekine et al. 2011). This suggests that the enhancement of cardiac performance is dependent on the survival of the transplanted cells. Treatments based on the injection of cell suspensions are limited by the loss of cells that occurs due to necrosis and outflow from the transplantation site. Zhang et al. utilized quantitative TUNEL analysis to determine the rate of necrosis and showed that most cardiomyocytes become necrotic several days after their transplantation as a cell suspension (Zhang et al. 2001). Studies have also examined the technical issues involved in the outflow of cells from the transplantation site. Terrovitis et al. found that only 17% of cardiac-derived stem cells were retained 1 h after their transplantation into a normal rat heart, but the engraftment rate was improved if the host heart was temporarily stopped while the cells were injected (Terrovitis et al. 2009). Hudson et al. performed intramyocardial injections of fluorescent microspheres rather than cells in a porcine cardiopulmonary bypass model and determined the retention rate to be only 10% irrespective of whether the heart was temporarily arrested (Hudson et al. 2007). The above findings indicate that cell transplantation by injection is associated with a low retention rate. Hofmann et al. carried out transcoronary grafting of a suspension of bone marrow-derived cells, and analysis by positron emission tomography revealed that only 1–3% of cells engrafted onto the heart, with many cells flowing to the liver and pancreas (Hofmann et al. 2005). Subsequently, Kutschka et al. demonstrated that the efflux of cells from the transplantation site was lower for cells mixed with collagen than for a cell suspension (Kutschka et al. 2007). This showed that it might be possible to reduce cell loss following the transplantation of a cell suspension. Nonetheless, cell survival is better for transplanted cell sheets than for injected cell suspensions

(Sekine et al. 2011), suggesting that the use of cell sheets can maximize the potential of transplanted cells.

The layering of cardiac cell sheets generates three-dimensional myocardial tissue with a high density of cells because the tissue consists only of cells and the ECM that they produce. An additional advantage of this construct is that it can be readily engrafted onto host tissue because ECM is present on the lower aspect of the cell sheet. For adherent cells like cardiomyocytes, adhesion to the ECM is an essential function for survival and proliferation. Consistent with this, many TUNEL-positive apoptotic cells are detected 24 h after the transplantation of a cell suspension. Apoptosis caused by defective cell adhesion is referred to as anoikis (Michel 2003). Adherent cells bind to a certain substrate via integrins, but when the adhesion is broken, the cells become suspended and initiate a signal to induce anoikis. The apoptotic signal is transmitted to the nucleus through several kinases (intracellular signaling molecules) such as focal adhesion kinase (FAK). Cells transplanted as a suspension are usually harvested from a culture dish using a protease to degrade proteins on the surface of the ECM or cell membrane. Therefore, the cells in a suspension are in a state in which their adhesion to the ECM has been disrupted. By contrast, a cell sheet used for transplantation is collected from a temperature-responsive culture dish as an intact sheet without any disruption of cell adhesion to the ECM. Since anoikis would be induced more easily in cell suspensions than in cell sheets, this would explain why significantly more TUNEL-positive apoptotic cells are evident in transplanted cell suspensions than in transplanted cell sheets.

Previous studies have demonstrated that cardiac cell sheets co-cultured with endothelial cells form a network of vascular endothelial cells that promote angiogenesis after transplantation (Sekiya et al. 2006). Twenty-four hours after transplantation onto an ischemic heart, cell sheets co-cultured with endothelial cells exhibit mature blood vessels surrounded by pericytes and luminalized endothelial cells, whereas cell suspensions show no evidence of mature blood vessels (Sekine et al. 2011). It is thought that the presence of a network of endothelial cells within the cell sheet facilitates the early maturation of blood vessels after transplantation. Furthermore, transplanted cell sheets show a dense arrangement of capillaries, whereas transplanted cell suspensions contain only a small number of sparsely distributed microvessels. Interestingly, new blood vessels in the graft are remodeled from both graft- and host-derived endothelial cells, and the graft-derived endothelial cells migrate to ischemic regions in the host myocardium to contribute to the development of blood vessel walls. The above results show that the inclusion of vascular cells within a cell sheet before transplantation can help to engineer tissue with a strong ability to promote angiogenesis and improve cardiac function (Sekine et al. 2008).

Based on these findings, in 2007 we set up a collaborative clinical research project with the Department of Cardiovascular and Respiratory Surgery at Osaka University to evaluate whether the transplantation of autologous skeletal myoblast sheets could be used as a therapy for severe dilated cardiomyopathy. Patients who had been fitted with a left ventricular assist device prior to treatment showed sufficient improvement in cardiac function 3 months after cell sheet transplantation to allow removal of the assist device and discharge from hospital (Sawa et al. 2012). In view of the promising results obtained in the above study, Terumo Co., Ltd. initiated a clinical trial in 2012 to evaluate the potential use of cell sheet transplantation in the treatment of ischemic heart disease, and regenerative therapy has been carried out in more than 35 cases to date.

6 Effects of Cell Sheet Transplantation in Infant Ischemic Hearts

Cell sheet-based therapy for severe heart failure in adults has achieved a certain degree of success in clinical trials. However, there have been no reports regarding the effects of cell sheet-based treatment on myocardial damage in infants. Therefore, we utilized a rat model of myocardial infarction to compare the therapeutic effects of cell sheet transplantation between infant and adult animals (Homma et al. 2017). Briefly, experimental myocardial infarction was induced in infant rats (2 weeks old) and adult rats (12 weeks old), triple-layered rat myoblast sheets were transplanted 1 week later, and the response to therapy was evaluated 2 weeks after transplantation. Cardiac catheterization studies demonstrated better improvement of cardiac function after cell sheet transplantation in infant rats than in adult rats. Moreover, histological evaluation showed that the hearts of infant rats exhibited greater wall thickness, less fibrosis, a higher number of dividing cardiomyocytes, and more neovascularization of the infarcted region than those of adult rats. The above results indicate that infant hearts have a greater regenerative ability than adult hearts and that this regenerative ability is stimulated by the implantation of cell sheets. It is thought that cardiomyocyte mitogenesis is one of the mechanisms contributing to the effects of cell sheet transplantation on infarcted hearts in infants, whereas this has not been reported in adult hearts. This implies that myocardial regeneration therapy might be particularly effective during infancy.

7 Vascularization of Engineered Myocardial Tissue for Scaling-Up of Tissue Size

Tissues constructed by conventional bioengineering techniques lack functional vascular networks, which results in oxygen and nutrient deficiency and accumulation of waste products. The development of technologies to generate a blood vessel network within a bioengineered construct has become an important focus of tissue engineering research, and various approaches have been attempted. A widely used method is to administer an angiogenesis-promoting growth factor either at the time of transplantation or more gradually by elution from a scaffold (Richardson et al. 2001). Although this technique can promote angiogenesis at the initial stage of transplantation, it generally does not sustain angiogenesis for a sufficiently long period of time to prevent necrosis of the inner region of the transplanted tissue due to ischemia. As a result, such use of angiogenesis-promoting growth factors is

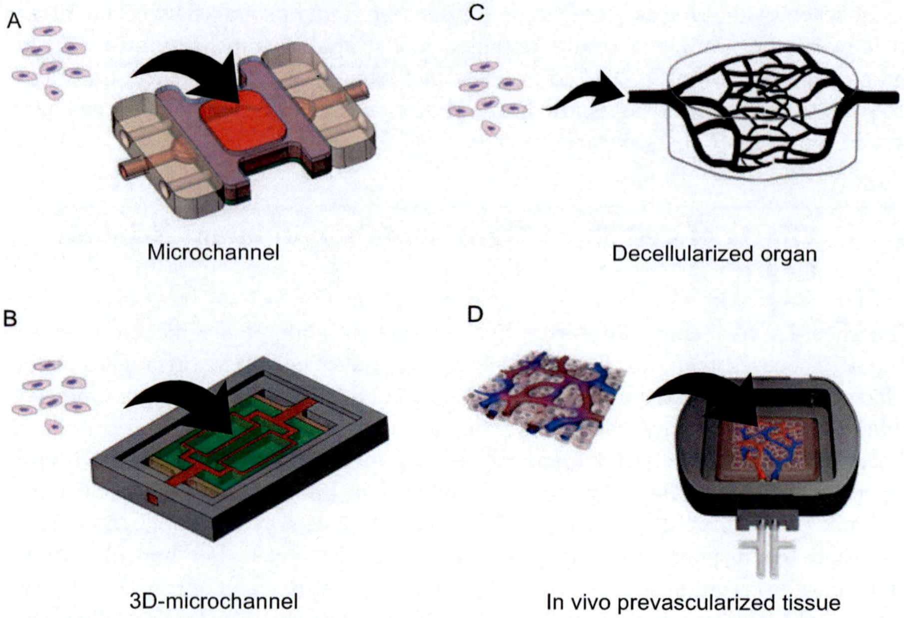

Fig. 3 Bioreactor system for application of vascular network fabrication

not suitable for the generation of thick tissues. Co-culturing with a cell type that is a constituent of blood vessels has received increasing attention in recent years as a method to promote the formation of a vascular network. As described above, we have shown that an endothelial cell network can be generated within a cell sheet containing co-cultured vascular endothelial cells and that this speeds up the development of a vascular network after transplantation (Sekiya et al. 2006). Although the co-cultured vascular cells directly contribute to blood vessel formation in the tissue after transplantation, they do not form continuous structures with lumens before transplantation. As a result, the thickness of the bioengineered three-dimensional tissue is still initially limited by ischemia after transplantation. Several methods have been advocated to overcome this limitation. We have developed a bioreactor system that initiates the formation of capillaries through the artificial construction of blood vessel-like structures in vitro that are perfused via their lumens. This technique involves the advance preparation of minute flow channels within a scaffold using microfabrication technology, which enables culture medium to be perfused through these channels during tissue culture (Fig. 3a) (Chouinard et al. 2009). An alternative approach utilizes a scaffold that imitates the three-dimensional structure of the microcirculation ("angiochip"), which is seeded with vascular endothelial cells from the luminal side to enable the construction of a vascular network during perfusion culture (Fig. 3b) (Zhang et al. 2016). Other studies have attempted to create vascularized tissues by decellularizing a living tissue or organ, seeding vascular endothelial cells within the original blood vessels, and performing perfusion culture (Fig. 3c) (Ott et al. 2008; Song et al. 2013; Ren et al. 2015).

An important aim of our research into regenerative therapy is to upscale the stacking of cell sheets to generate thick tissues. In order to achieve high-performance, the bioengineered tissue needs to be provided with a vascular network to improve the delivery of oxygen and nutrients and the removal of waste products. As a way of realizing this aim, we have devised a procedure that allows cardiac cell sheets to be sequentially transplanted every 24 h, which provides enough time for sufficient angiogenesis to occur from the host side of the layered cardiac cell sheets. Using this method, it was possible to create myocardial tissue with a thickness of about 1 mm that generated contractile force in vivo (Shimizu et al. 2006a). Following the success of the above experiments, we then tried to generate thick myocardial tissue by adding a vascular network to the construct in vitro (Fig. 4) (Sekine et al. 2013). First, we developed a vascular bed that would promote vascularization of the bioengineered tissue and a bioreactor for tissue perfusion, and we evaluated the formation of capillaries within cardiac cell sheets that were stacked on the vascular bed (Fig. 3d). The vascular bed was made from rat femoral muscle: the femoral muscle tissue was partially resected, reset to the original position, and then incubated in the host body for 1 week before complete resection with the femoral artery and vein. Cell sheets comprising cardiomyocytes co-cultured with endothelial cells were

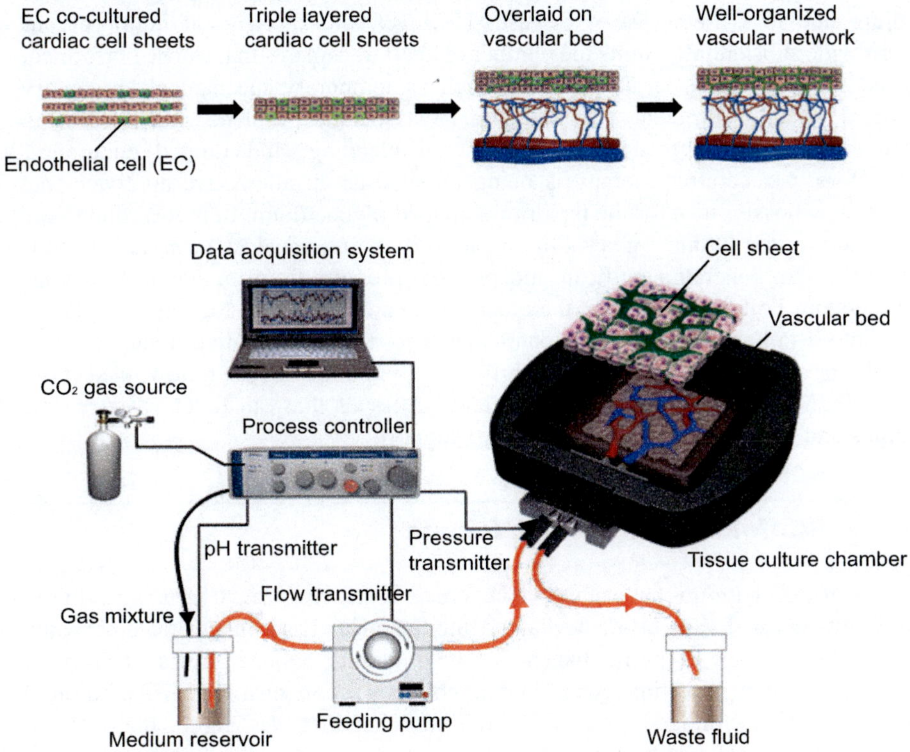

Fig. 4 Tissue perfusion using a bioreactor and stacking of cell sheets on a vascular bed

layered on the vascular bed, and tissue perfusion culture was performed using a solution supplemented with basic fibroblast growth factor (b-FGF), with inflow via the vascular bed artery and outflow via the vein. Connection of capillaries within the myocardial tissue to blood vessels of the vascular bed was confirmed after 3 days of perfusion culture, and the sequential stacking of triple-layered cell sheets on four occasions enabled the construction of myocardial tissue with a thickness of about 200 μm. Furthermore, when vascularized myocardial tissue prepared by the stepwise layering of cell sheets was transplanted into a rat and its vessels anastomosed to the cervical artery and vein of the host, the transplanted tissue was shown to engraft and retain its cellular function 2 weeks after transplantation.

In another study, cell sheets composed of cardiomyocytes co-cultured with endothelial cells were perfused via an artificial vascular bed consisting of a collagen gel containing microchannels. It was found that capillaries were constructed by the proliferation and migration of vascular endothelial cells within the microchannels and cell sheets (Sakaguchi et al. 2013).

8 Organ-Like Tissue Fabrication

Heart transplantation remains the best treatment for patients with severe cardiac failure due to ischemic heart disease. However, the shortage of donor organs worldwide substantially limits the number of heart transplants that can be performed. In addition, artificial cardiac systems such as temporary mechanical circulatory support or left ventricular assist devices have specific problems associated with thromboembolism, infection, gastrointestinal bleeding, and limited endurance. Therefore, regenerative therapy is being pursued as an alternative approach that offers new possibilities for the repair of damaged myocardium. In myocardial tissue engineering, the ultimate goal is the creation of a functional myocardial compartment that can generate significant independent pressure from its own spontaneous contraction. Following on from the creation of cardiac patches for direct implantation into a failing heart, the next challenge is to design organ-like tissues such as tubular or spherical structures that can function as pumps. In the following section, we introduce the bioengineering of organ-like tissues that can function as cardiac pumps and potentially provide circulatory support.

9 Scaffold-Based Cardiac Pumps

Several research groups have attempted to create organ-like structures using cultured cardiomyocytes. Evans et al. developed tubular heart tissue using rat embryonic heart cells containing type I collagen. A tube of aligned collagen fibers was formed using two counter-rotating cones and a polymerization chamber. Tube-cultured cardiomyocytes exhibited a level of differentiation that was higher than that of planar cultured cells and similar to that of neonatal ventricular myocytes in vivo (Evans et al. 2003). Yost et al. created a three-dimensional tubular collagen scaffold with a varying fibril angle from the inside to the outside and seeded it with neonatal rat cardiac cells on both the luminal and outer surfaces at weekly intervals (Yost et al. 2004; Franchini et al. 2007). Mechanical studies demonstrated that stiffness and viscosity were significantly increased in collagen tubes seeded with cardiac cells (Yost et al. 2004; Franchini et al. 2007). Birla et al. used a biodegradable hydrogel to produce a cell-based tubular cardiac structure that was capable of generating pressure. The constructs were made by culturing neonatal rat heart cells in a fibrin gel and then wrapping this around a silicon tube. The tubular structure was composed of layers of aligned cells and exhibited contractile function that generated an internal pressure of approximately 0.08 mmHg (Birla et al. 2008). Lee et al. described a cardiac organoid chamber prepared by mixing neonatal rat heart cells with a mixture of type I collagen and matrigel. The chamber construct exhibited structural and mechanical properties similar to those of a cardiac ventricle and generated an internal pressure of about 2 mmH_2O. The cardiac organoid also showed local variations in contractile function following cryoinjury, suggesting that it could be used as an in vitro model of myocardial infarction (Lee et al. 2008). Yildrim et al. developed bag-like heart tissue by mixing neonatal rat heart cells with type I collagen and matrigel in a globular mold. The sac-shaped heart tissue had structural and contractile properties comparable to those of natural myocardium. Two weeks after transplantation, an artificial graft covering the entire surface of the heart was shown to be in functional communication with the host myocardium (Yildirim et al. 2007). Ott et al. attempted to create a working heart through a process of decellularization followed by recellularization. First, detergent was perfused through the coronary arteries of a rat heart in order to decellularize it while preserving its extracellular matrix and vascular architecture. Then, this construct was recellularized with neonatal rat cardiac cells or rat aortic endothelial cells. The resulting bioengineered heart was capable of contracting and generating a left ventricular pressure of up to 2.4 mmHg after 8 days of culture. This use of such an approach could potentially solve the challenge of providing a blood supply to engrafted cells by maintaining the natural structure of the heart and inducing the formation of blood vessels (Ott et al. 2008).

10 Cell Sheet-Based Cardiac Pump

As an alternative to scaffold-based heart pumps, we have utilized a temperature-responsive culture dish to create beating cardiac tubes in vitro that could be used to provide circulatory support (Fig. 5). Three neonatal rat cardiac cell sheets were harvested from temperature-responsive culture surfaces, layered and then wrapped around a fibrin tube. The cardiac tube created in vitro exhibited spontaneous and synchronized pulsations at the macroscopic level and generated a measurable change in internal pressure in response to spontaneous contraction, with a mean internal pressure gradient of 0.11 mmHg. We also confirmed an inotropic effect of increased elevated extracellular Ca^{2+} concentration in the cardiac tube (Kubo et al. 2007).

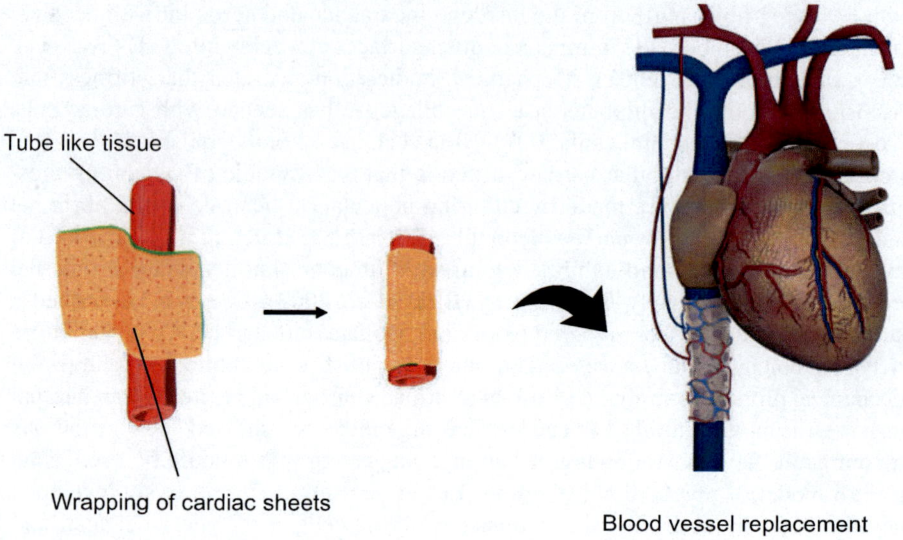

Fig. 5 Cell sheet-based myocardial tube for circulatory support

Subsequently, we created an implantable tubular myocardial structure by wrapping a neonatal rat cardiac cell sheet around a segment of thoracic aorta resected from a rat (Sekine et al. 2006a). Four weeks after transplantation of this construct into the abdominal aorta of an athymic rat, the artificial myocardial tube exhibited spontaneous and synchronous pulsations. The spontaneous contractile activity was confirmed by electrography to be independent of the host's heartbeat and generated a pressure within the graft of 5.9 mmHg. Histological examination and TEM showed that the myocardial tube was composed of tissue resembling that of the normal heart. The artificial tube was densely populated with stratified cells that stained positively for troponin T, indicating that they were cardiac cells. In addition, diffuse localization of connexin-43 throughout the graft tissue suggested that gap junctions had formed between the cells. TEM identified well-differentiated myocardial tissue with numerous mitochondria as well as myofilaments with sarcomeres. TEM also confirmed that functional microvessels containing red blood cells were present throughout the graft. When compared with a graft that was transplanted into the abdominal cavity of an animal, the myocardial tube used for aortic replacement contained significantly thicker tissue and higher expression levels of brain natriuretic peptide, alpha-myosin heavy chain, and beta-myosin heavy chain. These findings suggest that pulsatile blood flow through the lumen of the myocardial tube had stimulated the growth and hypertrophy of its cardiomyocytes. Previous studies have demonstrated that mechanical stress has the ability to induce myocardial hypertrophy during development and under pathological conditions. This phenomenon has been exploited in myocardial tissue engineering through the use of mechanical stretching to promote cardiomyocyte hypertrophy. Therefore, the application of mechanical loading, either in vitro or in vivo, appears to be an essential element in the generation of functional cardiac tissue.

We have also established that human tubular cardiac structures can improve hemodynamics after implantation in the rat inferior vena cava (Seta et al. 2017). Specifically, a triple-layered human cardiac cell sheet was wound around the rat inferior vena cava to form a tubular cardiac structure. Ultrasonography 4 weeks after transplantation demonstrated that the fabricated tubular cardiac structure was beating. Furthermore, catheterization and measurement of internal pressure revealed that the tubular myocardium generated a pulse pressure of 6.3 ± 1.6 mmHg at 4 weeks and 9.1 ± 3.2 mmHg at 8 weeks after transplantation. In addition, histological analysis confirmed that the transplanted human cardiac cell sheet had developed mature myofibrils and a rich vascular network. The creation of this cardiac tube represents the next step in myocardial tissue construction and a transition stage toward the production of an independently functioning structure with the potential to act as a bioengineered cardiac assist device.

11 Future Perspectives

The next exciting challenge in the field of regenerative therapy will be the design of organ-like tissues. Although the creation of organ-like tissues has only been implemented on a small scale, future solutions to current problems concerning cell sources and upscaling will facilitate the development of tissue-engineered cardiac assist devices or even replacement organs. The major obstacle in myocardial tissue engineering remains the poor delivery of oxygen to a three-dimensional structure, which limits the thickness of a construct to about 100 μm. Therefore, the production of thicker and more functional cardiac pumps will require new technologies to control blood vessel growth. Future attempts to promote vascular network growth and formation, using techniques such as growth factor administration, gene transfer, and co-culture with vascular progenitor cells, may contribute to the production of thicker tissues. Overcoming the limitations of passive diffusion should make it possible to create a powerful cardiac pump.

One possible technique for improving the contractile force of artificial cardiac pumps is to ensure the correct orientation of cardiomyocytes. Therefore, controlling cell orientation is also considered to be an essential element in the generation of a tissue-engineered pump with improved cardiac-like properties.

12 Conclusions

This chapter has described the advances made in regenerative medicine using cell sheets and outlined further possibilities. Regenerative medicine is expected to become a new therapeutic option for refractory diseases that are currently difficult to treat, and various approaches have been adopted in the development of potential therapeutic solutions. Cell sheets created using temperature-responsive culture dishes have several advantages over other currently available methods, including more efficient transplantation, the construction of tissues that are impossible to make with preexisting technology, and realization of cell stratification. In addition, cell sheet-based therapies have achieved highly promising results in human clinical trials.

To construct tissues and organs that can regulate the blood circulatory system, it will be necessary to overcome numerous challenges such as the development of cardiomyocytes that can be transplanted into humans and the upscaling of constructed tissues. It is anticipated that bioengineered cardiac structures will become an important treatment for diseases that cause severe heart failure, including congenital heart disease in children. We believe that this can be achieved through future research efforts and an interdisciplinary approach to technological development.

Acknowledgments This research was supported by JSPS KAKENHI grant number 19H04453. We thank OxMedComms (www.oxmedcomms.com) for writing assistance.

References

Birla RK, Dow DE, Huang YC, Migneco F, Khait L, Borschel GH, Dhawan V, Brown DL (2008) Methodology for the formation of functional, cell-based cardiac pressure generation constructs in vitro. In Vitro Cell Dev Biol Anim 44(8–9):340–350

Bursac N, Papadaki M, Cohen RJ, Schoen FJ, Eisenberg SR, Carrier R, Vunjak-Novakovic G, Freed LE (1999) Cardiac muscle tissue engineering: toward an in vitro model for electrophysiological studies. Am J Physiol 277(2):H433–H444

Caspi O, Huber I, Kehat I, Habib M, Arbel G, Gepstein A, Yankelson L, Aronson D, Beyar R, Gepstein L (2007) Transplantation of human embryonic stem cell-derived cardiomyocytes improves myocardial performance in infarcted rat hearts. J Am Coll Cardiol 50(19):1884–1893

Chouinard JA, Gagnon S, Couture MG, Levesque A, Vermette P (2009) Design and validation of a pulsatile perfusion bioreactor for 3D high cell density cultures. Biotechnol Bioeng 104(6): 1215–1223

Christman KL, Lee RJ (2006) Biomaterials for the treatment of myocardial infarction. J Am Coll Cardiol 48(5):907–913

Evans HJ, Sweet JK, Price RL, Yost M, Goodwin RL (2003) Novel 3D culture system for study of cardiac myocyte development. Am J Physiol Heart Circ Physiol 285(2):H570–H578

Franchini JL, Propst JT, Comer GR, Yost MJ (2007) Novel tissue engineered tubular heart tissue for in vitro pharmaceutical toxicity testing. Microsc Microanal 13(4):267–271

Haraguchi Y, Shimizu T, Yamato M, Kikuchi A, Okano T (2006) Electrical coupling of cardiomyocyte sheets occurs rapidly via functional gap junction formation. Biomaterials 27(27):4765–4774

Hata H, Matsumiya G, Miyagawa S, Kondoh H, Kawaguchi N, Matsuura N, Shimizu T, Okano T, Matsuda H, Sawa Y (2006) Grafted skeletal myoblast sheets attenuate myocardial remodeling in pacing-induced canine heart failure model. J Thorac Cardiovasc Surg 132(4):918–924

Hofmann M, Wollert KC, Meyer GP, Menke A, Arseniev L, Hertenstein B, Ganser A, Knapp WH, Drexler H (2005) Monitoring of bone marrow cell homing into the infarcted human myocardium. Circulation 111(17):2198–2202

Homma J, Sekine H, Matsuura K, Yamato M, Shimizu T (2017) Myoblast cell sheet transplantation enhances the endogenous regenerative abilities of infant hearts in rats with myocardial infarction. J Tissue Eng Regen Med 11(6):1897–1906

Hudson W, Collins MC, deFreitas D, Sun YS, Muller-Borer B, Kypson AP (2007) Beating and arrested intramyocardial injections are associated with significant mechanical loss: implications for cardiac cell transplantation. J Surg Res 142(2):263–267

Kondoh H, Sawa Y, Miyagawa S, Sakakida-Kitagawa S, Memon IA, Kawaguchi N, Matsuura N, Shimizu T, Okano T, Matsuda H (2006) Longer preservation of cardiac performance by sheet-shaped myoblast implantation in dilated cardiomyopathic hamsters. Cardiovasc Res 69(2): 466–475

Kubo H, Shimizu T, Yamato M, Fujimoto T, Okano T (2007) Creation of myocardial tubes using cardiomyocyte sheets and an in vitro cell sheet-wrapping device. Biomaterials 28(24): 3508–3516

Kushida A, Yamato M, Konno C, Kikuchi A, Sakurai Y, Okano T (1999) Decrease in culture temperature releases monolayer endothelial cell sheets together with deposited fibronectin matrix from temperature-responsive culture surfaces. J Biomed Mater Res 45(4):355–362

Kutschka I, Chen IY, Kofidis T, von Degenfeld G, Sheikh AY, Hendry SL, Hoyt G, Pearl J, Blau HM, Gambhir SS, Robbins RC (2007) In vivo optical bioluminescence imaging of collagen-supported cardiac cell grafts. J Heart Lung Transplant 26(3):273–280

Laflamme MA, Murry CE (2005) Regenerating the heart. Nat Biotechnol 23(7):845–856

Langer R, Vacanti JP (1993) Tissue engineering. Science 260(5110):920–926

Lee EJ, Kim DE, Azeloglu EU, Costa KD (2008) Engineered cardiac organoid chambers: toward a functional biological model ventricle. Tissue Eng Part A 14(2):215–225

Leor J, Aboulafia-Etzion S, Dar A, Shapiro L, Barbash IM, Battler A, Granot Y, Cohen S (2000) Bioengineered cardiac grafts: a new approach to repair the infarcted myocardium? Circulation 102(19 Suppl 3):III56–III61

Li RK, Jia ZQ, Weisel RD, Mickle DA, Choi A, Yau TM (1999) Survival and function of bioengineered cardiac grafts. Circulation 100(19 Suppl):II63–II69

Memon IA, Sawa Y, Fukushima N, Matsumiya G, Miyagawa S, Taketani S, Sakakida SK, Kondoh H, Aleshin AN, Shimizu T, Okano T, Matsuda H (2005) Repair of impaired myocardium by means of implantation of engineered autologous myoblast sheets. J Thorac Cardiovasc Surg 130(5):1333–1341

Menasche P, Hagege AA, Vilquin JT, Desnos M, Abergel E, Pouzet B, Bel A, Sarateanu S, Scorsin M, Schwartz K, Bruneval P, Benbunan M, Marolleau JP, Duboc D (2003) Autologous skeletal myoblast transplantation for severe postinfarction left ventricular dysfunction. J Am Coll Cardiol 41(7):1078–1083

Menasche P, Alfieri O, Janssens S, McKenna W, Reichenspurner H, Trinquart L, Vilquin JT, Marolleau JP, Seymour B, Larghero J, Lake S, Chatellier G, Solomon S, Desnos M, Hagege AA (2008) The Myoblast Autologous Grafting in Ischemic Cardiomyopathy (MAGIC) trial: first randomized placebo-controlled study of myoblast transplantation. Circulation 117(9):1189–1200

Michel JB (2003) Anoikis in the cardiovascular system: known and unknown extracellular mediators. Arterioscler Thromb Vasc Biol 23(12):2146–2154

Miyagawa S, Sawa Y, Sakakida S, Taketani S, Kondoh H, Memon IA, Imanishi Y, Shimizu T, Okano T, Matsuda H (2005) Tissue cardiomyoplasty using bioengineered contractile cardiomyocyte sheets to repair damaged myocardium: their integration with recipient myocardium. Transplantation 80(11):1586–1595

Miyagawa S, Saito A, Sakaguchi T, Yoshikawa Y, Yamauchi T, Imanishi Y, Kawaguchi N, Teramoto N, Matsuura N, Iida H, Shimizu T, Okano T, Sawa Y (2010) Impaired myocardium regeneration with skeletal cell sheets – a preclinical trial for tissue-engineered regeneration therapy. Transplantation 90(4):364–372

Miyahara Y, Nagaya N, Kataoka M, Yanagawa B, Tanaka K, Hao H, Ishino K, Ishida H, Shimizu T, Kangawa K, Sano S, Okano T, Kitamura S, Mori H (2006) Monolayered mesenchymal stem cells repair scarred myocardium after myocardial infarction. Nat Med 12(4):459–465

Nelson TJ, Martinez-Fernandez A, Yamada S, Perez-Terzic C, Ikeda Y, Terzic A (2009) Repair of acute myocardial infarction by human stemness factors induced pluripotent stem cells. Circulation 120(5):408–416

Okano T, Yamada N, Sakai H, Sakurai Y (1993) A novel recovery system for cultured cells using plasma-treated polystyrene dishes grafted with poly(*N*-isopropylacrylamide). J Biomed Mater Res 27(10):1243–1251

Ott HC, Matthiesen TS, Goh SK, Black LD, Kren SM, Netoff TI, Taylor DA (2008) Perfusion-decellularized matrix: using nature's platform to engineer a bioartificial heart. Nat Med 14(2):213–221

Ren X, Moser PT, Gilpin SE, Okamoto T, Wu T, Tapias LF, Mercier FE, Xiong L, Ghawi R, Scadden DT, Mathisen DJ, Ott HC (2015) Engineering pulmonary vasculature in decellularized rat and human lungs. Nat Biotechnol 33(10):1097–1102

Richardson TP, Peters MC, Ennett AB, Mooney DJ (2001) Polymeric system for dual growth factor delivery. Nat Biotechnol 19(11):1029–1034

Sakaguchi K, Shimizu T, Horaguchi S, Sekine H, Yamato M, Umezu M, Okano T (2013) In vitro engineering of vascularized tissue surrogates. Sci Rep 3:1316

Sawa Y, Miyagawa S, Sakaguchi T, Fujita T, Matsuyama A, Saito A, Shimizu T, Okano T (2012) Tissue engineered myoblast sheets improved cardiac function sufficiently to discontinue LVAS in a patient with DCM: report of a case. Surg Today 42(2):181–184

Sekine H, Shimizu T, Kosaka S, Kobayashi E, Okano T (2006a) Cardiomyocyte bridging between hearts and bioengineered myocardial tissues with mesenchymal transition of mesothelial cells. J Heart Lung Transplant 25(3):324–332

Sekine H, Shimizu T, Yang J, Kobayashi E, Okano T (2006b) Pulsatile myocardial tubes fabricated with cell sheet engineering. Circulation 114(1 Suppl):I87–I93

Sekine H, Shimizu T, Hobo K, Sekiya S, Yang J, Yamato M, Kurosawa H, Kobayashi E, Okano T (2008) Endothelial cell coculture within tissue-engineered cardiomyocyte sheets enhances neovascularization and improves cardiac function of ischemic hearts. Circulation 118(14 Suppl):S145–S152

Sekine H, Shimizu T, Dobashi I, Matsuura K, Hagiwara N, Takahashi M, Kobayashi E, Yamato M, Okano T (2011) Cardiac cell sheet transplantation improves damaged heart function via superior cell survival in comparison with dissociated cell injection. Tissue Eng Part A 17(23–24):2973–2980

Sekine H, Shimizu T, Sakaguchi K, Dobashi I, Wada M, Yamato M, Kobayashi E, Umezu M, Okano T (2013) In vitro fabrication of functional three-dimensional tissues with perfusable blood vessels. Nat Commun 4:1399

Sekiya S, Shimizu T, Yamato M, Kikuchi A, Okano T (2006) Bioengineered cardiac cell sheet grafts have intrinsic angiogenic potential. Biochem Biophys Res Commun 341(2):573–582

Seta H, Matsuura K, Sekine H, Yamazaki K, Shimizu T (2017) Tubular cardiac tissues derived from human induced pluripotent stem cells generate pulse pressure in vivo. Sci Rep 7:45499

Shiba Y, Gomibuchi T, Seto T, Wada Y, Ichimura H, Tanaka Y, Ogasawara T, Okada K, Shiba N, Sakamoto K, Ido D, Shiina T, Ohkura M, Nakai J, Uno N, Kazuki Y, Oshimura M, Minami I, Ikeda U (2016) Allogeneic transplantation of iPS cell-derived cardiomyocytes regenerates primate hearts. Nature 538(7625):388–391

Shimizu T, Yamato M, Isoi Y, Akutsu T, Setomaru T, Abe K, Kikuchi A, Umezu M, Okano T (2002) Fabrication of pulsatile cardiac tissue grafts using a novel 3-dimensional cell sheet manipulation technique and temperature-responsive cell culture surfaces. Circ Res 90(3):e40

Shimizu T, Sekine H, Isoi Y, Yamato M, Kikuchi A, Okano T (2006a) Long-term survival and growth of pulsatile myocardial tissue grafts engineered by the layering of cardiomyocyte sheets. Tissue Eng 12(3):499–507

Shimizu T, Sekine H, Yang J, Isoi Y, Yamato M, Kikuchi A, Kobayashi E, Okano T (2006b) Polysurgery of cell sheet grafts overcomes diffusion limits to produce thick, vascularized myocardial tissues. FASEB J 20(6):708–710

Song JJ, Guyette JP, Gilpin SE, Gonzalez G, Vacanti JP, Ott HC (2013) Regeneration and experimental orthotopic transplantation of a bioengineered kidney. Nat Med 19(5):646–651

Soonpaa MH, Koh GY, Klug MG, Field LJ (1994) Formation of nascent intercalated disks between grafted fetal cardiomyocytes and host myocardium. Science 264(5155):98–101

Terrovitis J, Lautamaki R, Bonios M, Fox J, Engles JM, Yu J, Leppo MK, Pomper MG, Wahl RL, Seidel J, Tsui BM, Bengel FM, Abraham MR, Marban E (2009) Noninvasive quantification and optimization of acute cell retention by in vivo positron emission tomography after intra-myocardial cardiac-derived stem cell delivery. J Am Coll Cardiol 54(17):1619–1626

Wilmut I, Leslie S, Martin NG, Peschanski M, Rao M, Trounson A, Turner D, Turner ML, Yamanaka S, Taylor CJ (2015) Development of a global network of induced pluripotent stem cell haplobanks. Regen Med 10(3):235–238

Yamada N, Okano T, Sakai H, Karikusa F, Sawasaki Y, Sakurai Y (1990) Thermo-responsive polymeric surfaces; control of attachment and detachment of cultured cells. Makromol Chem Rapid 11(11):571–576

Yang J, Yamato M, Kohno C, Nishimoto A, Sekine H, Fukai F, Okano T (2005) Cell sheet engineering: recreating tissues without biodegradable scaffolds. Biomaterials 26(33): 6415–6422

Yildirim Y, Naito H, Didie M, Karikkineth BC, Biermann D, Eschenhagen T, Zimmermann WH (2007) Development of a biological ventricular assist device: preliminary data from a small animal model. Circulation 116(11 Suppl):I16–I23

Yost MJ, Baicu CF, Stonerock CE, Goodwin RL, Price RL, Davis JM, Evans H, Watson PD, Gore CM, Sweet J, Creech L, Zile MR, Terracio L (2004) A novel tubular scaffold for cardiovascular tissue engineering. Tissue Eng 10(1–2):273–284

Zandonella C (2003) Tissue engineering: the beat goes on. Nature 421(6926):884–886

Zhang M, Methot D, Poppa V, Fujio Y, Walsh K, Murry CE (2001) Cardiomyocyte grafting for cardiac repair: graft cell death and anti-death strategies. J Mol Cell Cardiol 33(5):907–921

Zhang B, Montgomery M, Chamberlain MD, Ogawa S, Korolj A, Pahnke A, Wells LA, Masse S, Kim J, Reis L, Momen A, Nunes SS, Wheeler AR, Nanthakumar K, Keller G, Sefton MV, Radisic M (2016) Biodegradable scaffold with built-in vasculature for organ-on-a-chip engineering and direct surgical anastomosis. Nat Mater 15(6):669–678

Zimmermann WH, Melnychenko I, Wasmeier G, Didie M, Naito H, Nixdorff U, Hess A, Budinsky L, Brune K, Michaelis B, Dhein S, Schwoerer A, Ehmke H, Eschenhagen T (2006) Engineered heart tissue grafts improve systolic and diastolic function in infarcted rat hearts. Nat Med 12(4):452–458

Bioengineering of Trachea and Esophagus

Soichi Shibuya, Natalie Durkin, Matías Garrido, Paola Bonfanti, and Paolo De Coppi

Contents

Soichi Shibuya, Natalie Durkin and Matias Garrido Flores contributed equally with all other contributors.

S. Shibuya · N. Durkin
Stem Cell and Regenerative Medicine Section, Great Ormond Street Institute of Child Health, University College London, London, UK

M. Garrido · P. Bonfanti
Epithelial Stem Cell Biology and Regenerative Medicine Laboratory, The Francis Crick Institute, London, UK

P. De Coppi (✉)
Stem Cell and Regenerative Medicine Section, Great Ormond Street Institute of Child Health, University College London, London, UK

Specialist Neonatal and Paediatric Surgery, Great Ormond Street Hospital, London, UK
e-mail: p.decoppi@ucl.ac.uk; paolo.decoppi@gosh.nhs.uk

D. Eberli et al. (eds.), *Organ Tissue Engineering*, Reference Series in Biomedical Engineering, https://doi.org/10.1007/978-3-030-44211-8_18

Abstract

Tissue engineering offers huge potential as a novel strategy to treat complex congenital and acquired conditions of both the trachea and the esophagus where standard therapy has failed or current options for organ replacement fall short. Existing approaches have involved the use of scaffolds, cells, or a combination of both to reconstruct damaged organs, however increasing evidence suggests that hybrid techniques with exogenous cell delivery enhances tissue regeneration while reducing inflammation. While many cell lines have been used, increasing focus on the use of stem cells of mesenchymal origin holds promise, although the mechanism through which remodeling occurs remains unclear.

In vivo animal models have provided real insights into the use of these techniques for human therapy. While initial success in partial-thickness and patch defects models has been translated into clinical studies in humans, repair of circumferential defects remains altogether more challenging due to difficulties with stenosis, luminal collapse, and anastomotic leak. More work is required to establish safe and standardized methods for circumferential organ transplantation in this field prior to advancement to human candidates with open reporting of outcomes. Here, we review current approaches, evaluate the existing evidence, and discuss the future of tissue engineering of the trachea and esophagus.

1 Introduction

In the past decade, translation of tissue-engineered organs into clinical practice is more promising than ever owing to rapid progress in research focusing on cell biology and tissue regeneration. The major goal of bioengineering is to fabricate a biocompatible organ as a viable substitute for transplantation into patients in whom no other treatment options exist. This novel strategy has huge potential to save and improve the quality of life of patients suffering from either congenital or acquired disease. Simultaneously, however, it can pose ethical and economical concerns, particularly in clinical trials developed on a compassionate basis. In this chapter, we describe developing research on bioengineering of the trachea and the esophagus.

2 Tracheal Bioengineering

2.1 Introduction

The trachea is a vital organ which cannot be replaced by any other. Its dysfunction results in difficulty breathing, which is directly associated with mortality. A lack of alternative treatment for end-stage tracheal failure raises demands for novel treatment strategies, such as tracheal bioengineering. In adults, the most frequent cause of losing a healthy trachea is cancer treatment requiring bulk resection due to massive tumor invasion emerging from the esophagus or lung. If the length of resected trachea is greater than 50%, it is considered impossible to reconstruct by end-to-end anastomosis (Chiang et al. 2016). In children, the most significant indication for tracheal reconstruction is congenital tracheal malformation. This includes tracheal agenesis, tracheal stenosis, and severe tracheomalacia, all of which require intensive treatment immediately after birth and can result in devastating consequences despite painstaking medical efforts. Over the past decade, slide tracheoplasty has become the most reliable surgical treatment for tracheal stenosis, but despite significant improvement in mortality compared to other options it is technically demanding and can result in postoperative bronchomalacia, which is occasionally lethal (Butler et al. 2014). In addition, accidental events such as foreign body ingestion, especially of lithium button battery, and swallowing of caustic substances can devastate tracheal function. Due to the severity and urgency of these scenarios, compassionate application of innovative therapy is more likely to be adopted in the emergency setting than in chronic disease. Clinical trials in tracheal tissue engineering have been performed on a compassionate basis and have significantly contributed to the progress of this field. The first tracheal reconstruction was introduced in 1950 (Belsey 1950). A large defect in the trachea due to massive tumor resection was restored by the combination of steel wire scaffold and a free fascial graft from the thigh. Although eventual outcomes were not satisfactory, with collapse of the structure and granulation in the lumen, respiratory function was temporally maintained and epithelization on the luminal surface was observed. Current literature suggests that tracheal reconstruction using bioengineered grafts is complicated and potentially risky but not impossible if adequate tissue regeneration is induced by a biocompatible method (Delaere and Van Raemdonck 2016).

In the following sections, we discuss the fundamental knowledge, which is a prerequisite to formulating a bioengineering strategy, and the current achievements in clinical translation of the tissue-engineered trachea.

2.2 Anatomy and Embryology

The size of the trachea in an adult human is approximately 12–15 cm in length and 2–3 cm in diameter. The lumen is covered with a mucous membrane layer, composed of pseudostratified columnar epithelia resting on the basal lamina. The tracheal epithelium bears cilia on its apical surface, 0.25 nm hair-like projections which expel external particles through a polarized beating movement. Scattered throughout the cilia are mucus secreting goblet cells which help maintain the humidity in the airway and trap microorganisms. There are numerous blood and lymphatic vessels running throughout the lamina propria between the mucosa and the tracheal wall. The function of the vessels includes supply of oxygen and nutrition, control of intraluminal heat and removal of harmful substances trapped by epithelia. The wall of the tracheal tube is composed of elastic and collagen fibers, supported by horseshoe-shaped cartilages, which contributes to the contractility and stability. The tracheal cartilages cover only the ventral and lateral side of the trachea, leaving the dorsal side open, which allows the esophagus to bulge as food passes through. Muscle fibers running alongside the cartilage control air flow to the lungs by changing the length and the diameter of the trachea in synchronization with respiration. These muscles are controlled by the coordination of the specific autonomic nerves, namely, sympathetic fibers from the thoracic sympathetic trunk and parasympathetic fibers from the vagus nerves and their recurrent laryngeal branches. Furthermore, the muscle fibers also function as a barrier against harmful particles by triggering strong constriction when such substances are inhaled or aspirated, resulting in coughing.

Understanding normal embryonic development is important to establish a strategy for organ bioengineering. The development of airway begins in the fourth week of gestation in humans, where a laryngotracheal diverticulum emerges from the foregut endoderm just caudal to the level of the fourth pharyngeal pouches. The laryngotracheal diverticulum is a pouch-like structure located at the ventral side of the foregut, from which the respiratory bud sprouts and elongates caudally, giving rise to primordial lungs. Subsequently, the trachea becomes a distinctive structure from the original foregut tube, but the mechanism of separating the trachea from the esophagus is not well understood. There have been roughly two contrasting theories for explaining this process. One theory suggests that the respiratory system develops as a result of rapid outgrowth of the laryngotracheal diverticulum. This theory considers the tracheal primordium buds as a separate structure from the foregut during the subsequent stages of development. An alternative theory hypothesizes the growth of mesenchymal ridges, which appear between the laryngotracheal diverticulum and the foregut. Fusion of the ridges creates the tracheoesophageal septum which divides the foregut into the ventral and dorsal part, differentiating the trachea and the esophagus. Several studies using animal models have attempted to resolve the controversy and it is likely that the physical separation of the foregut tube plays a key role in the organogenesis, but this assumption is yet to be proven. After separation from the foregut, the tracheal primordium starts formulating the functional airway. The endodermal cells lining the tracheal lumen differentiate to the epithelium and the glands, while the mesodermal cells give rise to cartilage, connective tissue, and muscle.

2.3 Strategy for Tissue Engineering

The main role of the trachea is to allow filtered air to pass into the lungs while trapping and clearing harmful particles. This unique function is achieved by the characteristic cell layer covering the luminal surface of the tube. As such, the fundamental idea of tracheal bioengineering is to generate a conduit lined with ciliated mucosa capable of mucus secretion and ciliary movement. In addition, the graft needs to be airtight to prevent air leakage and robust enough to maintain its structure in the mediastinum. These requirements raise two major questions; how to create a three-dimensional (3D) tube-shaped scaffold and how to populate it with cells.

In order to facilitate clinical translation, biocompatibility of the graft is essential. Due to its function as a barrier between external airflow into the internal environment of body, the trachea is constantly exposed to foreign substances. As such, the immune system is more active than in other organs, making allogenic tracheal transplantation complicated; severe immune rejection frequently occurs, leading to a poor success rates. Any graft therefore needs to be immune tolerant, while retaining the immunological barrier function. The other crucial factor for successful graft implantation is adequate vascularization, especially if the graft covers a circumferential defect. Unlike lung, liver, or intestine, the trachea does not have a truncal blood supply, with a segmental vascular network providing nutritional and immune capabilities. This lack of major vessel connecting the trachea would not allow a graft to draw sufficient blood by vascular anastomosis. Therefore, techniques to promote vascularization, such as a pericardial patch, muscular patch, or prevascularization in vivo, is likely to be mandatory for preventing necrosis and collapse of the graft.

In summary, tracheal bioengineering for clinical application consists of three parts; obtaining biocompatible scaffolds, populating this with cells able to generate a functional mucosal layer, and establishing a proper blood supply.

2.4 Scaffolds

The major role of scaffolds is to provide a footing where cells can adhere and colonize. The scaffold must be mechanically robust to maintain structural integrity in order to resist extrinsic pressure induced by neck flexion and negative pressure during inspiration. Conversely, it needs to be flexible enough to fit in the narrow space without damaging neighboring vulnerable structures, such as the esophagus and aorta. Currently, materials commonly chosen as potential sources can be categorized into two types, synthetic and biologically derived scaffolds.

The advantage of synthetic scaffolds is the adjustability of the size according to recipient need and prompt availability when required. In early attempts, solid prostheses were directly used to reconstruct the trachea but this resulted in unacceptably high mortality associated with severe stenosis and fistula formation (Neville 1982; Toomes et al. 1985). These complications were secondary to severe inflammatory reactions of adjacent tissues causing granuloma and erosions. As such, solid prostheses are no longer used for clinical trials and many researchers began to use porous constructs, which can facilitate ingrowth and migration of indigenous cells. However, reconstruction using porous tubes has achieved limited success in the proximal airway while incidence of erosion and fistula formation remains high in the distal trachea (Maziak et al. 1996). Based on these results, the focus has shifted to biodegradable polymers, which gradually degrade in vivo, accommodating the surrounding environment. Biodegradable materials may especially benefit pediatric patients, allowing tissue growth beyond the initial size of the graft once this has completely degraded. Traditionally, Poly (glycolic acid) (PGA), poly(lactic acid) (PLA), Poly(lactic-co-glycolic acid) (PLGA), and polydioxanone (PDS) have been used as biodegradable scaffolds. PGA is the simplest linear polyester, with a high melting point, high solubility in water, and low solubility in organic solvents. Due to its hydrophilic nature, constructs made of PGA quickly uptake water and lose their mechanical strength. PLA has less hydrolytic properties than PGA and thus its degradation is slow. This characteristic is advantageous in terms of sustainability of mechanical strength, but can be problematic by remaining too long after implantation. PLGA is hybrid polymer of PGA and PLA, which enables tuning of the degradation parameters according to the ratio of glycolic acid and lactic acid. PDS is a unique material commonly used as a surgical suture. The tensile strength of PDS is lower than other synthetic materials; however it is more durable with a significantly slower degradability. All these materials have the advantage of being FDA-approved and are already in use as surgical sutures and implants, making their use in clinical translation straightforward. The major limitation of polymer structures is their hydrophobic nature and smooth surface, which may prevent adherence of cells to the scaffold. Recently, 3D printing technology combined with electro spinning methods enables nano-level surface modification of synthetic materials. In this state-of-the-art technique, polycaprolactone (PCL) has been the preferred choice due to its strength and long absorption time despite a low porosity (Jang et al. 2014; Park et al. 2019). Nevertheless, synthetic materials should be adopted only following thorough evaluation of their safety and biocompatibility in preclinical studies.

Biologically derived polymeric materials, such as collagen, fibrin, and hyaluronic acid sponges, have been used in a number of preclinical studies for partially repairing tracheal defects. However, grafts solely made from these materials lack the durability to maintain the mechanical strength required when implanted into a circumferential tracheal defect. Recently, scaffolds derived from natural organs by decellularization, in which indigenous cells are removed through perfusion or circulation of detergent and enzymatic solutions, have emerged as promising alternative (Conconi et al. 2005). By removing native cells, the expression of the major histocompatibility complex (MHC) becomes almost null, significantly decreasing immunoreactivity of non-autologous tissues. This technique, which is applicable to either donated human cadavers or size-matched animals, can convert xenogenic organs into natural scaffolds without the need for immunosuppression. Decellularized organs conserve structured extracellular matrix (ECM), which arguably retains cytokines and growth factors containing essential information for cell growth and maturation. Composition of the ECM is generally organ-specific, thus, using a decellularized trachea is likely the most efficient to establish proper differentiation of seeded cells. It is also reported that decellularized scaffolds tolerate cryopreservation for prolonged periods, indicating the possibility of an off-the-shelf availability, which is important for universal clinical translation and potential commercial distribution (Urbani et al. 2017). The downside of decellularized scaffolds is the inability to modify the mechanical properties. Donor organs tend to become less durable after decellularization; over-treatment should therefore be avoided as scaffolds lose mechanical strength with degradation of the ECM resulting in technical difficulty at anastomosis. Residual chemical reagents are also harmful, suppressing cell ingrowth and even causing cell death, necessitating vigorous washout during the procedure. Consequently, the decellularization protocol for the trachea tends to take long time (Aoki et al. 2019; Butler et al. 2017; Conconi et al. 2005). Conversely, however, insufficient treatment can leave native cells in scaffolds as tracheal chondrocytes are relatively stable and cartilage is hard to perfuse due to its density which could result in immune rejection on implantation. Therefore, optimization of the protocol as well as careful quality control is crucial to make decellularized scaffold to be satisfactory for clinical use.

2.5 Cells

In order to enhance biocompatibility, it is important to achieve sufficient growth of cells on the graft prior to transplantation. Use of recipient autologous cells is undoubtedly advantageous with potential to make the graft immune-tolerant. The dominant cell types in the native trachea are epithelial cells lining the luminal aspect and chondrocytes inside the cartilaginous matrices. If possible, constructing a muscle and nerve network is desirable to establish the physiological function of controlling air flow and the defense reaction to noxious aspiration. It is however challenging, because it not only demands a population of smooth muscle and nerve cells, but also connection of this with the autonomous nervous system.

Types of cells used in previous studies vary from mature cells to stem cells. Mature cell lineages, such as epithelial cells, chondrocytes, and skeletal muscle cells, have an advantage of already being differentiated and functional, whereas stem cells, such as mesenchymal stromal/stem cells (MSCs) and induced pluripotent stem cells (iPSCs), retain a high proliferative potential and the ability to differentiate into different types of cells. Although there is no consensus on the most efficient combination of cell types to be seeded, recapitulating the epithelial–mesenchymal interaction observed in embryonic development by co-seeding both endoderm- and mesoderm-derived cells is conceivably valuable. However, there is a lack of reliable method to appraise colonization of cells after seeding. Neocartilage formation, which is an area of persistent chondrocytes surrounded by remodeled ECM, is often reported as an index of fresh colonization of chondrocytes, but its association with functionality is hard to determine (Maughan et al. 2017; Park et al. 2019; Zhu et al. 2016).

The source of epithelial cells can be from biopsy of either nasal or bronchial mucosa. Owing to the good availability of autologous epithelial cells, seeding epithelial cells on the tracheal lumen has already been performed in several studies, demonstrating prevention of mucus plugging and distal airway infection. However, it is unclear whether seeded cells can survive long-term or just temporarily cover the surface of airway until recruited endogenous cells overlay the lumen. Clinical cases report that the engrafted bioengineered trachea eventually obtains a functional mucosal layer, but the process is slow and takes months to complete (Elliott et al. 2012). This suggests that the ultimate epithelial coverage was gained by migration of the recipient's native cells rather than ingrowth of seeded cells.

Chondrocytes, the other principal cell type, play an important role in maintaining integrity and elasticity of the graft by generating cartilage. Considering native tracheal cartilage is composed of hyaline cartilage, it is reasonable to isolate cells from body parts enriched with hyaline cartilage, such as the trachea itself, ribs, knees, and nasal septum. The ear contains elastic cartilage that has also been shown to be possible source of chondrocytes, which can ultimately produce hyaline cartilage (Fuchs et al. 2002). Obtaining a specimen from these anatomical sites, however, requires invasive procedures, which represents a significant hurdle for clinical translation. In an effort to overcome this problem, researchers have recently shifted to focus on pluripotent cells as a cell source of chondrocytes.

MSCs are widely used for tissue engineering, as they have the potential to differentiate into various mesenchymal lineages, including osteoblasts, chondroblasts, and myoblasts. Ease of accessibility is a significant advantage, as isolation is possible from adipose tissue, bone marrow, and amniotic fluid, which can be obtained without highly invasive procedures. In addition, many preclinical and clinical trials have demonstrated the safety of using MSCs in in vivo experiments. Furthermore, MSCs are not only useful as a source of chondrocytes, but also appear to be promotors of engraftment by interaction with the endogenous immune system (Elliott et al. 2017; Seguin et al. 2013). The variability of chondrogenic potential depending on the source of MSCs is, however, an ongoing issue, and how MSCs contribute to tissue regeneration is yet to be fully understood (Diekman et al. 2010).

There is increasing interest in the implication of patient-derived iPSC in bioengineering, which have the potential to be expanded indefinitely in an undifferentiated state retaining the pluripotency. Studies demonstrate differentiation of iPSCs into epithelial cells and chondrocytes in response to tissue-specific culture conditions, successfully generating hyaline cartilage in animal models (Diekman et al. 2012; Zhu et al. 2016). One of the major limitations of the use of iPSC is the time required to obtain pluripotency and to induce proper differentiation, which could be a practical problem on translation to clinical applications. For antenatally diagnosed congenital tracheal malformations, utilization of cells derived from amniotic fluid is likely to be a promising alternative. Amniotic fluid contains abundant cells including stem cells (AFSCs), which have been shown to differentiate into multiple cell lineages, and can be easily obtained from amniocentesis between prenatal diagnosis and delivery, allowing enough time for cell growth and preparation of a graft.

In addition to choice of cell, the technical method for seeding cells on the scaffold is important, as well as the maturation period prior to in vivo transplantation. Wide variation in seeding techniques has been reported ranging from superficial delivery to micro injection. The number of cells to be seeded is a further variable which may impact outcome. It is generally acknowledged that at least 1×10^6 cells/cm^2

epithelial cells are required to cover the intraluminal area of the scaffold. Nevertheless, the type of the scaffold may impact cell proliferation and colonization, so this must be optimized according to individual protocol (Butler et al. 2017). Some researchers have developed tailored bioreactors which can generate a continuous flow of culture media, which may facilitate cells to grow in vitro. Populated grafts are generally kept in a static condition before entering a dynamic culture for the purpose of allowing cells to adhere on the scaffold.

2.6 Animal Models

Bioengineered grafts are thought to contribute to organ restoration in two separate fashions. One is through direct colonization of implanted, exogenous cells, which remodel surrounding native tissues, and the other is through a paracrine effect via cytokines secreted by seeded cells or preserved in natural scaffolds, recruiting endogenous cells to the implanted site. The paracrine effect also potentially promotes remodeling and vascularization of the graft by stimulating immune cells in vivo. Animal studies have been performed attempting to understand how each mechanism contributes to tissue repair, ultimately aiming to maximize the effect of tissue-engineered grafts on tissue remodeling.

The first in vivo transplantation of a tissue-engineered trachea in an animal model was reported by Vacanti et al. (Vacanti et al. 1994). Seeding of chondrocytes obtained from the shoulder of newborn calves on a sheet of nonwoven PGA mesh produced cylindrical cartilages. After subcutaneous implantation in a mouse for 4 weeks, the grafts were used to substitute a circumferential defect created in immune deficient rats. Death of all animals from respiratory distress was assumed to be secondary to an inability to clear secretions, highlighting the necessity of functional epithelium. A fetal lamb model where the trachea was augmented by a tissue-engineered cartilage patch was subsequently developed as a possible fetal approach for congenital tracheal stenosis (Fuchs et al. 2002). In this model, chondrocytes were derived from either fetal ear or tracheal ring and seeded on PGA sheet, followed by in vitro maturation in bioreactors for 6–8 weeks. This had promising results, with fetal survival, engraftment of the patch, and epithelialization on the luminal side.

Partial tracheal resection models in rabbits have been widely utilized to evaluate the biocompatibility of newly invented materials (Grimmer et al. 2004; Shin et al. 2015). Advantages of this model include simplicity of the surgical technique and higher survival rate compared to rodent models. Park et al. fabricated a multilayered implant containing epithelial cells derived from nasal mucosa on the inner layer and chondrocytes derived from auricular cartilage on the outer layer by utilizing a 3D printing technique (Park et al. 2019). They repaired a semi-circumferential defect created in the rabbit trachea with the artificial patch and obtained a trachea-like ciliated epithelial layer 6 months after surgery. However, cartilage formation was limited even after 12 months, suggesting the patency of the airway was maintained by the original trachea rather than the patched neo-cartilage.

Pig models are useful for assessing the feasibility of circumferential transplantation prior to human clinical translation due to their robustness and anatomical comparability to humans. Go et al. compared outcomes of tracheal implantation in pigs using decellularized homologous tracheas with bone marrow-derived MSC (BM-MSC) and epithelial cell seeding (Go et al. 2010). Maintained graft integrity was seen in the cell-seeded group. The graft without MSCs resulted in severe luminal collapse and the graft without epithelium was significantly contaminated with bacterium and fungus, suggesting the importance of repopulation with both chondrocytes and epithelial cells.

2.7 Clinical Trials

Several clinical trials for substituting a severely impaired trachea with an engineered graft have been reported. Failing to achieve appropriate tissue regeneration results in stricture and collapse of the graft. Initially, pioneers struggled to overcome the period immediately after transplantation during which the graft is vulnerable to cell ischemia and external pressure. Pre-vascularization of grafts in a rich vascular environment, such as the omentum and pericardial membrane, has been shown to be an effective option to improve adaptation of grafts. Delaere et al. revascularized a cadaver allograft in the recipient's forearm using the radial artery and successfully transplanted this into a 4.5 cm tracheal defect with vascular anastomosis (Delaere et al. 2010). However, the time taken to obtain a sufficient blood supply means this process may not be feasible in emergency settings. In such cases, interposition of a pedicled wrap is likely to be the solution to shorten the pre-vascularization period. In terms of prevention of early disintegration, securing the lumen by placing a stent is arguably essential until tissue remodeling progresses and the structure of the graft becomes strong enough. Despite a number of trials, the best procedure to accomplish this goal is yet to be validated.

The first case of circumferential replacement of the proximal airway with a tissue-engineered graft was performed on an adult patient affected with severe bronchial stenosis and malacia due to post-tuberculous chronic bronchitis (Macchiarini et al. 2008). The scaffold was obtained by decellularization of the trachea retrieved from a human cadaver and repopulated with nasal epithelial cells and MSC-derived chondrocytes. The stenotic left main bronchus was replaced with the graft and vascularization promoted by interposition of an omental wrap. In this patient, signs of an establishing vascular supply were observed 4 days after implantation with subsequent restoration of the mucosal layer.

The first clinical application in a pediatric patient was led by a group in the UK in an 11-year-old-boy with long-segment congenital tracheal stenosis. This was initially repaired by an autologous patch tracheoplasty, but scarring of the patch and severe bronchomalacia required a stent placement, which resulted in creation of an aortotracheal fistula at 3 years old. Repair by implantation of a homologous graft was successful, but required ongoing stents for recurrent stenosis, which eventually lead to an acute hemorrhage due to stent erosion. Repeated failure of treatment and the devastating condition of the patient rendered him a candidate for tissue-engineering therapy. The scaffold was obtained by decellularization of the cadaveric trachea from a 30-year-old donor, matched to the recipient's trachea size. Because of the urgent clinical need, repopulation of the graft was performed during the surgery with hematopoietic stem cells derived from bone marrow aspiration. Human recombinant erythropoietin (hrEPO), G-CSF, and transforming growth factor β (TGF-β) were injected into the graft to promote angiogenesis, indigenous MSC recruitment, and chondrocyte differentiation. The omentum was mobilized and interposed between the transplanted trachea and the heart to prevent perforation and increase the vascularity of the graft. Bleeding from the luminal mucosa of the graft was observed 1 week postoperatively, suggesting an established blood supply to the mucosa. Although the intratracheal stent was periodically replaced after surgery, the graft eventually became stable allowing him to be stent free (Elliott et al. 2012; Hamilton et al. 2015). After some minor alterations, this resulted in a good manufacturing practice (GMP) compliant method to produce a decellularized tracheal scaffold with cell-seeding (Elliott et al. 2017). Sadly, however, the subsequent application of this method in a 15-year old girl with congenital tracheal stenosis resulted in a lethal acute airway obstruction postoperatively. The lack of a stent was deemed pivotal, and based on their experiences prolonged stenting posttransplantation is now advocated. This case emphasizes the requirement of a significant evidence base before further clinical use and full disclosure of negative as well as positive clinical outcomes.

2.8 Conclusion

For many decades, surgeons and scientists have striven for a solution to devastating tracheal disorders. Significant progress in cell biology and scaffold manufacture has realized clinical translation with long-term reports demonstrating adequate epithelialization and stabilization of grafts. These primary successes seemed a promising novel alternative for patients with limited other treatment options. Unfortunately, however, not all cases have been successful. This has led researchers back to the laboratory to find the solution; a deeper understanding of the mechanism of tissue regeneration will be key to help refine and improve existing techniques. Clinical lessons from these early successes and failures must also be harnessed. As the trachea is a truly vital organ, luminal collapse and occlusion debilitates the patient's condition immediately. As such, close monitoring for signs of stricture and blockage is crucial to prevent graft failure during the precarious postoperative period. Finally, trials of novel tissue-engineered therapies in humans must not be rushed and multidisciplinary discussion and ethical committees are essential prior to planning surgery, even when limited or no alternatives are available.

3 Esophageal Bioengineering

3.1 Introduction

Complex congenital and acquired esophageal pathologies may require esophageal substitution to restore anatomical continuity. This continues to present a significant challenge to both pediatric and adult surgeons as traditional surgical techniques carry significant morbidity. Tissue engineering techniques offer a promising alternative to treat these conditions, with different strategies required depending on the nature and depth of damage caused to the esophagus.

In the pediatric population, the primary indication for esophageal replacement is esophageal atresia. Faulty embryonic separation of the esophagus and trachea results in an abnormal connection between the two organs. In the majority of cases this results in a tracheoesophageal fistula (TEF) in which a primary anastomosis is usually possible. In approximately 10% of patients, however, limited distal esophagus exists (van der Zee et al. 2017). In this circumstance, primary anastomosis may not be feasible due to a large tissue deficit, resulting in the requirement for an esophageal substitute. Even patients undergoing primary anastomosis may require an esophageal substitute regardless in the eventuality of recurrent anastomotic leak or recurrent fistula, albeit infrequently. Although less common, severe esophageal strictures refractory to endoscopic intervention may also require consideration of esophageal replacement, secondary to caustic injury, significant gastro-esophageal reflux, ischemia, radiation exposure or anastomotic stricture.

Despite the advent of organ transplantation in the 1950s, current techniques for esophageal allografts are limited entirely by the segmental vascular supply to the esophagus from multiple arteries, rendering this extremely technically challenging in adults and impossible in neonates. The combination of this and other transplant-associated issues such as scarcity of organ availability, need for immunosuppression and infection risk precludes orthotopic donor transplantation from being a viable treatment option for esophageal replacement. The concept of esophageal replacement is long standing and initial attempts in children began in the first half of the twentieth century with jejunal, colonic, and gastric constructs. Over the last 50 years, while refinements have been made and outcomes have improved, techniques for surgical replacement remain essentially unchanged. All three techniques have respective weaknesses; gastric pull-ups are associated with worse reflux, colonic transpositions have a higher risk of redundancy and delayed emptying with an unknown malignancy risk, and jejunal interpositions have a significant risk of anastomotic leak and graft loss (Gallo et al. 2012). Additionally, in rare situations, failure of these replacements with no appropriate esophageal substitute precludes further reconstruction and renders the patient unable to feed orally.

While esophageal malignancy in children is extremely rare, esophageal cancer is the seventh most common cancer worldwide in adults with over 500,000 new cases in 2018 (World Cancer Research Fund; https://www.wcrf.org/dietandcancer/cancer-trends/esophageal-cancer-statistics). Where resection is possible, the standardized surgical option is esophagectomy with reconstruction. Gastric tubularization remains the primary reconstructive technique due to technical ease and safety, although colonic transpositions or free or pedicalized jejunal grafts are also used when use of the stomach is not possible. Regardless of technique, this is a major undertaking; even with the introduction of minimally invasive techniques, in a meta-analysis of over 14,000 patients, in-hospital-mortality was 4%. This reflects both the underlying health status of this population and gravity of the operation (Zhou et al. 2015). As such, overall and disease-free survival have been the primary aims with less focus on long-term functional outcomes despite over 70% of patients reporting long-term dysphagia in a recent a systematic review, with additional troublesome symptoms of reflux, dumping syndrome, and delayed gastric emptying (Irino et al. 2017). With ongoing improvements in oncological treatments and minimally invasive techniques, however, adult patients can expect to live longer with their esophageal substitute, meaning long-term functional results are increasingly as important as short-term outcomes.

Esophageal tissue engineering has huge potential to provide an alternative approach, critically without the loss of function of another gastrointestinal organ. Huge advances in this field have allowed for successful replacements of partial thickness esophageal defects in humans. While whole organ replacement is far from possible at present, animal models show increasing promise for imminent implementation of full thickness patch and circumferential replacements which could change the face of esophageal pathology.

3.2 Anatomy, Physiology, and Development

Bioengineering of the esophagus presents distinct challenges from that of the trachea. In adults, the esophagus is approximately 20–25 cm in length, corresponding to 8–10 cm in a neonate. It crosses three anatomical planes during its course to the gastro-esophageal junction; the neck, thorax, and abdomen. As such, it is intimately related to key mediastinal structures including the trachea, aorta, recurrent laryngeal, and vagus nerves, making approaches operatively challenging with high morbidity. Due to its length, the esophageal blood and lymphatic supply is segmental with the upper, middle, and lower thirds supplied by differing branches of regional vasculature.

While also acting primarily as a conduit, unlike the trachea the motility of the esophagus is essential. Gravity alone is insufficient to propel a food bolus from the oropharynx to the stomach and it is the complex interplay of neural and muscular mechanisms which enables coordinated peristalsis and enteral autonomy. To facilitate this, the esophageal wall is composed of four main layers; the mucosa, submucosa, muscularis externa, and adventitia. The *mucosa* comprises of three distinct regions. On the luminal aspect is stratified non-keratinized squamous epithelium. The presence of a highly proliferative basal epithelial layer can readily replenish the well-differentiated, flattened superficial layer when injury occurs, protecting the esophagus from both mechanical and chemical injury during passage of food. Deep to this, the lamina propria is composed of loose connective tissue, vessels, and sensory nerves with numerous cell types including fibroblasts, endothelial, and smooth muscle cells. Finally, the muscularis mucosae bestows the overall mucosal shape and contracts to move the luminal folds during the passage of food. The *submucosa* is composed of a dense layer of loose connective tissue, rich in collagen and elastin. The orientation of ECM fibers in this layer allow for the high circumferential distensibility of the esophagus without compromising longitudinal strength (Bonavina et al. 1995). Glands secrete mucous directly into the lumen via transmucosal ducts, allowing for lubrication of food. The submucosal nerve plexus regulates the secretion of these ducts, in addition to contraction of the muscularis mucosa. The orientation of inner circular and outer longitudinal layers of muscle in the *muscularis externa* gives rise to radial contractions and longitudinal shortening resulting in peristalsis and is coordinated by the myenteric nerve plexus, which lies between the two muscular layers. In the human esophagus, this changes from striated skeletal muscle to smooth muscle at approximately one third of the cranial to caudal distance. Peristalsis is therefore under voluntary control in the proximal one third, and autonomic control distally, however this distance is variable among species. Below the diaphragm, the musculature specializes to produce a physiological rather than anatomical lower esophageal sphincter at the gastro-esophageal junction. This reduces reflux of gastric contents due to a higher intrinsic spontaneous tone than that found in the musculature of the esophageal body, which relaxes on swallowing. The diaphragmatic crural muscle fibers surround the lower esophageal sphincter to act as a pinchcock, supplementing this anti-reflux mechanism.

3.3 Strategy and Challenges for Esophageal Tissue Engineering

Esophageal tissue engineering has traditionally involved three approaches; scaffold alone, cells alone, or a combination of scaffold and cells. The use of these techniques is dependent on clinical need; mucosal defects may only require epithelial cells for reconstruction whereas full thickness defects clearly require all layers of the esophageal wall. The location and size of the defect may also have implications for replacement requirements depending on the anatomy and function of that region. For example, a more proximal replacement should be skeletal rather than smooth muscle with less requirement for the epithelium to be reflux resistant. Similarly innervation, and therefore peristalsis of an esophageal patch, does not appear to be integral to global function whereas in a circumferential replacement this may be vital.

Evidence now suggests that successful full thickness esophageal tissue engineering requires three key components; a biocompatible scaffold, cells able to differentiate into epithelial and muscular tissues, and a vascular supply. The presence of these components alone, however, is not sufficient and organization to resemble a multi-striated tube is key for function. Significant challenges include avoiding stricture and coordinating peristalsis, with the absence of either resulting in dysphagia, regurgitation, or aspiration. Coordinated peristalsis requires communication of the enteric nervous system with a continuous bi-directional muscular layer. The biomechanical and elastic properties of the native esophagus must also be replicated to withstand the pressure of passage of a food bolus without perforation, for which appropriate choice of scaffold is essential. Finally, the ability to be resistant to chemical or mechanical injury from the passage of food and gastric acid requires an epithelial barrier, which also appears to be integral to protection against stricture.

The other main challenge for clinical translation is vascularization. The use of bioreactors prior to transplantation ensures transport of oxygen and nutrients to the scaffold to promote cell adherence, proliferation and migration, tissue maturation, and angiogenesis. Various "natural" bioreactors have been used including the greater omentum, latissimus dorsi, or thyroid gland flaps and results in animal models have been promising. Pre-vascularization with omental wrapping or in vitro bioreactor time clearly had a beneficial effect on tissue regeneration in circumferential cervical esophageal replacements with cellularized hybrid scaffolds compared to controls in a rat model (Kim et al. 2019). However, the effect of duration of pre-transplantation maturation still needs to be addressed with regard to mechanical strength and cell survival. Luc et al. attempted to identify the optimal duration of omental in vivo maturation with seeded and unseeded decellularized matrices in a murine model. While vascularization was present at 2 weeks, this was at the expense of a marked inflammatory infiltrate of mononuclear cells, suggesting this time period is too short. By 4 weeks, vascularization persisted with weaning of inflammatory infiltrate; however some degradation of the scaffold was seen and by 8 weeks, more than half of the matrix had degraded with no discernable layers (Luc et al. 2018). Omental maturation does not come without risks; it requires two operations rather than one and this has been reflected in lower weight gain in animal models with pre-vascularization compared to controls in porcine models (Luc et al. 2018). In addition, several studies found that epithelial cells specifically did not survive short omental pre-implantation periods (Nakase et al. 2008; Poghosyan et al. 2015). Further studies are required to understand whether there is a clinical advantage in pre-vascularization.

3.4 Scaffold

Minimum requirements for clinical translation require a scaffold to be biocompatible, non-immunogenic, and nontoxic. It must be porous enough to allow cell delivery, adherence, migration, and differentiation, while facilitating permeability of nutrients prior to neovascularization. Finally, it must retain elasticity while being mechanically robust. In addition to this, desirable qualities include biodegradability to prevent mismatch with patient growth and low inflammatory potential. The ideal scaffold would also be replicable, easy to store, and readily available in a variety of sizes.

Currently, as with the trachea, both synthetic and biologically derived scaffolds have been utilized for esophageal tissue engineering, in combination with, or without, cellular seeding. More recently, advances in hybrid or "intelligent matrices" may be a promising future alternative.

3.4.1 Synthetic Scaffolds

Nonabsorbable synthetic scaffolds have been used for esophageal replacement for over a century with initial unsuccessful attempts using materials including ivory, metal, rubber, and plastic tubes. Regeneration of full-thickness cervical esophageal defects in dogs was compared with one of the three materials; lyophilized dura mater (biological), polytetrafluoroethylene (PTFE), and polyethylene terephthalate (Dacron), both non-absorbable synthetic materials. While anastomotic leak was high in all three groups, significantly increased rates of foreign body reactions, delayed epithelialization, and circumferential stenoses were seen in the synthetic material group (Freud et al. 1999). Formulation of muscle generation was also absent, suggesting that while they provide mechanical support, they are unable to promote tissue regeneration. Results such as these paved the way for development of biodegradable synthetic polymers as absorbable scaffolds, which degrade in situ with absorption of by-products, leaving an entirely new biological layer. As with the trachea, similar materials have been used including PLGA, PLA, PCL, and PLLA, either in isolation or in combination with a protein coating.

The main advantages of absorbable synthetic scaffolds are the ability to customize them to the requirements of the patient with respect to size and biomechanical strength, with an "off-the-shelf" availability. The main limitation, however, remains biocompatibility; close attention must also be paid to the fine balance between early degradation resulting in loss of mechanical stability versus delayed degradation inhibiting tissue remodeling. Some studies have suggested acidic microenvironments on degradation of biodegradable scaffolds may result in a locally toxic environment, resulting in poorer cell adherence and contributing to a high stricture incidence (Ceonzo et al. 2006; Kohn et al. 2002). The final criticism of absorbable scaffolds for esophageal reconstruction is that they are predominantly used to "bridge the gap," rather than to reconstruct the complex native esophageal layers. One group attempted to address this issue with an electrospun polycarbonate polyurethane polymer (PCU) multilayered scaffold with broad pore layers on the luminal and exterior surface with an intervening small pore layer. This prevented mixing of different seeded cell populations while allowing for diffusion of nutrients and oxygen with promising results (Soliman et al. 2019). While this technology is still in proof of concept phase, other groups have shifted their focus to both natural and, more recently, hybrid alternatives, which are increasing in popularity.

3.4.2 Natural Scaffolds

For organ-specific tissue engineering, decellularized xenogenic matrices offer an attractive prospect; a ready-made, multilayered, three-dimensional scaffold with in situ polysaccharides, proteins encouraging cell repopulation, and tissue-appropriate cytokines to guide regeneration. Decellularized tissues commonly used for esophageal tissue engineering include small intestinal submucosa (SIS), urinary bladder (UBM), and esophageal matrices. Decellularized porcine esophagus appears to be a logical choice. It has similar anatomical dimensions to the human esophagus and a precedent exists for use of porcine decellularized matrix in humans with cardiac valves. In addition, using the organ-specific matrix may have advantages for effective tissue-specific recellularization as previously shown in the liver (Faulk et al. 2015). A reproducible technique for clinical grade porcine decellularized matrices for esophageal replacement has shown the scaffold remains structurally intact, with similar transverse tensile strength to that found in the native esophagus (Arakelian et al. 2019). While longitudinal strength appeared to be stiffer in decellularized specimens, this remained easily suturable in vivo, findings echoed by Luc et al. in 40 decellularizations of porcine esophagus (Luc et al. 2018).

3.4.3 Composite/Hybrid Scaffolds

The relative weaknesses of both biological and synthetic scaffolds have led to increased investigation into a combined approach with composite or "hybrid" scaffolds. Synthetic scaffolds can be coated with biological compounds to improve biocompatibility and cell adherence or biological scaffolds can be reinforced with synthetic materials to improve strength. PLGA scaffolds grafted with collagen and fibronectin demonstrated significant improvements in smooth muscle proliferation and epithelial regeneration, respectively, with no difference in the tensile strength of the scaffolds (Zhu et al. 2005; Zhu et al. 2007). Similarly, reinforcing a tubular SIS construct with electrospun PLGA nanofibers led to improved mechanical properties compared to SIS alone. This was also found to improve muscle cell alignment with limited inflammation in a subcutaneous rat model and allowed for delivery of implanted bioactive molecules of vascular endothelial growth factor (VEGF), which improved angiogenesis (Syed et al. 2014).

Nanoparticles will likely have an increasing role in the future of "intelligent matrices," either by delivering bioactive molecules to promote tissue regeneration as above or to improve structural integrity. When compared to decellularized esophagi alone, those conjugated with silver nanoparticles had improved structural stability and biocompatibility with reduced host immune response in a subcutaneous mouse model (Saleh et al. 2019). The addition of copper, known for its angiogenic properties, to SIS scaffolds promoted reepithelialization, revascularization, and muscular regeneration compared to SIS alone in a cervical patch esophagoplasty model (Tan et al. 2014). Finally, 3D printing represents an exciting development in the field of "scaffold-free" esophageal replacements. Takeoka et al. described a technique of multicellular spheroid culture, with subsequent harvesting and arrangement into a tubular formation by a 3D printer and fusion into a continuous layer after 1 week in a bioreactor. While questionable evidence of muscle development was seen and tensile strength was not comparable to native esophagus, this model was able to withstand esophago-gastric bypass transplantation in rats with no perforation at 30 days (Takeoka et al. 2019). Clearly more work is needed but this field may represent a promising alternative to current tissue engineering techniques.

3.5 Cells

While the precise role of exogenous cell delivery on tissue remodeling is as yet unknown, it appears to be beneficial. Whether biological or synthetic in nature, acellular scaffolds perform poorly in vivo compared to their cellularized counterparts with increased stricture rates and less muscle and epithelial regeneration (Nakase et al. 2008). Mature cells, stem cells, and muscle cell progenitors from mesenchymal and endodermal lineages have all been used to tackle scaffold cellularization. Autologous cells are preferential as these do not carry a risk of bacterial or viral transmission and do not require immunosuppression. Cells must be easy to harvest with minimal donor site morbidity and ideally have excellent proliferation and differentiation capacity once seeded. Cell selection is also dependent on the clinical need of the construct; partial thickness injury may only require one cell type for repair; for prevention of stricture formation after mucosectomy, for example, epithelial cells alone are required to regenerate the mucosa. When circumferential esophageal replacement is required, however, a combination of cells may be more effective for inducing regeneration in all layers.

3.5.1 Epithelial Cells

Epithelial cells are required as a physical barrier from mechanical stress, infection, and gastric acid. Regardless of scaffold type or presence of cells, ingrowth of endogenous epithelium appears to occur by 3 months; however this response appears to be slower and less comprehensive than in those with epithelial cell seeding (Wei et al. 2009). Interestingly, some studies have suggested decellularized scaffolds allow for a more mature, stratified epithelium than that seen on synthetic scaffolds (Beckstead et al. 2005).

Epithelial cells may be sourced from either buccal mucosal biopsy or endoscopic esophageal biopsy giving rise to oral muscosal epithelial cells (OMEC) or esophageal epithelial cells (EEC), respectively. Newer techniques for epithelial cell delivery include use of organoid units (Grikscheit et al. 2003; Spurrier et al. 2015) or epithelial cell sheets; culture of cells on thermo-responsive polymers allows the polymer to convert from a hydrophobic to hydrophilic state on temperature reduction, resulting in epithelial cell detachment without compromise to cell morphology or function (Yamato et al. 2001). This has now been extensively used with OMECs and is in clinical use for stricture prevention after submucosal dissection.

Smooth and Skeletal Muscle Progenitors

Attempts to repopulate the muscularis externa have included combinations of several different types of cell seeding. Autologous delivery of cells likely provides paracrine signals and growth factors, lowering the inflammatory response and inducing migration of host muscle cells from the edges of the implant. Saxena et al. obtained unidirectional, smooth-muscle-actin positive muscle fibers after seeding rat smooth muscle cells on collagen scaffolds in vitro (Saxena et al. 2009). In a canine model, esophageal decellularized matrix seeded with mature smooth muscle cells had muscular regeneration and significantly less inflammatory infiltrate at 3 weeks than unseeded controls (Marzaro et al. 2006). Main limitation of adult smooth muscle cells is their origin as they are obtained from either the vasculature or esophagus. In addition, they demonstrate slower expansion than other cell lines used for muscular replacement.

Skeletal muscle progenitor cells including myoblasts are easily obtained from skeletal muscle biopsies. Cultured human and porcine myoblasts on SIS produced multinucleated, desmin-positive skeletal muscle fibers with upregulation of late skeletal muscle markers including MyoD. When adapted for circumferential, full thickness use in a porcine model, evidence of circular muscle morphology was seen at 9 months (Poghosyan et al. 2016).

Mesoangioblasts (MABs)

Mesoangioblasts (MABs) are pericytes which are easy to culture, have prior approval for clinical use and have smooth and skeletal muscle differentiation potential. Seeding of MABs, fibroblasts, and neural crest cells on decellularized rat esophagi with omental maturation and subsequent EEC seeding resulted in an organized esophageal construct with a multi-stratified epithelium and mature smooth muscle layer (Urbani et al. 2018). Co-seeding with fibroblasts clearly improved the migratory potential of the MABs and cell distribution throughout the scaffold. The synergistic effect of co-seeding with fibroblasts does not appear to be a phenomenon exclusive to muscle progenitors; when EEC were co-seeded with fibroblasts, superior epithelial layer generation was seen compared to EEC alone (Miki et al. 1999).

Mesenchymal Stem/Stromal Cells (MSCs)

MSCs derived from bone marrow, adipose tissue, and amniotic fluid have all been used in in vivo esophageal models (Jensen et al. 2018; Luc et al. 2018; Tan et al. 2013; Wang et al. 2018). SIS seeded with BM-MSCs was compared to unseeded SIS in a canine patch esophagoplasty model; the presence of MSCs enhanced both epithelialization and muscularization with faster and more comprehensive generation of mucosal and muscular layers. Significant neovascularization and reduced inflammation in the BM MSC-group also suggests MSCs promote angiogenesis and tissue healing while reducing inflammation, properties previously affiliated with the role of MSCs in tissue repair (Tan et al. 2013). Pluripotency, ease of harvesting, and immunomodulatory effects continue to make MSCs an attractive option for cellular seeding.

3.5.2 Neural Progenitors

Cellular approaches have primarily focused on regeneration of epithelial or muscular layers; however neuronal structures are also required for esophageal function; the enteric nervous system is necessary for coordinated peristalsis and release of neuropeptides. Use of enteric nervous system progenitors has demonstrated promising results. When seeded with MABs and fibroblasts, neural crest cells distributed throughout the scaffold in a ring-like formation similar to that seen in the native esophagus. Subsequent differentiation into neurons and glial cells and connections with MABs represents promising evidence that this approach may result in functional neuromusculature when adapted in vivo (Urbani et al. 2018). Whether this seeding is required, however, remains to be seen. The presence of neural markers (S100B) was found in regenerated tissue despite the absence of neural cell seeding in in vivo models (Algarrahi et al. 2018; Jensen et al. 2018). Additionally, despite the absence of peristalsis on barium swallow or limited muscular regeneration on histology, some animal models were able to grow and feed normally (Urita et al. 2007; Yamamoto et al. 1999). Therefore, although ideal, peristalsis may not be essential for short segment circumferential esophageal replacements.

3.6 Clinical Trials

While strategies for mucosal replacement are well underway with some success in human studies, attempts at full thickness replacements are currently in early stages. Animal models have provided real insights into future techniques; however any potential construct for human use requires rigorous evaluation, particularly with regard to implant integration, immune response, survivability, and long-term functionality of the graft. This is especially critical when considering use in pediatric patients who require the replacement to provide a life-long, good functional outcome.

3.6.1 Partial Thickness Defects

Partial thickness defects of the esophagus are predominantly iatrogenic. Endoscopic mucosal resection and submucosal dissection (EMR/ESD) to treat superficial esophageal cancers reduces the need for esophagectomy. Unfortunately, however, they are associated with high stricture rates due to inflammation and scarring. Strictures significantly affect quality of life, requiring treatments such as dilatation, stenting, and application of steroids. In an animal model of mucosal loss by EMR, ulcer formation, inflammatory cell invasion, and collagen hyperplasia were observed in the first week with fibrosis of the submucosa by day 28. Recommendations to reduce stricturing included reduction of the inflammatory response, promotion of epithelial regeneration, and prevention of damage to the intrinsic muscle layer (Honda et al. 2011). Regenerative medicine approaches have attempted to address these via cell or scaffold delivery to areas of mucosal loss (Table 1).

Endoscopic injection of autologous cell suspensions has been used in animal models with success. Direct injection is simple and quick, however limited by the cell number isolated from the tissue and the viability rate posttransplantation. Various cell types have been used in EMR animal models. Autologous OMECs harvested from oral mucosal biopsy induced reepithelialization by 2 weeks compared to no regeneration in controls without injection (Sakurai et al. 2007). Injected adipose tissue-derived stromal cells showed a similar result in a canine model; improved dysphagia scores, reduced mucosal contraction, and improved angiogenesis were seen in the seeded group compared to controls at 8 weeks (Honda et al. 2011). Finally, injection of autologous keratinocytes from a split skin graft in a sheep model showed no evidence of stricture at 6 months (Zuercher et al. 2013). The contribution of the exogenous cells is unclear. While labeled OMECs were still present in the defect at 2 weeks in Sakurai's model, this is too early to determine if they contribute to long-term tissue remodeling and the rapid turnover of esophageal epithelium negates effective tracing over longer time periods. What appears clear, however, is that delivery of cells directly after EMR appears to promote early reepithelialization and prevent inflammation and fibrosis.

Table 1 Summary of papers which describes replacement of esophageal mucosa

Authors	Year	Cell/Material	Model	n = x	Control	Stricture	Histology: Epithelialization	Follow-up
Cell injection								
Sakurai et al.	2007	OMEC	Porcine	4	4	Reduced vs control	++	2 weeks
Honda et al.	2011	ADSC	Canine	5	5	Reduced dysphagia scores	Unreported	8 weeks
Zuercher et al.	2013	Skin keratinocytes	Ovine	9	Nil	Nil	Unreported	6 months
Cell sheet								
Ohki et al.	2006	OMEC	Canine	3	3	Nil	+++	4 weeks
Ohki et al.	2012	OMEC	Human	9	Nil	11% stricture	++*	4 weeks
Kanai et al.	2012	Epidermal	Porcine	4	4	Reduced vs control	+	2 weeks
Jonas et al.	2016	OMEC	Human	5	Nil	60% stricture	++ **	4 weeks
Perrod et al.	2017	ADSC	Porcine	6	6	Reduced vs control (17% vs 100%)	++ **	4 weeks
Yamaguchi et al.	2017	OMEC	Human	10	Nil	40% stricture	++*	105 weeks
Covered stents								
Nieponice et al.	2009	Unseeded porcine UBM	Canine	5	5	Reduced vs controls	+++	8 weeks
Badylak et al.	2011	Unseeded SIS	Human	5	Nil	100% requiring dilatation	+++	24 months
Barrett et al.	2014	Unseeded amniotic membrane	Porcine	10	10	100% by day 35, less severe in AM group	+	5 weeks
Han et al.	2017	Acellular dermal matrix	Porcine	7	7	Less severe in ACM group	+++	4 weeks

Oral mucosal epithelial cells (OMEC); Adipose tissue-derived stromal cells (ADSC); Urinary bladder matrix (UBM); Small Intestine Submucosa (SIS). Evaluation of the outcome: + patchy/incomplete; ++ continuous; +++ mature and stratified. * endoscopic assessment; ** confocal laser endomicroscopy

Endoscopically deployed acellular ECM grafts post EMR have had mixed results. Endoscopic placement of porcine UBM resulted in development of a stratified epithelium and weight gain compared to 100% stricture in controls in a dog model (Nieponice et al. 2009); however this initial success has not been replicated in subsequent models. A similar trial in humans using SIS with stent had poor results with strictures in all patients (Badylak et al. 2011). Median stricture-free survival was significantly better in a porcine model of human amniotic membrane (HAM) and stent compared to controls with no ECM or stent delivery after ESD. The lack of difference in survival between the "HAM/stent" and a "stent only" group, however, suggests that stricture reduction may be due to the presence of a stent, rather than the contribution of the acellular ECM graft (Barret et al. 2014). More recently, acellular dermal matrix (ADM) was applied with surgical clips in a porcine model and compared to a control group with no intervention. Despite the absence of stents in the study, the ADM group had less stenosis, a more complete reepithelialization and reduced inflammatory infiltrate compared to controls (Han et al. 2017). This suggests that ECM grafts alone may have a role; however the full extent of potential clinical application is yet to be seen.

Perhaps the most promising results for translational use have been with cell sheet technology. In 2006, tissue-engineered autologous OMEC sheets were transplanted in a canine model immediately after ESD. Those transplanted with OMEC sheets had significantly better results compared to controls with no OMEC; a stratified, mature epithelium resembling the native esophageal surface was seen in the OMEC group at 4 weeks compared with patchy, immature epithelium with ulcerated areas in all controls. Marking of cell sheets indicated that the epithelial cells were entirely derived from the exogenous OMECs, suggesting exogenous cells can contribute to tissue remodeling (Ohki et al. 2006). These results have been replicated with different cell types. Transplantation of allogenic adipose tissue-derived stromal cell sheets post-ESD had a lower stricture rate and fibrosis in a porcine model compared to controls (Perrod et al. 2016) and autologous epidermal sheets were as effective as OMEC sheets in preventing esophageal stricture after ESD in porcine models (Kanai et al. 2012). Proposed mechanisms for reduced stricture rate include provision of an epithelial barrier protecting the underlying submucosa and muscularis from mechanical damage and secretion of growth factors and cytokines to recruit host epithelial cells to proliferate and migrate into the wound.

Human feasibility and safety studies have now shown cell-sheet technology to be a reproducible and safe measure to promote early reepithelialization in ESD. The first transplantation of autologous OMEC sheets in humans had a median wound healing time of 3.5 weeks, with the exception of one case transecting the gastroesophageal junction, which took significantly longer and developed a refractory stricture (Kanai et al. 2012). A study of ten patients over 2 years demonstrated a 40% stricture rate with requirement for 1.5 median balloon dilatations and no significant complications or adverse reactions reported (Yamaguchi et al. 2017). Jonas et al. reported a similar time to reepithelialization, however a 60% stricture rate in five patients undergoing a similar technique. Possible explanations included a more distal esophageal ESD in patients with known reflux and fewer numbers of cell sheets used to cover the defect (Jonas et al. 2016). The lack of control arms in these human studies makes effectiveness compared to current standard treatments difficult to ascertain. This has paved the way for a multicenter, prospective, phase III clinical trial for which recruitment has now been completed.

Full Thickness Defects – Patch

Patch esophagoplasty, whereby a full-thickness patch of esophagus is replaced, may be a treatment option for localized perforation. Conservative management requires prolonged hospital stay, has a high failure rate, and often results in stricture formation as a sequelae of the healing process. Therapeutic endoscopic interventions including clipping, suturing, and stenting have high success rates in acute cases, however this is much lower in anastomotic leaks or chronic fistulae (Dasari et al. 2014; Mennigen et al. 2014). Where these techniques are not possible, due to failure or the size or complexity of the defect, surgery is required. Current surgical alternatives for esophageal reconstruction involve either a pedicled or free tissue flap requiring microvascular anastomoses with high morbidity (Lin et al. 2017; Sa et al. 2013). Surgical reconstruction of recalcitrant strictures refractory to endoscopic intervention is also increasingly used. In these circumstances, tissue-engineered patch esophagoplasty has the potential to allow preservation of the native esophagus and avoid complex esophageal reconstruction.

Natural scaffolds as models for patch esophagoplasty have had universally good results. When 50% patch defects were repaired with acellular SIS in dogs, survival was up to 15 months. While no evidence of stricture was seen in the patch group, all four dogs with circumferential replacements developed stricture (Badylak et al. 2000). A rat model had similar results; patch and circumferential defects were replaced with acellular SIS with survival of all animals in the patch group compared to none in the circumferential group due to obstruction, stricture, or leak (Lopes et al. 2006). Epithelialization was seen by 4 weeks with organized skeletal muscle bundles and neoinnervation in both models. Interestingly, in a patch esophagoplasty rat model with gastric acellular matrix, survival until 18 months was reported despite no evidence of muscle regeneration, suggesting peristalsis is not critical for esophageal function or survival in patch repair (Urita et al. 2007).

Despite evidence of tissue regeneration and good clinical outcomes in these models, the cell-seeded ECM approach has increased in popularity and with good reason. Cellularized patch defects appear to heal quicker compared to ECM patch esophagoplasty alone. Improved reepithelialization and skeletal muscle regeneration was seen in OMEC-seeded SIS compared to SIS alone in patch defects in a canine model, with a faster clinical recovery and better weight gain (Wei et al. 2009). This is not a result exclusive to epithelial cell seeding; smooth muscle and BM-MSC seeded ECM scaffolds also demonstrate earlier epithelialization and improved muscular regeneration compared to controls in patch esophagoplasty models (Marzaro et al. 2006; Tan et al. 2013).

Patch models on biodegradable synthetic scaffolds have had varied results. Unseeded poly-e-caprolactone (PCL) mesh used for abdominal esophageal patch replacement had a 25% 1-month mortality in 20 rabbits, with stricture in 13% of the surviving animals and pseudodiverticula in 60% (Diemer et al. 2015). This implies degradation of the mesh was too fast to allow for proper ingrowth of new tissue, a result previously found in vicryl mesh rabbit model with an 80% anastomotic leak rate (Jansen et al. 2004). Park et al. performed the only in vivo cell-seeded patch model using a synthetic scaffold in a rabbit model. At 3 weeks, the BM-MSC seeded fibrin-coated PCL group had epithelial and smooth muscle development compared to no regeneration in unseeded PCL controls, suggesting that cell seeding has the same effect of enhancing regeneration in synthetic scaffolds as in natural scaffolds (Park et al. 2016). In addition, no strictures or pseudodiverticulae were seen, albeit at a short end point. As such, the cell-seeded synthetic scaffold approach has now been adopted into circumferential defect models.

Novel synthetic scaffolds have included subcutaneous implantation of a silicone mould for 8 weeks to create a "biotube" of collagen-based tissue. After cervical transplantation, all animals survived until 12 weeks with normal epithelial covering and some evidence of organized skeletal muscular bundles on histology (Okuyama et al. 2018). However, success in this model must not be overstated as the defect was small (1 × 2 cm) and the true stricture rate is unknown due to short follow-up period. Algarrahi pioneered the use of bilayer silk fibroin grafts; a biodegradable scaffold with a porous foam compartment for host tissue integration and a buttressing film layer preventing leakage of luminal contents (Algarrahi et al. 2015). Superior muscular and neural regeneration compared to SIS in a rat patch esophagoplasty model resulted in its use in a subsequent porcine thoracic defect model. All animals survived to 3 months with peristalsis on fluoroscopy and no strictures observed. Although underdeveloped in comparison to native esophageal tissue, organized circular and longitudinal layers of muscularis externa were reported, perhaps representing the best evidence of muscular regeneration in a patch model to date. In addition, submucosal glands were observed and the presence of synaptophysin positive boutons and CD31+ endothelial cells confirmed the presence of neo-innervation and vascularization similar to native tissue (Algarrahi et al. 2018).

What is clear from animal studies is that patch defects are less susceptible to stricture formation than circumferential models, despite the use of natural or biological scaffolds, or cellular or acellular constructs (Table 2). Relatively good clinical outcomes despite a lack of comprehensive muscular regeneration also suggests that peristalsis in patch models may not be critical to outcome. Direct extrapolation of the success of these models into human studies must be done with caution as often defects were small, definitions of stricture and muscle generation varied, and follow-up relatively short. Additionally, limited data exists for the use of patch esophagoplasty in the thoracic or abdominal esophagus where the blood supply is not as well developed. A single series of four patients with esophageal strictures has been reported whereby conventional treatments had failed. Porcine UBM was used to repair three cervical and one distal defect of varying diameters. Despite being a heterogeneous group, this study had remarkable results; only one of four patients developed a postoperative stricture, allowing all patients to recover functionality and preserve their esophagus up to 16 months postoperatively (Nieponice et al. 2014). Clearly, tissue engineering represents a promising area for the rare situation where a

Table 2 Summary of papers which describes replacement of a patch of the esophagus

Authors	Year	Scaffold	Cells	Model	n = x	Control	Location	Clinical Stricture	Anastomotic Leak	Epithelium	Muscle	Maximum Follow up
Acellular												
Badylak et al.	2000	Porcine SIS and UBM	Nil	Canine	11	0	Cervical	Nil	Nil	++	+	15 months
Isch et al.	2001	Alloderm	Nil	Canine	12	0	Cervical	Nil	Nil	++	–	3 months
Jansen et al.	2004	Vicryl vs PVDF	Nil	Rabbit	5 vs 5	0	Abdominal	Nil	80% in vircyl group	+	+	3 months
Lopes et al.	2006	SIS	Nil	Rat	34	15	Cervical	Nil	Nil	+++	++	1-6 months
Urita et al.	2006	GAM	Nil	Rat	27	0	Abdominal	Nil	11%	+++	–	18 months
Aikawa et al.	2013	BAPP	Nil	Porcine	9	0	Thoracic	Nil	Nil	+++	+	12 weeks
Tan et al.	2014	Copper SIS	Nil	Canine	12	6 SIS	Cervical	Nil	Nil	+++	+ (better in CuSIS vs SIS alone)	8 weeks
Nieponice et al.	2014	Porcine UBM	Nil	Human	4	0	Cervical 3, thoracic 1	25%	25%	+++	Not assessed	12-16 months
Diemer et al.	2015	PCL mesh	Nil	Rabbit	15	0	Abdominal	13%	60% pseudodiverticula	+	+	4 weeks
Algarrahi et al.	2015	Silk fibroin	Nil	Rat	40	22 SIS	Abdominal	Nil	5% SIS	+++	++	2 months
Okuyama et al.	2018	IBTA biosheet	Nil	Canine	4	0	Cervical	Nil	Nil	+++	+	4,12 weeks
Algarrahi et al.	2018	Silk fibroin	Nil	Porcine	6	0	Thoracic	Nil	Nil	+++	++	3 months
Cellular												

(continued)

Table 2 (continued)

Authors	Year	Scaffold	Cells	Model	n = x	Control	Location	Clinical Stricture	Anastomotic Leak	Epithelium	Muscle	Maximum Follow up
Marzaro et al.	2006	DEM	SMC	Porcine (neonatal)	3	3 (unseeded)	Thoracic	Not assessed	Not assessed	Not assessed	+ (enhanced regeneration in SMC group)	3 weeks
Wei et al.	2009	SIS	OMEC	Canine	6	6 (unseeded)	Cervical	Nil	Nil	+++	+ (enhanced regeneration in OMEC group)	8 weeks
Tan et al.	2013	SIS	BM-MSC + myoblasts	Canine	6	6 (unseeded)	Cervical	Nil	Nil	+++	+ (enhanced regeneration in BM- MSC group)	12 weeks
Park et al.	2016	Fibrin coated PCL	BM-MSC	Rabbit	3	3 (unseeded)	Cervical	Nil	Nil	Not quantified	Epithelium and smooth muscle in seeded group only	3 weeks

Polyvinylidene fluoride (PVDF); Gastric acellular matrix (GAM); Bioabsorbable polymer patch (BAPP); Decellularized esophageal matrix (DEM); Poly-e-caprolactone (PCL); Autologous Smooth muscle cells (SMC); Bone marrow mesenchymal stem cells (BM-MSC). + islets/bundles of muscle cells, incomplete, disorientated; ++ two distinct layers of muscle patch esophagoplasty is required; however significant further investigation needs to be undertaken prior to further clinical use.

Full Thickness Defects – Circumferential

For conditions that require circumferential, full-thickness esophageal regeneration, the challenge is even more complex (Table 3). Initial attempts from a Japanese group using acellular natural scaffolds in canine models appeared positive. Cervical esophageal defects were replaced with double-layered collagen sponges with stent removal at 2, 3, and 4 weeks. Stent removal prior to 4 weeks resulted in stricture universally with no muscular regeneration by 6 months. In contrast, where the stent remained in situ for 4 weeks, epithelialization was established prior to stent removal, longitudinal and circular muscle regeneration was reported from the edges of the defect, and no strictures were seen by 12 months (Takimoto et al. 1998). The same group adapted this technique to an intrathoracic model with 4-week stent removal. While 89% survived until their defined end point, a degree of stenosis was present in all animals and muscle regeneration was less advanced; immature skeletal muscle myotubes were present, however they did not extend to the middle of the regenerated esophagus by 24 months (Yamamoto et al. 1999). They reflected that the difference in results between the cervical and thoracic esophagus may be due to the poor blood supply in the thoracic region. To address this, a final experiment with omental wrapping of the thoracic replacement was performed with stent for 8 weeks compared to controls with no wrap and 4-week stents. Surprisingly, those with omental wrapping performed worse than controls; delayed epithelial regeneration was seen with prolonged stent duration, and only 20% survived to their defined end point compared to 100% in controls (Yamamoto et al. 2000).

Subsequent circumferential reconstruction of the esophagus without cell seeding has invariably led to stricture formation and the absence of tissue remodeling. A collagen sponge scaffold with split-skin graft and latissimus dorsi muscle flap in a cervical rabbit model resulted in death of all 12 rabbits by day 16 due to aspiration (Saito et al. 2000) and SIS replacement in a cervical porcine model resulted in stricture in all but one of 14 animals by 24 days requiring early sacrifice (Doede et al. 2009). More recently, reconstruction with decellularized esophagus in a porcine 0% abdominal model also had high complication rates; 60% anastomotic leak and 40% stricture at 5 weeks. Although epithelialization was seen, there was little or no muscle regeneration with significant fibrosis and mononuclear infiltrate (Luc et al. 2018). The unifying feature in these experiments was the lack of stent, which had been used in earlier studies. Despite the use of silicone stents in an SIS intrathoracic piglet model, however, all six animals developed recurrent, symptomatic stricture post removal of stent at 4 weeks (Jönsson et al. 2011). Multiple lessons can be learnt from these early experiences. Both the presence of a stent and its duration in situ appear to be critical. Stricture is universal in both natural and synthetic acellular scaffolds after stent removal regardless of intact epithelium. This suggests the process is secondary to processes deeper in esophageal wall and it may be a stent is required until underlying muscular remodeling has occurred.

Table 3 Summary of papers which describes replacement of a circumferential esophagus

	Year	Scaffolds	Cells	Model	n = x	Control	Location
Acellular							
Takimoto et al.	1998	Collagen sponge +silicone	Nil	Canine	16	27 (early stent removal)	Cervical
Yamamoto et al.	1999	Collagen sponge +silicone	Nil	Canine	9	0	Thoracic
Yamamoto et al.	2000	Collagen sponge +silicone	Nil	Canine	5	9	Thoracic
Badylak et al.	2000	Porcine SIS/UBM	Nil	Canine	4	0	Cervical
Saito et al.	2000	Collgen sponge + STS	Nil	Rabbit	12	0	Cervical
Badylak et al.	2005	Porcine UBM	Nil	Canine	5	0	Cervical
Doede et al.	2009	SIS	Nil	Porcine (neonatal)	14	0	Cervical
Jönsson et al.	2011	SIS	Nil	Porcine (neonatal)	6	0	Thoracic
Luc et al.	2018	DEM	Nil	Porcine	3	3 no bioreactor	Abdominal
Cellular							
Nakase et al.	2008	Ham +PGA	Keratinocytes + fibroblasts	Canine	6	6 (acellular)	Thoracic
Poghosyan et al.	2015	SIS + HAM	Myoblasts + OMEC	Porcine	6	12 (acellular)	Cervical
Dua et al.	2016	Dermal matrix	PRP	Human	1	0	Cervical
Catry et al.	2017	SIS	BM-MSC	Mini-pig	10	10 (acellular)	Abdominal
Barron et al.	2018	PU	OMEC	Porcine	2	1 (acellular)	Thoracic
La Francesca et al.	2018	PU	a-MSC	Porcine	8	0	Thoracic

Stent	Bioreactor	Anastomotic Leak	Stricture	Epithelium	Muscle	Survival to endpoint	Maximum Follow-up
Silicone	Nil	0%	0% vs 81%	+++	++	100% vs 26%	12 months
Silicone	Nil	0%	100% (13-54% stenosis)	+++	+	89%	24 months
Silicone	Omental	20%	Not reported	++	Nil	20%	3 months
Nil	Nil	0%	100%	+	+	0%	15 months
Nil	Lat dorsi	0%	Insufficient survival	–	–	0%	16 days
Nil	Nil	0%	100%	–	–	0%	19 days
Nil	Nil	14%	93%	+	–	7%	4 weeks
Silicone	Nil	0%	100%	++	+	100%	17 weeks
Nil	Omental	60%	40%	+	–	83%	5 weeks
Nil	Omental	0%	33% vs 100%	+++	+	67% vs 0%	60 weeks
Polyflex	Omental	0% vs 42%%	67% vs 100%	+++	+	83% vs 8%	12 months
4 years	Nil	0%	100%	+++	Not available	100%	4 years
Polyflex	Omental	0%	100%	+++ (earlier in MSC)	+ (MSC only)	No difference between groups	3 months
Wallflex	Nil	0%	0%	+++	++	No defined endpoint	29 days
Wallflex	Nil	12.5%%	25% (no stent)	+++	++	No defined endpoint	19 months

(continued)

Table 3 (continued)

	Year	Scaffolds	Cells	Model	n = x	Control	Location
Jensen et al.	2018	PU	AF-MSC/ EEC	Porcine	- MSC,	1 (unseeded)	Thoracic
Kim et al.	2019	PU + PCL	a-MSC	Rat	21	14 (unseeded)	Cervical

Split thickness skin (STS); Human Amniotic Membrane (HAM); Polyglycolic Acid (PGA); Platelet-Rich Plasma (PRP); Polyurethane (PU); Adipose-derived mesenchymal stem cells (a-MSC); Amniotic fluid-derived mesenchymal stem cells (AF-MSC); Esophageal epithelial cells (EEC)

Stent	Bioreactor	Anastomotic Leak	Stricture	Epithelium	Muscle	Survival to endpoint	Maximum Follow-up
Biliary stent	Nil	0%	33%	+++	++ (AF > EEC > un seeded)	100%	6 months
Nil	Thyroid gland flap vs omental	Not quantified	Not quantified	++	+	0%	15 days

In parallel with results seen in patch defects, the approach in recent years has shifted to the use of pre-seeded scaffolds, which appear to result in lower stricture rates, enhanced epithelial and muscular regeneration, and a reduced inflammatory response. After 3 weeks of omental maturation, thoracic implantation of un-stented hybrid scaffolds seeded with OMEC and fibroblasts were compared to acellular controls. Within 3 weeks, stratified epithelialization was complete with polarized smooth muscle-like regeneration in contrast to acellular controls with incomplete epithelialization, significant inflammation, and complete stenosis. Stricture occurred in only one third of the cellular group, interestingly in the two animals with desquamation of the epithelial layer noted after bioreactor maturation (Nakase et al. 2008). Despite the absence of a stent, epithelial cell seeding was in some way protective of stricture formation. Both stents and epithelial cells are therefore potentially key components to stricture reduction and further work is required to understand if this effect is synergistic. While unable to show a difference in stricture rate or survival, BM-MSC on SIS also promoted epithelial regeneration compared to SIS alone in a porcine abdominal model. Mature epithelium was seen at 45 compared to 95 days in the unseeded control, with early initiation of muscle cell colonization, which was never found in the unseeded model (Catry et al. 2017). This suggests that, as in patch models, cell seeding of scaffolds accelerates both epithelial and muscle regeneration regardless of cell type. The effect of stenting on stricture rate in cell-seeded constructs also appears to be positive; all un-stented controls developed stricture compared to 50% of the stented group in a porcine cell-seeded biological scaffold cervical model. In addition, the un-stented group had a high anastomotic leak rate of 83% not seen in the stented group, highlighting additional clinical benefits of stenting (Poghosyan et al. 2015).

An alternative approach is the use of synthetic scaffolds as temporary templates to guide esophageal tissue regrowth rather than integration and degradation. Polyurethane electro-spun scaffolds seeded with autologous adipose-derived MSC were used to replace the thoracic esophagus in eight pigs. The extruded graft was removed at 3 weeks with stent exchange and platelet-rich plasma and MSC application. Subsequent stent exchange occurred every 3 weeks until 6 months. Epithelialization and organized smooth muscle were reported with symptom-free survival of two pigs at 18 and 19 months. The frequency of stent exchange in this model represents a major limitation if this approach were to be translated to humans (La Francesca et al. 2018). The same model was used to determine whether seeding with epithelial or mesenchymal cells resulted in better tissue regeneration, although numbers were small ($n = 4$). Scaffolds seeded with amniotic fluid-derived MSCs had improved muscular regeneration compared to EEC scaffolds with a bidirectional muscularis externa, however both groups were spatially disorganized. The unseeded control had incomplete epithelium at 6 months, no muscular regeneration, and significantly higher inflammatory infiltrate. Interestingly, no stricture was seen, although a stent was used for the duration of the study (Jensen et al. 2018).

Human experience is extremely limited due to the availability of alternative surgical options rather than the use of radical experimental techniques. A single case of a full thickness circumferential replacement was reported in 2016 on compassionate grounds after extensive esophageal injury failed all conventional treatment in a 24-year old patient. The 5-cm defect in the cervical esophagus was repaired using a self-expanding metal stent covered with Alloderm, an acellular dermal matrix, coated with autologous platelet-rich plasma adhesive gel. After initial issues with stent migration and embedding, the stent remained in situ for 3 years. One-year post removal, no evidence of stricture or fistula was seen. Biopsy-confirmed squamous epithelium, endoscopic ultrasonography confirmed "normal architecture" and high resolution manometry confirmed peristaltic contractile motility in the neo-esophagus (Dua et al. 2016). Most importantly, the patient had achieved oral enteral autonomy. With a lack of histology confirming contribution of muscular regeneration in this patient, it is difficult to know the tissue remodeling processes underlying this clinically successful outcome. These results do suggest, however, that exogenous cells are not essential for full thickness regeneration in the presence of a long-term stent.

3.7 Conclusions

Esophageal replacement remains a major challenge in both children and adults. Although replacement techniques exist, all have associated morbidity and none are able to fully replicate the function of the native esophagus. There is, therefore, a tangible clinical need for a tissue-engineered esophageal construct. Huge advances in tissue engineering over the last two decades have resulted in a number of viable options for replacement of partial thickness defects using regenerative medicine techniques. The complex anatomy and physiology of the esophagus means full thickness defects present a significantly more complex reconstructive challenge with stricture prevention, a bi-directional muscular layer and coordination of peristalsis the main obstacles.

While full esophageal replacement post-esophagectomy is a distant prospect, short circumferential defects as required in esophageal atresia are a very real possibility in the near future. Current evidence from animal models offers promising results with relative success of both natural and synthetic scaffolds. It is clear that cellularization of scaffolds enhances muscular and epithelial regeneration and reduces inflammation. Future approaches should focus on these techniques. However, further work to identify optimal cell combinations is required, with cell-labeling potentially offering a more thorough understanding of the contribution of exogenous cell delivery on tissue remodeling. Stents also appear to be essential to reducing stricture formation and improving clinical outcomes, however duration must be carefully considered for optimal results. Finally, increased work using intrathoracic rather than cervical animal models, with a particular focus on the merits of bioreactors and pre-vascularization, must be undertaken.

Animal models have provided real insights into future techniques for human therapies and demonstrate the huge potential therapeutic possibilities offered by regenerative medicine. Although further work is clearly needed prior to translation in humans, the implications of tissue engineering for both benign and malignant conditions of the esophagus is enormous.

4 Final Considerations

Recent innovation in both science and technology has encouraged the development of trailblazing research in the field of regenerative medicine, creating novel treatments for diseases in which limited or no alternative solutions exist. The promise of this cannot be underestimated; however, despite the significant progress achieved, there remains a long way to go to establish standardized and safe methods for bioengineered organ transplantation.

Research in regenerative medicine is often plagued with ethical controversy. Proof of biological mechanisms from preclinical studies is therefore imperative prior to proceeding with novel treatments in patients. Open reporting and standardization of definitions and outcomes between groups is key. The creation of an international register for preclinical trials with mandatory registration prior to publication of both positive and negative outcomes may also go some way to addressing this.

Compassionate clinical trials in patients should only be performed when all treatment options are exhausted and open discussions with both patients and ethical committees conclude the merit of the trial, particularly where the treatment may result in lethal adverse effects. Building valuable international collaborations, gathering expertise from varied specialties, and increasing the role of clinical scientists facilitating bench-to-bedside translation are crucial for the advancement of regenerative medicine techniques into viable and successful options for patients in the future.

Acknowledgments PDC is supported by Horizon 2020 grant INTENS 668294; PDC is supported by NIHR Professorship, NIHR UCL BRC-GOSH and the Oak Foundation. ND is supported by L. Spitz GOSHCC Clinical Research Training Fellowships. MG is supported by PhD scholarship Agencia Nacional de Investigación y Desarrollo(ANID)/DoctoradoBecasChile/2018 – 72190110. PB is supported by the Rosetrees Trust (M362; M362-F1; M553) and the Cystic Fibrosis Trust (SRC 006)

References

Aikawa M, Miyazawa M, Okamoto K, Okada K, Akimoto N, Sato H, Koyama I, Yamaguchi S, Ikada Y (2013) A bioabsorbable polymer patch for the treatment of esophageal defect in a porcine model. J Gastroenterol 48:822–829

Algarrahi K, Franck D, Ghezzi CE, Cristofaro V, Yang X, Sullivan MP, Chung YG, Affas S, Jennings R, Kaplan DL et al (2015) Acellular bi-layer silk fibroin scaffolds support functional tissue regeneration in a rat model of onlay esophagoplasty. Biomaterials 53:149–159

Algarrahi K, Franck D, Cristofaro V, Yang X, Savarino A, Affas S, Schäfer FM, Ghezzi C, Jennings R, Nedder A et al (2018) Bi-layer silk fibroin grafts support functional tissue regeneration in a porcine model of onlay esophagoplasty. J Tissue Eng Regen Med 12:e894–e904

Aoki FG, Varma R, Marin-Araujo AE, Lee H, Soleas JP, Li AH, Soon K, Romero D, Moriya HT, Haykal S et al (2019) De-epithelialization of porcine tracheal allografts as an approach for tracheal tissue engineering. Sci Rep 9:1–12

Arakelian L, Caille C, Faivre L, Corté L, Bruneval P, Shamdani S, Flageollet C, Albanese P, Domet T, Jarraya M et al (2019) A clinical-grade acellular matrix for esophageal replacement. J Tissue Eng Regen Med 13:2191–2203

Badylak S, Meurling S, Chen M, Spievack A, Simmons-Byrd A (2000) Resorbable bioscaffold for esophageal repair in a dog model. J Pediatr Surg 35:1097–1103

Badylak SF, Vorp DA, Spievack AR, Simmons-Byrd A, Hanke J, Freytes DO, Thapa A, Gilbert TW, Nieponice A (2005) Esophageal reconstruction with ECM and muscle tissue in a dog model. J Surg Res 128:87–97

Badylak SF, Hoppo T, Nieponice A, Gilbert TW, Davison JM, Jobe BA (2011) Esophageal preservation in five male patients after endoscopic inner-layer circumferential resection in the setting of superficial cancer: a regenerative medicine approach with a biologic scaffold. Tissue Eng - Part A 17:1643–1650

Barret M, Pratico CA, Camus M, Beuvon F, Jarraya M, Nicco C, Mangialavori L, Chaussade S, Batteux F, Prat F (2014) Amniotic membrane grafts for the prevention of esophageal stricture after circumferential endoscopic submucosal dissection. PLoS One 9:e100236

Barron MR, Blanco EW, Aho JM, Chakroff J, Johnson J, Cassivi SD, Carey WA, Wigle DA (2018) Full-thickness oesophageal regeneration in pig using a polyurethane mucosal cell seeded graft. J Tissue Eng Regen Med 12:175–185

Beckstead BL, Pan S, Bhrany AD, Bratt-Leal AM, Ratner BD, Giachelli CM (2005) Esophageal epithelial cell interaction with synthetic and natural scaffolds for tissue engineering. Biomaterials 26:6217–6228

Belsey R (1950) Resection and reconstruction of the intrathoracic trachea. Br J Surg 38:200–205

Bonavina L, Segalin A, Pavanello M, Faranda C, Cioffi U, Peracchia A (1995) Surgical treatment of esophageal stenosis caused by reflux. Ann Ital Chir 66:621–624

Butler CR, Speggiorin S, Rijnberg FM, Roebuck DJ, Muthialu N, Hewitt RJ, Elliott MJ (2014) Outcomes of slide tracheoplasty in 101 children: a 17-year single-center experience. J Thorac Cardiovasc Surg 147:1783–1790

Butler CR, Hynds RE, Crowley C, Gowers KHC, Partington L, Hamilton NJ, Carvalho C, Platé M, Samuel ER, Burns AJ et al (2017) Vacuum-assisted decellularization: an accelerated protocol to generate tissue-engineered human tracheal scaffolds. Biomaterials 124:95–105

Catry J, Luong-Nguyen M, Arakelian L, Poghosyan T, Bruneval P, Domet T, Michaud L, Sfeir R, Gottrand F, Larghero J et al (2017) Circumferential esophageal replacement by a tissue-engineered substitute using mesenchymal stem cells: an experimental study in mini pigs. Cell Transplant 26:1831–1839

Ceonzo K, Gaynor A, Shaffer L, Kojima K, Vacanti CA, Stahl GL (2006) Polyglycolic acid-induced inflammation: role of hydrolysis and resulting complement activation. Tissue Eng 12:301–308

Chiang T, Pepper V, Best C, Onwuka E, Breuer CK (2016) Clinical translation of tissue engineered trachea grafts. Ann Otol Rhinol Laryngol 125:873–885

Conconi MT, De Coppi P, Di Liddo R, Vigolo S, Zanon GF, Parnigotto PP, Nussdorfer GG (2005) Tracheal matrices, obtained by a detergent-enzymatic method, support in vitro the adhesion of chondrocytes and tracheal epithelial cells. Transpl Int 18:727–734

Dasari BVM, Neely D, Kennedy A, Spence G, Rice P, Mackle E, Epanomeritakis E (2014) The role of esophageal stents in the management of esophageal anastomotic leaks and benign esophageal perforations. Ann Surg 259:852–860

Delaere P, Van Raemdonck D (2016) Tracheal replacement. J Thorac Dis 8:S186–S196

Delaere P, Vrancks J, Verleden G, de Leyn P, van Raemdonck D (2010) Tracheal allotransplantation after withdrawal of immunosuppressive therapy. Bull Acad Natl Med 194:138–183

Diekman BO, Rowland CR, Lennon DP, Caplan AI and Guilak F (2010) Chondrogenesis of adult stem cells from adipose tissue and bone marrow: induction by growth factors and cartilage-derived matrix. In *Tissue Engineering - Part A*, pp. 523–533. Mary Ann Liebert Inc.

Diekman BO, Christoforou N, Willard VP, Sun H, Sanchez-Adams J, Leong KW, Guilak F (2012) Cartilage tissue engineering using differentiated and purified induced pluripotent stem cells. Proc Natl Acad Sci U S A 109:19172–19177

Diemer P, Markoew S, Le DQS, Qvist N (2015) Poly-ε-caprolactone mesh as a scaffold for in vivo tissue engineering in rabbit esophagus. Dis Esophagus 28:240–245

Doede T, Bondartschuk M, Joerck C, Schulze E, Goernig M (2009) Unsuccessful alloplastic esophageal replacement with porcine small intestinal submucosa. Artif Organs 33:328–333

Dua KS, Hogan WJ, Aadam AA, Gasparri M (2016) In-vivo oesophageal regeneration in a human being by use of a non-biological scaffold and extracellular matrix. Lancet 388:55–61

Elliott MJ, De Coppi P, Speggiorin S, Roebuck D, Butler CR, Samuel E, Crowley C, McLaren C, Fierens A, Vondrys D et al (2012) Stem-cell-based, tissue engineered tracheal replacement in a child: a 2-year follow-up study. Lancet 380:994–1000

Elliott MJ, Butler CR, Varanou-Jenkins A, Partington L, Carvalho C, Samuel E, Crowley C, Lange P, Hamilton NJ, Hynds RE et al (2017) Tracheal replacement therapy with a stem cell-seeded graft: lessons from compassionate use application of a GMP-compliant tissue-engineered medicine. Stem Cells Transl Med 6:1458–1464

Faulk DM, Wildemann JD, Badylak SF (2015) Decellularization and cell seeding of whole liver biologic scaffolds composed of extracellular matrix. J Clin Exp Hepatol 5:69–80

Freud E, Efrati I, Kidron D, Finally R, Mares AJ (1999) Comparative experimental study of esophageal wall regeneration after prosthetic replacement. J Biomed Mater Res 45:84–91

Fuchs JR, Terada S, Ochoa ER, Vacanti JP, Fauza DO (2002) Fetal tissue engineering: in utero tracheal augmentation in an ovine model. J Pediatr Surg 37:1000–1006

Gallo G, Zwaveling S, Groen H, Van Der Zee D, Hulscher J (2012) Long-gap esophageal atresia: a meta-analysis of jejunal interposition, colon interposition, and gastric pull-up. Eur J Pediatr Surg 22:420–425

Go T, Jungebluth P, Baiguero S, Asnaghi A, Martorell J, Ostertag H, Mantero S, Birchall M, Bader A, Macchiarini P (2010) Both epithelial cells and mesenchymal stem cell-derived chondrocytes contribute to the survival of tissue-engineered airway transplants in pigs. J Thorac Cardiovasc Surg 139:437–443

Grikscheit T, Ochoa ER, Srinivasan A, Gaissert H, Vacanti JP (2003) Tissue-engineered esophagus: experimental substitution by onlay patch or interposition. J Thorac Cardiovasc Surg 126:537–544

Grimmer JF, Gunnlaugsson CB, Alsberg E, Murphy HS, Kong HJ, Mooney DJ, Weatherly RA (2004) Tracheal reconstruction using tissue-engineered cartilage. Arch Otolaryngol - Head Neck Surg 130:1191–1196

Hamilton NJ, Kanani M, Roebuck DJ, Hewitt RJ, Cetto R, Culme-Seymour EJ, Toll E, Bates AJ, Comerford AP, McLaren CA et al (2015) Tissue-engineered tracheal replacement in a child: a 4-year follow-up study. Am J Transplant 15:2750–2757

Han Y, Guo J, Sun S, Wu W, Wang S, Ge N, Liu X, Wang G, Wang S (2017) Acellular dermal matrix for esophageal stricture prevention after endoscopic submucosal dissection in a porcine model. Gastrointest Endosc 86:1160–1167

Honda M, Hori Y, Nakada A, Uji M, Nishizawa Y, Yamamoto K, Kobayashi T, Shimada H, Kida N, Sato T et al (2011) Use of adipose tissue-derived stromal cells for prevention of esophageal stricture after circumferential EMR in a canine model. Gastrointest Endosc 73:777–784

Irino T, Tsekrekos A, Coppola A, Scandavini CM, Shetye A, Lundell L, Rouvelas I (2017) Long-term functional outcomes after replacement of the esophagus with gastric, colonic, or jejunal conduits: a systematic literature review. Dis Esophagus 30

Jang YS, Jang CH, Cho YB, Kim M, Kim GH (2014) Tracheal regeneration using poly-caprolactone/collagen-nanofiber coated with umbilical cord serum after partial resection. Int J Pediatr Otorhinolaryngol 78:2237–2243

Jansen PL, Klinge U, Anurov M, Titkova S, Mertens PR, Jansen M (2004) Surgical mesh as a scaffold for tissue regeneration in the esophagus. Eur Surg Res 36:104–111

Jensen T, Wanczyk H, Sharma I, Mitchell A, Sayej WN, Finck C (2018) Polyurethane scaffolds seeded with autologous cells can regenerate long esophageal gaps: an esophageal atresia treatment model. J Pediatr Surg

Jonas E, Sjöqvist S, Elbe P, Kanai N, Enger J, Haas SL, Mohkles-Barakat A, Okano T, Takagi R, Ohki T et al (2016) Transplantation of tissue-engineered cell sheets for stricture prevention after endoscopic submucosal dissection of the oesophagus. United Eur Gastroenterol J 4:741–753
Jönsson L, Gatzinsky V, Jennische E, Johansson C, Nannmark U, Friberg LG (2011) Piglet model for studying esophageal regrowth after resection and interposition of a silicone stented small intestinal submucosa tube. Eur Surg Res 46:169–179
Kanai N, Yamato M, Ohki T, Yamamoto M, Okano T (2012) Fabricated autologous epidermal cell sheets for the prevention of esophageal stricture after circumferential ESD in a porcine model. Gastrointest Endosc 76:873–881
Kim IG, Wu Y, Park SA, Cho H, Choi JJ, Kwon SK, Shin JW, Chung EJ (2019) Tissue-engineered esophagus via bioreactor cultivation for circumferential esophageal reconstruction. Tissue Eng - Part A 25:1478–1492
Kohn DH, Sarmadi M, Helman JI, Krebsbach PH (2002) Effects of pH on human bone marrow stromal cells in vitro: implications for tissue engineering of bone. J Biomed Mater Res 60:292–299
La Francesca S, Aho JM, Barron MR, Blanco EW, Soliman S, Kalenjian L, Hanson AD, Todorova E, Marsh M, Burnette KL et al (2018) Long-term regeneration and remodeling of the pig esophagus after circumferential resection using a retrievable synthetic scaffold carrying autologous cells. Sci Rep 8:4123
Lin YC, Hsiao JR, Lee YC (2017) Patch esophagoplasty with a free proximal lateral leg flap for focal stricture of the cervical esophagus: a case report. Microsurgery 37:426–430
Lopes MF, Cabrita A, Ilharco J, Pessa P, Paiva-Carvalho J, Pires A, Patrício J (2006) Esophageal replacement in rat using porcine intestinal submucosa as a patch or a tube-shaped graft. Dis Esophagus 19:254–259
Luc G, Charles G, Gronnier C, Cabau M, Kalisky C, Meulle M, Bareille R, Roques S, Couraud L, Rannou J et al (2018) Decellularized and matured esophageal scaffold for circumferential esophagus replacement: proof of concept in a pig model. Biomaterials 175:1–18
Macchiarini P, Jungebluth P, Go T, Asnaghi MA, Rees LE, Cogan TA, Dodson A, Martorell J, Bellini S, Parnigotto PP et al (2008) Clinical transplantation of a tissue-engineered airway. Lancet 372:2023–2030
Marzaro M, Vigolo S, Oselladore B, Conconi MT, Ribatti D, Giuliani S, Nico B, Perrino G, Nussdorfer GG, Parnigotto PP (2006) In vitro and in vivo proposal of an artificial esophagus. J Biomed Mater Res - Part A 77:795–801
Maughan EF, Hynds RE, Proctor TJ, Janes SM, Elliott M, Birchall MA, Lowdell MW, De Coppi P (2017) Autologous cell seeding in tracheal tissue engineering. Curr Stem Cell Reports 3:279–289
Maziak DE, Todd TRJ, Keshavjee SH, Winton TL, Van Nostrand P, Pearson FG, Wood DE, Grillo HC (1996) Adenoid cystic carcinoma of the airway: thirty-two-year experience. J Thorac Cardiovasc Surg 112:1522–1532
Mennigen R, Senninger N, Laukoetter MG (2014) Novel treatment options for perforations of the upper gastrointestinal tract: endoscopic vacuum therapy and over-the-scope clips. World J Gastroenterol 20:7767–7776
Miki H, Ando N, Ozavva S, Sato M, Hayashi K, Kitajima M (1999) An artificial esophagus constructed of cultured human esophageal epithelial cells, fibroblasts, polyglycolic acid mesh, and collagen. ASAIO J 45:502–508
Nakase Y, Nakamura T, Kin S, Nakashima S, Yoshikawa T, Kuriu Y, Sakakura C, Yamagishi H, Hamuro J, Ikada Y et al (2008) Intrathoracic esophageal replacement by in situ tissue-engineered esophagus. J Thorac Cardiovasc Surg 136:850–859
Neville WE (1982) Reconstruction of the trachea and stem bronchi with Neville prosthesis. Int Surg 67:229–234
Nieponice A, McGrath K, Qureshi I, Beckman EJ, Luketich JD, Gilbert TW, Badylak SF (2009) An extracellular matrix scaffold for esophageal stricture prevention after circumferential EMR. Gastrointest Endosc 69:289–296
Nieponice A, Ciotola FF, Nachman F, Jobe BA, Hoppo T, Londono R, Badylak S, Badaloni AE (2014) Patch Esophagoplasty: esophageal reconstruction using biologic scaffolds. Ann Thorac Surg 97:283–288

Ohki T, Yamato M, Murakami D, Takagi R, Yang J, Namiki H, Okano T, Takasaki K (2006) Treatment of oesophageal ulcerations using endoscopic transplantation of tissue-engineered autologous oral mucosal epithelial cell sheets in a canine model. Gut 55:1704–1710

Okuyama H, Umeda S, Takama Y, Terasawa T, Nakayama Y (2018) Patch esophagoplasty using an in-body-tissue-engineered collagenous connective tissue membrane. J Pediatr Surg 53:223–226

Park SY, Choi JW, Park JK, Song EH, Park SA, Kim YS, Shin YS, Kim CH (2016) Tissue-engineered artificial oesophagus patch using three-dimensionally printed polycaprolactone with mesenchymal stem cells: a preliminary report. Interact Cardiovasc Thorac Surg 22:712–717

Park JH, Yoon JK, Lee JB, Shin YM, Lee KW, Bae SW, Lee JH, Yu JJ, Jung CR, Youn YN et al (2019) Experimental tracheal replacement using 3-dimensional bioprinted artificial trachea with autologous epithelial cells and chondrocytes. Sci Rep 9:2103

Perrod G, Rahmi G, Pidial L, Camilleri S, Bellucci A, Casanova A, Viel T, Tavitian B, Cellier C, Clement O (2016) Cell sheet transplantation for esophageal stricture prevention after endoscopic submucosal dissection in a porcine model. PLoS One 11

Poghosyan T, Sfeir R, Michaud L, Bruneval P, Domet T, Vanneaux V, Luong-Nguyen M, Gaujoux S, Gottrand F, Larghero J et al (2015) Circumferential esophageal replacement using a tube-shaped tissue-engineered substitute: an experimental study in minipigs. Surg (United States) 158:266–277

Poghosyan T, Catry J, Luong-Nguyen M, Bruneval P, Domet T, Arakelian L, Sfeir R, Michaud L, Vanneaux V, Gottrand F et al (2016) Esophageal tissue engineering: current status and perspectives. J Chir Viscerale 153:21–30

Sa YJ, Du Kim Y, Kim CK, Park JK, Moon SW (2013) Recurrent cervical esophageal stenosis after colon conduit failure: use of myocutaneous flap. World J Gastroenterol 19:307–310

Saito M, Sakamoto T, Fujimaki M, Tsukada K, Honda T, Nozaki M (2000) Experimental study of an artificial esophagus using a collagen sponge, a latissimus dorsi muscle flap, and split-thickness skin. Surg Today 30:606–613

Sakurai T, Miyazaki S, Miyata G, Satomi S, Hori Y (2007) Autologous buccal keratinocyte implantation for the prevention of stenosis after EMR of the esophagus. Gastrointest Endosc 66:167–173

Saleh T, Ahmed E, Yu L, Kwak HH, Kang BJ, Park KM, Choi KY, Kim BM, Kang KS, Woo HM (2019) Characterization of silver nanoparticle-modified decellularized rat esophagus for esophageal tissue engineering: structural properties and biocompatibility. J Biosci Bioeng 128:613–621

Saxena AK, Kofler K, Ainödhofer H, Höllwarth ME (2009) Esophagus tissue engineering: hybrid approach with esophageal epithelium and unidirectional smooth muscle tissue component generation in vitro. J Gastrointest Surg 13:1037–1043

Seguin A, Baccari S, Holder-Espinasse M, Bruneval P, Carpentier A, Taylor DA and Martinod E (2013) Tracheal regeneration: evidence of bone marrow mesenchymal stem cell involvement. J Thorac Cardiovasc Surg 145: 1297-1304.e2

Shin YS, Choi JW, Park JK, Kim YS, Yang SS, Min BH, Kim CH (2015) Tissue-engineered tracheal reconstruction using mesenchymal stem cells seeded on a porcine cartilage powder scaffold. Ann Biomed Eng 43:1003–1013

Soliman S, Laurent J, Kalenjian L, Burnette K, Hedberg B, La Francesca S (2019) A multilayer scaffold design with spatial arrangement of cells to modulate esophageal tissue growth. J Biomed Mater Res - Part B Appl Biomater 107:324–331

Spurrier RG, Speer AL, Hou X, El-Nachef WN, Grikscheit TC (2015) Murine and human tissue-engineered esophagus form from sufficient stem/progenitor cells and do not require micro-designed biomaterials. Tissue Eng - Part A 21:906–915

Syed O, Walters NJ, Day RM, Kim HW, Knowles JC (2014) Evaluation of decellularization protocols for production of tubular small intestine submucosa scaffolds for use in oesophageal tissue engineering. Acta Biomater 10:5043–5054

Takeoka Y, Matsumoto K, Taniguchi D, Tsuchiya T, Machino R, Moriyama M, Oyama S, Tetsuo T, Taura Y, Takagi K et al (2019) Regeneration of esophagus using a scaffoldfree biomimetic structure created with biothree- dimensional printing. PLoS One 14:e0211339

Takimoto Y, Nakamura T, Yamamoto Y, Kiyotani T, Teramachi M, Shimizu Y (1998) The experimental replacement of a cervical esophageal segment with an artificial prosthesis with the use of collagen matrix and a silicone stent. J Thorac Cardiovasc Surg 116:98–106

Tan B, Wei RQ, Tan MY, Luo JC, Deng L, Chen XH, Hou JL, Li XQ, Yang ZM, Xie HQ (2013) Tissue engineered esophagus by mesenchymal stem cell seeding for esophageal repair in a canine model. J Surg Res 182:40–48

Tan B, Wang M, Chen X, Hou JL, Chen XH, Wang Y, Li-Ling J, Xie HQ (2014) Tissue engineered esophagus by copper-small intestinal submucosa graft for esophageal repair in a canine model. Sci China Life Sci 57:248–255

Toomes H, Mickisch G, Vogt-Moykopf I (1985) Experiences with prosthetic reconstruction of the trachea and bifurcation. Thorax 40:32–37

Urbani L, Maghsoudlou P, Milan A, Menikou M, Hagen CK, Totonelli G, Camilli C, Eaton S, Burns A, Olivo A et al (2017) Long-term cryopreservation of decellularised oesophagi for tissue engineering clinical application. PLoS One 12:e0179341

Urbani L, Camilli C, Phylactopoulos DE, Crowley C, Natarajan D, Scottoni F, Maghsoudlou P, McCann CJ, Pellegata AF, Urciuolo A et al (2018) Multi-stage bioengineering of a layered oesophagus with in vitro expanded muscle and epithelial adult progenitors. Nat Commun 9: 4286

Urita Y, Komuro H, Chen G, Shinya M, Kaneko S, Kaneko M, Ushida T (2007) Regeneration of the esophagus using gastric acellular matrix: an experimental study in a rat model. Pediatr Surg Int 23:21–26

Vacanti CA, Paige KT, Kim WS, Sakata J, Upton J, Vacanti JP (1994) Experimental tracheal replacement using tissue-engineered cartilage. J Pediatr Surg 29:201–205

van der Zee DC, Bagolan P, Faure C, Gottrand F, Jennings R, Laberge JM, Martinez Ferro MH, Parmentier B, Sfeir R, Teague W (2017) Position paper of INoEA working group on long-gap esophageal atresia: for better care. Front Pediatr 5:63–63

Wang F, Maeda Y, Zachar V, Ansari T, Emmersen J (2018) Regeneration of the oesophageal muscle layer from oesophagus acellular matrix scaffold using adipose-derived stem cells. Biochem Biophys Res Commun 503:271–277

Wei RQ, Tan B, Tan MY, Luo JC, Deng L, Chen XH, Li XQ, Zuo X, Zhi W, Yang P et al (2009) Grafts of porcine small intestinal submucosa with cultured autologous oral mucosal epithelial cells for esophageal repair in a canine model. Exp Biol Med 234:453–461

Yamaguchi N, Isomoto H, Kobayashi S, Kanai N, Kanetaka K, Sakai Y, Kasai Y, Takagi R, Ohki T, Fukuda H et al (2017) Oral epithelial cell sheets engraftment for esophageal strictures after endoscopic submucosal dissection of squamous cell carcinoma and airplane transportation. Sci Rep 7:17460

Yamamoto Y, Nakamura T, Shimizu Y, Matsumoto K, Takimoto Y, Kiyotani T, Sekine T, Ueda H, Liu Y, Tamura N (1999) Intrathoracic esophageal replacement in the dog with the use of an artificial esophagus composed of a collagen sponge with a double-layered silicone tube. J Thorac Cardiovasc Surg 118:276–286

Yamamoto Y, Nakamura T, Shimizu Y, Matsumoto K, Takimoto Y, Liu Y, Ueda H, Sekine T, Tamura N (2000) Intrathoracic esophageal replacement with a collagen sponge-silicone double layer tube: evaluation of omental-pedicle wrapping and prolonged placement of an inner stent. ASAIO J 46:734–739

Yamato M, Utsumi M, Kushida A, Konno C, Kikuchi A, Okano T (2001) Thermo-responsive culture dishes allow the intact harvest of multilayered keratinocyte sheets without dispase by reducing temperature. Tissue Eng 7:473–480

Zhou C, Zhang L, Wang H, Ma X, Shi B, Chen W, He J, Wang K, Liu P, Ren Y (2015) Superiority of minimally invasive oesophagectomy in reducing in-hospital mortality of patients with resectable oesophageal cancer: a meta-analysis. PLoS One 10:1371

Zhu Y, Chan-Park MB, Chian KS (2005) The growth improvement of porcine esophageal smooth muscle cells on collagen-grafted poly(DL-lactide-co-glycolide) membrane. J Biomed Mater Res - Part B Appl Biomater 75:193–199

Zhu Y, Leong MF, Ong WF, Chan-Park MB, Chian KS (2007) Esophageal epithelium regeneration on fibronectin grafted poly(l-lactide-co-caprolactone) (PLLC) nanofiber scaffold. Biomaterials 28:861–868

Zhu Y, Wu X, Liang Y, Gu H, Song K, Zou X, Zhou G (2016) Repair of cartilage defects in osteoarthritis rats with induced pluripotent stem cell derived chondrocytes. BMC Biotechnol 16:1–11

Zuercher BF, George M, Escher A, Piotet E, Ikonomidis C, Andrejevic SB, Monnier P (2013) Stricture prevention after extended circumferential endoscopic mucosal resection by injecting autologous keratinocytes in the sheep esophagus. Surg Endosc 27:1022–1028

Part II

Digestive and Exocrine Systems

Liver Tissue Engineering

Iris Pla-Palacín, Natalia Sánchez-Romero, Sara Morini, Daniela Rubio-Soto, and Pedro M. Baptista

Contents

Iris Pla-Palacín and Natalia Sánchez-Romero contributed equally with all other contributors.

I. Pla-Palacín · D. Rubio-Soto
Instituto de Investigación Sanitaria Aragón (IIS Aragón), Zaragoza, Spain

N. Sánchez-Romero
Instituto de Investigación Sanitaria Aragón (IIS Aragón), Zaragoza, Spain

Cytes Biotechnologies, Barcelona, Spain

S. Morini
Instituto de Investigación Sanitaria Aragón (IIS Aragón), Zaragoza, Spain

Department of Bioengineering and iBB-Institute for Bioengineering and Biosciences, Instituto Superior Técnico, Universidade de Lisboa, Lisbon, Portugal

P. M. Baptista (✉)
Instituto de Investigación Sanitaria Aragón (IIS Aragón), Zaragoza, Spain

Centro de Investigación Biomédica en Red de Enfermedades Hepáticas y Digestivas (CIBERehd), Zaragoza, Spain

Instituto de Investigación Sanitaria de la Fundación Jiménez Díaz, Madrid, Spain

Departamento de Ingeniería Biomédica y Aeroespacial, Universidad Carlos III de Madrid, Madrid, Spain

Fundación ARAID, Zaragoza, Spain
e-mail: pmbaptista@iisaragon.es

D. Eberli et al. (eds.), *Organ Tissue Engineering*, Reference Series in Biomedical Engineering, https://doi.org/10.1007/978-3-030-44211-8_2

Abstract

Up to date, liver transplantation is the only definitive cure for patients with end-stage liver disease. However, donor organs are limited, and several patients succumb to liver failure before a suitable donor is found. Liver tissue engineering and regenerative medicine are promising new technologies that can help reduce the burden of liver shortage by increasing the number of organs available for transplantation. In this chapter, we focus on various aspects of liver bioengineering, describing current liver diseases and the most relevant liver models used in liver tissue engineering, with particular attention to liver decellularization and recellularization processes. Also, we highlight the importance of bioreactor systems to allow a dynamic growth of all the cells seeded in liver scaffolds. Furthermore, we describe new cell culture systems that help us to pave the way from the bench to the bedside, as well as the regulatory and technological challenges related to liver bioengineering.

1 Introduction

The liver is the second largest organ in the human body. With an approximate weight of 1.2–1.5 kg in an adult human, the liver constitutes the most significant gland of the organism.

It is involved in many vital functions like the regulation of the energetic metabolism, processing all the nutrients; in the synthesis of essential proteins and enzymes needed during the digestion and normal organism behavior; in the maintenance of hormone balance; and in the detoxification and elimination of compounds for a correct incorporation of amino acids, carbohydrates, lipids and vitamins, and their storage.

The liver is considered a metabolic organ and an exocrine gland, too, because of its role in metabolic and bloodstream secretion functions. Furthermore, because of its capacity for bile production and secretion, it is also considered as an exocrine gland.

This organ has the particularity of receiving a double blood contribution: 80% enters through the portal vein, with low oxygenation and providing the liver with all the substances drained from the abdominal organs. The remaining 20% enters through the hepatic artery, carrying well-oxygenated blood. Arterial and venous blood converge both at the hepatic sinusoid, which allows the transvascular interchange between the blood and parenchymal cells.

It is anatomically divided into eight independent segments, and its functional morphologic unit is the hepatic lobule, organized surrounding the central vein (CV). Hepatic cords and hepatic sinusoids are surrounding this CV. In the apex of the hepatic lobule is located the portal trial, formed by a hepatic artery, a portal vein, and a bile duct. Repetitions of this functional unit form the hepatic tissue.

Furthermore, many types of cells coexist in the liver tissue: the parenchymal cells, which are the hepatocytes, constitute 80% of the tissue. Non-parenchymal cells constitute the remaining 20% and include liver sinusoidal endothelial cells (LSEC), ductular cells (DC), Kupffer cells (KC), stellate cells (HSC), and Pit cells.

1.1 Brief Description of Acute, Inborn Errors of Metabolism and End-Stage Liver Disease Worldwide

Liver diseases are any conditions that lead to liver inflammation or damage, affecting thus its normal function. Many factors like infections, exposure to drugs or toxic compounds, autoimmune processes, or genetic malignancies can cause these. The effects include inflammation, scarring, blood clotting, bile duct obstructions, and liver failure. The following table (Table 1) summarizes most liver diseases:

Liver disease causes approximately two million deaths per year worldwide: one million originated by cirrhosis complications and the remaining million because of viral hepatitis or hepatocellular carcinoma (Mokdad et al. 2014). Cirrhosis is the 11th most common cause of death worldwide, with 1.16 million global deaths per year. Liver cancer represents the 16th most common cause of death, with 788,000 global deaths (Asrani et al. 2019).

Some global data analyzed from 1990 to 2010 suggest that Latin America, the Caribbean, Middle East, and North Africa have the highest percentage of deaths because of liver disease. Egypt, Moldova, and Mongolia have the highest cirrhosis mortality rate in the world. India represents an 18,3% of cirrhosis death in the world, followed by China with 11%. Central Asia, the Russian Federation, and Europe's mortality are increasing.

Almost any chronic liver disease leads to cirrhosis. Globally, the leading causes are hepatitis B and C virus and alcohol. Other causes can be immune liver diseases, drugs, cholestatic disease, among others. The causes vary between different countries: in Western and industrialized countries, alcohol and nonalcoholic fatty liver disease are the leading causes, whereas hepatitis B is the main cause in developing countries (Asrani et al. 2019; Lim and Kim 2008; Lozano et al. 2012).

Acute liver failure (ALF) is a rare disorder with a high mortality rate (Bernal et al. 2010). It can be produced by viral causes, predominantly in developed countries; in

Table 1 Characteristics and causing agents of different hepatic diseases. (Source: Adapted from (AACC 2019))

Liver disease	Characteristics	Causing agents
Acute liver failure	A rapid decrease in liver function	Drugs, toxins, liver diseases
Alcoholic liver disease	Liver damage which can lead to fatty liver, alcoholic hepatitis, or cirrhosis	Abusive alcohol intakes
Autoimmune liver disease	Body's immune system attacks liver cells	Primary biliary cirrhosis, autoimmune hepatitis
Biliary obstruction	Bile duct blocking	Trauma, tumors, inflammation
Cirrhosis	Liver scarring	Chronic hepatitis, alcoholism, chronic bile duct obstruction
Genetic diseases	Gene mutation	Hemochromatosis, Wilson disease
Hepatitis	Acute or chronic liver inflammation	Virus, alcohol, drugs, toxins
Infections	Can originate liver damage and/or bile duct blockage	Virus, parasites
Liver cancer	Abnormal growth of liver cells	Cirrhosis, chronic hepatitis, virus

the USA and Western Europe, drug-induced injury is the leading cause, and there are many other cases with no clear origin. ALF can be associated with multiorgan failure.

Nonalcoholic fatty liver disease (NAFLD) is an increasing health problem associated with diabetes and obesity that affects one-third of adults in developed countries (Cohen et al. 2011).

This clinicopathological condition comprises a broad spectrum of liver damage, ranging from simple steatosis to steatohepatitis, fibrosis, and cirrhosis in patients who do not abuse alcohol. This illness starts with an abnormal accumulation of triglyceride droplets within the hepatocytes, leading to hepatic steatosis. This stage is usually self-limited, but it can progress to nonalcoholic steatohepatitis (NASH) due to hepatic injury because of hepatocyte death due to ballooning, inflammatory infiltrate, or collagen deposition (fibrosis). NASH can then develop a cirrhotic stage, where hepatocytes are replaced by scar tissue, decreasing hepatic function and altering blood flow. There are several degrees of fibrosis, being the last step the progression of the cirrhotic stage to hepatocellular carcinoma (HCC).

Hepatocellular carcinoma is usually the result of a complication of liver diseases, commonly liver cirrhosis (Fattovich et al. 2004). The WHO estimates that in the 2000, primary liver cancer is supposed the fifth most prevalent illness in men and the ninth in women, representing about the 5,6% of all human cancers (Parkin et al. 2001), with 564,000 new cases per year and expected to increase by 2020 (Bosch et al. 2004).

The Barcelona-Clínic Liver Cancer (BCLC) has classified HCC in liver cirrhotic patients into five different stages: stage 0 (very early stage), stage A (early stage), stage B (intermediate stage), stage C (advanced stage), and stage D (terminal stage), and nowadays it is utilized by many centers.

1.2 Liver Transplant Worldwide

According to the data published in the *Newsletter Transplant*, handled by the NTO (National Transplant Organization) in collaboration with the World Health Organization, Spain revalidates its world leadership for the 26 consecutive years, with a rate of 47 donors per million population (p.m.p). This country provides 19,2% of organ donations in the EU and 6,4% of the registrations worldwide (34,096). Spain also maintains the leadership in transplants, with 113,4 p.m.p, above the USA, which has 109,7 p.m.p. In 2018, there were 135,860 organs transplanted worldwide. This number means an increase of 7,2% to the previous year (which was 126,670). From these, 89,823 were kidney transplants, 30,352 liver transplants, 7,626 heart transplants, 5,497 lung transplants, 2,342 where pancreas transplants, and 220 intestine transplants. By the end of 2018, there were 56,399 European people on the waiting list. According to these data, ten patients died per day waiting for an available organ in the EU.

The donation rate has increased in the last years worldwide: there are 31,7 donors p.m.p in the USA; Australia reached 20,8 p.m.p and Canada 21,9 p.m.p. Russia figures its rate in only 4 donors p.m.p, and Latin America reached 9,5 donors p.m.p.

The following table (Table 2) recapitulates data from a liver transplant in 2017 (data obtained from the Newsletter Transplant):

Table 3 summarizes the number of patients in the waiting list (WL) for a liver transplant and the number of patients who died in 2017 (data obtained from the Newsletter Transplant):

1.3 Bottlenecks and Limitations

As mentioned above, liver transplantation means the only effective treatment option for patients suffering from end-stage liver disease, acute hepatic failure, and hepatocellular carcinoma. Although short-term graft and recipient survival outcomes have improved, thanks to advancements in the surgical technique, perioperative management, and immunosuppressive therapy, there are still two main limitations in liver transplantation: the first one and the most important is organ shortage. That is

Table 2 Number of liver transplant worldwide

Country	Number of transplanted livers
Australia	282
Canada	585
European Union	7.984
Latin America	3.288
New Zealand	55
Russian Federation	438
Saudi Arabia	226
USA	8.082

Table 3 The number of patients waiting for a liver transplant and *exitus* number of total patients in 2017 across the world

Country	Number of patients in the WL	*Exitus* number
Australia	482	11
Canada	–	74
European Union	16.064	990
Latin America	8.112	1.087
New Zealand	–	–
Russian Federation	1.666	141
Saudi Arabia	658	–
USA	24.178	1306

why significant efforts are being developed to increase the existing scarce donor pool. This effort has derived in the use of liver allografts from donors after cardiac death in combination with extended criteria donors. The goal is to get a better selection of donors, selecting not only the adequate ones but has also helped in the development of mechanical perfusion strategies (Jadlowiec and Taner 2016). The second limitation is the long-term recipient's complications after liver transplantation. Infections, allograft failure, cardiovascular events, or renal failure are the most common causes of later mortality after liver transplantation, which derive from long-term immunosuppression (Watt et al. 2010). Substantial efforts are being developed to improve the recipient's long-term outcomes.

1.4 Solutions Developed Throughout the Years to Increase Organ Availability

Due to organ shortage, it became necessary to find alternative therapies to treat liver failure. Extracorporeal liver support systems for patients suffering from liver failure consist of temporary relief developed to speed up liver recovery from injury or as a bridge to transplantation. There are two types of devices for temporal support: artificial livers (AL), which use nonliving components to remove toxins accumulated in the blood or plasma due to liver failure, using membrane separation associated with columns or sorbents. On the other hand, bioartificial liver (BAL) provides not only liver detoxification but also synthetic functions by combining chemical procedures and bioreactors containing cells to maintain the liver function (Carpentier et al. 2009). Appropriate cell choice for BAL devices is still under investigation.

Primary human hepatocytes represent the ideal source for replacing liver function, but they still are difficult to maintain in vitro. Primary porcine hepatocytes, on the contrary, enable the availability of large quantities of cells but with the immunologic and infectious problems that they carry. This is the reason why human or human-derived cells are more desirable than animal cells. Other cell types investigated are immortalized cells [like C8-B (Cai et al. 2000), HepZ (Werner et al. 2000), HH25 (Kono et al. 1995). . .], which are easily cultivated in vitro and maintain liver-specific functions. Their problem still resides in their potential transmission of oncogenic substances to the patient (Allen et al. 2001; Carpentier et al. 2009). Up to date, BAL is still under investigation. Although some of them have FDA authorization for clinical development (like HepaMate, ELAD, or Excorp Medical), there is still a lot to do regarding their capacity to provide and replace liver functions. Significant issues have to be overcome, like cost, cell availability, or maintenance of cell viability that have delayed their appearance on the clinic.

Split livers represent another choice. This technique means a way to increase the number of cadaveric donor organs for children and adults: it provides a left lateral (to be transplanted into a child) and a right-extended liver graft (to be transplanted into one adult). Furthermore, the outcomes showed comparable results as those in whole organ liver transplantation (Broering et al. 2004). Nevertheless, the main problem with split livers resides in that it is a sophisticated variant of liver transplantation that requires a very high level of technical and logistical skills and an extensive knowledge of possible anatomic variations. It also needs a reliable judgment on graft quality and also an optimal graft-recipient size match (in order to avoid the small-size liver syndrome). However, in the latest years, new technologies like organ and tissue engineering have emerged not in order to maintain liver function while injury but for the creation of new organs in vitro able to be transplanted.

Another potential alternative to liver transplantation is allogeneic hepatocyte transplantation (Iansante et al. 2018; Laconi et al. 1998) to restore hepatic function once engrafted in the recipient's liver. Hepatocytes can be isolated from different sources: whole donor livers unsuitable for transplant, from liver segments after split liver transplantation, or from neonatal and fetal livers (which could provide very high hepatocyte quality). The advantages of these techniques include the fact that it is less invasive and less expensive than surgery, and it can be performed repeatedly if necessary. Cryopreserved cells are also available when necessary, and because this procedure requires few cell quantities, different recipients could beneficiate from the same donor organ. On the contrary, the limitations reside in the difficulty of isolating and maintaining high-quality hepatocytes (Ibars et al. 2016; Stephenne et al. 2007). Liver cell engraftment is usually poor (approximately 0,1–0,3%) (Wang et al. 2002), and allogeneic rejection may be the leading cause of failure in cell graft function.

Mesenchymal stem cells (MSC) are another important source to consider. They are an attractive source due to their capacity to proliferate and differentiate in vitro and their anti-inflammatory and immune status (Iansante et al. 2018). Furthermore, there are several tissues containing hMSC: bone marrow, adipose tissue, and umbilical cord. Some authors have recently isolated MSCs from the liver, called liver-derived human MSCs, which can be transdifferentiated toward hepatocyte-like cells (Najimi et al. 2007). However, their role is still under investigation in liver-recipient repopulation and providing satisfactory hepatic function.

Tissue-engineered whole livers are becoming very promising too. By combining tissue engineering techniques and biology, autologous livers might be able to be created in vitro and transplanted, eliminating thus the problem of an organ donor, waiting list, and rejection.

2 Tissue Engineering

2.1 First Approaches Used for Liver Tissue Engineering

Tissue engineering is one of the most promising fields in regenerative medicine. As described in 1993 by Robert Langer and Joseph Vacanti, it is the conjugation of biomaterials (synthetic or naturally derived) with cells, in order to generate tissue constructs that can be implanted into patients to substitute a lost function and maintain or gain new functions (Langer and Vacanti 1993). The current paradigm is suitable for the engineering of thin constructs like the bladder, skin, or blood vessels. In the specific case of the liver, the 3D architecture and dense cellular mass require novel tissue engineering approaches and the development of vascularized biomaterials, in order to support thick tissue masses and be readily transplantable. Additionally, to the vascular support for large tissue masses, hepatocyte function maintenance represents the ultimate aim in any organ engineering or regenerative medicine strategy for liver disease. Hepatocytes are known to be attachment-dependent cells and lose rather quickly their specific functions without optimal media and ECM (extracellular matrix) composition and cell-cell contacts. Also, the function and differentiation of liver cells are influenced by the 3D organ architecture (Mooney et al. 1992).

In the last two decades, multiple strategies for the culture of adult hepatocytes in combination with several types of 3D, highly porous polymeric matrices, have been attempted (Fiegel et al. 2008; Kim et al. 2000b; Lin et al. 2004; Linke et al. 2007; Tong et al. 1990). However, in the absence of vasculature, restriction in cell growth and function is joint due to the limitations in nutrient and oxygen diffusion. Some of these problems are being now partially overcome with the development of bioreactors that provide continuous perfusion of culture media and gases, allowing a 3D culture configuration and hepatocyte function maintenance (Gerlach et al. 1994; Torok et al. 2001, 2006).

The tissue engineering concept has several advantages over the injection of cell suspensions into solid organs. The matrices provide sufficient volume for the transplantation of an adequate cell mass up to whole-organ equivalents. Transplantation efficiency could readily be improved by optimizing the microarchitecture and composition of the matrices, as well as by attaching growth factors and extracellular matrix molecules to the polymeric scaffold, helping to recreate the hepatic microenvironment (Mooney et al. 1992). The use of naturally derived matrices has also proved to be very helpful in hepatocyte culture (Lin et al. 2004). These matrices, besides preserving some of the microarchitecture features of the tissues that they are derived from, also retain bioactive signals (e.g., cell-adhesion peptides and growth factors) required for the retention of tissue-specific gene expression (Kim et al. 2000a; Voytik-Harbin et al. 1997). Additionally, cell transplantation into polymeric matrices is, in contrast to cell injection into tissues and organs, a reversible procedure since the cell-matrix constructs may be removed if necessary.

Finally, heterotopic hepatocyte transplantation in matrices has already been demonstrated in long-term studies (Johnson et al. 1994; Kaufmann et al. 1999).

Nonetheless, initial engraftment rates are suboptimal. One of the reasons for this is the absolute requirement of the transplanted hepatocytes for hepatotrophic factors that the liver regularly receives through its portal circulation (Starzl et al. 1973). Thus, the development of a tissue-engineered liver construct capable of being orthotopically transplanted is essential.

2.2 Whole-Liver Scaffolds (Decellularization)

One other alternative is based on the decellularization process to generate a whole-organ scaffold. These have to be later recellularized using different cell types to get a functional bioengineered organ to be transplanted into recipient animals (Fig. 1). The strategy of decellularization of extracellular matrices has induced the development of other disciplines such as cell biology, tissue engineering, and regenerative medicine further than the implementation of more simple naturally derived biomaterials. The resultant scaffolds can be considered to be cell-derived matrices, which are a compound of natural proteins, macromolecules, and other components such as associated growth factors. They can recapitulate the native ECM, thanks to the conservation of its organization and composition. As these scaffolds are derived from the ECM, they provide biological and mechanical support, and above all, a template of the organ, so that cells can migrate, attach, proliferate, organize, and

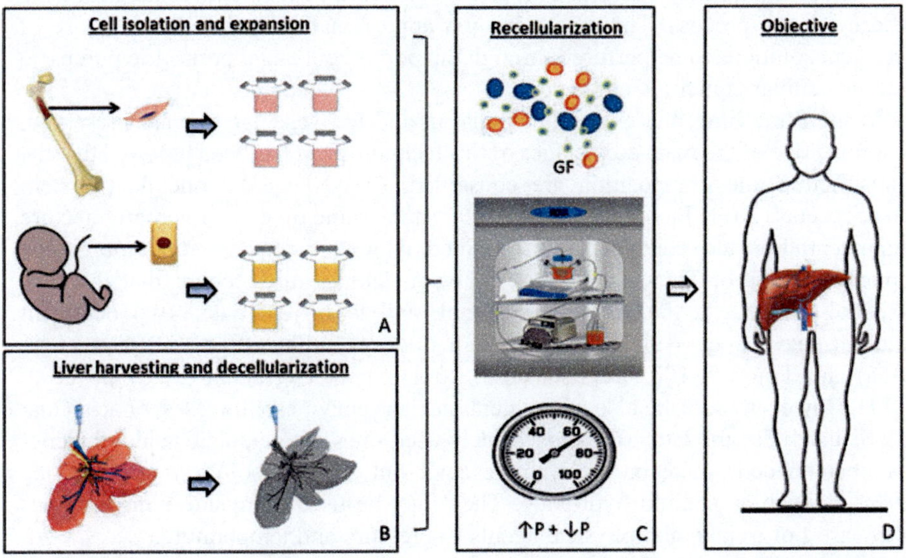

Fig. 1 Components of liver bioengineering. The process needs two components: cells, which are isolated and long-scale in vitro expanded (**a**), and a scaffold, which is decellularized (**b**) and recellularized with the isolated cells assembled in a bioreactor system, in combination with growth factors and different pressure conditions (**c**). The goal is the creation of a whole human bioengineered liver able to be transplanted into patients who require a new organ (**d**)

differentiate at the same time that it provides paracrine factor production (Robb et al. 2018).

Currently, decellularized tissues have some applications in the clinic in different areas. It is the case of orthopedics, dentistry, surgery, and even in the cardiovascular field (Parmaksiz et al. 2016). Nonetheless, the replacement of solid organs as the liver is not a clinical reality, yet as they are considerably larger than thinner tissues, their three-dimensional structure is much more complex, and they need a robust vascular network so that it can stay functional. In the field of whole-liver engineering, human, rat, and pig livers have been used as the primary sources to obtain scaffolds. However, some studies show that the spleen can also be used (Xiang et al. 2015). The generation of bioengineered livers able to replicate its native function has some limitations, including the formation of a patent vascular network and the vast expansion of cells to get the required number so that a minimal degree of function can be achieved. The first one is needed so that the blood flow becomes normal because the ECM without cellular components stimulates the formation of blood clots since it is considerably thrombogenic. However, the advances that have been made in the recellularization process give rise to an increase in vascular patency. Although this technology is expected to be used in the long term to treat patients suffering from end-stage liver failure, in recent years, some advances have been made in the field of whole-liver engineering. Researches have described different protocols that can be used to decellularize an organ in order to obtain a specific tissue-derived matrix. The strategies used are based on physical, chemical, mechanical, or enzymatic methods and even on a combination of them to maximize the efficacy of the process. The most common approximation is based on the use of detergent solutions to be perfused through the portal vein using peristaltic pumps so that the cellular content is removed.

At the same time, the extracellular matrix and the vascular tree are preserved, becoming one of the main advantages of this technology. It has been reported that the ultrastructure and composition are conserved, as well as the bile duct system (Baptista et al. 2011; Fukumitsu et al. 2011). At the same time, the microarchitecture and essential bioactive signals are maintained, which are considered complex to be replicated in vitro. This fact is quite relevant, taking into account that they are essential for cells to be viable, functional, and to differentiate. Two detergent solutions have been widely used: sodium dodecyl sulfate (SDS) (Buhler et al. 2015) and Triton X-100 (Baptista et al. 2011; Barakat et al. 2012; Crapo et al. 2011). However, there are also other agents such as enzymes, proteases, or acids that can be used. For the latter, the most widely used ones are peracetic acid, ethylenediaminetetraacetic acid, or deoxycholic acid, but alkaline solutions can also be perfused, such as sodium hydroxide. They may be used alongside with enzymes as DNase I or trypsin and physical agents as pressure and temperature.

On the one hand, for the first one, high hydrostatic pressure can be used or even supercritical CO_2. On the other hand, when the temperature is used, freeze-thaw cycles may be implemented. In the protocols published so far, the reagents mentioned above are used in multiple concentrations and even in different combinations (Faulk et al. 2015; Wang et al. 2017). Researchers have also reported the existence of diverse strategies to perfuse the detergent solutions including the use of the portal vein (Buhler et al. 2015), the portal vein together with the hepatic artery (Baptista et al. 2011; Barakat et al. 2012; Ko et al. 2015), and even the inferior *venae cavae* (Shupe et al. 2010). At the same time, both the hepatic artery and the portal vein are used in the recellularization procedure (Faulk et al. 2015; Wang et al. 2017).

Uygun and his co-workers have been able to decellularize ischemic rat livers implementing the perfusion of SDS through the portal vein using different concentrations (0.01%, 0.1%, and 1%) for 24 h each of them (Uygun et al. 2010). As a result, they obtained a scaffold without cells characterized by being translucent and by maintaining the main physical structure of the organ. On the other hand, Baptista and colleagues have reported the use of a protocol that depends on the size of the liver and its structure (Baptista et al. 2011). In this case, different animal models have been studied, including mice, rabbits, and pigs. In this case, SDS is not used, but it is based on the perfusion of a detergent solution together with other agents. Briefly, the authors identified high efficacy when 1% Triton X-100 with 0.1% ammonium hydroxide was perfused through the portal vein after perfusing a volume of deionized water equal to 40 times the liver. The authors reported that the matrices obtained were transparent, with the vascular tree visible. Finally, other authors, as De Kock et al., have described protocols in which both detergent solutions are used. They were able to decellularize a rat liver in just an hour using first 1% Triton-X 100 half of the time and 1% SDS during the second half an hour, with perfusion through the portal vein. This strategy gave rise to translucent liver scaffolds with no cellular component and able to endure fluid flows, thanks to the preservation of its architecture (De Kock et al. 2011).

2.3 Cellular Components, Their Source, and Role in Generating Hepatic Tissue

Once the decellularization procedure has been performed, to obtain a functional organ, it is necessary to seed different types of cells in the generated scaffold. To do so, they have to be sterilized and incorporated into a designed bioreactor system to be recellularized. As it was mentioned before, obtaining the required number of cells from each required type is one of the main limitations of this strategy.

As in the case of decellularization, there are different protocols reported and already available for the recellularization. Soto-Gutiérrez and collaborators compared three of them by studying direct parenchymal injection, multistep infusion, and continuous perfusion (Soto-Gutierrez et al. 2011). The first one, as its name indicates, is based on injecting the cells directly into the hepatic lobes of the scaffold, while the continuous perfusion strategy requires perfusing the obtained cells in suspension together with the culture medium using a bioreactor system to control its flow rate across the liver scaffold. Both the multistep infusion and the continuous perfusion involve delivering cells in small but multiple batches separated by intervals of time equally separated. The conclusion of this study indicated that the best results were achieved when the multistep infusion strategy was used as seeding conditions.

On the other hand, cell number is another critical parameter to be taken into account (Fukumitsu et al. 2011). In order to mimic liver functionality and morphology, its spatial distribution has to be recreated with the appropriate concentration and cellular type. Different cells can be used in the field of tissue engineering and regenerative medicine, including primary hepatocytes, which can be human or porcine, human hepatic cell lines, and immortalized hepatocytes, as well as pluripotent stem cells or fetal progenitors.

Regarding the cell types that have to be used to generate relevant hepatic tissue in vitro, the gold standard is primary human liver cells, isolated from liver tissue or whole livers that did not fulfill the conditions to be used in transplantation. Their relevance comes from their source, the human liver, and because consequently, they give rise to the functionality of this organ in vivo. Thus, in the toxicological and pharmacological research, their use can provide predictive results. However, multiple donors are required what causes differences in the characteristics of the cells used, such as the age, sex, or liver damage of each one of them. These variations give rise to some deviations in the experimental results that make it difficult to standardize the generated models. Indeed, the mentioned differences determine the success of the isolation process, as well as other factors as cell isolation conditions or intraoperative factors. During the mentioned procedure, enzymatic digestion may be performed, requiring the use of collagenase and other enzymes, and plastic culture dishes are often used. However, these conditions change some characteristics of these cells, like morphology or gene expression, due to a process known as dedifferentiation that represents a practical limitation. Some studies have been carried out to identify the best strategy to overcome it improving the survival of these cells and their functionality. The alternatives evaluated include the modification of the signals from the microenvironment by including soluble factors in the culture medium. The cell media can be modified with hormones, growth factors, or vitamins, in order to modulate the functionality of hepatocytes (Guillouzo 1998; Kidambi et al. 2009; Miyazaki et al. 1985).

The similarities between pigs and humans make it possible to use primary porcine hepatocytes as the availability of human sources is limited. Conversely, some disadvantages need to be considered. The xenogeneic character of these cells makes it possible to transfer zoonotic diseases and to identify an incompatibility between proteins. Subsequently, more appropriate cell sources have to be identified. To overcome the limitations mentioned above, human hepatocyte cell lines have been considered as an alternative. They have been immortalized using different approximations as transfection using simian virus 40 T antigen (Li et al. 2005). However, the potential to give rise to tumorigenic effects has not been assessed sufficiently in these studies. The immortalized cell lines derived from the liver that are more widely used include the HepaRG, HepG2, and Hep3B, among others (Guguen-Guillouzo et al. 2010). The first one, derived from a human hepatoma, conserves the expression of liver-specific functions, membrane transporters, and nuclear receptors (Aninat et al. 2006), presenting a stable karyotype and high proliferative capacity. Even more critical, this cell line has revealed that it can provide consistent and reproducible data as a result of different experiments. Still, its expression of liver-specific functions is lower than the one that can be identified for primary hepatocytes (Marion et al. 2010).

Another approximation that has been evaluated is the co-culture of primary human hepatocytes with non-parenchymal cells, such as Kupffer cells, liver endothelial cells, and hepatic stellate cells. The reason to do so is that researchers have identified that these cells have a critical role for liver function and in processes like acute inflammation or chronic liver diseases. As an example, some investigations have tried to obtain co-cultures with Kupffer cells to evaluate the reaction of hepatocytes in a pro-inflammatory environment (Nguyen et al. 2015).

Stem cells have also been considered to generate hepatic tissue as, in theory, they represent an unlimited cell source. Autologous liver stem cells are the leading type of cells used, but others are also being studied, such as induced pluripotent stem cells (iPSCs), mesenchymal stem cells, and humanized hepatocytes (Agmon and Christman 2016). The main advantage of iPSCs is their autologous character and that they are considered to be limitless being able to repopulate a scaffold with the appropriate size. Furthermore, these cells reduce the limitation that can be identified when others are used, such as ethical considerations when embryonic stem cells are used. Hepatocyte-like cells can also be obtained over-expressing transcription factors (Huang et al. 2011) specific from the hepatocyte lineage, starting with adult cells giving rise to induced hepatocyte-like cells or iHep. Further research is required to improve this technology, but the results obtained so far are promising.

Other cell sources have also been used, such as umbilical vein endothelial cells, both from a human and porcine source. In 2017, Mao and his co-workers were able to achieve complete coverage of the vascular network with appropriate patency in a porcine model using these cells (Mao 2017).

The number of cells that need to be used to generate functional hepatic tissue is not the only important factor. The cellular type is also a crucial parameter as it determines the flow-seeding conditions that have to be used. Then, the most appropriate speed has to be identified to optimize cellular engraftment. Hence, there are still relevant issues to be addressed to obtain functional organs before it becomes a clinical reality.

2.4 Bioreactors for Liver Tissue Engineering

Bioreactors are extensively used in tissue engineering for the development of tissue-engineered constructs. A bioreactor represents the central essence of any biochemical process in which plant, microbial, enzymes, or mammalian cell systems are used for the manufacture of a broad range of biological products. The primary function of a bioreactor is to provide a controlled in vitro environment for tissue generation and growth while mimicking the mechanochemical regulation that cells and tissues experience in vivo in their native environment. Ideally, and in the context of tissue engineering, bioreactors serve as a system able to allow the uniform seeding of cells to a scaffold, control the physiological conditions in the cell culture medium (i.e., nutrients, oxygen levels, temperature, pH), supply sufficient metabolites, and provide physiologically relevant signals in the form of mechanical loads (Altman et al. 2002; Freed and Vunjak-Novakovic 2000). Bioreactors are broadly used almost at any step of the generation of a tissue-engineered liver construct. Before their large-scale expansion or seeding in a scaffold, primary cells are isolated from autologous or allogeneic liver tissue or whole organ and cultured under static conditions in T-flasks or Petri dishes to let them recover from the isolation and purification stress and to obtain the adequate large mass of cells required for the creation of the final desired bioproduct.

An appropriate design of the bioreactor system is crucial to create better tissue-engineered products, such as artificial organs for transplantation. Each decision made in the design of the bioreactor may strongly impact the overall process performance. Theoretically, any bioreactor useful for tissue engineering purposes has to be robust enough to maintain the cultured cells or tissues for long periods, structured to deliver nutrients, growth factors, peptides, and gases into the culture system in order to encounter and sustain the needs of the cultured cells or tissues, easily cleanable, and operator friendly. From the performance point of view, the most inescapable hindrance regarding the biological design of a liver bioreactor is meeting the basic metabolic requirements of the cells or tissue.

Oxygen tension is finely controlled in the liver, varying along the sinusoid from 85 μM periportal to approximately 45 μM pericentral. Hepatocytes consume oxygen at 10- to 100-fold the rates of most cells, so there is a desperate need to balance oxygen consumption with oxygen delivery to guarantee an optimal hepatocellular function (Ebrahimkhani et al. 2014). Each hepatocyte contains over 1500 mitochondria, which consume oxygen at a rate of 0.3–0.9 nmol/sec/million cells (Nahmias et al. 2007), while an average rate of oxygen utilization by many other cells is about 2–40 picomol/sec/million cells (Wagner et al. 2011). This issue represents a considerable challenge in 3D cultures, as the oxygen gradient across a layer of five cell diameters, which represents a distance of approximately 120 μm, ranges in the liver from normoxic to hypoxic (Ebrahimkhani et al. 2014).

Another crucial step in the development of liver constructs for liver replacement is represented by the seeding of a vast amount of liver cells uniformly on or throughout a scaffold. Furthermore, the adherent cells require appropriate amounts of nutrients, oxygen, peptides, and growth factors to survive and proliferate and adequate chemo/physical signals to organize and differentiate and to generate cellular structures and metabolic zonation resembling that of the liver. On the one hand, a high initial cell seeding might favor tissue formation (Dvir-Ginzberg et al. 2003), but on the other hand, it would require a high amount of organ tissue for the isolation of the cells and more time to obtain an adequate number of cells to seed. Moreover, a uniform initial cell distribution on the scaffold is needed to avoid spatial variation in nutrients, oxygen, and metabolite concentrations that would condition the survival and metabolism of cells at different positions in the scaffold (Catapano and Gerlach 2007).

When designing a bioreactor for liver engineering, it is imperative to consider the diffusion distance existing between cells and the blood. Indeed, liver cells are sensitive to waste metabolites and have essential nutrient requirements that need a small diffusion distance (i.e., within a few hundred microns in the sinusoids) (Catapano and Gerlach 2007). It is an enormous challenge to provide a system to supply essential substrates (e.g., oxygen, glucose, and amino acids) or clear waste metabolites (e.g., CO_2, ammonia, urea, lactate) from liver cells in large 3D constructs and a central prerequisite to promote cell growth, differentiation, and long-term survival (Catapano 1996; Martin et al. 2004; Martin and Vermette 2005).

Plenty of efforts have been made to develop bioreactor designs adapted for liver cell culture. Hollow fiber cell culture bioreactors have been developed in the 1970s and used until today for the generation of high concentration of cell-derived products, such as monoclonal antibodies, recombinant proteins, growth factors, viruses, and virus-like particles. Nowadays, these reactors include control of pH, oxygen, fluid dynamics, and medium exchange, allowing the production of high numbers of bioproducts. The traditional hollow fiber bioreactor is a three-dimensional cell-culturing system based on small, semipermeable capillary membranes arranged in a parallel array. These membranes are bundled and housed within tubular polycarbonate shells to create the bioreactor cartridges containing two compartments: the intracapillary space (IC) within the hollow fibers and the extracapillary space (EC) surrounding the IC. The culture medium is pumped through the IC space outside the fibers while delivers molecular components with the cells cultured and expanded into the EC space. Here, the exchanges occur in a manner resembling some features of capillary blood-tissue exchanges in vivo, being the membranes the core of the bioreactor itself. As these membranes are permeable, it is possible to control the molecular exchange properties between the nutrient fed and tissue compartment through the selection of membrane permeability properties. These bioreactors have been applied to liver cell culture, practically from the beginning. Wolf C. et al. firstly describe that a rat hepatoma cell line could carry out bilirubin conjugation in hollow fiber culture with either 10 kDa or 50 kDa cutoff membranes (Wolf and Munkelt 1975), with speculation that the rates of conjugation may be limited by the transport of the bilirubin carrier protein albumin across the membrane. Next, Jauregui H. et al. found that primary rat cells cultured up to 45 days in a 7-cm-long hollow fiber reactor with 520–0.3 mm diameter fibers maintained almost half the diazepam-metabolizing capacity after 10 days provided the perfusion medium was equilibrated with 30% oxygen atmosphere (Jauregui et al. 1994). Due to the similarity between hollow fiber cell culture bioreactors and blood dialysis units, liver culture in hollow fiber bioreactors appeared as a potentially promising approach for extracorporeal liver support to provide metabolic capacity in synthesis of urea and metabolism of toxic metabolites like bilirubin (Demetriou et al. 2004; Jauregui et al. 1994; Rozga et al. 1993, 1994; Struecker et al. 2014).

Stirred-tank bioreactors have mainly been used in the large-scale culture of mammalian cells to create a suitable microenvironment, where variables are tightly controlled, such as pH and oxygen levels. With the appropriate design of impellers, the fluid mechanic microenvironment in these systems can provide relatively uniform low-shear mixing, and the commercial availability of reactor systems makes them accessible for general use (Morini et al. 2018). When a stirred-tank bioreactor is used for growing adherent cells, specific microcarriers are used as a supporting surface for cell attachment. Alternatively, adherent cells are grown as aggregates in suspension.

The ability of hepatocytes to form aggregates when seeded on low-adhesive surfaces, and for aggregates to maintain liver-specific function better compared to cells in monolayer cultures, has been described and demonstrated (Powers 1997). The utility of stirred-tank bioreactors for efficient formation of relatively uniform, functional spheroidal aggregates of primary rat hepatocytes that exhibit polarization of bile canaliculi was already illustrated over 20 years ago (Wu et al. 1996). Recently, Alves and colleagues established and tested a perfused bioreactor system for the long-term maintenance of primary cultures of human hepatocyte spheroids from three different donors. They found that using this method, the generated hepatocyte spheroids reproducibly recapitulated in vivo hepatic functions and structure, despite inter-donor variability. They hypothesized that these reproducible time-course profiles were made possible because of the tight control of critical environmental variables at physiological values. Moreover, they observed that the spheroid's inner structure resembled the liver architecture, with functional bile canaliculi-like structures and liver-specific markers (such as CYP450 expression and phase II and III drug-metabolizing enzyme gene expression and transport activity). This system constitutes an ideal long-term culture platform for analyzing hepatic function for drug development tests (Tostoes et al. 2012).

Commercially available microcarriers with different sizes and composition can also be used for large-scale cell culture for many different anchorage-dependent cell types. Microcarriers differ in their porosity, specific gravity, optical properties, presence of animal components, and surface chemistry and can be made of different materials (i.e., glass, polystyrene plastic, collagen, alginate, dextran) which can influence cellular behavior, phenotype, morphology, and proliferation. The main advantages of microcarrier technology reside in providing a larger surface-area-to-volume ratio for the growth of anchorage-dependent cells in a suspension culture system, ease of scale-up, and the ability to monitor and control various physiological conditions when used in a bioreactor. Several attempts have been made to take advantage of microcarrier technology for liver tissue engineering purposes. Gao Y. et al. used Cytodex-3 microcarriers to cultivate high-density human liver cell line Cl-1 to improve the cultivation efficiency and yield and evaluated specific functions of liver cells periodically. They observed that the human liver cell line Cl-1 can be cultivated to a high density on these microcarriers and has better biological functions, such as albumin and urea synthesis and diazepam transformation, along with the prolonging of the cultivation (Gao et al. 1999). Zhang L. and colleagues reported that rat hepatocytes cultivated on chitosan microcarriers cross-linked by oxidized lactose retained the spherical shape as they have in vivo. Moreover, liver-specific functions such as albumin secretion and glucose metabolism were stably maintained for 7 days in culture, and the metabolic activity of hepatocytes cultured on these specific microcarriers was higher than those of hepatocytes cultured on chitosan microcarriers cross-linked by glutaraldehyde and on Cytodex-3 (Zhang et al. 2003).

2.5 In Vivo Results of Liver Tissue-Engineered Constructs

Hepatic tissue engineering using primary hepatocytes has been considered a new exciting approach for the treatment of different classes of liver diseases. Hepatocyte-based therapies have been experimentally and clinically explored and generate new interest in the bioengineering of alternative liver systems in vivo based on the manipulation of in vitro-modified hepatocyte cultures (Griffith and Naughton 2002; Ohashi et al. 2001; Strom and Fisher 2003). Ohashi and co-workers developed a method to engineer a uniformly continuous sheet of hepatic tissue using isolated primary hepatocytes cultured on temperature-responsive surfaces. Sheets of hepatic tissue transplanted into the subcutaneous space resulted in efficient engraftment to the surrounding cells, with the formation of two-dimensional hepatic tissues that stably persisted for longer than 200 days. The engineered tissues also showed several characteristics of liver-specific functionality. They described this technology as simple, minimally invasive, and free of potentially immunogenic biodegradable scaffolds (Ohashi et al. 2007).

Clinical trials on hepatocyte transplantation have demonstrated long-term safety, but donor hepatocyte engraftment and restoration of failing host livers have not been adequate to reduce the need for organ transplantation (Dhawan et al. 2010). One alternative could be represented by the use of hydrogels reproducing the biochemistry of tissue-specific ECM proteins. ECM hydrogels were derived from decellularized rat livers and employed for both 2D-plate coatings and in vivo hepatocyte transplantation. Primary rat hepatocytes cultured on a liver ECM hydrogel-coated substrate exhibited higher viability and improved hepatic functions compared to cells cultured on a non-coated or collagen type I-coated substrate. Besides, liver ECM hydrogels engineered with rat hepatocytes maintained the hepatic phenotype and function after in vivo transplantation (Lee et al. 2014).

3 Future Perspectives

In the last years, the advances and improvements regarding the technology of liver bioengineering have been used to understand the hepatic regeneration mechanism better, and it has facilitated the development of different in vitro models near-physiological ex vivo culture system. In this section, different in vitro models showing a more precise control over the liver cell microenvironment will be discussed.

3.1 Liver Organoids

In the middle of the last century, different authors began to use the term "organoids" (Vendrely 1950), and this nomenclature started to appear in various publications in the 1960s. The advances in the organoid field over the last decade are the consequence of years of work to understand more profoundly the role of progenitor cells, self-organization of dissociated tissues, and ECM biology.

Usually, the term organoids are used to refer to a range of 3D culture systems that resemble the modeled organ to varying extents (Prior et al. 2019). Specifically, an organoid is an in vitro 3D cellular cluster derived from tissue-resident stem/progenitor cells, embryonic stem cells (ESCs), or induced pluripotent stem cells (iPSCs) capable of self-renewal and self-organization that recapitulate the functionality of the tissue of origin (Huch and Koo 2015; Lancaster and Knoblich 2014; Prior et al. 2019). Considering this definition, the organoid culture requires the isolation of stem/progenitor cells, either pluripotent stem cells or tissue-resident cells from embryonic stages of adult tissues, which are cultured by using specific medium supplemented with growth factors that recreate the signals to give rise to the specific tissue. Also, the organoid cultures require a specialized physical environment, and it usually involves culturing cells in suspension, on an air-liquid interface, or embedded in a suitable ECM such as Matrigel. Thus, when the isolated cells are cultured with the correct growth medium and a specific physical environment, they can follow intrinsic developmental or homeostatic/repair programs to proliferate and self-organize into 3D organoid structures, allowing the use of these models in different areas of knowledge. For example, the use of organoids has high relevance in disease modeling, as alpha-1 antitrypsin (A1AT) deficiency, an example of a monogenic liver disease that affects the liver parenchyma. Recently, the group led by Hans Clever showed that organoids derived from adult liver tissue from patients with A1AT deficiency had been successfully developed and recapitulated critical aspects of this disease: accumulation of protein aggregates and reduced ability to block elastase activity. They also showed signs of ER stress, such as phosphorylation of eIF2α, among other aspects, showing similar characteristics to what had been observed in the original biopsy from patients. Thus, organoids from A1AT-deficiency patients can be expanded in vitro and mimic the in vivo pathology recapitulating critical features of the disease in vitro and enabling further understanding of the disease processes (Huch et al. 2015). Another area of interest where liver organoids have been exploited is for personalized medicine use, as HCC.

Nuciforo et al. reported the generation of long-term organoid cultures from tumor needle biopsies of HCC patients with various etiologies and tumor stages. These HCC organoids showed a preserved morphology, as well as the expression pattern of HCC tumor markers, and preserved the genetic heterogeneity of the originating tumors. This study showed that liver cancer organoids could be used to test sensitivity to sorafenib, providing a tool for developing tailored therapies (Nuciforo et al. 2018).

After the examples described above, it is easy to understand the relevance of liver organoids. It is focused on its applications since this type of culture shows an improved understanding of hepatic regenerative pathways and the development of in vitro systems to mimic both the expansion and differentiation of hepatocytes. As a consequence of the relevance in its applications, the organoids were named "Method of the year 2017" by *Nature Methods* showing the interest and promise of this rapidly expanding field to provide new experimentally tractable and physiologically relevant models of organ development, human pathologies, and paving the way for therapeutic applications (Methods 2018; Prior et al. 2019).

3.2 Liver Buds

Cell-based therapy has been proposed as a useful alternative to orthotopic liver transplantation. Organ bud progenitor cells play essential roles in organ development. These cells have a remarkable capacity for rapid cell growth and differentiation into multi-lineage cells.

A liver bud is defined as the primordial cellular diverticulum of the embryonic foregut endoderm that gives rise to the parenchyma of the liver. Hepatic progenitor cells (HPCs), or liver bud progenitors, are specified from foregut endoderm at embryonic day (E)9.5 in mice. These is followed by a massive HPC expansion with a 10,000-fold increase in population, doubling from E9.5 to E13.5 in mice (Koike et al. 2014; Takebe et al. 2013). This tremendous growth is regulated by signals secreted from neighboring mesenchyme such as hepatocyte growth factors (Matsumoto et al. 2001), bone morphogenetic proteins (Rossi et al. 2001), and fibroblast growth factors (Serls et al. 2005) and by transcriptional networks that act intrinsically in the HPCs, such as Tbx3 (Suzuki et al. 2008), Smad2/3 (Weinstein et al. 2001), and beta-catenin (Micsenyi et al. 2004). However, the mechanism regulating this intensive and transient amplification in developing liver bud is mainly unknown.

Organoid technology represents a revolutionary paradigm toward therapy but is not yet applied in humans, due to lack of reproducibility and scalability. Several attempts have been made to overcome these limitations. Takebe and colleagues created a scalable organ bud production platform from human-induced pluripotent stem cells (iPSC). First of all, they identified three progenitor populations that can effectively generate liver buds in a highly reproducible manner: hepatic endoderm, endothelium, and septum mesenchyme. Furthermore, they achieved human scalability by developing an Omni-well-array culture platform for mass-producing homogeneous and miniaturized liver buds on a clinically relevant large scale. Vascularized and functional liver tissues generated entirely from iPSCs significantly improved subsequent hepatic functionalization potentiated by stage-matched developmental progenitor interactions, enabling functional rescue against acute liver failure via transplantation (Takebe et al. 2017).

Yanagi and co-workers have made another attempt. They reported a novel transplantation method for liver buds that were grown in vivo involving orthotopic transplantation on the transected parenchyma of the liver, which showed prolonged engraftment and marked growth in comparison with heterotopic transplantation. Furthermore, they demonstrated a method for rapidly fabricating scalable liver-like tissue fusing hundreds of liver bud-like spheroids using a 3D bioprinter. The ex vivo-fabricated human liver-like tissue exhibited self-tissue organization and engraftment on the liver of nude rats (Yanagi et al. 2017).

At the moment, no one succeeded in assembling human liver buds containing HSCs and LSECs. Recently, Li J. et al. described a reproducible, easy-to-follow, and comprehensive self-assembly protocol to generate 3D human liver buds from naïve MSCs, MSC-derived hepatocytes, and HSC- and LSEC-like cells. By optimizing the ratio between these different cell lineages, the cell mixture self-assembled into 3D human liver buds within 72 h in vitro and exhibited similar characteristics with early stage murine liver buds. In a murine model of acute liver failure, the mesenteric transplantation of self-assembled human liver buds effectively rescued animal death and triggered hepatic amelioration effects that were better than the ones observed after splenic transplantation of human hepatocytes or naïve MSCs. Besides, transplanted human liver buds underwent maturation during injury alleviation, after which they exhibited a gene expression profile signature similar to one of adult human livers (Li et al. 2018).

3.3 3D Bioprinting

Three-dimensional bioprinting is an innovative technology based on the successive addition of small quantities of biomaterials and cells which are deposited layer by layer. It comprises different technologies to obtain artificial constructs with a high precision level (Murphy and Atala 2014).

The biomaterials used can be natural polymers such as collagen and fibrin, synthetic polymers, and even decellularized matrices. Indeed, two different approaches can be followed when 3D bioprinting technology is used (Tiruvannamalai-Annamalai et al. 2014). On the one hand, the traditional approach of top down, which is based on seeding the cells into a porous scaffold to promote cell proliferation and scaffold degradation, gives rise to the engineered tissue. The other modular approach, or bottom-up, cells, together with biomaterials, are printed using the tridimensional technology to obtain the engineered tissue. The first one has different limitations that are solved in the second one, such as diffusion limitations and slow vascularization. However, most purists consider that just the modular approach represents three-dimensional bioprinting. The number of applications of 3D printing in medicine has increased over the past years. It can be used today to obtain customized implants, pre-surgery models, prostheses, exoskeletons, and even in drug discovery, delivery, and dosage forms (Klein et al. 2013). Hence, the uses of this technology belong to two different categories: the production of tissues and organs and research in the pharmaceutical area. For the latter, 3D bioprinting has been used to determine if a particular drug could work as a treatment for a specific patient as a screening technology (Banks 2013). Some strategies can be used to overcome some of the limitations of the process itself. For example, cells can also be printed using liver spheroids to protect them from the effect of shear stress as it can be generated along with the procedure (Bhise et al. 2016).

Currently, 3D bioprinting has the potential to produce complex organs, such as the liver, with a high density of cells (Zhang et al. 2017). Nevertheless, there are some limitations still to be solved. It is the case of achieving the appropriate vascular network. However, it can considerably simplify the process compared with traditional approaches. Due to the described characteristics of this technology, some companies have decided to invest in it, like Organovo™.

They have been able to generate 3D vascularized liver constructs using different types of cells, including HSC, LSEC, and hepatocytes. They have identified high cellular viability, as well as robustness for drug metabolism, resembling native hepatic lobules. At the end of 2016, this company announced that it would investigate the use of 3D bioprinting technology to obtain human liver tissue that could be transplanted directly to human patients (Delivery 2016). Their plan was focused on obtaining a clinical solution for acute-on-chronic liver failure and pediatric errors in metabolism. The preclinical studies that have been carried out demonstrated that after 60 days of implantation, there was sufficient vascularization, as well as engraftment and functionality. The next step was to test this approach to humans. However, due to the difficulty of a long-term function of non-vascularized bioprinted tissues after transplantation in larger animal models, the company investors declared that they would not proceed further, and the company stopped its research activities in May 2019.

4 Regulatory Landscape for Tissue-Engineered Livers

In legal terms, the regulatory landscape of liver tissue engineering and regenerative medicine, in general, is uninterruptedly moving toward a more favorable situation, allowing the creation and commercialization of new innovative products that can improve the quality of life of patients in need. These types of products are challenging due to the novelty, complexity, and technical specificity, so it is essential to understand the regulations that guarantee the quality and safety of these novel products. Next, three different regulatory agencies will be described: the Food and Drug Administration (FDA) in the USA, the European Medicines Agency (EMA) in Europe, and the Ministry of Health, Labour, and Welfare (MHLW) in Japan. All these organizations have similar objectives, but their systems of operation are different, and the approval of one of them does not imply the endorsement by the other (Bertram et al. 2013).

4.1 Food and Drug Administration (FDA)

In the USA, the FDA's Center for Biologics Evaluation and Research is the regulatory agency responsible for guaranteeing the safety, purity, potency, and effectiveness of many biologically derived products. All these functions are performed by the six centers and the several offices in which the FDA is divided.

The field of tissue engineering is englobed under the term "tissue-engineered medical products" (TEMP). This new terminology was defined in a standard document of the American Society for Testing and Materials, and it was included in the FDA-recognized consensus standards database (ASTM F2312-11). The term TEMP can consist of a variety of different constituents as cells, scaffolds, device, or any combination of these, and the FDA classifies these products as combination products. The US Congress recognized the existence of combination products when it enacted the Safe Medical Device Act of 1990, and it was defined in the 21 Code of Federal Regulation 1271 Part C 210/211/820 (CFR; FDA). The offices involved in the evaluation of TEMPs include the Office of Combination Products (OCP), the Office of Regulatory Affairs (ORA), and the Office of Orphan Products (OOP). Notably, the Office of Combination Products (OCP) determines the primary mode of action (PMOA). PMOA establishes its regulatory and product development framework and assigns it to the proper center to lead the review of that product, with the other two centers providing input (Montagne et al. 2011). The PMOA is such an essential concept that the FDA published a docket in August 2005 entitled *Definition of Primary Mode of Action of a Combination Product*. The PMOA is defined as "the single model of action of a combination product that provides the most important therapeutic effect of the combination product" (Food and Drug Administration 2005).

4.2 European Medicines Agency (EMA)

The EMA is the agency of the European Union in charge of the evaluation and supervision of medicinal products.

Under the term, advanced therapy medicinal products (ATMP) are englobed in the field of liver tissue engineering. ATMP is defined as being a somatic cell therapy medicinal product (SCTMP), a tissue-engineered product (TEP), a gene therapy medicinal product (GTMP), or a combined ATPM (EMA 2018).

Considering the progress of liver tissue engineering in particular and regenerative medicine in general, in 2007, the European Parliament and Council of the European Union (EU) issued an amendment to Directive 2001/83/EC and Regulation No. 776/2004 to include regulatory provisions for ATMPs defined in Regulation EC No 1394/2007. This regulation established that when a product contained viable cells or tissues, the pharmacological, immunological, or metabolic action of those cells or tissues has to be considered as the principal mode of action of the product (EPC 2007). To offer high-level expertise to assess the quality, safety, and efficacy of ATMPs, the EMA established a multidisciplinary committee called the committee for advanced therapies (CAT). This committee is responsible for reviewing applications for marketing authorization for advanced therapy medicinal products (EMA 2017). The application for the authorization of an ATMP is supervised down in Article 6 of Regulation No 726/2004 (EMA 2004), and among other requirement, the application should also include the description of the product design method by the Annex 1 to Directive 2001/83/EC (EPC 2001).

4.3 Ministry of Health, Labour, and Welfare (MHLW)

The Ministry of Health, Labour, and Welfare (MHLW) is the regulatory body that oversees food and drugs in Japan, which includes creating and implementing safety standards for medical devices and drugs (JPMA 2015).

Japan established that liver tissue engineering under a regulatory landscape is considered as cellular and tissue-based product. In Japan, the approved cell/tissue-engineered products were regulated as medical devices by adapting existing legislation under clause 2 of the Pharmaceutical Affairs Law (PAL) (Yano et al. 2015). These types of products are intended to be used for reconstruction, repair, or formulation of a structure or function of the body and treatment or prevention of disease or to be inducted into human cells for gene therapy. The underlying technical requirements to assure the quality and safety of these cellular and tissue-based products are specified in Notification No. 0912006 of the PFSB dated September 12, 2008.

In conjunction with the MHLW, the Pharmaceuticals and Medical Device Agency (PMDA) is an independent agency that is responsible for the in-country representation, certification processes, licensing, and quality assurance systems (Jokura et al. 2018). The PMDA consist of 25 offices, and the office of cellular and tissue-based products is the one in charge to confirm clinical trial notifications and adverse drug reactions, and it also conducts reviews required for approval, reexaminations, and reevaluations of regenerative medical products (cellular and tissue-based products and gene therapy products), preliminary reviews for approval or verification based on the Cartagena Protocol, and quality review of antibody preparations (JPMA 2017).

5 Conclusions

This chapter is focused on state-of-the-art strategies for liver tissue engineering. Through the different sections, it has been explained the current need of available organs for human liver transplantation, emphasizing the importance of achieving an optimal method of liver decellularization and recellularization to create new organs, as well as the development of innovative and efficient models for the study of the liver trying to recreate the physiological aspects found in vivo. The regulatory landscape is also discussed to have a global idea concerning the creation and commercialization of tissue-engineered products that can improve the quality of life of patients in need.

Even though regenerative medicine has exponentially increased in the last years, giving new hopes to the development of effective treatments for liver diseases, further work has to be performed in order to obtain functional therapies and models that represent the liver to its fullest. Critical components to consider in the liver tissue engineering field include the proper technology to obtain a high-quality scaffold after decellularization, as well as the identification, selection, and large-scale production of the most suitable cell sources for the most effective scaffold seeding.

Optimal recellularization is also a crucial step to generate a clinically functional organ. To accomplish this goal, bioreactors are an excellent tool to control all the process parameters. Nevertheless, their configuration is critical for the control of vital parameters such as oxygen, nutrient supply to cells, the control of biochemical concentrations, and gradients. However, it is essential to highlight that large scaffolds are complex and highly depend on cell metabolism and bioreactor configurations. Up to date, a complete recellularization using all the necessary liver cell types and the generation of a fully functional liver has not been accomplished.

In the last years, the technological advances in liver bioengineering have been used to understand the hepatic regeneration mechanism better, and it has facilitated the development of different in vitro models and ex vivo culture systems. Through this chapter, we have described relevant sophisticated engineering tools, as liver organoids, liver buds, and 3D bioprinting technology. All these in vitro models show a more precise control over the liver cell microenvironment, and they allow us to enhance our knowledge of disease development and progression, demonstrating faithful recapitulation of disease pathways in vitro.

The rapid advances and improvements achieved in tissue engineering require rigorous control over the laws and regulations of each country. For these reasons, the regulatory landscape of some regulatory agencies as the FDA, EMA, and MHLW has been intensely discussed.

In conclusion, liver tissue bioengineering requires a multidisciplinary approach, with biologists, clinicians, and bioengineers working closely to understand further how liver cells can self-organize to build the liver, one of the most complex organs in our body, in order to get a transplantable liver for those in need.

Acknowledgments This work was supported by Instituto de Salud Carlos III, through a predoctoral fellowship i-PFIS IFI15/00158 (I.P-P), by MINECO through a Torres Quevedo Grant fellowship PTQ-17-09267 (N.S-R), and by a predoctoral fellowship from Fundação para a Ciência e a Tecnologia Portugal through PD/BD/144057/2015 (S.M) and PMB and was supported with the project PI15/00563 and PI18/00529 from Instituto de Salud Carlos III, Spain, with project CIBEREHD EHD16PI02 from CIBERehd, Spain, and with the project LMP226_18 from the Government of Aragon, Spain.

References

AACC (2019). https://labtestsonline.org/conditions/liver-disease

Agmon G, Christman KL (2016) Controlling stem cell behavior with decellularized extracellular matrix scaffolds. Curr Opin Solid State Mater Sci 20:193–201

Allen JW, Hassanein T, Bhatia SN (2001) Advances in bioartificial liver devices. Hepatology 34:447–455

Altman GH, Lu HH, Horan RL, Calabro T, Ryder D, Kaplan DL, Stark P, Martin I, Richmond JC, Vunjak-Novakovic G (2002) Advanced bioreactor with controlled application of multi-dimensional strain for tissue engineering. J Biomech Eng 124:742–749

Aninat C, Piton A, Glaise D, Le Charpentier T, Langouet S, Morel F, Guguen-Guillouzo C, Guillouzo A (2006) Expression of cytochromes P450, conjugating enzymes, and nuclear receptors in human hepatoma HepaRG cells. Drug Metab Dispos 34:75–83

Asrani SK, Devarbhavi H, Eaton J, Kamath PS (2019) Burden of liver diseases in the world. J Hepatol 70:151–171

ASTM F2312-11 (2020) Standard Terminology Relating to Tissue Engineered Medical Products. https://www.astm.org/Standards/F2312.html.

Banks J (2013) Adding value in additive manufacturing: researchers in the United Kingdom and Europe look to 3D printing for customization. IEEE Pulse 4:22–26

Baptista PM, Siddiqui MM, Lozier G, Rodriguez SR, Atala A, Soker S (2011) The use of whole organ decellularization for the generation of a vascularized liver organoid. Hepatology 53:604–617

Barakat O, Abbasi S, Rodriguez G, Rios J, Wood RP, Ozaki C, Holley LS, Gauthier PK (2012) Use of decellularized porcine liver for engineering humanized liver organ. J Surg Res 173:e11–e25

Bernal W, Auzinger G, Dhawan A, Wendon J (2010) Acute liver failure. Lancet 376:190–201

Bertram T, Hellman KB, Bayon Y, Ellison S, Wilburn S (2013) The regulatory imperative: international perspective. Tissue Eng Part B Rev 19:191–193

Bhise NS, Manoharan V, Massa S, Tamayol A, Ghaderi M, Miscuglio M, Lang Q, Shrike Zhang Y, Shin SR, Calzone G et al (2016) A liver-on-a-chip platform with bioprinted hepatic spheroids. Biofabrication 8:014101

Bosch FX, Ribes J, Diaz M, Cleries R (2004) Primary liver cancer: worldwide incidence and trends. Gastroenterology 127:S5–S16

Broering DC, Schulte am Esch J, Fischer L, Rogiers X (2004) Split liver transplantation. HPB (Oxford) 6:76–82

Buhler NE, Schulze-Osthoff K, Konigsrainer A, Schenk M (2015) Controlled processing of a full-sized porcine liver to a decellularized matrix in 24 h. J Biosci Bioeng 119:609–613

Cai J, Ito M, Westerman KA, Kobayashi N, Leboulch P, Fox IJ (2000) Construction of a non-tumorigenic rat hepatocyte cell line for transplantation: reversal of hepatocyte immortalization by site-specific excision of the SV40 T antigen. J Hepatol 33:701–708

Carpentier B, Gautier A, Legallais C (2009) Artificial and bioartificial liver devices: present and future. Gut 58:1690–1702

Catapano G (1996) Mass transfer limitations to the performance of membrane bioartificial liver support devices. Int J Artif Organs 19:18–35

Catapano G, Gerlach JC (2007) Bioreactors for liver tissue engineering. In: Ashammakhi N, Reis R, Chiellini E (eds) Topics in tissue engineering. Vol 3. pp 1–41

CFR F. Code of Federal Regulations Title 21 Parts 210 and 211. https://www.accessdata.fda.gov/scripts/cdrh/cfdocs/cfcfr/cfrsearch.cfm

Cohen JC, Horton JD, Hobbs HH (2011) Human fatty liver disease: old questions and new insights. Science 332:1519–1523

Crapo PM, Gilbert TW, Badylak SF (2011) An overview of tissue and whole organ decellularization processes. Biomaterials 32:3233–3243

De Kock J, Ceelen L, De Spiegelaere W, Casteleyn C, Claes P, Vanhaecke T, Rogiers V (2011) Simple and quick method for whole-liver decellularization: a novel in vitro three-dimensional bioengineering tool? Arch Toxicol 85:607–612

Delivery DDa (2016) Organovo introduces 3D bioprinted human liver as leading therapeutic tissue in preclinical development. https://ir.organovo.com/news-releases/news-release-details/organovo-introduces-3dbioprinted-human-liver-leading/

Demetriou AA, Brown RS Jr, Busuttil RW, Fair J, McGuire BM, Rosenthal P, Am Esch JS 2nd, Lerut J, Nyberg SL, Salizzoni M et al (2004) Prospective, randomized, multicenter, controlled trial of a bioartificial liver in treating acute liver failure. Ann Surg 239:660–667. discussion 667–670

Dhawan A, Puppi J, Hughes RD, Mitry RR (2010) Human hepatocyte transplantation: current experience and future challenges. Nat Rev Gastroenterol Hepatol 7:288–298

Dvir-Ginzberg M, Gamlieli-Bonshtein I, Agbaria R, Cohen S (2003) Liver tissue engineering within alginate scaffolds: effects of cell-seeding density on hepatocyte viability, morphology, and function. Tissue Eng 9:757–766

Ebrahimkhani MR, Neiman JA, Raredon MS, Hughes DJ, Griffith LG (2014) Bioreactor technologies to support liver function in vitro. Adv Drug Deliv Rev 69–70:132–157

EMA (2004) Regulation No 726/2004 of the European Parliament and of the Council of 31 March 2004 laying down community procedures for the authorization and supervision of medicinal products for human and veterinary use establishing a European Medicines Agency

EMA (2017) EMA reflection paper on classification of advanced therapy medicinal products, EMA/CAT 600280/2010 rev.1

EMA (2018) Advanced therapy medicinal products (ATMPs) and ATMP Regulation. https://www.ema.europa.eu/en/documents/presentation/presentation-advanced-therapy-medicinal-products-atmps-atmp-regulation-patrick-celis_en.pdf
EPC (2001) Directive 2001/83/EC of the European Parliament and of the council of 6 November 2001 on the Community code relating to medicinal products for human use
EPC (2007) Regulation (EC) No 1394/2007 of the European Parliament and of the council of 13 November 2007 on advanced therapy medicinal products and amending Directive 2001/83/EC and Regulation (EC) No 726/2004
Fattovich G, Stroffolini T, Zagni I, Donato F (2004) Hepatocellular carcinoma in cirrhosis: incidence and risk factors. Gastroenterology 127:S35–S50
Faulk DM, Wildemann JD, Badylak SF (2015) Decellularization and cell seeding of whole liver biologic scaffolds composed of extracellular matrix. J Clin Exp Hepatol 5:69–80
FDA, H.P. FDA, USA. Regulation of human cells, tissues, and cellular and tissue-based products (HCT/Ps). 21 CFR, Parts 1270 and 1271
Fiegel HC, Kaufmann PM, Bruns H, Kluth D, Horch RE, Vacanti JP, Kneser U (2008) Hepatic tissue engineering: from transplantation to customized cell-based liver directed therapies from the laboratory. J Cell Mol Med 12:56–66
Food, and Drug Administration, H.H.S (2005) Definition of primary mode of action of a combination product. Final rule. Fed Regist 70:49848–49862
Freed LE, Vunjak-Novakovic G (2000) Tissue engineering bioreactors, 2nd edn. Academic, London
Fukumitsu K, Yagi H, Soto-Gutierrez A (2011) Bioengineering in organ transplantation: targeting the liver. Transplant Proc 43:2137–2138
Gao Y, Xu XP, Hu HZ, Yang JZ (1999) Cultivation of human liver cell lines with microcarriers acting as biological materials of bioartificial liver. World J Gastroenterol 5:221–224
Gerlach J, Unger J, Hole O, Encke J, Muller C, Neuhaus P (1994) Bioreactor for long-term maintenance of differentiated hepatic cell functions. ALTEX 11:207–215
Griffith LG, Naughton G (2002) Tissue engineering – current challenges and expanding opportunities. Science 295:1009–1014
Guguen-Guillouzo C, Corlu A, Guillouzo A (2010) Stem cell-derived hepatocytes and their use in toxicology. Toxicology 270:3–9
Guillouzo A (1998) Liver cell models in in vitro toxicology. Environ Health Perspect 106(Suppl 2):511–532
Huang P, He Z, Ji S, Sun H, Xiang D, Liu C, Hu Y, Wang X, Hui L (2011) Induction of functional hepatocyte-like cells from mouse fibroblasts by defined factors. Nature 475:386–389
Huch M, Koo BK (2015) Modeling mouse and human development using organoid cultures. Development 142:3113–3125
Huch M, Gehart H, van Boxtel R, Hamer K, Blokzijl F, Verstegen MM, Ellis E, van Wenum M, Fuchs SA, de Ligt J et al (2015) Long-term culture of genome-stable bipotent stem cells from adult human liver. Cell 160:299–312
Iansante V, Mitry RR, Filippi C, Fitzpatrick E, Dhawan A (2018) Human hepatocyte transplantation for liver disease: current status and future perspectives. Pediatr Res 83:232–240
Ibars EP, Cortes M, Tolosa L, Gomez-Lechon MJ, Lopez S, Castell JV, Mir J (2016) Hepatocyte transplantation program: lessons learned and future strategies. World J Gastroenterol 22:874–886
Jadlowiec CC, Taner T (2016) Liver transplantation: current status and challenges. World J Gastroenterol 22:4438–4445
Jauregui HO, Naik S, Santangini H, Pan J, Trenkler D, Mullon C (1994) Primary cultures of rat hepatocytes in hollow fiber chambers. In Vitro Cell Dev Biol Anim 30A:23–29
Johnson LB, Aiken J, Mooney D, Schloo BL, Griffith-Cima L, Langer R, Vacanti JP (1994) The mesentery as a laminated vascular bed for hepatocyte transplantation. Cell Transplant 3:273–281

Jokura Y, Yano K, Yamato M (2018) Comparison of the new Japanese legislation for expedited approval of regenerative medicine products with the existing systems in the USA and European Union. J Tissue Eng Regen Med 12:e1056–e1062

JPMA (2015) Quality and safety assurance for drug products or medicinal devices with processed (homogeneous) human-derived cells or tissue. http://www.jpma.or.jp/english/parj/pdf/2015.pdf

JPMA (2017) Pharmaceutical administration and regulations in Japan. http://www.jpma.or.jp/english/parj/pdf/2017.pdf

Kaufmann PM, Kneser U, Fiegel HC, Kluth D, Herbst H, Rogiers X (1999) Long-term hepatocyte transplantation using three-dimensional matrices. Transplant Proc 31:1928–1929

Kidambi S, Yarmush RS, Novik E, Chao P, Yarmush ML, Nahmias Y (2009) Oxygen-mediated enhancement of primary hepatocyte metabolism, functional polarization, gene expression, and drug clearance. Proc Natl Acad Sci U S A 106:15714–15719

Kim BS, Baez CE, Atala A (2000a) Biomaterials for tissue engineering. World J Urol 18:2–9

Kim SS, Sundback CA, Kaihara S, Benvenuto MS, Kim BS, Mooney DJ, Vacanti JP (2000b) Dynamic seeding and in vitro culture of hepatocytes in a flow perfusion system. Tissue Eng 6:39–44

Klein GT, Lu Y, Wang MY (2013) 3D printing and neurosurgery – ready for prime time? World Neurosurg 80:233–235

Ko IK, Peng L, Peloso A, Smith CJ, Dhal A, Deegan DB, Zimmerman C, Clouse C, Zhao W, Shupe TD et al (2015) Bioengineered transplantable porcine livers with re-endothelialized vasculature. Biomaterials 40:72–79

Koike H, Ueno Y, Naito T, Shiina T, Nakata S, Ouchi R, Obana Y, Sekine K, Zheng YW, Takebe T et al (2014) Ring1B promotes hepatic stem/progenitor cell expansion through simultaneous suppression of Cdkn1a and Cdkn2a in mice. Hepatology 60:323–333

Kono Y, Yang S, Letarte M, Roberts EA (1995) Establishment of a human hepatocyte line derived from primary culture in a collagen gel sandwich culture system. Exp Cell Res 221:478–485

Laconi E, Oren R, Mukhopadhyay DK, Hurston E, Laconi S, Pani P, Dabeva MD, Shafritz DA (1998) Long-term, near-total liver replacement by transplantation of isolated hepatocytes in rats treated with retrorsine. Am J Pathol 153:319–329

Lancaster MA, Knoblich JA (2014) Organogenesis in a dish: modeling development and disease using organoid technologies. Science 345:1247125

Langer R, Vacanti J (1993) Tissue engineering. Science 260:920–926

Lee JS, Shin J, Park HM, Kim YG, Kim BG, Oh JW, Cho SW (2014) Liver extracellular matrix providing dual functions of two-dimensional substrate coating and three-dimensional injectable hydrogel platform for liver tissue engineering. Biomacromolecules 15:206–218

Li J, Li LJ, Cao HC, Sheng GP, Yu HY, Xu W, Sheng JF (2005) Establishment of highly differentiated immortalized human hepatocyte line with simian virus 40 large tumor antigen for liver based cell therapy. ASAIO J 51:262–268

Li J, Xing F, Chen F, He L, So KF, Liu Y, Xiao J (2018) Functional 3D human liver bud assembled from MSC-derived multiple liver cell lineages. Cell Transplant:963689718780332

Lim YS, Kim WR (2008) The global impact of hepatic fibrosis and end-stage liver disease. Clin Liver Dis 12:733–746. vii

Lin P, Chan WC, Badylak SF, Bhatia SN (2004) Assessing porcine liver-derived biomatrix for hepatic tissue engineering. Tissue Eng 10:1046–1053

Linke K, Schanz J, Hansmann J, Walles T, Brunner H, Mertsching H (2007) Engineered liver-like tissue on a capillarized matrix for applied research. Tissue Eng 13:2699–2707

Lozano R, Naghavi M, Foreman K, Lim S, Shibuya K, Aboyans V, Abraham J, Adair T, Aggarwal R, Ahn SY et al (2012) Global and regional mortality from 235 causes of death for 20 age groups in 1990 and 2010: a systematic analysis for the Global Burden of Disease Study 2010. Lancet 380:2095–2128

Mao SA (2017) Sustained in vivo perfusion of a re-endothelialized tissue engineered porcine liver. Int J Transplant Res Med 3:2572–4045

Marion MJ, Hantz O, Durantel D (2010) The HepaRG cell line: biological properties and relevance as a tool for cell biology, drug metabolism, and virology studies. Methods Mol Biol 640:261–272

Martin Y, Vermette P (2005) Bioreactors for tissue mass culture: design, characterization, and recent advances. Biomaterials 26:7481–7503

Martin I, Wendt D, Heberer M (2004) The role of bioreactors in tissue engineering. Trends Biotechnol 22:80–86

Matsumoto K, Yoshitomi H, Rossant J, Zaret KS (2001) Liver organogenesis promoted by endothelial cells prior to vascular function. Science 294:559–563

Methods N (2018) Method of the Year 2017: organoids. Nat Methods 15:1

Micsenyi A, Tan X, Sneddon T, Luo JH, Michalopoulos GK, Monga SP (2004) Beta-catenin is temporally regulated during normal liver development. Gastroenterology 126:1134–1146

Miyazaki M, Handa Y, Oda M, Yabe T, Miyano K, Sato J (1985) Long-term survival of functional hepatocytes from adult rat in the presence of phenobarbital in primary culture. Exp Cell Res 159:176–190

Mokdad AA, Lopez AD, Shahraz S, Lozano R, Mokdad AH, Stanaway J, Murray CJ, Naghavi M (2014) Liver cirrhosis mortality in 187 countries between 1980 and 2010: a systematic analysis. BMC Med 12:145

Montagne K, Huang HY, Ohara K, Matsumoto K, Mizuno A, Ohta K, Sakai Y (2011) Use of liposome encapsulated hemoglobin as an oxygen carrier for fetal and adult rat liver cell culture. J Biosci Bioeng 112:485–490

Mooney D, Hansen L, Vacanti J, Langer R, Farmer S, Ingber D (1992) Switching from differentiation to growth in hepatocytes: control by extracellular matrix. J Cell Physiol 151:497–505

Morini S, Sánchez-Romero N, Plá IP, de la Maza Arnal PS, Almeida M, Verscheijden L, Almeida JI, Luè A, Llorente S, Almeida H, Royo Dachary P, García Gil A, Serrano-Aulló T, Baptista PM (2018) Liver tissue engineering. In: Cabral J, da Silva CL (eds) Bioreactors for stem cell expansion and differentiation. Taylor & Francis, Boca Raton

Murphy SV, Atala A (2014) 3D bioprinting of tissues and organs. Nat Biotechnol 32:773–785

Nahmias Y, Berthiaume F, Yarmush ML (2007) Integration of technologies for hepatic tissue engineering. Adv Biochem Eng Biotechnol 103:309–329

Najimi M, Khuu DN, Lysy PA, Jazouli N, Abarca J, Sempoux C, Sokal EM (2007) Adult-derived human liver mesenchymal-like cells as a potential progenitor reservoir of hepatocytes? Cell Transplant 16:717–728

Nguyen TV, Ukairo O, Khetani SR, McVay M, Kanchagar C, Seghezzi W, Ayanoglu G, Irrechukwu O, Evers R (2015) Establishment of a hepatocyte-kupffer cell coculture model for assessment of proinflammatory cytokine effects on metabolizing enzymes and drug transporters. Drug Metab Dispos 43:774–785

Nuciforo S, Fofana I, Matter MS, Blumer T, Calabrese D, Boldanova T, Piscuoglio S, Wieland S, Ringnalda F, Schwank G et al (2018) Organoid models of human liver cancers derived from tumor needle biopsies. Cell Rep 24:1363–1376

Ohashi K, Park F, Kay MA (2001) Hepatocyte transplantation: clinical and experimental application. J Mol Med (Berl) 79:617–630

Ohashi K, Yokoyama T, Yamato M, Kuge H, Kanehiro H, Tsutsumi M, Amanuma T, Iwata H, Yang J, Okano T et al (2007) Engineering functional two- and three-dimensional liver systems in vivo using hepatic tissue sheets. Nat Med 13:880–885

Parkin DM, Bray F, Ferlay J, Pisani P (2001) Estimating the world cancer burden: Globocan 2000. Int J Cancer 94:153–156

Parmaksiz M, Dogan A, Odabas S, Elcin AE, Elcin YM (2016) Clinical applications of decellularized extracellular matrices for tissue engineering and regenerative medicine. Biomed Mater 11:022003

Powers MJ (1997) Cell-substratum adhesion strength as a determinant of hepatocyte aggregate morphology. Biotechnol Bioeng 53(4):415–26. https://doi.org/10.1002/(SICI)1097-0290(19970220)53:4<415::AIDBIT10>3.0.CO;2-F.

Prior N, Inacio P, Huch M (2019) Liver organoids: from basic research to therapeutic applications. Gut 68(12):2228–2237 https://doi.org/10.1136/gutjnl-2019-319256

Robb KP, Shridhar A, Flynn LE (2018) Decellularized matrices as cell-instructive scaffolds to guide tissue-specific regeneration. ACS Biomater Sci Eng 4:3627–3643

Rossi JM, Dunn NR, Hogan BL, Zaret KS (2001) Distinct mesodermal signals, including BMPs from the septum transversum mesenchyme, are required in combination for hepatogenesis from the endoderm. Genes Dev 15:1998–2009

Rozga J, Williams F, Ro MS, Neuzil DF, Giorgio TD, Backfisch G, Moscioni AD, Hakim R, Demetriou AA (1993) Development of a bioartificial liver: properties and function of a hollow-fiber module inoculated with liver cells. Hepatology 17:258–265

Rozga J, Morsiani E, Lepage E, Moscioni AD, Demetriou AA, Giorgio T (1994) Isolated hepatocytes in a bioartificial liver: a single group view and experience. Biotechnol Bioeng 43:645–653

Serls AE, Doherty S, Parvatiyar P, Wells JM, Deutsch GH (2005) Different thresholds of fibroblast growth factors pattern the ventral foregut into liver and lung. Development 132:35–47

Shupe T, Williams M, Brown A, Willenberg B, Petersen BE (2010) Method for the decellularization of intact rat liver. Organogenesis 6:134–136

Soto-Gutierrez A, Zhang L, Medberry C, Fukumitsu K, Faulk D, Jiang H, Reing J, Gramignoli R, Komori J, Ross M et al (2011) A whole-organ regenerative medicine approach for liver replacement. Tissue Eng Part C Methods 17:677–686

Starzl TE, Francavilla A, Halgrimson CG, Francavilla FR, Porter KA, Brown TH, Putnam CW (1973) The origin, hormonal nature, and action of hepatotrophic substances in portal venous blood. Surg Gynecol Obstet 137:179–199

Stephenne X, Najimi M, Ngoc DK, Smets F, Hue L, Guigas B, Sokal EM (2007) Cryopreservation of human hepatocytes alters the mitochondrial respiratory chain complex 1. Cell Transplant 16:409–419

Strom S, Fisher R (2003) Hepatocyte transplantation: new possibilities for therapy. Gastroenterology 124:568–571

Struecker B, Raschzok N, Sauer IM (2014) Liver support strategies: cutting-edge technologies. Nat Rev Gastroenterol Hepatol 11:166–176

Suzuki A, Sekiya S, Buscher D, Izpisua Belmonte JC, Taniguchi H (2008) Tbx3 controls the fate of hepatic progenitor cells in liver development by suppressing p19ARF expression. Development 135:1589–1595

Takebe T, Sekine K, Enomura M, Koike H, Kimura M, Ogaeri T, Zhang RR, Ueno Y, Zheng YW, Koike N et al (2013) Vascularized and functional human liver from an iPSC-derived organ bud transplant. Nature 499:481–484

Takebe T, Sekine K, Kimura M, Yoshizawa E, Ayano S, Koido M, Funayama S, Nakanishi N, Hisai T, Kobayashi T et al (2017) Massive and reproducible production of liver buds entirely from human pluripotent stem cells. Cell Rep 21:2661–2670

Tiruvannamalai-Annamalai R, Armant DR, Matthew HW (2014) A glycosaminoglycan based, modular tissue scaffold system for rapid assembly of perfusable, high cell density, engineered tissues. PLoS One 9:e84287

Tong JZ, Bernard O, Alvarez F (1990) Long-term culture of rat liver cell spheroids in hormonally defined media. Exp Cell Res 189:87–92

Torok E, Pollok JM, Ma PX, Kaufmann PM, Dandri M, Petersen J, Burda MR, Kluth D, Perner F, Rogiers X (2001) Optimization of hepatocyte spheroid formation for hepatic tissue engineering on three-dimensional biodegradable polymer within a flow bioreactor prior to implantation. Cells Tissues Organs 169:34–41

Torok E, Vogel C, Lutgehetmann M, Ma PX, Dandri M, Petersen J, Burda MR, Siebert K, Dullmann J, Rogiers X et al (2006) Morphological and functional analysis of rat hepatocyte spheroids generated on poly(L-lactic acid) polymer in a pulsatile flow bioreactor. Tissue Eng 12:1881–1890

Tostoes RM, Leite SB, Serra M, Jensen J, Bjorquist P, Carrondo MJ, Brito C, Alves PM (2012) Human liver cell spheroids in extended perfusion bioreactor culture for repeated-dose drug testing. Hepatology 55:1227–1236

Uygun BE, Soto-Gutierrez A, Yagi H, Izamis ML, Guzzardi MA, Shulman C, Milwid J, Kobayashi N, Tilles A, Berthiaume F et al (2010) Organ reengineering through development of a transplantable recellularized liver graft using decellularized liver matrix. Nat Med 16:814–820

Vendrely C (1950) Contribution to the cytochemical study of nucleic acids of some cell organoids. Arch Anat Histol Embryol 33:113–155

Voytik-Harbin SL, Brightman AO, Kraine MR, Waisner B, Badylak SF (1997) Identification of extractable growth factors from small intestinal submucosa. J Cell Biochem 67:478–491

Wagner BA, Venkataraman S, Buettner GR (2011) The rate of oxygen utilization by cells. Free Radic Biol Med 51:700–712

Wang LJ, Chen YM, George D, Smets F, Sokal EM, Bremer EG, Soriano HE (2002) Engraftment assessment in human and mouse liver tissue after sex-mismatched liver cell transplantation by real-time quantitative PCR for Y chromosome sequences. Liver Transpl 8:822–828

Wang Y, Nicolas CT, Chen HS, Ross JJ, De Lorenzo SB, Nyberg SL (2017) Recent advances in decellularization and recellularization for tissue-engineered liver grafts. Cells Tissues Organs 203:203–214

Watt KD, Pedersen RA, Kremers WK, Heimbach JK, Charlton MR (2010) Evolution of causes and risk factors for mortality post-liver transplant: results of the NIDDK long-term follow-up study. Am J Transplant 10:1420–1427

Weinstein M, Monga SP, Liu Y, Brodie SG, Tang Y, Li C, Mishra L, Deng CX (2001) Smad proteins and hepatocyte growth factor control parallel regulatory pathways that converge on beta1-integrin to promote normal liver development. Mol Cell Biol 21:5122–5131

Werner A, Duvar S, Muthing J, Buntemeyer H, Lunsdorf H, Strauss M, Lehmann J (2000) Cultivation of immortalized human hepatocytes HepZ on macroporous CultiSpher G microcarriers. Biotechnol Bioeng 68:59–70

Wolf CF, Munkelt BE (1975) Bilirubin conjugation by an artificial liver composed of cultured cells and synthetic capillaries. Trans Am Soc Artif Inter Organs 21:16–27

Wu FJ, Friend JR, Hsiao CC, Zilliox MJ, Ko WJ, Cerra FB, Hu WS (1996) Efficient assembly of rat hepatocyte spheroids for tissue engineering applications. Biotechnol Bioeng 50:404–415

Xiang JX, Zheng XL, Gao R, Wu WQ, Zhu XL, Li JH, Lv Y (2015) Liver regeneration using decellularized splenic scaffold: a novel approach in tissue engineering. Hepatobiliary Pancreat Dis Int 14:502–508

Yanagi Y, Nakayama K, Taguchi T, Enosawa S, Tamura T, Yoshimaru K, Matsuura T, Hayashida M, Kohashi K, Oda Y et al (2017) In vivo and ex vivo methods of growing a liver bud through tissue connection. Sci Rep 7:14085

Yano K, Watanab N, Tsuyuki K, Ikawa T, Kasanuki H, Yamato M (2015) Regulatory approval for autologous human cells and tissue products in the United States, the European Union, and Japan. JSRM 1:45–56

Zhang L, Pan J, Li J, Wu W, Yu Y (2003) Studies on the preparation of chitosan microcarriers cross-linked by oxidized lactose and culture of primary hepatocytes. Artif Cells Blood Substit Immobil Biotechnol 31:293–301

Zhang YS, Yue K, Aleman J, Moghaddam KM, Bakht SM, Yang J, Jia W, Dell'Erba V, Assawes P, Shin SR et al (2017) 3D bioprinting for tissue and organ fabrication. Ann Biomed Eng 45:148–163

Bioprinting Strategies to Engineer Functional Salivary Gland Organoids

Christabella Adine and João Ferreira

Contents

Abstract

The generation of three-dimensional (3D) organoids can allow researchers to resemble in vitro distinct types of cellular compartments within specific organs, including exocrine glands. The development of salivary gland organoids can entail the use of biomaterial substrates (usually hydrogel- or Matrigel-based) or can be substrate-free to allow cells to produce their own extracellular matrix molecules. Our research strategies focus on the latter and use innovative biofabrication platforms via 3D bioassembly of salivary gland primary cells or

C. Adine
Department of Biomedical Engineering, Faculty of Engineering, National University of Singapore, Singapore, Singapore
e-mail: bieca@nus.edu.sg

J. Ferreira (✉)
Exocrine Gland Biology and Regeneration Research Group, Faculty of Dentistry, Chulalongkorn University, Bangkok, Thailand

Faculty of Dentistry, National University of Singapore, Singapore, Singapore
e-mail: Joao.F@chula.ac.th

D. Eberli et al. (eds.), *Organ Tissue Engineering*, Reference Series in Biomedical Engineering, https://doi.org/10.1007/978-3-030-44211-8_5

oral stem cells using magnetic nanoparticles. These magnetic nanoparticles can tag these cells and spatially arrange these either at the bottom of a culture well (magnetic 3D bioprinting) or at the air-media interface (magnetic 3D levitation) with specific magnetic fields. In this chapter, we discuss how epithelial compartments can be formed within these organoids with different 3D culture platforms and the importance of the presence of a neuronal network to make them achieve a functional status. A neuronal network that responds to parasympathetic and sympathetic neurotransmitters is crucial for testing saliva secretion arising from the epithelia compartment. These neuroepithelial organoids will support the discovery and cytotoxicity screening of novel drugs for dry mouth syndrome or xerostomia. The advantages and limitations of both magnetic bioassembly platforms will be discussed to optimize these salivary gland-like organoids.

1 Introduction

Exocrine glands, such as mammary, prostate, thyroid, pancreas, lacrimal, and salivary glands, are organs frequently targeted by cancer or autoimmune disorders (Stewart and Wild 2018; Lombaert et al. 2017). Functional damage in these glands usually occurs after conventional cancer therapies are delivered, thereby impairing the health of patients. In the salivary gland (SG), following radiotherapy for head and neck cancers, up to 60% of the secretory epithelia can be irreversibly damaged, and saliva secretion is severely compromised in those cases. No pharmacological or medical interventions exist to fully rescue the saliva production in these cancer patients. Immunosuppression-free gland transplants are currently not available for gland replacement. Organoid bioprinting has been emerging as a promising 3D biofabrication strategy toward (1) the development an in vitro and “off-the-shelf” gland transplant from patient’s own cryopreserved cells (Mironov et al. 2009a; Groll et al. 2016); or (2) the discovery of new saliva stimulants (sialogogues) or drugs in in vitro organoid platforms (Adine et al. 2018; Ferreira et al. 2019). To avoid immunological incompatibilities and potential cytotoxicity from scaffold biomaterials in the long-term, our group generated a scaffold-free culture platform to promptly produce 3D functional salivary gland-like organoids using magnetic nanoparticles (Adine et al. 2018; Ferreira et al. 2019). This biofabrication process induces cellular spatial arrangement in 3D and scalable organoid production within 24h. In addition, this process facilitates organoid handling and transfer as well as high-throughput analysis and imaging. A magnetic nanoparticle solution, consisting of gold, iron oxide, and poly-L-lysine, is used in this 3D bioassembly process. These nanoparticles support cell proliferation and metabolism, without increasing deleterious processes such as inflammation and oxidative stress (Tseng et al. 2015). In addition, negligible immune response after transplantation is reported in animal models (Lin et al. 2016).

Thus, in this chapter, we will discuss the development secretory neuroepithelial organoids similar to the native SG using stem cells and two 3D bioassembly platforms in vitro. The overall objective was to generate a scalable SG organoid with innervation and bio-functional properties upon neurostimulation to overcome current limitations in stem cell therapies for dry mouth or xerostomia.

2 Organotypic Three-Dimensional (3D) Cell Culture System

2.1 Overview of 3D Cell Culture Systems

In this last decade, the organotypic three-dimensional (3D) cell culture system has been utilized as a promising strategy for the ex vivo generation of robust 3D organoids or miniature glands for transplantation in patients suffering from xerostomia (Ferreira et al. 2016). Also, 3D cell culture models serve as an auspicious and feasible model for the study of the molecular and cellular mechanisms underlying diseases as well as for drug discovery (Langhans 2018; Hagemann et al. 2017).

The 3D cell culture system is rapidly prevailing over the two-dimensional (2D) culture system and in vivo models, because better mimics physiologic conditions and maintains the cellular and tissue function by establishing appropriate cell-signaling pathways and extracellular matrix interactions (Campbell and Watson 2009). These unique properties enable it to overcome many of the shortcomings associated with both in vivo animal experimentation and two-dimensional (2D) tissue culture. In vivo transgenic animal models, while undoubtedly representing the gold standard for basic molecular studies and diagnostic work, are expensive to produce and are subject to stringent regulation, resulting in studies utilizing low replicate numbers, with consequences for scientific rigor (Campbell and Watson 2009).

2.2 Spheroid 3D Culture Systems

Scaffold-based tissue engineering faces some challenges including (i) immunogenicity, (ii) acute and long-term inflammatory response resulting from the host response to the scaffold and its biodegradation products, (iii) mechanical mismatch with the surrounding tissue, (iv) difficulties in incorporating high numbers of cells uniformly distributed within the scaffold, and (v) limitations in introducing multiple cell types with positional specificity (Jakab et al. 2010). In contrast, in scaffold-free spheroids culture, cellular cross talk proceeds naturally (Jakab et al. 2010).

Table 1 summarizes the most common 3D spheroid culture techniques currently employed. All these techniques have their own advantages and challenges (Lombaert et al. 2017; Ferreira et al. 2016; Achilli et al. 2012; Fennema et al. 2013). A few of 3D bioengineering systems have effectively shown the capability of primary SG epithelial cells to self-assemble and display polarization properties on engineered scaffolds incorporating SG basement membrane molecules such as Matrigel as well as other natural polymers (chitosan) and synthetic polymers (PEG, PLGA). However, some of these matrices (Matrigel) contain xenogeneic materials and thus limit its application in vivo (Ferreira et al. 2016). Other polymers may degrade too fast before epithelial cells

Table 1 Advantages and challenges of 3D spheroid culture systems. HTS: high-throughput screening

Techniques		Advantages	Limitations
Hanging drop		Compatible with HTS Requires a few step	Cell aggregation depends on cell seeding numbers and cell line Spheroid size not uniform for particular cell lines Sample handling Slow spheroid formation (>24 h)
Nonadherent surface methods		User-friendly in 1-step Allow HTS Able to scale up	Nonuniform spheroid size Spheroid formation depends on cell seeding numbers Slow spheroid formation
Suspension culture			
	Spinner flask	Enable massive production	Shear forces can modify cell function Nonuniform spheroid size
	Bioreactor	Reduces shear force on cells	Nonuniform spheroid size Requires specialized equipment
Scaffolds-based			
	Natural biomaterials	The maximum resemblance to the in vivo conditions	May contain xenogeneic substrates Batch-to-batch variation
	Hydrogels (synthetic/granular)	Reproducible mechanical and physical properties Comprises micron-sized pores Microinvasive injectability	Unable to generate larger, clinically viable tissue or organ replacement
3D bioassembly (magnetic bioprinting)		Rapid spheroid formation (24 h) Spheroids form intrinsic ECM Millimeter size range spheroid	Potentially increases oxidative stress with high concentration of magnetic particles
3D Bioprinting			
	Scaffold-assisted Spheroid-based	Enabling the reproduction of complex in vivo structures Useful for disease modeling and drug screening	Requires specialized equipment Cost, cell density capacity, cell viability Increased resolution, printing speed, biocompatibility, and scaling-up Regulatory and safety issues Lacks an universal bioink
Microfluidic and microchip methods		Control of the fluid shear stresses and the concentration of soluble factors	Requires specialized equipment Difficulty collecting cells for analysis

self-assemble and tight junctions can form (Sfakis et al. 2018). In contrast, scaffold-free bioengineered platforms can facilitate a rapid cell clustering in 3D as well as cell-cell interactions to allow the generation of a tightly packed epithelium to finally become a functional organ. Scaffold-free or spheroid culture can be efficiently produced through seeding cells in nonadherent surfaces and suspension culture. However, these methods will form nonuniform spheroids, and the size of the spheroids cannot be controlled for proper mass transfer and nutrition (Ferreira et al. 2016; Achilli et al. 2012; Fennema et al. 2013).

In contrast, cell patterning techniques offer uniformly formed spheroids. Cell patterning arranges cells into desired patterns mimicking the real tissue by applied external guiding or manipulation. In general, cell patterning can be classified as passively and actively direct cell patterning. Passive cell patterning requires a cell-adhesive ECM protein-coated onto regions defined by photoresistance or applied by a micro-fabricated stamp, leaving other uncoated regions unfavorable for cell adhesion. The seeded cells attach and occupy ECM-coated regions and hence form a clean-cut pattern (Fukuda et al. 2006; Fu et al. 2011).

Despite promising results, the effectiveness of passive cell patterning is limited by the natural cell adhesion process, which is slow, irregular, and divergent from cell type to cell type. In contrast, active cell patterning employs techniques that apply force to actively direct cells to desired positions (Table 1).

Similar to the in vivo environment, cells in a 3D spheroid are free to assemble without constraint. They establish 3D gradients of nutrients, metabolites, and cell signals as well as barriers to the transport of molecules (Achilli et al. 2012). In addition to gradients, spheroids create an in vivo-like microenvironment by forming more complex cell-to-cell interactions and cell-to-ECM adhesions. They maximize cell-to-cell contact, form numerous adhesions between surface adhesion molecules, and even form direct cell couplings such as gap junctions (Fennema et al. 2013).

Thus, spheroids from adult stem cells bioprinted onto the 3D printed scaffold are a fascinating approach for future clinical trials, since they can form larger, complex, and functional autologous tissues and mini-organs (Fennema et al. 2013).

In summary, there are several advantages of using spheroids as building blocks for organoid development (Achilli et al. 2012; Fennema et al. 2013; Parfenov et al. 2018a; Mironov et al. 2009b). First, spheroids have the highest theoretically possible cell density equivalent to native tissue. Second, spheroids have a compact rounded shape, which is ideally suitable for their handling, manipulation, transfer, processing, and bioprinting. Third, they have a complex internal structure, multicellular composition and lumens and can be pre-vascularized. Finally, when the spheroids are closely placed and touching each other, they inherently begin to fuse and producing complex 3D tissue constructs. This phenomenon was found during fetal development and organogenesis, and it is a fundamental principle of bioprinting and biofabrication technologies (Mironov et al. 2009b).

2.3 Potential Cell Sources for Salivary Gland Tissue Engineering

The ideal cell source for SG tissue engineering should entail cell types such as primary, progenitor, or stem cells isolated from autologous tissue (Lombaert et al. 2017). These cells can be isolated from the patient's SG prior to radiotherapy (RT), grown and expanded in the laboratory, and transplanted into the gland after RT has been completed. Autologous SG cells would be accepted by the host with no risk of immune rejection. Although studies on SG epithelial transplants from digested tissue have demonstrated potential, most reveal a partial capability of transplanted cells to restore SG function. Isolated primary SG cells rapidly dedifferentiate and undergo structural changes as a result of the loss of extracellular and intercellular communications (Nelson et al. 2013). Unfortunately, harvesting such cells generally causes great patient discomfort and morbidity, and thus such cell lines are not widely available. Furthermore, these cells do not grow efficiently in vitro (Nelson et al. 2013) as they become apoptotic when dissociated into single-cell suspensions and difficult to achieve acinar formation (Szlavik et al. 2008). Differentiation of human submandibular gland (HSG) intercalated duct-like cell lines toward the acinar phenotype is possible on a substrate containing laminin-1 or Matrigel (Nelson et al. 2013). However, the lack of tight junctions (TJs) crucial for the development of polarized epithelia and unidirectional secretion has become a major limitation of this cell line. Moreover, transfection of tight junction proteins did not improve HSG polarity and form weak epithelial resistance (Aframian et al. 2002). Therefore, this cell line is not suitable for tissue engineering of SG epithelia. Currently, no primary cell line fully recapitulates the morphological and functional features of the native salivary acinar cells (Hegde et al. 2014).

Other potential cell sources are embryonic stem cells (ESC) and induced pluripotent stem cells (iPSC). These are considered pluripotent stem cells because they can differentiate into all cell types from all three germinal layers. A few studies have differentiated ESC into SG tissues. Mouse ESC was shown to successfully differentiate into SG-like cells through direct co-culture with human SG cells (Kawakami et al. 2013) or in the presence of mouse fetal rudiments by induction of Sox9 and Foxc1 transcription factors (Tanaka et al. 2018). However, the translation of these ESC mouse studies into human ESC is still far for becoming a reality due to the tumorigenicity of these cells lines and rudiments. Moreover, multiple studies have reported the generation of iPSC from various dental tissues, but very few targeted the SG (Hynes et al. 2015). In the scarce literature, mouse iPSC has been shown to support SG differentiation (Ono et al. 2015) and is capable to partially regenerate damaged SG (Alaa El-Din et al. 2018). iPSC differentiation into SG-like cells still warrants further investigation. Despite the preliminary promising results, both stem cells (ESC, iPSC) have technical and moral obstacles. In addition, these cells are not easy to control, due to their genomic instability and propensity to form tumor (Hynes et al. 2015).

In contrast, adult stem cells become a very attractive cell source because they are widely available and have multi-lineage differentiation capacity. Adult stem cells (somatic stem cells) are undifferentiated cells in the body after a multiplication process during its development when the cells will eventually cease to exist or are damaged. Stem cells are generally considered to be more "primitive" and are the precursors of progenitor cells, which are more lineage-committed, have less capacity to self-renew, and maybe organ-specific. Multiple groups have been evaluating the potential of different stem cell sources for SG repair or regeneration which will be discussed in the following section. Furthermore, there are four clinical trials evaluating the effectiveness of adult stem cells to alleviate xerostomia, listed in Table 2.

Table 2 Types of stem cells used in NIH-reported clinical trials for xerostomia conditions

MSC location	Origin	Xerostomia condition	Clinical trial
Bone marrow	Autologous	Radiotherapy-induced	NCT03743155
Adipose	Autologous	Radiotherapy-induced	NCT02513238
Umbilical cord	Allogeneic	Sjögren's syndrome	NCT00953485
Lip and cheek	Autologous	Sjögren's syndrome	NCT00023491

2.3.1 Salivary Gland Stem/Progenitor Cells

As stem cells might be organ-specific, it is therefore obvious to explore the potential of stem cells derived from the SG to repair or regenerate the damaged gland. Earlier on, the SG organ was thought to have a slow cellular turnover, and therefore the presence of stem cells was seen unlikely. However, ductal ligation experiments in rat parotid and submandibular glands identified proliferating cells with regenerative capabilities. These cells emerged upon deligation and were labeled stem/progenitor cells, and gland atrophy was reversed with the appearance of new acinar cells (Burford-Mason et al. 1993; Takahashi et al. 2004). These and other studies have shown that SG stem/progenitors cells are present among the differentiated cells, and therefore, it is necessary to have an effective isolation methodology to test their regenerative potential.

Nanduri et al. (Nanduri et al. 2013a) have shown that isolation with the c-Kit marker was enough to restore morphology and functionality of irradiated salivary glands. Salisphere derived c-Kit+ cells expressing CD24 and/or CD49f markers, successfully restored tissue homeostasis in the submandibular gland of irradiated mice. However, the population of c-Kit+ is very limited (<1%) in the human SG and greatly diminishes in aging patients (Pringle et al. 2016).

Furthermore, Jeong et al. (Jeong et al. 2013) develop a method to efficiently isolate SG progenitor without cell surface marker selection but repeated subculturing over five passages. These cells exhibited MSC-like characteristics and successfully differentiated into amylase-secreting cells in vitro. More importantly, these cells were able to ameliorate hyposalivation and rescued SG morphology in irradiated rat SG.

In addition, an interesting study by Mishima et al. (Mishima et al. 2012) has identified and isolated side population (SP) cells from mouse SG via cell sorting. This cell population was able to recover the radiotherapy-induced hypofunction of salivary and lacrimal glands, although cell reconstitution was not observed in vivo.

Table 3 summarizes studies of SG-derived stem/progenitor cells (SGSC) isolation and evaluation of their regenerative capacity, especially for radiotherapy-induced xerostomia. Although studies have shown that origin-specific stem cells are favorable for treatment of the tissue origin, SGSC has its own limitations. The treatment is heavily dependent on residual endogenous acinar cell, and therefore it is not suitable for severe cases. Hence it is worthwhile to explore the use of other sources of adult stem cells.

2.3.2 Bone Marrow-Derived Mesenchymal Stem Cells (BM-MSC)

Several groups have reported the potential of bone marrow-derived cells (BMC) for salivary gland regeneration (Sumita et al. 2011; Lombaert et al. 2006; Lin et al.

Table 3 Studies on SG-derived stem cells/progenitor (SGSC) in vitro culture systems with in vivo SG regenerative outcomes

Refs.	Cell source	Selected cell population	SG markers	Culture system	In vivo methods/ outcomes
Lombaert et al. (2008a)	C57BL/6 mouse	c-Kit^{+}	Sca-1 c-Kit Musashi-1	Salisphere	IG: 300 cells* ↑SFR, GW, AC, Amy ↑proliferation
Nanduri et al. (2011)		CD133^{+}	c-Kit CD133 CD24 CD29 CD49f	Salisphere	IG: 1x10^{4} cells* ↑SFR, GW, AC, Amy ↑proliferation
Nanduri et al. (2014)		CD24^{hi}/CD49f^{hi} Old vs. young	–	Salisphere	IG: 5000 cells* ↑SFR, AC
Nanduri et al. (2013b)		c-Kit^{+}/ CD24^{+}/ CD49f^{+}	–	Salisphere	IG: 400 cells* ↑SFR, AC ↓fibrosis
Xiao et al. (2014)		Lin^{-}/CD24^{+}/ c-Kit^{+}/Sca1^{+}	c-Kit KRT5 KRT14 Vimentin	Salisphere	IG: 300–1000 cells* ↑SFR, AC
Maimets et al. (2016)		Wnt3a+ R-spondin 1 stimulation	Ductal epithelia EpCam^{+}	Salisphere	IG: 10-1x10^{4} cells* ↑SFR, AC
Mishima et al. (2012)	C57BL/6 mouse	SP (FACS) clusterin stimulation	CD31	2D	IG: 1x10^{4} cells ↑SFR
Jeong et al. (2013)	Human	Long-term culture in collagen type I substrate	CD44 CD49f CD90 CD105	2D	IG: 5x10^{5} cells ↑SFR, GW, AC ↓apoptosis
Yi et al. (2016)	Human	Modified subfractionation culture method	LGR5 MSC-like markers	2D	IG: 1x10^{5} cells ↑SFR, Amy, engraft
Pringle et al. (2016)	Human	c-Kit^{+}	SG-like markers	Salisphere	IG: 1200 cells* ↑SFR

SP, side population; FACS, fluorescence-activated cell sorting; MSC, mesenchymal stem cells; RIX, radiotherapy-induced xerostomia; IG, intraglandular injection, SFR; salivary flow rate, GW; gland weight, AC; acinar cells, SAC; saliva secreting acinar cells, # in vivo rescue in the secondary recipient. NA, not applicable. *cells dissociated from salisphere

2011). These studies used very heterogeneous stem cell populations, which may produce different functional effects when transplanted in vivo. In order to address this, Lim et al. utilized a subfractionation culturing method to obtain highly homogenous MSC cell lines (Lim et al. 2013a). The intraglandular injection of these MSC restored the morphology and function of the submandibular gland after radiation injury. The paracrine effect appears to induce functional restoration through the survival of epithelial secretory cells, mobilization of host progenitor cells, and vascularization.

To further investigate this matter, Tran et al. (Tran et al. 2013) evaluated the paracrine effect of BM-MSC. These researchers extracted and lysed the crude bone marrow cells and its soluble intracellular content (termed BM soup) and then injected into irradiated mice. Injection of BM soup was as effective as the transplantation of BM cells in restoring the function of irradiated salivary glands, highlighting the importance of paracrine effects in salivary gland regeneration. The utilization of cell extract treatment strategies can definitely overcome the issues related to immune rejection after allogeneic transplantation of MSC or other adult stem/progenitor cells.

Table 4 displays the studies evaluating BMC and BM-MSC for SG regeneration.

Multiple research groups have also identified several reprogramming factors that can lead to transdifferentiation to SG-like epithelial cell types and/or SG regeneration. These reprogramming factors include ankyrin repeat domain-containing protein 56 (ANKRD56), high mobility group protein 20B (HMG20B), transcription factor E2a (TCF3), pancreas-specific transcription factor 1a (PTF1α), muscle, intestine, and stomach expression-1 (MIST-1), achaete-scute complex homolog 3 (ASCL3), bone morphogenetic protein 6 (BMP-6), interleukin-6 (IL-6), and/or CCAAT enhancer-binding protein beta (CEBPB).

2.3.3 Adipose-Derived Stem Cells

Although BM-derived cells hold a relevant potential for SG tissue regeneration, autologous or allogeneic cell harvesting from BM or transplantation may not be possible for a variety of reasons (compromised immunity, abnormal hematopoiesis, hip morbidity, lack of patient acceptance, etc.). Alternatively, researchers explored other sources of mesenchymal stem cells (MSC). Studies have shown that one can have a high yield of MSC from adult adipose tissues, and fat tissue biopsies are easier to perform when compared to bone marrow biopsies and produce less morbidity (Konno et al. 2013).

Therefore, several studies using adipose-derived stem cells are currently registered in clinical trial databases (Table 2). Direct administration of adipose-derived stromal cells into the SG can improve saliva production by stimulating vascularization within salivary gland tissues, since transplanted ASC differentiate into endothelial cells as well as ductal epithelial cells (Kojima et al. 2011). However, the secretory acinar compartment did not display prominent signs of regeneration. Later,

Table 4 Studies evaluating the therapeutic potential of bone marrow-derived cell transplantation for xerostomia

Refs.	Cell type	Disease model	Transplantation method	In vivo outcomes
Lombaert et al. (2006)	BMC	RIX	IV: Cells+G-CSF	↑GW, AC
Lombaert et al. (2008b)	BMC	RIX	IV: $3x10^6$ cells + Flt-3 L/SCF/G-CSF	↑VA, proliferation
Yuan et al. (2013)	BMC	RIX	Tail vein: $5x10^6$ cells	↑AQP5, KRT19, αSMA
Sumita et al. (2011)	BMC	RIX	Tail vein: 1x × 10^7 cells, 2x/wk. for 6 wks	↑GW, AC, VA, proliferation ↓apoptosis
Lim et al. (2013a)	BM-MSC[a]	RIX	IG: $1x10^5$ cells	↑AC, VA, proliferation ↓apoptosis
Lin et al. (2011)	BM-MSC[b]	RIX	IG: $1x10^6$ cells	↑GW, AC, Amy
Elsaadany et al. (2017)	BM-MSC	RIX	Tail vein: $1x10^7$ cells	↑VA ↓apoptosis, mucositis
Xu et al. (2012)	BM-MSC	SS	Tail vein: $1x10^5$ cells	↑SFR ↓inflammation
Khalili et al. (2010, 2012, 2014)	BM-MSC	SS	Tail vein: 1 × 10^7 cells, 2x/wk. for 2 or 6 wks	↑SFR ↓inflammation
Almansoori et al. (2019)	BM-MSC	Allogeneic transplantation	Marginal ear vein: $2x10^6$ cells/kg wt	↓atrophy ↑SG survival rate

Legend: [a]purified
[b]BM-MSC is labeled with iron oxide. BMC, bone marrow-derived cells; BM-MSC, bone marrow mesenchymal stem cell; SCID, Severe Combined Immunodeficiency' RIX, radiotherapy-induced xerostomia; SS, Sjögren's syndrome; IG, intraglandular; GW, gland weight; AC, acinar cells; SFR, saliva flow rates; VA, vascularization; NA, SFR was not evaluated

Lim et al. (Lim et al. 2013b) administered systemically human adipose-derived MSC (hAd-MSC) to subdue radiotherapy-induced SG damage. This strategy improved salivary gland function and epithelial proliferation, and suppressed cellular apoptosis. Furthermore, like BM-MSC studies, a limited number of adipose-derived MSC transdifferentiate into epithelial secretory cells. These findings may suggest that MSC, regardless of their source, have a paracrine effect rather than a transdifferentiation effect.

A phase 1/2 clinical trial evaluating the safety and efficacy of autologous hAd-MSC in 15 subjects with radiotherapy-induced xerostomia was recently reported (NCT02513238) (Gronhoj et al. 2018). At 4 months, this clinical trial reported 50% improvement in saliva secretion and no major adverse events. Thus, ASC treatment was safe and exhibited promising efficacy in a short-term follow-up on both objective and patient-reported outcomes. It would be interesting to assess if these clinical outcomes can be maintained or improved beyond 4 months.

2.3.4 Dental Pulp Stem Cells

Dental pulp stem cells (DPSC) arise from the neural crest-derived mesenchyme surrounding the oral primitive ectoderm during craniofacial morphogenesis. These cells possess several MSC-like markers and share a common mesenchymal embryonic origin with the salivary gland stroma (Adine et al. 2018). Therefore, DPSC can be part of regeneration strategies for the SG epithelia (severely damaged in radiotherapy-induced xerostomia), the stroma compartment, and the neuronal network (neurons are derived from neural crest mesenchymal lineages).

In 2D cultures, Janebodin et al. (Janebodin and Reyes 2012) co-cultured DPSC with a human submandibular gland cell line (HSG) in Matrigel which increased the number of secretory epithelial cells in vitro. In vivo subcutaneous co-transplantation of HSG and DPSC with hyaluronic acid (HA) substrate led to an increase of SG epithelial tissue, innervation, and vascularization. Yamamura et al. (Yamamura et al. 2013) successfully differentiated DPSC into endothelial cells (DPECs) and demonstrated that radiotherapy-induced SG hypofunction was partially reverted following in vivo transplantation of DPECs.

Recently, our laboratory was able to differentiate human DPSC in 3D culture platform toward SG-like cells that express epithelial acinar and ductal markers as well as neuronal markers in the presence of FGF10 (Adine et al. 2018). This 3D cellular assembly created the first innervated SG epithelial organoid via 3D bioprinting.

3 Incorporating Different Biomaterials for Salivary Gland Bioengineering

A recent advancement in SG regenerative medicine showed that a bioengineered gland made from embryonic epithelium and mesenchyme can be transplanted into an adult mouse to produce a whole functional SG (Ogawa et al. 2013). This bioengineered SG was composed of a variety of progenitor and stem cells, including cell from epithelial, mesenchymal, endothelial, and neuronal origins. More interestingly, the SG reconnected with the existing ductal system and possessed functional activity. The new SG was able to secrete saliva, protect the oral cavity from bacteria, and restore swallowing functions.

Hence, future research may translate these bioengineering strategies to animal models with salivary glands that have more morphological and functional similarities to the human SG. Further studies may also focus on the usage of stem cells or adult salivary progenitors with high expansion capabilities in 3D scaffolds in order to form a bioengineered construct that grows into a functional gland in the adult microenvironment.

Salivary gland tissue engineering requires three essential components: (1) the stem/progenitor cells that retain epithelial progenitor biomarkers typical of the native salivary gland (SG); (2) the extracellular matrix (ECM) proteins that can orchestrate the differentiation of progenitor cells into functional structures; and (3) a biocompatible and biodegradable three-dimensional (3D) scaffold that can hold these components together to recreate the microenvironment found in the native SG (Aframian and Palmon 2008).

Since dynamic cell-ECM interactions are essential in processes such as epithelial ductal formation/branching, a recent strategy has been to engineer scaffolds that structurally and functionally resemble native ECM architecture. Three-dimensional (3D) collagen matrices have been used for homing salisphere stem/progenitor cells which form epithelial ductal structures with mucin-positive acini, indicating their capability to differentiate in response to the ECM environment (Nanduri et al. 2014). Various biomaterials such as collagen type I, Matrigel, and other animal-derived products have been showing promising results in the differentiation and organization of human SG cells (Nanduri et al. 2014; Maria et al. 2011; Pradhan et al. 2009); nevertheless, these biomaterials are not human-compatible. Thus, tissue engineering-based research is gearing toward the creation of xeno-free biomaterials, which can eventually be transplanted into humans.

In 2010, researchers have started to utilize the soft hyaluronic acid (HA) hydrogels, which are human-compatible, as biocompatible substrates for SG tissue engineering (Pradhan et al. 2010). When encapsulated in HA hydrogels, human SG cells can grow into organized spheroid structures that merge and proliferate to form larger acini-like structures with a central lumen and are maintained for long-term in these gels in vitro (Pradhan et al. 2010). These in vitro 3D acini-like structures also secrete α-amylase, express β-adrenergic and muscarinic receptors that activate protein transport, and induce calcium oscillations upon treatment with cholinergic stimulants. Furthermore, these 3D spheroids continue to secrete α-amylase when hydrogels were implanted in vivo in an athymic rat model (Pradhan-Bhatt et al. 2013). However, these latter 3D structures have reversed polarity suggesting that further environmental cues from the ECM and the myoepithelial cells may be needed to reverse inside out acini and correct the polarity. Culture of salivary gland progenitor cells on human perlecan domain IV peptide has been shown to support the formation of 3D acini-like salivary units that express α-amylase (Pradhan et al. 2009). It will be useful to incorporate the perlecan IV domain peptide into biomaterial scaffolds to mediate differentiation, correct polarity, and directional secretion of the 3D salivary gland cell cultures in the future.

Other research groups have used poly(lactic)-co-glycolic acid (PLGA), an FDA approved constituent in implantable dental and orthopedic devices, as a synthetic material to show that it can support the attachment, proliferation, and survival of salivary gland epithelial cells (Jean-Gilles et al. 2010). The same group further shows that nanofiber PLGA scaffolds can support development and morphogenesis of intact fetal SMG organ cultures and promote natural self-organization of dissociated SMG cells into branched SG-like structures (Sequeira et al. 2012). However, adult SG cells grown on flat polymeric substrates fail to form a complex 3D branching structure and are unable to assemble tight junctions that are needed for unidirectional flow of saliva. To overcome this, recent studies generated lithographically based micropatterning curved "craters" that mimic the physical structure of the basement membrane, which have increased both the surface area and allowed apicobasal polarization and differentiation of salivary gland epithelial cells

(Soscia et al. 2013). An increase in aquaporin-5, a water channel protein marking acinar differentiation, was also detected in SG cells cultured on higher curvature scaffolds. Further studies with PLGA nanofibers coupled with laminin-111 and chitosan showed that laminin-111 promotes the formation of mature epithelial tight junctions and apicobasal polarization, and, on the other hand, the chitosan antagonizes this phenomenon (Pradhan-Bhatt et al. 2013).

Taken together, current cell-based therapies and tissue engineering studies have provided a promising outlook to regenerate SG and restore the saliva secretory function. However, in order to test these techniques in humans, several hurdles need to be surpassed. To overcome these hurdles, further research steps should include (1) the elimination of xenogeneic elements from transplants for feasible human use to comply with good manufacturing practices; (2) a thorough assessment of histocompatibility barriers; (3) an evaluation of long-term transplant survival and saliva secretion in larger animal models with a better SG human resemblance (i.e., pigs); and (4) an assessment of tumor sensitivity to bioengineered transplants in SG cancer models.

4 3D Bioprinting

One of the major challenges for tissue engineering is to produce a large volume of tissues or organs and to scale them up for clinical applications. The use of 3D bioprinting in regenerative medicine is to address this need for tissues and organs suitable for transplantation. Bioprinting and bioassembly constitute the two major biofabrication setups. Bioprinting allows the precise spatial organization (patterning) of living cells, biomaterials, and bioactive molecules layer by layer, whereas bioassembly facilitates the automated assembly of cell-containing building blocks via cell-driven self-organization or through the preparation of hybrid cell-material building blocks (Moroni et al. 2018).

Only magnetic-based bioprinting and bioassembly have been recently tested for the generation of secretory epithelia in salivary glands (Adine et al. 2018; Ferreira et al. 2019).

4.1 Magnetic-Based Bioprinting

There are two technologies for bioprinting utilizing magnetic forces. The first mode involved suspending cells in paramagnetic liquid via the addition of gadolinium (Gd^{3+})-based solution which termed as a label-free approach (Turker and Arslan-Yildiz 2018; Abdel Fattah et al. 2016). When a magnetic field is applied, the magnetized fluid is attracted to regions of high-magnetic field gradient, displacing the cells toward regions with a low gradient, a process called as diamagnetophoresis (Abdel Fattah et al. 2016). Since cell patterning through diamagnetophoresis can be controlled and it is nozzle-free, this technique is convenient to rapidly print multi-cellular spheroids. The technology has been reported to assemble epithelial breast cancer cell lines and can be modified to print multiple cell types (Turker and Arslan-Yildiz 2018). Despite promising results, the major concern over this technology relates to the usage of cytotoxic paramagnetic suspending media. The high concentrations of Gd^{3+} could be potentially toxic for tissue spheroids, and a certain risk exists for osmotic pressure imbalance due to excessive use of ions in the paramagnetic medium (Parfenov et al. 2018b). Further, this label-free levitation magnetic assembly technology is currently only able to fabricate soft small living blocks due to the size of the magnets underneath each well. Therefore, it is necessary to use either superconducting magnets or apply a higher concentration of the paramagnetic solution to fabricate a larger size of tissue-engineered constructs (Parfenov et al. 2018b).

The second magnetic-based bioprinting technology requires cell labeling with magnetic nanoparticles. Souza et al. (Souza et al. 2010) developed a magnetizing solution consisting of poly-L-lysine, iron oxide, and gold nanoparticles named NanoShuttle™ for 3D cell culture formation (Souza et al. 2010; Tseng et al. 2013; Haisler et al. 2013; Timm et al. 2013). Upon magnetic nanoparticle (MNPs) uptake, the cells can be easily directed using mild magnetic forces. Spatial patterning of the 3D cell assembly into the desired morphology is controlled through varying the shape or configuration of the magnetic field. Furthermore, the culture platform can consistently control the size of the spheroids by tuning the following variables: MNP's concentration, number of cells, and the magnet size. Upon removal of the magnetic field, spheroids rapidly contract in size due to cell rearrangement to reach an equilibrium size and simultaneously produce their own extracellular matrix. This system has already been validated in various cell lines (Souza et al. 2010; Tseng et al. 2013; Haisler et al. 2013; Timm et al. 2013). Furthermore, the developed magnetic 3D cell culture platform has been used in disease modeling of epithelial tissues such as breast cancer (Jaganathan et al. 2014). Given the spheroids contrast with media due to their brown color imparted by the MNPs, the change in size can be imaged at programmed intervals using a mobile device (Timm et al. 2013). As a result, spheroids can be imaged simultaneously in a multi-well plate, thereby improving imaging throughput and efficiency. Due to magnetized spheroids, adding and removing solutions is made easy by the use of magnetic forces to hold down spheroids during aspiration, limiting spheroid loss. Spheroids can also be picked up and transferred between vessels using magnetic tools such as the MagPen™. NanoShuttle™ and magnetic field have been shown to have no compromising effect on cell proliferation, viability, and metabolism as reported in previous publications. Therefore, this technology offers reproducible spheroid formation, scalable, ease of imaging, no specialized equipment or media required, biocompatible, and ready for automation (Timm et al. 2013).

Epithelial cells from the salivary gland are known tightly become packed in the presence of tight junction proteins and exhibit polarization properties (Varghese et al. 2019). However, the majority of the bioengineering studies do not support the development of a neuronal network within the epithelial tissues. Innervation is deemed crucial to maintain and repair SG epithelial cells after radiotherapy damage (Knox et al. 2010). In collaboration with Souza and colleagues, our research team has recently produced innervated and bio-functional SG-like epithelial organoids using the NanoShuttle™ MNPs and its magnetic 3D bioprinting (M3DB) platform (Fig. 1) (Adine et al. 2018). Our first study indicated that M3DB supported the metabolic activity of human dental pulp stem cells (hDPSC) and did not inflicted any cytotoxicity to the cells. Further, M3DB was capable of maintaining hDPSC stemness during a 3-day expansion culture to assemble the cells. Next, our team showed that retinoic acid and FGF10 induced SG epithelial differentiation and created a neuronal network in 11 days of culture. The final 3D bioprinted tissue had not only amylase-secreting secretory epithelia but also functional and responsive neurons in vitro, mimicking the normal SG. Such innervated secretory epithelia secreted salivary amylase and were responsive to both cholinergic (parasympathetic) and β-adrenergic (sympathetic) neurotransmitters. Furthermore, our study showed the bioengineered organoids exhibited tight junction and transepithelial resistance, required for unidirectional saliva flow. This 3D tissue was therefore termed as SG-like organoid as per the above SG features. To determine the regenerative potential, an injured mouse SG model was established after RT damage was induced. SG-like organoids were able to restore branching morphogenesis following transplantation into damaged SG, and neuron integration was established with the native SG neuronal network (Adine et al. 2018). The millimeter-level organoids generated in our study are remarkable but can pose technical challenges in terms of nutrient and oxygen diffusion and upon analyzing or imaging biological mechanisms as a 3D SG in vitro model. Confocal and multiphoton microscopy with multiple detector systems through the spheroid allowed us to provide more biological assessments into the inner cellular core of the organoid. Several proliferative cell compartments were present in both the center and periphery of SG-like organoids after transplantation. With about 1 to 1.2 millimeter in size, vascularization in the organoid is necessary for nutrient and oxygen exchange. Despite this, an apicobasal polarization was challenging to determine due to tightly packed epithelial cells, and vascularization was limited in the developed epithelial organoids. We reported that our organoids have a limited vascular network partially because of the low hematopoietic population of DPSC capable of generating CD31+ cells.

Moreover, our work showed that neurons, as well as SOX2+ cells, are also diminished in the irradiated SG, suggesting that tissue degeneration is due to loss of progenitors and their regulators (Adine et al. 2018). Emmerson et al. (Emmerson et al. 2018) also found that SOX2+ stem cell population is essential for acinar cell maintenance and is capable of replenishing acini after radiotherapy-induced damage. The maintenance of parasympathetic nerves was required to host the SOX2+ population. In our study, SG-like organoid exhibited significant increase in the SOX2+ population in mice ex vivo submandibular glands (Adine et al. 2018). Therefore, we hypothesize that the rescue of epithelial bud growth was due to the survival of SOX2+ cells and the presence of a neuronal (parasympathetic) network in the organoid. In future clinical studies, autologous human DPSC may be isolated from the patient's impacted tooth, differentiated into SG-like organoids in vitro, and transplanted back to regenerate or repair the damaged SG.

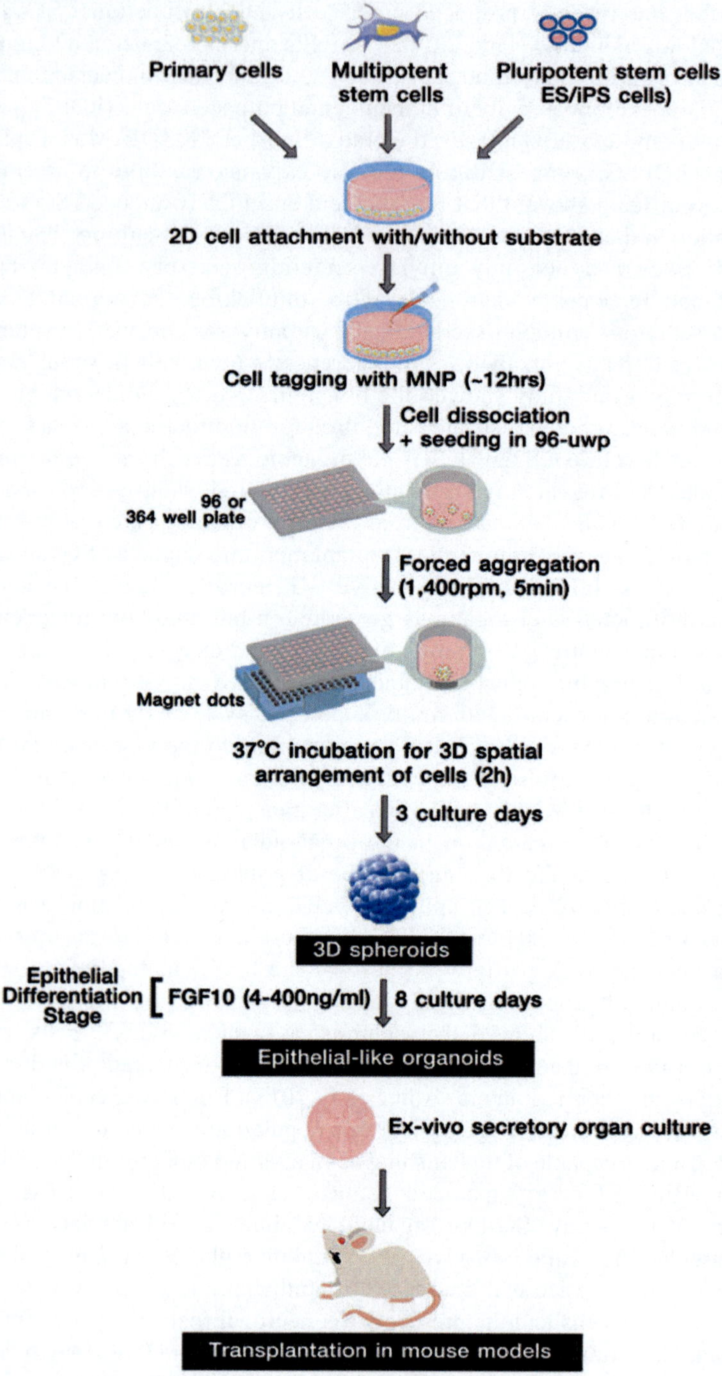

Fig. 1 Flow diagram with the methodological steps to generate epithelial-like SG organoids with magnetic-based bioprinting

4.2 Magnetic Levitation and 3D Bioassembly

Starting in 2016, our research team developed a user-friendly and reproducible magnetic-based levitation methodology (M3DL) with NanoShuttle™ MNPs for the generation of organoids mimicking native SG cellular compartments and the neuroepithelial communication and secretory function (Ferreira et al. 2019). The M3DL platform used porcine SG primary cells, which supported the formation of epithelial tissue spheres from 24 hours up to 7 days with viable and proliferating cells at both the center and periphery. A 200–250 μm diameter aggregate can be developed after 7 days with a consistent sphere-like shape. More importantly, M3DL-formed organoids did exhibit SG-like secretory functions upon muscarinic stimulation because intracellular calcium activity and salivary amylase secretion were swiftly amplified. This rapid stimulation of the organoid's secretory activity is an advantage because it is crucial to obtain a clinically relevant saliva secretion in the shortest time possible in patients with dry mouth (Ferreira et al. 2019). These functional mini-glands can potentially be translated to humanized models and be applied in the repair or regeneration of damaged secretory SG epithelia in hyposalivation and dry mouth medical conditions. There is also ongoing works to use these mini-glands as an early in vitro screening tool for drug toxicity or to create in vitro dry mouth models induced by radiotherapy.

In addition, future magnetic levitation studies may need to address phenotypic differences between porcine and human primary cells extracted using various SG cell isolation methodologies, as well as evaluate the effectiveness of these cells in generating functional organoids.

4.3 Scaffold-Assisted Bioprinting

Despite promising progresses, scaffold-assisted bioprinting faces several challenges when laying biomaterials together with cells in a 3D spatial arrangement. This technology is biomaterial-dependent and energy-intensive, making it poorly compatible with some of the intended biological applications. The biomaterial needs to be supportive of both cells and the challenges of the bioprinting process. Thus, a universal biomaterial is required because often each cell type needs to be embedded in a different hydrogel (Kang et al. 2016). More importantly, the common printing methods are intrinsically stressful to live cells, by exposing them to high shear stress, overheating, and/or toxic compounds generated even from initially cell-friendly materials (Nair et al. 2009). If the printing process slows down, then printing large amount of cells may become time-consuming. Also, when epithelial cells from exocrine glands are encapsulated within the bioink, the cells need to both dissolve their biomaterial scaffold and to proliferate to the point where they must contact and produce tight junctions for epithelial polarization and proper fluid secretion. These are among the reasons and challenges of using scaffold-assisted bioprinting techniques in SG epithelial regeneration and organoid development.

At the moment, the future of SG biofabrication may benefit more from the use of a scaffold-free 3D bioprinting or bioassembly approaches until better bioinks are developed to promote epithelial viability and proper epithelial spatial arrangement and polarization. The ideal properties for a SG epithelial-supportive bioink would include printability (viscosity, gelation method, rheology), biocompatibility and material biomimicry, fast degradation kinetics, and nontoxic byproducts, with mechanical/structural properties similar to SG basement membrane molecules (e.g., perlecan). The other important desirable features of a bioink include industrial scalability, availability, short post-printing time for maturation, permeability to oxygen and nutrient transport, and ability to eliminate metabolic waste generated by cells. Hopefully, the emerging 4D bioprinting strategies may be able to overcome the challenges related with the lack of a universal bioink compatible with epithelial organs like the SG.

5 Conclusions

The use of tissue engineering and bioprinting approaches of human tissues or organs for transplantation has developed substantially over the past decade. Many studies have demonstrated promising in vitro and in vivo results to overcome the challenges of achieving innervation, vascularization, and functionalization in bioprinted tissues or organs. Bioengineering the epithelia of the salivary gland is now possible with scaffold-free or with different proteoglycans/hydrogels (e.g., hyaluronic acid, perlecan domain IV, hyaluronic acid, etc.), and innervation can be achieved with a combination of specific cell lines, growth factors (e.g., FGF10, Neurturin, etc.), and bioprinting techniques. However, bioprinted tissues are still restricted to millimeter size constructs with only immature vascular networks. Combining different bioprinting techniques can potentially facilitate the biofabrication of vascularized tissues together with the support of endothelial cells, novel biomaterials, and growth factors. The pathway from organ bioprinting to implantation into a human must be performed within a reasonable amount of time; thus standardized protocols involving patient-specific design, fabrication techniques, maturation processes, surgical operations, and postoperative care are essential for customized functional organ fabrication.

Acknowledgments Our research work was supported by these grants and institutions: Special Task Force for Activating Research (STAR) for Exocrine Gland Biology and Regeneration Research Group, Ratchadaphiseksomphot Endowment Fund, Chulalongkorn University, Grant Number: STF 6202432001-1; Chulalongkorn University Faculty of Dentistry Special Funding 2018, Grant Number: DRF62018.

References

Abdel Fattah AR et al (2016) In situ 3D label-free contactless bioprinting of cells through Diamagnetophoresis. ACS Biomater Sci Eng 2(12):2133–2138

Achilli TM, Meyer J, Morgan JR (2012) Advances in the formation, use and understanding of multi-cellular spheroids. Expert Opin Biol Ther 12(10):1347–1360

Adine C, Ng KK, Rungarunlert S et al (2018) Engineering innervated secretory epithelial organoids by magnetic three-dimensional bioprinting for stimulating epithelial growth in salivary glands. Biomaterials 180:52–66

Aframian DJ, Palmon A (2008) Current status of the development of an artificial salivary gland. Tissue Eng Part B Rev 14(2):187–198

Aframian DJ et al (2002) Absence of tight junction formation in an allogeneic graft cell line used for developing an engineered artificial salivary gland. Tissue Eng 8(5):871–878

Alaa El-Din Y et al (2018) Potential therapeutic effects of induced pluripotent stem cells on induced salivary gland cancer in experimental rats. Biotech Histochem 94:1–8

Almansoori AA et al (2019) Mesenchymal stem cell therapy in submandibular salivary gland Allotransplantation: experimental study. Transplantation 103:1111

Burford-Mason AP et al (1993) Immunohistochemical analysis of the proliferative capacity of duct and acinar cells during ligation-induced atrophy and subsequent regeneration of rat parotid gland. J Oral Pathol Med 22(10):440–446

Campbell JJ, Watson CJ (2009) Three-dimensional culture models of mammary gland. Organogenesis 5(2):43–49

Elsaadany B et al (2017) Effect of transplantation of bone marrow derived mesenchymal stem cells and platelets rich plasma on experimental model of radiation induced Oral mucosal injury in albino rats. Int J Dent 2017:8634540

Emmerson E et al (2018) Salivary glands regenerate after radiation injury through SOX2-mediated secretory cell replacement. EMBO Mol Med 10(3)

Fennema E et al (2013) Spheroid culture as a tool for creating 3D complex tissues. Trends Biotechnol 31(2):108–115

Ferreira JN et al (2016) Three-dimensional bioprinting nanotechnologies towards clinical application of stem cells and their Secretome in salivary gland regeneration. Stem Cells Int 2016:7564689

Ferreira JN, Hasan R, Urkasemsin G et al (2019) A magnetic three-dimensional levitated primary cell culture system for the development of secretory salivary gland-like organoids. J Tissue Eng Regen Med 13(3):495–508

Fu CY et al (2011) A simple cell patterning method using magnetic particle-containing photosensitive poly (ethylene glycol) hydrogel blocks: a technical note. Tissue Eng Part C Methods 17(8):871–877

Fukuda J et al (2006) Micromolding of photocrosslinkable chitosan hydrogel for spheroid microarray and co-cultures. Biomaterials 27(30):5259–5267

Groll J, Boland T, Blunk T et al (2016) Biofabrication: reappraising the definition of an evolving field. Biofabrication 8(1):013001

Gronhoj C et al (2018) Safety and efficacy of mesenchymal stem cells for radiation-induced xerostomia: a randomized, placebo-controlled phase 1/2 trial (MESRIX). Int J Radiat Oncol Biol Phys 101(3):581–592

Hagemann J et al (2017) Spheroid-based 3D cell cultures enable personalized therapy testing and drug discovery in head and neck Cancer. Anticancer Res 37(5):2201–2210

Haisler WL et al (2013) Three-dimensional cell culturing by magnetic levitation. Nat Protoc 8(10):1940–1949

Hegde S, Anuradha A, Sodhi A (2014) Reviving the salivary cells – pathway to new horizon. IJSS Case Reports Rev 1(2):27–29

Hynes K et al (2015) Induced pluripotent stem cells: a new frontier for stem cells in dentistry. J Dent Res 94:1508

Jaganathan H et al (2014) Three-dimensional in vitro co-culture model of breast tumor using magnetic levitation. Sci Rep 4:6468

Jakab K et al (2010) Tissue engineering by self-assembly and bio-printing of living cells. Biofabrication 2(2):022001

Janebodin K, Reyes M (2012) Neural crest-derived dental pulp stem cells function as Ectomesenchyme to support salivary gland tissue formation. Dentistry 2

Jean-Gilles R, Soscia D, Sequeira S et al (2010) Novel modeling approach to generate a polymeric nanofiber scaffold for salivary gland cells. J Nanotechnol Eng Med 1(3):31008

Jeong J et al (2013) Human salivary gland stem cells ameliorate hyposalivation of radiation-damaged rat salivary glands. Exp Mol Med 45:e58
Kang HW et al (2016) A 3D bioprinting system to produce human-scale tissue constructs with structural integrity. Nat Biotechnol 34(3):312–319
Kawakami M et al (2013) Functional transplantation of salivary gland cells differentiated from mouse early ES cells in vitro. Hum Cell 26(2):80–90
Khalili S et al (2010) Bone marrow cells are a source of undifferentiated cells to prevent Sjogren's syndrome and to preserve salivary glands function in the non-obese diabetic mice. Int J Biochem Cell Biol 42(11):1893–1899
Khalili S et al (2012) Mesenchymal stromal cells improve salivary function and reduce lymphocytic infiltrates in mice with Sjogren's-like disease. PLoS One 7(6):e38615
Khalili S et al (2014) Treatment for salivary gland hypofunction at both initial and advanced stages of Sjogren-like disease: a comparative study of bone marrow therapy versus spleen cell therapy with a 1-year monitoring period. Cytotherapy 16(3):412–423
Knox SM et al (2010) Parasympathetic innervation maintains epithelial progenitor cells during salivary organogenesis. Science 329(5999):1645–1647
Kojima T et al (2011) Regeneration of radiation damaged salivary glands with adipose-derived stromal cells. Laryngoscope 121(9):1864–1869
Konno M et al (2013) Adipose-derived mesenchymal stem cells and regenerative medicine. Develop Growth Differ 55(3):309–318
Langhans SA (2018) Three-dimensional in vitro cell culture models in drug discovery and drug repositioning. Front Pharmacol 9:6
Lim JY et al (2013a) Intraglandular transplantation of bone marrow-derived clonal mesenchymal stem cells for amelioration of post-irradiation salivary gland damage. Oral Oncol 49(2):136–143
Lim JY et al (2013b) Systemic transplantation of human adipose tissue-derived mesenchymal stem cells for the regeneration of irradiation-induced salivary gland damage. PLoS One 8(8):e71167
Lin CY et al (2011) Cell therapy for salivary gland regeneration. J Dent Res 90(3):341–346
Lin H, Dhanani N, Tseng H et al (2016) Nanoparticle improved stem cell therapy for erectile dysfunction in a rat model of cavernous nerve injury. J Urol 195(3):788–795
Lombaert IM et al (2006) Mobilization of bone marrow stem cells by granulocyte colony-stimulating factor ameliorates radiation-induced damage to salivary glands. Clin Cancer Res 12(6):1804–1812
Lombaert IM et al (2008a) Rescue of salivary gland function after stem cell transplantation in irradiated glands. PLoS One 3(4):e2063
Lombaert IM et al (2008b) Cytokine treatment improves parenchymal and vascular damage of salivary glands after irradiation. Clin Cancer Res 14(23):7741–7750
Lombaert I, Movahednia MM, Adine C, Ferreira JN et al (2017) Concise review: salivary gland regeneration: therapeutic approaches from stem cells to tissue organoids. Stem Cells 35(1):97–105
Maimets M et al (2016) Long-term in vitro expansion of salivary gland stem cells driven by Wnt signals. Stem Cell Rep 6(1):150–162
Maria OM, Zeitouni A, Gologan O et al (2011) Matrigel improves functional properties of primary human salivary gland cells. Tissue Eng Part A 17(9–10):1229–1238
Mironov V, Visconti RP, Kasyanov V et al (2009a) Organ printing: tissue spheroids as building blocks. Biomaterials 30(12):2164–2174
Mironov V et al (2009b) Organ printing: tissue spheroids as building blocks. Biomaterials 30(12):2164–2174
Mishima K et al (2012) Transplantation of side population cells restores the function of damaged exocrine glands through clusterin. Stem Cells 30(9):1925–1937
Moroni L et al (2018) Biofabrication: a guide to technology and terminology. Trends Biotechnol 36(4):384–402
Nair K et al (2009) Characterization of cell viability during bioprinting processes. Biotechnol J 4(8):1168–1177
Nanduri LS et al (2011) Regeneration of irradiated salivary glands with stem cell marker expressing cells. Radiother Oncol 99(3):367–372

Nanduri LS et al (2013a) Salisphere derived c-Kit+ cell transplantation restores tissue homeostasis in irradiated salivary gland. Radiother Oncol 108(3):458–463

Nanduri LS et al (2013b) Salisphere derived c-Kit+ cell transplantation restores tissue homeostasis in irradiated salivary gland. Radiother Oncol 108(3):458–463

Nanduri LS et al (2014) Purification and ex vivo expansion of fully functional salivary gland stem cells. Stem Cell Rep 3(6):957–964

Nelson J, Manzella K, Baker OJ (2013) Current cell models for bioengineering a salivary gland: a mini-review of emerging technologies. Oral Dis 19(3):236–244

Ogawa M, Oshima M, Imamura A et al (2013) Functional salivary gland regeneration by transplantation of a bioengineered organ germ. Nat Commun 4:2498

Ono H et al (2015) Regenerating salivary glands in the microenvironment of induced pluripotent stem cells. Biomed Res Int 2015:293570

Parfenov VA et al (2018a) Scaffold-free, label-free and nozzle-free biofabrication technology using magnetic levitational assembly. Biofabrication 10(3):034104

Parfenov VA et al (2018b) Scaffold-free, label-free and nozzle-free biofabrication technology using magnetic levitational assembly. Biofabrication 10(3):034104

Pradhan S, Zhang C, Jia X et al (2009) Perlecan domain IV peptide stimulates salivary gland cell assembly in vitro. Tissue Eng Part A 15(11):3309–3320

Pradhan S, Liu C, Zhang C et al (2010) Lumen formation in three-dimensional cultures of salivary acinar cells. Otolaryng Head Neck 142(2):191–195

Pradhan-Bhatt S, Harrington DA, Duncan RL et al (2013) Implantable three-dimensional salivary spheroid assemblies demonstrate fluid and protein secretory responses to neurotransmitters. Tissue Eng Part A 19(13–14):1610–1620

Pringle S et al (2016) Human salivary gland stem cells functionally restore radiation damaged salivary glands. Stem Cells 34(3):640–652

Sequeira SJ, Soscia DA, Oztan B et al (2012) The regulation of focal adhesion complex formation and salivary gland epithelial cell organization by nanofibrous plga scaffolds. Biomaterials 33(11):3175–3186

Sfakis L et al (2018) Mesenchymal cells affect salivary epithelial cell morphology on PGS/PLGA Core/Shell nanofibers. Int J Mol Sci 19(4)

Soscia DA, Sequeira SJ, Schramm RA et al (2013) Salivary gland cell differentiation and organization on micropatterned PLGA nanofiber craters. Biomaterials 34(28):6773–6784

Souza GR et al (2010) Three-dimensional tissue culture based on magnetic cell levitation. Nat Nanotechnol 5(4):291–296

Stewart BW, Wild CW (eds) (2018) World Cancer Report 2018. International Agency for Research on Cancer, France, pp 16–53

Sumita Y et al (2011) Bone marrow-derived cells rescue salivary gland function in mice with head and neck irradiation. Int J Biochem Cell Biol 43(1):80–87

Szlavik V et al (2008) Differentiation of primary human submandibular gland cells cultured on basement membrane extract. Tissue Eng Part A 14(11):1915–1926

Takahashi S et al (2004) Cell death and cell proliferation in the regeneration of atrophied rat submandibular glands after duct ligation. J Oral Pathol Med 33(1):23–29

Tanaka J et al (2018) Generation of orthotopically functional salivary gland from embryonic stem cells. Nat Commun 9(1):4216

Timm DM et al (2013) A high-throughput three-dimensional cell migration assay for toxicity screening with mobile device-based macroscopic image analysis. Sci Rep 3:3000

Tran SD et al (2013) Paracrine effects of bone marrow soup restore organ function, regeneration, and repair in salivary glands damaged by irradiation. PLoS One 8(4):e61632

Tseng H et al (2013) Assembly of a three-dimensional multitype bronchiole coculture model using magnetic levitation. Tissue Eng Part C Methods 19(9):665–675

Tseng H, Gage JA, Shen T et al (2015) A spheroid toxicity assay using magnetic 3D bioprinting and real-time mobile device-based imaging. Sci Rep 5:13987

Turker E, Arslan-Yildiz A (2018) Recent advances in magnetic levitation: a biological approach from diagnostics to tissue engineering. ACS Biomater Sci Eng 4(3):787–799

Varghese JJ et al (2019) Salivary gland cell aggregates are derived from self-organization of acinar lineage cells. Arch Oral Biol 97:122–130

Xiao N et al (2014) Neurotrophic factor GDNF promotes survival of salivary stem cells. J Clin Invest 124(8):3364–3377

Xu J et al (2012) Allogeneic mesenchymal stem cell treatment alleviates experimental and clinical Sjogren syndrome. Blood 120(15):3142–3151

Yamamura Y et al (2013) Treatment of salivary gland hypofunction by transplantation with dental pulp cells. Arch Oral Biol 58(8):935–942

Yi T et al (2016) Single cell clones purified from human parotid glands display features of multipotent epithelial mesenchymal stem cells. Sci Rep 6:36303

Yuan Y-W et al (2013) The ability of transplanted bone marrow-derived cells to differentiate into parenchymal cells of salivary glands. J Hard Tissue Biol 22(4):433–438

Tissue-Engineered Thymus

Gauri Kulkarni and John D. Jackson

Contents

Abstract

Impaired thymus function due to aging or other clinical conditions may have a number of consequences for the immune system such as an enhanced predisposition to infection and autoimmunity, slow response to vaccines with age, and

G. Kulkarni (✉) · J. D. Jackson (✉)
Wake Forest Institute for Regenerative Medicine, Wake Forest School of Medicine, Medical Center Boulevard, Winston Salem, NC, USA
e-mail: gkulkarn@wakehealth.edu; jojackso@wakehealth.edu

D. Eberli et al. (eds.), *Organ Tissue Engineering*, Reference Series in Biomedical Engineering, https://doi.org/10.1007/978-3-030-44211-8_11

possible risk of cancer development. Current approaches for exogenous thymus regeneration focus on the modulation of growth factors and hormones secreted by thymic epithelial cells. Bioengineering approach to create and use transplantable thymus tissue can offer effective regenerative strategy. This chapter aims to describe the cellular architecture and function of primary lymphoid organ thymus and discuss the current and potential bioengineering approaches to regenerate the thymus. Continued research to understand the mechanisms that regulate thymic recovery after injuries and strategies that can boost its endogenous repair is essential for furthering current and developing new promising regenerative technologies.

1 Introduction

Throughout our lives, we need an internal system to fight off infections. Our immune system protects us from pathogens and abnormal cells such as cancer cells using a specialized type of white blood cell, the T cell, which is produced in the thymus. The thymus is a primary lymphoid organ located in the mediastinum (the midline above the heart) and is composed of two identical lobes (Fig. 1a; Safieddine and Keshavjee 2011). During the neonatal and pre-adolescent periods, the thymus is at its largest and most active phase, after which it gradually decreases in size and function

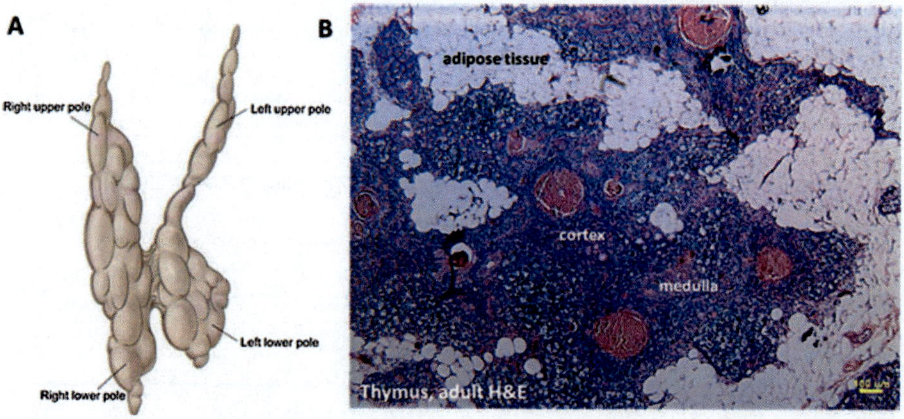

Fig. 1 (**a**) Schematic representation of human thymus anatomy. (**b**) Histology of adult human thymus stained with hematoxylin and eosin. Thymus is enclosed by a thin connective tissue capsule. Numerous septa extend into the thymus subdividing the two lobes; two lobes divided into numerous lobules (about 0.5–2 mm in diameter). Blood vessels enter and leave the thymus via the connective tissue septa. Each lobulus is divided into thymus cortex – a darker peripheral zone – and thymus medulla, a lighter central zone. Medullary tissue is continuous from lobule to lobule throughout each lobe. After puberty much of the parenchyma of the thymus is replaced by adipose tissue (adipocytes), particularly cortical lymphoid tissue. This process is called involution which is under the control of both sexual hormones and stress hormones. (Figure 1a is reprinted from Safieddine and Keshavjee (2011), with permission from Elsevier)

(von Gaudecker 1991) and is replaced by fat (Fig. 1b; Jackson et al. unpublished). The thymus is responsible for the production of all the T cells which is involved in cellular immunity (Pearse 2006). The epithelial cells of the thymus support the proliferation and maturation of T cells as well as smaller numbers of other lymphoid cells. The specificity of T cells that are released into the systemic circulation is also under thymic control.

Early phases of thymus development are characterized by rapid growth. The thymus is one of the first organs to degenerate postnatally, beginning early in life. This process called thymus involution (Steinmann et al. 1985) significantly impacts the immune system and diminished its capacity through a reduced output of naive T cells (Rudd et al. 2011). Structural changes that happen within the thymus include loss of tissue structure, fibro-adipogenetic transformation, and decline of naive T-cell export. The consequential changes in the peripheral T-cell compartment are believed to be the clinical signs of immunosenescence. Thymus involution differs from the normal aging process in other organs because of its early initiation of atrophy that continues during the life of the individual. Furthermore, the involution process also impairs the ability of the thymus to regenerate from acute damage. Analysis of thymus from various vertebrates including avian, amphibian, and teleost reveals that it undergoes age-associated involution, suggesting that this process is an evolutionary conserved event which appears to occur early in life (Shanley et al. 2009). One evolution theory called the disposable soma theory which is based on the allocation of resources by an organism between maintenance and repair and other functions to maximize Darwinian fitness appears to be the most compatible with regression of the thymus (Kirkwood 1977).

The thymus is particularly sensitive to endogenous and exogenous insults such as infection, chemotherapies or radiation therapies, shock, sex steroids, and graft-versus-host disease (GVHD). Of prime concern are chemotherapy and radiation therapy. For example, the process of hematopoietic stem cell transplantation (HSCT) can acutely damage the thymus through the treatment of tumors via chemotherapy, radiotherapy, or antibody therapy. While the early thymus has a remarkable capacity for endogenous regeneration, this capacity declines considerably with age. Age-related thymic involution may not represent a problem in healthy individuals; however, it becomes a significant clinical issue for sustaining immune competence after insult or injury that leads to immune depletion. For example, insufficient recovery of thymic activity has been directly linked to adverse clinical outcome and opportunistic infections in recipients of allo-HSCT therapy (Wils et al. 2011). In some circumstances, such as a genetic condition called DiGeorge syndrome, the thymus is either very small or absent altogether. This leads to severe immunodeficiency and autoimmune diseases (Davies 2013).

The adult thymus has a limited capacity for regeneration (endogenous thymus regeneration). This endogenous repair may not be sufficient to fully restore the function of the thymus, particularly in the face of injuries or age-related degeneration. In such cases, exogenous interventions will be required to regenerate or even replace lost thymic function. Such strategies include use of hormonal modulation such as sex steroid inhibition or growth hormone administration; use of cytokines

(IL-7 and IL-22); and growth factors (like keratinocyte growth factor (KGF)) (Chaudhry et al. 2016), Other novel approaches being tested include use of precursor T cells (Dudakov and van den Brink 2011). Regenerative medicine is showing great promise in repair of a variety of tissues and organs in the body. Bioengineering of the thymus is one of them. This is an exciting intersection of regenerative medicine, tissue engineering, and immune biology, which may be beneficial in the treatment for disorders of the immune system, such as DiGeorge syndrome and age-related immunodeficiency. This chapter will describe the current state of bioengineering of the thymus.

2 Structure and Function of the Thymus

2.1 Structure and Histology

The thymus is an epithelial organ. The mammalian thymus is located in the pericardial mediastinum, anterior to the major vessels of the heart, and ventral to the base of the heart and aortic arch, with variable extension of one or both lobes into the cervical region in the rat (Haley 2003). In the guinea pig, it is located in the neck region (Dijkstra and Sminia 1990). The thymus consists of two distinct lobes connected by a connective tissue isthmus (Fig. 1a; Safieddine and Keshavjee 2011). It is also histologically consistent across different species. Each lobe of the bi-lobed thymus can be divided into morphologically distinct regions referred to as the cortex and the medulla, both of which are separated by an intermediate cortico-medullary zone (von Gaudecker 1991). In adults, thymus gets involuted due to invasion of excessive adipose tissue (Fig. 1b).

Capsule: The capsule is a thin connective tissue layer that surrounds each lobe. The capsule can further be divided into an outer and inner layer of collagen and reticular fibers, between which occasional clusters of lymphocytes can be found. The inner capsule layer gives rise to interlobular septum that partially subdivides the thymus into inter-connecting lobules of variable size and orientation. Fine trabeculae also extend from the capsule or the septum into the center of the lobules.

Epithelial Stroma: A network of epithelial reticular cells form the bulk of the supporting framework in the thymus (Banks 1993), while epithelium-free areas (known as EFAs or holes) present in the sub-capsular area extending deep into the cortex (Bruijntjes et al. 1993; Elmore 2006). Epithelial cells in the sub-capsular region form a layer one or two cells deep. The cell morphology is usually thin and sheet-like (also in the outer cortex and areas surrounding blood vessels), but elsewhere, they assume a stellate appearance. These epithelial cells form an open framework that predominantly has T lymphocytes, but smaller populations of plasma cells, B lymphocytes, and other cells such as neuroendocrine cells can also be seen.

Epithelium-Free Area (EFA): The epithelium-free areas (EFAs) in the thymus are lymphocyte-rich regions, devoid of stromal elements, non-vascularized, and with unknown function (Bruijntjes et al. 1993). It is postulated that they may be lymphocyte reservoirs (Van Ewijk 1984), proliferation sites of lymphocytes (Duijvestijn et al. 1981), or a specific intra-thymic pathway for T lymphocytes (Bruijntjes et al. 1993). These EFAs are located in the sub-capsular region, and serial sections show that they run from the sub-capsular area to deep in the cortex, often bordering the medulla (Bruijntjes et al. 1993).

It is hypothesized that the "EFAs" offer T lymphocytes a separate intra-thymic pathway where immature lymphocytes can move between the cortex and medulla and avoid contact with the stromal elements that are associated with selection (Bruijntjes et al. 1993; Elmore 2006).

Cortex: The cortex region mostly contains small, immature, and densely packed thymocytes, in addition to a transient bone marrow-derived population of predominantly phagocytic macrophages. Mitotically active thymocytes can be found in the sub-capsular cortex. Also seen from the outer cortex to the cortico-medullary junction are a gradient of small, less mitotically active cells. In the sub-capsular region and cortex region, large apoptotic bodies can also be seen. These apoptotic bodies are derived from short-lived, rapidly dividing lymphocytes that undergo apoptosis. The developing thymocytes are in contact with cortical epithelial cells that direct their growth and maturation.

Medulla: The medulla forms a significant portion of the thymus and can form small buds that reach deep into the cortex or even to the capsule in some places. It has a more elaborate cellular composition, consisting of mature T cells, epithelial cells, B lymphocytes, dendritic cells (bone marrow-derived, non-phagocytic cells), Hassall's corpuscles (mainly in primates and humans), and mixed macrophages. The medullary T lymphocytes are larger, paler-staining and have more cytoplasm than cortical lymphocytes. Hassall's corpuscles are rare in rodent species when compared with humans and primates. Component cells are polygonal with a large nucleus and pale, occasionally granular cytoplasm with variable cystic degeneration and dystrophic calcification (Pearse 2006). Neuroendocrine cells occur in low numbers. Their physiological function is not understood, but they can give rise to carcinoid tumors. Mast cells and eosinophils are also variably present.

Cortico-medullary Junction: The cortico-medullary junction is the region between the cortex and the medulla and is characterized by presence of abundant blood vessels. Mature and immature T lymphocytes are present, along with variable numbers of plasma cells and perivascular B lymphocytes. The number of plasma and B lymphocytes increases with increasing age of the individual.

2.2 Function

The thymus functions as a primary lymphoid organ. T-cell precursors (hematopoietic precursors) enter the thymus at the cortico-medullary junction and then undergo stages of highly ordered differentiation. During intra-thymic migration, the thymocytes (developing T cells) proliferate and differentiate and also express interleukin receptors and differentiation antigens (Van Ewijk et al. 1988). This process occurs in four different subdivisions of the thymus cortex, referred to as regions 1–4 (Fig. 2).

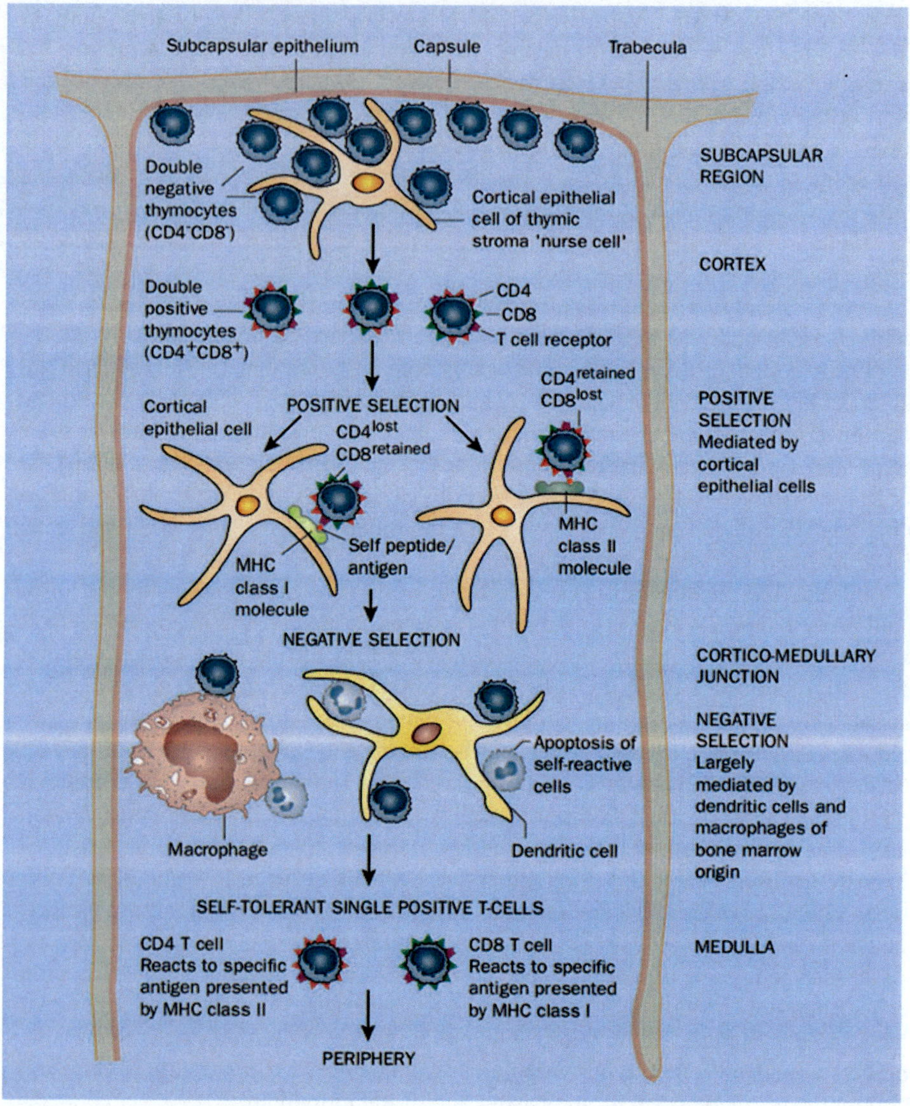

Fig. 2 The process of formation of functional T cells from hematopoietic precursors within the thymus. Circulating T-lymphoid progenitor cells migrate into the thymic parenchyma through the vasculatures around the cortico-medullary junction in the sub-capsular region of postnatal thymus. T-cell differentiation is characterized by ordered expression of various CD surface molecules. Cells bearing a T-cell receptor that recognizes self MHC are positively selected in the cortex and pass into the cortico-medullary junction. Here, T cells that react with self-antigens undergo apoptosis by the process of negative selection. Double-positive (CD4+CED8+) thymocytes differentiate into single-positive cells either CD4+CD8− or CD4−CD8+ and exit the thymus into the blood and lymphoid tissues. (Image sourced from Parkin and Cohen 2001)

Region 1 is the cortico-medullary junction, which contains the T-cell precursors, called CD4−CD8− double-negative 1 (DN1) cells at this stage. In region 2, these cells differentiate to the double-negative 2 (DN2) stage and undergo a proliferative expansion. In region 3, they transform into double-negative 3 (DN3) cells, where T-cell lineage commitment occurs, and the cells undergo T-cell receptor (TCR) β-chain rearrangement. The transition from double-negative to double-positive (DP) CD4+CD8+ status happens in region 4. The DP cells migrate through the cortex and into the medulla where they now differentiate into either CD4+ or CD8+ single-positive (SP) cells. During the cortico-medullary transition, these cells are also subjected to positive and negative selection. While positive selection happens mainly in the cortex and involves cortical TECs, negative selection occurs mainly in the medulla, where thymic dendritic cells (DCs) and medullary TECs mediate this process. Finally, after the differentiation and maturation within the thymic microenvironment, the mature SP cells are released into the circulation as functional naive T cells (Abbas 2004; Brown et al. 2002; Tizard 2004) (Fig. 2).

The thymic epithelial cells (TECs) are involved in positive and negative selection, while the dendritic cells (DC) are more efficient than epithelial cells in mediating negative selection. Thymic nurse cells, which are specialized epithelial cells present in the sub-capsular region, are important in early T-cell differentiation. Sub-capsular and medullary epithelial cells produce the thymic hormones, such as Thymosin that is involved in T-cell maturation. Other factors such as thymic humoral factor, serum thymic factor, and thymopoietin are also produced in the thymus which enhance lymphocyte responsiveness (Banks 1993).

3 Embryology and Development

In mammals, the endoderm of the third and fourth pharangeal pouches and surrounding mesenchyme give rise to the thymus (Dijkstra and Sminia 1990). As development progresses, the thymus (along with the parathyroid and thyroid) migrates caudally. They separate around day 15 when the thymus migrates into the thorax. After migration is complete, the epithelial cells organize into a loose meshwork separated by the developing vasculature. Lymphocyte precursors from developing hematopoietic populate the developing thymus, which now is a lymphoepithelial organ. Immediately after birth, the thymus grows considerably in response to stimulation from postnatal antigen and the demand for large numbers of mature T cells. It is now known that genetic factors can also influence its age of onset, rate of development, and the magnitude of immunological function. In a newborn child, the thymus is more of a tri-lobular structure than bi-lobular, whereas with advancement of age, the shape and then relative size of the thymus decrease progressively compared to the surrounding tissues (Fig. 1b).

The epithelial microenvironment of the thymus develops from endodermal cells of the third pharyngeal pouch, which is located in close apposition to the parathyroid primordium. The establishment of the thymic fate of the pharyngeal epithelium is believed to be controlled by the FOXN1 gene expression (Bleul et al. 2006; Garfin et al. 2013). The mechanisms by which FOXN1 expression is initially induced in the pharyngeal endoderm are not clear, but are thought to involve interaction between multiple signaling and growth factors, including sonic hedgehog (SHH), bone morphogenetic proteins (BMPs), and the WNT signaling pathways (Osada et al. 2006; Rodewald 2008; Manley and Condie 2010). Bipotent thymic epithelial cell (TEC) progenitors develop from the endoderm of the third pharyngeal pouch and express the transcription factor FOXN1. Based on functional studies, there is evidence for the presence of bipotent progenitors that gives rise to cortical TECs (cTECs) and medullary TECs (mTECs) through compartment-specific intermediate precursor cells (Rodewald et al. 2001; Bleul et al. 2006; Rossi et al. 2006) (Fig. 3). There have been attempts to derive this cell type using the in vitro differentiation of embryonic stem and induced pluripotent stem (iPS) cells. Cortical TECs-like environments have been generated in vivo and in vitro by transgenic reprogramming using specific factors, such as CXC chemokine ligand 12 (CXCL12), Delta-like ligand 4 (DLL4), and Notch ligand 1.

The thymus generates and supplies native T cells and a broad TCR repertoire to the body. The process of T-cell development is tightly regulated and involves a bidirectional crosstalk between the developing thymocytes and the thymus stroma (Takahama 2006). Within the thymus stroma, the thymic epithelial cells (TECs), fibroblasts, endothelial cells, and dendritic cells guide the differentiation of bone marrow-derived T-cell progenitors that ultimately lead to the formation of mature T cells (CD4+ or CD8+) expressing an MHC-restricted, antigen-specific TCR. The events that occur during this process involve the following: (i) immature thymocytes derived from blood-borne progenitor cells accumulate beneath the thymus capsule and then move inward through the cortex and the cortico-medullary junction (CMJ) and finally toward the medulla; (ii) thymocytes undergo multiple rounds of

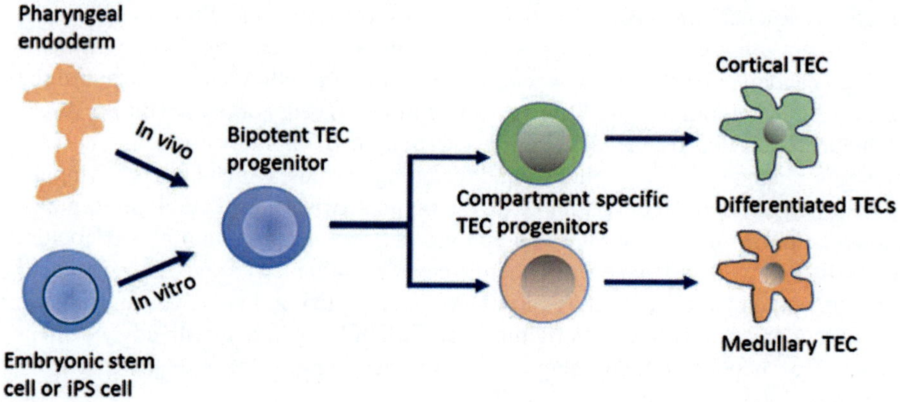

Fig. 3 Developmental lineage of thymic epithelial cell (TEC). Third pharyngeal pouch gives rise to bipotent thymic epithelial cell (TEC) progenitors that express the transcription factor forkhead box protein N1 (FOXN1). These progenitors generate mature cortical TECs (cTECs) and medullary TECs (mTECs) via compartment-specific progenitor cells

proliferation and differentiation as they progressively encounter a variety of specialized stromal cells along this pathway; and finally (iii) the thymocytes mature into T lymphocytes. The differential expression of chemokines and chemokine receptors on the surface of stromal cells and thymocytes, respectively, are now known to be involved in the trafficking of thymocytes (Norment and Bevan 2000). The mature T cells are attracted to the endothelium of blood vessels in the CMJ and medullary areas, from where the self-tolerant, MHC-restricted T cells are finally exported into the circulation (Rossi and Zlotnik 2000; Ueno et al. 2002).

4 Thymus Dysfunction and Damage

The thymus is extremely sensitive to insults such as toxins, stress hormones, and drugs and can be easily damaged. It can regenerate on its own to a certain extent, but this ability is severely reduced with increasing age. The functional capacity of the thymus is gradually reduced due to age-related degeneration, referred to as thymus involution. Damage to the thymus can result from inborn conditions, infection, or cytoablative therapies or because of graft-versus-host disease following allogeneic transplantation. Some of these conditions are discussed here with the aim to understand the molecules, cells, and processes that can be targeted for repair and regeneration of thymus structure and function.

4.1 Thymus Involution

Unlike other tissues or organs of the body, the thymus exhibits a unique pattern of aging. First, thymus involution starts much earlier, beginning in early childhood and accelerating at puberty (Steinmann et al. 1985). It is estimated that after the initial phase, approximately 3% of the thymus tissue is lost per year until middle age and approximately 1% per year subsequently. This process is called thymus involution. In addition to individual variations, a sexual dimorphism in the rate of involution exists, and it is found to be greater in males than in females (Gui et al. 2012). Some of the etiological factors that can affect the pattern of involution include genetic polymorphisms and hormones (sex steroids such as androgens) (Hsu et al. 2003; Gui et al. 2012). The greatest rate of age-related thymus involution is observed at puberty, and a decline in growth hormone (GH) concentration with age has been shown to play a role in thymus involution (Boehm et al. 2013; Griffith et al. 2012). Another factor playing a key role in aging of the thymus is damage from oxygen free radicals, particularly accumulated damage later in life. This is supported by studies showing that reducing metabolic activity (through caloric restriction or modulation of IGF signaling) can reduce involution of the thymus (Yang et al. 2009; Vallejo et al. 2009). Other evidence that supports free radical-related damage is the lack of enzyme catalase in thymus stromal cells, making these cells more sensitive to damage by oxidative by-products (Griffith et al. 2015).

Immunosenescence is a term used to describe the array of defects seen in adaptive and innate immunity with advancing age (Goronzy and Weyand 2013). This impairs the aging body's ability to respond to vaccinations and pathogens, resulting in higher mortality from diseases in comparison to younger individuals. Immunosenescence can also lead to a loss of tumor immune surveillance and higher incidence of autoimmune disease in the elderly. Involution of the thymus is a part of immunosenescence where a decrease in cellularity and a loss of tissue organization lead to profound age-related defects in both quality and quantity of T cell produced (lynch et al. 2009). The involution of the thymus due to aging is a major reason for the decreased production of naive T cells and reduced immunity. Changes in the thymus microenvironment here include the loss of thymic epithelial cell (TEC) numbers and function, fibro-adipogenetic transformation, and a decline in crosstalk between developing thymocytes and thymic stromal cells. This crosstalk is very important for thymus homeostasis. This communication is however disrupted during aging, resulting in depletion of cells both in epithelial and lymphoid compartments.

The process of fibro-adipogenesis transformation needs a special mention with respect to thymus involution. In an aging thymus, the TECs decline, while the thymic adipocyte and fibroblast numbers increase. An expansion of lipid-bearing cells within the thymic medulla follows the adipocyte accumulation, while further adipocyte infiltration is observed in several regions such as the capsular region, sub-capsular cortex, and the perivascular space (Fig. 1b) (Yang et al. 2009; Dixit 2010). Dedifferentiation of thymic epithelial cells triggers EMT (epithelial to mesenchymal transition) first, and then the resulting fibroblast cells undergo the conventional route of differentiation program toward adipocyte lineage (Yang et al. 2009; Youm et al. 2009). This loss of TECs and rise of TEC-derived adipocytes and fibroblasts directly compromises thymopoiesis (differentiation of thymocytes into mature T lymphocytes) during aging. Also, the age-related decrease in cellular communications within the cortical and medullary thymic microenvironments results in the decreased T-cell receptor rearrangement, failure of the selection processes, and loss of self versus non-self recognition (Aspinall 1997). From a clinical perspective, elderly individuals (or individuals with highly reduced thymus function) will not only show decreased immune responsiveness but will also have severely delayed immune recovery following cytoablative treatments that can lead to increased morbidity and mortality because of opportunistic infections (Mackall et al. 1995). Additionally, age-induced alterations in the bone marrow (BM) and intrinsic defects in BM-derived hematopoietic stem cells (HSCs) can affect their capacity to generate functional lymphoid progenitors. This can further contribute to the thymus involution.

One consequence of thymus involution is the development of autoimmune disease in elderly individuals. As the T-cell negative selection is lost with age, the ability to mediate self-tolerance is impaired, and the chances of auto-reactive T cells being released into the periphery become greater (Prelog 2006; Coder et al. 2015). However, it is important to note that despite this degenerative process, some residual thymus function does persist into old age. This residual function suggests that instead of becoming a vestigial organ later in life, the thymus continues to function at a lower efficiency. This lower function can easily be lost in the face of the acute insults that are more common in the elderly population. This compounded by the reduced ability of the aged thymus to endogenously regenerate presents several clinical challenges. Therefore, strategies to bioengineer or regenerate the thymus are particularly relevant to the elderly population.

4.2 Infection

The thymus was previously thought to be an immune-privileged organ. However, increasing scientific evidence over the past decade has shown that like other organs, the thymus can get infected by bacteria, fungi, parasites, and viruses (Savino 2006). Damage to the thymus can be caused either directly by the pathogen infecting the organ or by the systemic effects that result from an infection. In a direction infection scenario, lympho-stromal disruption can negatively affect the export of newly generated naive T cells, which can further impair the immune responses against the pathogen (Nunes-Alves et al. 2013). Additionally, infection of the thymus can accelerate its involution. The thymus can be directly infected by pathogens. The human immunodeficiency virus (HIV) infection is a classic example, where the virus infects the T-cell progenitors and thymocytes (CD4+ SP) and induces apoptosis of uninfected thymocytes through secretion of viral products (Douek et al. 1998; Dion et al. 2004; Fang et al. 2008). The HIV also infects dendritic cells and TECs, leading to degradation of the thymus microenvironment (Stanley et al. 1993; Rozmyslowicz et al. 2010). Similarly, infection of the thymus by cytomegalovirus (CMV) and parasites such as *Trypanosoma cruzi* disrupts its environment by increasing deposition of laminin and fibronectin and increases expression of chemokine ligands (such as CCL4 and CXCL12) by stromal cells. Such changes also promote premature release of DP thymocytes from the thymus and further enhance thymic involution (Mendes-da-Cruz et al. 2006).

Factors produced by the systemic pathogens can also damage the thymus. One such notable factor is the bacterial lipopolysaccharide (LPS), which is released from bacteria such as *Escherichia coli*. Acute thymic atrophy can be seen due to LPS, which can lead to loss of DP thymocytes (Hick et al. 2006). Stress response caused by other infections can trigger a surge in glucocorticoid levels, which can in turn lead to apoptosis of thymocytes (Gruver and Sempowski 2008). Damage to the thymus via depletion of the DP population can happen due to other inflammatory mediators that are produced during infection. Such mediators include interferon-γ released by *Salmonella* infection (Deobagkar-Lele et al. 2013) and tumor necrosis factor-α (TNF-α) released by *Trypanosoma cruzi* (Pérez et al. 2007). Thymus atrophy seen across a broad range of pathogen infections might be a virulence strategy employed by pathogens to survive within the host by deliberately destabilizing the immune responses. In contrast, studies from mycobacterial infection of the thymus suggest that infection in the absence of atrophy can result in generation of naive T cells that are tolerant to the pathogen (Nobrega et al. 2010). Taken together, it seems that the thymic atrophy seen as a result of infection could be a deliberate strategy employed by the host so that generation of an immune repertoire tolerant to certain pathogens can be avoided. Further investigation in this direction will be necessary not only to understand this phenomenon but to also develop potential therapies that can rejuvenate a damaged thymus.

4.3 Cytoablative Therapies

Cytoablative therapies (including chemotherapies and radiation therapies) that are primarily directed against malignant cells in a patient can also target the hematopoietic system (Mackall 1999). Alkylating agents (such as cyclophosphamide) have been shown to deplete all thymocyte subsets within the thymus (Goldberg et al. 2010), while chemotherapy or radiation can also damage the thymic stroma (Fletcher et al. 2009). Thymic epithelial cells (TECs) located within the medulla (mTECs) are particularly vulnerable to the effects of chemotherapy (Fletcher et al. 2009). Since these cells are known to play a key role in negative selection, it is possible that this depletion of mTECs can have profound negative implications for development of tolerance to self-antigens following such treatments. Following chemotherapy, recovery of thymic function in younger patients can be augmented by peripheral expansion of T cells; however, similar recovery can take years in older patients (Storek et al. 1995). Similar to chemotherapy, radiation treatment can cause acute thymic damage, and the double-positive (DP) thymocyte population are particularly sensitive to this damage (Williams et al. 2009; Gentil Dit Maurin et al. 2015). Considerable reduction in TEC numbers is also seen due to stromal damage, while notably, innate lymphoid cells (ILCs) and endothelial cells have been found to be relatively resistant to radiation exposure (Dudakov et al. 2012; Zhang et al. 2014). These two cell types play an important role in endogenous thymus regeneration.

During the postnatal development of the thymus, increased levels of steroid hormones such as the glucocorticoids induce apoptosis of double-positive (DP) thymocytes that leads to thymus involution (Dooley and Liston 2012). Interestingly, some glucocorticoids are also produced by the thymus epithelium itself (Vacchio et al. 1994), which can inhibit T-cell receptor (TCR)-mediated deletion of DP thymocytes. This modulates positive or negative selection of T-cell repertoire (Mittelstadt et al. 2012). Sex steroid hormones such as estrogen, progesterone, and testosterone also affect the state of the thymus. They can reduce the lymphocyte pool within the thymus and directly induce apoptosis of thymocytes (Patiño et al. 2000). Increased levels of these hormones result in thymus involution, primarily through their effects on non-hematopoietic stromal cells (Olsen et al. 1991). The rate of thymus involution increases rapidly during puberty (Steinmann et al. 1985), and acute transient involution of the thymus can also be seen during pregnancy (Dixit et al. 2003). Thymic epithelial cells (TECs), which express a functional androgen receptor, are important targets of androgens. Recently it has been shown that androgens directly inhibit thymopoietic factors in TECs (Williams et al. 2008; Velardi et al. 2014), while other sex steroids show their effect on HSCs and the bone marrow microenvironment. Other effects of sex steroids on thymus include reduction in the number of available T-cell progenitors and decreased lymphoid differentiation (Dudakov et al. 2009).

4.4 Graft-Versus-Host Disease

The graft-versus-host disease (GVHD) arises as a complication of allogeneic hematopoietic stem cell transplantation (HSCT). There are three distinct phases of this disease that include damage to the tissue resulting from conditioning therapy, allo-reactive donor T-cell activation by host antigen-presenting cells, and soluble effectors-mediated target tissue damage (Ferrara et al. 2009; Blazar et al. 2012). Traditionally, GVHD has been viewed as a disease of the skin, liver, and gut. However, recent studies have provided evidence that it also targets the thymus via allo-reactive T cells (Krenger et al. 2000; Krenger and Holländer 2008). In the thymus, GVHD-related changes to both the stromal and lymphocyte compartments can lead to acute involution. Adverse effects include loss of TECs, loss of cortical and medullary thymocytes, and disruption of the gland architecture. This leads to distorted TCR repertoire and, consequently, decreased output of naive T cells (Lapp et al. 1985; Przybylski et al. 2007). Recent studies using mouse models have shown that TECs are not only direct targets for allo-reactive T cells, but these cells can also act as antigen-presenting cells to prime allo-reactive T cells (Hauri-Hohl et al. 2007). Other events that happen during this process include impairment of thymocyte negative selection and Treg development, which not only results in thymic damage but also the development of chronic GVHD and autoimmunity (Hollander et al. 2010; Wu et al. 2013; Dertschnig et al. 2015). It is likely that during sub-clinical GVHD, the damage to the thymus has already started, which may not be detected in the clinical setting. However, it can begin to have detrimental consequences for T-cell reconstitution. Further complicating this situation could be the use of corticosteroids to treat GVHD. Corticosteroids are popular first line of treatment for GVHD and are also known to induce thymic involution (Blazar et al. 2012). Therefore, alternate therapeutic approaches will be required to treat GVHD of the thymus.

5 Endogenous Thymus Regeneration

It was previously thought that the thymus, once damaged, does not have the capacity to regenerate itself (endogenous regeneration). However, now it is known that this is not the case. Endogenous thymus regeneration is actually a crucial function that allows for renewal of immune competence after the efficiency of the immune system is affected due to disease, infection, stress, chemotherapy/radiation therapy, etc. While most of these insults target the CD4+CD8+ DP thymocytes (T-cell progenitors), TECs are also targets of chemotherapy and irradiation therapy. The number of TECs in the thymus also shows variation during the lifetime of an individual, where they initially increase and then gradually decrease over time. At the molecular level, downregulation of transforming growth factor-β receptor 2 (TGFβR2) and increased expression of forkhead box protein N1 (FOXN1) contribute to the life-long regulation of TEC numbers. The TEC numbers decline after an insult to the thymus, followed by recovery where endogenous regeneration induces an increase of TEC progenitors. During this stage, a premature depletion of the progenitor cell pool can also happen where the TEC numbers fall more rapidly than normal, and the recovery is probably transient. Therapeutic interventions, such as fibroblast growth factor 7 (FGF7) treatment, sex hormone ablation, etc., can lead to a boost in thymic function. However, long-term studies will be needed to determine if this boost is due to increased TEC progenitors or increase in the rate of progenitor cell differentiation.

The thymus has a remarkable capacity to regenerate itself after injury. While the mechanisms underlying this regeneration remain poorly understood, studies in the past few years have revealed several pathways. One of the first studies that identified a potential pathway of endogenous thymus regeneration used mouse models, where fibroblast growth factor 7 (FGF7) was found to be crucial for regeneration after total body irradiation (TBI) injury (Alpdogan et al. 2006). Another study identified a second pathway of endogenous thymus regeneration that involved innate lymphoid cells (ILCs) in the thymus (Dudakov et al. 2012). Here, it was found that production of IL-22 by intra-thymic ILCs directly leads to survival and proliferation of TECs, which in turn supports thymopoiesis (differentiation of thymocytes into mature T lymphocytes) through regeneration of the supporting microenvironment. Another study showed a correlation between increased expressions of Foxn1 (a molecule important for thymic maintenance and regeneration) and increased expression of IL-22 after damage (Pan et al. 2014). Therefore, activation of Foxn1 after damage could represent another potent pathway of endogenous thymus regeneration. Other molecules that might be involved in endogenous thymus regeneration are those that also play important roles in steady-state thymopoiesis. These include CCL25, CXCL12, DLL4, Notch ligand, IL-7, SCF, and VEGF.

In addition to the acute damage, thymus involution drastically reduces the regenerative capacity of the thymus. Endogenous regeneration is a critical process to maintain immune competence following insults. However, chronic infection and repeated rounds of chemotherapy or radiation therapy, compounded by age-related involution, can be a significant clinical challenge that can lead to decreased capacity for immune surveillance, reduced response to vaccines, and increase in opportunistic infections. According to some studies, BMP4 has been found to drive thymic regeneration after injury (Tsai et al. 2003). This happens primarily through the upregulation of FOXN1 (the key TEC transcription factor) and its downstream targets. It is now known that in addition to TEC maintenance, FOXN1 likely contributes to age-related thymic involution when its expression declines (Chen et al. 2009; Nowell et al. 2011; Rode et al. 2015). The regulation of the Notch ligand DLL4 has also been shown to be relevant for steady-state thymus function and also for T-cell development and thymus size (Velardi et al. 2014).

6 Exogenous Thymus Regeneration

Regeneration of the immune function remains a prominent unmet need in many clinical situations. Intense research to understand the mechanisms that regulate thymic recovery after injuries and strategies that can boost its endogenous repair have resulted in advancement of knowledge that will help further exogenous regeneration strategies. At present, thymus transplantation is used as a therapeutic intervention conditions where the thymus is either absent (as in DiGeorge syndrome) or non-functional. Thymus transplantation in other types of patients is risky because the implant would be attacked by the patient's own T cells. For transplantation, thymus fragments from very young patients that are obtained as a by-product of heart surgery are generally used.

6.1 Use of Cytokines and Growth Factors

Exogenous regeneration strategies for thymus regeneration include application of cytokines and growth factors such as interleukin-7 (IL-7), interleukin-22 (IL-22), growth hormones (GH), insulin-like growth factor-1 (IGF-1), keratinocyte growth factor (KGF), and sex steroid inhibitors (SSI). In the thymus, IL-7 is primarily produced by cTECs and to a lesser extent by fibroblasts. In the thymus, stimulation of the IL-7 receptor (IL-7R) promotes proliferation, differentiation, and survival of the developing thymocytes. Due to its critical role in promoting pro-survival signals to peripheral lymphocytes and in thymopoiesis, IL-7 has been extensively studied for its potential to enhance recovery after immune insults (Mackall et al. 2011). Studies have also shown that exogenous administration of IL-7 enhances thymopoiesis and increases the proliferation of mature peripheral T cells (Alpdogan and van den Brink 2005; van Lent et al. 2009). A phase I clinical trial involving the use of recombinant human IL-7 (hIL-7) in patients that had allogeneic hematopoietic stem cell transplantation (HSCT) post-T-cell depletion showed increased CD3, CD4, and CD8 counts after hIL-7 treatment, along with a broader TCR beta repertoire diversity compared to untreated patients (Perales et al. 2012; clinical trial No. NCT00684008).

IL-22 has a role in mediating endogenous recovery of thymus function after acute damage (Dudakov et al. 2012). In mice studies where the animals were subjected to sub-lethal irradiation, exogenous administration of IL-22 promoted accelerated thymic recovery. It is now known that IL-22 can act directly on TECs, promoting their proliferation and survival, which ultimately leads to the rejuvenation of thymopoiesis. These findings suggest that exogenous application of IL-22 represents a novel therapeutic strategy for thymus regeneration and immune function recovery. Currently, clinical trials are being conducted to evaluate the safety and pharmacokinetic profile of IL-22 treatment in healthy volunteers, as well as a phase IIa study where the safety of human recombinant IL-22 (hrIL-22) in combination with systemic corticosteroids is being evaluated in the treatment of gastrointestinal acute GVHD in patients receiving HSCT (Study No. NCT02406651). If there are positive outcomes from such studies, IL-22 may be part of an immune-boosting therapy approach.

Several neuroendocrine hormones have been shown to have important effects on the immune system. This includes growth hormone (GH) and insulin-like growth factor-1 (IGF-1). Using GH-deficient mice (having defects in T-cell development), the effects of GH on thymus function have been extensively investigated. Also, in pre-clinical studies using mouse models, exogenous administration of recombinant GH was shown to promote thymus regrowth (Murphy et al. 1992). Results from other studies show that use of GH can increase TCR diversity, promote better thymic cellularity, and improve recovery of the hematopoietic compartment in aged and immune-compromised animals (Chen et al. 2003; Redelman et al. 2008). The mode of action of GH with respect to thymus regeneration includes enhanced proliferation of TECs and trafficking of common lymphoid progenitors (CLPs) into the thymus (Knyszynski et al. 1992; Tsuji et al. 1994). Insulin-like growth factor-1 (IGF-1) is known as a principal mediator of the biological effects of GH (Montecino-Rodriguez et al. 1998; Alpdogan et al. 2003). In the thymus, IGF-1 has been implicated as a positive thymic regulator based on early observations that age-related declines in thymic function paralleled declines in plasma concentrations of IGF-1. IGF-1 receptor (IGF-1R) is expressed on thymocytes and peripheral T cells (Chu et al. 2008). Since, GH and IGF-1 are connected through the same pathway (Chu et al. 2008), it is expected that using IGF-1 for exogenous thymus regeneration may have similar effects as that for GH.

Keratinocyte growth factor (KGF) is another growth hormone that is being used for exogenous thymus regeneration. In the thymus, a splice variant of the fibroblast growth factor receptor 2 (FgfR2-IIIb) expressed on TECs is known to bind KGF. Considering this association, exogenous administration of recombinant KGF has been tested for enhancing thymus regeneration, particularly with reference to HSCT. Results from such studies show that KGF can protect TECs from different types of thymic injuries (Min et al. 2002; Rossi et al. 2002). TEC proliferation leading to an increase in thymocyte expansion and enhanced T-cell export was also observed. Pre-clinical studies in non-human primates have been conducted, including one where KGF treatment after an autologous HSCT in adult rhesus macaques improved thymopoiesis, enhanced naive T-cell recovery, and also accelerated hematopoietic recovery after the transplant (Wils et al. 2012). However, more studies in non-human primates and human subjects will be needed to determine the therapeutic efficacy and safety of this approach. Palifermin (recombinant human keratinocyte growth factor) is an FDA-approved drug that is used for the prevention of mucositis in recipients of high-dose chemotherapy. However, more clinical studies will be needed to evaluate the efficacy of KGF in directly enhancing T-cell reconstitution and restoring thymus function in immune-compromised patients (Min et al. 2002).

Sex steroid inhibition (SSI) using chemical or surgical approach is a well-described method to promote thymus growth (Sutherland et al. 2005; Hince et al. 2008). Using mouse models, several studies have demonstrated thymus enlargement and accelerated thymic recovery after immune insults (Sutherland et al. 2008; Goldberg et al. 2009, 2010). In the past years, several pharmacological treatments that have been developed to suppress the release of androgens from the gonads or block the androgen receptor on the target cells are now being investigated for their potential to reversibly inhibit sex steroid hormones and promote thymic regrowth and immune reconstitution. The exact mechanisms underlying SSI-based thymus regenerative effects are still not completely understood. Recent studies have shown that SSI can directly promote the expression of Notch ligand DLL4 (Velardi et al. 2014) and CCL25 (Williams et al. 2008). The function of hematopoietic stem and progenitor cells is also promoted using this treatment (Khong et al. 2015). Since, transient ablation of sex steroids represents a promising regenerative strategy for the immune system, further work will be needed to understand the different effector mechanisms so that clinically relevant therapies can be developed. Currently, two ongoing clinical trials are testing the effects of leuprolide acetate (an LHRH agonist) with palifermin (recombinant KGF) in promoting immune recovery after allogenic HSCT transplants in human patients (Study No. NCT01746849 and NCT01338987).

Another molecule that can have a key role in exogenous thymus regeneration is the RANK ligand. The RANK ligand is a TNF family member, which has emerged as an important regulator of epithelial cell growth and differentiation in different tissues (Duheron et al. 2011). It has been shown that ex vivo administration of RANKL substantially enhances the cellularity of cTEC and mTEC subsets within the thymus. It also supports T-cell progenitor homing, TEC recovery, and de novo thymopoiesis and thus represents a new therapeutic strategy to boost thymic regeneration. More importantly, RANKL administration was also shown to be efficient for T-cell regeneration and TEC recovery in older individuals with highly diminished thymic function due to natural involution (Gray et al. 2006; Tomimori et al. 2009; Chinn et al. 2012).

6.2 Use of Stem Cells, Progenitor Cells, and Induced Pluripotent Stem Cells

Stem cells and progenitor cells are considered as a promising therapeutic option for thymus repair and regeneration. Regeneration of a functional thymus was demonstrated when thymus stem or progenitor cells from a donor mouse were transplanted into a recipient mouse that lacked a thymus (Fan et al. 2015). Although, this approach is very promising, many challenges still remain that include growing these cells in the lab, expanding enough cells for transplantation, and studying their regenerative capacity in humans. These challenges will have to be overcome if thymic stem or progenitor cells are being considered for clinical use. Use of pluripotent stem cells (such as the induced pluripotent stem cells or iPSCs) is another promising avenue for exogenous thymus regeneration. Researchers have successfully created thymus-like cells from mouse and human pluripotent cells (Lai and Jin 2009; Sun et al. 2013). In order to overcome the challenge of immune rejection and create de novo human thymus tissue for transplantation, use of patient's own cells to create iPSCs is a practical option.

The generation of thymus-derived T cells can take several months after hematopoietic stem cell transplantation (HSCT), and this process can be delayed even further by situations such as aging and GVHD (Bosch et al. 2012). The use of precursor T (pre-T) cell infusion during HSCT is one of the therapeutic approaches to accelerate immune recovery in these immune-compromised patients. For cell therapy, pre-T cells can be generated using ex vivo co-culture of HSCs with OP9 cells that have been ectopically transduced with two critical factors required for thymocyte proliferation, commitment, and differentiation, such as Delta-like 1 (Dll1) and Delta-like 4 (Dll4) (De Smedt et al. 2004; La Motte-Mohs et al. 2005; Holmes and Zúñiga-Pflücker 2009). The OP9-DL1 cells are a bone marrow-derived stromal cell line that ectopically expresses the Notch ligand, Delta-like 1 (Dll1). This simple and efficient culture system for ensuring proliferation, differentiation, and commitment of T-lineage cells from different sources of stem cells is promising. It can be used to shorten the duration of immunodeficiency in patients that have a dysfunctional thymus. The mechanism by which pre-T cells support immune system recovery is also being explored. Studies have shown that pre-T cells help T-cell reconstitution by providing a ready source of T-cell progenitors. A recently study also reported that the pre-T cells can also have a secondary effect, where through crosstalk with other cellular components, the pre-T cells can enhance thymic stromal function long after these transplanted cells have transited through the thymus (Awong et al. 2013). Another advantage of using pre-T cells is that they can be used as an "off-the-shelf" therapeutic strategy that can be administered across MHC barriers. They can also be used as vehicles for chimeric antigen receptors during immunotherapy.

Pluripotent stem cells hold great promise in the field of regenerative medicine. The selective generation of thymic epithelial progenitor cells (TEPCs) from ESCs or iPSCs in vitro for transplantation may have important implications for treatment of T-cell immune deficiencies in clinical settings. When TEPCs that have been derived from mESC were placed in vivo, they are self-renewed, develop into TECs, and reconstitute the normal thymic architecture. After bone marrow transplantation, these mESC-derived TEPCs caused an increase in the number of functional naive T cells and enhanced thymocyte regeneration (Sun et al. 2013). In a different study, transplantation of mESC-derived TEPCs resulted in the efficient generation of naive T cells in both young and old recipients following allogeneic BM transplantation (Lai et al. 2011). Hence, in a clinical setting, use of autologous pluripotent stem cell-derived TEPC grafts may lead to broad applications for restoring immune damage resulting from dysfunction of the thymus.

6.3 Targeting Cellular Pathways Within the Thymus

Understanding and exploiting the pathways underlying endogenous thymic regeneration can help us develop novel therapeutic strategies for exogenous thymus regeneration. One such insight can be obtained from study of endothelial cells (ECs). It is now known that ECs can have an active role in tissue repair via production of angiocrine factors (Rafii et al. 2016). Angiocrine growth factors are molecules produced by tissue-/organ-specific blood vessel endothelial cells that can stimulate organ-specific repair activities in diseased or damaged tissue/organs. In one study regeneration within the thymus was observed using Akt-activated ex vivo propagated ECs (Zhang et al. 2004). While it is clear that ECs can promote TEC function and regeneration after damage, further studies will be required to investigate the role that TECs play in guiding EC function with respect to thymus regeneration.

FoxN1 is a transcription factor mainly expressed in the thymic epithelium and skin. FOXN1 plays a critical role in the maturation of the thymus and skin epithelial cells essential for its development (Nehls et al. 1994). In human patients, mutations in the FoxN1 gene result in an athymic condition where thymopoiesis is completely blocked, leading to severe primary T-cell immune deficiency (Frank et al. 1999). This severe immune deficiency can often lead to death in early childhood because of severe infections. Studies have shown that expression of FoxN1 can drive differentiation of the whole TEC network from TEPC and also lead to successful colonization of thymic rudiment by bone marrow-derived precursors (Nowell et al. 2011). This knowledge can be used to develop therapeutic interventions using FoxN1 expression as a driving force for thymus regeneration.

Age-induced thymic involution is multifactorial. Therefore, targeting thymus regeneration using isolated pathways would be transitory and incomplete. One approach to achieve sustained rejuvenation of the thymus would be to target both TECs and lymphoid cells, along with other means such as activating endogenous thymic epithelial progenitor cells (TEPCs), sustained expression of forkhead box N1 (a critical TEC transcription factor), and generating TEPCs from pluripotent stem cells. Thymus epithelial stem/progenitor cells from early embryonic thymus are able to differentiate into the full range of thymic epithelial subsets (Rossi et al. 2006; Baik et al. 2013). Recently, an adult thymic epithelial progenitor population within the TEC subset has also been identified and characterized with respect to their self-renewal, colony-forming potential and generation of mature cTEC and mTEC (Wong et al. 2014). This may allow for the development of new therapies targeting thymus rejuvenation (Ventevogel et al. 2013).

Multiple strategies have been proposed and evaluated to boost thymus regeneration. However, many of these have failed to generate compelling effectiveness in clinical studies. The reason for this can be attributed to the fact that thymus requires a reciprocal cellular communication between the hematopoietic and stromal compartments for its development, maintenance, and function. This fact needs to be incorporated when designing and administering therapies aimed at thymus rejuvenation and regeneration. One approach would be a combination strategy such as administering KGF along with pre-T cells to have a synergistic effect (Zakrzewski et al. 2006) or combining sex steroid ablation (SSA) with p53 inhibition (Kelly et al. 2008).

6.4 External Modulation of T-Cell Response Within the Thymus

The role of the thymus in the establishment of the T-cell receptor repertoire is considered as an important area of intervention related to thymus regeneration. Also, a long-pursued goal in the field of transplantation has been to develop strategies to induce immune tolerance through the direct introduction of antigens into the thymus (Posselt et al. 1990). Intra-thymic injection of antigens using lentiviral vectors is one of the methods being used to modulate specific T-cell responses exogenously (Gottrand et al. 2012). Even though a majority of such studies about external modulation of T-cell response have been conducted in mouse models, there is an immense translational potential in the context of organ transplantation to induce tolerance this way. Recently, a double-positive stage of thymocyte development was identified, which could be a potential target for the modulation of FOXP3+ regulatory T cells (Nunes-Cabaco et al. 2011). As a strategy to overcome the delay in immune reconstitution following HSCT (a major cause or morbidity and mortality in such cases), intra-thymus delivery of thymocyte progenitor cells is also being explored. In one study using mouse models, it was shown that delivery of semi-allogeneic progenitor cells within the thymus results in sustained T-cell development, even across histocompatibility barriers (de Barros et al. 2013).

6.5 Thymus Transplantation

Thymus transplantation has been long used as a treatment option in cases of severe immunodeficiency that result from either mutation in the genes essential for TEC differentiation (such as FOXN1) or absence of thymus in conditions such as DiGeorge syndrome. Thymus transplantation aims for more complete reconstitution to produce naive T cells that show a broad T-cell receptor (TCR) repertoire. Postnatal thymic tissue can be readily available because it is routinely removed from infants undergoing open heart surgery through a median sternotomy. A clinical study showed approximately 75% long-term survival in 60 patients under 2 years with a complete DiGeorge anomaly after transplantation of postnatal allogeneic cultured thymus tissue (Markert et al. 2010). There was evidence of thymopoiesis, and a diverse repertoire of naive circulating T-cell responses was seen in survivors. Xenotransplantation has also been considered as an option for thymus regeneration (Zhou et al. 2012). Combined use of hematopoietic stem cell transplantation (HSCT) along with thymus transplantation is being evaluated to facilitate long-term reconstitution of immune function in cases with severe combined immune deficiencies (SCIDs) (Hu and Yang 2012).

6.6 Artificial Thymopoietic Environments

Reconstitution of certain features of the thymus environments by artificial means has also been attempted as an alternative to de novo generation of TEC progenitors. These artificial thymopoietic environments have been generated both in vitro (Beaudette-Zlatanova et al. 2011) and in vivo (Calderón and Boehm 2012). In such studies, the Delta-like ligand 4 (DLL4) was found to be a key determinant of T-cell lineage induction (Hozumi et al. 2008). In in vivo reconstitution experiments, the CXC chemokine ligand 12 (CXCL12) was found to be conducive to the generation of T cells (CD4+CD8+) (Calderón and Boehm 2012). It was also found that in individuals where the lympho-stromal crosstalk was absent during the early stages of development, functional thymopoietic environments can be successfully established at later stages of life using exogenous application of selected cytokines or growth factors, including FOXN1 (Bleul et al. 2006; Corbeaux et al. 2010).

7 Thymus Bioengineering

The advancements in regenerative medicine technologies, particularly tissue engineering, have enabled development of many tissue and organs outside the body. We know that the capacity to replace thymus function therapeutically will be beneficial in a variety of clinical settings, including improving recovery following bone marrow transplantation and supporting tolerance to transplanted cells, tissue, and organs. A fully functional bioengineered thymus can help achieve these goals. This would also be a practical therapeutic option for patients that have complete athymia due to congenital immune deficiencies, such as that seen in DiGeorge syndrome. In the past decade, several efforts have been made to characterize the complex microenvironment of the thymus and also to rebuild the complex 3D structure of this organ in vitro. To obtain full functionality of bioengineered thymus, not only will the spatial organization of the native thymus need to be recreated, but full TEC functionality will be required in each compartment. This would include cTEC and mTEC. For in vivo use, particularly in human patients, bioengineering of a thymus structure would require using artificial matrices. A particularly important area of study has been the identification and use of biomaterials to create an artificial 3D matrix that can support cell-to-cell interactions of thymus-derived cells. An example of such study is the demonstration that thymic stromal cells can support T-cell development from precursor hematopoietic cells in artificial 3D matrices (Pinto et al. 2013).

As for the cell source, the normal thymus tissue-derived cells would still represent the best source for tissue engineering. However, this would require large numbers of cells to be used in the bioengineered construct (Fig. 4). The in vitro expansion of adult thymic epithelial cells (TECs) is challenging because they tend to lose their epithelial phenotype and show diminished growth. Other cell sources for thymus tissue engineering would include endogenous thymic epithelial progenitor cells (TEPCs) or TEPC derived from differentiation of embryonic stem cells or iPSs (Sun et al. 2013; Bredenkamp et al. 2014; Su et al. 2015) and transdifferentiation of some cell lineages into TEC-like cells. A novel approach for generation of new TECs was carried out by direct reprogramming of mouse embryonic fibroblasts (MEFs) using the forced expression of FOXN1 [Bredenkamp et al. 2014]. These TECs (referred to as iTECs) were shown to promote T-cell development both in vitro and in vivo after transplantation into nude mice.

The scaffold forms a very important part of many tissue engineering strategies. Use of decellularized tissue scaffolds ensures that the bioengineered tissue or organ retains the structural and ECM compositions of the native tissue (Fig. 5; Jackson et al. unpublished). The same strategy has been used for the thymus. In a study by Fan et al. (2015), three-dimensional thymus organoids were created using decellularized

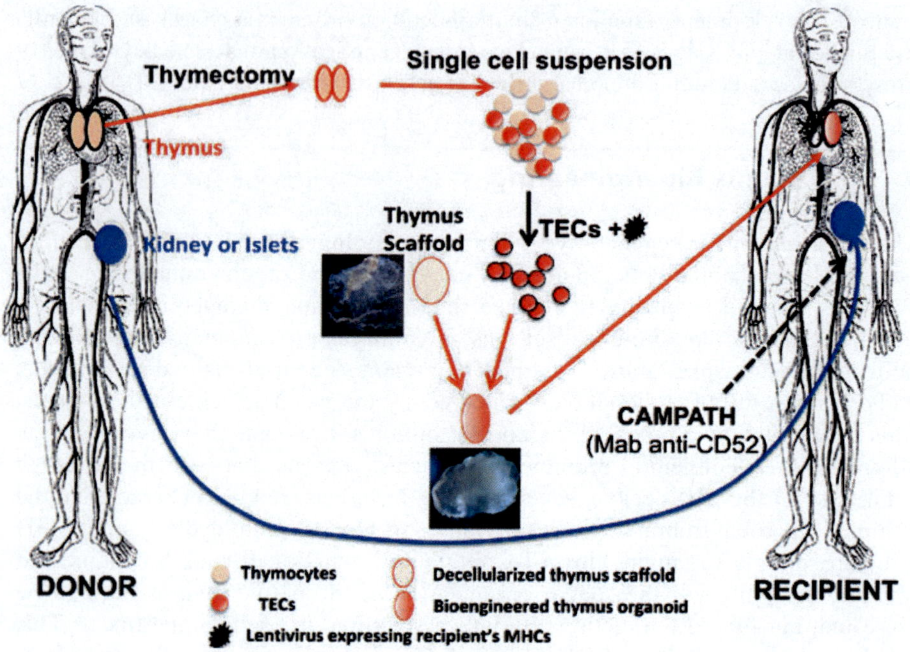

Fig. 4 Using thymus bioengineering to introduce donor-specific allograft tolerance. Thymus gland is obtained by thymectomy from a cadaver donor along with the organ to be transplanted (like kidney or pancreatic islets). The thymic epithelial cells (TECs) will be isolated from dissociated thymus and transduced with lentivirus particles expressing the recipient's MHC molecules. These cells will be used to reconstruct thymus organoids with surrogate thymus scaffold. Shown here are representative images of decellularized thymus scaffold (mouse) and a bioengineered thymus organoid (mouse). This bioengineered thymus organoid will be transplanted into the recipient at the time of organ transplantation. The transplant will be received by a recipient who is pre-conditioned with Campath (anti-CD52) or ATG (anti-thymocyte immunoglobulin). TEC, Thymic epithelial cells; MHCs, major histocompatibility complex; CAMPATH, anti-CD52 monoclonal antibody. (Image used with permission from Tajima et al. 2016)

scaffolds from mouse thymus and thymic epithelial cell. When transplanted in vivo, these organoids supported thymopoiesis and also homing of lymphocyte progenitor cells. Transplanting these thymus organoids also resulted in development of immune tolerance to skin allografts in these mice, showing that it is feasible to restore certain thymus functions using bioengineered thymus organoids. The clinical implications of this approach are also being investigated. Bioengineered thymus organoids are also being used for understanding and modulating thymic negative selection. Negative selection is a key checkpoint in the T-cell development pathway where T-cell clones with high affinity to self MHC complexes (pMHCs) and self-peptides are eliminated. This is also referred to as clonal deletion, and both thymic antigen-presenting cells (such as B cells, macrophages, and thymic dendritic cells) and medullary thymic epithelial cells (mTECs) contribute to this process. Using islet cell autoantigen 69 (ICA69) as a model antigen, it was demonstrated that ICA69-reactive T cells can escape clonal deletion when thymus organoids constructed with

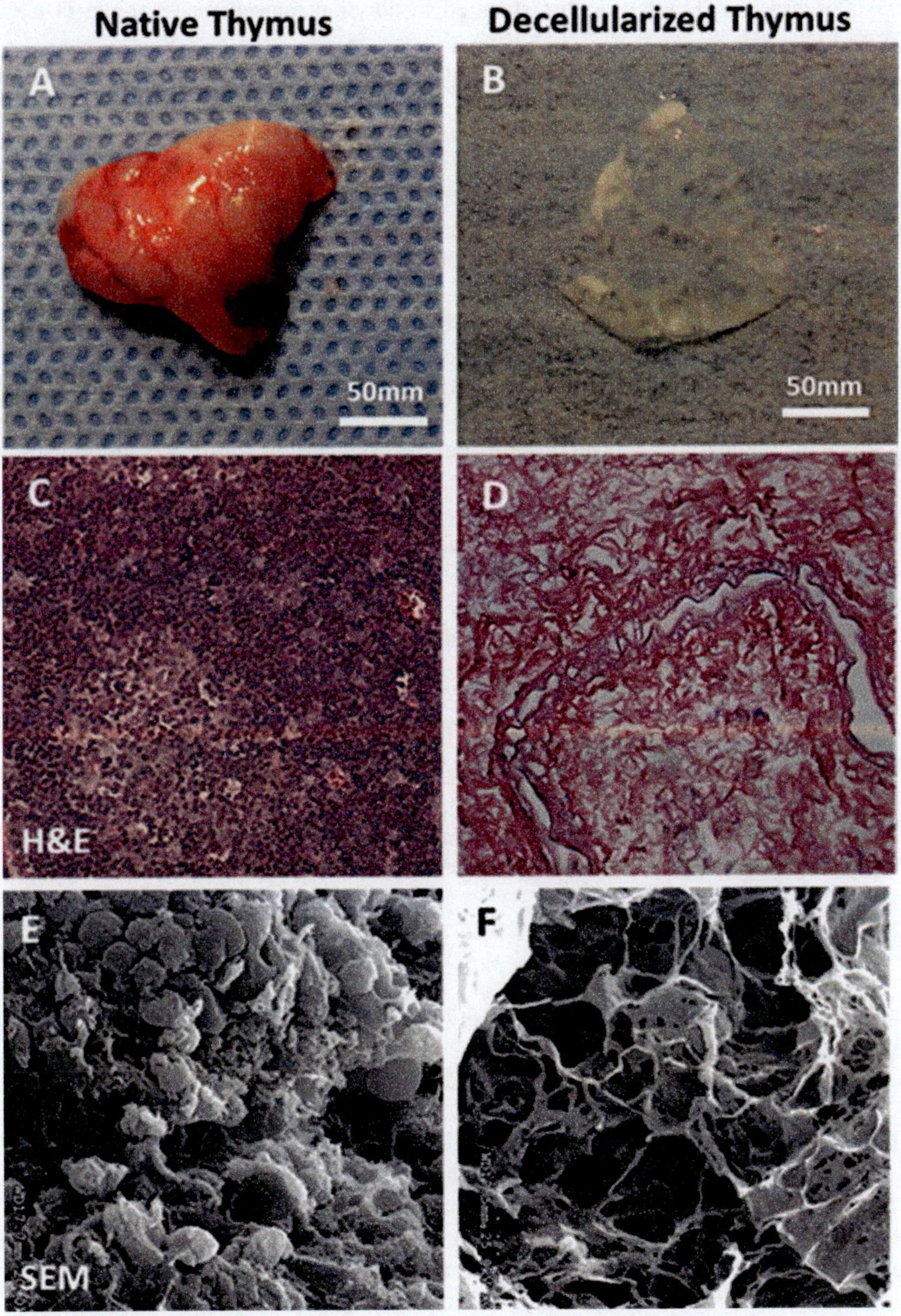

Fig. 5 Rat thymus before and after decellularization (**a**, **b**) represents thymus anatomy before and after decellularization process. (**c**, **d**) The histology using hematoxylin and eosin staining suggests the removal of cells from the decellularized thymus scaffold. (**e**, **f**) SEM images of native and decellularized thymus showing structural preservation thymus. Extracellular matrix scaffold after decellularization (Jackson et al. unpublished)

ICA69-deficient mTECs were transplanted in athymic nude mice (Pradhan et al. 2018).

Using thymus bioengineering, it might be possible to modulate the adaptive immune system of recipients so that donor-specific immune tolerance of allogeneic grafts can be achieved without using any mmunosuppressive drugs (Fan et al. 2015). In one such strategy described by Tajima et al. (2016), the thymus gland is harvested along with the needed (transplantable) organ from a cadaver donor and dissociated into single cell suspension. Then, TECs and other thymic mesenchymal cells are enriched and used to populate the decellularized thymus scaffolds (Fig. 4). These scaffolds can be of either allogeneic or even xenogeneic origins as the ECM components normally are not antigenic. Using lentiviral vectors, the recipient's MHC molecules can be introduced and expressed in donor TECs, followed by culturing of the TEC-populated scaffold (bioengineered thymus organoids) to enable colonization and expansion of the cells. The graft (like kidney or pancreatic islets, etc.) is transplanted into the patient preconditioned with either anti-CD52 monoclonal antibody (CAMPATH) or anti-thymocyte immunoglobulin (ATG) so as to deplete the T cells (cause immunosuppression) within the graft (Fig. 4). Now the bioengineered thymus organoids are engrafted into the thoracic cavity of the recipient, which will facilitate the generation of a new repertoire of tolerant naive T cells. These new T cells would not respond to the donor cells, but exert adaptive immune function, resulting in the long-term survival of the transplant without the need for immunosuppression. In spite of the abovementioned progress, engineering a full thymus organ de novo still remains a challenge. This is mainly because of the highly complex structural and functional compartmentalization of this organ. Additionally, reconstitution of all the cell types that contribute to the main thymus function (differentiation and maturation of T cells) further adds to the complexity of the system. To fabricate such a complex structure having multiple cell types, use of advanced biofabrication technologies such as 3D bioprinting can offer a solution.

8 Current Challenges

Due to the nature and function of the thymus, both endogenous and exogenous regeneration strategies face many challenges. The thymic stroma is essential for the survival and function of TECs. However, recapitulating this unique 3D architecture in vitro has been a major challenge. Additionally, growing adult TECs in culture is difficult. This is in contrast to epithelial cells from other organs that easily form 2D sheets on artificial substrates. It is now known that the expression of genes critical for TEC proliferation, specification, and function is dependent on the 3D configuration of TECs within the thymus stroma (Novell et al. 2011). Therefore, when TECs are separated from their natural microenvironment and cultured in vitro, most of them either express markers of terminally differentiated epithelial cells (Saunders et al. 1995) or even transdifferentiate into skin cells (Bonfanti et al. 2010). Although we now have 2D culture systems to propagate TECs in vitro, suitable 3D culture systems will still be needed to generate sufficient TECs for therapeutics and tissue engineering. Toward this goal, TECs seeding onto Matrigel or other collagen-based 3D matrices have been explored but with limited success. An encouraging development worth mentioning is the observation that when artificial 3D matrices with mTECs were supplemented with dermal fibroblasts, some of the key molecular factors required for negative selection within the thymus (such as TRAs) were expressed in the mTECs (Pinto et al. 2013).

A big challenge toward exogenous thymus regeneration using cell or bioengineered approaches is the limited number of TECs that can be harvested from the adult thymus. Additionally, as mentioned above, their culture and expansion ex vivo are also a challenge. The proliferation of thymic epithelial progenitor cells (TEPCs) drops drastically within the first week after birth, and the TEC repertoire starts to decrease at around first year after birth in human. Preventing this early loss of TEPCs/TECs and successfully expanding them in culture remains a challenging task. When TECs are isolated and cultured as single cells, they lose cell-cell contact and geometrical organization when placed in an artificial environment. In fact, in an intact thymus, the mTECs and cTECs are compartmentalized to support specific stages of thymocyte development. To recreate this in a bioengineered environment (such as organoid, synthetic scaffold) can be challenging. However, use of 3D biofabrication tools (such as 3D bioprinting) can be utilized to achieve this compartmentalization. Another way to achieve aggregation of TECs within a specific region of an artificial matrix is by using immobilized TEC-specific antibodies, similar to what Tajima et al. (2015) have used. They successfully generated mini aggregates of TECs by incorporating TEC-specific antibodies in a polypeptide-based, self-assembling hydrogel system. These 3D aggregates of TECs that could support T lymphogenesis were transplanted into athymic nude mice. Similar results can be achieved by using 3D bioprinting to create a bioengineered thymus having TEC compartmentalization that is similar to native thymus.

Currently, transplantation of thymus tissue is primarily used as an investigational treatment for infants who have congenital anomalies, such as the DiGeorge syndrome. In such cases, donor thymus tissues are obtained from infants undergoing unrelated surgeries (such as treatment of congenital heart diseases). Tissues obtained from donors under 9 months of age are normally used for transplantation. Therefore, in the absence of a bioengineered thymus tissue or whole organ, treatment of patients with congenital anomalies or the ones who have lost most of their thymus function is a big challenge. Aging-associated progressive degeneration of the thymus and immune function (immune senescence) leads to increased susceptibility to opportunistic infections, incidence of cancer, and autoimmunity, among other issues. Preventing or reversing this immunosenescence in aged individuals so that they can lead a healthy, disease-free life is an idealistic but practical outlook for medicine. However, achieving this is challenging. This problem is significant considering the fact that more than 20% of US population is projected to be aged 65 or older by the year 2030. Also, much of the current research using stem cells as a potential tool to regenerate and repair the thymus has been performed in mice. Translating these findings to humans will be a big challenge both for researchers and clinicians.

9 Conclusions

Thymic involution occurs with age and is central to the decline in immune system function. Current approaches for exogenous thymus regeneration focus on the modulation of growth factors and hormones secreted by TECs. However, such efforts can only lead to transient and partial thymus recovery. For complete recovery, new strategies such as TEPC activation, TEC recovery using induction of FoxN1 expression, de novo generation of TEPC from stem cells, etc. can be used. Targeting pathways like BMP4 to modulate expression of key transcription factors necessary for TEC and thymocyte development such as FoxN1 or Dll4 has also opened up new possibilities for exogenous thymus regeneration. Another promising approach for boosting immune functions in patients whose thymus has been severely damaged is through the administration of RANKL that can be beneficial at several levels, including thymus homing of T-cell progenitors, TEC regeneration, de novo thymopoiesis, etc. Further, use of bioengineering approach using decellularized thymus scaffolds or functional thymus organoids can not only support regeneration of a functional T-cell compartment but also help to introduce donor-specific immune tolerance.

A major challenge to recreate the thymus microenvironment is the complex composition and organization of its extracellular matrix (ECM). Using biological scaffolds generated from decellularized can be an effective strategy to reproduce its in vivo-like ECM microenvironment. As we obtain greater understanding of the molecular mechanisms involved in thymus development, function, and involution, we move forward in developing new therapeutic tools for the repair, regeneration, or replacement of this critical organ. For treating aging-associated progressive immunosenescence, successful creation and use of transplantable thymus tissue using bioengineering approaches seems like the best option. Another solution would be to transplant artificial thymus organoids that have been engineered with host MHC-expressing TEPCs (derived either from genetically modified hESCs or iPSCs) so as to rejuvenate T-cell immunity and also treat aged-related immune disorders. The challenge for the future of thymus regeneration will be to continue the development of current promising technologies and develop new ones. Those technologies with highest potential should then be subjected to rigorous evaluation in a clinical environment.

Despite many encouraging improvements, use of current methods (including pharmacological interventions) has not been able to achieve physiologically relevant and long-term improvements in decreased thymus function. Advances in cell biology, stem cell technology, and cellular engineering are making it possible to use tissue-resident epithelial stem cells, stem cell-derived TECs, and even in vitro generated thymus tissue for thymus regeneration and/or replacement. For using thymus tissue-resident epithelial stem cells, a sufficient lack of knowledge about the compartment-specific intermediate cells and identity of bipotent TEC progenitors are the major roadblocks for endogenous thymus regeneration. A better understanding about the behavior of these cells during tissue maintenance and after injury would be valuable to devise strategies where activation of quiescent endogenous stem cells or introduction of reprogrammed cells with thymopoietic fate can achieve clinically significant effects for restoring failing thymus function.

The bone morphogenetic protein 4 (BMP4) has been shown to be critical for promoting TEC regeneration after injury. After peaking early after damage, BMP4 expression by ECs is seen to return to baseline levels by day 21. Additionally, BMP4 is also known to inhibit the differentiation of T cells. Taken together, these mechanisms point toward potential therapeutic strategies for regenerating the immune system in patients whose thymus has been irrevocably damaged. This direction needs to be further explored. The proliferation of TECs can be limited by TGF-β signaling, while the rate of TEC proliferation increases due to FGF7 signaling both during the steady state and after insult to the thymus. Contrary to this, an increase in IL-22 signaling seems to have little effect on steady-state thymopoiesis. Therefore, a long-term strategy to induce and support thymus recovery would be to provide exogenous FGF7 and IL-22 while blocking the effects of TGF-β.

The presence of a thymus is characteristic of vertebrate immune systems, with commonalities and differences seen over the course of evolution. Therefore, a comparative analysis of thymopoiesis between evolutionarily distant organisms can probably provide important insights into the factors and lesser-known pathways involved in thymus generation and regeneration. Such a comparative analysis can also provide important insights into developing allogenic or even xenogeneic transplant therapies (Yamada et al. 2012). The ECM environment of a decellularized thymus can provide the appropriate microenvironment for the long-term survival of adult TECs and can be used to create thymus organoids and tissues in vitro. Since there is no cellular component in a decellularized tissue scaffold, allogeneic or even xenogeneic rejection should not be a concern. Additionally, since decellularized thymus scaffold from mouse can be stored at 4oC for up to 1 month before use (Fan et al. 2015), it is expected that decellularized thymus scaffold from humans can be handled the same way, providing an off-the-shelf option for creating thymus organoids and tissues for regenerative applications.

A major challenge in the success of solid organ transplantation is the induction of donor-specific immune tolerance (Sachs et al. 2011). Even though the use of latest immunosuppressive drugs has caused the rates of acute rejection to decrease significantly over the past decade, the long-term survival statistics of allograft have not improved correspondingly. Use of modern immunomodulatory protocols (depletion of mature T cells, promotion of regulatory T cells, blockage of costimulatory molecules, etc.) is not helping much. One strategy to overcome this limitation is by introducing allogenic donor TECs into bioengineered thymus organoids so that the need for immunosuppressive drugs to maintain allograft survival is reduced or eliminated. Use of gene editing and delivery technologies can also be valuable in terms of thymus regeneration and bioengineering. One way of using this would be to create patient-specific thymus organoids where the antigen presentation properties of TECs are fine-tuned. This can be used either to create self-tolerance to specific tissues in treating autoimmune disorders or in case of clinical transplantation to have long-term donor-specific immune unresponsiveness. In the future, it is conceivable that patients with thymus dysfunction can be transplanted with bioengineered thymus organoids or tissues that are constructed from either cryopreserved donor TECs or TEC-like cells derived from the patient's own cells (e.g., from iPSCs). It has been seen in DiGeorge syndrome (DGS) patients treated with bioengineered thymus organoids that around 70% of them show successful thymopoiesis and many of them can avoid immunoglobulin replacement therapy. This indicates the clinical feasibility of the approach for regenerating adaptive immunity.

As in other bioengineered solid tissues or organs, vascularization will be a critical component in the success of thymus bioengineering. Vascularization will be needed not only to prevent ischemia-induced cell death but also to transport cytokines, growth factors, hormones, and cells to and from this bioengineered organ to support normal immune function. While decellularized scaffold can be used for preserving the ECM of the vascular framework, it has been seen that there is a significant loss of factors that support angiogenesis. Studies in NOD scid gamma humanized mice have shown that human TECs and VEGF-expressing thymic mesenchyme can recruit human hematopoietic progenitor cells and promote their differentiation into T cells (Chung et al. 2014), while in other studies, VEGF produced by the thymus stroma was critical for its growth and induction of angiogenesis (Cuddihy et al. 2009). Therefore, incorporating angiogenesis-inducing factors within biomaterials of in vitro generated thymus tissues can be an effective way to promote vascularization post-implantation. There is no doubt that thymus bioengineering will form a critical component of strategies that will be used for restoration or replacement of lost thymus function and immune competence. However, more research and clinical studies will be needed for this to be translated into the clinic.

Acknowledgments We would like to thank Dr. Prafulla Chandra (Wake Forest Institute for Regenerative Medicine) for the editorial help.

References

Abbas A (2004) Diseases of immunity. In: Kumar V, Abbas A, Fausto N (eds) Robbins and Cotran, pathologic basis of disease. W. B. Saunders, Philadelphia, pp 193–267

Alpdogan O, van den Brink M (2005) IL-7 and IL-15: therapeutic cytokines for immunodeficiency. Trends Immunol 26:56

Alpdogan O et al (2003) Insulin-like growth factor-I enhances lymphoid and myeloid reconstitution after allogeneic bone marrow transplantation. Transplantation 75:1977–1983

Alpdogan O et al (2006) Keratinocyte growth factor (KGF) is required for postnatal thymic regeneration. Blood 107:2453–2460

Aspinall R (1997) Age-associated thymic atrophy in the mouse is due to a deficiency affecting rearrangement of the TCR during intrathymic T cell development. J Immunol 158:3037–3045

Awong G et al (2013) Human proT-cells generated in vitro facilitate hematopoietic stem cell-derived T-lymphopoiesis in vivo and restore thymic architecture. Blood 122:4210–4219

Baik S, Jenkinson EJ, Lane PJ, Anderson G, Jenkinson WE (2013) Generation of both cortical and Aire + medullary thymic epithelial compartments from CD205 + progenitors. Eur J Immunol 43:589–594

Banks W (1993) Applied veterinary histology. Mosby, St. Louis

Beaudette-Zlatanova BC et al (2011) A human thymic epithelial cell culture system for the promotion of lymphopoiesis from hematopoietic stem cells. Exp Hematol 39:570–579

Blazar BR, Murphy WJ, Abedi M (2012) Advances in graft-versus-host disease biology and therapy. Nat Rev Immunol 12:443–458

Bleul CC et al (2006) Formation of a functional thymus initiated by a postnatal epithelial progenitor cell. Nature 441:992–996

Boehm T, Swann JB (2013) Thymus involution and regeneration: two sides of the same coin? Nat Rev Immunol 13:831–838

Bonfanti P, Claudinot S, Amici AW, Farley A, Blackburn CC, Barrandon Y (2010) Microenvironmental reprogramming of thymic epithelial cells to skin multipotent stem cells. Nature 466 (7309):978–982

Bosch M, Khan FM, Storek J (2012) Immune reconstitution after hematopoietic cell transplantation. Curr Opin Hematol 19:324–335

Bredenkamp N, Ulyanchenko S, O'Neill KE, Manley NR, Vaidya HJ, Blackburn CC (2014) An organized and functional thymus generated from FOXN1-reprogrammed fibroblasts. Nat Cell Biol 16:902–908

Brown T Jr, Suter M, Slauson D (2002) Immunopathology. In: Slauson D, Cooper B (eds) Mechanisms of disease. A textbook of comparative general pathology. Mosby, St. Louis, pp 247–297

Bruijntjes JP, Kuper CF, Robinson JE, Schuurman HJ (1993) Epithelium-free area in the thymic cortex of rats. Dev Immunol 3:113–122

Calderón L, Boehm T (2012) Synergistic, context-dependent and hierarchical functions of epithelial components in thymic microenvironments. Cell 149:159–172

Chaudhry MS, Velardi E, Dudakov JA, van den Brink MR (2016) Thymus: the next (re)generation. Immunol Rev 271(1):56–71

Chen BJ, Cui X, Sempowski GD, Chao NJ (2003) Growth hormone accelerates immune recovery following allogeneic T-cell-depleted bone marrow transplantation in mice. Exp Hematol 31:953–958

Chen L, Xiao S, Manley NR (2009) Foxn1 is required to maintain the postnatal thymic microenvironment in a dosage-sensitive manner. Blood 113:567–574

Chinn IK, Blackburn CC, Manley NR, Sempowski GD (2012) Changes in primary lymphoid organs with aging. Semin Immunol 24:309–320

Chu YW et al (2008) Exogenous insulin-like growth factor 1 enhances thymopoiesis predominantly through thymic epithelial cell expansion. Blood 112:2836–2846

Chung B, Montel-Hagen A, Ge S, Blumberg G, Kim K, Klein S et al (2014) Engineering the human thymic microenvironment to support thymopoiesis in vivo. Stem Cells 32(9):2386–2396

Coder BD, Wang H, Ruan L, Su D-M (2015) Thymic involution perturbs negative selection leading to autoreactive T cells that induce chronic inflammation. J Immunol 194:5825–5837

Corbeaux T et al (2010) Thymopoiesis in mice depends on a Foxn1 positive thymic epithelial cell lineage. Proc Natl Acad Sci U S A 107:16613–16618

Cuddihy AR, Ge S, Zhu J, Jang J, Chidgey A, Thurston G et al (2009) VEGF-mediated cross-talk within the neonatal murine thymus. Blood 113(12):2723–2731

Davies EG (2013) Immunodeficiency in DiGeorge syndrome and options for treating cases with complete athymia. Front Immunol 4:322

de Barros SC, Vicente R, Chebli K, Jacquet C, Zimmermann VS, Taylor N (2013) Intrathymic progenitor cell transplantation across histocompatibility barriers results in the persistence of early thymic progenitors and T-cell differentiation. Blood 121(11):2144–2153

De Smedt M, Hoebeke I, Plum J (2004) Human bone marrow CD34+ progenitor cells mature to T cells on OP9-DL1 stromal cell line without thymus microenvironment. Blood Cells Mol Dis 33:227–232

Deobagkar-Lele M, Chacko SK, Victor ES, Kadthur JC, Nandi D (2013) Interferon-γ- and glucocorticoid-mediated pathways synergize to enhance death of CD4(+) CD8(+) thymocytes during *Salmonella enterica* serovar typhimurium infection. Immunology 138:307–321

Dertschnig S, Hauri-Hohl MM, Vollmer M, Holländer GA, Krenger W (2015) Impaired thymic expression of tissue-restricted antigens licenses the de novo generation of autoreactive CD4+ T cells in acute GVHD. Blood 125:2720–2723

Dijkstra C, Sminia T (1990) Normal anatomy, histology, immunohistology, ultrastructure, rat. In: Jones T, Ward J, Mohr U, Hunt R (eds) Hematopoietic system. Monographs on pathology of laboratory animals. Springer, Berlin, pp 249–256

Dion M-L, Poulin J-F, Bordi R et al (2004) HIV infection rapidly induces and maintains a substantial suppression of thymocyte proliferation. Immunity 21:757–768

Dixit VD (2010) Thymic fatness and approaches to enhance thymopoietic fitness in aging. Curr Opin Immunol 22:521–528

Dixit VD, Sridaran R, Edmonsond MA, Taub D, Thompson WE (2003) Gonadotropin-releasing hormone attenuates pregnancy-associated thymic involution and modulates the expression of antiproliferative gene product prohibitin. Endocrinology 144:1496–1505

Dooley J, Liston A (2012) Molecular control over thymic involution: from cytokines and micro-RNA to aging and adipose tissue. Eur J Immunol 42:1073–1079

Douek DC, McFarland RD, Keiser PH et al (1998) Changes in thymic function with age and during the treatment of HIV infection. Nature 396:690–695

Dudakov JA, van den Brink MR (2011) Greater than the sum of their parts: combination strategies for immune regeneration following allogeneic hematopoietic stem cell transplantation. Best Pract Res Clin Haematol 24(3):467–476

Dudakov JA, Goldberg GL, Reiseger JJ, Vlahos K, Chidgey AP, Boyd RL (2009) Sex steroid ablation enhances hematopoietic recovery following cytotoxic antineoplastic therapy in aged mice. J Immunol 183:7084–7094

Dudakov JA et al (2012) Interleukin-22 drives endogenous thymic regeneration in mice. Science 336:91–95

Duheron V, Hess E, Duval M, Decossas M, Castaneda B, Klopper JE, Amoasii L, Barbaroux JB, Williams IR, Yagita H et al (2011) Receptor activator of NFkappaB (RANK) stimulates the proliferation of epithelial cells of the epidermo-pilosebaceous unit. Proc Natl Acad Sci U S A 108:5342–5347

Duijvestijn AM, Hoefsmit ECM (1981) Ultrastructure of the rat thymus: the micro-environment of T-lymphocyte maturation. Cell Tissue Res 218(2):279–292

Elmore S (2006) Enhanced histopathology evaluation of thymus. Toxicol Pathol 34:656–665

Fan Y, Tajima A, Goh SK, Geng X, Gualtierotti G, Grupillo M, Coppola A, Bertera S, Rudert WA, Banerjee I, Bottino R, Trucco M (2015) Bioengineering thymus organoids to restore thymic function and induce donor-specific immune tolerance to allografts. Mol Ther 23(7):1262–1277

Fang RHT, Colantonio AD, Uittenbogaart CH (2008) The role of the thymus in HIV infection: a 10 year perspective. AIDS 22:171–184

Ferrara JLM, Levine JE, Reddy P, Holler E (2009) Graft-versus-host disease. Lancet 373:1550–1561

Fletcher AL et al (2009) Ablation and regeneration of tolerance-inducing medullary thymic epithelial cells after cyclosporine, cyclophosphamide, and dexamethasone treatment. J Immunol 183:823–831

Frank J, Pignata C, Panteleyev AA, Prowse DM, Baden H, Weiner L et al (1999) Exposing the human nude phenotype. Nature 398:473–474

Garfin PM et al (2013) Inactivation of the RB family prevents thymus involution and promotes thymic function by direct control of Foxn1 expression. J Exp Med 210:1087–1097

Gentil Dit Maurin A, Lemercier C, Collin-Faure V, Marche PN, Jouvin-Marche E, Candéias SM (2015) Developmental regulation of p53-dependent radiation-induced thymocyte apoptosis in mice. Clin Exp Immunol 179:30–38

Goldberg GL et al (2009) Luteinizing hormone-releasing hormone enhances T cell recovery following allogeneic bone marrow transplantation. J Immunol 182:5846–5854

Goldberg GL et al (2010) Sex steroid ablation enhances immune reconstitution following cytotoxic antineoplastic therapy in young mice. J Immunol 184:6014–6024

Goronzy JJ, Weyand CM (2013) Understanding immunosenescence to improve responses to vaccines. Nat Immunol 14:428–436

Gottrand G, Taleb K, Ragon I et al (2012) Intrathymic injection of lentiviral vector curtails the immune response in the periphery of normal mice. J Gene Med 14:90–99

Gray DH, Seach N, Ueno T, Milton MK, Liston A, Lew AM, Goodnow CC, Boyd RL (2006) Developmental kinetics, turnover, and stimulatory capacity of thymic epithelial cells. Blood 108:3777–3785

Griffith AV, Fallahi M, Venables T, Petrie HT (2012) Persistent degenerative changes in thymic organ function revealed by an inducible model of organ regrowth. Aging Cell 11:169–177

Griffith AV et al (2015) Metabolic damage and premature thymus aging caused by stromal catalase deficiency. Cell Rep 12:1071–1079

Gruver AL, Sempowski GD (2008) Cytokines, leptin, and stress-induced thymic atrophy. J Leukoc Biol 84:915–923

Gui J, Mustachio LM, Su D-M, Craig RW (2012) Thymus size and age-related thymic involution: early programming, sexual dimorphism, progenitors and stroma. Aging Dis 3:280–290

Haley PJ (2003) Species differences in the structure and function of the immune system. Toxicology 188(1):49–71

Hauri-Hohl MM et al (2007) Donor T-cell alloreactivity against host thymic epithelium limits T-cell development after bone marrow transplantation. Blood 109:4080–4088

Hick RW, Gruver AL, Ventevogel MS, Haynes BF, Sempowski GD (2006) Leptin selectively augments thymopoiesis in leptin deficiency and lipopolysaccharide-induced thymic atrophy. J Immunol 177:169–176

Hince M, Sakkal S, Vlahos K, Dudakov J, Boyd R, Chidgey A (2008) The role of sex steroids and gonadectomy in the control of thymic involution. Cell Immunol 252:122–138

Hollander GA, Krenger W, Blazar BR (2010) Emerging strategies to boost thymic function. Curr Opin Pharmacol 10:443–453

Holmes R, Zúñiga-Pflücker JC (2009) The OP9-DL1 system: generation of T-lymphocytes from embryonic or hematopoietic stem cells in vitro. Cold Spring Harb Protoc. https://doi.org/10.1101/pdb.prot5156

Hozumi K et al (2008) Delta-like 4 is indispensable in thymic environment specific for T cell development. J Exp Med 205:2507–2513

Hsu H-C et al (2003) Age-related thymic involution in C57BL/6J × DBA/2J recombinant-inbred mice maps to mouse chromosomes 9 and 10. Genes Immun 4:402–410

Hu Z, Yang Y-G (2012) Human lymphohematopoietic reconstitution and immune function in immunodeficient mice receiving cotransplantation of human thymic tissue and CD34(þ) cells. Cell Mol Immunol 9:232–236

Parkin J, Cohen B (2001) An overview of the immune system. Lancet 357(9270):1777–1789

Kelly RM, Highfill SL, Panoskaltsis-Mortari A, Taylor PA, Boyd RL, Hollander GA, Blazar BR (2008) Keratinocyte growth factor and androgen blockade work in concert to protect against conditioning regimen-induced thymic epithelial damage and enhance T-cell reconstitution after murine bone marrow transplantation. Blood 111:5734–5744

Khong DM et al (2015) Enhanced hematopoietic stem cell function mediates immune regeneration following sex steroid blockade. Stem Cell Rep 4:445–458

Kirkwood TB (1977) Evolution of ageing. Nature 270:301–304

Knyszynski A, Adler-Kunin S, Globerson A (1992) Effects of growth hormone on thymocyte development from progenitor cells in the bone marrow. Brain Behav Immun 6:327–340

Krenger W, Holländer GA (2008) The immunopathology of thymic GVHD. Semin Immunopathol 30:439–456

Krenger W, Rossi S, Hollander GA (2000) Apoptosis of thymocytes during acute graft-versus-host disease is independent of glucocorticoids. Transplantation 69:2190–2193

La Motte-Mohs RN, Herer E, Zuniga-Pflucker JC (2005) Induction of T-cell development from human cord blood hematopoietic stem cells by Delta-like 1 in vitro. Blood 105:1431–1439

Lai L, Jin J (2009) Generation of thymic epithelial cell progenitors by mouse embryonic stem cells. Stem Cells 27(12):3012–3020

Lai L, Cui C, Jin J, Hao Z, Zheng Q, Ying M et al (2011) Mouse embryonic stem cell-derived thymic epithelial cell progenitors enhance T-cell reconstitution after allogeneic bone marrow transplantation. Blood 118:3410–3418

Lapp WS, Ghayur T, Mendes M, Seddik M, Seemayer TA (1985) The functional and histological basis for graft-versus-host-induced immunosuppression. Immunol Rev 88:107–133

Lynch HE, Goldberg GL, Chidgey A, van den Brink MRM, Boyd R, Sempowski GD (2009) Thymic involution and immune reconstitution. Trends Immunol 30:366–373

Mackall CL (1999) T-cell immunodeficiency following cytotoxic antineoplastic therapy: a review. Oncologist 4:370–378

Mackall CL, Fleisher TA, Brown MR, Andrich MP, Chen CC, Feuerstein IM et al (1995) Age, thymopoiesis, and CD4+ T-lymphocyte regeneration after intensive chemotherapy. N Engl J Med 332:143–149

Mackall CL, Fry TJ, Gress RE (2011) Harnessing the biology of IL-7 for therapeutic application. Nat Rev Immunol 11:330–342

Manley NR, Condie BG (2010) Transcriptional regulation of thymus organogenesis and thymic epithelial cell differentiation. Prog Mol Biol Transl Sci 92:103–120

Markert ML, Devlin BH, Mccarthy EA (2010) Thymus transplantation. Clin Immunol 135:236–246

Mendes-da-Cruz DA, Silva JS, Cotta-de-Almeida V, Savino W (2006) Altered thymocyte migration during experimental acute Trypanosoma cruzi infection: combined role of fibronectin and the chemokines CXCL12 and CCL4. Eur J Immunol 36:1486–1493

Min D et al (2002) Protection from thymic epithelial cell injury by keratinocyte growth factor: a new approach to improve thymic and peripheral T-cell reconstitution after bone marrow transplantation. Blood 99:4592–4600

Mittelstadt PR, Monteiro JP, Ashwell JD (2012) Thymocyte responsiveness to endogenous glucocorticoids is required for immunological fitness. J Clin Invest 122:2384–2394

Montecino-Rodriguez E, Clark R, Dorshkind K (1998) Effects of insulin-like growth factor administration and bone marrow transplantation on thymopoiesis in aged mice. Endocrinology 139:4120–4126

Murphy WJ, Durum SK, Anver MR, Longo DL (1992) Immunologic and hematologic effects of neuroendocrine hormones. Studies on DW/J dwarf mice. J Immunol 148:3799–3805

Nehls M, Pfeifer D, Schorpp M, Hedrich H, Boehm T (1994) New member of the winged-helix protein family disrupted in mouse and rat nude mutations. Nature 372:103–107

Nobrega C et al (2010) Dissemination of mycobacteria to the thymus renders newly generated T cells tolerant to the invading pathogen. J Immunol 184:351–358

Norment AM, Bevan MJ (2000) Role of chemokines in thymocyte development. Semin Immunol 12:445–455

Nowell CS, Bredenkamp N, Tetelin S, Jin X, Tischner C, Vaidya H et al (2011) Foxn1 regulates lineage progression in cortical and medullary thymic epithelial cells but is dispensable for medullary sublineage divergence. PLoS Genet 7(11)

Nunes-Alves C, Nobrega C, Behar SM, Correia-Neves M (2013) Tolerance has its limits: how the thymus copes with infection. Trends Immunol 34:502–510

Nunes-Cabaco H, Caramalho I, Sepulveda N, Sousa AE (2011) Differentiation of human thymic regulatory T cells at the double positive stage. Eur J Immunol 41:3604–3614

Olsen NJ, Watson MB, Henderson GS, Kovacs WJ (1991) Androgen deprivation induces phenotypic and functional changes in the thymus of adult male mice. Endocrinology 129:2471–2476

Osada M et al (2006) The Wnt signaling antagonist Kremen1 is required for development of thymic architecture. Clin Dev Immunol 13:299–319

Pan B et al (2014) Acute ablation of DP thymocytes induces up-regulation of IL-22 and Foxn1 in TECs. Clin Immunol 150:101–108

Patiño JAG, Marino MW, Ivanov VN, Nikolich-Zugich J (2000) Sex steroids induce apoptosis of CD8+CD4+ double-positive thymocytes via TNF-alpha. Eur J Immunol 30:2586–2592

Pearse G (2006) Normal structure, function and histology of the thymus. Toxicol Pathol 34(5):504–514

Perales MA et al (2012) Recombinant human interleukin-7 (CYT107) promotes T-cell recovery after allogeneic stem cell transplantation. Blood 120:4882–4891

Pérez AR et al (2007) Thymus atrophy during Trypanosoma cruzi infection is caused by an immuno-endocrine imbalance. Brain Behav Immun 21:890–900

Pinto S, Schmidt K, Egle S, Stark HJ, Boukamp P, Kyewski B (2013) An organotypic coculture model supporting proliferation and differentiation of medullary thymic epithelial cells and promiscuous gene expression. J Immunol 190:1085–1093

Posselt AM, Barker CF, Tomaszewski JE et al (1990) Induction of donor-specific unresponsiveness by intrathymic islet transplantation. Science 249:1293–1295

Pradhan I, Tajima A, Bertera S, Trucco M, Fan Y (2018) Modulating thymic negative selection with bioengineered thymus organoids. In: Soboloff J, Kappes DJ (eds) Signaling mechanisms regulating T cell diversity and function. CRC Press/Taylor & Francis, Boca Raton. Chapter 4

Prelog M (2006) Aging of the immune system: a risk factor for autoimmunity? Autoimmun Rev 5:136–139

Przybylski GK, Kreuzer K-A, Siegert W, Schmidt CA (2007) No recovery of T-cell receptor excision circles (TRECs) after non-myeloablative allogeneic hematopoietic stem cell transplantation is correlated with the onset of GvHD. J Appl Genet 48:397–404

Rafii S, Butler JM, Ding BS (2016) Angiocrine functions of organ-specific endothelial cells. Nature 529:316–325

Redelman D, Welniak LA, Taub D, Murphy WJ (2008) Neuroendocrine hormones such as growth hormone and prolactin are integral members of the immunological cytokine network. Cell Immunol 252:111–121

Rode I, Martins VC, Küblbeck G, Maltry N, Tessmer C, Rodewald HR (2015) Foxn1 protein expression in the developing, aging, and regenerating thymus. J Immunol 195(12):5678–5687

Rodewald HR (2008) Thymus organogenesis. Annu Rev Immunol 26:355–388

Rodewald HR, Paul S, Haller C, Bluethmann H, Blum C (2001) Thymus medulla consisting of epithelial islets each derived from a single progenitor. Nature 414:763–768

Rossi D, Zlotnik A (2000) The biology of chemokines and their receptors. Annu Rev Immunol 18:217–242

Rossi S et al (2002) Keratinocyte growth factor preserves normal thymopoiesis and thymic microenvironment during experimental graft-versus-host disease. Blood 100:682–691

Rossi SW, Jenkinson WE, Anderson G, Jenkinson EJ (2006) Clonal analysis reveals a common progenitor for thymic cortical and medullary epithelium. Nature 441:988–991

Rozmyslowicz T, Murphy SL, Conover DO, Gaulton GN (2010) HIV-1 infection inhibits cytokine production in human thymic macrophages. Exp Hematol 38:1157–1166

Rudd BD et al (2011) Nonrandom attrition of the naive CD8+ T cell pool with aging governed by T cell receptor: pMHC interactions. Proc Natl Acad Sci U S A 108:13694–13699

Sachs DH, Sykes M, Kawai T, Cosimi AB (2011) Immuno-intervention for the induction of transplantation tolerance through mixed chimerism. Semin Immunol 23:165–173

Safieddine N, Keshavjee S (2011) Anatomy of the thymus gland. Thorac Surg Clin 21(2):191–5,viii

Saunders DJ, Georgiou HM, Wu L, Shortman K (1995) Induction of limited growth and differentiation of early thymic precursor cells by thymic epithelial cell lines. Immunol Lett 47(1–2):45–51

Savino W (2006) The thymus is a common target organ in infectious diseases. PLoS Pathog 2:e62

Shanley DP, Aw D, Manley NR, Palmer DB (2009) An evolutionary perspective on the mechanisms of immunosenescence. Trends Immunol 30(7):374–381

Stanley SK et al (1993) Human immunodeficiency virus infection of the human thymus and disruption of the thymic microenvironment in the SCID-hu mouse. J Exp Med 178:1151–1163

Steinmann GG, Klaus B, Müller-Hermelink HK (1985) The involution of the ageing human thymic epithelium is independent of puberty. A morphometric study. Scand J Immunol 22:563–575

Storek J, Witherspoon RP, Storb R (1995) T cell reconstitution after bone marrow transplantation into adult patients does not resemble T cell development in early life. Bone Marrow Transplant 16:413–425

Su M et al (2015) Efficient in vitro generation of functional thymic epithelial progenitors from human embryonic stem cells. Sci Rep 5:9882

Sun X, Xu J, Lu H, Liu W, Miao Z, Sui X, Liu H, Su L, Du W, He Q, Chen F, Shi Y, Deng H (2013) Directed differentiation of human embryonic stem cells into thymic epithelial progenitor-like cells reconstitutes the thymic microenvironment in vivo. Cell Stem Cell 13(2):230–236

Sutherland JS et al (2005) Activation of thymic regeneration in mice and humans following androgen blockade. J Immunol 175:2741–2753

Sutherland JS et al (2008) Enhanced immune system regeneration in humans following allogeneic or autologous hemopoietic stem cell transplantation by temporary sex steroid blockade. Clin Cancer Res 14:1138–1149

Tajima A, Liu W, Pradhan I et al (2015) Bioengineering mini functional thymic units with EAK16-II/EAKIIH6 self-assembling hydrogel. Clin Immunol 160(1):82–89

Tajima A, Pradhan I, Trucco M, Fan Y (2016) Restoration of thymus function with bioengineered thymus organoids. Curr Stem Cell Rep 2(2):128–139

Takahama Y (2006) Journey through the thymus: stromal guides for T-cell development and selection. Nat Rev Immunol 6:127–135

Tizard I (2004) Veterinary immunology. An introduction, 7th edn. WB Saunders, Philadelphia

Tomimori Y, Mori K, Koide M, Nakamichi Y, Ninomiya T, Udagawa N, Yasuda H (2009) Evaluation of pharmaceuticals with a novel 50-hour animal model of bone loss. J Bone Miner Res 24:1194–1205

Tsai PT, Lee RA, Wu H (2003) BMP4 acts upstream of FGF in modulating thymic stroma and regulating thymopoiesis. Blood 102:3947–3953

Tsuji Y, Kinoshita Y, Hato F, Tominaga K, Yoshida K (1994) The in vitro proliferation of thymus epithelial cells stimulated with growth hormone and insulin-like growth factor-I. Cell Mol Biol (Noisy-le-Grand) 40:1135–1142

Ueno T et al (2002) Role for CCR7 ligands in the emigration of newly generated T lymphocytes from the neonatal thymus. Immunity 16:205–218

Vacchio MS, Papadopoulos V, Ashwell JD (1994) Steroid production in the thymus: implications for thymocyte selection. J Exp Med 179:1835–1846

Vallejo AN, Michel JJ, Bale LK, Lemster BH, Borghesi L, Conover CA (2009) Resistance to age-dependent thymic atrophy in long-lived mice that are deficient in pregnancy-associated plasma protein A. Proc Natl Acad Sci U S A 106:11252–11257

Van Ewijk (1984) Immunohistology of lymphoid and non-lymphoid cells in the thymus in relation to T lymphocyte differentiation. American Journal of Anatomy 170(3):311–330

van Ewijk W, Brekelmans PJ, Jacobs R, Wisse E (1988) Lymphoid microenvironments in the thymus and lymph node. Scanning Microsc 2:2129–2140

van Lent AU et al (2009) IL-7 enhances thymic human T cell development in "human immune system" Rag2-/-IL-2Rgammac-/- mice without affecting peripheral T cell homeostasis. J Immunol 183:7645–7655

Velardi E et al (2014) Sex steroid blockade enhances thymopoiesis by modulating Notch signaling. J Exp Med 211:2341–2349

Ventevogel MS, Sempowski GD (2013) Thymic rejuvenation and aging. Curr Opin Immunol 25 (4):516–522

von Gaudecker B (1991) Functional histology of the human thymus. Anat Embryol 183:1–15. https://doi.org/10.1007/BF00185830

Williams KM et al (2008) CCL25 increases thymopoiesis after androgen withdrawal. Blood 112:3255–3263

Williams KM, Mella H, Lucas PJ, Williams JA, Telford W, Gress RE (2009) Single cell analysis of complex thymus stromal cell populations: rapid thymic epithelia preparation characterizes radiation injury. Clin Transl Sci 2:279–285

Wils EJ, van der Holt B, Broers AE, Posthumus-van Sluijs SJ, Gratama JW, Braakman E, Cornelissen JJ (2011) Insufficient recovery of thymopoiesis predicts for opportunistic infections in allogeneic hematopoietic stem cell transplant recipients. Haematologica 96:1846–1854

Wils EJ et al (2012) Keratinocyte growth factor and stem cell factor to improve thymopoiesis after autologous CD34+ cell transplantation in rhesus macaques. Biol Blood Marrow Transplant 18:55–65

Wong K, Liser NL, Barsanti M, Lim JM, Hammet MV, Khong DM, Siatskas C, Gray DH, Boyd RL, Chidgey AP (2014) Multilineage potential and self-renewal define an epithelial progenitor cell population in the adult thymus. Cell Rep 8:1198–1209

Wu T, Young JS, Johnston H, Ni X, Deng R, Racine J, Wang M, Wang A, Todorov I, Wang J, Zeng D (2013) Thymic damage, impaired negative selection, and development of chronic graft-versus-host disease caused by donor CD4+ and CD8+ T cells. J Immunol 191(1):488–499

Yamada K, Scalea J (2012) Thymic transplantation in pig-to-nonhuman primates for the induction of tolerance across xenogeneic barriers. Methods Mol Biol 885:191–212

Yang H, Youm YH, Dixit VD (2009) Inhibition of thymic adipogenesis by caloric restriction is coupled with reduction in age-related thymic involution. J Immunol 183:3040–3052

Youm YH, Yang H, Sun Y, Smith RG, Manley NR, Vandanmagsar B, Dixit VD (2009) Deficient ghrelin receptor-mediated signaling compromises thymic stromal cell microenvironment by accelerating thymic adiposity. J Biol Chem 284:7068–7077

Zakrzewski JL, Kochman AA, Lu SX et al (2006) Adoptive transfer of T-cell precursors enhances T-cell reconstitution after allogeneic hematopoietic stem cell transplantation. Nat Med 12(9):1039–1047

Zhang F et al (2004) Adenovirus E4 gene promotes selective endothelial cell survival and angiogenesis via activation of the vascular endothelial-cadherin/Akt signaling pathway. J Biol Chem 279:11760–11766

Zhang SL et al (2014) Chemokine treatment rescues profound T-lineage progenitor homing defect after bone marrow transplant conditioning in mice. Blood 124:296–304

Zhou J, Wang X, Luo G et al (2012) Partial tolerance induced by transplantation of spatially separated thymuses: a cue for T cell retolerization in thymus grafts. Scand J Immunol 75:401–408

Part III

Excretory and Respiratory Systems

Tissue-Engineered Renal Tissue

Diana Lim, Anthony Atala, and James J. Yoo

Contents

Abstract

It is estimated that 5–10 million people die annually from kidney disease. Astounding as these numbers are, it is predicted that these numbers will only grow especially in countries that lack the infrastructure to support its treatments. In spite of this increasing prevalence, current treatments for renal failure are still limited to dialysis and donor organ transplantation. Although these treatments are indeed lifesaving, dialysis does not fully restore kidney function, and a worldwide shortage of transplantable kidneys poses a great problem. To address these issues, the field of regenerative medicine has aimed for many years to develop treatments in the areas of tissue engineering, cell therapies, and gene therapies. In this chapter, we will discuss developments in one such strategy of engineering a cell-based renal tissue construct capable of restoring all kidney function with a focus on possible cell sources, materials, and vascularization strategies.

D. Lim · A. Atala · J. J. Yoo (✉)
Wake Forest Institute for Regenerative Medicine, Wake Forest School of Medicine, Medical Center Boulevard, Winston-Salem, NC, USA
e-mail: dlim@wakehealth.edu; aatala@wakehealth.edu; jyoo@wakehealth.edu

D. Eberli et al. (eds.), *Organ Tissue Engineering*, Reference Series in Biomedical Engineering, https://doi.org/10.1007/978-3-030-44211-8_12

1 Introduction

It has been 74 years since the first hemodialysis machine was made using the humble beginnings of a few sausage casings and an automobile water pump (Morrissey 2012). This novel effort to artificially replace native tissue function brought about a profound shift in the paradigm of clinical care for patients with renal disease and also ignited the development of other organ replacement therapies (Morrissey 2012). Hemodialysis now keeps over two million patients alive worldwide and, after all of these years, remains the only viable long-term extracorporeal organ replacement therapy (United States Renal Data System 2017). However, dialysis is only a maintenance therapy and not a curative treatment for renal failure as it provides clearance of uremic toxins but does not provide the vital endocrine, metabolic, and homeostatic functions of the native kidney. The lack of key endocrine functions such as the production of erythropoietin, a hormone necessary for the formation of red blood cells, puts dialysis patients at a high risk of developing severe anemia (Chen et al. 2019; Huml and Sehgal 2019; Johansen et al. 2005). The fatigue of severe anemia in conjunction with multiple required dialysis treatments per week prevents many patients from continuing in the workforce to maintain their livelihood (Wang et al. 2016). Providing a treatment that can recover all lost kidney functions would improve patients' lives and lift a significant societal, economic burden (Wang et al. 2016).

Currently, one such curative treatment exists – donor organ transplantation. However, with increasing numbers of end-stage renal disease (ESRD) worldwide, many patients die while waiting for a donor kidney (United States Renal Data System 2017). It is estimated that less than 10% of the global need for kidney transplantation is met (United States Renal Data System 2017). Furthermore, even if a patient were able to receive a transplant, they would necessarily be put at significant risk for complications from life-long immunosuppression (Sen et al. 2019). In light of these facts, it is becoming increasingly apparent that new therapies are necessary. The pressing need for new treatments for ESRD is further highlighted when evaluating demographic shifts in the global ESRD patient population (Harambat and Ekulu 2016; Kumar and Jha 2016; Xie et al. 2018). By 2030, over 70% of patients with ESRD are expected to be inhabitants of developing countries that do not have the infrastructure to support dialysis, organ transplantation, or post-procedure care (Luyckx et al. 2018). If the current healthcare discrepancies for renal failure treatment are not bridged, or new therapies are not found, it is evident that a fast-approaching global healthcare crisis will be unavoidable.

This chapter will highlight some of the advancements in the study of cell-based renal tissue engineering – currently, one of the three main approaches (tissue engineering, cell therapy, and gene therapy) to providing alternative treatments for renal failure.

2 Renal Tissue Engineering: Foundational Knowledge

The field of renal tissue engineering provides many promising strategies toward alternative treatments that aim to disrupt the clinical climate of care for patients with renal failure. Tissue engineering has been defined as creating new tissue for the therapeutic reconstruction of the human body, by the deliberate and controlled stimulation of selected target cells through a systematic combination of molecular and mechanical signals (Williams 2006). One of the basic strategies for cell-based tissue engineering is to support the in vitro development of 3D renal structures from a cell source by expanding these cells and seeding them on a construct that consists of an extracellular matrix, soluble growth factors, and scaffolding for subsequent implantation (Fig. 1) (Guimaraes-Souza et al. 2012; Chung et al. 2015; Chang and Davies 2012). This strategy is one of the most promising for treating renal failure because organs with this degree of damage may lack adequate numbers of viable cells to recover injury and, due to high fibrosis, lack a niche that facilitates regeneration. A summary of the cell-based tissue engineering approaches discussed in the chapter are listed in Table 1.

The major challenges of renal tissue engineering are the complexities of the structures that need to be recapitulated. It is very difficult to artificially induce the growth, differentiation, and organization of the over 26 types of cells from different embryological lineages that make up this organ and its functional unit, the nephron (Sagrinati et al. 2008). The nephron is mainly responsible for its filtration function that excretes waste products and water as urine and controls blood pH, electrolyte concentrations, and blood volume. It is within this filtration function that the importance of the precise placement and organization of the cells of the nephron becomes apparent. Blood is filtered in the glomerulus through a specialized three-layer filtration system consisting of fenestrated capillary endothelium, basement membrane, and podocyte processes that create a size and charge barrier that produces a plasma ultrafiltrate devoid of cells and proteins. This ultrafiltrate passes through a series of regions along the nephron tube, each with unique membrane transporters, channels, and mechanical features that provide unique filtration

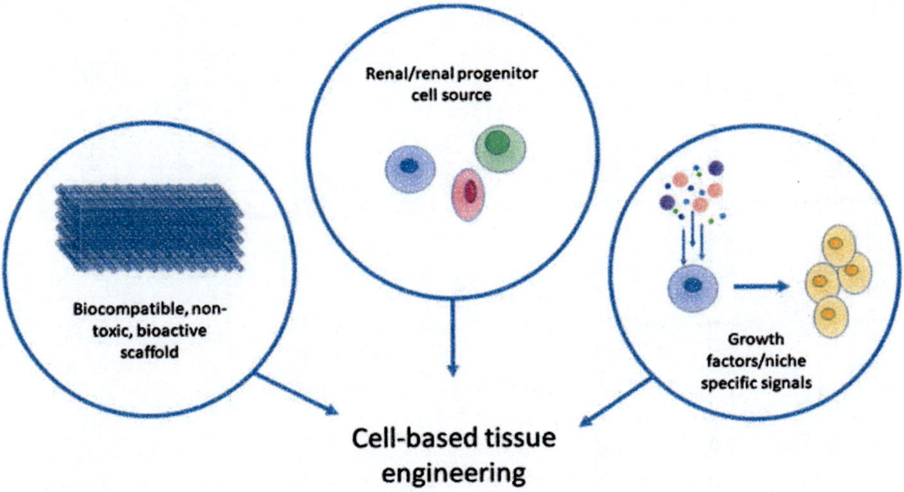

Fig. 1 Schematic of cell-based tissue engineering strategies. Biocompatible, sometimes bioactive, scaffolds are seeded with the primary renal, renal progenitor, stem cells, and exposed to regenerative niche factors to create renal tissue structures

Table 1 Cell-based tissue engineering approaches producing renal structures

Starting cell type	Scaffold material	Strategy	End product
Rabbit renal cells (Atala et al. 1995)	Polyglycolic acid	In vitro culture followed by scaffold seeding and subcutaneous implantation into athymic mice	Formation and organization of nephron like structures
Murine renal cells (Fung et al. 1996)	Polyglycolic acid	In vitro culture, single cell suspension, scaffold seeding, and subcutaneous implantation into immune-competent mice (Fung et al. 1996)	Successful reconstitution of different nephron segments including proximal tubules, distal tubules, loops of Henle, and collecting ducts (Fung et al. 1996)
Murine renal cells (Yoo et al. 1996)	Polycarbonate	In vitro culture, scaffold seeding onto tubular scaffold connected at both ends to silastic catheters connected to a reservoir and implanted subcutaneously into athymic mice (Yoo et al. 1996)	Organization into structures positive for proximal tubule, loop of Henle, distal tubule, and glomerular markers. Produced yellow reservoir fluid with significant urea and creatinine content (Yoo et al. 1996)
Renal metanephric cell clones (Lanza et al. 2002)	Polyglycolic acid	Nucleus of adult bovine fibroblasts transplanted into enucleated bovine oocytes. Cells isolated from metanephros and expanded in vitro and seeded onto cylindrical polycarbonate scaffolds coated in collagen and connected to silastic catheters connected to a reservoir. Implanted subcutaneously to develop, than implanted orthotopically (Lanza et al. 2002)	Highly vascularized glomerular and tubule like structures that functionally produced more urine than allogeneic and unseeded constructs (Lanza et al. 2002)
Fetal rat kidney cells (Kim et al. 2007)	Polyglycolic acid scaffold and fibrin gel matrix	Cells isolated from different gestational stages seeded onto a scaffold for transplantation into the omentum (Kim et al. 2007)	Early-stage cells resulted in formation of renal and nonrenal tissues whereas later stages developed nephron structures only (Kim et al. 2007)
Glomerular epithelial and mesangial cells (Wang and Takezawa 2005)	Collagen/vitrigel	Coculture of glomerular epithelial and mesangial cells on both surfaces of collagen vitrigel in vitro (Wang and Takezawa 2005)	Creation of 3D glomerular organoid with polarity of epithelial layer

Mixed neonatal rat renal cells (Lü et al. 2012)	Collagen/matrigel	Cells seeded in collagen I supplemented with Matrigel in casting mold to exert static stretch when renal constructs contracted (Lü et al. 2012). In vitro culture	Reconstitution of renal tubular and glomeruli-like structures with different appearances at varying developmental stages (Lü et al. 2012)
Fetal kidney tissue (Rosines et al. 2007)	Hyaluronic acid	Metanephric kidneys and ureteric buds were isolated from 13-day rat embryos and cultured in transwell filters followed by treatment with hyaluronic acid (Rosines et al. 2007)	HA found to simultaneously modulate ureteric bud branching morphogenesis, induce MET, and promote differentiation of both metanephric mesenchyme and ureteric bud (Rosines et al. 2007)
Pluripotent murine embryonic stem cells (Ross et al. 2009)	Decellularized rat kidney scaffold	Cells were seeded into scaffolds antegrade through the artery and retrograde through the ureter and cultured in vitro (Ross et al. 2009)	Glomerular, vascular, and tubular structures were populated and cells lost their embryonic appearance and began differentiating, demonstrating the extracellular matrix of decellularized scaffolds can direct regeneration of kidneys (Ross et al. 2009)
Fetal kidney explants (Nakayama et al. 2010)	Decellularized rhesus monkey kidney scaffold	Fetal kidney explants were layered onto fetal, infant, juvenile, and adult age-matched decellularized scaffolds and cultured in vitro (Nakayama et al. 2010)	Scaffold was able to support attachment and migration of nephron progenitor populations to recellularize the scaffold (Nakayama et al. 2010)
Epithelial and endothelial cells (Song et al. 2013)	Decellularized rat kidneys	Scaffolds seeded with cells, then perfused in whole organ bioreactor. The grafts were also transplanted orthotopically in rats (Song et al. 2013)	Grafts produce rudimentary urine in vitro when vascular bed reperfused (Song et al. 2013). Orthotopic transplant led to urine production via ureteral conduit in vivo (Song et al. 2013)

functions. For example, the proximal tubule resorbs glucose through an elegant system utilizing a pre-established sodium gradient to co-transport glucose into cells. Then, using another glucose transporter, moves glucose from the cells back into circulation. The sodium gradient is also used to resorb all amino acids and 70% of the water in this portion. The filtrate continues through the loop of Henle that has a highly water permeable descending portion and a solute permeable ascending portion that creates a countercurrent multiplication system responsible for concentrating the filtrate. The concentrated filtrate then passes through the thick ascending loop with tight junctions that prevent the resorption of water, so only solutes are modified there, followed by the distal convoluted tubule that resorbs salt, and finally a collecting duct responsible for the secretion of acid before it moves into the renal calyx and to the bladder for excretion. In addition to these filtration functions, nephron regions also hold specific cells responsible for endocrine and metabolic functions such as the production of erythropoietin (EPO) discussed earlier, the production of 1-alpha-hydroxylase for vitamin D metabolism and activation, and the production of renin that participates in the renin-angiotensin-aldosterone axis to increase blood pressure in response to renal hypotension (Donnelly 2003; Banerjee and Jha 2019; Obert and Frazier 2019).

With the current resolution of bioengineering techniques, the complex and specific placement of cells that provide nephron functions is currently impossible to recreate artificially. To work around this barrier, researchers have focused on harnessing the pre-programmed ability of various renally differentiated cells to self-organize into the nephron structure within a bioengineered construct.

3 Potential Cell Sources for Scaffold Seeding

Early studies of the cell sources capable of forming renal structures within a scaffold for tissue engineering naturally focused on primary renal cells. One such successful study by Atala et al. expanded in vitro a heterogeneous population of donor rabbit kidney cells that included proximal tubule, distal tubule, and glomerular cells and seeded them onto biodegradable scaffolds for subcutaneous implantation into athymic mice (Atala et al. 1995). Histological evaluation of these scaffolds showed the progressive formation and organization of nephron like structures (Atala et al. 1995). However, it was unclear whether these structures were forming de novo from the cells themselves or undigested pre-formed nephron structures, leading to the following study that created a single-cell suspension of murine kidney cells for scaffold seeding and implantation into immune-competent hosts (Atala et al. 1995; Fung et al. 1996). This study showed that renal tubule-like structures were able to form from the isolated cells themselves, with histological studies indicating the successful reconstitution of different nephron segments including the proximal tubules, distal tubules, loops of Henle, and collecting ducts (Fung et al. 1996). Branching from these studies, Yoo et al. evaluated whether the structures that resulted from these methods were functional in vivo (Yoo et al. 1996). Murine renal cells were seeded onto a tubular scaffold connected to a reservoir via a catheter and implanted this structure subcutaneously into athymic mice (Yoo et al. 1996). Histological analysis showed that the seeded cells self-organized into proximal tubule marker, loop of Henle marker, and distal tubule marker positive structures connected to a glomerular-like structure (Yoo et al. 1996). Evaluation of the reservoir fluid revealed a urine-like fluid with significant uric acid and creatinine concentrations – indicating that renal structures formed from isolated and expanded primary renal cells seeded onto scaffolds are capable of filtering and excreting a urine-like fluid (Yoo et al. 1996).

More recent studies by George et al. have established cell culture methods that enable the expansion of primary renal cells from normal and damaged primary renal human tissues (Guimaraes-Souza et al. 2012). In this study, viable cells were isolated from normal and damaged kidneys and expanded in vitro (Guimaraes-Souza et al. 2012). Primary renal cells isolated from both normal and diseased kidneys showed similar phenotypic characteristics and growth kinetics (Guimaraes-Souza et al. 2012). Characterization of cells cultured from damaged kidneys showed a composition of roughly 80% proximal tubular cells, 11% distal tubular cells, and 2% podocytes – comparable to normal kidney cell isolation (Guimaraes-Souza et al. 2012). These studies demonstrate the feasibility of using primary renal cells isolated from patients' damaged kidneys in future studies to form autologous engineered kidney constructs (Guimaraes-Souza et al. 2012). The main benefit of creating autologous constructs is that immunogenicity is diminished as the patients' own cells with homologous HLA surface markers would be used – eliminating the complications of and need for life-long immunosuppression (George et al. 2016).

Lanza et al. took a different approach to provide an autologous cell source for tissue engineering (Lanza et al. 2002). This group seeded cloned renal metanephric cells onto cylindrical polycarbonate membranes coated in collagen and connected them to catheters ending in a reservoir for subcutaneous implantation – similar to studies done by Yoo et al. using primary cells discussed earlier (Yoo et al. 1996; Lanza et al. 2002). The histocompatible clonal cells were created by taking the nucleus of adult bovine fibroblasts and transplanting them into enucleated bovine ooctyes (Lanza et al. 2002). Renal cells were then isolated from 56-day-old cloned metanephros and expanded in vitro (Lanza et al. 2002). A total of 31 constructs (19 clones cells, 6 allogenic cells, and 6 no cells) were left to develop subcutaneously for 12 weeks and subsequently implanted into the same steer that the cells were derived from (Lanza et al. 2002). When constructs were harvested for analysis, they showed high vascularization and assembly into glomerular and tubule-like structures (Lanza et al. 2002). The cloned cell constructs also functionally produced more urine than allogenic and unseeded constructs (Lanza et al. 2002). In hypersensitivity tests, the cloned constructs showed no rejection response (Lanza et al. 2002). It is important to note that this study used methods that generated an early-stage bovine embryo to isolate renal cells from; therefore, this strategy cannot be applied to humans for ethical reasons (Lanza et al. 2002). However, this study demonstrates the feasibility of using different autologous stem cell sources, specifically differentiated into metanephric cells to seed scaffolds to create functional renal structures in future studies.

Previous studies have also established methods of isolating primary purified functional cell types from the kidney that may be used in future tissue engineering applications. One such study by Van der Hauwaert et al. purified proximal tubular epithelial cells from human renal tissue through methods of mechanical and enzymatic dissociation followed by flow cytometry to select CD10 and CD13 double-positive cells (Van der Hauwaert et al. 2013). This population of CD10+CD13+ cells dependably proliferated in culture and maintained proximal tubular cell markers over the long term (Van der Hauwaert et al. 2013). Another study on specific functional renal cell isolation by Maxwell et al. identified the restricted location of erythropoietin producing cells in the interstitium of the cortex and outer medulla of the kidney using transgenic mice with marker tagged erythropoietin gene loci (Maxwell et al. 1993). These studies found that the cells in these regions producing EPO were fibroblast-like type 1 interstitial cells (Maxwell et al. 1993). Based on these findings, Yamaleyeva et al. isolated fibroblast-like type 1 interstitial cells from cadaveric kidneys to enrich human primary renal cell cultures with EPO+ cells to evaluate enrichment effects on rat chronic kidney disease models (Yamaleyeva et al. 2012). The intrarenal delivery of these EPO+ cell-enriched primary renal cells resulted in better functional recovery and provided renoprotective, anti-fibrotic, and anti-inflammatory effects compared to controls (Yamaleyeva et al. 2012). These studies demonstrated the functional use of these isolated primary cell populations and potential benefits when presented together with other cell groups for renal tissue engineering applications.

The use of primary renal cells discussed above is the most well-established for future use in engineering renal constructs. However, other cell sources have also shown success in producing renal structures and present some advantages to using primary renal cells. For example, the use of stem cells as a source of renal cells may be more beneficial for the commercialization of tissue-engineered constructs because they have the potential to be cultured for longer periods of time and yield larger amounts of renal cells for successful seeding of scaffolds.

A potential autologous stem cell source is the adult renal progenitor cell population. The identification of this cell type as a stem cell is still a controversial topic. Though the ability of the kidney to regenerate during times of acute injury has been documented for many years – indicating the presence of a regenerative cell source – there remains much debate about whether the source of tissue recovery is a self-renewing stem cell population or an adult cell induced to exhibit stem-like behaviors during times of injury. In either case, the regenerative cells have specific surface markers that can be utilized to isolate these cell types for tissue engineering purposes. Different populations of renal progenitor cells have been identified throughout the nephron and papilla of the kidney (Li and Wingert 2013). One population is a subset of CD133+CD24+ parietal epithelial cells in the Bowman's capsule that has the potential to differentiate into podocytes and tubular epithelium, as well as adipocytes and osteocytes (Bussolati et al. 2005; Sagrinati et al. 2006). Following studies by Ronconi et al. found that within this population, CD133+CD24+PDX- cells were able to differentiate into podocytes and tubules, whereas CD133+CD24 +PDX+ cells were only able to differentiate into podocytes with the former obviously more successful in providing functional recovery of reduced proteinuria and glomerular damage after injection in Adriamycin-induced nephropathy mouse models (Ronconi et al. 2009). Another population of cells in the renal papilla was identified as possibly stem-like in studies by Oliver et al. tracking low-cycling cells in the adult kidney using nucleotide bromodeoxyuridine (BrdU) labeling (Oliver et al. 2004). Ward et al. and Patschan et al. further reported that CD133/1+nestin+ cells isolated from the adult human papilla and cortex were capable of undergoing tubulogenesis, and in metanephric organ culture, CD133/1+nestin+ cells integrated into mouse-derived tubules in in vitro cell culture (Ward et al. 2011; Patschan et al. 2007). Other populations of renal progenitor cells have been found to be localized in the proximal tubule and distal convoluted tubule with a CD133+CD24+CD106- expression profile. Injection of these cells into SCDI mice with acute kidney injury showed engraftment of these cells, formation of new tubule cells, and improved renal function (Angelotti et al. 2012).

Another approach to finding a renal cell source has been to use embryonic stem cell (ESC) populations that will be able to produce the different renal cell types needed for kidney formation. Though the use of ESCs is not feasible for human studies and clinical applications, the findings of these studies can inform the direction of future tissue engineering studies using other progenitor cell populations and are therefore discussed below. The identification of which progenitor cell populations will be able to reconstitute functional renal cells is heavily dependent on an understanding of kidney embryology, which we will briefly discuss below.

Viewed simplistically, the primitive streak that engenders all body structures at the beginning of embryological development forms a polarity. The primitive posterior streak develops into the intermediate mesoderm (IM) that further differentiates into an anterior and posterior IM (Mugford et al. 2008). The anterior IM develops into the nephric duct that yields the ureteric bud, and the posterior IM develops into the metanephric mesenchyme that yields the cap mesenchyme (Basta et al. 2014). The cap mesenchyme is a self-renewing renal stem cell population that condenses to undergo a mesenchymal-to-epithelial transition after receiving a series of signals from the ureteric bud to form an epithelialized renal vesicle that undergoes proliferation, morphogenesis, and maturation to eventually become the intricately segmented nephron structure that contains discrete glomerular, proximal tubule, and distal tubule regions (Rothenpieler and Dressler 1993; Little and McMahon 2012; Herzlinger et al. 1992). Simultaneously, the ureteric bud receives signals from the condensing renal vesicles to undergo branching to eventually develop the collecting duct system that the developing nephron connects to (Little and McMahon 2012; Kurtzeborn et al. 2018). Vascularization of the developing kidney occurs as endothelial precursors differentiated from initial kidney anlagen and ingrowth from external sources migrate to a groove called the vascular cleft on the lower aspect of the renal vesicle, most likely driven by VEGF-VEGFR signaling, to begin forming the glomerular capillary tuft (Abrahamson 2009) (Figs. 2 and 3).

Additionally, clonal analysis tracing renal progenitors from E13.5 to P1 during kidney development showed tubules had polyclonal origin – indicating that renal vesicles in different stages of nephron branching contain different populations of

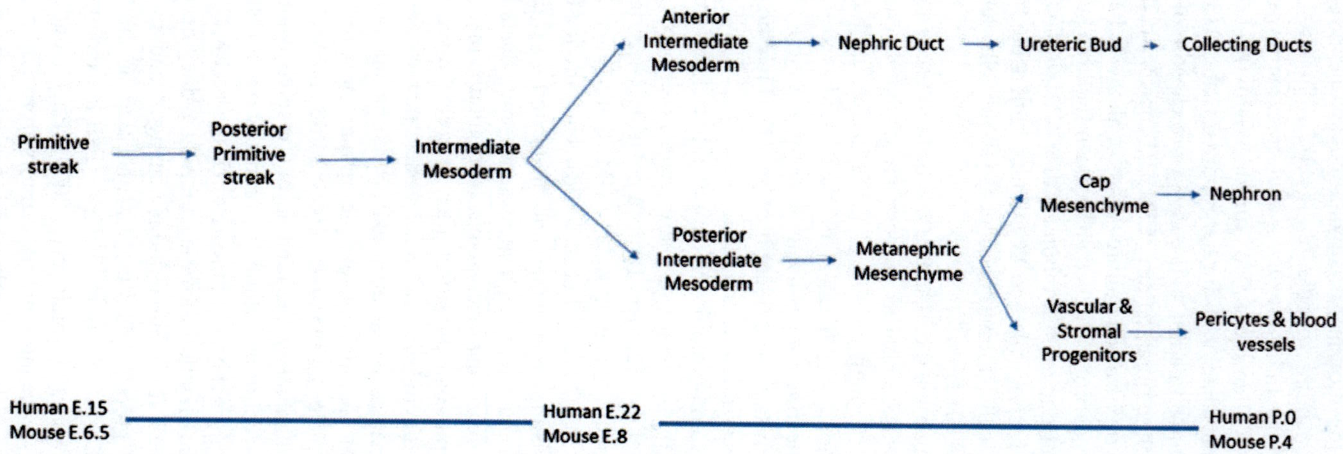

Fig. 2 Developmental stages of kidney structures. Primitive streak cells develop an anterior-posterior polarity. The posterior primitive streak develops into intermediate mesoderm. The intermediate mesoderm itself then develops an anterior-posterior polarity. The anterior intermediate mesoderm develops into the Wolffian/mesonephric duct – subsequently forming a ureteric bud that begins to branch after receiving signals from the cap mesenchyme to become the collecting duct system. The posterior intermediate mesoderm develops into the metanephric mesenchyme that develops into the nephron progenitors of the cap mesenchyme. The metanephric mesenchyme also forms the vascular and stromal progenitors of the kidney

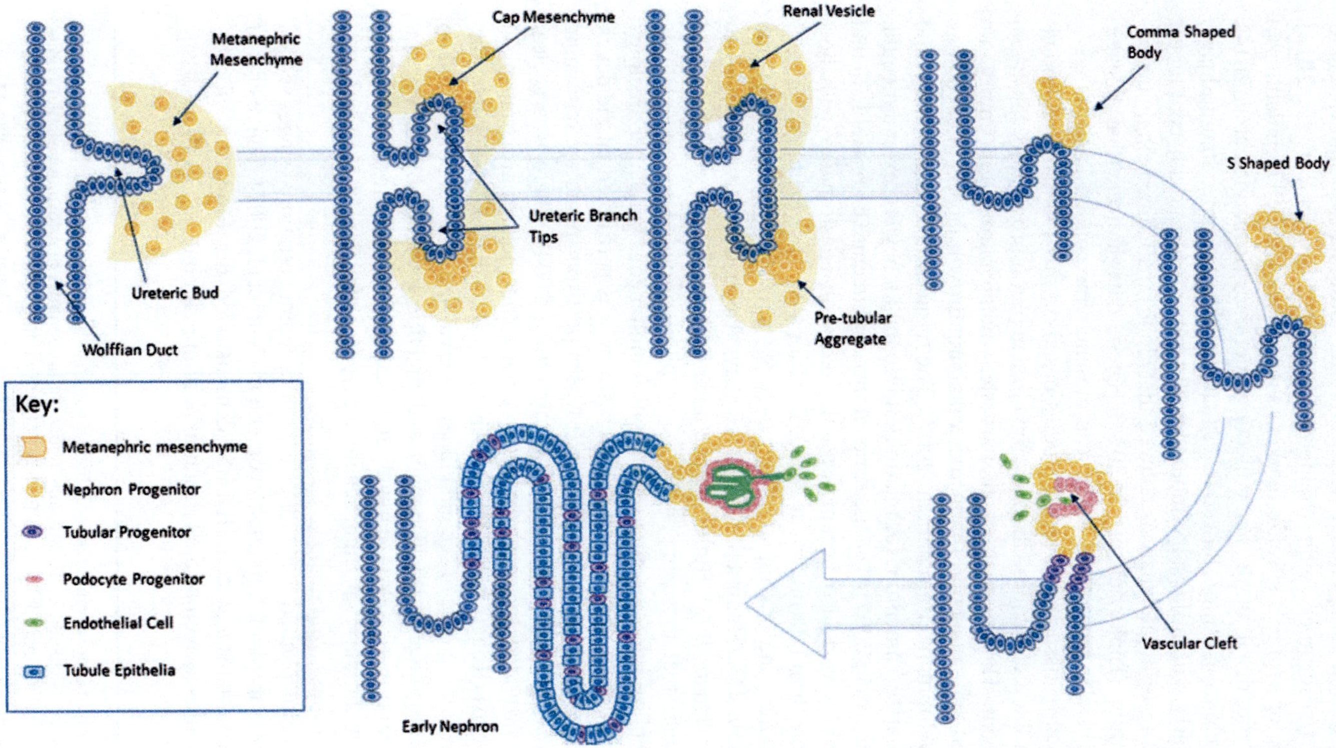

Fig. 3 Nephron formation. The mature human kidney begins to form from the metanephros after a series of rudimentary kidney-like structures called the pronephros and mesonephros. In the development of the native kidney, the Wolffian/mesonephric duct forms an outgrowth called the ureteric bud. This outpouching extends into the metanephric mesenchyme, and the two cell populations, the ureteric bud and the metanephric mesenchyme, exchange signals that promote the branching of the ureteric bud and the condensation of metanephric mesenchyme cells around the ureteric bud tips to become the cap mesenchyme. With further growth signals, the cap mesenchyme cells form a pre-tubular aggregate and subsequently form a renal vesicle. The renal vesicle contains the

progenitor cells (Rinkevich et al. 2014). To tag different clonal populations to genetically trace individual epithelial cells within the adult kidney, Rinkevich et al. crossed Actin CrER mice with "rainbow" mice (Rinkevich et al. 2014). Immunostaining for segment-specific markers showed specific cell populations of the proximal tubule, distal tubule, and collecting duct fates (Rinkevich et al. 2014). Clonal expansion during injury showed that specific clones expanded longitudinally and perpendicularly within the same tubule but did not extend into adjacent tubule segments, and long-term fate analysis showed that clones maintained single epithelial lineage identity (Rinkevich et al. 2014). These data indicate that lineage-restricted renal vesicles are created during cap mesenchyme condensation.

Some of the early tissue engineering studies based on this foundational knowledge focused on the metanephros that develops into nephron progenitors for in situ organ development. Hammerman et al. harvested metanephroi from embryonic tissue then placed the metanephroi in media containing growth factors to encourage further renal development (Hammerman 2007). The tissue was then implanted into the highly vascularized omentum for maturation into the kidney cortex and medulla structures (Hammerman 2007). The tissue developed architecture comparable to that of the native kidney and also became vascularized with anastomoses with host arteries in the omentum (Hammerman 2007). Functionally, these developed tissues were able to ultrafilter inulin with a GFR between 6% and 11%, produce EPO, and metabolize vitamin D3 (Hammerman 2007). They also had a survival of up to 32 weeks (Hammerman 2007). Building upon these findings, Rogers et al. evaluated the survival benefit that these implanted metanephroi provided by transplanting allogenic metanephroi into omentum of rats with simultaneous single nephrectomy (Rogers and Hammerman 2004). Ureteroureterostomy was performed on the developing kidney 3 weeks post-implant, and at 20 weeks post-transplant, the remaining native kidney was removed (Rogers and Hammerman 2004). There was a clear survival benefit provided by rats that had received the metanephroi transplant compared to controls (Rogers and Hammerman 2004). Rats that had their ureteroureterostomy severed at the time of second kidney removal had survival comparable to controls – indicating clear survival benefit and function provided by the implant (Rogers and Hammerman 2004). Marshall et al. expanded upon these studies and demonstrated that the survival time benefit was proportional to the transplanted metanephroi renal mass (Marshall et al. 2007).

Kim et al. also transplanted fetal kidney cells at different gestational stages seeded onto a scaffold for transplantation into the omentum (Kim et al. 2007). The results of using early gestational stage cells (E14.5) demonstrated the formation of early glomeruli and tubules as well as the formation of nonrenal bone and cartilage tissue

Fig. 3 (continued) nephron progenitor cells that form a comma-shaped body and an S-shaped body. The distal portion of the S-shaped body develops podocyte progenitors that form a vascular cleft – attracting endothelial cells most likely via VEGF-VEGFr signaling. The vascular cleft eventually develops into the glomerulus, and the extending tubular progenitors develop the specialized tubular epithelia

(Kim et al. 2007). Engineered tissues using cells from later gestational stages (E20.5) developed into nephron structures without differentiation into bone and cartilage tissues (Kim et al. 2007). However, constructs using these later-stage cells showed fewer nephron structures than those using earlier stage cells for seeding (Kim et al. 2007). This study highlights the importance of selecting the correct timing for harvest when using fetal kidney cells, or fetal kidney like cells, for renal tissue engineering purposes.

What we can gain from this research is that metanephroi or metanephroi-like cells hold much potential to be grown into functional kidney tissue in situ. Though these results are profound and the potential of using cells from embryological origin to more fully understand renal tissue formation is great, as mentioned earlier, the use of embryonic tissues or embryonic stem cells for practical clinical applications is hindered by ethical barriers and governmental policies. ESC research is permitted in limited locations worldwide. However, the findings from embryonic stem cell studies discussed above can inform the direction of future studies using induced pluripotent stem cells (iPSCs), a promising alternative to ESCs. iPSCs are cells from skin fibroblast, blood cell, or even mature kidney origin that can be treated with a series of four transcriptional factors, often referred to as the Yamanaka factors, that force the de-differentiation of these cells into an ESC like state capable of spontaneous self-renewal (Zhou et al. 2012). Several groups have developed methods of driving these de-differentiated human iPSCs into renal cells. These groups have all utilized an understanding of key stepwise developmental cues provided in different stages of kidney embryogenesis to provide culture conditions necessary to artificially guide renal development.

Takasato et al. are a group leading these iPSCs to renal cell conversion efforts that successfully upregulated wingless/integrated signaling (Wnt signaling) by exposing iPSCs to CHIR 99021, a glycogen synthase kinase beta 3 inhibitor, followed by fibroblast growth factor 9 (FGF9) to produce renal progenitor cells (Takasato et al. 2014, 2016a; Takasato and Little 2016). Longer periods of Wnt signaling were found to encourage the intermediate mesoderm to generate metanephric mesoderm and nephron progenitors, whereas shorter periods of Wnt signaling encouraged the formation of ureteric bud progenitors (Takasato et al. 2016b). Other groups found that retinoic acid exposure pushed intermediate mesoderm cells toward ureteric bud lineage, whereas blockade of retinoic acid pushes IM to form metanephric mesenchyme and nephron progenitors (Takasato et al. 2016b). These initial studies also showed that 3D aggregate culture of these renal precursors into organoids made from iPSCs was able to develop renal structures that mimic tissue architecture. This strategy may potentially be used in the future to seed renal scaffolds to yield more stable renal structures or be used in other micro tissue engineering efforts to model kidney disease or test drugs. However, these organoid studies are still in the early stages. One concern that needs to be addressed before using organoids for these purposes is that there are some discrepancies between renal structures formed within organoids and those formed within native tissue (Eremina et al. 2003; Bernstein et al. 1981). For example, podocytes created within organoids do not generate the full repertoire of mature proteins seen in normal development (Eremina et al. 2003).

Furthermore, tubules developed in organoids structurally resemble proximal and distal tubule segments, but again, do not express the full repertoire of RNAs characterizing these cells in vivo (Wu et al. 2018). It is currently unclear whether these discrepancies are due to the need for further maturation of these organoids or due to a discrepancy in cell types formed (Morizane et al. 2017; Morizane and Bonventre 2017). Other points of concern for using organoid structures for renal tissue engineering are that several cell types typically not thought to play a role in renal embryogenesis are found within organoids – possibly indicating that iPSC differentiation protocols need better control (Morizane et al. 2017). Human kidney organoids do not recapitulate the normal kidney's organization into the cortex, outer medulla, and inner medulla.

Other potential cell sources that may be used for renal tissue engineering are external adult stem cell populations like amniotic fluid stem cells (AFSCs) and mesenchymal stem cells (MSCs) that are readily available commercially and can also be isolated autologously.

AFSCs have been successfully isolated to maintain self-renewal capacity and have been shown to have some beneficial renoprotective and regenerative outcomes in cell therapy treatments in renal insufficiency models (Hauser et al. 2010; Rota et al. 2012; Sedrakyan et al. 2012, 2017; Baulier et al. 2014). More studies need to be done on the mechanism of this effect as there is currently no data demonstrating the direct differentiation of AFSCs into renal structures for use in tissue engineering.

Renal-derived MSCs have been discovered to have a more nephrogenic gene expression profile and have been shown to differentiate into renal epithelial cells, EPO+ fibroblasts, and juxtaglomerular cells containing renin granules (Chen et al. 2008). This is an attractive cell source for tissue engineering, but their populations are exhausted over the course of renal disease progression, making them unsuitable as an autologous cell source. Nonrenal MSCs from bone marrow, adipose tissue, placenta, cord, and blood origins have been studied in depth using renal insufficiency models. These studies demonstrate that MSC delivery during renal insufficiency provides a renoprotective and regenerative effect (Morigi et al. 2004; Choi et al. 2009). However, the consensus has been that these effects are mainly a result of MSCs' paracrine and endocrine effects that support the proliferation of local cells, with evidence of a very small percentage of direct differentiation into renal structures themselves. It may be the case that in natural injury settings MSCs may mainly work through paracrine and endocrine effects, but Yokoo et al. profoundly demonstrated the potential of bone marrow-derived human MSCs to directly differentiate into renal structures when put into a nephrogenic environment – confirming that they possessed the machinery to be differentiated into renal structures only postulated in previous studies (Yokoo et al. 2005). This study injected bone marrow-derived human MSCs pre-transfected with GDNF and labeled with LacZ into the nephrogenic region of a rat embryo that naturally provided all of the necessary developmental signals for nephrogenesis (Yokoo et al. 2006). These constructs were transplanted into the omentum for development and were found to become highly vascularized and integrated into omental tissue with functional EPO production in response to anemia (Yokoo et al. 2006; Yokoo et al. 2008).

4 Tissue Construct Template Materials

In addition to determining adequate cell populations for seeding, identifying efficient scaffolds and 3D culture conditions for the creation of a suitable niche to guide renal regeneration is crucial to the success of renal tissue engineering strategies. A scaffold has been defined as a biomaterial structure that serves as a substrate and guide for tissue repair and regeneration (Williams 2006). To this end, three main types of scaffolding systems have been explored: natural polymers, synthetic polymers, and biologic tissue matrices or decellularized organ scaffolds.

Natural polymers are biocompatible with characteristics of native tissue that are able to support and guide cell behaviors like cell activation, migration, proliferation, and differentiation. Collagen, HA, chitosan, fibrin, alginate, and agarose are some that have been used in renal tissue engineering studies. Of these, collagen and HA have been extensively studied (Manuela Gomes et al. 2008).

Collagen has been used extensively for tissue engineering applications over the past decades because it is biodegradable, biocompatible, easily available, and highly versatile (Parenteau-Bareil et al. 2010). Collagen and collagen-based hydrogels have been specifically used in many renal tissue engineering studies that have demonstrated the successful generation of renal structures from seeded cell sources. Wang et al. demonstrated the successful recapitulation of native glomerular tissue using a collagen vitrigel-based membrane scaffold when seeded with primary glomerular epithelial and mesangial cells isolated from renal tissues (Wang and Takezawa 2005). On a larger scale, Lü et al. also used a collagen-based scaffold to provide adequate scaffolding for a mixed population of neonatal rat renal cells to self-assemble into renal tissues that contained tubule and glomeruli-like structures (Lü et al. 2012). Revisiting a study mentioned above when discussing cell sources, Guimaraes-Souza et al.'s study of using primary renal cells isolated from normal and diseased kidneys for tissue engineering purposes demonstrated not only proper renal structure formation but also the development of functional capabilities in renal cells grown in a collagen-based scaffold (Guimaraes-Souza et al. 2012; George et al. 2016). Furthermore, this functional tissue construct, when implanted subcutaneously in mice, was viable for up to 6 weeks (Guimaraes-Souza et al. 2012; George et al. 2016). Taken together, these studies and others using collagen-based scaffolds identify this material as one supportive of functional renal structure formation.

Hyaluronic acid (HA) is a glycosaminoglycan that has also been used in several gel-based scaffolds for tissue engineering purposes because it is highly biocompatible and has cross-linking capabilities (Chircov et al. 2018). Hyaluronic acid is of specific interest in renal tissue engineering because a high level of synthesis and degradation naturally occurs during kidney development (Rosines et al. 2007). Rosines et al. studied the role of hyaluronic acid concentration during renal organogenesis and found specific benefits of using this ECM material for renal tissue engineering (Rosines et al. 2007). This group found that hyaluronic acid has the ability to modulate ureteric bud branching, promote mesenchymal-to-epithelial transformation for nephron formation, and promote differentiation of both the metanephric mesenchyme and ureteric bud (Rosines et al. 2007). More specifically, they found that hyaluronic acid stimulated ureteric bud branch morphogenesis at low concentrations (0.1%) and low molecular weights (6.55 kDa) but inhibited branching at high concentrations (3.75%) and molecular weight (234.4 kDa) (Rosines et al. 2007). Furthermore, hyaluronic acid at every molecular weight tested promoted collecting duct differentiation, and a variety of molecular weights strongly promoted mesenchymal to epithelial transformation for nephron differentiation in a concentration-dependent manner (Rosines et al. 2007). These findings suggest that hyaluronic acid may be a good scaffold material for renal tissue engineering that not only supports growth but also potentially provides a means of controlling and guiding specific morphogenesis through the use of different concentrations.

Though using these natural polymers as biomaterials is very promising for renal TE, a major shortcoming is that it is limited in architectural control and mechanical properties, and there is fast degradation (Bryant and Anseth 2001). For this reason, synthetic scaffolds have also been studied for renal TE. Synthetic polymers like polylactic acid, polyglycolic acid, and polylactic-co-glycolic acid have been studied for the purposes of tissue engineering because they are nontoxic and biodegradable, have a well-controlled degradation profile, and allow for simple fabrication techniques. In early studies, Yoo et al. demonstrated the feasibility of using these polymers in vitro and in vivo (Yoo et al. 1996). These studies, mentioned above, used a porous tubular polycarbonate scaffold seeded with proximal tubule epithelial cells connected to a reservoir via a catheter that supported the formation of functional renal structures in vitro (Yoo et al. 1996). This system was subsequently subcutaneously implanted and resulted in the formation of a urine-like fluid collected in the reservoir (Yoo et al. 1996). These studies profoundly demonstrated that polycarbonate scaffolds could feasibly be used for future tissue engineering studies as they supported not only the proper formation of renal structures but also the function of these engineered tissues. Synthetic scaffolds to support renal structure formation was also utilized by Kim et al.'s studies evaluating the importance of gestational age of fetal kidney cells for use in renal tissue engineering also used a synthetic polymer, porous PGA, and successfully showed that this scaffold material supported the development of nephron structures (Kim et al. 2007).

Although the formation of renal structures is supported in these scaffolds, the shortcomings of these synthetic materials are that there are only a few materials compatible with 3D printing, and more importantly, the materials are inherently unable to fully recapitulate a regenerative niche because of their inert qualities that do not allow for the participation in the regenerative process. It is thought that to recapitulate the regenerative environment of the kidney fully, an inert material is not the ideal source (Williams and Challenges 2019).

A promising alternative source of biomaterials that adequately recapitulates renal regenerative niches are decellularized scaffolds. The basic strategy of using a decellularized scaffold is to remove native cells using detergents to leave behind an extracellular matrix that maintains the organ's original architecture. This scaffold is subsequently repopulated with cells, possibly from the different cell sources discussed above. The use of a decellularized scaffold as a template for tissue regeneration is attractive because they have collagen-rich matrices that degrade more slowly than those created with natural polymers. They are also able to actively participate in the tissue regeneration process unlike inert synthetic scaffolds because they can be efficiently remodeled by extracellular matrix proteins produced by seeded cells (Williams and Challenges 2019). Another critical advantage of using this strategy is that the native spatial architecture of the kidney is currently impossible to recapitulate using natural and synthetic polymers. This strategy is still in its early stages with many promising approaches that still need to be tested for the successful development of functional kidney structures and for vascular integrity and reperfusion over long-term periods. Early studies in using decellularized scaffolds have only been performed in short-term experiments that show limited success due to the relative immaturity and seeding efficiency of cells. Longer-term studies are postulated to have better results in the future.

Ross et al. performed initial studies that harvested and decellularized rat kidneys using detergents and seeded them with murine ESCs via arterial or ureteral perfusion (Ross et al. 2009). The scaffold after decellularization was evaluated by scanning electron microscopy and immunohistochemistry to reveal an intact microstructure and a type 4 collagen and laminin-rich network (Ross et al. 2009). Histological analyses of the scaffold showed that the perfused ESCs populated glomerular, tubular, and vascular regions of the scaffold (Ross et al. 2009). A similar method of decellularization and reperfusion was used by Nakayama et al. to decellularize rhesus monkey kidneys at different stages of development (Nakayama et al. 2010). These scaffolds were evaluated with immunohistochemistry to assess the preservation of extracellular matrix proteins important to kidneys, including heparan sulfate, fibronectin, type I collagen, type IV collagen, and lamanin (Nakayama et al. 2010). Their findings demonstrated that the decellularized kidneys provided sufficient structural properties that supported the migration of seeded cells in an age-dependent manner by proving the spatial and organizational influences on human ESC migration and differentiation (Nakayama et al. 2010).

Song et al. were another group that utilized a decellularized scaffold for renal tissue engineering. This group decellularized cadaveric rat kidneys that were seeded first with endothelial cells sources from human umbilical vein via perfusion through the renal artery, followed by seeding with a mixture of neonatal kidney cells of different epithelial phenotypes via the ureter (Song et al. 2013). The kidney vasculature was successfully lined with endothelial cells after 3–5 days in culture and epithelial repopulation of the scaffold was seen as early as 4 days in culture (Song et al. 2013). Histological evaluation showed that the mixture of neonatal kidney cells engrafted based on their phenotype to specific regions on the extracellular matrix (Song et al. 2013). For example, the tubular epithelial cells organized into tubular structures and podocytes, specifically seeded glomeruli (Song et al. 2013). Functionally, the seeded kidneys had a 10% clearance of creatinine and retained albumin and resorbed glucose better than decellularized controls (Song et al. 2013). However, the seeded kidneys showed less effective creatinine clearance and urine excretion function than cultured cadaveric kidneys (Song et al. 2013). These seeded kidney constructs were then orthotopically transplanted after a single nephrectomy in rats (Song et al. 2013). The transplants demonstrated successful reperfusion and urine production – though again less than native controls (Song et al. 2013). These studies are significant in that they demonstrate the feasibility of using a decellularized scaffold for renal tissue engineering as functional renal structures resulted after seeding. However, there is a need for future studies to optimize this method to provide a better niche that allows for improved functionality of the renal structures before this strategy becomes clinically viable. For sustainable use, the fabrication of acellular renal scaffolds on a clinical scale with efficient recellularization strategies also still needs to be developed.

Furthermore, successful vascularization and perfusion remain a major challenge in being able to use decellularized scaffolds as a tissue engineering template. The seeded scaffolds have to be perfused successfully with blood for nutrient and oxygen delivery as well as waste removal. However, if functional complete endothelial seeding is not achieved before perfusion, thrombosis is likely to occur. To address this issue, Novitskaya et al. pre-treated vascular lumen of porcine decellularized kidneys with CD31 antibody conjugation to enhance the retention of subsequently seeded endothelial cells (Novitskaya et al. 2014). This strategy improved the vascular patency of the scaffold post-implantation (Novitskaya et al. 2014).

Vascularization and reperfusion is not only a major challenge of using decellularized scaffolds but of tissue engineering strategies as a whole when considering organizing the discussed cell sources and scaffold materials to create sustainable constructs. Some of these strategies are discussed below.

5 Vascularization

Vascularization of bioengineered organs has presented significant challenges in large organ development. Thus far, construct design has been limited by the fact that cells typically need to be within 100–200 μm of an oxygen and nutrient source for survival (Jain et al. 2005). For sufficient survival of cells without vascular adjustment, constructs need to be small, thin, or porous to rely on diffusion from the host's vascular supply (Rouwkema and Khademhosseini 2016). Furthermore, the rate of spontaneous vascular infiltration is often limited to several tenths of micrometers a day – meaning that it would take several weeks for host vasculature to naturally vascularize an implant several millimeters thick (Rouwkema and Khademhosseini 2016). To work around these challenges for larger organs, several approaches have been taken, including scaffold functionalization, cell-based techniques, modular assembly, bioreactor designs, in vivo systems, and microelectromechanical systems-based microfluidics approaches. Scaffold functionalization is achieved by loading scaffolds with pro-angiogenic factors like vascular endothelial growth factor (VEGF) and/or adjusting the porosity or channeling of a scaffold (Zisch et al. 2003a; Radisic et al. 2006; Zhang and Suggs 2007; Rosa et al. 2017; van Rensburg et al. 2017; Zisch et al. 2003b; Andrae et al. 2008; Caballero Aguilar et al. 2019). Cell-based techniques coculture scaffolds with either endothelial cells or growth factor-producing cells (Wenger et al. 2004, 2005; Borges et al. 2003; Rouwkema et al. 2006; Yang et al. 2007). Modular assembly combines microtissues seeded with desired organ cell types coated with endothelial cells into a macrotissue structure or combines small single- or multi-channel hydrogels to form a larger structure (Kelm et al. 2006; McGuigan and Sefton 2006; Chrobak et al. 2006). These techniques increase angiogenesis, but the vascular organization and connectivity with host vasculature once implanted is slow and not biomimetic, often resulting in vessels that were disorganized, unstable, leaky, and hemorrhagic. Bioreactor designs mechanically force media or nutrients through a porous scaffold to maintain viability in vitro so constructs can develop before being placed in vivo (Fassina et al. 2018; Birru et al. 2019). In vivo systems approaches involve multistep surgical procedures in which constructs are initially placed in vivo next to arterial loop systems to develop vasculature and then removed and placed in the functional anatomical location (Shimizu et al. 2006). Though effective, this approach unfavorably required multiple surgeries and can lead to complications. Lastly, microelectromechanical systems-based synthetic or biodegradable microfluidics approaches build vascular networks to be placed into tissue constructs using either stacking of single-layer microfluidic networks followed by bonding to form a 3D structure with plasma etching or lithographic techniques or by using direct-write assembly and computational modeling techniques to generate 3D microfluidic devices using a mixture of sacrificial and non-sacrificial materials (Hasan et al. 2014; Chandra and Atala 2019; He et al. 2014; Lovett et al. 2009; Therriault et al. 2003; Liu et al. 2019; Zhang et al. 2018). Previous studies using these microfluidics approaches were successful in creating vascular networks, but the design complexity was limited and did not recapitulate native vasculature (Lovett et al. 2009).

Huling et al. have recently taken a novel microfluidics approach to generate a more biomimetic vascular network. This group developed a technique that involves using a sacrificial material (polycaprolactone, PCL) that is perfused into the arteries of native kidney tissue (Huling et al. 2019). After this material solidifies, the surrounding tissue is digested, leaving behind a cast of just the kidney vasculature (Huling et al. 2019). This cast is then coated with collagen and subsequently removed, leaving behind a hollow collagen microvascular scaffold that is then seeded with endothelial cells and placed in a kidney cell-laden hydrogel (Huling et al. 2019). The results show that this process successfully generates a biomimetic vascular network; however, achieving functional anastomosis between the implant and host vasculature in a timely manner remains a challenge to be addressed in the future.

6 Conclusions

Many advancements have been made in the study of renal cell-based tissue engineering in the past few decades. Cell-based studies have discovered several cell sources capable of recapitulating renal structures when cultured within a three-dimensional scaffolding system. Furthermore, these studies have utilized and assessed the success of a variety of biomaterials for renal tissue engineering purposes.

However, several challenges still remain before renal tissue engineering can be used therapeutically. It is becoming increasingly apparent that foundational knowledge regarding the natural microenvironments for renal regeneration is lacking. How the mechanics of scaffold materials themselves affect renal structure formation is still an area of understanding that needs to be developed. Furthermore, the exact signaling pathways and duration of factor secretion during times of kidney injury and repair need to be better understood for adequate recapitulation of the regenerative niche. Safe and effective scaffold seeding and implantation techniques also need to be further developed with a special focus on vascularization to allow for better construct survival as well.

Despite this, or perhaps due to, these challenges renal tissue engineering presents an exciting field of future study. From foundational research to clinical studies, the study of the multifaceted approaches to making this technology feasible for therapeutic use holds the potential to profoundly impact the lives of patients worldwide and change the clinical climate.

References

Abrahamson DR (2009) Development of kidney glomerular endothelial cells and their role in basement membrane assembly. Organogenesis 5:275–287. https://doi.org/10.4161/org.7577

Andrae J, Gallini R, Betsholtz C (2008) Role of platelet-derived growth factors in physiology and medicine. Genes Dev 22:1276–1312. https://doi.org/10.1101/gad.1653708

Angelotti ML et al (2012) Characterization of renal progenitors committed toward tubular lineage and their regenerative potential in renal tubular injury. Stem Cells 30:1714–1725. https://doi.org/10.1002/stem.1130

Atala A et al (1995) Renal cell growth in vivo after attachment to biodegradable polymer scaffold. J Urol 153:4

Banerjee D, Jha V (2019) Vitamin D and cardiovascular complications of CKD: what's next? Clin J Am Soc Nephrol 14:932–934. https://doi.org/10.2215/CJN.12581018

Basta JM, Robbins L, Kiefer SM, Dorsett D, Rauchman M (2014) Sall1 balances self-renewal and differentiation of renal progenitor cells. Development 141:1047–1058. https://doi.org/10.1242/dev.095851

Baulier E et al (2014) Amniotic fluid-derived mesenchymal stem cells prevent fibrosis and preserve renal function in a preclinical porcine model of kidney transplantation. Stem Cells Transl Med 3:809–820. https://doi.org/10.5966/sctm.2013-0186

Bernstein J, Cheng F, Roszka J (1981) Glomerular differentiation in metanephric culture. Lab Investig 45:183–190

Birru B, Mekala NK, Parcha SR (2019) Mechanistic role of perfusion culture on bone regeneration. J Biosci 44:23

Borges J et al (2003) Engineered adipose tissue supplied by functional microvessels. Tissue Eng 9:1263–1270. https://doi.org/10.1089/10763270360728170

Bryant SJ, Anseth KS (2001) The effects of scaffold thickness on tissue engineered cartilage in photocrosslinked poly(ethylene oxide) hydrogels. Biomaterials 22:619–626

Bussolati B et al (2005) Isolation of renal progenitor cells from adult human kidney. Am J Pathol 166:545–555. https://doi.org/10.1016/S0002-9440(10)62276-6

Caballero Aguilar LM, Silva SM, Moulton SE (2019) Growth factor delivery: defining the next generation platforms for tissue engineering. J Control Release 306:40–58. https://doi.org/10.1016/j.jconrel.2019.05.028

Chandra P, Atala A (2019) Engineering blood vessels and vascularized tissues: technology trends and potential clinical applications. Clin Sci (Lond) 133:1115–1135. https://doi.org/10.1042/CS20180155

Chang CH, Davies JA (2012) An improved method of renal tissue engineering, by combining renal dissociation and reaggregation with a low-volume culture technique, results in development of engineered kidneys complete with loops of Henle. Nephron Exp Nephrol 121:e79–e85. https://doi.org/10.1159/000345514

Chen J et al (2008) Kidney-derived mesenchymal stem cells contribute to vasculogenesis, angiogenesis and endothelial repair. Kidney Int 74:879–889. https://doi.org/10.1038/ki.2008.304

Chen N et al (2019) Roxadustat treatment for anemia in patients undergoing long-term dialysis. N Engl J Med. https://doi.org/10.1056/NEJMoa1901713

Chircov C, Grumezescu AM, Bejenaru LE (2018) Hyaluronic acid-based scaffolds for tissue engineering. Romanian J Morphol Embryol 59:71–76

Choi S et al (2009) The role of mesenchymal stem cells in the functional improvement of chronic renal failure. Stem Cells Dev 18:521–529. https://doi.org/10.1089/scd.2008.0097

Chrobak KM, Potter DR, Tien J (2006) Formation of perfused, functional microvascular tubes in vitro. Microvasc Res 71:185–196. https://doi.org/10.1016/j.mvr.2006.02.005

Chung HC, Ko IK, Atala A, Yoo JJ (2015) Cell-based therapy for kidney disease. Korean J Urol 56:412–421. https://doi.org/10.4111/kju.2015.56.6.412

Donnelly S (2003) Why is erythropoietin made in the kidney? The kidney functions as a 'critmeter' to regulate the hematocrit. Adv Exp Med Biol 543:73–87. https://doi.org/10.1007/978-1-4419-8997-0_6

Eremina V et al (2003) Glomerular-specific alterations of VEGF-A expression lead to distinct congenital and acquired renal diseases. J Clin Invest 111:707–716. https://doi.org/10.1172/JCI17423

Fassina L et al (2018) Biomaterials and biophysical stimuli for bone regeneration. J Biol Regul Homeost Agents 32:41–49

Fung LCT et al (1996) Reconstitution of poor EGFr-poor renal epithelial cells into tubular structures on biodegradable polymer scaffold. Pediatrics 98:605

George SK et al (2016) Potential use of autologous renal cells from diseased kidneys for the treatment of renal failure. PLoS One 11:e0164997. https://doi.org/10.1371/journal.pone.0164997

Guimaraes-Souza NK, Yamaleyeva LM, AbouShwareb T, Atala A, Yoo JJ (2012) In vitro reconstitution of human kidney structures for renal cell therapy. Nephrol Dial Transplant 27:3082–3090. https://doi.org/10.1093/ndt/gfr785

Hammerman MR (2007) Transplantation of renal primordia: renal organogenesis. Pediatr Nephrol 22:1991–1998. https://doi.org/10.1007/s00467-007-0554-7

Harambat J, Ekulu PM (2016) Inequalities in access to pediatric ESRD care: a global health challenge. Pediatr Nephrol 31:353–358. https://doi.org/10.1007/s00467-015-3263-7

Hasan A et al (2014) Microfluidic techniques for development of 3D vascularized tissue. Biomaterials 35:7308–7325. https://doi.org/10.1016/j.biomaterials.2014.04.091

Hauser PV et al (2010) Stem cells derived from human amniotic fluid contribute to acute kidney injury recovery. Am J Pathol 177:2011–2021. https://doi.org/10.2353/ajpath.2010.091245

He J, Zhu L, Liu Y, Li D, Jin Z (2014) Sequential assembly of 3D perfusable microfluidic hydrogels. J Mater Sci Mater Med 25:2491–2500. https://doi.org/10.1007/s10856-014-5270-9

Herzlinger D, Koseki C, Mikawa T, al-Awqati Q (1992) Metanephric mesenchyme contains multipotent stem cells whose fate is restricted after induction. Development 114:565–572

Huling J et al (2019) Kidney regeneration with biomimetic vascular scaffolds based on vascular corrosion casts. Acta Biomater. https://doi.org/10.1016/j.actbio.2019.04.001

Huml AM, Sehgal AR (2019) Hemodialysis quality metrics in the first year following a failed kidney transplant. Am J Nephrol:1–7. https://doi.org/10.1159/000501605

Jain RK, Au P, Tam J, Duda DG, Fukumura D (2005) Engineering vascularized tissue. Nat Biotechnol 23:821–823. https://doi.org/10.1038/nbt0705-821

Johansen KL, Doyle J, Sakkas GK, Kent-Braun JA (2005) Neural and metabolic mechanisms of excessive muscle fatigue in maintenance hemodialysis patients. Am J Physiol Regul Integr Comp Physiol 289:R805–R813. https://doi.org/10.1152/ajpregu.00187.2005

Kelm JM et al (2006) Design of custom-shaped vascularized tissues using microtissue spheroids as minimal building units. Tissue Eng 12:2151–2160. https://doi.org/10.1089/ten.2006.12.2151

Kim SS et al (2007) Kidney tissue reconstruction by fetal kidney cell transplantation: effect of gestation stage of fetal kidney cells. Stem Cells 25:1393–1401. https://doi.org/10.1634/stemcells.2006-0183

Kumar V, Jha V (2016) End-stage renal disease care in South Asia: demographics, economics, and opportunities. Clin Nephrol 86(2016):23–26. https://doi.org/10.5414/CNP86S103

Kurtzeborn K, Cebrian C, Kuure S (2018) Regulation of renal differentiation by trophic factors. Front Physiol 9:1588. https://doi.org/10.3389/fphys.2018.01588

Lanza RP et al (2002) Generation of histocompatible tissues using nuclear transplantation. Nat Biotechnol 20:689–696. https://doi.org/10.1038/nbt703

Li Y, Wingert RA (2013) Regenerative medicine for the kidney: stem cell prospects, challenges. Clin Transl Med 2:11. https://doi.org/10.1186/2001-1326-2-11

Little MH, McMahon AP (2012) Mammalian kidney development: principles, progress, and projections. Cold Spring Harb Perspect Biol 4. https://doi.org/10.1101/cshperspect.a008300

Liu X et al (2019) Fabrication and characterization of reconstituted silk microgels for the storage and release of small molecules. Macromol Rapid Commun 40:e1800898. https://doi.org/10.1002/marc.201800898

Lovett M, Lee K, Edwards A, Kaplan DL (2009) Vascularization strategies for tissue engineering. Tissue Eng Part B Rev 15:353–370. https://doi.org/10.1089/ten.TEB.2009.0085

Lü SH et al (2012) Self-assembly of renal cells into engineered renal tissues in collagen/Matrigel scaffold in vitro. J Tissue Eng Regen Med 6:786–792. https://doi.org/10.1002/term.484

Luyckx VA, Tonelli M, Stanifer JW (2018) The global burden of kidney disease and the sustainable development goals. Bull World Health Organ 96:414D–422D. https://doi.org/10.2471/BLT.17.206441

Manuela Gomes HA, Malafaya P, Silva S, Oliveira J, Silva G, Sousa R, Mano J, Reis R (2008) Natural polymers in tissue engineering applications. In: Tissue engineering. Academic, London, pp 145–192

Marshall D, Dilworth MR, Clancy M, Bravery CA, Ashton N (2007) Increasing renal mass improves survival in anephric rats following metanephros transplantation. Exp Physiol 92:263–271. https://doi.org/10.1113/expphysiol.2006.036319

Maxwell PH et al (1993) Identification of the renal erythropoietin-producing cells using transgenic mice. Kidney Int 44:1149–1162. https://doi.org/10.1038/ki.1993.362

McGuigan AP, Sefton MV (2006) Vascularized organoid engineered by modular assembly enables blood perfusion. Proc Natl Acad Sci U S A 103:11461–11466. https://doi.org/10.1073/pnas.0602740103

Morigi M et al (2004) Mesenchymal stem cells are renotropic, helping to repair the kidney and improve function in acute renal failure. J Am Soc Nephrol 15:1794–1804. https://doi.org/10.1097/01.asn.0000128974.07460.34

Morizane R, Bonventre JV (2017) Generation of nephron progenitor cells and kidney organoids from human pluripotent stem cells. Nat Protoc 12:195–207. https://doi.org/10.1038/nprot.2016.170

Morizane R, Miyoshi T, Bonventre JV (2017) Concise review: kidney generation with human pluripotent stem cells. Stem Cells 35:2209–2217. https://doi.org/10.1002/stem.2699

Morrissey M (2012) Willem J Kolff (1911-2009): physician, inventor and pioneer: father of artificial organs. J Med Biogr 20:136–138. https://doi.org/10.1258/jmb.2012.012030

Mugford JW, Sipilä P, McMahon JA, McMahon AP (2008) Osr1 expression demarcates a multipotent population of intermediate mesoderm that undergoes progressive restriction to an Osr1-dependent nephron progenitor compartment within the mammalian kidney. Dev Biol 324:88–98. https://doi.org/10.1016/j.ydbio.2008.09.010

Nakayama KH, Batchelder CA, Lee CI, Tarantal AF (2010) Decellularized rhesus monkey kidney as a three-dimensional scaffold for renal tissue engineering. Tissue Eng Part A 16:2207–2216. https://doi.org/10.1089/ten.tea.2009.0602

Novitskaya T et al (2014) A PTBA small molecule enhances recovery and reduces postinjury fibrosis after aristolochic acid-induced kidney injury. Am J Physiol Renal Physiol 306:F496–F504. https://doi.org/10.1152/ajprenal.00534.2013

Obert LA, Frazier KS (2019) Intrarenal renin-angiotensin system involvement in the pathogenesis of chronic progressive nephropathy-bridging the informational gap between disciplines. Toxicol Pathol. https://doi.org/10.1177/0192623319861367

Oliver JA, Maarouf O, Cheema FH, Martens TP, Al-Awqati Q (2004) The renal papilla is a niche for adult kidney stem cells. J Clin Invest 114:795–804. https://doi.org/10.1172/JCI20921

Parenteau-Bareil R, Gauvin R, Berthod F (2010) Collagen-based biomaterials for tissue engineering applications. Materials (Basel) 3:1863–1887

Patschan D et al (2007) Normal distribution and medullary-to-cortical shift of Nestin-expressing cells in acute renal ischemia. Kidney Int 71:744–754. https://doi.org/10.1038/sj.ki.5002102

Radisic M et al (2006) Biomimetic approach to cardiac tissue engineering: oxygen carriers and channeled scaffolds. Tissue Eng 12:2077–2091. https://doi.org/10.1089/ten.2006.12.2077

Rinkevich Y et al (2014) In vivo clonal analysis reveals lineage-restricted progenitor characteristics in mammalian kidney development, maintenance, and regeneration. Cell Rep 7:1270–1283. https://doi.org/10.1016/j.celrep.2014.04.018

Rogers SA, Hammerman MR (2004) Prolongation of life in anephric rats following de novo renal organogenesis. Organogenesis 1:22–25. https://doi.org/10.4161/org.1.1.1009

Ronconi E et al (2009) Regeneration of glomerular podocytes by human renal progenitors. J Am Soc Nephrol 20:322–332. https://doi.org/10.1681/ASN.2008070709

Rosa AR et al (2017) Development of VEGF-loaded PLGA matrices in association with mesenchymal stem cells for tissue engineering. Braz J Med Biol Res 50:e5648. https://doi.org/10.1590/1414-431X20175648

Rosines E, Schmidt HJ, Nigam SK (2007) The effect of hyaluronic acid size and concentration on branching morphogenesis and tubule differentiation in developing kidney culture systems: potential applications to engineering of renal tissues. Biomaterials 28:4806–4817. https://doi.org/10.1016/j.biomaterials.2007.07.034

Ross EA et al (2009) Embryonic stem cells proliferate and differentiate when seeded into kidney scaffolds. J Am Soc Nephrol 20:2338–2347. https://doi.org/10.1681/ASN.2008111196

Rota C et al (2012) Human amniotic fluid stem cell preconditioning improves their regenerative potential. Stem Cells Dev 21:1911–1923. https://doi.org/10.1089/scd.2011.0333

Rothenpieler UW, Dressler GR (1993) Pax-2 is required for mesenchyme-to-epithelium conversion during kidney development. Development 119:711–720

Rouwkema J, Khademhosseini A (2016) Vascularization and angiogenesis in tissue engineering: beyond creating static networks. Trends Biotechnol 34:733–745. https://doi.org/10.1016/j.tibtech.2016.03.002

Rouwkema J, de Boer J, Van Blitterswijk CA (2006) Endothelial cells assemble into a 3-dimensional prevascular network in a bone tissue engineering construct. Tissue Eng 12:2685–2693. https://doi.org/10.1089/ten.2006.12.2685

Sagrinati C et al (2006) Isolation and characterization of multipotent progenitor cells from the Bowman's capsule of adult human kidneys. J Am Soc Nephrol 17:2443–2456. https://doi.org/10.1681/ASN.2006010089

Sagrinati C, Ronconi E, Lazzeri E, Lasagni L, Romagnani P (2008) Stem-cell approaches for kidney repair: choosing the right cells. Trends Mol Med 14:277–285. https://doi.org/10.1016/j.molmed.2008.05.005

Sedrakyan S et al (2012) Injection of amniotic fluid stem cells delays progression of renal fibrosis. J Am Soc Nephrol 23:661–673. https://doi.org/10.1681/ASN.2011030243

Sedrakyan S et al (2017) Amniotic fluid stem cell-derived vesicles protect from VEGF-induced endothelial damage. Sci Rep 7:16875. https://doi.org/10.1038/s41598-017-17061-2

Sen A, Callisen H, Libricz S, Patel B (2019) Complications of solid organ transplantation: cardiovascular, neurologic, renal, and gastrointestinal. Crit Care Clin 35:169–186. https://doi.org/10.1016/j.ccc.2018.08.011

Shimizu T et al (2006) Polysurgery of cell sheet grafts overcomes diffusion limits to produce thick, vascularized myocardial tissues. FASEB J 20:708–710. https://doi.org/10.1096/fj.05-4715fje

Song JJ et al (2013) Regeneration and experimental orthotopic transplantation of a bioengineered kidney. Nat Med 19:646–651. https://doi.org/10.1038/nm.3154

Takasato M, Little MH (2016) A strategy for generating kidney organoids: recapitulating the development in human pluripotent stem cells. Dev Biol 420:210–220. https://doi.org/10.1016/j.ydbio.2016.08.024

Takasato M et al (2014) Directing human embryonic stem cell differentiation towards a renal lineage generates a self-organizing kidney. Nat Cell Biol 16:118–126. https://doi.org/10.1038/ncb2894

Takasato M, Er PX, Chiu HS, Little MH (2016a) Generation of kidney organoids from human pluripotent stem cells. Nat Protoc 11:1681–1692. https://doi.org/10.1038/nprot.2016.098

Takasato M et al (2016b) Kidney organoids from human iPS cells contain multiple lineages and model human nephrogenesis. Nature 536:238. https://doi.org/10.1038/nature17982

Therriault D, White SR, Lewis JA (2003) Chaotic mixing in three-dimensional microvascular networks fabricated by direct-write assembly. Nat Mater 2:265–271. https://doi.org/10.1038/nmat863

United States Renal Data System (2017) USRDS annual data report: epidemiology of kidney disease in the United States. National Institutes of Health, National Institute of Diabetes and Digestive and Kidney Diseases

Van der Hauwaert C et al (2013) Isolation and characterization of a primary proximal tubular epithelial cell model from human kidney by CD10/CD13 double labeling. PLoS One 8:e66750. https://doi.org/10.1371/journal.pone.0066750

van Rensburg AJ et al (2017) Improved vascularization of porous scaffolds through growth factor delivery from heparinized polyethylene glycol hydrogels. Acta Biomater 49:89–100. https://doi.org/10.1016/j.actbio.2016.11.036

Wang PC, Takezawa T (2005) Reconstruction of renal glomerular tissue using collagen vitrigel scaffold. J Biosci Bioeng 99:529–540. https://doi.org/10.1263/jbb.99.529

Wang V, Vilme H, Maciejewski ML, Boulware LE (2016) The economic burden of chronic kidney disease and end-stage renal disease. Semin Nephrol 36:319–330. https://doi.org/10.1016/j.semnephrol.2016.05.008

Ward HH et al (2011) Adult human CD133/1(+) kidney cells isolated from papilla integrate into developing kidney tubules. Biochim Biophys Acta 1812:1344–1357. https://doi.org/10.1016/j.bbadis.2011.01.010

Wenger A et al (2004) Modulation of in vitro angiogenesis in a three-dimensional spheroidal coculture model for bone tissue engineering. Tissue Eng 10:1536–1547. https://doi.org/10.1089/ten.2004.10.1536

Wenger A et al (2005) Development and characterization of a spheroidal coculture model of endothelial cells and fibroblasts for improving angiogenesis in tissue engineering. Cells Tissues Organs 181:80–88. https://doi.org/10.1159/000091097

Williams DF (2006) To engineer is to create: the link between engineering and regeneration. Trends Biotechnol 24:4–8. https://doi.org/10.1016/j.tibtech.2005.10.006

Williams D, Challenges F (2019) With the development of biomaterials for sustainable tissue engineering. Front Bioeng Biotechnol 7:127. https://doi.org/10.3389/fbioe.2019.00127

Wu H et al (2018) Comparative analysis and refinement of human PSC-derived kidney organoid differentiation with single-cell transcriptomics. Cell Stem Cell 23:869–881.e868. https://doi.org/10.1016/j.stem.2018.10.010

Xie Y et al (2018) Analysis of the Global Burden of Disease study highlights the global, regional, and national trends of chronic kidney disease epidemiology from 1990 to 2016. Kidney Int 94:567–581. https://doi.org/10.1016/j.kint.2018.04.011

Yamaleyeva LM et al (2012) Cell therapy with human renal cell cultures containing erythropoietin-positive cells improves chronic kidney injury. Stem Cells Transl Med 1:373–383. https://doi.org/10.5966/sctm.2011-0048

Yang J et al (2007) Effects of myocardial transplantation of marrow mesenchymal stem cells transfected with vascular endothelial growth factor for the improvement of heart function and angiogenesis after myocardial infarction. Cardiology 107:17–29. https://doi.org/10.1159/000093609

Yokoo T et al (2005) Human mesenchymal stem cells in rodent whole-embryo culture are reprogrammed to contribute to kidney tissues. Proc Natl Acad Sci U S A 102:3296–3300. https://doi.org/10.1073/pnas.0406878102

Yokoo T et al (2006) Xenobiotic kidney organogenesis from human mesenchymal stem cells using a growing rodent embryo. J Am Soc Nephrol 17:1026–1034. https://doi.org/10.1681/ASN.2005101043

Yokoo T et al (2008) Generation of a transplantable erythropoietin-producer derived from human mesenchymal stem cells. Transplantation 85:1654–1658. https://doi.org/10.1097/TP.0b013e318173a35d

Yoo JJ et al (1996) Creation of functional kidney structures with excretion of kidney like fluid in vivo. Pediatrics 98S:605

Zhang G, Suggs LJ (2007) Matrices and scaffolds for drug delivery in vascular tissue engineering. Adv Drug Deliv Rev 59:360–373. https://doi.org/10.1016/j.addr.2007.03.018

Zhang B et al (2018) Microfabrication of AngioChip, a biodegradable polymer scaffold with microfluidic vasculature. Nat Protoc 13:1793–1813. https://doi.org/10.1038/s41596-018-0015-8

Zhou T et al (2012) Generation of human induced pluripotent stem cells from urine samples. Nat Protoc 7:2080–2089. https://doi.org/10.1038/nprot.2012.115

Zisch AH, Lutolf MP, Hubbell JA (2003a) Biopolymeric delivery matrices for angiogenic growth factors. Cardiovasc Pathol 12:295–310

Zisch AH et al (2003b) Cell-demanded release of VEGF from synthetic, biointeractive cell ingrowth matrices for vascularized tissue growth. FASEB J 17:2260–2262. https://doi.org/10.1096/fj.02-1041fje

Engineering of the Bladder and Urethra

Xian Lin Yi, Diana Lim, Anthony Atala, and James J. Yoo

Contents

Abstract

Many conditions can damage the lower urinary tract tissues, including trauma, inflammation, cancer, and congenital anomaly. Unfortunately, reconstruction of the human bladder and urethra remains a great urological challenge. This is due to the limited availability of tissue substitutes that can be used for reconstruction.

X. L. Yi
Wake Forest Institute for Regenerative Medicine, Wake Forest School of Medicine, Winston Salem, NC, USA

Department of Urology, Tumor Hospital of GuangXi Medical University, Nanning, China

D. Lim · A. Atala · J. J. Yoo (✉)
Wake Forest Institute for Regenerative Medicine, Wake Forest School of Medicine, Medical Center Boulevard, Winston-Salem, NC, USA
e-mail: dlim@wakehealth.edu; aatala@wakehealth.edu; jyoo@wakehealth.edu

D. Eberli et al. (eds.), *Organ Tissue Engineering*, Reference Series in Biomedical Engineering, https://doi.org/10.1007/978-3-030-44211-8_1

Consequently, the use of intestinal tissue has remained as the gold standard for bladder reconstruction and repair, despite the associated complications, such as mucus production, electrolyte imbalances, recurrent infections, and malignancies. Similarly, the option for urethral tissue reconstruction is also limited. The autologous buccal mucosa is the most widely used material currently; however, donor site morbidity and stricture recurrence are continued problems. Tissue engineering has been introduced as a promising solution to repair and reconstruct lower urinary tract tissues, including the bladder and urethra. Clinical translation of tissue-engineered products has made significant progress in developing tangible therapies and inspiring the next generation of medical science over recent decades. Cell-based tissue engineering approaches have been employed to treat bladder and urethral pathologies in patients, demonstrating that multicellular tissues and organs with complex functions can be built for clinical use. This chapter covers the recent advancements in tissue engineering in the lower urinary tract tissues. Specifically, we discuss the strategic approaches, components used, supporting technologies, and tissue applications of the bladder and urethra.

1 Introduction

The lower urinary tract is mainly comprised of the bladder and urethra. The bladder is a hollow organ that stores and expels urine through the urethra (Balsara and Li 2017). Anatomically, the bladder is composed of two main parts: a highly specialized urothelium and a compliant detrusor. The bladder wall consists of a urothelial cell-lined lumen surrounded by connective tissue and smooth muscle layers (Ajalloueian et al. 2018). The detrusor muscle, composed of smooth muscle fibers, relaxes to accommodate urine deposition and contracts to expel urine when full. This elastic tissue property permits the storage of a large volume of urine. The urothelium, consisting of transitional epithelial cells, covers the bladder lumen, which acts as a barrier to preventing urine and waste product absorption. The connective tissue layer present between the urothelium and detrusor muscle is lamina propria, which is involved in modulating bladder tissue responses. The lamina propria contains fibroblasts, interstitial cells, and adipocytes, rich in blood vessels, afferent, and efferent nerve endings (Andersson 2014).

The lower portion of the bladder or neck is connected to the urethra, consisting of fibromuscular tubular tissue. The urethra serves as a passage for urine expelled from the bladder and semen from the ejaculatory ducts. The male urethra is 18–20 cm in length and 8–9 mm in diameter and is divided into prostatic, membranous, bulbar, and penile urethra (Partin et al. 2015). The most proximal portion is the prostatic urethra, which commences from the bladder and is entirely enclosed within the prostate. The mid-portion of the prostatic urethra is the opening of the paired ejaculatory ducts. The membranous portion extends from the apex of the prostate to the bulbar urethra. Pseudostratified columnar epithelium covers the membranous urethra and the external urethral sphincter and urogenital diaphragm surrounds it.

From the bulb of the penis to the urethral meatus is the penile urethra (Abbas et al. 2019). In adult females, the urethra is 3–4 cm in length and is embedded behind the symphysis pubis. The urethral wall divided into four layers from innermost to outer are (1) epithelium, (2) sub-mucosa, (3) fascial layer, and (4) two layers of smooth muscles (Gabrich et al. 2007). Squamous epithelial cells line most of the female urethra except for a small group of transitional epithelial cells in the proximal urethra (Liu et al. 2016).

There exist many conditions that can damage the lower urinary tract tissues. These include trauma, inflammation, cancer, and congenital anomaly (Lazzeri et al. 2016). Over the years, reconstructing the lower urinary tract has been performed to repair damaged or diseased tissues using various tissue substitutes, ranging from autologous to allogenic to manufactured synthetic sources. While reconstructive efforts have had variable successes over the years, challenges remain to be addressed on the horizon. Reconstruction of the bladder remains one of the most significant challenges in urology. Since the first cystectomy for bladder cancer in 1887, pursuing the most suitable replacement tissue substitutes for the bladder has been an ongoing challenge (Poole Wilson and Barnard 1971). Consequently, the use of intestinal tissue has remained as the gold standard for bladder reconstruction and repair, despite the associated complications, such as mucus production, electrolyte imbalances, recurrent infections, and malignancies (Adamowicz et al. 2019a; Pi et al. 2018). Similarly, the urethral tissue reconstruction due to trauma or congenital malformation such as hypospadias is also a surgical challenge (Keays and Dave 2017). The autologous buccal mucosa is the most widely used material; however, donor site morbidity and stricture recurrence are continued problems (Caldamone et al. 1998; Markiewicz et al. 2008).

Tissue engineering has been introduced as a promising solution to repair and reconstruct lower urinary tract tissues, including the bladder and urethra. Tissue engineering is a field that applies the principles of cell biology, biomaterials science, and engineering to develop biological substitutes for repairing damaged tissues and restore normal function (Griffith and Naughton 2002; Mangir 2019a). In 2006, the first clinical trial for bladder augmentation using tissue engineering was reported, a milestone in the field (Atala et al. 2006a). Using cell-seeded biodegradable grafts proved that multicellular structure organs with complex functions could be built for clinical use. The urethra is another tissue repaired using tissue engineering techniques in patients with stricture diseases (Raya-Rivera et al. 2011). Clinical translation of tissue-engineered products has made significant progress in developing tangible therapies and inspiring the next generation of medical science over recent decades. Increasing numbers of FDA-approved tissue products are being tested in the clinic, indicating that tissue engineering has become a promising alternative approach for tissue and organ replacement (Griffith and Naughton 2002). This chapter covers the recent advancements in tissue engineering applications in the bladder and urethra. We discuss the strategic approaches, components used, supporting technologies, and tissue applications of the bladder and urethra.

2 Engineering Strategies for Bladder and Urethra

In 1917, the initial application of a free fascia tissue graft for canine bladder augment was reported. Since then, numerous graft materials have been tried in bladder reconstructive studies (Neuhof 1917). In 1950, bladder reconstruction using a plastic mold was performed in dogs (Bohne and Urwiller 1957), and then in humans (Sanchez et al. 1958). The efforts to find ideal materials for bladder tissue reconstruction have been continued. Reconstruction of the human bladder remains one of the greatest challenges in urology. Currently, intestinal tissue is being used as a gold standard for bladder repair. However, because of gastrointestinal tissue incompatibility with urine, a series of complications occur, including mucus production, recurrent urinary tract infections, electrolyte imbalances, and malignancy. As such, investigators have sought to develop alternative methods, including auto-augmentation, tissue expansion, ureterocystoplasty, and regenerative medicine with cell transplantation (Partin 2015).

Regenerative medicine strategies include cell-based therapies, the use of scaffolds seeded with cells, and biomaterials alone (Partin 2015). An ideal neobladder for application in patients should have a large storage capacity, high compliance, low pressure for continence, and voluntary emptying without residual urine (Koie et al. 2018). The desired tissue substitute for urinary bladders' engineering should be nonmigratory, non-antigenic, volume stable, and safe (Partin 2015). In respect of functionalization strategies, bioactive factors could be incorporated into biomaterials and scaffolds. For example, pharmacological factors released by electrospun biomaterials to stimulate new extracellular matrix formation and angiogenesis have been suggested (Mangır et al. 2017; Mangır et al. 2016).

Although the urethra is an adjacent tissue structure connected to the bladder, its engineering approaches depend primarily on the type of tissue substitutes used, such as degradable biomaterials. The ultimate goal is to reconstruct normal functioning urethra that includes the structural and biological properties similar to native tissue. Urethral stricture is a common condition caused by injury, infection, and congenital anomaly, characterized by narrowing or obstruction of urine passage. Management of complicated, lengthy urethral strictures is considered a difficult and challenging task due to frequent recurrence rates. Many tissue substitutes have been used for urethral repair applications, such as skin and oral mucosa grafts; however, these materials have not been entirely satisfactory (Mangir 2019a). Recent success in human urethra repair using tissue engineering techniques provides hopes to patients in need (Raya-Rivera et al. 2011).

Urethral tissue engineering strategies require the universal requirements such as proper mechanical compliance, barrier function to avoid urine leakage to the underlying layer, and specific considerations for different urethral segments such as elastic properties to accommodate structural changes during the erection of the penis. Furthermore, the engineered tissue compliance is vital in adapting pressure during micturition, where the bladder pressure increases up to 50–60 cm H_2O with the urine flow rate of approximately 20–30 mL/s (Mijailovich et al. 2007). Therefore, urethral tissue with passive viscoelastic properties must prevent the escalation of pressure in the bladder that could reflux and damage the upper urinary tract (Yalla and Burros 1974). It is known that even a very short stricture can affect renal function. The most common cause of renal transplantation before puberty is posterior urethral valves in children where normal urine flow is disrupted (Husmann and Rathbun 2006).

To achieve urethral tissue's functional properties, efforts have focused on developing tissue constructs that replicate the native tissue configuration, consisting of smooth muscle and epithelial layers. The epithelial lining is an important tissue component that prevents leakage of cytotoxic urine into the body, thus avoiding the possible occurrence of inflammation, fibrosis, and stricture (Akkad 2007). For the cell-based tissue constructs, the viability of cells within the scaffold can be affected by exposure to urine (Rajasekaran et al. 2006). Therefore, an indwelling catheter is usually placed after surgery to divert urine to provide adequate time for the cells to mature and cover the urethral lumen to perform the urothelial barrier function. *The* cell sheet engineering technique, which allows for cellular sheet fabrication, has been applied (Kajbafzadeh et al. 2017).

3 Vascularization

Establishing a vascular network is critical in achieving successful tissue engineering therapies. Vascularization involves the sprouting of capillaries from existing blood vessels in vivo. This process involves complex interactions among endothelial and non-endothelial cells, growth factors, enzymes, and adhesion molecules. Many approaches have been introduced to promote vascularization, including the use of biological factors, such as growth factors and endothelial cells to stimulate angiogenesis. Other strategies include cell sheet engineering, scaffold functionalization, arteriovenous loops, spheroid coculture, modular assembly, and bioprinting. Toward this goal, one study designed a scaffold that promotes attachment and proliferation of the endothelial cell, thus facilitating the capillary formation. The addition of growth factors into the microenvironment and incorporating a modified angiogenic cell source was also attempted (Mangir et al. 2019b). Recently, a phase 2 clinical study of a bioengineered human acellular vessel (HAV) was performed and evaluated. A biodegradable mesh scaffold seeded with human vascular cells was implanted as a hemodialysis conduit in patients with end-stage renal disease. After 16 to 200 weeks follow-up, human acellular vessels were transformed into functional multilayered living tissues by endothelial, myogenic, and progenitor cells of the host. The new vessels were functional in transporting blood and self-healing after cannulation injury (Dahl et al. 2011a). While this approach may not directly apply to vascularize implanted tissues, it is hopeful that this technology may be modified and further developed to benefit tissue implants in the future. Attempts to enhance vascularization have been tried in bladder urethral tissue applications. Rapid angiogenesis is crucial for tissue regeneration, particularly for large-sized grafts. Overcoming the diffusion limitation is key to success, and numerous investigators have sought to solve this challenge (Rouwkema et al. 2008).

Establishing vascularization of full-thickness engineered soft tissue constructs with multilayer epithelium is not easily achieved, leading to cell necrosis and shedding. Shukui Zhou et al. have monitored the tissue-engineered urethra construct after subcutaneous implantation (Zhou et al. 2017). The investigators observed that new capillaries infiltrated into the implant and maintained an intact epithelial layer. However, blood capillaries were not fully matured. This study indicates that the implanted tissue can recruit new blood vessels that support the survival of cells. Another study used a rat-tail collagen-based acellular collagen scaffold to repair urethral defects in rabbits (Pinnagoda et al. 2016). Repopulation of urothelial and muscle cells was observed on all grafts with vessel formation at 1 month. To further facilitate vascularization, one group conjugated vascular endothelial growth factor (VEGF) to collagen (Jia et al. 2015). In this study, a modified tubularized collagen scaffold was used to repair a 5-cm long anterior urethral defect in dogs. The results demonstrated that the thickness of the epithelial layers was maintained with increased formation of blood vessels in the VEGF conjugated scaffold group.

4 Bioreactors

Bioreactors have become a standard supporting technology for tissue engineering. They serve many functions, from tissue preconditioning and maturation of tissue constructs to large-scale cell cultures (Martin et al. 2004). Bioreactors are complex engineering simulation biosystems that can control environmental factors (pH, dO_2, and T°) to recapitulate physiological body conditions (de Bournonville et al. 2018). Bioreactor systems have been used for bladder and urethral tissue applications. For example, organization and maturation of engineered bladder tissue construct can be simulated by providing mechanical filling and emptying control using an in vitro bioreactor (Haberstroh et al. 2002). In vivo bioreactors can improve vascularization of the bioengineered bladder, decrease the loss of contractility, and prevent fibrosis (Horst et al. 2013a).

Precise mechanical and nutrient environmental controls are also crucial in bladder bioreactors for enhancing cell proliferation and differentiation (Farhat and Yeger 2008). The goal is to achieve an impermeable epithelial surface, well-differentiated urothelial and smooth muscle cells, and excellent compliant extracellular matrices (Serrano-Aroca et al. 2018a). It is reported that mechanical stimulations positively affect the growth and development of newly developed tissues; however, high shear stress inhibits the proliferation of endothelial and smooth muscle cells (Stock and Vacanti 2001). In one study, Faisal M. Shaikh et al. have developed a pulsatile bioreactor that exerts dynamic uniform cyclical pressure over a urinary bladder matrix (UBM) scaffold (Shaikh et al. 2010). When cyclical mechanical pressure was applied on a cell-seeded urinary bladder matrix scaffold, significantly greater growth and urothelium viability were observed in the bioreactor compared with the conventional static conditions. A commercialized bioreactor system (BOSE BioDynamic®) was applied to urothelial and smooth muscle cells seeded collagen scaffolds. Mechanical stimulation of smooth muscle cells promoted cell growth, improved cell alignment, and distribution. However, dynamic mechanical stimulation did not affect the proliferation and differentiation of urothelial cells.

Mechanical stimulation in urethral engineering is beneficial in maturing the engineered tissue constructs. Magnan et al. investigated the self-assembly technique for urethral tissue constructs using an in vitro bioreactor (Magnan et al. 2009). Their results showed that the engineered tissue grafts displayed sufficient resistance to the pressure applied by the bioreactor. Cattan et al. have combined the self-assembled scaffold with mechanical stimulation with in vitro bioreactors to produce tissue-engineered tubular graft (Cattan et al. 2011). Their results showed that mechanical stimulation leads to maturation and formation of a stratified urothelium.

5 Biomaterials for Bladder and Urethra

Biomaterials are a major component of tissue engineering. Ideal biomaterials for scaffold should be biocompatible, non-antigenic, and biodegradable. The scaffolds provide structural support and a suitable microenvironment for cells to attach and proliferate (Chan and Leong 2008). Therefore, scaffolds should be designed and fabricated according to the specifics of tissue being engineered. In this regard, the bladder scaffold should provide storage of adequate urine volume and the ability to contract for physiologic voiding. Thus, many tissues and biomaterials of diverse nature have been examined for tissue engineering of the bladder. The bladder scaffold made by natural or synthetic biomaterials is the key element supporting the bladder function (Oberpenning et al. 1999a). The scaffold for the bladder should endure dynamic mechanical and chemical stimuli under normal physiological filling and emptying. The proliferation, migration, and differentiation of the seeded cells will be affected by the matrix (Oberpenning et al. 1999a). Therefore, the biomaterials used for bladder applications should be biocompatible and possess sufficient mechanical and chemical properties. Furthermore, the luminal surface should allow for uniform urothelial cell attachment to prevent urine leakage into the sub-urothelial tissue, and the outer side should harbor the smooth muscle cells for contraction, innervation, and vascularization (Horst et al. 2013b).

Similar to the scaffolds for bladder, the ideal biomaterial for the urethra should possess adequate mechanical support to prevent premature collapse before tissue regeneration is achieved (Orabi et al. 2013a). One study shows that a self-assembled engineered cell sheet has many of these favorable characteristics. It is reported that a well-stratified urothelial cell layer formed on self-assembled collagen sheets can withstand a high burst pressure in a porcine model (Bouhout et al. 2010). The self-assembled collagen sheets made from human adipose-derived stromal cells also showed similar mechanical and architectural characteristics (Rousseau et al. 2015). Nondegradable materials were also considered for urinary tract applications; however, they resulted in calcification, chronic hematuria, stone formation, fistulae, and graft contracture (Bouhout et al. 2010; Feil et al. 2006). Degradable biopolymers, on the other hand, allowed for tissue ingrowth, resulting in becoming tissues closer to the native normal urethra.

Degradable polymers for urethral tissue engineering stem from natural sources or by synthesis. Natural biomaterials such as acellular matrices and natural polymers are known to promote cell growth and differentiation and enhance angiogenesis (Ouellet et al. 2011). Currently, the most widely used polymers for urinary tract tissue applications include linear polyesters, copolymers, poly-lactic acid (PLA), poly-glycolic acid (PGA), poly lactic-co-glycolide (PLGA), and polycaprolactone (PCL). Synthetic polymers have the advantage of providing reproducible mechanical characteristics and controlled degradation.

6 Decellularized Matrices

Two classes of biomaterials, acellular matrices(Chen et al. 1999a) and biopolymers (Oberpenning et al. 1999a), have been used as scaffolds in bladder regeneration. These biomaterials can be configured as three-dimensional porous structures (Dhandayuthapani et al. 2011a), in which the seeded scaffold construct can serve as a template for the growth of new tissue (Ajalloueian et al. 2018). Tissue-derived decellularized matrices have been favorably used for bladder and urethra applications due to their biocompatibility and similarity in composition and inner architecture to native tissue (Ahmed 2019). Allogenic acellular matrices, such as the bladder submucosa matrix, offer excellent trophic support, thus stimulating tissue growth, facilitating cell–material interactions, maintaining the functional phenotype (Chun et al. 2007). Decellularized bladder tissue-derived scaffold is usually obtained from porcine due to its similarity to human bladders in anatomy and biology (Crapo et al. 2011; Zhang et al. 2000). It has been shown that the decellularized matrix facilitates cellular adhesion, proliferation, and maturation. However, some disadvantages include potential process-induced damage to the extracellular matrix ultrastructure and mechanical properties. In addition, the host innate immune response may be induced by the degradation of extracellular matrix scaffolds (Taylor et al. 2018).

Extracellular spaces contain a network of extracellular matrix (ECM) proteins and polysaccharides (Alberts et al. 2002). The ECM is an intricate crosslinked network structure of proteins, growth factors, small molecules, and glycosaminoglycans, and the commonly used tissue ECM for urinary tract applications include bladder acellular matrix (BAM), small intestinal submucosa (SIS), pericardium, fascia lata, amniotic membrane, and tunica vaginalis (Ramuta and Kreft 2018). ECM's main constituents are collagens, fibronectin, elastin, laminins, proteoglycans, glycoproteins, and glycosaminoglycans (Theocharis et al. 2016). As an example, an ECM scaffold was generated by cells seeded on polyglycolic acid scaffolds to deposit ECM, followed by a decellularization process to produce a vascular graft. The FDA approved this product for clinical use (Dahl et al. 2011b; Niklason et al. 1999). The acellular vascular graft or human acellular vessel (HAV) maintained mechanical integrity when implanted in patients after up to 4 years follow-up (Kirkton et al. 2019). In one study involving urethral repair, ECM scaffolds were shown to play a signaling role and maintained cellular homeostasis. More importantly, ECM is found to regulate the urethra's mechanical compliance of the urethra by dilating its lumen during micturition and stretch during erections (Shirozu et al. 1995; Ohel et al. 1995).

The typical and frequently used decellularized xenogeneic or allogeneic matrices are porcine small intestine submucosa (SIS) and human bladder acellular matrix (BAM). As a collagen-based biomaterial, SIS is processed by mechanically removing the mucosal, muscular, and serosal layers from the inner and outer surfaces, yielding a collagen-rich membrane of approximately 0.15–0.25 mm in thickness, mostly consisting of the submucosal tissue (Badylak et al. 1989). In the 1980s, Badylak et al. first applied SIS for bladder augmentation in dogs (Badylak et al. 1989). In 2005, Lu et al. (Lu et al. 2005) seeded cells onto SIS scaffolds. In their study, histological results indicated that seeded muscle-derived cells migrated throughout the graft after 20 days, and initial remodeling of the SIS occurred within 10 days. However, moderate-to-heavy adhesion and graft shrinkage was reported in both the cell-seeded and unseeded SIS by Zhang and colleagues (Zhang et al. 2006a). Many studies have demonstrated that the urothelium regeneration was achieved quicker than the muscle layer (Kropp et al. 1996; Caione et al. 2012). In another report, Schaefer et al. suggested that unseeded small intestine submucosa is not well suited for human enterocystoplasty because of the insufficient increase in bladder compliance (Schaefer et al. 2013). In other studies, SIS was shown to promote cellular growth and angiogenesis. The results of SIS was comparable to skin and mucosal grafts in animal experiments of urethral replacement (Fiala et al. 2007; Palminteri et al. 2012). SIS also showed favorable outcomes when used as a corporal body graft (Hayn et al. 2009) and as a seeded graft in a urethroplasty (Zhang et al. 2016). It was found that SIS decreased the risk of urine extravasation and reduced early irritation when used for urethral replacement (Yoo et al. 1998). However, the data of clinical experiments with SIS were not entirely satisfactory, mainly caused by infection (Orabi et al. 2013b).

Bladder acellular matrix (BAM) is an attractive biomaterial for tissue engineering (Zhang et al. 2006b). It is an ECM with an intact microarchitecture, first described in 1975 (Brown et al. 2005). One of the first applications of BAM was to support the regeneration of the rat bladder wall (Probst et al. 1997). A study evaluated the potential of the bladder submucosa matrix for dog bladder augmentation (Yoo et al. 1998). In this study, one side of the scaffold was seeded with autologous urothelial cells and the opposite side with smooth muscle cells followed by augmentation cystoplasty. The result showed that a 99% increase in capacity was observed in the cell-seeded group, whereas the unseeded group showed only a 30% increase. This study concluded that cell-seeded allogenic bladder submucosa is an excellent biomaterial for bladder augmentation. This research paved the way for the subsequent human bladder augmentations (Atala et al. 2006b). In another study, BAM seeded with marrow-derived mesenchymal stem cells was transplanted in rats after partial cystectomy. Histological analyses showed increased muscle regeneration and the bladder capacity nearly recovered completely for up to 6 months (Coutu et al. 2014).

BAM and SIS have been widely used for urethral tissue repair experimentally and clinically (Davis et al. 2018a). In a rabbit model of urethroplasty, cell seeded BAM scaffold demonstrated a normal urethral architecture when implanted for 4 weeks, and achieved near natural urethra after 6 months (Chun et al. 2015; Gu et al. 2012). In another rabbit study, autologous urethral tissue was repaired with combined BAM and suggested that the use of BAM may be a viable option for long-segment urethral stricture (Chun et al. 2015). However, residual immunogenic components are a safety concern of BAM, despite the stringent decellularization process (Roth and Kropp 2009). Other tissue-derived scaffolds have been developed for urinary tract applications. In one study, decellularized trachea was applied as a tissue-engineered neo-urinary conduit (Anirudha et al. 2018). In this study, a decellularized rabbit trachea was seeded with human smooth muscle cells and urothelial cells. The results showed that decellularized trachea possesses appropriate biomechanical properties and structural integrity. Another promising natural scaffold biomaterial proposed for bladder applications is tissue-engineered pericardium, which has achieved favorable results in the test of small-sized scaffolds (Kajbafzadeh et al. 2011).

Tissue-derived matrices are biocompatible, durable, effective, and are easily remodeled (Horst et al. 2013c). However, seeding smooth muscle cells is difficult to infiltrate into the scaffold because of the high density of these materials (Atala 2011). They are also prone to fibrosis and exhibit poor mechanical and dimensional stability. Therefore, artificial tissue-engineered biomaterials or a combination of synthetic and biological materials may be a choice for use in bladder reconstructive surgery (Zhang et al. 2006b).

7 Natural Polymers

Natural polymers have been frequently used for many tissue engineering studies due to their biocompatibility, less toxicity, and superior interactions with cells compared with synthetic polymers (Dhandayuthapani et al. 2011b). Collagen, gelatin, fibrinogen, and elastin are more commonly used scaffold materials for bladder and urethra applications. Collagen is an ECM-derived natural material and plays a crucial role in promoting cell attachment and maintaining structural integrity, thus improves tissue regeneration (Hubbell 2003) and decreases inflammatory and antigenic responses (Furthmayr and Timpl 1976). Many collagen-based materials have been tried for urethral reconstruction such as collagen gels, bladder-derived acellular submucosa and acellular urethral submucosa (Partin et al. 2015). In one study, rat-tail collagen was used to generate acellular collagen tubular scaffolds to repair a 2 cm-long rabbit urethral defect (Pinnagoda et al. 2016). The collagen tube possessed good surgical handling and did not require synthetic polymer support to maintain structural integrity. Results demonstrated that spontaneous repopulation of urothelial and muscle cells was observed on all grafts. The cellular organization increased with time, although fistula and stenosis occurred in 20% of the animals postoperatively.

As a natural polymer, silk fibroin (SF) is derived from Bombyx mori cocoons. SF is a biocompatible polymer that can degrade into peptides and amino acids (Wang et al. 2008). SF has been used for urethral applications. In one study, urothelial cell-seeded silk fibroin grafts were functioning without causing strictures after implanted for 6 months in urethroplasty (Xie et al. 2013). Bilayer silk fibroin (BLSF) scaffolds have also been evaluated for urethral repair in a rabbit model of onlay urethroplasty. Histological analyses demonstrated the formation of innervated and vascularized neotissues at graft sites (Algarrahi et al. 2018). The bilayer silk fibroin (BLSF) matrices were recently investigated in pigs with acute partial bladder outlet obstruction. The animals underwent augmentation cystoplasty with BLSF grafts. The bladder capacity and compliance in the graft augmented group significantly increased by 79% and 171%, respectively, compared to the baseline values after 3 months. The study shows that BLSF scaffolds can improve the capacity and compliance of bladder and promote the formation of neotissues (Saif et al. 2019). Another important natural polymer is hyaluronic acid (HA), which has been applied in wound healing and 3D bioprinting applications (Burdick and Prestwich 2011). The US Food and Drug Administration (FDA) also approved another natural polymer, alginate, used for bladder tissue engineering after modification with cell-adherent peptides (Lee and Mooney 2012).

8 Synthetic Biomaterials

Over the past decades, many synthetic and organic materials have been used for bladder regeneration. Synthetic materials used for bladder and urethral tissue applications include polyvinyl sponge, Teflon, collagen matrices, polylactic acid (PLA), poly (glycolic acid) (PGA), Polycaprolactone (PCL) (Pinnagoda et al. 2016), copoly(lactic/glycolic) acid (PLGA) (Yao et al. 2013; El-Taji et al. 2015), poly(ε-caprolactone)/poly (L-lactic acid) (PCL/PLLA)(Shakhssalim et al. 2013), Poly (Carbonate-Urethane) Urea (PCUU), Nanocellulose (Bacakova et al. 2019), poly(1,8-octanediol-co-cirtic acid) (POC)(Liu et al. 2016), and silicone (Serrano-Aroca et al. 2018b). These polymers can be fabricated reproducibly, with controllable mechanical and degradation properties, porosity, and structures (Horst et al. 2013c). Electrospun nanofibers with core-shell structure have been applied in rat bladder augmentation. Domed scaffold fabricated by coaxial electrospinning of poly(l-lactide)/poly(e-caprolactone) (PLCL) and Hyaluronic acid (HA) mesh possessed good cytocompatibility and was convenient for smooth muscle tissue proliferation (Chunxiang et al. 2019). In another study, Poly (Carbonate-Urethane) Urea (PCUU) scaffold for bladder engineering was evaluated. The biodegradable elastomers showed cytocompatibility, increased porosity, and stretch, which can aid smooth muscle cells infiltration. Bladder augmentation with Poly (Carbonate-Urethane) Urea scaffolds can increase the bladder capacity and voiding volume, and enhance survival of the rat disease model (Sivaraman et al. 2019). Synthetic matrices seeded with cells have been applied for urethral reconstruction. A widely used synthetic material for urethral repair is polylactide-co-glycolide (PLGA) today (Adamowicz et al. 2019b).

While there are advantages to using synthetic materials, the disadvantages include urinary stone formation and mechanical failure for nondegradable polymers. In contrast, degradable materials often result in fibrosis, scarring, graft contracture, and a long-term reduction of voiding volume (Partin 2015). Other disadvantages of synthetic materials are their biological inertness, foreign body reaction, the lack of trophic support for cell activity, and the natural barrier function of urothelium (Horst et al. 2019). It has been reported that pre-seeding the scaffolds with cells is shown to relieve graft shrinkage (Zhang et al. 2006b; Filippo et al. 2002). However, this process requires time-consuming preparatory steps of cell harvesting, isolation, seeding, and proliferation on scaffolds. Despite the laborious procedures, the preferred option for bladder engineering seems to be the use of autologous cells in combination with synthetic biomaterials (Horst et al. 2019).

To be able to use synthetic polymers for cell-based applications, the surface of polymers must be in favor of cell attachment. A novel surface-modification of poly-L-lactic acid (PLLA) scaffold was produced to support urethral cell adhesion and proliferation (Fu et al. 2011). In this study, a hollow cylinder was wrapped with PLLA filaments, followed by chitosan and fibronectin treatment. Rabbit uroepithelial cells were adhered to and cultivated on the scaffold. The scaffold has more than 90% porosity, and its mesh was fully populated by uroepithelial cells after 3–7 days. In another study, electrospun Poly(l-lactide)/Poly(ethylene glycol) (PLLA/PEG) hybrid scaffolds were produced to support urethral tissue regeneration (Lv et al. 2016). The scaffolds were then seeded with human amniotic mesenchymal cells (hAMSCs) for repairing rabbit urethral defects. The results showed that hAMSCs-loaded PLLA/PEG scaffolds produced multilayered urothelium with similar characteristics to the native urethra.

Cellulose is known to be the most abundant biopolymer on earth. Nanocellulose is the cellulose of nanostructures with a feature that, at least in one dimension, is not exceeding 100 nm (Bacakova et al. 2019). A three-dimensional porous bacterial cellulose (BC) scaffold seeded with lingual keratinocytes was used to repair rabbit ventral urethral defects. The results demonstrated that all urethras maintained a wide caliber in 3D porous BC seeded with cells. The smooth muscle content and endothelia density were significantly increased at 3 months. This study shows that the BC scaffolds seeded with cells can enhance urethral tissue regeneration (Huang et al. 2015).

9 Smart and Hybrid Polymers

Smart biomimetic materials are a significant advancement in the field of tissue engineering. Some smart biomaterials are designed to respond reversibly to temperature, pH scale, light, or ionic strength (Aguilar and San Román 2019). The responses of these polymers may include gelation, reversible adsorption on a surface, collapse of a hydrogel, and alteration between hydrophilic and hydrophobic states. Smart biomaterials have been studied in many areas, including drug delivery, medical devices, and tissue engineering applications (Kowalski et al. 2018). Furthermore, combinations of natural and synthetic polymers can be designed to produce hybrid biomaterials with specific properties for tissue engineering. Such properties may include controlling porosity, mechanical strength, cell affinity, biocompatibility, and biodegradability (Chun et al. 2015).

A group of electroconductive polymers has demonstrated promise for engineering electrically active tissues. Electroactive polyurethane polymers have manifested enormous potential in bladder tissue engineering, especially in regenerating muscular components and innervation (Hardy et al. 2015; Wu et al. 2016). Electroconductive polymers may be applied with electrical stimulation for advanced maturation of engineered urethra (StÖlting et al. 2016). In one study, a smart acellular collagen scaffold with growth factors was used in sheep bladder regeneration. The results showed that the presence of growth factors, VEGF, FGF2, and HB-EGF improved bladder regeneration (Roelofs et al. 2016).

The largest class of smart polymers used is thermo-responsive polymers. A reversibly alterable phase or volume transition can occur following a change of temperature in these materials. A typically used biomaterial for this purpose is poly (N-isopropylacrylamide) (PNIPAM). The lower critical solution temperature (LCST) of PNIPAM is around 32 °C, not far from the human body temperature (Kim and Matsunaga 2017). Accordingly, PNIPAM solution can be changed from a hydrophobic to hydrophilic surface following the change of temperature in water. When the temperature is above the lower critical solution temperature, the polymer chains of PNIPAM are precipitated, making it hydrophobic, while below LCST, polymer chains hydrated and made PNIPAM hydrophilic (Kim and Matsunaga 2017). This characteristic is beneficial in producing three-layered urethral tissue constructs. It was reported that viable urothelial cell sheets could be generated using the temperature-responsive cell culture method (Shiroyanagi et al. 2003).

A smart bilayered scaffold seeded with keratinocytes and muscle cells was also tested in canine urethral reconstruction. The bilayer scaffold consisted of a microporous network of silk fibroin (SF) and a nanoporous bacterial cellulose (BC) scaffold with a porosity of 85%. The result showed that the smart SF-BC structures could enhance adhesion and proliferation of lingual keratinocytes and lingual muscle cells. At 3 months after graft implantation, the urethra reconstructed with the seeded SF-BC scaffold showed superior structure formation compared to the unseeded control scaffold. The nanoporous network offers excellent support for epithelial cells, while the microporous scaffolds sustain the growth and penetration of smooth muscle cells (Lv et al. 2018). Shape-memory polymers (SMPs) have also been tested in vascular and bone tissue engineering. The characteristics of SMPs are suitable for urethral tissue engineering, as urethra is subject to expansion during erection and recoils during the flaccid state. Smart acellular collagen scaffold with growth factors have been applied in sheep bladder regeneration studies (Roelofs et al. 2016).

10 Cells for Engineering Bladder and Urethra

Cells are an integral part of tissue engineering. While cells from various sources have been used for tissue regeneration studies, autologous cells are considered the preferred source for bladder and urethral applications in patients currently (Rashidbenam et al. 2019). Experimentally, both somatic and stem cells have been used for bladder and urethra tissue studies, spanning from autologous, allogeneic, and xenogeneic sources.

10.1 Somatic Cells

One tissue engineering strategy that has been shown to be effective in reconstructing bladder and urethra tissues involves the use of autologous cells from the bladder (Zhang and Frey 2003). The donor tissue is dissociated to isolate urothelial and muscle cells. The cells are then culture expanded, attached to a supporting matrix followed by implantation back into the same host (Atala et al. 1992; Li et al. 2008; Feng et al. 2011; Atala 2009). This approach has shown to provide sufficient numbers of cells for bladder and urethral tissue applications clinically. While the ideal cell source for tissue application is from the target organ of interest, there are instances where obtaining a tissue biopsy from the host may not be feasible. This is true when the organ or tissue to be reconstructed cannot provide healthy cells, such as bladder cancer. In such a case, cells, either somatic or stem cells, should be obtained from other tissue sources.

Epithelial cells for urethral tissue engineering could originate from different organs, such as the bladder, penile foreskin, and buccal mucosa. In addition, autologous urothelial cells can also be isolated from urine (Nagele et al. 2008). In one study, the penile foreskin was used as a source of autologous epidermal cells. These cells were seeded on acellular collagen matrix and used as a tissue implant for urethral applications in animal models with positive results (Fu et al. 2007). Autologous oral keratinocytes was previously obtained from a buccal mucosa biopsy for urethroplasty (Barbagli et al. 2018). The sources for smooth muscle cells include the bladder, blood vessel, and gastrointestinal tissues. The culture of both cell types is well established and their expansion capacity far exceeds the cell numbers necessary for reconstruction.

10.2 Stem Cells

Stem cells have become an attractive cell source for reconstructive tissue engineering due to their ability to differentiate into multiple cell types, and if guided appropriately, they can transform into target cells of interest. For this reason, stem cells have been used extensively in urinary tract application studies (Serrano-Aroca et al. 2018a). One stem cell type that is widely investigated is adult mesenchymal stem cells (MSCs), which can be isolated from multiple body sources such as bone marrow and adipose tissue (Adipose-derived Stem cells; ASCs). Mesenchymal stem cells have been shown to be safe and effective in many tissue applications, including the urethra (Stangel-Wojcikiewicz et al. 2014; Deng et al. 2014). Moreover, these cells possess immunomodulatory properties, thus eliminating the need for autologous cells for tissue applications (Davis et al. 2018b).

Adipose-derived stem cells have been widely used due to the ease of harvest, isolation, and expansion as compared to other sources of MSCs. They are also immunomodulatory and can secrete various growth factors to promote angiogenesis and neurogenesis (Kingham et al. 2014). As such, ASCs have been favored as a cell source for bladder regeneration studies (Wang et al. 2019). These cells have been shown to differentiate into smooth muscle (Salem et al. 2013) and urothelial-like phenotype (Zhang et al. 2013), thus avoiding the use of native cells from bladder cancer patients. In one study, bioengineered bladder patches constructed from multilayered adipose-derived stem cell sheets were successfully used for bladder regeneration (Wang et al. 2019). The effectiveness of ASCs has also been demonstrated in urethral applications. Davis NF et al. have reviewed 11 in vivo preclinical studies that investigated various stem-cell therapies for reconstructing the urethra (Davis et al. 2018b). Among them, ten studies showed the data on the patency in the stem cell-seeded bio-scaffolds; 100% patency was reported in nine studies, and the remaining one was 75% patency rate. These studies indicate that stem cell-seeded urethral scaffolds can recruit endogenous cells for tissue regeneration, and this effect is believed to be moderated by angiogenic trophic factors via a paracrine mechanism of stem cells (Caplan and Dennis 2006). In another study, Zhou S et al. used the myoblast-induced adipose-derived stem cell sheets to form muscular layers (Zhou et al. 2017). The adipose-derived stem cells were isolated from fat tissue in the groin of beagle dogs. The cells were cultured in myoblast differentiated culture medium and then stimulated with vitamin C to form ECM. The structure and function of neo-urethras were similar to normal urethra after 3 months. Human ASCs were also used with self-assembled collagen sheets. The mechanical and architectural characteristics were shown to be similar to that of human fibroblasts (Rousseau et al. 2015).

It has been demonstrated that bone marrow-derived stem cells (BMSCs) have the potential to differentiate into bladder muscle (Shukla et al. 2008a) and urothelial cells (Anumanthan et al. 2008). The differentiated cells resembled bladder smooth muscle cells and promoted muscle bundle formation (Zhang et al. 2005). A study used the bladder acellular matrix seeded with BMSCs for bladder tissue engineering. The results showed that the BMSC-seeded bladder acellular matrix improved nearly 100% of normal bladder capacity for up to 6 months after partial cystectomy (Coutu et al. 2014). In another study, a research team performed bladder augmentation after hemicystectomy in six dogs using the BMSC-seeded small intestine submucosa (SIS) (Zhang et al. 2005). Upon retrieval, solid smooth muscle bundle formation was observed throughout the cell-seeded grafts. In the unseeded SIS scaffolds, SMC regeneration was seen only at the graft edges adjacent to native tissue, suggesting the effects of BMSCs. In a porcine model, Shukla and colleagues also showed that the BMSCs were able to differentiate into smooth muscle in the pig (Shukla et al. 2008b). Recently, bladder tissue consisting of full three layers was successfully generated using differentiated human BMSCs (Gabr et al. 2018). BMSCs were cocultured separately with urothelial and muscle cell-derived conditioned medium to obtain the corresponding biological characterization. These cells were then seeded on the decellularized bladder adventitia scaffold. Histological examination of the recellularized tissue revealed the full three layers of bladder tissue (Shukla et al.

2008b). The advantage of synergistic effect of allogeneic BMSCs and endothelial progenitor cells (EPCs) was also investigated for treatment in a long segment urethral defect. These cells were seeded onto decellularized human amniotic scaffolds (dHAS) to repair a 3 cm-long circumferential urethral defect in dogs. The animals showed unhindered urination and capacious urethral caliber similar to the normal group in dHAS+BMSCs+EPCs and dHAS+EPCs groups (Chen et al. 2018).

Induced pluripotent stem cells (iPSCs), generated by somatic cell reprograming, have recently gained enormous attention as a non-controversial cell source of embryonic stem cells (Takahashi and Yamanaka 2006). These cells possess self-renewal properties and the ability to differentiate into various specialized cell types (Brivanlou et al. 2003). Due to the enormous potential of utility, iPSCs have been extensively investigated in numerous areas. Likewise, iPSCs have been studied in urology. These cells have been shown to induce and differentiate into urothelial cells and smooth muscle cells (Osborn et al. 2014; Zhe et al. 2016). However, further studies are necessary to use iPSCs for tissue applications.

Amniotic fluid and placenta stem cells are fetal-derived cells that have great potential for development. These cells do not present with ethical controversies that embryonic stem cells have but exhibit similar cellular characteristics. The cells are multipotent, highly replicative, and can differentiate into all three germ layers (De Coppi et al. 2007a; De Coppi et al. 2007b). These cells have been studied in many cell-based applications. In one study, human amniotic mesenchymal cells (hAMSCs) were seeded on Poly(l-lactide)/Poly(ethylene glycol) scaffolds and implanted to repair urethral defects in rabbits. The hAMSCs-seeded scaffold showed that epithelial cells covered the defect and formed multilayer mucosa membranes similar to normal urethra after implanted for 12 weeks (Lv et al. 2016).

Another stem cell type gaining much attention in urology is urine-derived stem cells (USCs) due to their existence in urine (Zhang et al. 2008). These cells are found to possess stem cell characteristics, robust proliferative, and differentiation potential. More importantly, these cells are easily obtained from voided urine, and the isolation procedures are simple and low costing. USCs have been shown to differentiate into many cell types, including urothelial, smooth muscle, vascular endothelial, osteogenic, neurogenic, and adipogenic cells (Bharadwaj et al. 2011; Bodin et al. 2010). USCs have been used in urological cell-based applications for urethral and bladder reconstructions (Bodin et al. 2010; Wu et al. 2011; Bharadwaj et al. 2013).

11 Applications of Bladder and Urethra

For more than 100 years, gastrointestinal segments have been used for bladder replacement or repair in the clinic, despite the associated complications (Partin 2015; Anderson and McKiernan 2018). Tissue engineering approaches were initially applied to address the reconstructive challenges associated with pathologic bladder and urethra conditions. Early studies have tested different scaffolds as cell carriers. PGA scaffold coated with PLGA were examined to determine whether the synthetic biomaterial could serve as a comparable biomaterial as acellular scaffolds. Human bladder urothelial and muscle cells were attached to both collagen matrices derived for bladder submucosa and nonwoven meshes of PGA in vitro (Atala et al. 1993; Cilento et al. 1994). When implanted into animals, these constructs were shown to survive and reorganize into newly formed multilayered structures that exhibit spatial orientation and vascularized in vivo. In a subsequent study, autologous urothelial and smooth muscle cells were seeded on PGA scaffold, and implanted the tissue-engineered neobladder in a canine subtotal cystectomy model. The tissue-engineered neo-organs, consisting of a bladder-shaped biodegradable polymer and autologous urothelial and muscle cells showed a normal bladder capacity and compliance and histologic findings comparable to normal bladder tissue. This was the first report demonstrating the successful reconstruction of the tissue-engineered bladder in a large animal model (Oberpenning et al. 1999b).

The successful preclinical study led to a clinical study using the tissue-engineered bladder construct. Autologous tissue-engineered bladder substitutes were used to treat patients with poorly compliant neurogenic bladders due to congenital anomaly (Atala et al. 2006a). The scaffolds made of collagen and polyglycolic acid (PGA) were fabricated for augmentation cystoplasty. The postoperative urodynamic studies on patients implanted with the engineered bladders showed varying degrees of contractility, capacity, and compliance. The margin between the composite matrix-based engineered segments and the native bladders was grossly indistinguishable during the cystoscopic evaluations. Patients treated with the engineered bladder showed no metabolic consequences or formed urinary calculi. Their levels of mucus production were normal, and renal function was preserved. The biopsies taken from the engineered bladder region showed adequate structural architecture and phenotype. This was the first study in patients demonstrating that engineered bladder tissues can be used in patients requiring bladder reconstruction.

Engineered tissues have also been applied in the urethra. The initial proof of concept studies examining the natural matrix as a biomaterial for urethra reconstruction was conducted on small segment urethral defects in New Zealand white male rabbits (Chen et al. 1999b). Porcine bladder matrices were used to repair a urethral defect in an onlay fashion. Urethrography showed that the reconstructed urethras maintained their diameter over 2 months. Histological evaluation of the retrieved constructs showed vessel formation and a transitional epithelial cell layer by 2 weeks. The smooth muscle layer was formed by 6 months post-implantation.

This basic strategy was used to guide a small clinical study (Atala et al. 1999). Four patients with age ranging from 4 to 20 years with a history of failed surgeries to correct hypospadias underwent reconstructive urethral repair using an acellular collagen matrix derived from cadaveric bladder. The acellular collagen matrix was trimmed to a rectangular shape of the appropriate size ranging 5–15 cm and attached to the urethral plate in an onlay fashion. At 1 year following surgery, retrograde cystourethrography indicated no narrowing of the engineered urethras. Three-year follow-up examinations showed all patients' engineered urethras functioned normally. One patient who received a 15 cm reconstruction developed a subglanular fistula easily repaired by standard methods. Encouraged by the clinical cohort's outcomes, a more extensive study was subsequently conducted (El-Kassaby et al.

2003). A total of 28 patients aged 22–61 with urethral stricture was included in the study. Acellular bladder submucosa was used to repair urethral defects ranging from 1.5 cm to 16 cm in an onlay fashion. Postoperative urethrography showed a patent urethra of wide caliber in all 24 of the successful reconstructions. Average urine flow rates increased from 6 +/− 1.57 ml/sec before the procedure to 9 +/− 1.3 ml/sec postoperatively. Patency of the entire reconstructed sections was confirmed by cystoscopy, and biopsies confirmed the restoration of normal urethral tissue.

While acellular collagen scaffolds were successful for onlay reconstructive procedures in which an intact urethral plate was maintained, studies showed that the use of the material for more complex reconstruction requiring a tubularized urethra resulted in stricture formation. Later studies indicated that, in recellularization of the acellular biomaterial, only implants occurred over short distances from healthy tissue (Dorin et al. 2008). It was found that fibrosis occurred in any biomaterial that was farther than 5 mm from the graft's edge. Onlay reconstruction with acellular matrix was successful because the procedure creates a urethra in which only 10 mm of the circumference of the urethra is comprised of the biomaterial. This allows recellularization to proceed 5 mm from either side of the onlay. Consequently, a study was conducted in rabbits to determine if autologous engineered urethral constructs composed of acellular porcine bladder submucosa seeded with autologous bladder cells could be used to repair complex urethral defects (De Filippo et al. 2002). Porcine bladder matrix was tubularized and seeded with autologous rabbit epithelial cells into the lumen of the scaffolds and smooth muscle cells on the exterior. The seeded grafts of 1 cm in length were used to repair a defect in 12 animals. Normal urethral caliber was maintained in all rabbits that received cell-seeded constructs. Histological analysis of the retrieved constructs demonstrated a normal transitional cell layer surrounded by muscle fiber bundles that became more organized over 6 months following implantation. Organ bath studies performed on retrieved engineered scaffolds showed an appropriate contractile response to chemical and electrical stimuli. This study demonstrated the utility of cell-seeded tubularized engineered urethras for the reconstruction of urethral defects.

A subsequent rabbit study was conducted that demonstrated the feasibility of using cellularized engineered urethral constructs to reconstruct extended urethral defects involving the entire circumference of the urethra (De Filippo et al. 2015). In this study, tubularized acellular bladder submucosa was also used to repair 3 cm long urethral defects in rabbits. All animals that received cellularized engineered constructs had normal urethral caliber across the length of the reconstructed segments. Grossly, the cellularized engineered constructs developed the appearance of normal urethra over time. The lumen of the constructs contained well-organized epithelium, and muscle fiber bundles developed over 6 months following implantation. These results indicate that tubularized collagen-based scaffolds seeded with autologous epithelial and smooth muscle cells can be used to bioengineer a construct to reconstruct a urethral defect much longer than can be accomplished with an acellular biomaterial.

These studies have led to a clinical study in five male pediatric patients with urethral strictures. (Raya-Rivera et al. 2011). Autologous urothelial and smooth muscle cells were isolated from a 1 × 1 cm bladder biopsy. Cells seeded on tubularized scaffolds were used to repair strictures.

No complications occurred during any of the surgeries. Cystourethroscopy confirmed patent urethras in all patients. Voiding cystourethrograms showed the maintenance of wide caliber urethras in all patients over the period of follow-up. Urethral biopsies showed that the engineered grafts had developed a normal-appearing architecture by 3 months after implantation. This clinical study shows that tubularized urethras can be engineered and remain functional in a clinical setting for up to 6 years, and that the engineered urethras can be used in patients who need complex urethral reconstruction.

12 Conclusions

Recent advances in tissue engineering have provided potential solutions to the existing challenges related to surgical reconstruction. Over the past decades, enormous progress has been made in the area of bladder and urethral reconstruction, experimentally and clinically. This chapter discussed the recent development and the progress of cell-based engineering approaches, covering foundational research to clinical studies. Research related to cells and biomaterials were discussed with a focus on bladder and urethral applications. The development of new and innovative technologies integrating into the tissue building processes has allowed for rapid translation. In addition, support technologies such as bioreactors have further accelerated the development of effective therapies. Continued advancement is expected in all areas leading to the production of viable tissue and organ substitutes through innovation. Although this chapter did not cover 3D bioprinting, it is anticipated that this technology will play a significant role in the tissue engineering field. Studies utilizing the 3D bioprinting technology to build improved lower urinary tract tissues are already being investigated with great interest. The success of tissue engineering will likely depend on innovation.

References

Abbas TO, Yalcin HC, Pennisi CP (2019) From acellular matrices to smart polymers: degradable scaffolds that are transforming the shape of urethral tissue engineering. Int J Mol Sci 20(7):1763

Adamowicz J et al (2019a) Reconstructive urology and tissue engineering: converging developmental paths. J Tissue Eng Regen Med 13(3):522–533

Adamowicz J et al (2019b) Reconstructive urology and tissue engineering: converging developmental paths. J Tissue Eng Regen Med 13(3):522–533

Aguilar MR, San Román J (2019) Chapter 1 - introduction to smart polymers and their applications. In: Aguilar MR, Román JS (eds) Smart polymers and their applications, 2nd edn. Woodhead Publishing, Sawston, pp 1–11

Ahmed S et al (2019) New generation of bioreactors that advance extracellular matrix modelling and tissue engineering. Biotechnol Lett 41(1):1–25

Ajalloueian F et al (2018) Bladder biomechanics and the use of scaffolds for regenerative medicine in the urinary bladder. Nat Rev Urol 15:155

Akkad T et al (2007) Influence of intravesical potassium on pelvic floor activity in women with recurrent urinary tract infections: comparative urodynamics might lead to enhanced detection of dysfunctional voiding. BJU Int 100(5):1071–1074
Alberts B et al (2002) Molecular biology of the cell, vol 91, 4th edn. Garland Science, New York, p 401
Algarrahi K et al (2018) Repair of injured urethras with silk fibroin scaffolds in a rabbit model of onlay urethroplasty. J Surg Res 229:192–199
Anderson CB, McKiernan JM (2018) Surgical complications of urinary diversion. Urol Clin North Am 45(1):79–90
Andersson KE, McCloskey KDJN (2014) Lamina propria: the functional center of the bladder? Neurourol Urodyn 33(1):9–16
Anirudha S et al (2018) Tissue-engineered neo-urinary conduit from decellularized trachea. Tissue Eng Part A 24(19–20):1456–1467
Anumanthan G et al (2008) Directed differentiation of bone marrow derived mesenchymal stem cells into bladder urothelium. J Urol 180(4 Suppl):1778–1783
Atala A (2009) Engineering organs. Curr Opin Biotechnol 20(5):575–592
Atala A (2011) Tissue engineering of human bladder. Br Med Bull 97(1):81–104
Atala A et al (1992) Formation of urothelial structures in vivo from dissociated cells attached to biodegradable polymer scaffolds in vitro. J Urol 148(2 Pt 2):658–662
Atala A et al (1993) Implantation in vivo and retrieval of artificial structures consisting of rabbit and human urothelium and human bladder muscle. J Urol 150(2 Pt 2):608–612
Atala A, Guzman L, Retik AB (1999) A novel inert collagen matrix for hypospadias repair. J Urol 162(3 Pt 2):1148–1151
Atala A et al (2006a) Tissue-engineered autologous bladders for patients needing cystoplasty. Lancet 367(9518):1241–1246
Atala A et al (2006b) Tissue-engineered autologous bladders for patients needing cystoplasty. Lancet 367(9518):1241–1246
Partin AW, Kavoussi LR, Peters CA (2015) Campbell-Walsh urology. In: Atala A (ed) Principles of tissue engineering, vol 1, 11th edn. Elsevier Health Sciences
Bacakova L et al (2019) Versatile application of Nanocellulose: from industry to skin tissue engineering and wound healing. Nanomaterials (Basel) 9(2):164
Badylak SF et al (1989) Small intestinal submucosa as a large diameter vascular graft in the dog. J Surg Res 47(1):74–80
Balsara ZR, Li X (2017) Sleeping beauty: awakening urothelium from its slumber. Am J Physiol Renal Physiol 312(4):F732–F743
Barbagli G et al (2018) Anterior Urethroplasty using a new tissue engineered Oral mucosa graft: surgical techniques and outcomes. J Urol 200(2):448–456
Bharadwaj S et al (2011) Characterization of urine-derived stem cells obtained from upper urinary tract for use in cell-based urological tissue engineering. Tissue Eng Part A
Bharadwaj S et al (2013) Multi-potential differentiation of human urine-derived stem cells: potential for therapeutic applications in urology. Stem Cells
Bodin A et al (2010) Tissue-engineered conduit using urine-derived stem cells seeded bacterial cellulose polymer in urinary reconstruction and diversion. Biomaterials 31(34):8889–8901
Bohne AW, Urwiller KJTJou (1957) Experience with urinary bladder regeneration. J Urol 77(5):725–732
Bouhout S et al (2010) In vitro reconstruction of an autologous, watertight, and resistant Vesical equivalent. Tissue Eng Part A 16(5):1539–1548
Brivanlou AH et al (2003) Stem cells. Setting standards for human embryonic stem cells. Science 300(5621):913–916
Brown RA et al (2005) Ultrarapid engineering of biomimetic materials and tissues: fabrication of Nano- and Microstructures by plastic compression. Adv Funct Mat 15(11):1762–1770
Burdick JA, Prestwich GD (2011) Hyaluronic acid hydrogels for biomedical applications. Adv Mater 23(12):H41–H56
Caione P et al (2012) Bladder augmentation using acellular collagen biomatrix: a pilot experience in exstrophic patients. Pediatr Surg Int 28(4):421–428

Caldamone AA et al (1998) Buccal mucosal grafts for urethral reconstruction. Urology 51(5, Supplement 1):15–19
Caplan AI, Dennis JE (2006) Mesenchymal stem cells as trophic mediators. J Cell Biochem 98(5):1076–1084
Cattan V et al (2011) Mechanical stimuli-induced urothelial differentiation in a human tissue-engineered tubular genitourinary graft. Eur Urol 60(6):1291–1298
Chan BP, Leong KW (2008) Scaffolding in tissue engineering: general approaches and tissue-specific considerations. Eur Spine J 17(Suppl 4):467–479
Chen F, Yoo JJ, Atala AJU (1999a) Acellular collagen matrix as a possible "off the shelf" biomaterial for urethral repair. Urology 54(3):407–410
Chen F, Yoo JJ, Atala A (1999b) Acellular collagen matrix as a possible "off the shelf" biomaterial for urethral repair. Urology 54(3):407–410
Chen C et al (2018) Transplantation of amniotic scaffold-seeded mesenchymal stem cells and/or endothelial progenitor cells from bone marrow to efficiently repair 3-cm circumferential urethral defect in model dogs. Tissue Eng Part A 24(1–2):47–56
Chun SY et al (2007) Identification and characterization of bioactive factors in bladder submucosa matrix. Biomaterials 28(29):4251–4256
Chun SY et al (2015) Urethroplasty using autologous urethral tissue-embedded acellular porcine bladder submucosa matrix grafts for the management of long-segment urethral stricture in a rabbit model. J Korean Med Sci 30(3):301–307
Chunxiang F et al (2019) Electrospun nanofibers with core–shell structure for treatment of bladder regeneration. Tissue Eng Part A 25:1289–1299
Cilento BG et al (1994) Phenotypic and cytogenetic characterization of human bladder urothelia expanded in vitro. J Urol 152(2 Pt 2):665–670
Coutu DL et al (2014) Tissue engineering of rat bladder using marrow-derived mesenchymal stem cells and bladder acellular matrix. PLoS One 9(12):e111966
Crapo PM, Gilbert TW, Badylak SFJB (2011) An overview of tissue and whole organ decellularization processes. Biomaterials 32(12):3233–3243
Dahl SLM et al (2011a) Readily available tissue-engineered vascular grafts. Sci Transl Med 3(68):68ra9-68ra9
Dahl SL et al (2011b) Readily available tissue-engineered vascular grafts. Sci Transl Med 3(68):68ra9-68ra9
Davis NF et al (2018a) Tissue engineered extracellular matrices (ECMs) in urology: evolution and future directions. Surgeon 16(1):55–65
Davis NF et al (2018b) The role of stem cells for reconstructing the lower urinary tracts. Curr Stem Cell Res Ther 13(6):458–465
de Bournonville S et al (2018) 233 - a modular, standalone perfusion bioreactor for robust, monitored and controlled tissue engineering. Cytotherapy 20(5 Supplement):S81–S82
De Coppi P et al (2007a) Isolation of amniotic stem cell lines with potential for therapy. Nat Biotechnol 25(1):100–106
De Coppi P et al (2007b) Amniotic fluid and bone marrow derived mesenchymal stem cells can be converted to smooth muscle cells in the cryo-injured rat bladder and prevent compensatory hypertrophy of surviving smooth muscle cells. J Urol 177(1):369–376
De Filippo RE, Yoo JJ, Atala A (2002) Urethral replacement using cell seeded tubularized collagen matrices. J Urol 168(4 Pt 2):1789–1792; discussion 1792-3
De Filippo RE et al (2015) Penile urethra replacement with autologous cell-seeded tubularized collagen matrices. J Tissue Eng Regen Med 9(3):257–264
Deng K et al (2014) Mesenchymal stem cells and their secretome partially restore nerve and urethral function in a dual muscle and nerve injury stress urinary incontinence model. Am J Physiol Renal Physiol 308(2):F92–F100
Dhandayuthapani B et al (2011a, 2011) Polymeric scaffolds in tissue engineering application: a review. Int J Polym Sci
Dhandayuthapani B et al (2011b) Polymeric scaffolds in tissue engineering application: a review. Int J Polym Sci

Dorin RP et al (2008) Tubularized urethral replacement with unseeded matrices: what is the maximum distance for normal tissue regeneration? World J Urol 26(4):323–326

El-Kassaby AW et al (2003) Urethral stricture repair with an off-the-shelf collagen matrix. J Urol 169(1):170–173; discussion 173

El-Taji OM, Khattak AQ, Hussain SA (2015) Bladder reconstruction: the past, present and future. Oncol Lett 10(1):3–10

Farhat WA, Yeger HJWJoU (2008) Does mechanical stimulation have any role in urinary bladder tissue engineering? World J Urol 26(4):301–305

Feil G et al (2006) Investigations of Urothelial cells seeded on commercially available small intestine submucosa. Eur Urol 50(6):1330–1337

Feng C et al (2011) Reconstruction of three-dimensional neourethra using lingual keratinocytes and corporal smooth muscle cells seeded acellular corporal spongiosum. Tissue Eng Part A

Fiala R et al (2007) Porcine small intestinal submucosa graft for repair of anterior urethral strictures. Eur Urol 51(6):1702–1708

Filippo RED, Yoo JJ, Atala A (2002) Urethral replacement using cell seeded tubularized collagen matrices. J Urol 168(4 Part 2):1789–1793

Fu Q et al (2007) Urethral replacement using epidermal cell-seeded tubular acellular bladder collagen matrix. BJU Int 99(5):1162–1165

Fu W-J et al (2011) A surface-modified biodegradable urethral scaffold seeded with urethral epithelial cells. Chin Med J 124(19):3087–3092

Furthmayr H, Timpl R (1976) Immunochemistry of collagens and procollagens. In: Hall DA, Jackson DS (eds) International review of connective tissue research. Elsevier, pp 61–99

Gabr H et al (2018) 234 - tissue engineering for bladder reconstruction: exploring bioscaffolds. Cytotherapy 20(5, Supplement):S82

Gabrich PN et al (2007) Avaliação das medidas do comprimento peniano de crianças e adolescentes. J Pediatr 83:441–446

Griffith LG, Naughton G (2002) Tissue engineering–current challenges and expanding opportunities. Science 295(5557):1009–1014

Gu GL et al (2012) Tubularized urethral replacement using tissue-engineered peritoneum-like tissue in a rabbit model. Urol Int 89(3):358–364

Haberstroh KM et al (2002) A novel in-vitro system for the simultaneous exposure of bladder smooth muscle cells to mechanical strain and sustained hydrostatic pressure. J Biomech Eng 124(2):208–213

Hardy JG et al (2015) Electroactive tissue scaffolds with aligned pores as instructive platforms for biomimetic tissue engineering. Bioengineering 2(1):15–34

Hayn MH, Bellinger MF, Schneck FX (2009) Small intestine submucosa as a corporal body graft in the repair of severe Chordee. Urology 73(2):277–279

Horst M et al (2013a) Engineering functional bladder tissues. J Tissue Eng Regen Med 7(7):515–522

Horst M et al (2013b) Engineering functional bladder tissues. J Tissue Eng Regen Med 7(7):515–522

Horst M et al (2013c) A bilayered hybrid microfibrous PLGA–acellular matrix scaffold for hollow organ tissue engineering. Biomaterials 34(5):1537–1545

Horst M et al (2019) Tissue engineering in pediatric bladder reconstruction – the road to success. Front Pediatr 7:91

Huang J-W et al (2015) Urethral reconstruction with a 3D porous bacterial cellulose scaffold seeded with lingual keratinocytes in a rabbit model. Biomed Mater 10(5):055005

Hubbell JA (2003) Materials as morphogenetic guides in tissue engineering. Curr Opin Biotechnol 14(5):551–558

Husmann DA, Rathbun SR (2006) Long-term followup of visual internal urethrotomy for management of short (less than 1 cm) penile urethral strictures following hypospadias repair. J Urol 176(4, Supplement):1738–1741

Jia W et al (2015) Urethral tissue regeneration using collagen scaffold modified with collagen binding VEGF in a beagle model. Biomaterials 69:45–55

Kajbafzadeh A-M et al (2011) In-vivo autologous bladder muscular wall regeneration: application of tissue-engineered pericardium in a model of bladder as a bioreactor. J Pediatr Urol 7(3):317–323

Kajbafzadeh A-M et al (2017) Future prospects for human tissue engineered urethra transplantation: decellularization and recellularization-based urethra regeneration. Reprod Tissue Eng 45(7):1795–1806

Keays MA, Dave S (2017) Current hypospadias management: diagnosis, surgical management, and long-term patient-centred outcomes. Can Urol Assoc J 11(1-2Suppl1):S48–S53

Kim Y-J, Matsunaga YT (2017) Thermo-responsive polymers and their application as smart biomaterials. J Mater Chem B 5(23):4307–4321

Kingham PJ et al (2014) Stimulating the neurotrophic and angiogenic properties of human adipose-derived stem cells enhances nerve repair. Stem Cells Dev 23(7):741–754

Kirkton RD et al (2019) Bioengineered human acellular vessels recellularize and evolve into living blood vessels after human implantation. Sci Transl Med 11(485):eaau6934

Koie T et al (2018) Robotic cross-folded U-configuration intracorporeal ileal neobladder for muscle-invasive bladder cancer: initial experience and functional outcomes. Int J Med Robot Comp Assisted Surg 14(6):e1955

Kowalski PS et al (2018) Smart biomaterials: recent advances and future directions. ACS Biomater Sci Eng 4(11):3809–3817

Kropp BP et al (1996) Regenerative urinary bladder augmentation using small intestinal submucosa: urodynamic and histopathologic assessment in long-term canine bladder augmentations. Investig Urol 155(6):2098–2104

Lazzeri M et al (2016) Incidence, causes, and complications of urethral stricture disease. Eur Urol Suppl 15(1):2–6

Lee KY, Mooney DJ (2012) Alginate: properties and biomedical applications. Prog Polym Sci 37(1):106–126

Li C et al (2008) Urethral reconstruction using oral keratinocyte seeded bladder acellular matrix grafts. J Urol 180(4):1538–1542

Liu JS et al (2016) Bone marrow stem/progenitor cells attenuate the inflammatory milieu following substitution Urethroplasty. Sci Rep 6:35638

Lu S-H et al (2005) Biaxial mechanical properties of muscle-derived cell seeded small intestinal submucosa for bladder wall reconstitution. Biomaterials 26(4):443–449

Lv X et al (2016) Electrospun poly(l-lactide)/poly(ethylene glycol) scaffolds seeded with human amniotic mesenchymal stem cells for urethral epithelium repair. Int J Mol Sci 17(8):1262

Lv X et al (2018) A smart bilayered scaffold supporting keratinocytes and muscle cells in micro/nano-scale for urethral reconstruction. Theranostics 8(11):3153–3163

Magnan M et al (2009) Tissue engineering of a genitourinary tubular tissue graft resistant to suturing and high internal pressures. Tissue Eng Part A 15(1):197–202

Mangir N et al (2019a) Current state of urethral tissue engineering. Curr Opin Urol 29(4):385–393

Mangır N et al (2016) Production of ascorbic acid releasing biomaterials for pelvic floor repair. Acta Biomater 29:188–197

Mangır N et al (2017) Oestradiol-releasing biodegradable mesh stimulates collagen production and angiogenesis: an approach to improving biomaterial integration in pelvic floor repair. Eur Urol Focus 5(2):280–289

Mangir N et al (2019b) Current state of urethral tissue engineering. Curr Opin Urol 29(4):385–393

Markiewicz MR et al (2008) Morbidity associated with oral mucosa harvest for urological reconstruction: an overview. J Oral Maxillofac Surg 66(4):739–744

Martin I, Wendt D, Heberer M (2004) The role of bioreactors in tissue engineering. Trends Biotechnol 22(2):80–86

Mijailovich SM et al (2007) Effect of urethral compliance on the steady state p-Q relationships assessed with a mechanical analog of the male lower urinary tract. Neurourol Urodyn 26(2):234–246

Nagele U et al (2008) In vitro investigations of tissue-engineered multilayered Urothelium established from bladder washings. Eur Urol 54(6):1414–1422

Neuhof H (1917) Facial transplantation into visceral defects: an experimental and clinical study. Surg Gynecol Obstet 25:383–427
Niklason L et al (1999) Functional arteries grown in vitro. Science 284(5413):489–493
Oberpenning F et al (1999a) De novo reconstitution of a functional mammalian urinary bladder by tissue engineering. Nat Biotechnol 17(2):149
Oberpenning F et al (1999b) De novo reconstitution of a functional mammalian urinary bladder by tissue engineering. Nat Biotechnol 17(2):149–155
Ohel G, Haddad S, Samueloff A (1995) Fetal urine production and micturition and fetal behavioral state. Am J Perinatol 12(2):91–92
Orabi H et al (2013a, 2013) Tissue engineering of urinary bladder and urethra: advances from bench to patients. Sci World J:154564–154564
Orabi H et al (2013b) The use of small intestinal submucosa graft for hypospadias repair: pilot study. Arab J Urol 11(4):415–420
Osborn SL et al (2014) Induction of human embryonic and induced pluripotent stem cells into urothelium. Stem Cells Transl Med 3(5):610–619
Ouellet G et al (2011) Production of an optimized tissue-engineered pig connective tissue for the reconstruction of the urinary tract. Tissue Eng Part A 17(11–12):1625–1633
Palminteri E et al (2012) Long-term results of small intestinal submucosa graft in bulbar urethral reconstruction. Urology 79(3):695–701
Partin, A.W., et al., Campbell-Walsh urology, 11th Edition. Principles of tissue engineering, ed. A. Atala. Vol. 1. 2015: Elsevier Health Sciences. 482–498
Pi Q et al (2018) Digitally tunable microfluidic bioprinting of multilayered Cannular tissues. Adv Mater 30(43):e1706913
Pinnagoda K et al (2016) Engineered acellular collagen scaffold for endogenous cell guidance, a novel approach in urethral regeneration. Acta Biomater 43:208–217
Poole Wilson DS, Barnard RJBjou (1971) Total cystectomy for bladder tumors. Br J Urol 43(1):16–23
Probst M et al (1997) Reproduction of functional smooth muscle tissue and partial bladder replacement. Br J Urol 79(4):505–515
Rajasekaran M, Stein P, Parsons CLJIjou (2006) Toxic factors in human urine that injure urothelium. Int J Urol 13(4):409–414
Ramuta TŽ, Kreft ME (2018) Human amniotic membrane and amniotic membrane–derived cells: how far are we from their use in regenerative and reconstructive urology? Cell Transplant 27(1):77–92
Rashidbenam Z et al (2019) Overview of urethral reconstruction by tissue engineering: current strategies, clinical status and future direction. Tissue Eng Regen Med
Raya-Rivera A et al (2011) Tissue-engineered autologous urethras for patients who need reconstruction: an observational study. Lancet 377(9772):1175–1182
Roelofs LAJ, Oosterwijk E et al (2016) Bladder regeneration using a smart Acellular collagen scaffold with growth factors VEGF, FGF2 and HB-EGF. Tissue Eng Part A 22(1–2):83–92
Roth CC, Kropp BPJCUR (2009) Recent advances in urologic tissue engineering. Curr Urol Rep 10(2):119–125
Rousseau A et al (2015) Adipose-derived stromal cells for the reconstruction of a human vesical equivalent. J Tissue Eng Regen Med 9(11):E135–E143
Rouwkema J, Rivron NC, van Blitterswijk CA (2008) Vascularization in tissue engineering. Trends Biotechnol 26(8):434–441
Saif A et al (2019) Augmentation cystoplasty of diseased porcine bladders with bi-layer silk fibroin grafts. Tissue Eng Part A 25(11–12):855–866
Salem SA et al (2013) Human adipose tissue derived stem cells as a source of smooth muscle cells in the regeneration of muscular layer of urinary bladder wall. Malay J Med Sci MJMS 20(4):80–87
Sanchez RP et al (1958) Vesical regeneration in the human after total cystectomy and implantation of a plastic mould. Br J Urol 30(2):180–188

Schaefer M et al (2013) Bladder augmentation with small intestinal submucosa leads to unsatisfactory long-term results. J Pediatr Urol 9(6, Part A):878–883
Serrano-Aroca Á, Vera-Donoso CD, Moreno-Manzano V (2018a) Bioengineering approaches for bladder regeneration. Int J Mol Sci 19(6):1796
Serrano-Aroca A, Vera-Donoso CD, Moreno-Manzano V (2018b) Bioengineering approaches for bladder regeneration. Int J Mol Sci 19(6)
Shaikh FM et al (2010) New pulsatile hydrostatic pressure bioreactor for vascular tissue-engineered constructs. Artif Organs 34(2):153–158
Shakhssalim N et al (2013) Bladder smooth muscle cells interaction and proliferation on PCL/PLLA electrospun Nanofibrous scaffold. Int J Artif Organs 36(2):113–120
Shiroyanagi Y et al (2003) Transplantable Urothelial cell sheets harvested noninvasively from temperature-responsive culture surfaces by reducing temperature. Tissue Eng 9(5):1005–1012
Shirozu H et al (1995) Penile tumescence in the human fetus at term — a preliminary report. Early Hum Dev 41(3):159–166
Shukla D et al (2008a) Bone marrow stem cells for urologic tissue engineering. World J Urol 26(4):341–349
Shukla D et al (2008b) Bone marrow stem cells for urologic tissue engineering. World J Urol 26(4):341
Sivaraman S et al (2019) Evaluation of poly (carbonate-urethane) urea (PCUU) scaffolds for urinary bladder tissue engineering. Ann Biomed Eng 47(3):891–901
Stangel-Wojcikiewicz K et al (2014) Autologous muscle-derived cells for the treatment of female stress urinary incontinence: a 2-year follow-up of a polish investigation. Neurourol Urodyn 33(3):324–330
Stock UA, Vacanti JP (2001) Cardiovascular physiology during fetal development and implications for tissue engineering. Tissue Eng 7(1):1–7
StÖlting MNL et al (2016) Magnetic stimulation supports muscle and nerve regeneration after trauma in mice. Muscle Nerve 53(4):598–607
Takahashi K, Yamanaka S (2006) Induction of pluripotent stem cells from mouse embryonic and adult fibroblast cultures by defined factors. Cell 126(4):663–676
Taylor DA et al (2018) Decellularized matrices in regenerative medicine. Acta Biomater 74:74–89
Theocharis AD et al (2016) Extracellular matrix structure. Adv Drug Deliv Rev 97:4–27
Wang Y et al (2008) In vivo degradation of three-dimensional silk fibroin scaffolds. Biomaterials 29(24):3415–3428
Wang Y et al (2019) Bioengineered bladder patches constructed from multilayered adipose-derived stem cell sheets for bladder regeneration. Acta Biomater 85:131–141
Wu S et al (2011) Human urine-derived stem cells seeded in a modified 3D porous small intestinal submucosa scaffold for urethral tissue engineering. Biomaterials 32(5):1317–1326
Wu Y et al (2016) Electroactive biodegradable polyurethane significantly enhanced Schwann cells myelin gene expression and neurotrophin secretion for peripheral nerve tissue engineering. Biomaterials 87:18–31
Xie M et al (2013) Evaluation of stretched electrospun silk fibroin matrices seeded with urothelial cells for urethra reconstruction. J Surg Res 184(2):774–781
Yalla SV, Burros HM (1974) Conduit and regional compliance of female canine urethra: observations during voiding and passive urethral infusion. Urology 4(2):155–161
Yao C et al (2013) Nanostructured polyurethane-poly-lactic-co-glycolic acid scaffolds increase bladder tissue regeneration: an in vivo study. Int J Nanomedicine 8:3285–3296
Yoo JJ et al (1998) Bladder augmentation using allogenic bladder submucosa seeded with cells. Urology 51(2):221–225
Zhang YY, Frey P (2003) Growth of cultured human urothelial cells into stratified urothelial sheet suitable for autografts. Adv Exp Med Biol 539(Pt B):907–920
Zhang Y et al (2000) Coculture of bladder urothelial and smooth muscle cells on small intestinal submucosa: potential applications for tissue engineering technology. J Urol 164(3 Part 2): 928–935

Zhang Y et al (2005) Growth of bone marrow stromal cells on small intestinal submucosa: an alternative cell source for tissue engineered bladder. BJU Int 96(7):1120–1125
Zhang Y et al (2006a) Challenges in a larger bladder replacement with cell-seeded and unseeded small intestinal submucosa grafts in a subtotal cystectomy model. BJU Int 98(5):1100–1105
Zhang Y et al (2006b) Challenges in a larger bladder replacement with cell-seeded and unseeded small intestinal submucosa grafts in a subtotal cystectomy model. BJU Int 98(5):1100–1105
Zhang Y et al (2008) Urine derived cells are a potential source for urological tissue reconstruction. J Urol 180(5):2226–2233
Zhang M et al (2013) Differentiation of human adipose-derived stem cells co-cultured with Urothelium cell line toward a Urothelium-like phenotype in a nude murine model. Urology 81(2):465.e15–465.e22
Zhang L et al (2016) Development of a cell-seeded modified small intestinal submucosa for urethroplasty. Heliyon 2(3):e00087
Zhe W et al (2016) Smooth muscle precursor cells derived from human pluripotent stem cells for treatment of stress urinary incontinence. Stem Cells Dev 25(6):453–461
Zhou S et al (2017) Fabrication of tissue-engineered bionic urethra using cell sheet technology and labeling by ultrasmall Superparamagnetic Iron oxide for full-thickness urethral reconstruction. Theranostics 7(9):2509–2523

Tissue-Engineered Ovary

Monica M. Laronda

Contents

M. M. Laronda (✉)
Stanley Manne Children's Research Institute, Ann & Robert H. Lurie Children's Hospital of Chicago, Chicago, IL, USA

Department of Pediatrics, Feinberg School of Medicine, Northwestern University, Chicago, IL, USA
e-mail: mlaronda@luriechildrens.org; m-laronda@northwestern.edu

D. Eberli et al. (eds.), *Organ Tissue Engineering*, Reference Series in Biomedical Engineering, https://doi.org/10.1007/978-3-030-44211-8_13

Abstract

An engineered ovary could benefit a variety of patients with ovarian insufficiency including those with an inherited condition, *de novo* mutation, or those whose treatment regimens could reduce ovarian function. This chapter highlights the impact of developing an engineered ovary, highlights the important features of an ovary to be recreated, and reviews the current state of fertility preservation. Encapsulating hydrogels and scaffolds have been used to restore ovarian hormones and fertility in animal models and to monitor in vitro growth of human ovarian cells. Additional advancements to these technologies will rely on an understanding of ovarian biology in order to recapitulate and restore long-term and physiological function. Considerations for the different microenvironments within the cortical and medullar compartments and cell types that support oocyte maturation, including granulosa and heterogeneous stromal cells, will be essential for recreating appropriate folliculogenesis. Mechanical cues and facilitation of paracrine and endocrine signals to reach target cells can also be tailored within engineered environments. These studies may one day improve the current fertility restoration options or even provide options for patients that currently lack options.

1 Introduction

1.1 Causes of Premature Ovarian Insufficiency

Premature ovarian insufficiency (POI) is defined by the cessation of normal ovarian function prior to age 40 and it occurs in approximately 1% of women in the USA (Development 2016). It is characterized by a decrease in ovarian hormones, increase in pituitary gonadotropins, and amenorrhea. Primary POI can occur as a result of an inherited condition, *de novo* mutation or from insults to the ovarian tissue (Care 2014). Cancer patients are particularly at risk for POI due to their chemotherapy and radiation treatments (Wallace et al. 2005a, 2011; Duffy and Allen 2009; Jadoul et al. 2010). Survivors of childhood cancer are more likely to have trouble getting pregnant than their siblings (Barton et al. 2013), and 1 in 6 female childhood cancer survivors are expected to suffer from POI (Chow et al. 2016). In the USA, more than 11,000 children aged 0 to 14 years will be diagnosed with cancer this year alone, resulting in potentially 1800 new cases of gonadal insufficiency due to cancer treatments this year (Chow et al. 2016).

With the increased survival rate aided by these treatments, there has been an additional focus toward quality of life after cancer. A quality of life measure for some patients includes the option of having biological children. In a small Swedish Cancer Registry survey of participants ages 18 to 45, 72.8% of women and 83.0% of men, who reported a pretreatment desire to have children, later reported a definite or possible desire for children 3 to 7 years posttreatment (Armuand et al. 2014). Even with an average of 10 years posttreatment, women who did not have children were significantly distressed about their infertility and interrupted childbearing years (Canada and Schover 2012).

Ten to twenty percent of patients can attribute their POI to a heritable trait (Cordts et al. 2011). Mutations that affect ovarian function can arise within genes that are essential for ovarian follicle development and expansion or within the endocrine system that controls gonadotropin-dependent growth and ovulation. These conditions may result in differences in sex development (DSD). The X chromosome is a significant player in this disorder, as X monosomy, X trisomy, and X chromosome rearrangements on the long arm of the chromosome alter expression or dosing of genes that are critical for ovarian development and account for up to 12% of POI cases (Shelling et al. 2000; Goswami 2005; Qin et al. 2014). Mutations in X chromosome genes, including bone morphogenic protein 15 (BMP15) (Chand et al. 2006; Dixit et al. 2006; Pasquale et al. 2006; Rossetti et al. 2009), fragile X mental retardation (FMR1 and FMR2) (Allingham-Hawkins et al. 1999; Murray et al. 1999; Allen et al. 2007; Espeche et al. 2017), have also been found to induce POI. Other examples of genes that have been identified as having mutations in patients with gonadal dysgenesis or POI include gonadotropin receptors, luteinizing hormone receptor (LHR) (Toledo et al. 1996; Puett et al. 2010), and follicle stimulating hormone receptor (FSHR) (Aittomäki et al. 1995), hormone receptors estrogen receptor alpha (ESR1) (Cordts et al. 2012), steroidogenic pathway genes, steroidogenic factor (SF1) (Lourenço et al. 2009) and aromatase (CYP19A1) (Lin et al. 2007) and a transcription factor necessary for granulosa cell development, forkhead box L2 (FOXL2) (Beysen et al. 2004).

1.2 Recognition of the Problem

The Oncofertility Consortium is a collection of institutions around the world that seek to provide fertility preservation and restoration options for patients undergoing treatments for their cancer or hematologic disease (Woodruff 2007). These and other fertility preservation clinics offer preservation options for those with an 80% or more chance of undergoing premature gonadal insufficiency. POI risk and age of onset can be calculated for some treatments based on models of ovarian reserve rates of decline (Faddy and Gosden 1996; Wallace et al. 2005b; Wallace and Kelsey 2010). For example, it is estimated that a 12-year-old child who undergoes 10 Gy of pelvic radiation may experience POI at 19.5 years old due to the treatment-induced depletion of her primordial follicles (Wallace et al. 2005b; Wallace and Kelsey 2010). For adult women, fertility preservation can often mean undergoing oocyte harvest and egg or embryo freezing. However, some women, like those with estrogen-responsive cancers, cannot undergo the supraphysiologic treatments required for oocyte harvest. Additionally, prepubertal girls do not yet produce eggs and their only option to preserve their fertility is to freeze ovarian tissue. Ovarian tissue cryopreservation (OTC) requires removal of ovarian tissue that is processed into strips of tissue for cryopreservation through slow-freeze or vitrification techniques. Transplantation of this tissue has resulted in restoring hormones to induce or maintain cyclical hormones and has resulted in over 130 reported live births (Donnez and Dolmans 2017; Pacheco and Oktay 2017; Corkum et al. 2019). While oncology patients comprise the majority of fertility preservation cases, clinics have begun to offer services to DSD patients that may have some immature gametes within their gonadal tissue (Finlayson et al. 2017; Johnson et al. 2017; Finney et al. 2019).

1.3 Beyond Fertility

Ovarian hormones are responsible for the development of secondary sex characteristics that arise during puberty and are essential for female adult anatomy and physiology. Puberty is the transition from an adolescent to a mature being and encompasses physiological, physical, and psychological changes (Laronda and Woodruff 2017). It is initiated by gonadotropin-releasing hormone (GnRH) neurons in the hypothalamus that control the release of follicle stimulating hormone (FSH) and luteinizing hormone (LH) from the pituitary. FSH and LH elicit responses from the gonads to produce opposing factors, such as estradiol, progesterone, inhibin, and testosterone. This cross talk is known as the hypothalamic-pituitary-gonadal axis, and it occurs in a cyclical manner. Cessation of normal ovarian function has effects that reach beyond reproductive capacity and health. Traditional male and female sex hormones regulate growth hormone (GH) both directly, at the hypothalamus, and indirectly, by regulating the downstream tissue response to GH (Juul et al. 1994; Veldhuis et al. 2000). Women/girls with POI are at increased risk for comorbidities that include cognitive, metabolic, osteogenic, skin, and cardiovascular diseases that could be life threatening (Cooper and Sandler 1998; Ossewaarde et al. 2005; Shuster et al. 2010; Muka et al. 2016; Chemaitilly et al. 2017).

Life expectancy is reduced in women with POI. Long-term studies by the Mayo Clinic found a significant increase in mortality rates in women with surgical menopause that were mainly due to cardiovascular disease, osteoporosis, and bone fractures (Rocca et al. 2016). They also reported an increase in dementia, Parkinsonism, and reduced psychological well-being (Rocca et al. 2007; Shuster et al. 2010). Additional studies in Dutch, Norwegian, South Korean, and Japanese cohorts revealed a life expectancy of approximately 2 years shorter in women with ovarian insufficiency from various causes (Jacobsen et al. 1999; Ossewaarde et al. 2005; Amagai et al. 2006; Hong et al. 2007).

Current hormone replacement therapies (HRTs) are neither developmentally dynamic, responsive, comprehensive, nor ideal long-term solutions. HRTs may be used to initiate puberty, mitigate comorbidities in postmenopausal women, or promote gender-affirming characteristics. Current HRT protocols utilize patches that are semiquantitatively cut to approximate appropriate and increasing doses at night to mimic the female pubertal transition that occurs with at first low, then increasing levels of nocturnal estradiol (Ankarberg-Lindgren et al. 2014, 2019). The goal is to provide hormones that promote growth, secondary sex characteristics (pubic hair and breasts), and cognitive function that mimics their peers. However, current therapies and trials with available options are not sufficient (Matthews et al. 2017).

While they may improve sex characteristics, they fail, for instance, to fully restore linear growth (Leung et al. 2004). The ovary produces hormones other than estrogen and progesterone, which are absent in pharmaceutical regimens and may be key to mimicking normal puberty and sustained health. Inhibin A increases bone mass and strength, independent of sex steroids, by stimulating osteoblast activity in mice with gonads disconnected from vessels or removed altogether (Perrien et al. 2007). Inhibin from gonadal sources also modulate hemoglobin accumulation and erythropoiesis in human bone marrow cells (Yu et al. 1987; Meunier et al. 1988). Additionally, anti-Müllerian hormone (AMH), which is secreted exclusively by granulosa cells of growing follicles, can inhibit abnormal growth of tissues derived from the Müllerian duct and potentially prevent endometriosis, adenomyosis, and uterine cancer(Barbie et al. 2003; Chung et al. 2015). Further, each mode of HRT administration offers different pros and cons, which may vary based on patient's age and diagnosis. For example, transdermal administration of estradiol results in a more physiological breakdown of estradiol, estrone, and bioestrogen concentrations, with greater suppression of pituitary gonadotropins over synthetic oral formulations (Taboada et al. 2011). Yet in some studies, oral estrogen replacement therapies, but not transdermal therapies, reduce myocardial infarction and cardiovascular mortality in postmenopausal women by decreasing Insulin-like Growth Factor-1 (IGF-1) and increasing C-reactive protein, and promoting endothelium-dependent vasodilation and antiatherogenic changes (Vehkavaara et al. 2000; Vongpatanasin et al. 2003).

Women suffer from ovarian insufficiency symptoms for decades after menopause. The age of menopause has stayed relatively constant; however, women's life expectancy has extended. Therefore, women are living longer with ovarian insufficiency and its comorbidities. Increased bone resorption and impaired bone remodeling is accelerated during menopause, making women susceptible to developing osteoporosis. The "timing hypothesis" for estrogen therapy in women theorizes that a prompt restoration of estradiol/progesterone would be beneficial to maintain sexual, cardiovascular, bone, and cognitive health and to significantly lower mortality; whereas delayed therapy may increase risk of disease (Lobo et al. 2016). A cell-based therapy that can respond to decreases in the patient's serum ovarian hormone levels could potentially remove the speculation of appropriate timing and provide the full hormonal milieu in response to internal stimuli.

Ovarian tissue transplant after OTC has been shown to restore endocrine function in one pediatric cancer patient so that she could transition through puberty with her own hormones (Ernst et al. 2013). However, the function of this graft ceased after 19 months, due in part to the limited supply of hormone-producing cells within the tissue. The lifespans of ovarian tissue transplants are limited. The current average lifespan of transplanted cortical tissue, as monitored by menses and normal levels of gonadotropin hormones, is between 2 and 5 years (Donnez and Dolmans 2013; Stoop et al. 2014; Pacheco and Oktay 2017). Upon transplant of cryopreserved tissue, the primordial oocyte pool undergoes increased activation that depletes the reserve and reduces the hormone and egg production of the tissue (Gavish et al. 2018). This has been measured in women with ovarian transplants as an initial significant increase in AMH, at 140–280 days after transplant followed by a sharp decline at approximately 280 days after transplant (Silber 2015). This accelerated loss may be attributed to disruption of their ECM microenvironment (Kawamura et al. 2013). More critically, some ovarian cortical tissue examined from patients with leukemia, breast, gastric, uterine, and cervical cancers contained metastatic disease (Dolmans et al. 2010; Bastings et al. 2013; Laronda et al. 2015), which increases the risk of reseeding cancer with transplantation.

While great strides have been made to recognize causes of POI and the physical and psychological consequences of this condition, more needs be done to understand several aspects of restoring ovarian function. This includes investigating basic ovarian biology and how the microenvironment influences folliculogenesis, how these environments and the resulting gametes may differ depending on age or pubertal status during isolation and transplantation, optimizing the procurement, processing, cryopreservation, recovery, and transplantation of the ovarian tissue and cells, and engineering the conditions that are amenable to appropriate folliculogenesis with long-term production of fertilizable gametes and secretion of the full complement of cyclical hormones.

2 Current State of Restoring Ovarian Function

2.1 Ovarian Tissue Cryopreservation (OTC) and Transplantation

Protocols for cryopreserving human ovarian tissue were developed using slow-freeze techniques and transplanted into immunocompromized mice (Oktay et al. 1998, 2000). With the administration of long-term FSH, the transplanted cortical strips produced estradiol in ovariectomized mice and supported large antral follicle growth (Oktay et al. 1998). The first case where human ovarian tissue was cryopreserved and autologously transplanted to restore ovarian hormones was performed in New York, USA, in 2000. The tissue was placed in the pelvic peritoneum and antral follicle formation was observed without stimulation. Ovulation with endometrium thickening was observed following 10,000 IU of human chorionic gonadotropin (Oktay and Karlikaya 2000). The first live birth that was reported following transplantation of cryopreserved ovarian tissue was in Brussels, Belgium, in 2004 (Donnez et al. 2004). This participant had stage IV Hodgkin's lymphoma and chose to cryopreserve her tissue prior to chemotherapy and radiotherapy treatments at 27 years old. Within months she had undergone POI and by the age of 32, she had all of her cryopreserved tissue orthotopically transplanted. This tissue restored hormone function and supported a spontaneous pregnancy and live birth (Donnez et al. 2004). To date, there have been over 130 reported live births from tissue that was cryopreserved with an average age of 30 years old at the point of cryopreservation (Donnez and Dolmans 2017; Pacheco and Oktay 2017; Corkum et al. 2019). An assessment of 60 ovarian tissue transplants performed in Belgium, Denmark, and Spain revealed that 93% restored estradiol and suppressed FSH, which was detected by 3.5–6.5 months posttransplant (Donnez et al. 2013). Hormone production lasted up to 12 years with an average of 2–5 years of sustained FSH suppression in these reports (Donnez and Dolmans 2017; Pacheco and Oktay 2017). The average age of women seeking transplantation of their ovarian tissue is 34 years old and 20–30% of these procedures have restored fertility (Donnez and Dolmans 2017; Pacheco and Oktay 2017; Corkum et al. 2019). Of these successful births, 62% were achieved spontaneously while the remainder were achieved using additional assisted reproductive technologies, such as in vitro fertilization (Corkum et al. 2019).

There is one report of a live birth using tissue that was cryopreserved when the patient was peripubertal (Demeestere et al. 2015) and a report in the lay press of a child born following transplantation of tissue that had been cryopreserved when the woman was prepubertal (Donnelly 2016). While prior to these reports it was unclear if ovarian oocytes from prepubertal patients were viable, it was at least clear from animal studies that primordial follicles from young patients could mature into eggs following xenotransplantation (Lotz et al. 2014). There have been 13 total transplants resulting in 8 live births from patients who underwent OTC procedures at 20 years old or younger (Corkum et al. 2019). It may be too soon to fully realize the implications of OTC for pediatric patients, as they may not request the tissue for several years or even decades after preservation.

2.2 Location of Transplantation

Ovarian tissue can successfully restore function when transplanted in numerous locations, such as the peritoneal window, ovarian medulla, beneath the ovarian cortex, and subcortical ovarian pocket (Donnez et al. 2013). Orthotopic transplantation can be achieved when one of the ovaries or partial ovarian tissue remains. The ovarian tissue pieces can be quilted to create a larger piece for transplant handling (Silber et al. 2008; Sánchez-Serrano et al. 2010). Alternatively, pieces can be adhered with fibrin glue or slipped into pockets created just below the ovarian surface epithelium (Meirow et al. 2005; Donnez et al. 2011). If there is no remaining ovarian tissue, or the tissue is damaged and less than ideal for transplantation, a peritoneal pocket may be created on the anterior leaf of the broad ligament and fixed with fibrin glue (Donnez et al. 2012). Additional techniques using heterotopic sites are also options. A location that is not intraabdominal could be used to monitor new techniques or materials, follicular growth, or provide a potentially safer location if there is any small risk of reintroducing cancer cells with the ovarian tissue transplant. The forearm was used as an easily accessible location to restore ovarian hormones in women that had undergone bilateral oophorectomies for squamous cell carcinoma of the cervix with familial history of ovarian cancer and recurrent benign serous cystadenoma (Oktay et al. 2003). Both patients' grafts were functional for >20 months and ultrasound had indicated several antral follicles (Oktay et al. 2003).

2.3 In Situ Options for Fertility

Prior to complete engineering of a safe bioprosthetic ovary, additional fertility restoration techniques need to be developed, especially for those cancer survivors who may have metastatic cells within their ovarian tissue and are at risk of reintroducing the disease with an ovarian transplant (Meirow et al. 2008; Rosendahl et al. 2010; Donnez et al. 2011; Laronda et al. 2015). In vitro maturation is one option that could provide a fertilizable egg that could produce an embryo and be implanted into the mother's or surrogate's uterus. Several labs have grown human primordial follicles within ovarian cortical pieces to achieve large antral follicles (Picton et al. 2008; Laronda et al. 2014; Jakus et al. 2017) and there was one indication of a polar body, representing resumption of meiosis (McLaughlin et al. 2018). Additionally, 20% (4/20) of isolated human secondary follicles cultured in a supportive biomaterial were matured to MII eggs (Xiao et al. 2015b). It would be ideal to restore both fertility and hormone function without the risk of reintroducing disease. Therefore, techniques for isolating primordial follicles and transplanting them within a supportive matrix or scaffolding are also being explored.

3 Important Features in an Ovary to Be Recreated

Follicles are the functional units of the ovary with a centralized oocyte, or potential egg cell, and singular or multiple layers of support cells. The two functional components of the ovary are linked as ablating the follicle would deplete both the sources of fertility and hormone production. Primordial follicles are located in the cortical region of the ovary, most within 500 μm from the ovarian capsules in large mammals (Fig. 1). Maturing follicles grow toward the center of the ovary and have an additional support cell called theca cells. As folliculogenesis occurs, the ~30 μm primordial follicle grows ~600 times that size (~20 mm) prior to ovulation. This dynamic change is under the control of endocrine, paracrine, and structural controls. Communication between the meiotically arrested oocyte and its nurse cells is essential to maintain the health and support the growth and maturation of the oocyte.

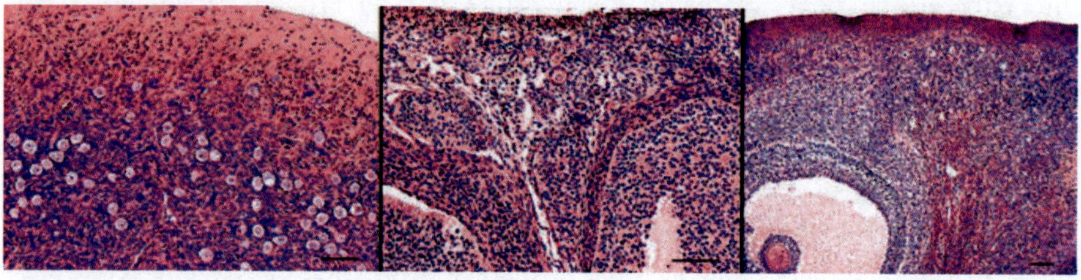

Fig. 1 Histological images of human, porcine, and bovine ovaries (left to right). The ovarian surface epithelium can be seen at the top of each image and primordial follicles are visible in all examples. Antral space (porcine, middle) and antral follicle with an oocyte (bovine, right) is visible deeper within the ovarian tissue. Scale bars, 100 μm

The granulosa and theca cells produce the ovarian hormones in response to gonadotropins from the pituitary, which is primed during puberty to support the maturation of the egg and trigger ovulation. The oocyte also controls the stage of the granulosa cells that can proliferate and respond to pituitary hormones. This orchestration is essential for the cyclical output of hormones and fertilizable eggs.

3.1 Oocytes

Gonadal tissues arise from the celomic epithelium that develops as a protrusion of cells from the Müllerian duct. The primordial germ cells are a selection of cells that remain in the extraembryonic ectoderm and migrate to the gonadal ridge. The battle between defined testis and ovarian characteristics occurs within the supportive cells, pre-Sertoli cells in the testis and pre-granulosa cells in the ovary. One main driver of this decision is the presence or absence of a functional SRY, which drives testis determination. In the ovary, the primordial germ cells, now referred to as gonocytes, are surrounded by pre-granulosa cells to create nests of synchronously dividing germ cells. After initiation of meiosis and recombination of homologous chromosomes, the oocytes arrests in the diplotene I stage (Pan and Li 2019). The support cells invade the nests and primordial follicles are formed with a single centralized oocyte and squamous granulosa cells, which occurs between 15 and 22 weeks of gestation in humans (Maheshwari and Fowler 2008). This meiotic arrest is maintained by high levels of cAMP produced by the surrounding granulosa cells (Pan and Li 2019). The primordial follicle pool represents the finite source of potential egg cells and finite source of ovarian hormones for that individual.

3.2 Support Cells and Folliculogenesis

The number of primordial follicles can dictate the length of time an ovary can function. The peak number of follicles is an average of 3×10^5(95% range 0.35–25.3×10^5) in the fifth month of gestation and this number declines continuously after birth (Wallace and Kelsey 2010). There are two waves of increased apoptosis that reduces the number of oocytes within the ovary: during nest breakdown and during cyclical primordial follicle recruitment. Once a primordial follicle is recruited to grow it will continue and either mature and ovulate, in a postpubertal woman, or undergo atresia which occurs in ~90% of the activated follicles undergoing atresia (Baker 1963). Once the reserve is reduced to approximately 1000 follicles, the ovary ceases to produce enough hormones to continue a normal menstrual cycle, which naturally occurs as menopause in adult women around 50 years old (Wallace and Kelsey 2010). It takes approximately 85 days for a human preantral follicle to become preovulatory size (Gougeon 1986). Activation occurs downstream of the serine/threonine kinase, AKT, and phosphatidylinositol3-kinase (AKT/PI3K) pathway which induces phosphorylation of the forkhead transcription factor (FOXO3) and translocation from the nucleus to the cytoplasm and is described in mice and cows (John et al. 2008; Andrade et al. 2017). However, this pathway may not be important for primate oocyte activation as FOXO3 is absent in rhesus macaque ovaries and there have been no sequence variations in human FOXO3 in women with POI or primary amenorrhea (Gallardo et al. 2008; Ting and Zelinski 2017).

The granulosa cells play an important role in supporting the quiescent oocyte and can trigger activation.Transzonal projections (TZP) are filipodia-like structures that stem from the granulosa cells and to the oocyte. These projections are formed after the primordial to primary transition and remain until the oocyte is released as an egg (El-Hayek et al. 2018). Early attempts to culture these cell aggregates in vitro have made the importance of maintaining these connections obvious. Ovarian follicles cultured with a thick collagen gel culture maintained their spherical shape and intercellular connections between the cells and the oocytes, which led to greater efficiency in in vitro follicle growth and maintenance of healthy oocytes over traditional 2D cultures (Gomes et al. 1999). Follicles are now often grown in encapsulation materials. Hormone production and meiotic maturation were compared with different encapsulation culture conditions, including alginate, collagen I, collagen IV, fibronectin, and laminin (Kreeger et al. 2006). Murine follicle growth and estradiol production was significantly improved with alginate by maintaining the 3D organization of the follicle (Kreeger et al. 2006).

There is a morphological change from squamous to polarized, cuboidal that occurs in granulosa cells during follicle activation and results in a cell that is four times narrower and six time taller (Hirshfield 1991; Picton 2001; Silva-Buttkus et al. 2008). Once an oocyte is surrounded by a full layer of cuboidal cells, the follicle is termed a primary. This initial cuboidalization is coupled with TZP formation. The cuboidal granulosa cells proliferate more than the cells with flattened morphology (Gougeon and Busso 2000; Stubbs et al. 2007) and at this point granulosa cell proliferation occurs in a layering or stacking manner to achieve secondary or multilayered secondary developmental status (Silva-Buttkus et al. 2008). Beginning at the primary stage, interstitial stromal cells are recruited to the outside of the basal laminae that envelop the granulosa cells to form a flattened layer of theca cells. This is also the point in which FSHR mRNA can be detected in granulosa cells in cows (Xu et al. 1995; Bao and Garverick 1998), sheep (Tisdall et al. 1995), and rats (Presl et al. 1974). While the basal laminae exclude blood vessels and nerves from the oocytes and granulosa cell layers (Rodgers et al. 2003), there is a recruitment of vessels through angiogenic factors released by the granulosa cells and recruitment of lymphatic fluid that fills and creates space and pressure between two developing layers of granulosa cells. The granulosa cells that remain in contact with the oocyte are called cumulus cells while the cells that remain at the periphery of the follicle are called mural granulosa cells.

3.3 Hormones

Theca cells express luteinizing hormone receptor (LHR) and produce androstenedione in response to LH signaling. Theca cells express steroidogenic acute regulatory protein (STAR), which shuttles cholesterol to the mitochondria, cytochrome P450 subfamily members (CYP11A1, CYP17), and a steroid dehydrogenase (HSD3B2) to produce this androgen in an LH dose–dependent manner (Smyth et al. 1995). Aromatase (CYP19A1) is expressed in granulosa cells in an FSH dose–dependent manner and converts androstenedione to estradiol. There are several growing follicles at a given time in the ovary but only approximately 10 gain gonadotropin growth dependence and continue to grow under peak levels of FSH (Fauser and van Heusden 1997). It is necessary for estradiol levels to be high enough to trigger an LH pulse in the late follicular phase of the cycle. The production of estradiol and inhibin B also suppress FSH in a negative feedback loop. This feedback that shortens the duration of FSH appears important for a single follicle to escape atresia and become the dominant follicle that becomes increasingly sensitive to FSH and LH stimulation (Wallach and Hodgen 1982; Pache et al. 1990; van Santbrink et al. 1995; Schipper et al. 1998). Granulosa cells from mature follicles express LHR (Sullivan et al. 1999; Blockeel et al. 2009). The LH surge triggers ovulation of a mature egg from the dominant follicle and the remaining follicular cells develop the corpus luteum (CL). The CL contains small luteal cells (formerly theca cells) that produce the androgen precursors required for large luteal cells (formerly granulosa cells) to aromatize into progesterone. The CL gland is maintained if the egg becomes fertilized and pregnancy occurs; otherwise, eosinophils, T lymphocytes, and macrophages are recruited to the CL (Kirsch et al. 1981; Murdoch 1987).

In addition to the induction and inhibition of gonadotropins and preparation of the reproductive tract to accept an embryo, ovarian hormones play important roles in development and maintenance of other systems. In addition to estradiol and progesterone, those hormones then can be supplemented with hormone replacement therapies, ovaries produce testosterone, inhibins, activin, anti-Müllerian hormone, and insulin-like growth factor and relaxin, among others. Almost half of postmenopausal women develop metabolic syndrome and hypertriglyceridemia (Chedraui et al. 2007; Ali et al. 2014). The increase in metabolic syndrome severity during the menopausal transition was not changed by hormone replacement therapy use (Gurka et al. 2016).

3.4 Vascularization

A vascular plexus infiltrates the ovary at the hilum and stems from the abdominal aorta. Because the larger growing follicles surpass the diffusion limit, growing follicles are located near blood vessels in the murine ovary (Feng et al. 2017). Ovarian pericytes that express vascular endothelial growth factor (VEGF) migrate to the theca cell layers of growing follicles in bovine, ovine, and porcine ovaries (Reynolds and Redmer 1998). Neovascularization is also necessary for antrum and CL formation. Conversely, hypoxia may mediate the dormant state of the primordial follicles. Murine-induced pluripotent stem cell (iPSC)-derived oocytes were cocultured with murine fetal gonad tissue to develop primordial follicles (Shimamoto et al. 2019). Under control conditions, the follicles within the reconstituted ovary undergo activation as opposed to the in vivo condition where the ovarian follicle pool is maintained by primordial follicles remaining in a quiescent state. This is alleviated under hypoxic conditions with 5% oxygen at a greater extent than the addition of exogenous FOXO3a (Shimamoto et al. 2019).

The ovary produces both water- and fat-soluble hormones, in the form of peptide and steroid hormones. Water-soluble hormones can move freely through the circulation to target tissues but are generally repelled by cell membranes and require receptors to induce cellular responses. Steroid hormones can travel through cellular membranes and generally function at a nuclear receptor. They rely on carrier proteins such as steroid hormone–binding globulins to travel through the circulation. General proximity to important target organs may be beneficial for steroid hormones produced by the ovary. There are increased concentrations of estradiol and progesterone in fallopian tube and uterine vessels than systemic blood circulation indicating localized control of these hormones within the circulation (Cicinelli et al. 2004).

3.5 Compartmentalization and Physical Features

The ovary is surrounded by a capsule or layer of ovarian surface epithelium (OSE) and is divided into two main, visibly distinct compartments, the cortex and medulla. The cortical region contains mostly quiescent primordial follicles while the medulla region is more vascularized and contains growing follicles. The compartmentalization of the ovary is dictated by the extracellular matrix (ECM) composition, organization and density and can be seen in some ovarian tissue cross-sections (Fig. 2). Additionally, a recent proteomic analysis of a porcine ovary revealed significant differences in expression of matrisome proteins, ECM, and associated proteins, across the cortical and medullary compartments (Henning et al. 2019). Collagen is concentrated in the cortex of mouse ovaries (Bochner et al. 2015; Henning et al. 2019) and is present in more organized sheets in bovine ovaries (Laronda et al. 2015). While it has never been directly tested, it is hypothesized that the cortical region of the ovary is stiffer than the medullary region (Woodruff and Shea 2011). Many polycystic ovarian syndrome (PCOS) patients are anovulatory and have increased numbers of primordial follicles, which may be due, in part, to the increased rigidity of their ovaries as measured through magnetic resonance elastography (Wood et al. 2015). Alternatively, a reduction in relative stiffness and improper development of ovarian tissue, as in streak ovaries within Turner syndrome patients, may cause the increased rate of primordial follicle activation and depletion (Reindollar 2011). These observations in human patients are supported in experimental models.

Compression of murine ovarian tissue is required for primordial follicles to remain quiescent. Disruption of the FOXO3a gene in mice accelerates primordial follicle growth and addition of FOXO3a inhibits it (Castrillon et al. 2003; Liu et al. 2007). Digestion of the ovary with collagenase IV and trypsin released the primordial follicles of the ECM-associated stress as indicated by loosening of actin stress fibers and a progression in granulosa cell morphology from squamous to cuboidal (Nagamatsu et al. 2019). Additionally, FOXO3a was exported from the nucleus to the cytoplasm, an additional indication of follicle activation in primordial to growing

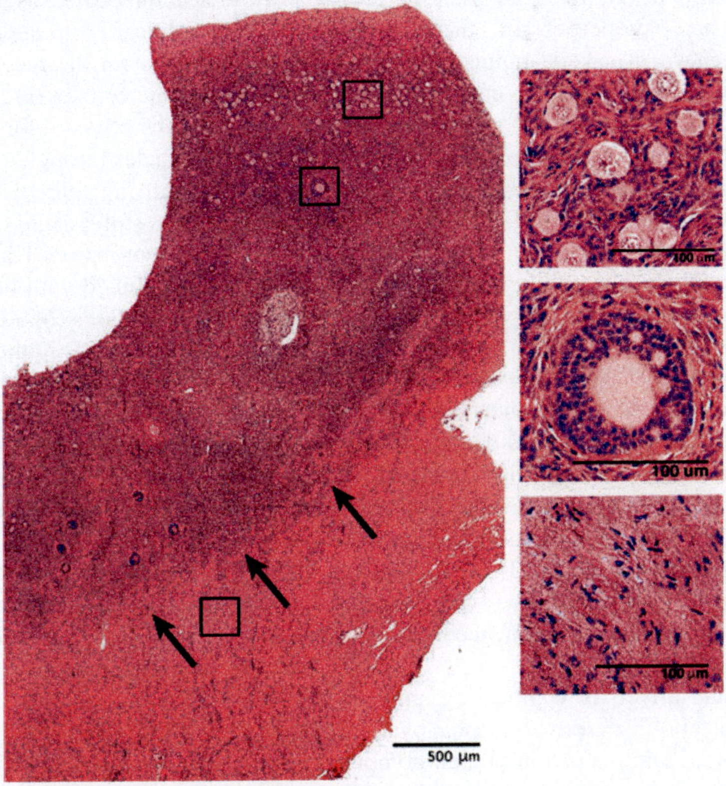

Fig. 2 Histological images of human ovarian tissue biopsy. Primordial follicle clusters are visible closer to the ovarian surface epithelium (top square). Small growing follicles are visible a little deeper from the surface (middle square of a secondary follicle). The cortical to medullary transition is visible (black arrows) and differences in stromal cell density is visible in the medullary region (bottom square). This de-identified tissue biopsy was obtained from a 7.09 year old OTC patient with sickle cell anemia that had not obtained any previous potential gonadotoxic treatment. Black squares correspond the region that is magnified to the right. Scale bars, 500, 100 μm

follicles in mice. However, growing these enzyme-treated ovaries in a pressure chamber reversed this phenotype and the primordial follicles were maintained in the cortex of the ovary (Nagamatsu et al. 2019).

4 Materials Used for Engineered Ovaries

4.1 Overarching Consideration for Engineering an Ovary

The connections between the oocyte and supportive granulosa cells are essential for the health and growth of the female gamete. Therefore, an environment that maintains the spheroid shape of the follicle is essential. Additionally, follicles require space to grow or require a material that breaks down to accommodate this growth. The physical properties of these biomaterials are critical for supporting folliculogenesis at several points, as described above to maintain the mechanical equilibrium of the follicle an enable oocyte–granulosa cell connections (Fig. 3). In particular, the activation of the quiescent follicle is controlled by physical forces and mechanotransductive cues. In the formation of secondary follicles through an organized cuboidalization, stacking and packing is dependent on the follicular basal laminae, theca cell layer, and directionality of mitotic divisions (Silva-Buttkus et al. 2008). The antral space formation, directionality of microenvironment resistance established by localized matrix degradation and thinning, combined with building osmotic pressure facilitates ovulation. Both matrix metalloproteinases (MMPs) and their tissue inhibitors (TIMPs) are expressed by theca and stromal cells in the ovary (García et al. 1997; McCaffery et al. 2000).

Additionally, the overall mechanical properties of the encapsulation or scaffolding material needs to be considered. Ovaries that are too rigid or not rigid enough may contribute to an inappropriate rate at which primordial follicles are activated, causing them to be too slow or too fast, respectively. Ovarian wedge resections and ovarian drilling have been used to facilitate follicle growth in anovulatory PCOS patients (Farquhar et al. 2012). These examples highlight the necessity of dynamic reciprocity of ovarian follicles with their microenvironment, and that the follicles must be appropriately influenced and be able to modify and influence its microenvironment.

One main driver for engineering an ovary is to benefit patient populations that have metastatic disease within their ovarian cortical tissue and, therefore, are not candidates for reimplantation of their cryopreserved tissue as it naturally exists. The functional unit of the ovary, the follicle, that produces both the female gamete and produces and responds to hormones is self-encapsulated within a basal lamina that is not penetrated by white blood cells, vessels, or nerve processes until ovulation (Rodgers et al. 2003). Follicles could be isolated from ovarian tissue containing cancer cells and washing steps have been developed using both human and mouse ovaries that demonstrate this (Kniazeva et al. 2016; Soares et al. 2017). However, the stromal cell population is necessary for normal folliculogenesis as it facilitates intra- and extraovarian communication and provides a source of theca cells and supports recruitment of vessels and immune cells.

4.2 Restoration of Ovarian Hormones

Current hormone replacement therapies (HRTs) are neither developmentally dynamic, responsive, comprehensive, nor ideal long-term solutions. HRTs may be used to initiate puberty, mitigate comorbidities in postmenopausal women, or promote gender-affirming characteristics. In the right engineered environment cellular therapies could provide continuous function that responds to the body's stimuli. Mice that were ovariectomized before puberty were able to go through pubertal transitions with a transplant of ovarian cells on a decellularized scaffold by

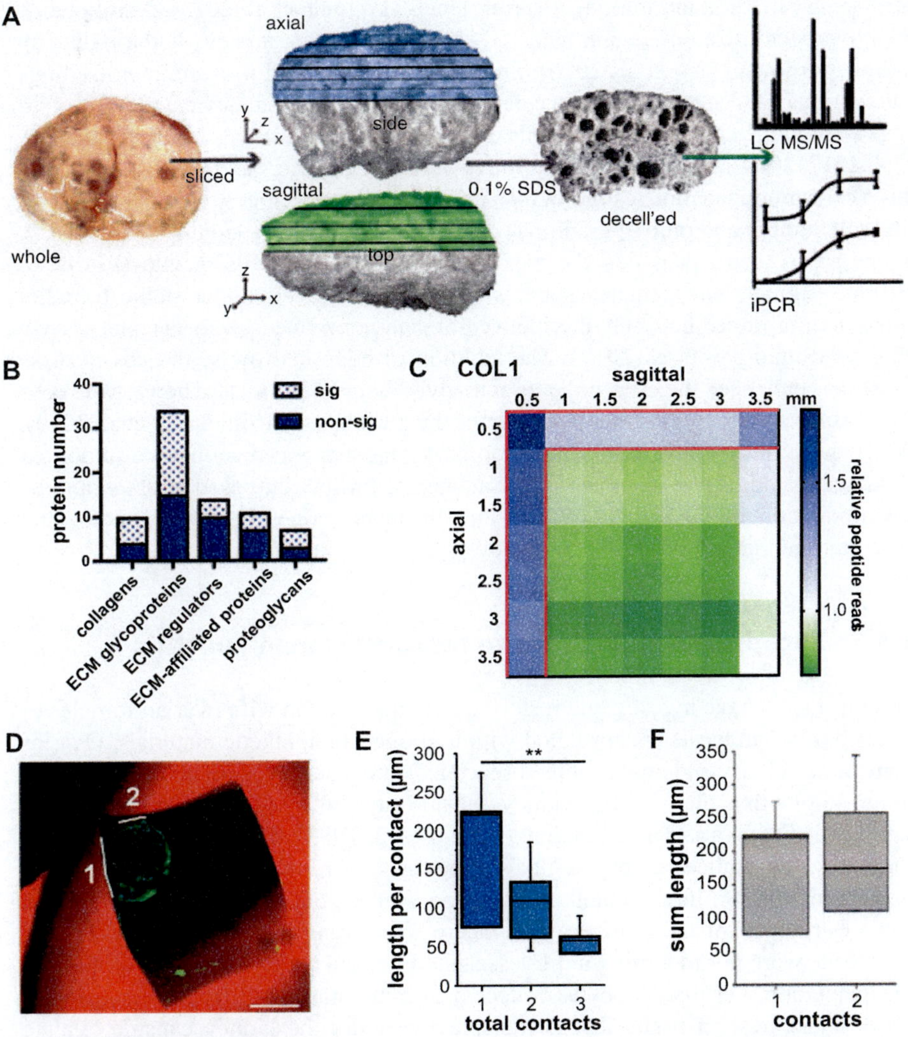

Fig. 3 Evaluation of biochemical cues and mechanical equilibrium of ovarian follicles. (**a**) Schematic of processing the porcine ovary with a tissue slicer, prior to decellularization then proteomics analysis and iPCR validation. Ovaries were sliced axially and sagittally. SDS, sodium dodecyl sulfate; "decell'ed," decellularized; LC MS/MS, liquid chromatography tandem mass spectrometry; iPCR, immuno PCR. (**b**) Number of matrisome proteins by category significantly differentially expressed across slices ("sig") or not ("non-sig"). (c) COL1 expression at sagittal and axial intersections, pink outline represents cortical layer. (Modified: Henning et al. 2019 Sci Reports). (d) Image slice that designates follicle (calcein, green) strut contact, rhodaminestained (red), at 2 contacts, scale ¼ 100 μm. (e) As the number of strut contacts increased, the length of follicle adhesion along one strut decreased, P ¼ 0.0029. (f) Side contact lengths summed. No significant difference between 1 and 2 side contact, P ¼ 0.59. (Modified from Laronda et al. (2017) and Henning et al. (2019), both licensed under CC BY 4.0)

increasing estradiol and inhibin A serum levels (Laronda et al. 2015). Additionally, adult ovariectomized mice that received encapsulated ovarian cells had a reduction in bone turnover (Guo et al. 2010). Studies were performed in ovariectomized rats using primary theca and granulosa cells encapsulated in alginate where the granulosa cells were in the center and theca cells on the periphery (Liu et al. 2013b; Sittadjody et al. 2017, 2019). Granulosa cells cultured with microcarriers and encapsulated with theca cells produced more estradiol than granulosa cells separately (Liu et al. 2013b). The cell number and ratio of granulosa cells to theca cells revealed that a ratio of 1:2, granulosa to theca, provided the maximum estradiol synthesis in vito (Liu et al. 2013a). Alginate encapsulated spheres reduced FSH levels under stable estradiol levels and improved body fat, uterine weight, and bone density versus controls over 90 days (Sittadjody et al. 2017). The addition of bone marrow stem cells to these constructs enhanced the estradiol output in vivo, though it is unclear how, as the cells did not affect viability or vascularization of the graft in vivo (Sittadjody et al. 2019). While these findings are important proofs of concepts and demonstrate important sustained estradiol levels, an additional advancement for cell-based HRTs would be the production of cyclical hormones and those hormones besides estradiol and progesterone that are produced by the ovary.

4.3 Encapsulation Methods to Restore Ovarian Function

Several studies have investigated restoring ovarian function with ovarian follicles or pieces of ovarian tissue encapsulated with hydrogels or synthetic materials. Ovaries from 8- to 11-day-old mice were dissociated and encapsulated in collagen gels before implanting them in the kidney capsule of adult ovariectomized recipients with confirmed ovarian insufficiency (Felicio et al. 1983). The transplants restored hormones, as indicated by vaginal openings, increased uterine weight and reemergence of cornified epithelium in vaginal smears, but seemed to be maintained in the estrus part of the murine cycle. There was a range of growing follicles but theca cells were not identified until vessels had formed approximately 1 week after transplantation. Fertilized oocytes collected from the transplant formed blastocysts. While intact ovarian tissue that is transplanted under the kidney capsule ovulate (Felicio et al. 1983), the dissociated ovary encapsulated in collagen did not (Telfer et al. 1990).

Similar experiments with fibrin clots were performed using primordial follicles and other ovarian cells from 6- to 8-day-old mice (Gosden 1990). They were transplanted into the ovarian bursa of mice that had 0.5 Gy of radiation, to eliminate follicles, or in ovariectomized mice. Follicle loss was recorded from both extrusion of the follicles from the clot and necrosis of follicles in the center. The grafts that were transplanted into a bursa that contained an x-radiated ovary formed into the tissue. Folliculogenesis appeared to occur in the transplant in a similar pattern, with primordial follicles residing in the cortical region. CLs with and without anovulated oocytes were found and pups were born from both transplant groups, though donor eggs were not distinguishable from recipient eggs (Gosden 1990). The use of primordial follicles from cryopreserved ovarian tissue was an important addition toward translation of these experiments. Recovered primordial follicles were encapsulated in a fibrin clot and transplanted into the ovarian bursa of adult ovariectomized mice. Eight out of eighteen mice demonstrated estrogenic activity and four mice produced live pups after natural mating (Carroll and Gosden 1993).

Other studies used fibrin, fibrin-collagen, and fibrin-alginate gels containing primordial follicles near the periphery of the beads to test the efficacy of different formulations of materials in a model of orthotopic transplants in ovariectomized mice. The fibrin-encapsulated follicles in the ovarian bursa had a twofold increase in survival over other encapsulation groups and all groups contained a majority of primordial follicles after 9 days. Mice containing a fibrin transplant or fibrin transplant with VEGF were mated. Two mice (out of six) with transplants with VEGF produced pups that were of the coat color of the donor (Kniazeva et al. 2016).

Alginate is used successfully in in vitro follicle growth of mammalian follicles (Pangas et al. 2003; Xu et al. 2006; Hornick et al. 2013; Skory et al. 2015; Xiao et al. 2015a, b). However, a step that removes the grown follicle from the alginate is required to accommodate full growth and ovulation. Follicles were isolated from 12-day-old mice and encapsulated in 0.5% alginate and transplanted into the ovarian bursa or subcutaneous pockets of adult ovariectomized mice to demonstrate an immune-isolated method (Rios et al. 2018). Those in the subcutaneous location survived at a great rate than those in the ovarian bursa. Oocytes from antral follicles were retrieved from both experimental sites and matured in vitro to produce eggs with normal spindle morphology. Those matured from the subcutaneous site could mature to the four-cell embryo (Rios et al. 2018).

Synthetic poly(ethylene glycol) or (PEG)-based hydrogels have also been used to test their function as a substrate for an engineered ovary. Nondegradable, proteolytically degradable, and dual PEG hydrogels were investigated using ovarian tissue pieces with subcutaneous transplants. No antral follicles developed in nondegradable PEG, but were found in the other groups; there was no vessel infiltration. Recipients with degradable PEG had reduced FSH levels and cornified vaginal epithelium but this was delayed in the nondegradable and dual PEG (Day et al. 2018). Hydrogels made of PEG with RGD (integrin-binding peptide) allowed for vessel infiltration and large antral follicles and CLs were visible while maintaining primordial follicles (Kim et al. 2016).

4.4 Scaffold Development for Ovarian Restoration

The main goals of an engineered ovary are to restore long-term fertility and ovarian hormone support. These results will depend on the number of primordial follicles that can be transplanted and maintained as a pool of potential gametes. Therefore, an encapsulation method may be difficult to scale; additionally, alternative methods for developing a transplantable material that lends itself totailored compartmentalized microenvironments that mimic human ovarian compartments need to be explored. To this end scaffolds made of decellularized ovaries, 3D printed gelatin, and

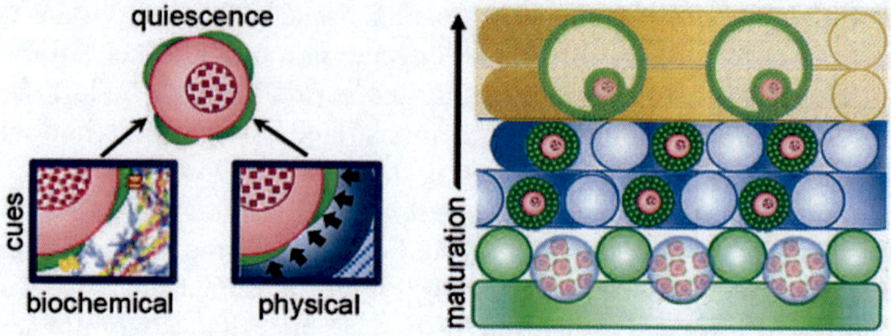

Fig. 4 Schematic for building a bioprosthetic ovary. The quiescent primordial follicle is maintained by both biochemical and physical cues. These cues can be incorporated into an engineered scaffold that supports the changing needs of the maturing ovarian follicle through growth, maturation, and ovulation

electrospun poly(epsilon caprolactone) (PCL) have been investigated (Laronda et al. 2015, 2017; Liverani et al. 2019). Bovine ovaries were decellularized with sodium dodecyl sulfate (SDS) and reseeded with follicular cells from young mice. While few follicles were present in the single-cell preparation, there were follicles identified that did form and grew within the transplant after two weeks under the kidney capsule (Laronda et al. 2015). Because it is inefficient to seed follicles into pores that were previously created by other follicles, 3D printing was explored as a way to manufacture pores that appropriately support and maintain the essential oocyte–support cell connections. Several architectural designs were created and tested in vitro before downselecting to an advancing angle with an architecture that allowed the follicles to be caught within the pore but also allowed for expansion during follicle growth. The bioprosthetic ovaries made of a 3D printed gelatin scaffold and isolated small follicles were transplanted into the bursa of ovariectomized mice. The scaffolds enabled vessel infiltration, without the addition of VEGF, restored AMH, and inhibin A, and supported ovulation and live birth of offspring that were genetically distinct from the recipient mice (Laronda et al. 2017). An improved version of the bioprosthetic ovary scaffold would consider both biochemical and physical cues (Fig. 4). These cues may change as a gradient or consider the dynamic reciprocity of the maturing ovarian follicles to enable follicle expansion, and recruitment of vessels.

4.5 In Vitro Studies with Human Follicles

A process for isolating primordial follicles from human tissue has been established, though remains inefficient in follicle yield (Laronda et al. 2014; Chiti et al. 2017a). Isolated human follicles in plasma clots or ovarian cortical tissue pieces were xenotransplanted into the ovarian bursa of adult intact mice. Both experiments resulted in some growth after seven days (Dolmans et al. 2007). At five months, the xenotransplant with isolated follicles grew to secondary and antral follicles, with a few primordial follicles remaining (Dolmans et al. 2008). Another set of experiments where isolated human preantral follicles were encapsulated in fibrin xenografted into a mouse for seven days also examined the addition of hyaluronic acid to the fibrin clot. More primordial follicles remained in the hyaluronic acid group and no difference in proportion of the follicle stages, unlike the fibrin alone clot which supported growth to secondary follicles (Paulini et al. 2016). Fibrin clots were additionally investigated for their fiber thickness, rigidity, and ability to hold follicles short term. Clots with increased concentrations of fibrin and thrombin resulted in fibers of similar thickness to human ovarian cortical tissue and rigidity of 3–10 kPa. There was significant loss of follicles in all formulations tested (Chiti et al. 2017b).

Human ovarian tissue was decellularized with 0.1% SDS and tested as a potential scaffold for human and mouse follicles (Laronda et al. 2015; Pors et al. 2019). These scaffolds supported mature granulosa cells isolated from follicular fluid and ovarian stroma cells in culture. Preantral human follicles in Matrigel were seeded on top of the human decellularized scaffold and xenotransplanted into mice. There was significant follicle loss after three weeks and the only remaining follicles were primary. There was better recovery and signs of folliculogenesis from transplants of preantral murine follicles in Matrigel that were cultured on top of decellularized human medullary scaffolds, were cultured in between two decellularized scaffolds, or within a pocket without Matrigel (Pors et al. 2019).

5 Conclusions

An ideal engineered ovary will support continued fertility and hormone restoration long term. For this to occur, the engineered material must consider the necessity of the spherical shape of the ovarian follicle, the differences in rigidity, and biological signals available within different anatomical compartments that influence folliculogenesis in order to maintain a quiescent pool adjacent to the recruited and growing follicles. Considerations should be made to the location of the transplant and potential downstream use, i.e., isolation of stimulated eggs for retrieval in a heterotopic site versus an orthotopic site that may allow for ovulation through the fallopian tubes. Sources of stromal cell populations, that are required for theca cell development and modulation of signals, and vessel infiltration should also be considered.

Immunoprotective materials may be an important feature of the engineered ovary if the cellular of scaffolding materials elicit an immune response. However, the main driver of developing an engineered ovary is to restore both hormone function and fertility and implies that biological offspring is the desired outcome. Additionally, it is unclear if full cyclicity would be restored with these immunoprotective materials, because they prevent vessel infiltration, something that is required for normal ovarian peptides hormones and the complete establishment of the hypothalamic-pituitary-gonadal axis, in addition to preventing ischemia. Additionally, immune cells are a natural part of CL formation. To facilitate the ovarian biology and alleviate these potential immune responses we are tasked with isolating ovarian follicles and cells for an autologous transplant.

There could be several autologous stromal cell sources. If engineered ovaries are able to last longer as transplants in hormone and fertility restoration than unaltered ovarian cortical tissue pieces that have been cryopreserved, then an engineered ovary could be recreated using the isolated follicles and the stromal cells from that ovary. Additionally, gonadal tissue from patients with DSD may also have some stromal cells that could be used. For those patients who have potentially metastatic disease, the source of stromal material that may be safe to transplant could be from those same patients following completion of the cancer-eradicating treatment or through expansion of populations from complimentary organs. For those patients that do not have the option to use their own source of cells to restore ovarian function, future stem cell therapies may be a welcome advancement.

The current protocols for differentiating stem cells into gametes produced PGCs and require the nurturing environment of their support cells to undergo meiosis for both mouse (Imamura et al. 2013; Hikabe et al. 2016; Ishikura et al. 2016; Zhou et al. 2016; Shimamoto et al. 2019) and human stem cells (Yamashiro et al. 2018). These are positive advancements that have the potential to benefit all patients desiring biological offspring. It is essential to have a clear understanding of how each of these new technologies, from scaffold materials and design to source of cells, will influence the gamete quality as they will ultimately affect the next generation born of these technologies.

Acknowledgments Processing and embedding and routine staining services were provided by the Microscopy and Histology Group at the Stanley Manne Children's Research Institute, affiliated with Ann & Robert H. Lurie Children's Hospital of Chicago. MML is the Warren and Eloise Batts Scholar and a Burroughs Wellcome Fund Career Award at the Scientific Interface Recipient.

References

Aittomäki K, Lucena JD, Pakarinen P et al (1995) Mutation in the follicle-stimulating hormone receptor gene causes hereditary hypergonadotropic ovarian failure. Cell 82:959–968. https://doi.org/10.1016/0092-8674(95)90275-9

Ali SB, Belfki-Benali H, Aounallah-Skhiri H et al (2014) Menopause and metabolic syndrome in Tunisian women. Biomed Res Int 2014:1–7. https://doi.org/10.1155/2014/457131

Allen EG, Sullivan AK, Marcus M et al (2007) Examination of reproductive aging milestones among women who carry the FMR1 premutation. Hum Reprod 22:2142–2152. https://doi.org/10.1093/humrep/dem148

Allingham-Hawkins DJ, Babul-Hirji R, Chitayat D et al (1999) Fragile X premutation is a significant risk factor for premature ovarian failure: the international collaborative POF in fragile X study – preliminary data. Am J Med Genet 83:322–325. https://doi.org/10.1002/(sici)1096-8628(19990402)83:4<322::aid-ajmg17>3.0.co;2-b

Amagai Y, Ishikawa S, Gotoh T et al (2006) Age at menopause and mortality in Japan: the Jichi Medical School Cohort Study. J Epidemiol 16:161–166

Andrade GM, da Silveira JC, Perrini C et al (2017) The role of the PI3K-Akt signaling pathway in the developmental competence of bovine oocytes. PLoS One 12:e0185045. https://doi.org/10.1371/journal.pone.0185045

Ankarberg-Lindgren C, Kriström B, Norjavaara E (2014) Physiological estrogen replacement therapy for puberty induction in girls: a clinical observational study. Horm Res Paediatr 81:239–244. https://doi.org/10.1159/000356922

Ankarberg-Lindgren C, Gawlik A, Kriström B et al (2019) Estradiol matrix patches for pubertal induction: stability of cut pieces at different temperatures. Endocr Connect 1:360–366. https://doi.org/10.1530/ec-19-0025

Armuand GM, Wettergren L, Rodriguez-Wallberg KA, Lampic C (2014) Desire for children, difficulties achieving a pregnancy, and infertility distress 3 to 7 years after cancer diagnosis. Support Care Cancer 22:2805–2812. https://doi.org/10.1007/s00520-014-2279-z

Baker TG (1963) A quantitative and cytological study of germ cells in human ovaries. Proc R Soc Lond Ser B Biol Sci 158:417–433. https://doi.org/10.1098/rspb.1963.0055

Bao B, Garverick HA (1998) Expression of steroidogenic enzyme and gonadotropin receptor genes in bovine follicles during ovarian follicular waves: a review. J Anim Sci 76:1903–1921. https://doi.org/10.2527/1998.7671903x

Barbie TU, Barbie DA, MacLaughlin DT et al (2003) Mullerian Inhibiting Substance inhibits cervical cancer cell growth via a pathway involving p130 and p107. Proc Natl Acad Sci 100:15601–15606. https://doi.org/10.1073/pnas.2636900100

Barton SE, Najita JS, Ginsburg ES et al (2013) Infertility, infertility treatment, and achievement of pregnancy in female survivors of childhood cancer: a report from the Childhood Cancer Survivor Study cohort. Lancet Oncol 14:873–881. https://doi.org/10.1016/s1470-2045(13)70251-1

Bastings L, Beerendonk CCM, Westphal JR et al (2013) Autotransplantation of cryopreserved ovarian tissue in cancer survivors and the risk of reintroducing malignancy: a systematic review. Hum Reprod Update 19:483–506. https://doi.org/10.1093/humupd/dmt020

Beysen D, Vandesompele J, Messiaen L et al (2004) The human FOXL2 mutation database. Hum Mutat 24:189–193. https://doi.org/10.1002/humu.20079

Blockeel C, Vos MD, Verpoest W et al (2009) Can 200 IU of hCG replace recombinant FSH in the late follicular phase in a GnRH-antagonist cycle? A pilot study. Hum Reprod 24:2910–2916. https://doi.org/10.1093/humrep/dep253

Bochner F, Fellus-Alyagor L, Kalchenko V et al (2015) A novel intravital imaging window for longitudinal microscopy of the mouse ovary. Sci Rep-uk 5:1:12. https://doi.org/10.1038/srep12446

Canada AL, Schover LR (2012) The psychosocial impact of interrupted childbearing in long-term female cancer survivors. Psycho Oncol 21:134–143. https://doi.org/10.1002/pon.1875

Care C on AH (2014) Committee opinion no. 605: primary ovarian insufficiency in adolescents and young women. Obstet Gynecol 124(1):193–197. https://doi.org/10.1097/01.AOG.0000451757.51964.98

Carroll J, Gosden RG (1993) Transplantation of frozen-thawed mouse primordial follicles. Hum Reprod Oxf Engl 8:1163–1167. https://doi.org/10.1093/oxfordjournals.humrep.a138221

Castrillon DH, Miao L, Kollipara R et al (2003) Suppression of ovarian follicle activation in mice by the transcription factor foxo3a. Science 301:215–218. https://doi.org/10.1126/science.1086336

Chand AL, Ponnampalam AP, Harris SE et al (2006) Mutational analysis of BMP15 and GDF9 as candidate genes for premature ovarian failure. Fertil Steril 86:1009–1012. https://doi.org/10.1016/j.fertnstert.2006.02.107

Chedraui P, Hidalgo L, Chavez D et al (2007) Quality of life among postmenopausal ecuadorian women participating in a metabolic syndrome screening program. Maturitas 56:45–53. https://doi.org/10.1016/j.maturitas.2006.05.008

Chemaitilly W, Li Z, Krasin MJ et al (2017) Premature ovarian insufficiency in childhood cancer survivors: a report from the St. Jude lifetime cohort. J Clin Endocrinol Metab 102:2242–2250. https://doi.org/10.1210/jc.2016-3723

Chiti MC, Dolmans M-M, Hobeika M et al (2017a) A modified and tailored human follicle isolation procedure improves follicle recovery and survival. J Ovarian Res 10:1–9. https://doi.org/10.1186/s13048-017-0366-8

Chiti MC, Dolmans M-M, Mortiaux L et al (2017b) A novel fibrin-based artificial ovary prototype resembling human ovarian tissue in terms of architecture and rigidity. J Assist Reprod Gen 32:1–8. https://doi.org/10.1007/s10815-017-1091-3

Chow EJ, Stratton KL, Leisenring WM et al (2016) Pregnancy after chemotherapy in male and female survivors of childhood cancer treated between 1970 and 1999: a report from the childhood cancer survivor study cohort. Lancet Oncol 17:567–576. https://doi.org/10.1016/s1470-2045(16)00086-3

Chung YJ, Kim HJ, Park SH et al (2015) Transcriptome analysis reveals that Müllerian inhibiting substance regulates signaling pathways that contribute to endometrial carcinogenesis. Int J Oncol 46:2039–2046. https://doi.org/10.3892/ijo.2015.2920

Cicinelli E, Einer-Jensen N, Cignarelli M et al (2004) Preferential transfer of endogenous ovarian steroid hormones to the uterus during both the follicular and luteal phases. Hum Reprod 19:2001–2004. https://doi.org/10.1093/humrep/deh404

Cooper GS, Sandler DP (1998) Age at natural menopause and mortality. Ann Epidemiol 8:229–235

Cordts EB, Christofolini DM, dos AA S et al (2011) Genetic aspects of premature ovarian failure: a literature review. Arch Gynecol Obstet 283:635–643. https://doi.org/10.1007/s00404-010-1815-4

Cordts EB, Santos AA, Peluso C et al (2012) Risk of premature ovarian failure is associated to the PvuII polymorphism at estrogen receptor gene ESR1. J Assist Reprod Gen 29:1421–1425. https://doi.org/10.1007/s10815-012-9884-x

Corkum KS, Rhee DS, Wafford QE et al (2019) Fertility and hormone preservation and restoration for female children and adolescents receiving gonadotoxic cancer treatments: a systematic review. J Pediatr Surg. https://doi.org/10.1016/j.jpedsurg.2018.12.021. ahead of print

Day JR, David A, Cichon AL et al (2018) Immunoisolatingpoly(ethylene glycol) based capsules support ovarian tissue survival to restore endocrine function. J Biomed Mater Res A 106:1381–1389. https://doi.org/10.1002/jbm.a.36338

Demeestere I, Simon P, Dedeken L et al (2015) Live birth after autograft of ovarian tissue cryopreserved during childhood: Figure 1. Hum Reprod 30:2107–2109. https://doi.org/10.1093/humrep/dev128

Development N-EKSNI of CH and H (2016) Primary Ovarian Insufficiency (POI)_ Condition Information_NICHD – Eunice Kennedy Shriver National Institute of Child Health and Human Development.pdf

Dixit H, Rao LK, Padmalatha VV et al (2006) Missense mutations in the BMP15 gene are associated with ovarian failure. Hum Genet 119:408–415. https://doi.org/10.1007/s00439-006-0150-0

Dolmans M-M, Martinez-Madrid B, Gadisseux E et al (2007) Short-term transplantation of isolated human ovarian follicles and cortical tissue into nude mice. Reproduction 134:253–262. https://doi.org/10.1530/rep-07-0131

Dolmans M-M, Yuan WY, Camboni A et al (2008) Development of antral follicles after xenografting of isolated small human preantral follicles. Reprod Biomed Online 16:705–711. https://doi.org/10.1016/s1472-6483(10)60485-3

Dolmans M-M, Marinescu C, Saussoy P et al (2010) Reimplantation of cryopreserved ovarian tissue from patients with acute lymphoblastic leukemia is potentially unsafe. Blood 116:2908–2914. https://doi.org/10.1182/blood-2010-01-265751

Donnelly L (2016) Woman gives birth to baby using ovary frozen in her childhood in 'world first.' The Telegraph

Donnez J, Dolmans M-M (2013) Fertility preservation in women. Nat Rev Endocrinol 9:735–749. https://doi.org/10.1038/nrendo.2013.205

Donnez J, Dolmans M-M (2017) Fertility preservation in women. New Engl J Med 377:1657–1665. https://doi.org/10.1056/nejmra1614676

Donnez JJ, Dolmans M-M, Demylle DD et al (2004) Livebirth after orthotopic transplantation of cryopreserved ovarian tissue. Lancet 364:1405–1410. https://doi.org/10.1016/s0140-6736(04)17222-x

Donnez J, Silber S, Andersen CY et al (2011) Children born after autotransplantation of cryopreserved ovarian tissue. A review of 13 live births. Ann Med 43:437–450. https://doi.org/10.3109/07853890.2010.546807

Donnez J, Jadoul P, Pirard C et al (2012) Live birth after transplantation of frozen-thawed ovarian tissue after bilateral oophorectomy for benign disease. Fertil Steril 98:720–725. https://doi.org/10.1016/j.fertnstert.2012.05.017

Donnez J, Dolmans M-M, Pellicer A et al (2013) Restoration of ovarian activity and pregnancy after transplantation of cryopreserved ovarian tissue: a review of 60 cases of reimplantation. Fertil Steril 99:1503–1513. https://doi.org/10.1016/j.fertnstert.2013.03.030

Duffy C, Allen S (2009) Medical and psychosocial aspects of fertility after cancer. Cancer J 15:27–33. https://doi.org/10.1097/ppo.0b013e3181976602

El-Hayek S, Yang Q, Abbassi L et al (2018) Mammalian oocytes locally remodel follicular architecture to provide the foundation for germline-soma communication. Curr Biol 28:1124–1131.e3. https://doi.org/10.1016/j.cub.2018.02.039

Ernst E, Kjærsgaard M, Birkebæk NH et al (2013) Case report: stimulation of puberty in a girl with chemo- and radiation therapy induced ovarian failure by transplantation of a small part of her frozen/thawed ovarian tissue. Eur J Cancer 49:911–914. https://doi.org/10.1016/j.ejca.2012.09.028

Espeche LD, Chiauzzi V, Ferder I et al (2017) Distribution of FMR1 and FMR2 repeats in argentinean patients with primary ovarian insufficiency. Genes-Basel 8:194. https://doi.org/10.3390/genes8080194

Faddy MJ, Gosden RG (1996) A model conforming the decline in follicle numbers to the age of menopause in women. Hum Reprod 11:1484–1486

Farquhar C, Brown J, Marjoribanks J (2012) Laparoscopic drilling by diathermy or laser for ovulation induction in anovulatory polycystic ovary syndrome. Cochrane Database Syst Rev 6:CD001122. https://doi.org/10.1002/14651858.cd001122.pub4

Fauser BCJM, van Heusden AM (1997) Manipulation of human ovarian function: physiological concepts and clinical consequences*. Endocr Rev 18:71–106. https://doi.org/10.1210/edrv.18.1.0290

Felicio LS, Nelson JF, Gosden RG, Finch CE (1983) Restoration of ovulatory cycles by young ovarian grafts in aging mice: potentiation by long-term ovariectomy decreases with age. Proc Natl Acad Sci 80:6076–6080. https://doi.org/10.1073/pnas.80.19.6076

Feng Y, Cui P, Lu X et al (2017) CLARITY reveals dynamics of ovarian follicular architecture and vasculature in three-dimensions. Sci Rep-uk 7:1:13. https://doi.org/10.1038/srep44810

Finlayson C, Fritsch MK, Johnson EK et al (2017) Presence of germ cells in disorders of sex development: implications for fertility potential and preservation. J Urol 197:937–943. https://doi.org/10.1016/j.juro.2016.08.108

Finney EL, Johnson EK, Chen D et al (2019) Gonadal tissue cryopreservation for a girl with partial androgen insensitivity syndrome. J Endocr Soc 3:887–891. https://doi.org/10.1210/js.2019-00023

Gallardo TD, John GB, Bradshaw K et al (2008) Sequence variation at the human FOXO3 locus: a study of premature ovarian failure and primary amenorrhea. Hum Reprod 23:216–221. https://doi.org/10.1093/humrep/dem255

García R, Ballesteros LM, Hernández-Pérez O et al (1997) Metalloproteinase activity during growth, maturation and atresia in the ovarian follicles of the goat. Anim Reprod Sci 47:211–228. https://doi.org/10.1016/s0378-4320(96)01637-5

Gavish Z, Spector I, Peer G et al (2018) Follicle activation is a significant and immediate cause of follicle loss after ovarian tissue transplantation. J Assist Reprod Gen 35:1–9. https://doi.org/10.1007/s10815-017-1079-z

Gomes JE, Correia SC, Gouveia-Oliveira A et al (1999) Three-dimensional environments preserve extracellular matrix compartments of ovarian follicles and increase FSH-dependent growth. Mol Reprod Dev 54:163–172. https://doi.org/10.1002/(sici)1098-2795(199910)54:2<163::aid-mrd8>3.0.co;2-4

Gosden RG (1990) Restitution of fertility in sterilized mice by transferring primordial ovarian follicles. Hum Reprod Oxf Engl 5:117–122. https://doi.org/10.1093/oxfordjournals.humrep.a137053

Goswami D (2005) Premature ovarian failure. Hum Reprod Update 11:391–410. https://doi.org/10.1093/humupd/dmi012

Gougeon A (1986) Dynamics of follicular growth in the human: a model from preliminary results. Hum Reprod 1:81–87. https://doi.org/10.1093/oxfordjournals.humrep.a136365

Gougeon A, Busso D (2000) Morphologic and functional determinants of primordial and primary follicles in the monkey ovary. Mol Cell Endocrinol 163:33–42. https://doi.org/10.1016/s0303-7207(00)00220-3

Guo X, Zhou J, Xu Q et al (2010) Prevention of osteoporosis in mice after ovariectomy via allograft of microencapsulated ovarian cells. Anat Rec 293:200–207. https://doi.org/10.1002/ar.21036

Gurka MJ, Vishnu A, Santen RJ, DeBoer MD (2016) Progression of metabolic syndrome severity during the menopausal transition. J Am Heart Assoc 5. https://doi.org/10.1161/jaha.116.003609

Henning NF, LeDuc RD, Even KA, Laronda MM (2019) Proteomic analyses of decellularized porcine ovaries identified new matrisome proteins and spatial differences across and within ovarian compartments. Sci Rep 9. https://doi.org/10.1038/s41598-019-56454-3

Hikabe O, Hamazaki N, Nagamatsu G et al (2016) Reconstitution in vitro of the entire cycle of the mouse female germ line. Nature 539:1:17. https://doi.org/10.1038/nature20104

Hirshfield AN (1991) Development of follicles in the mammalian ovary. Int Rev Cytol 124:43–101. https://doi.org/10.1016/s0074-7696(08)61524-7

Hong JS, Yi S-W, Kang HC et al (2007) Age at menopause and cause-specific mortality in South Korean women: kangwha cohort study. Maturitas 56:411–419. https://doi.org/10.1016/j.maturitas.2006.11.004

Hornick JE, Duncan FE, Shea LD, Woodruff TK (2013) Multiple follicle culture supports primary follicle growth through paracrine-acting signals. Reproduction 145:19–32. https://doi.org/10.1530/rep-12-0233

Imamura M, Hikabe O, Lin ZY-C, Okano H (2013) Generation of germ cells in vitro in the era of induced pluripotent stem cells. Mol Reprod Dev 81:2–19. https://doi.org/10.1002/mrd.22259

Ishikura Y, Yabuta Y, Ohta H et al (2016) In vitro derivation and propagation of spermatogonial stem cell activity from mouse pluripotent stem cells. Cell Rep 17:2789–2804. https://doi.org/10.1016/j.celrep.2016.11.026

Jacobsen BK, Knutsen SF, Fraser GE (1999) Age at natural menopause and total mortality and mortality from ischemic heart disease the adventist health study. J Clin Epidemiol 52:303–307. https://doi.org/10.1016/s0895-4356(98)00170-x

Jadoul P, Dolmans M-M, Donnez J (2010) Fertility preservation in girls during childhood: is it feasible, efficient and safe and to whom should it be proposed? Hum Reprod Update 16:617–630. https://doi.org/10.1093/humupd/dmq010

Jakus AE, Laronda MM, Rashedi AS et al (2017) "Tissue Papers" from organ-specific decellularized extracellular matrices. Adv Funct Mater 27:1700992–1700914. https://doi.org/10.1002/adfm.201700992

John GB, Gallardo TD, Shirley LJ, Castrillon DH (2008) Foxo3 is a PI3K-dependent molecular switch controlling the initiation of oocyte growth. Dev Biol 321:197–204. https://doi.org/10.1016/j.ydbio.2008.06.017

Johnson EK, Finlayson C, Rowell EE et al (2017) Fertility preservation for pediatric patients: current state and future possibilities. J Urol 198:1–25. https://doi.org/10.1016/j.juro.2016.09.159

Juul A, Bang P, Hertel NT et al (1994) Serum insulin-like growth factor-I in 1030 healthy children, adolescents, and adults: relation to age, sex, stage of puberty, testicular size, and body mass index. J Clin Endocrinol Metab 78:744–752. https://doi.org/10.1210/jcem.78.3.8126152

Kawamura K, Cheng Y, Suzuki N et al (2013) Hippo signaling disruption and Akt stimulation of ovarian follicles for infertility treatment. Proc Natl Acad Sci 110:17474–17479. https://doi.org/10.1073/pnas.1312830110

Kim J, Perez AS, Claflin J et al (2016) Synthetic hydrogel supports the function and regeneration of artificial ovarian tissue in mice. NPJ Regen Med 1:16010–16018. https://doi.org/10.1038/npjregenmed.2016.10

Kirsch TM, Friedman AC, Vogel RL, Flickinger GL (1981) Macrophages in corpora lutea of mice: characterization and effects on steroid secretion. Biol Reprod 25:629–638. https://doi.org/10.1095/biolreprod25.3.629

Kniazeva E, Hardy AN, Boukaidi SA et al (2016) Primordial follicle transplantation within designer biomaterial grafts produce live births in a mouse infertility model. Sci Rep-uk 5:srep17709. https://doi.org/10.1038/srep17709

Kreeger PK, Deck JW, Woodruff TK, Shea LD (2006) The in vitro regulation of ovarian follicle development using alginate-extracellular matrix gels. Biomaterials 27:714–723. https://doi.org/10.1016/j.biomaterials.2005.06.016

Laronda MM, Woodruff TK (2017) Opportunities for enabling puberty. Pediatr Adolesc Oncofertil 140:99–118. https://doi.org/10.1007/978-3-319-32973-4_7

Laronda MM, Duncan FE, Hornick JE et al (2014) Alginate encapsulation supports the growth and differentiation of human primordial follicles within ovarian cortical tissue. J Assist Reprod Gen 31:1013–1028. https://doi.org/10.1007/s10815-014-0252-x

Laronda MM, Jakus AE, Whelan KA et al (2015) Initiation of puberty in mice following decellularized ovary transplant. Biomaterials 50:20–29. https://doi.org/10.1016/j.biomaterials.2015.01.051

Laronda MM, Rutz AL, Xiao S et al (2017) A bioprosthetic ovary created using 3D printed microporous scaffolds restores ovarian function in sterilized mice. Nat Commun 8:1–10. https://doi.org/10.1038/ncomms15261

Leung K-C, Johannsson G, Leong GM, Ho KKY (2004) Estrogen regulation of growth hormone action. Endocr Rev 25:693–721. https://doi.org/10.1210/er.2003-0035

Lin L, Ercan O, Raza J et al (2007) Variable phenotypes associated with aromatase (CYP19) insufficiency in humans. J Clin Endocrinol Metab 92:982–990. https://doi.org/10.1210/jc.2006-1181

Liu L, Rajareddy S, Reddy P et al (2007) Infertility caused by retardation of follicular development in mice with oocyte-specific expression of Foxo3a. Development 134:199–209. https://doi.org/10.1242/dev.02667

Liu C, Luan X, He Y et al (2013a) Endogenous release of female hormones from co-microencapsulated rat granulosa and theca cells. Biomed Microdevices 16:209–216. https://doi.org/10.1007/s10544-013-9824-2

Liu C, Xia X, Miao W et al (2013b) An ovarian cell microcapsule system simulating follicle structure for providing endogenous female hormones. Int J Pharm 455:312–319. https://doi.org/10.1016/j.ijpharm.2013.07.004

Liverani L, Raffel N, Fattahi A et al (2019) Electrospun patterned porous scaffolds for the support of ovarian follicles growth: a feasibility study. Sci Rep-uk 9:1–14. https://doi.org/10.1038/s41598-018-37640-1

Lobo RA, Pickar JH, Stevenson JC et al (2016) Back to the future: hormone replacement therapy as part of a prevention strategy for women at the onset of menopause. Atherosclerosis 254:282–290. https://doi.org/10.1016/j.atherosclerosis.2016.10.005

Lotz L, Liebenthron J, Nichols-Burns SM et al (2014) Spontaneous antral follicle formation and metaphase II oocyte from a non-stimulated prepubertal ovarian tissue xenotransplant. Reprod Biol Endocrin 12:41. https://doi.org/10.1186/1477-7827-12-41

Lourenço D, Brauner R, Lin L et al (2009) Mutations in NR5A1 associated with ovarian insufficiency. New Engl J Med 360:1200–1210. https://doi.org/10.1056/nejmoa0806228

Maheshwari A, Fowler PA (2008) Primordial follicular assembly in humans – revisited. Zygote 16:285–296. https://doi.org/10.1017/s0967199408004802

Matthews D, Bath L, Högler W et al (2017) Hormone supplementation for pubertal induction in girls. Arch Dis Child 102:975. https://doi.org/10.1136/archdischild-2016-311372

McCaffery FH, Leask R, Riley SC, Telfer EE (2000) Culture of bovine preantral follicles in a serum-free system: markers for assessment of growth and development. Biol Reprod 63:267–273. https://doi.org/10.1095/biolreprod63.1.267

McLaughlin M, Albertini DF, Wallace WHB et al (2018) Metaphase II oocytes from human unilaminar follicles grown in a multi-step culture system. Mol Hum Reprod 19:52–58. https://doi.org/10.1093/molehr/gay002

Meirow D, Levron J, Eldar-Geva T et al (2005) Pregnancy after transplantation of cryopreserved ovarian tissue in a patient with ovarian failure after chemotherapy. New Engl J Med 353:318–321. https://doi.org/10.1056/nejmc055237

Meirow D, Hardan I, Dor J et al (2008) Searching for evidence of disease and malignant cell contamination in ovarian tissue stored from hematologic cancer patients. Hum Reprod 23:1007–1013. https://doi.org/10.1093/humrep/den055

Meunier H, Rivier C, Evans RM, Vale W (1988) Gonadal and extragonadal expression of inhibin alpha, beta A, and beta B subunits in various tissues predicts diverse functions. Proc Natl Acad Sci U S A 85:247–251

Muka T, Oliver-Williams C, Kunutsor S et al (2016) Association of age at onset of menopause and time since onset of menopause with cardiovascular outcomes, intermediate vascular traits, and all-cause mortality: a systematic review and meta-analysis. Jama Cardiol 1:767–776. https://doi.org/10.1001/jamacardio.2016.2415

Murdoch WJ (1987) Treatment of sheep with prostaglandin F2α enhances production of a luteal chemoattractant for eosinophils. Am J Reprod Im Mic 15:52–56. https://doi.org/10.1111/j.1600-0897.1987.tb00152.x

Murray A, Webb J, Dennis N et al (1999) Microdeletions in FMR2 may be a significant cause of premature ovarian failure. J Med Genet 36:767. https://doi.org/10.1136/jmg.36.10.767

Nagamatsu G, Shimamoto S, Hamazaki N et al (2019) Mechanical stress accompanied with nuclear rotation is involved in the dormant state of mouse oocytes. Sci Adv 5:eaav9960. https://doi.org/10.1126/sciadv.aav9960

Oktay K, Karlikaya G (2000) Ovarian function after transplantation of frozen, banked autologous ovarian tissue. New Engl J Med 342:1919–1919. https://doi.org/10.1056/nejm200006223422516

Oktay K, Newton H, Mullan J, Gosden RG (1998) Development of human primordial follicles to antral stages in SCID/hpg mice stimulated with follicle stimulating hormone. Hum Reprod 13:1133–1138. https://doi.org/10.1093/humrep/13.5.1133

Oktay K, Newton H, Gosden RG (2000) Transplantation of cryopreserved human ovarian tissue results in follicle growth initiation in SCID mice. Fertil Steril 73:599–603. https://doi.org/10.1016/s0015-0282(99)00548-8

Oktay KK, Buyuk EE, Rosenwaks ZZ, Rucinski JJ (2003) A technique for transplantation of ovarian cortical strips to the forearm. Fertil Steril 80:193–198. https://doi.org/10.1016/s0015-0282(03)00568-5

Ossewaarde ME, Bots ML, Verbeek ALM et al (2005) Age at menopause, cause-specific mortality and total life expectancy. Epidemiology 16:556–562. https://doi.org/10.1097/01.ede.0000165392.35273.d4

Pache TD, Wladimiroff JW, de Jong FH et al (1990) Growth patterns of nondominant ovarian follicles during the normal menstrual cycle. Fertil Steril 54:638–642. https://doi.org/10.1016/s0015-0282(16)53821-7

Pacheco F, Oktay K (2017) Current success and efficiency of autologous ovarian transplantation: a meta-analysis. Reprod Sci 24:1111–1120. https://doi.org/10.1177/1933719117702251

Pan B, Li J (2019) The art of oocyte meiotic arrest regulation. Reprod Biol Endocrin 17:8. https://doi.org/10.1186/s12958-018-0445-8

Pangas SA, Saudye H, Shea LD, Woodruff TK (2003) Novel approach for the three-dimensional culture of granulosa cell-oocyte complexes. Tissue Eng 9:1013–1021. https://doi.org/10.1089/107632703322495655

Pasquale ED, Rossetti R, Marozzi A et al (2006) Identification of new variants of human BMP15 gene in a large cohort of women with premature ovarian failure. J Clin Endocrinol Metab 91:1976–1979. https://doi.org/10.1210/jc.2005-2650

Paulini F, Vilela JMV, Chiti MC et al (2016) Survival and growth of human preantral follicles after cryopreservation of ovarian tissue, follicle isolation and short-term xenografting. Reprod Biomed Online 33:425–432. https://doi.org/10.1016/j.rbmo.2016.05.003

Perrien DS, Akel NS, Edwards PK et al (2007) Inhibin A is an endocrine stimulator of bone mass and strength. Endocrinology 148:1654–1665. https://doi.org/10.1210/en.2006-0848

Picton HM (2001) Activation of follicle development: the primordial follicle. Theriogenology 55:1193–1210. https://doi.org/10.1016/s0093-691x(01)00478-2

Picton HM, Harris SE, Muruvi W, Chambers EL (2008) The in vitro growth and maturation of follicles. Reproduction 136:703–715. https://doi.org/10.1530/rep-08-0290

Pors SE, Ramløse M, Nikiforov D et al (2019) Initial steps in reconstruction of the human ovary: survival of pre-antral stage follicles in a decellularized human ovarian scaffold. Hum Reprod Oxf Engl. https://doi.org/10.1093/humrep/dez077

Presl J, Pospísil J, Figarová V, Krabec Z (1974) Stage-dependent changes in binding of iodinated FSH during ovarian follicle maturation in rats. Endocrinologia 8:291–298

Puett D, Angelova K, da Costa MR et al (2010) The luteinizing hormone receptor: insights into structure–function relationships and hormone-receptor-mediated changes in gene expression in ovarian cancer cells. Mol Cell Endocrinol 329:47–55. https://doi.org/10.1016/j.mce.2010.04.025

Qin Y, Jiao X, Dalgleish R et al (2014) Novel variants in the SOHLH2 gene are implicated in human premature ovarian failure. Fertil Steril 101:1104–1109.e6. https://doi.org/10.1016/j.fertnstert.2014.01.001

Reindollar RH (2011) Turner syndrome: contemporary thoughts and reproductive issues. Semin Reprod Med 29:342–352. https://doi.org/10.1055/s-0031-1280919

Reynolds LP, Redmer DA (1998) Expression of the angiogenic factors, basic fibroblast growth factor and vascular endothelial growth factor, in the ovary. J Anim Sci 76:1671. https://doi.org/10.2527/1998.7661671x

Rios PD, Kniazeva E, Lee HC et al (2018) Retrievable hydrogels for ovarian follicle transplantation and oocyte collection. Biotechnol Bioeng 115:1–25. https://doi.org/10.1002/bit.26721

Rocca WA, Bower JH, Maraganore DM et al (2007) Increased risk of cognitive impairment or dementia in women who underwent oophorectomy before menopause. Neurology 69:1074–1083. https://doi.org/10.1212/01.wnl.0000276984.19542.e6

Rocca WA, Gazzuola-Rocca L, Smith CY et al (2016) Accelerated accumulation of multimorbidity after bilateral oophorectomy: a population-based cohort study. Mayo Clin Proc 91:1577–1589. https://doi.org/10.1016/j.mayocp.2016.08.002

Rodgers RJ, Irving-Rodgers HF, Russell DL (2003) Extracellular matrix of the developing ovarian follicle. Reproduction (Cambridge, England) 126:415–424

Rosendahl M, Andersen MT, Ralfkiær E et al (2010) Evidence of residual disease in cryopreserved ovarian cortex from female patients with leukemia. Fertil Steril 94:2186–2190. https://doi.org/10.1016/j.fertnstert.2009.11.032

Rossetti R, Pasquale ED, Marozzi A et al (2009) BMP15 mutations associated with primary ovarian insufficiency cause a defective production of bioactive protein. Hum Mutat 30:804–810. https://doi.org/10.1002/humu.20961

Sánchez-Serrano M, Crespo J, Mirabet V et al (2010) Twins born after transplantation of ovarian cortical tissue and oocyte vitrification. Fertil Steril 93:268.e11–268.e13. https://doi.org/10.1016/j.fertnstert.2009.09.046

Schipper I, de Jong FH, Fauser BC (1998) Lack of correlation between maximum early follicular phase serum follicle stimulating hormone concentrations and menstrual cycle characteristics in women under the age of 35 years. Hum Reprod 13:1442–1448. https://doi.org/10.1093/humrep/13.6.1442

Shelling AN, Burton KA, Chand AL et al (2000) Inhibin: a candidate gene for premature ovarian failure. Hum Reprod 15:2644–2649. https://doi.org/10.1093/humrep/15.12.2644
Shimamoto S, Nishimura Y, Nagamatsu G et al (2019) Hypoxia induces the dormant state in oocytes through expression of Foxo3. Proc Natl Acad Sci 116:12321–12326. https://doi.org/10.1073/pnas.1817223116
Shuster LT, Rhodes DJ, Gostout BS et al (2010) Premature menopause or early menopause: long-term health consequences. Maturitas 65:161–166. https://doi.org/10.1016/j.maturitas.2009.08.003
Silber S (2015) Unifying theory of adult resting follicle recruitment and fetal oocyte arrest. Reprod Biomed Online 31:1–4. https://doi.org/10.1016/j.rbmo.2015.06.022
Silber SJ, DeRosa M, Pineda J et al (2008) A series of monozygotic twins discordant for ovarian failure: ovary transplantation (cortical versus microvascular) and cryopreservation. Hum Reprod 23:1531–1537. https://doi.org/10.1093/humrep/den032
Silva-Buttkus PD, Jayasooriya GS, Mora JM et al (2008) Effect of cell shape and packing density on granulosa cell proliferation and formation of multiple layers during early follicle development in the ovary. J Cell Sci 121:3890–3900. https://doi.org/10.1242/jcs.036400
Sittadjody S, Saul JM, McQuilling JP et al (2017) In vivo transplantation of 3D encapsulated ovarian constructs in rats corrects abnormalities of ovarian failure. Nat Commun 8:1–13. https://doi.org/10.1038/s41467-017-01851-3
Sittadjody S, Enck KM, Wells A et al (2019) Encapsulation of mesenchymal stem cells in 3D ovarian cell constructs promotes stable and long-term hormone secretion with improved physiological outcomes in a syngeneic rat model. Ann Biomed Eng:1–13. https://doi.org/10.1007/s10439-019-02334-w
Skory RM, Xu Y, Shea LD, Woodruff TK (2015) Engineering the ovarian cycle using in vitro follicle culture. Hum Reprod 30:1386–1395. https://doi.org/10.1093/humrep/dev052
Smyth CD, Miro F, Howles CM, Hillier SG (1995) Effect of luteinizing hormone on follicle stimulating hormone-activated paracrine signalling in rat ovary. Hum Reprod 10:33–39. https://doi.org/10.1093/humrep/10.1.33
Soares M, Saussoy P, Maskens M et al (2017) Eliminating malignant cells from cryopreserved ovarian tissue is possible in leukaemia patients. Br J Haematol 28:255–259. https://doi.org/10.1111/bjh.14657
Stoop D, Cobo A, Silber S (2014) Fertility preservation for age-related fertility decline. Lancet 384:1311–1319. https://doi.org/10.1016/s0140-6736(14)61261-7
Stubbs SA, Stark J, Dilworth SM et al (2007) Abnormal preantral folliculogenesis in polycystic ovaries is associated with increased granulosa cell division. J Clin Endocrinol Metab 92:4418–4426. https://doi.org/10.1210/jc.2007-0729
Sullivan MW, Stewart-Akers A, Krasnow JS et al (1999) Ovarian responses in women to recombinant follicle-stimulating hormone and luteinizing hormone (LH): a role for LH in the final stages of follicular maturation 1. J Clin Endocrinol Metab 84:228–232. https://doi.org/10.1210/jcem.84.1.5389
Taboada M, Santen R, Lima J et al (2011) Pharmacokinetics and pharmacodynamics of oral and transdermal 17β estradiol in girls with turner syndrome. J Clin Endocrinol Metab 96:3502–3510. https://doi.org/10.1210/jc.2011-1449
Telfer E, Torrance C, Gosden RG (1990) Morphological study of cultured preantral ovarian follicles of mice after transplantation under the kidney capsule. Reproduction 89:565–571. https://doi.org/10.1530/jrf.0.0890565
Ting AY, Zelinski MB (2017) Characterization of FOXO1, 3 and 4 transcription factors in ovaries of fetal, prepubertal and adult rhesus macaques. Biol Reprod 96:1052–1059. https://doi.org/10.1093/biolre/iox034
Tisdall DJ, Watanabe K, Hudson NL et al (1995) FSH receptor gene expression during ovarian follicle development in sheep. J Mol Endocrinol 15:273–281. https://doi.org/10.1677/jme.0.0150273

Toledo SP, Brunner HG, Kraaij R et al (1996) An inactivating mutation of the luteinizing hormone receptor causes amenorrhea in a 46,XX female. J Clin Endocrinol Metab 81:3850–3854. https://doi.org/10.1210/jcem.81.11.8923827

van Santbrink EJP, Hop WC, van Dessel TJHM et al (1995) Decremental follicle-stimulating hormone and dominant follicle development during the normal menstrual cycle* * Supported by the StichtingVoortplantingsgeneeskunde Rotterdam and the StichtingMenselijkeVoortplanting, Rotterdam, The Netherlands. Fertil Steril 64:37–43. https://doi.org/10.1016/s0015-0282(16)57652-3

Vehkavaara S, Hakala-Ala-Pietilä T, Virkamäki A et al (2000) Differential effects of oral and transdermal estrogen replacement therapy on endothelial function in postmenopausal women. Circulation 102:2687–2693

Veldhuis JD, Roemmich JN, Rogol AD (2000) Gender and sexual maturation-dependent contrasts in the neuroregulation of growth hormone secretion in prepubertal and late adolescent males and females – a general clinical research center-based study. J Clin Endocrinol Metab 85:2385–2394. https://doi.org/10.1210/jcem.85.7.6697

Vongpatanasin W, Tuncel M, Wang Z et al (2003) Differential effects of oral versus transdermal estrogen replacement therapy on C-reactive protein in postmenopausal women. J Am Coll Cardiol 41:1358–1363. https://doi.org/10.1016/s0735-1097(03)00156-6

Wallace WHB, Kelsey TW (2010) Human ovarian reserve from conception to the menopause. PLoS One 5:e8772–e8779. https://doi.org/10.1371/journal.pone.0008772

Wallace WHB, Anderson RA, Irvine DS (2005a) Fertility preservation for young patients with cancer: who is at risk and what can be offered? Lancet Oncol 6:10–10. https://doi.org/10.1016/s1470-2045(05)70092-9

Wallace WHB, Thomson AB, Saran F, Kelsey TW (2005b) Predicting age of ovarian failure after radiation to a field that includes the ovaries. Int J Radiat Oncol Biol Phys 62:738–744. https://doi.org/10.1016/j.ijrobp.2004.11.038

Wallace WHB, Critchley HOD, Anderson RA (2011) Optimizing reproductive outcome in children and young people with cancer. J Clin Oncol 30:3–5. https://doi.org/10.1200/jco.2011.38.3877

Wallach EE, Hodgen GD (1982) The dominant ovarian follicle**the Ayerst lecture presented at the thirty-eighth annual meeting of the American Fertility Society, Las Vegas, Nevada, March 23, 1982. Fertil Steril 38:281–300. https://doi.org/10.1016/s0015-0282(16)46509-x

Wood CD, Vijayvergia M, Miller FH et al (2015) Multi-modal magnetic resonance elastography for noninvasive assessment of ovarian tissue rigidity in vivo. Acta Biomater 13:295–300. https://doi.org/10.1016/j.actbio.2014.11.022

Woodruff TK (2007) The emergence of a new interdiscipline: oncofertility. Oncofert Fertil Preserv Cancer Surviv 138:3–11. https://doi.org/10.1007/978-0-387-72293-1

Woodruff TK, Shea LD (2011) A new hypothesis regarding ovarian follicle development: ovarian rigidity as a regulator of selection and health. J Assist Reprod Gen 28:3–6. https://doi.org/10.1007/s10815-010-9478-4

Xiao S, Duncan FE, Bai L et al (2015a) Size-specific follicle selection improves mouse oocyte reproductive outcomes. Reproduction 150:183–192. https://doi.org/10.1530/rep-15-0175

Xiao S, Zhang J, Romero MM et al (2015b) In vitro follicle growth supports human oocyte meiotic maturation. Sci Rep-uk 5:17323–17325. https://doi.org/10.1038/srep17323

Xu Z, Garverick HA, Smith GW et al (1995) Expression of follicle-stimulating hormone and luteinizing hormone receptor messenger ribonucleic acids in bovine follicles during the first follicular wave. Biol Reprod 53:951–957. https://doi.org/10.1095/biolreprod53.4.951

Xu M, Kreeger PK, Shea LD, Woodruff TK (2006) Tissue-engineered follicles produce live, fertile offspring. Tissue Eng 12:2739–2746. https://doi.org/10.1089/ten.2006.12.2739

Yamashiro C, Sasaki K, Yabuta Y et al (2018) Generation of human oogonia from induced pluripotent stem cells in vitro. Science 362:356–360. https://doi.org/10.1126/science.aat1674

Yu J, Shao LE, Lemas V et al (1987) Importance of FSH-releasing protein and inhibin in erythrodifferentiation. Nature 330:765–767. https://doi.org/10.1038/330765a0

Zhou Q, Wang M, Yuan Y et al (2016) Complete meiosis from embryonic stem cell-derived germ cells in vitro. Cell Stem Cell 18:1–26. https://doi.org/10.1016/j.stem.2016.01.017

Tissue-Engineered Approaches for Penile Reconstruction

Heung Jae Park

Contents

Abstract

Organs in our body are not simple in composition. Penis also is composed of delicate complex tissues. Moreover penis is distinctive for its dual actions, voiding and sexual functions, and has an impact on the psychological aspect for some people. There are lots of reasons for penile reconstruction, including various penile diseases, penile traumas, and congenital penile anomalies. Other niche fields for penile reconstruction would be transsexual surgeries and penile augmentations. As other tissues, penile tissue has little substitutes and shortage of supply for reconstructive surgery. And allogenic transplantation as kidney would be impossible except for the exceptional situations because of the ethical issue. Tissue engineering would be one of the solutions for the sufficient supply of penile tissue on demand.

H. J. Park (✉)
Department of Urology, Kangbuk Samsung Hospital, Sungkyunkwan University School of Medicine, Seoul, Republic of Korea

D. Eberli et al. (eds.), *Organ Tissue Engineering*, Reference Series in Biomedical Engineering, https://doi.org/10.1007/978-3-030-44211-8_14

This chapter deals with how the penile reconstruction using tissue engineering has been started and developed. Clinical trials in penile tissue engineering will be introduced. Also, the new approaches and investigations for better physiologic reconstruction of penis in the future including stem cell applications, cell therapy and gene therapy, new scaffolds, and 3D printing technics will be discussed.

1 Introduction

1.1 Anatomy and Physiology of Penis

Penis is the complex organ composed of various tissues. Grossly penis can be divided into two parts in structure and function. Anatomical structure of penis is well known (Drake et al. 2009; Dwyer et al. 2011) (Fig. 1). The penis contains three cylindrical structures. A pair of tissue located at the dorsum of penis is called as corpus cavernosum and these structures are responsible for penile erection. They are covered with a thick fibroelastic structure called tunica albuginea (TA). Corporal tissue is composed of trabeculae with various sizes of pores resembling sponge architecture. These sinusoidal structures are composed of connective tissue, mainly several types of collagen and elastin (Goldstein et al. 1985). Smooth muscle cells (SMCs) and endothelial cells (ECs) are intermingled with these cavernosal sinusoidal matrices. Once arterial blood flows into corpus cavernosum, erection starts. This inflow of blood is pooled in the irregular size sinuses of corpus cavernosum resulting in expanding volume of cavernosal tissue. This tissue expansion is restricted by the surrounding thick, partially elastic TA. As the arterial inflow slowing down for back pressure of corpus cavernosum, the venules penetrating TA, that bridges in and out

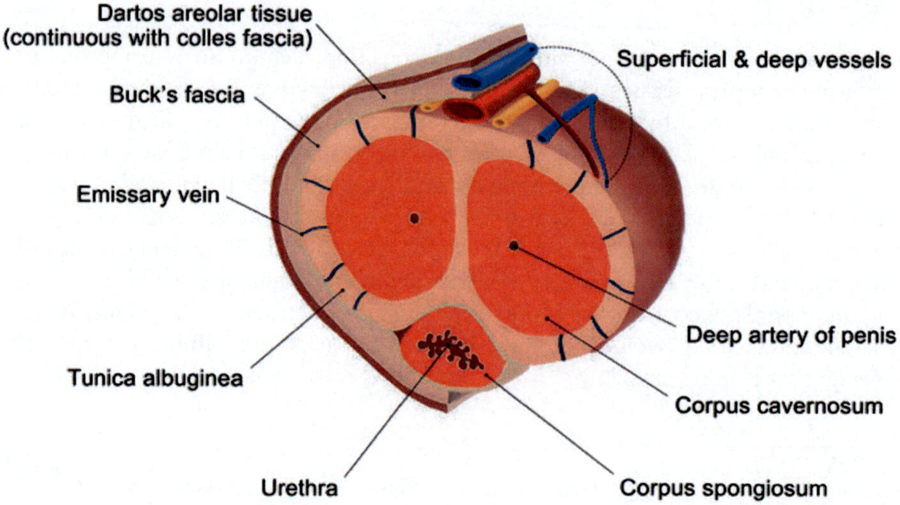

Fig. 1 Cross-sectional anatomy of penis

of corpus cavernosum, and draining blood pooled inside the corpus cavernosum, are compressed by increasing intracavernosal pressure inside the TA. This results in stasis of blood inside corpus cavernosum and maintenance of high pressured cavernosal tissue (Lue 2016). The stiffness that originates from corpus cavernosum is erection.

Another cylindrical structure is the corpus spongiosum. It is located at the ventral side of the penis and surrounds the urethra. The urethra is composed of several muscle layers and acts as a common pathway of urine and ejaculate. The corpus spongiosum is also covered with TA. Penile subcutaneous tissue is composed of Buck's fascia, Colle's fascia, and small arteries, veins, and nerves covering the TA (Awad et al. 2011; Hsieh et al. 2012).

Histology of corpus cavernosum shows that this structure is able to shrink and dilate under appropriate conditions, and sustaining erection is possible with pooling of blood in the sinusoidal structure of corpus cavernosum with the help of TA. The key mechanism of erectile process is coordination of various nerves and vessels. Aside of erectile process, ejaculatory process happens in a millisecond, also with the cooperation of autonomic nerves, sensory and motor nerves, and muscles (Lue 2016).

Though penile reconstruction includes reconstruction of urethra as well as corporal tissue, reconstruction of urethra and corpus spongiosum is dealt in another chapter of this book.

1.2 History of Tissue-Engineered Penile Reconstruction

There are instances when penile reconstruction is needed. These are pathologic conditions to deteriorate the penile functions or ruin penile cosmetic aspects. These penile diseases include Peyronie's disease (PD), urethral stricture, penile fracture, burns, infections, penile cancer, penile trauma as penetrating injuries and blunt force trauma, bites, and congenital anomalies like hypospadias, epispadias, buried penis, and so on (Lumen et al. 2015). Nowadays there is another specific demand for penile reconstruction, as penile augmentation or transsexual surgery. Damaged penis usually loses its function as erection and urination. A lot of pathologic conditions of penis could not be restored with conservative approaches, therefore depending on severity of disease, surgical reconstruction would be the appropriate solution (Williams et al. 2016).

Even though various surgical treatment modalities have been introduced to restore penis since the 1930s, the results had not fulfilled the expectations of doctors and patients. The primary goal of penile reconstruction was structural restoration. Since the Russian surgeon Nikolai Borgoras tried penile restoration using a tubed abdominal pedicled flap combined with rib cartilage in 1936 (Bogoras 1936; Goodwin and Scott 1952), clinicians have utilized various well-vascularized musculofascial tissues such as the radial forearm flap, thigh flap, free fibular flap, osteofasciocutaneous flaps, or the anterolateral thigh flap as well as autologous skin, muscle, vessels, facial tissue from various parts of body in penile reconstruction

(Felici and Felici 2006; Mutaf et al. 2006; Salgado et al. 2011). Materials, allogenic nongenital tissues including skin, muscle, and xenogenic tissues from various locations were also applied as the substitution of penile reconstruction due to deficit autologous genial tissue (Palese and Burnett 2001). Nevertheless the results of surgical procedures revealed risks of graft failure due to tissue necrosis, infection, and donor site problems (Horton and Dean 1990). Adverse immunologic reactions were another problem in non-autologous graft materials. Usually these surgical procedures have been performed in multiple stages depending on the cases and methods. The first one-staged penile reconstruction was performed in 1984 (Chang and Hwang 1984).

A novel technic called "tissue engineering" was introduced in 1988 to create the appropriate tissue using cells and biomaterials. For the purpose of properly functioning tissue or organ reconstruction, tissue engineering would be the appropriate treatment modality compared to the traditional methods in obtaining sufficient and efficient biologic substitutes (Machluf and Atala 1998). In 1999, the concept of "regenerative medicine" has been proclaimed. The terminology Regenerative Medicine include wider range of field than Tissue Engineering, including cell therapy, nuclear, or gene therapy adding to tissue engineering.

As another approach for penile reconstruction, penile transplantation has been tried three times until now. In spite of the ethical issues in 2006, penile allotransplantation was performed without serious physical postoperative complications of the transplanted penis. Instead, the result was dismal for the psychological problem of recipient (Caplan et al. 2017; Hu et al. 2006).

There are two main factors of tissue engineering: cells and biomaterials as scaffolds. Researchers have investigated for the selection and culture of proper cells, methods to harvest enough cells, inventing appropriate biocompatible scaffolds to restore specific organs.

Studies have been started for the proper cell selection and effective methods of cellular proliferation composing corpus cavernosum. Cell proliferation technic like cell culture was not familiar and not studied systematically in the clinical field until the 1980s (Atala 2012). Major cell composite of corpus cavernosum is smooth muscle cells (SMCs) and endothelial cells (ECs). Researchers have successfully set up technics to culture and harvest purified SMCs and ECs (Carson et al. 1989).

Another wing of necessary component is biocompatible biomaterials. Ideal biomaterials would support the growth of cells, act as a scaffold of aimed structure, and avoid immune reactions to the seeded cells and implanted host. Biomaterials are classified into two categories. The first one is synthetic materials. Ideal properties of synthetic biomaterials would be biocompatible, nontoxic, nonantigenic, non-teratocarcinogenic. A lot of synthetic polymers were introduced as polyglycolic acid (PGA), polylactic acid (PLA), poly (lactic-co-glycolic acid): PLGA, poly (caprolactone: PCL), and so on. These synthetic materials have advantages such as production in large scale, reproduction, and availability of mixing components as enhancing or inhibiting factors and are also capable in quality control on specific demands. On the other hand, some synthetic polymers have limitations of their own as PLGA is not elastic enough for a scaffold in contractile tissues (Boland et al. 2006).

Another category would be natural materials like acellular matrices (ACM) from native tissue and decellularized bladder submucosa or small intestinal submucosa (SIS), collagen, alginate, hyaluronic acid (HA), elastin, and fibronectin derivatives manufactured from allogenic or xenogenic sources. The merits of natural materials are lower immune reaction and natural 3D architecture in case of ACM. Collagen has been one of the most popular substances of them. It can be derived from human and animals, and has been approved for clinical application from the US Food and Drug Administration (FDA). Collagen has been adopted in the production of various textures for its properties of low immunogenicity and low inflammatory responses. Another advantage of collagen is that it is one of the raw materials relatively easy in molding to certain shapes and structures. Also collagen is one of the main component of the ACM (Williams et al. 2016). About the safety issue, FDA approved collagen and PLA, PGA, PLGA as the materials applicable to humans in 1981. This was followed by approval of HA in 2003, Poly-L-lactic acid (PLLA) in 2004, Polymethylmethacrylate (PMMA), and calcium hydroxyapatite (CaHA) in 2006. Autologous fibroblast was approved in 2011. Studies have developed to combine natural materials with synthetic polymers as hybrid biomaterial on demand. One of the examples would be mixture of hybridizing a high molecular weight Poly (ε-caprolactone) (PCL) and type I collagen. This composite biomaterial was designed for the endurable material under the environment of high flow and pressure as in physiologic vascular status. This composite biomaterial showed improved mechanical properties as longer stability and provided better condition for vascular cells compared to scaffold made from single material (Lee et al. 2008). A variety of biomaterials have been tried for penile reconstruction using tissue engineering, synthetic materials as PLGA, natural materials as ACM, collagen, etc.

2 Corpus Cavernosum

Researches for reconstructing penile tissue using tissue engineering have progressed by the following steps: (1) establishing method and process of culture and expansion of necessary cells, SMCs and ECs, (2) animal studies of cell-biomaterial complexes using cells combined with synthetic biomaterial scaffolds, (3) animal studies of cell-corporal ACM complexes, and (4) replacing entire corporal tissue with cell-corporal ACM complexes in animal for the recovery of penile function.

The first step was constructing a tissue similar to corpus cavernosum from the cell-scaffold complex. In the 1980s technics of cell culture for implantation was not familiar and infrequently performed by biologists and clinicians. Human corporal SMCs, one of the major cells composing corpus cavernosum, was isolated from human corporal tissue using explant technic, then cultured and expanded in the cell culture chambers under the sterile condition. The PGA polymer scaffolds were manufactured as porous architecture for its durability and hydrolytic properties. These cultured SMCs were combined to the biodegradable PGA polymers. The SMCs-polymer complexes have been implanted at the subcutaneous space of rats. In 1998, Kershen et al. reported the results of a study about engineered tissue in vivo using animals. The implanted cell-polymer complexes were harvested at 7, 14, and 24 days after surgery. Histologic analysis showed retrieved SMCs-PGA complexes formed smooth muscle cell layers and showed penetrating vessels into SMCs-PGA complexes from surrounding tissue, which was necessary for cell viability and development of cells into tissue. Alpha smooth muscle actin, the smooth muscle specific marker, was detected by Western blotting and immunocytochemical staining in these neo-tissues formed from cell-polymer complexes. The PGA scaffolds degraded with time in the rats as designed. The cultured cells in vitro combined with synthetic biodegradable polymer scaffolds developed into smooth muscular tissue in vivo. This was the first study on reconstruction of corpus cavernosal tissue using cultured human corporal SMCs seeded on biodegradable PGA polymers (Kershen et al. 2002).

Oxygen and nutrients are necessary in cell survival and reconstituting tissue from cells. These elements are delivered by blood. The sufficient vasculatures would be important to promote the reconstitution of tissue or organ from individual cells and to avoid necrosis and fibrosis of implants or grafts (de Vocht et al. 2018; Novosel et al. 2011). Surgical procedures result in disconnection of vascular supplies at the site of incision or implantations from surrounding native environments. The organic implants as cells or tissue need ingrowth of vessels from surrounding tissues into implanted tissues or autologous vascular formation from implants by themselves. One of the huddles in reviving cellular implants was to build abundant vasculatures around the cell-containing implants. ECs are the basic component of vasculatures and are necessary for vascular formation and it is another major composite of corpus cavernosum together with SMCs. Investigation for coexistence of SMCs and ECs in the implant complexes was attempted. Human SMCs and ECs (ECV 304) were cultured individually, and then seeded on PGA polymer scaffolds in stepwise manner. These SMCs- and ECs-seeded PGA polymer complexes were implanted in the skin of athymic nude mice. The basic process of these methods including cell preparation and seeding, are similar to various other applications of tissue engineering (Lanza 2002; Yoo et al. 1998b). Park et al. (1999) reported the construction of vascularized corporal tissue in vivo by combining human SMCs and ECs to PGA scaffolds. The implanted cell-PGA complexes were retrieved at 1–42 days after surgery and they showed well-organized smooth muscle architecture surrounded by the accumulation of endothelium lining the porous luminal structures and plentiful of neo-vascularities around the implanted complexes from 5 days of implantation. This study revealed that ECs had facilitated the ingrowth of vasculature from the surrounding native tissue and new capillary formation by itself, resulting in assisting reconstruction of corporal tissue with abundant vasculatures. This was the first study of the possibility in combination of different types of cells on a scaffold at a time and the way to supply enough oxygen and nutrients through neo-vasculatures using cultured ECs in vivo.

Studies stepped ahead to (1) introduction of new three-dimensional matrix for penile reconstruction, as ACM, (2) improved cellular distribution and attachment technics as dynamic seeding method applying bioreactors, and (3) developing method of total replacement of penis by cell–matrix complexes in vivo.

There was a need for more physiologic, biocompatible scaffold specific for corpus cavernosum. Manufacturing scaffold similar to the native cavernosal sinusoids was an obstacle. Pores of sponge-like corporal sinusoids vary in size and are irregular in distribution. Fabricated synthetic polymer scaffolds were not suitable for the natural corpus cavernosal architecture. Corpus cavernosum is composed of elastic fibers, collagen (mainly type I and part of type III) and SMCs, ECs (Lue 2016). Three-dimensional scaffold similar to corpus cavernosum was investigated. Decellularization method is removing any cells from tissue and results in acellular backbone structures as end product. Acellular Matrix (ACM) made of allogenic tissue has the merits of same meticulous microarchitecture as native properties, less immune responses, and contains some growth factors or cytokines that help in cellular growth compared to conventional synthetic biomaterials (Hoganson et al. 2010). In previous studies, acellular matrices were processed using animal bladder and urethra by cell lysis technic, then these ACMs were applied in animal for bladder and urethral reconstructions (Atala et al. 1999). Falke et al. (2003) processed donor rabbit corporal tissues to get corporal ACM using cell decellularization technic, which was the same method used in previous bladder or urethral tissue preparations. These products had the same architecture of corporal sinusoids avoid of cells as native corpus. Acellularity of these matrices was confirmed by histology and electron microscopy. They have cultured human SCMs, ECs, and then implanted 80 of ACM-human SMCs and ECs complexes in athymic nude mice. These implants were harvested from 3 days to 8 weeks after implantation. Retrieved implants were analyzed histologically, physiologically including organ bath study, scanning electron microscopy (SEM), and by molecular analyses including Western blot, RT PCR. The results showed the presence of neo-vascularities in the sinusoidal spaces of ACM with increased organization of smooth muscle and collagen over time. Analysis showed the characteristics of human SMCs and ECs in immunohistochemical studies (alpha-actin and factor VIII, each) and Western blot, and revealed expression of muscarinic acetylcholine receptor subtype m4 mRNA on RT-PCR in 8 weeks implants. In 4 weeks implants, ACM-cells complexes showed cell-covered sinusoidal architecture. Appropriate collagen composition was found with hydroxyproline quantification. Contractility was detected in harvested ACM-cells complexes on organ bath electrical field stimulation also. This study meant human SMCs- and ECs- seeded corporal ACM complexes constructed well-vascularized corporal structures similar to the native corporal body. ACM worked as a better substitute than conventional corporal scaffolds as cell delivery vehicle in vivo (Falke et al. 2003).

Kwon et al. (2002) replaced segments of rabbit corporal body with the cells-ACM complexes made in the same processes as mentioned above. They implanted 18 of ACM-SMCs and ECs complexes along with 8 of ACMs without cell as control into the rabbit penis. Rabbit corpus cavernosum was excised segmentally in 7 mm length leaving urethra intact, then cells-ACM complexes were interposed at the excised cavernosal sites. The rabbits sacrificed after 3, 6 months of implantation. Rabbit cavernosal structure and intracavernosal pressure were evaluated before retrieving cells-ACM complexes. They showed continuous visualization of cavernosum, this meant intact patent corpus cavernosal cylindrical architecture despite of corporal replacement on cavernosography and improved intracavernosal pressure up to 50% of normal compared to null in control group (only ACM implantation) on cavernosometry. Histological analysis showed corporal sinusoids were covered with SMCs and ECs compared to fibrosis and calcifications in ACM-only group. Levels of eNOS and nNOS of cell-ACM complex group were similar to that of normal rabbit in Western blot analysis compared to decreased nitric oxide synthase activity in ACM without cell group. The rabbits implanted with the cells–matrix complexes mated 3 month after implantation and the sperms were detected from female rabbit after mating. This result meant tissue-engineered penile implant worked more physiologically, about 50% of normal and even showed ability to achieve the sufficient erection, vaginal penetration, and ejaculation. This study showed successful tissue-engineered autologous corporal tissue in structural and functional aspects, implying potentials for applying this technic into clinical penile reconstruction.

To improve the maturation of implanted cell–matrix complex, cellular distribution in the matrix is important. The methods were investigated for improving delivery, attachment, and growth of cells in the new environment of scaffolds (Hasan et al. 2014).

In normal instance, penis erects spontaneously for several times during sleep physiologically (nocturnal tumescence) without any sexual stimulations. Nocturnal tumescence is known as preventing fibrotic degeneration of cavernosal tissue and inducing angiogenesis in corpus by regular oxygenation of corpus cavernosum. Insufficient supply of oxygen and nutrients result in fibrosis and degeneration of tissue. Abundant vasculature around and inside the reconstituting structure helps cells to mature into organized tissue. Appropriate distribution of ECs into the scaffold together with the ECs on scaffold's surface would be ideal to form vascular structures than ECs residing on surface of scaffold only. Similar condition as nocturnal tumescence was designed during cell culture and cell seeding on biomaterials in vitro with the help of bioreactor (Eberli et al. 2008). Previous results of Kwon et al. (2002)'s trial was not satisfactory as intracavernosal pressure was less than normal in reconstructed corporal tissue in rabbit. It was assumed that one of the reasons for deficit intracavernosal pressure might be insufficient number of functional cells in the implanted ACM-cells complexes (Persson et al. 1989). Previous implants showed that cell densities in engineered reconstructed tissues were lesser than cells in normal corporal sinusoids. Increasing cell densities in the implants, especially SMCs, would improve intracavernosal pressure to the physiologic level. It was postulated that applying the hydrodynamic forces on solution containing cells would be advantageous in distributing cells into the irregular, narrow spaces by the form of cells in solution. Since 1990s, there have been studies about cell seeding technics and improved results were stated in the use of hydrodynamic forces, dynamic seeding, and bioreactor system (Burg et al. 2000; Xiao et al. 1999).

So Eberli et al. (2008) tried to improve the quantity and quality of cell viability and cell attachment on biomaterials for corpus cavernosum using bioreactor system consisting of a closed glass beaker and a magnetic stirrer with hydrodynamic forces. ECs and ACMs obtained from rabbits were processed as routine. EBM2 solution containing rabbit ECs were saturated on ACMs in static manner or dynamic method. For dynamic seeding group, ECs-ACM complexes were cultured using bioreactors stirred in 40 rpm for 48 h. Each cell-ACM complexes were implanted in athymic mice subcutaneously. These implants were harvested and analyzed after 14 and 28 days of surgery. Specimens of dynamically seeded group showed better cell viability and better cell attachment compared to that of static group. Retrieved tissues of dynamically seeded group revealed enhanced tissue organization, higher cellular density, intact cellular lining of the sinusoids, and complete coverage of the sinusoidal space on immunohistochemical, histologic evaluations. MTT assay represents overall metabolic activity of 3D tissue, not a quantifying tool for counting cell number. It measures mitochondrial activity that would imply active metabolizing cells. Dynamic seeding group showed greater metabolic activity compared to static seeding group in MTT assay. Superior in cellular density was measured with DNA assay and also showed superior in the dynamically seeded ACMs than static group (71% versus 39% compared to the superior in cellular density of normal corpus cavernosum). Results of histology and SEM revealed structural superiority in the dynamic seeding group also. These results implied dynamic seeding using bioreactor could result in enhancement of cellular distribution, cellular differentiation improving tissue restoration.

Following Eberli et al. (2008)'s trial of dynamic seeding of ECs only, Falke et al. (2003) made the cell-seeded ACMs by dynamic seeding of dual cells, as SMs and ECs, and replaced short segment of rabbit corpus cavernosum with these ACM-autologous cells complexes. After the successful results of above trials, Chen et al. (2010) replaced total corpora with engineered tissue in the rabbits applying dynamic cell seeding. The basic procedure was similar to the previous studies using cells and biomaterials (Fig. 2) SMCs and ECs were prepared from biopsied rabbit corpus tissue. They processed ACMs from donor rabbit corpus by decellularization. Then they seeded these cells on ACMs in a manner of multistep static/dynamic seeding method for the purpose of homogenous cellular distribution in ACMs and to enhance cellular attachment. These cells-ACM complexes were implanted into excised penile space of 12 rabbits. Nonsurgical rabbit group and rabbits operated without implantation group were as control and comparison groups. Evaluations were made in 1, 3, 6 months after implantation. Cavernosography and cavernosometry showed normal shape of gross corpus cavernosum and normal intracavernosal pressure in cell–matrix complex group compared to nonsurgical group. On physiologic test using organ bath studies of muscle contractors, relaxants, and electrical stimulation, they could observe the contraction of implanted ACM-cell complexes by electric stimulates or pharmaceutics and these implied that penile nerve innervated into the implanted engineered corpus. Similar to previous trial in vivo, mating, ejaculation was observed in these engineered tissue implanted rabbits and gave birth to descendants. This research showed the development of penile tissue engineering achieved the tissue construction performing similar functions as normal animal (Chen et al. 2010). Even bearing problem of allogenic corporal ACM supply (Caplan et al. 2017), this technic implied potential of reconstructing penile tissue. Also this study triggered the need for the technics to

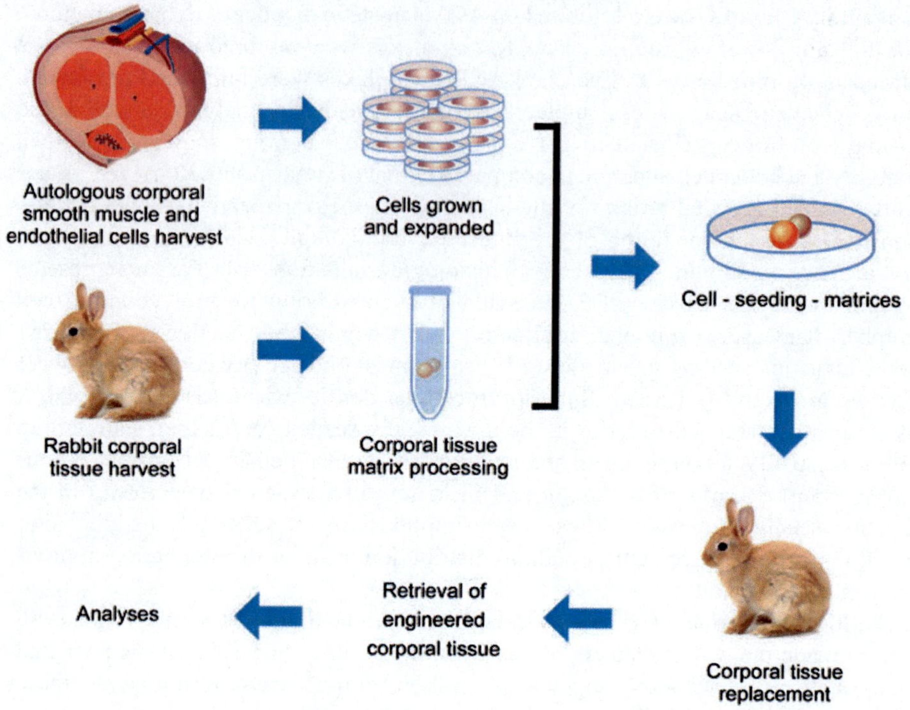

Fig. 2 Basic process of in vitro and in vivo study of cavernosal reconstruction

fabricate meticulous synthetic polymer scaffolds identical to natural corporal matrix along with ACMs.

There was a report of corporal ACM processed from human corpus cavernosum. They obtained human tissue under guidance of institutional review board approval and got informed consent from transgender patients before sex reassignment surgery. They decellularized human corpus cavernosum, then implanted human corporal ACM into the peritoneal cavity of Sprague–Dawley rats. Histologic evaluation revealed intact corporal sinusoidal structure with increasing vasculature around, ingrowth of SMCs and ECs from 1 month of implantation. The implanted ACM connected to the microcirculation of host rat and inward spread of host cells into ACM was observed. These results might imply that there was natural regenerating ability in the body, as the concept of natural bioreactor (Kajbafzadeh et al. 2017).

There have been clinical applications in several organs made by tissue engineering already. Clinical applications on bladder and urethra have been reported. Atala et al. (2006) applied tissue-engineered autologous bladder as cystoplasty clinically. They also reconstructed human urethra utilizing engineered tissue (Raya-Rivera et al. 2011).

In the long run, researchers are expecting that clinical trial of tissue-engineered penile tissue could be approved by FDA. Definitive rabbit study must be the first step for providing safety data. This study includes scaffolds preparation, cell isolation, cell characterization, cell seeding, and characterization of seeded scaffold, implantation, and characterization in vivo. After confirming these studies, permission of human clinical trial from FDA would be expected. A Phase 1 clinical trial using autologous corporal cells was approved by the FDA and is being initiated to demonstrate safety and feasibility.

A lot of suggestions are proposed for the construction of the tissue-engineered penis closer to physiologic status. To create corporal tissue from cells-scaffold complexes efficiently, we expect the developments in several aspects, proper distribution of cells with help of bioreactors or 3D bioprinting, methods of collaborating pharmaceutical compounds, cytokines or growth factors aiding tissue maturation from seeded cells, methods of enriching vasculature into the cell-scaffold complexes, introduction of modified cells including stem cells (SCs), or applying DNA or gene therapies.

3 Penile Prosthesis

Another issue of penile reconstruction is penile prosthesis. Inserting penile prosthesis is the definitive treatment of erectile dysfunction (Levine et al. 2016). Initially, the concept of penile prosthesis was for the structural buttress in reconstructing damaged penis aside of erectile functions. Bone or cartilages as rib covered with autologous musculofascial tissue was applied to form penis-like construct for substitution of damaged penis. Following the introduction of inflatable penile prosthesis (Scott et al. 1973), a variety of penile prosthesis has developed around 1980s to treat erectile dysfunction. Currently available penile prosthesis is classified as malleable type and inflatable type. Penile prosthesis implantation: past, present, and future (Simmons and Montague 2008). Malleable penile prosthesis is semirigid and simply bendable with the spring compartment in the prosthesis. The merits of malleable type are its lower incidence of mechanical failure and it is less expensive than inflatable prosthesis. But implanted penis is not natural and it is difficult to conceal for its constant semirigid property. Complications as erosion and protrusion of implantation might occur. Inflatable prosthesis includes two-pieces, three-pieces, and positionable prosthesis. These provide better tactile sensation as normal penis and these can be deflated when not using. But mechanical dysfunction is more frequent than malleable type and is more expensive. Major composition of this conventional prosthesis is silicon. Prosthesis made of silicon is widely used but still have some drawbacks as biocompatibility, infection, erosion, mechanical problems like autoinflation, leakage, and pump malfunction (Muench 2013; Nukui et al. 1997). Yoo et al. (1998a) reported tissue-engineered penile prosthesis. They created penile prosthetic material using cartilage cells. Harvesting articular cartilage tissue from calf, chondrocytes were isolated from these cartilages, and then cells were culture-expanded. These culture-multiplied chondrocytes were seeded on rod-shaped PGA polymer scaffolds. These chondrocyte-PGA polymer complexes were implanted at the back of athymic mice along with polymer without cell implant group as control. Chondrocyte-PGA polymer complexes have formed in milky solid cartilaginous rod compared to shrinkage of hydrolyzed polymers without cell (Yoo et al. 1998a). As a next study, cartilage was harvested from rabbit ear, isolated chondrocytes, and expanded in culture. These autologous chondrocytes were seeded on polymer rods and implanted in the rabbit corpus compared to implanting polymers without cell. One month later, implants were retrieved and cell-seeded polymers formed milky cartilage rods. On the other hand, implants of polymer without cell degraded in 2 months after surgery. The cell–polymer complex implanted rabbits could copulate and impregnate female rabbits (Yoo et al. 1999). Investigation of mechanical properties of engineered cartilage rod was taken for the possibility in clinical application. Cartilage was obtained from human ear and chondrocytes were isolated. Cultured chondrocytes were seeded on polymer rods in the same manner as the previous experiments and harvested after 2 months. Mechanical properties were evaluated in harvested implants. These cartilaginous tissues were flexible, elastic, and endurable to high degrees of compressive pressure that might have had potential for pressure durability in vaginal penetration during intercourse. These mechanical properties were comparable to silicone prosthesis used in clinical field (Kim et al. 2002). These results showed tissue-engineered cartilage tissue would have potential in application to malleable prosthesis.

In general, biomaterials originated from natural tissue might have merits of tissue friendly properties as lower immune reactions. So these studies opened the possibility of developing penile prosthesis with ideal mechanical properties and biocompatibility: as semirigid penile prosthesis made of cartilage equipped with spring components inside, bi-layered inflatable prosthesis as outer biocompatible tissue layer covering inner synthetic biomaterial as silicon regarding the virtue of the materials, or designing single layered prosthesis made of material combining natural and synthetic biomaterials. The technologies of 3D bioprinting could assist the precise design of engineered penile prosthesis.

4 Tunica Albuginea

Tunica albuginea (TA) is a relatively firm, elastic tissue covering corpus cavernosum. TA is a two-layered structure of outer longitudinal and inner circular layer with 1.5–3 mm thickness depending on its location. The main composition of TA is known as type I and type III collagen fibers nesting on elastin fibers (Wein et al. 2016). In case of penile injuries involving TA as penile fracture, old age, or Peyronie's disease (PD), malfunction of TA result in the so-called venogenic impotence. Penile fracture is the laceration of TA arising from exacerbating penile bending by external forces. It is one of the emergent conditions to be operated promptly.

PD is a pathologic condition of formation of fibrous plaque on TA. The fibrous plaque shows abundant abnormal collagen and elastin tissue (Davis Jr. 1997). The cause is uncertain. Blunt trauma as excessive bending of penis intolerable to elastic limit of TA would be suspected. It is estimated that TGF-β1 plays a major role in pathogenesis of Peyronie's plaque (Hinz 2015). The incidence of PD is estimated to be about 1–13%. PD could develop penile deformities as curvature, penile shortening, and pain during erection and even erectile dysfunction. Spontaneous resolution of penile deformity in PD occurs only in 3.2–12% of patients. The Impact of Peyronie's Disease on the Patient: Gaps in Our Current Understanding (Goldstein et al. 2016).

Kinds of conservative medical treatments have been tried for PD: cold massage, compression, oral medications as vitamin E, potassium aminobenzoate, tamoxifen, colchicine, analgesics, anti-inflammatories, vasodilators, intralesional injections as steroid, calcium antagonists, vasodilators, interferon or collagenase enzymes (Abdel Raheem et al. 2017), botulinum toxin (Munoz-Rangel et al. 2015) vaccum device, extracorporeal shockwave therapy (ESWT), and combination of some of these (Tan et al. 2014). Results of current medical therapies have not been successful. Surgical correction is the effective treatment modality of PD (Guillot-Tantay et al. 2014). Definite treatment of PD would be elimination of fibrous plaque on TA. Surgeons have tried to correct PD using incision and plication of plaque and complete excision of plaque followed by grafting with various materials depending on size and location of plaque (Perovic and Djordjevic 2001).

Incision without removing plaque usually resulted in shortening of penile length. Patch grafts after excision of plaque has been tried for compromising shortening of penis. Several materials have been applied for a tunical graft, usually made by autologous, allogenic natural biomaterials. These were tunica vaginalis, SIS, veins allogenic pericardium, dermal graft, and muscle fasciae, depending on the defect size of the excised plaque (Chun et al. 2001; Kadioglu et al. 2007; Montorsi et al. 2003). However these materials also had some drawbacks as donor site complications, shrinkage of graft, and availability of material supplies, hence alternatives are needed. Surgeons needed efficient, biocompatible, and off-the-shelf graft materials.

Porcine vesical ACM was evaluated for graft material of TA. ACMs were obtained from porcine bladder by decellularization. Histologic analysis using SEM and RT-PCR assay about growth factors that would assist tissue regeneration were made. SEM showed fibrous collagen tissue of various sizes of pores. Vascular endothelial growth factor (VEGF) receptor, FGF-1 receptor and neuregulin mRNA were detected in ACMs by RT-PCR. These ACMs were implanted to TA of rabbits and the ACM implants were harvested after 2 months. Histologic analysis revealed grafted ACM was linked to neighboring native TA without serious inflammatory reaction. There was minimal contracture in grafted ACM and restored ACM-graft tissue on corpus cavernosum was not different from native TA histologically after 6 months of implantation. It implied the possibility of porcine vesical ACM as a graft material of TA (Joo et al. 2006).

Schultheiss et al. (2004) reported efficacy of dynamic seeding of fibroblasts in vitro compared to static seeding technic in constructing tunical collagen tissue as for graft material of TA. They biopsied porcine fascia and isolated fibroblasts. Collagen matrices were prepared using decellularization technic. Cells were seeded on these matrices. In dynamic seeding group, the fibroblasts–matrix complexes were cultured in a bioreactor under the multiaxial stress for 3 weeks. Static cultures were done in other group of cell–matrix complexes as control. Results after 7 days showed better cell array, better cell infiltration into the matrix, and formation of multilayered tissue superior to static culture group in dynamic group. Also dynamically cultured fibroblasts produced more extracellular matrix proteins than that of static culture group.

TGF-β1 is known as enhancing Peyronie's plaque. Most of the animal models of Peyronie's disease were inducted using TGF-β1 treatment. Castiglione et al. applied stem cell (SC) therapy in PD rats. They made the rat model of PD with TGF-β1 treatment and then injected adipose tissue–derived stem cells (ADSCs) at Peyronie's plaque intralesionally at the active phase of PD along with no injection of ADSCs as control group. Cavernosometry and histologic analysis was made. Compared to sham-operated group and control group, which showed increased abnormal collagen III and elastin proteins on histologic analysis, ADSC injected rats revealed increased intracavernosal pressure and prevention of fibrotic changes of PD plaques. This was the first trial of SC injection therapy on PD animal (Castiglione et al. 2013; Shindel 2013).

Lander et al. (2016) reported the results of clinical trial using adipose stromal vascular fraction (SVF) combined with ESWT in PD patients. They evaluated the effect and safety of intralesional injection of SVF into fibrous plaque followed by ESWT in 11 PD patients. The results were decrease of plaque size, improved penile curvature, and erectile function.

5 Penile Augmentation

Penile augmentation is another issue regarding reconstruction of penis. Concept of penile surgery as augmentation is not familiar yet. This procedure is still on debate as a controversial issue, so is not a recommended medical procedure by most of the medical authorities. Uncertain indications for surgery, absence of standardization of procedures, lack of guidelines, and lack of evidence-based studies could be the reasons (Park 2016).

At present, the only positive consensus for this procedure would be in the cases of congenital problems of penis. There are congenital problems as micropenis, penile hypoplasia, or hermaphroditism (Gillies 1948; Horton et al. 1987; Johnston 1974; Snyder 1964; Woo et al. 1996). Similar situation happens in traumatic total loss of penis and genial thermal burn (Hotchkiss et al. 1956).

Deformity or decrease in length or girth of penis occurs after prostatectomy, pelvic surgeries, PD, and other penile diseases. Men who underwent radical prostatectomy and possibly radiation therapy and hormonal treatment are susceptible to penile shortening (Kabalin et al. 1990). Some of these patients feel depression, anxiety, locker room syndrome, and represent erectile dysfunction (Dillon et al. 2008; Pietropinto 1986; Wylie and Eardley 2007).

Another example would be as the penile dysmorphophobia (Spyropoulos et al. 2005). This medical problem occurs when a man is dissatisfied with his penile size despite anatomically normal developed penis. Penile augmentation is helpful in some patients psychologically. Also a lot of evidence has been found historically that there is a tendency to relate the size of the penis to the self-esteem or potency of sexual power, the so-called phallocentrism (Cheng et al. 2007; Lever et al. 2006; Thompson et al. 1999; Wylie and Eardley 2007). Because of this myth, various methods were tried to enlarge penis for ages as traction, massage on penis, foreign body insertion, herbal medicines, and snake poison, which proved to be not only ineffective but also harmful.

Medical treatments were not able to achieve satisfactory results, and surgical methods were introduced (Nugteren et al. 2010). Development of penile augmentation technic has close relationship with penile reconstruction, PD, graft or flap procedures, and materials used in tissue engineering. Documents revealed attempts made to augment penile girth using fat graft since 1893 by Neuber. Also there have been various surgical procedures for penile lengthening procedures. Experiencing unexpected complications such as infection and graft failures, through trials and errors, technics and materials have developed (Vardi et al. 2008; Wassermann and Greenwald 1995). Various materials have been applied for penile augmentation as bone, ivory, metal, silicon, artificial oil forms of Vaseline, paraffin, silicon, collagen and fat, dermis, fascia, cartilage, pericardium, and so on. Some of them resulted in serious complications and they are not used anymore (De Siati et al. 2013; Moon et al. 2003; Park 1991; Sukop et al. 2013).

Nowadays synthetic biocompatible materials or autologous, allograft, and xenograft natural materials are used in augmentation penoplasty, as are similar to the materials used in tissue engineering (Austoni et al. 2002; Perovic and Djordjevic 2000).

Examples of synthetic biomaterials used in penile augmentation are calcium hydroxyapatite, polyacrylamide, silicone, and PLGA. HA. Silicon and PLGA are used as fillers also.

Examples of xenogenic materials are dermis, SIS, pericardium of bovine or porcine origin, which have stable structures and lower immunologic reactions on host. There are several commercial xenogenic products as porcine dermal collagen tissue, Permacol® (Covidien, Mansfield, USA). Lyoplant® (B. Braun Aesculap, Tuttlingen, Germany) is a collagen implant made from acellular bovine pericardium. MegaDerm Ultra® (L&C BIO Inc., Seongnamsi, Korea) is derived from porcine dermis. InteXen® (American Medical Systems, Minnetonka, MN, USA) is made from porcine acellular dermal matrix. Xenogenic materials have been used in various clinical fields. Acellular dermal matrix from fetal bovine dermis, SurgiMend® (TEI Biosciences Inc., Boston, USA), has been used for hernia repair. Porcine dermal ACM is applied clinically for penile girth augmentation (Alei et al. 2012). Xenogenic type I collagen was applied clinically in penile augmentation also (Kim 2013). Permacol has been applied as a graft material for TA in PD patients also.

Allogenic materials as acellular dermal matrix are processed from human source, as ACM (Alloderm: LifeCell Inc., Branchburg, USA) from human cadaver skin, MegaDerm® (L&C BIO, Korea), and SureDerm® (Hans Biomed, Korea). AlloDerm® has been applied for treatment in burn, breast reconstruction, hernia repair, and other reconstructive procedures (Gaertner et al. 2007), and this was applied as grafting material in penile girth enhancement along with other allogenic materials

(Solomon et al. 2013). Collagen products are xenogenic or allogenic. Allogenic collagen material is derived from human fibroblasts. One of the xenogenic collagen product approved by the US FDA in 2008 is Evolence® made from swine tendon. Xenogenic product showed higher allergic reaction than allogenic one but the adverse events were usually mild and uncommon.

Compared to penile lengthening that mostly depends on surgical skills and tissue rearrangement as suspensory ligament release, Z plasty, V-Y advanced plasty, or penile disassembly technic, a variety of materials are interposed in augmentation of glans and penile girth. The materials used at present include allogenic materials as Alloderm®, xenogenic materials as Lyoplants®, autologous fat, dermal fat graft, fascia, pedicled graft, and fillers made of silicon, PLGA, HA, etc. HA is a natural component of human skin, thus host immune reaction is far rare. Kwak et al. (2011) reported clinical results of penile girth enhancement using injectable hyaluronic acid gel. HA gel injection has been used to glans augmentation clinically (Kim et al. 2003).

Complications were infection, edema, ulceration, skin necrosis, absorption of implants, seroma, epidermal cyst, uneven skin, penile deformity including asymmetry or curvature, calcification, and so on. The most serious complication of penile augmentation would be infection. Solomon et al. reported that patients using allograft materials showed 42% postoperative infection rate (Plaza and Lautenschlager 2010; Solomon et al. 2013).

Female to male transsexual surgery needs neo-penis construction (Laub et al. 1989). Various pedicled flaps have been tried in phalloplasty (Bettocchi et al. 2005). Tubed groin flap with hydraulic inflation device (Puckett and Montie 1978), island tensor fascia lata flap (Santanelli and Scuderi 2000) radial artery-based forearm free flap (Garaffa et al. 2010) had been used for phalloplasty in transsexual surgery.

6 Recent Novel Technics

6.1 Cell Therapy and Stem Cell Therapy in ED

The position of the penis is easily accessible for injection therapy. In 1979, Green et al. proposed clinical application of cell therapy as cultured human cells for grafting. One of the considerations would be hemodynamic change of the cavernosal tissue. Continuous circulation results in washing the injected materials out from corporal tissue. The method of anchoring cells inside corpus cavernosum for the appropriate duration would be the task.

Various cells have been introduced to reconstruct penile tissue using tissue engineering technic for the physiologically compatible to native tissue. Mature adult cells have several drawbacks as inability to differentiate into different cell types and low proliferative capacity in culture and expansion (Atala 1998). To overcome the immunologic issues and to produce enough tissue easily, stem cells (SCs) were investigated as an available source of off-shelf tissue production.

The merits of SC are potential of “homing” differentiation as differentiating into target tissue, self-renewal, capacity of differentiating potentials to various cell types, capability of extensive proliferation, and secreting various cytokines (Chamberlain et al. 2007). Martin (1981) reported about the existence of SC in mouse embryo. At first it was thought that SCs resided only in the embryonic tissues, so researches in human SCs were difficult for ethical issue. Nevertheless Thomson et al. (1998) reported about the human embryonic stem cells (ESCs). ESCs are pluripotent that can be differentiated into all cell types of the three germinal layers through embryonic developments. ECSs are known as provoking immune reaction and having teratogenic potential. Researches about human ESCs have barrier for the ethical issue also. There are multipotent stem cells as adult SCs which differentiate confined to the same native germinal cell types. The merit of adult SCs would be nonimmune reactive and low risk of oncogenic potential. One of the variant SCs is stromal vascular fraction cells (SVFCs). SVFCs are derived from autologous adipose tissue, which contains Adipose-derived stem cells (ADSCs), endothelial progentior cells (EPCs), and various other cells (Bora and Majumdar 2017).

SCs are divided into three categories based on their differentiation capacity as totipotential, pluripotential, and multipotential SCs (Mahla 2016). Also SCs are classified by their origin. 1) ESC derived from the inner cell mass of a blastocyst. ESC is pluripotent; 2) SCs from amniotic fluid, placenta, and umbilical cord blood. They are multipotent and have partially differentiated cells and express both markers of embryonic and adult SCs (Williams et al. 2016); 3) Adult stem cell (ASC) derived from various adult tissue as neural crest SCs (NCSCs), epithelial SCs, mesenchymal SCs (MSCs) from bone marrow, muscle (MDSCs), or adipose tissue (ADSCs), urine, and so on. MSCs have paracrine effect as releasing various cytokines and growth factors. ASC is multipotent (Alwaal et al. 2015; Gimble et al. 2007; Volarevic et al. 2011). One of the SCs frequently adopted in researches is ADSCs. The source of ADSCs is adipose tissue, which is relatively convenient to obtain by subcutaneous liposuction which is a less invasive procedure compared to harvesting other adipose tissues (Zuk et al. 2001).

Novel concepts were reported about transforming multipotent ASCs into pluripotent SCs. Takahashi and Yamanaka (2006) reported about reprogrammed mouse fibroblast into induced pluripotent state, called as iPS. These cells showed immortal growth, proliferated into embryonic bodies in vitro and teratoma in vivo.

There are various sources of SCs in our body including penis (Lin et al. 2015). These SCs repair the injured tissue and keep homeostatic tissue environment (Xin et al. 2016). These innate SCs reside with supporting cells and extracellular matrix. They are connected and communicate with each other through signal molecules. This microenvironment is called SC niche (Moore and Lemischka 2006). Mobilization of innate SCs to target tissue and activation of innate SCs would be helpful in regeneration of tissue, apart from extrinsic supply of SCs as injecting cells. Several molecular factors have been known to help mobilizing SCs into blood stream for target tissue, as granulocyte colony-stimulating factor (Deng et al. 2011). Activation of p38 pathway is important in cellular differentiation of SCs (Jones et al. 2005). Chemicals as icariside II activate p38 pathway. Icariside II has been tried on cavernosal nerve injury rat model and diabetic rat model, resulting in improved erectile function (Qiu et al. 2013; Xu et al. 2015). Investigators mentioned about the effect of ESWT in regeneration and healing process. It has been proposed that ESWT enhance activation of SCs, angiogenesis, reduce inflammatory reaction, and proliferate and differentiate cells (Aicher et al. 2006; Mittermayr et al. 2012).

There have been lots of preclinical studies applying SCs for treatment of ED and for regeneration of corporal tissue using animal models. Most of the reports showed established efficacy of SCs (Orabi et al. 2012; Soebadi et al. 2017). Song et al. (2009) reported SMCs derived from the other parts of the body could be used in penile corporal tissue regeneration. They implanted SMCs derived from the human umbilical artery-seeded ACM in athymic mice subcutaneously. The SMCs from umbilical artery, not identical to corporal SMCs, grew in tissue similar to native corpus cavernosum histologically. Laks et al. (2015) investigated the effectiveness of tissue regeneration depending on the routes of SCs delivery. They derived the MSCs from rabbit bone marrow and delivered these cells by intravenous infusion or by implantation in the form of ACM-MSCs complexes in rabbits along with no cell seeding group. Cavernosography showed that dye-filled entire corpus cavernosum of rabbits treated with intravenous infusion of MSCs and cell-ACM complexes implanted rabbits compared to filling defects of corporal tissue in control group rabbits. Cavernosometry also revealed higher intracavernosal pressure in both cell-delivered groups than control group. Implants of MSC with ACM showed organization into partial sinusoidal structure (Laks et al. 2015). In another study, MSCs derived from rabbit was seeded on ACM and then implanted at the excised corporal space of rabbits along with the control group as implants without cells. MSCs seeded ACM grew into tissues similar to natural corpus sinusoids architecture and enriched smooth muscle and increased NO synthetic activity compared to ACM without cell-implanted group in 6 months after implantation. These results meant the implanted MSCs differentiated into SMCs and ECs successfully in vivo (Ji et al. 2011).

Investigators explored the effects of SCs in treating plaque of PD. Simple SCs, or combination with interferon or gene-transfected SCs were injected into the fibrous plaque of PD in rats. These also showed changes of composition of collagen and elastin proteins, increase of NOSs and cGMP, growth factors, and improved erectile function (Levy et al. 2015; Song et al. 2008). Gokce et al. reported series of studies of SC therapy for Peyronie's plaque and ED. They have injected ADSCs alone or genetically modified ADSCs with human interferon alpha-2b intratunically in rats. TGF-β1, known as the provoking factor of fibrotic plaque on TA, was injected to these rats. Decreased collagen deposition and less fibrotic plaque formation as well as better intracavernosal pressure were noted in both ADSCs treated groups after 6 weeks. This result showed the ADSCs were effective in preventing formation of Peyronie's plaque and prevented decrease of erectile function related to PD (Gokce et al. 2015; Gokce et al. 2016). Another study was using a combination of ADSCs and SIS. Researchers prepared SIS seeded with ADSCs and implanted into rat TA. Compared to implanting SIS alone, SIS-ADSCs implants showed elevation of NOS activation, inhibition of fibrosis, improved angiogenesis, and resulted in preservation of corporal tissue and improved erectile function (Ma et al. 2012).

Castiglione et al. (2019) employed SVF injected intratunically to prevent fibrosis of PD in rat model.

Researchers used established diabetic rat model, cavernous nerve injury model, PD model, and old rats as similar to clinical ED etiologies for disease-specific investigations. ADSCs alone or combined with insulin or interferon alpha-2b, BMSCs alone or combined with extracorporeal shock wave therapy (ESWT), human urine-derived SCs c growth factors, MDSCs, have been studied on diabetic rat models (Qiu et al. 2013; Ryu et al. 2016; Sun et al. 2012; Zhou et al. 2016). ESWT has been known as stimulants for angiogenesis and endogenous SC activation (Qiu et al. 2013; Vardi et al. 2012) (Table 1). Jeon et al. (2016) reported improved erectile function treated with human ADSCs with ESWT in cavernosal nerve injury rat model. Shan et al. (2017) reported the improved erectile function with treatment of ESWT plus bone marrow MSCs in diabetes mellitus (DM) rat model. SVF from human breast adipose tissue was applied for diabetic ED rats. This xenogenic SVF resulted in increase of SMCs, ECs, neo-vascularization, eNOS, and nerve regeneration (Das et al. 2014).

For application to the nerve-damaged ED patients after radical prostatectomy or lower abdominal surgery as for colon cancer, cavernosal nerve injured models were used (Mangir and Turkeri 2017). ADSCs, BMSCs, MSCs, urine-derived SCs, and SVF alone or combined with various growth factors, PDE5I, graft, or ESWT were treated in animal models (Alwaal et al. 2015). Miyamoto et al. implanted CD133+ cells derived from human bone marrow in cavernosal nerve injury rat model. The results revealed regeneration of excised cavernosal nerve and increase of nerve-derived growth factors and cytokines, and this might be applied in ED patients after prostatectomy (Miyamoto et al. 2014). There were reports about combined transplantation of MSC and endothelial progenitor cells for restoration of injured cavernous nerve (Fang et al. 2018). Similar to ED after prostatectomy, radiation around genital area might result in ED. ADSCs were treated in radiation-induced ED rat model. The result showed increased SMCs and nNOS in corporal tissue, and improved erectile function (Qiu et al. 2012) (Fig. 3).

Intracavernosal injection of SCs, instead of intralesional application on exact location of lesion, also restored remote defected nerve function. Intracavernosal injection of ADSCs improved erectile function in rat model by modulating pelvic ganglion. This result implied recruitment of injected SCs in pelvic ganglion (Fandel et al. 2012). Another report by Matsuda et al. was that intravenous injection of bone marrow–derived MSCs improved erectile function in rat cavernosal nerve injury model (Matsuda et al. 2018).

Table 1 Summary of papers about ESWT

	Mediator	Effect
Nishida et al. (2004)	VEGF, VEGF-R	Angiogenesis
Aicher et al. (2006)	Stem cell-derived factor-1 (SDF-1)	Stem cell activation & recruitment
Ha et al. (2013)	Erk 1/2, nitric oxide synthase	Angiogenesis, immune-reaction
Weihs et al. (2014)	ATP, Erk 1/2	Cell division, stem cell activation

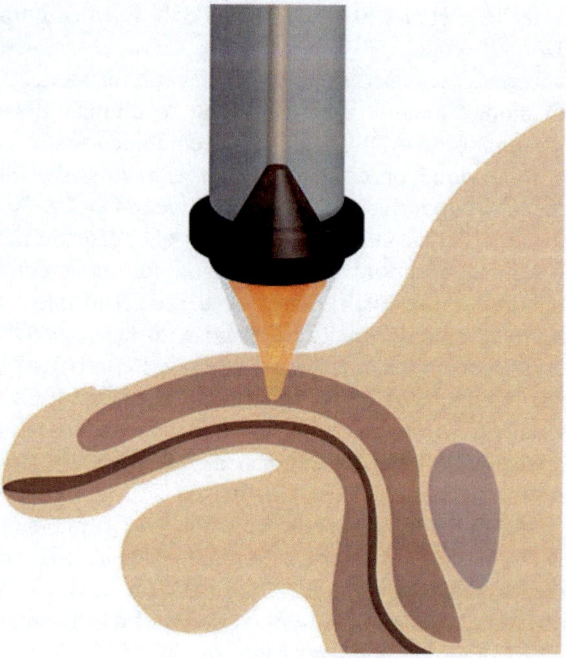

Fig. 3 Effect of ESWT. VEGF↑ → angiogenesis. Activate & recruit circulating or resident stem cells. Activate nerve cells.

Most of the studies showed improved histologic composition of corpus cavernosum; cavernosal nerve regeneration; increased cavernosal eNOS, nNOS, and vasculogenic growth factors; intracavernosal pressure; and restoration of erectile function (Albersen et al. 2010; Jeon et al. 2016; Kendirci et al. 2010; Ryu et al. 2014; Song et al. 2014). The most obvious etiologic factor of erectile dysfunction is aging. Similar studies were performed with SCs in old rats and the results were successful also (Bivalacqua et al. 2007; Liu et al. 2017).

SCs have been applied in glans reconstructing experiment. Egydio et al. (2015) processed human glans matrix by decellularization. MSCs were obtained from rats and these cells were seeded on human glans ACM in static manner. After 2 weeks in vitro culture, cells maintained their integrity and viability up to 2 weeks.

Several clinical applications of SC therapy have been already reported (Capogrosso et al. 2018). The first clinical trial might have started in 2010. Bahk et al. (2010) injected a total of 1.5 × 107 human umbilical cord blood SCs into corpus cavernosum of type 2 diabetic ED patients. No immunosuppressive measures were taken in any of the patients. Morning erections were regained in three participants within 1 month, and for all except one by the third month, and maintained for more than 6 months. Rigidity increased as a result of SCs alone, but was insufficient for penetration. With the addition of PDE5 inhibitor before coitus, two achieved penetration and experienced orgasm, and maintained for more than 6 months. Blood glucose levels decreased by 2 weeks, and medication dosages were reduced in all but one subject for 4–7 months. Glycosylated hemoglobin levels improved after treatment for up to 3–4 months (Bahk et al. 2010). Afterward several open label, phase I clinical trials were made. Haahr et al. (2018) reported the trial of intracavernosal injection of autologous adipose-derived regenerative cells in 17 patients with erectile dysfunction following radical prostatectomy as a phase I clinical trial, and they presented 12 months follow-up data of this trial in 2018. This clinical data showed 8 out of 15 (53%) ED patients after radical prostatectomy showed improvement of erectile function in self-evaluating questionnaire, International Index of Erectile Function (IIEF) – 5 score, enough to intercourse after 12 months of SCs injection. There was no report of serious side effects (Haahr et al. 2018). Another pilot clinical trial was conducted in 2016. Autologous bone marrow–derived mononucleated cells were injected into corpus cavernosum to patients of ED undergone prostatectomy. They showed improved erectile function on IIEF symptom score and IIEF-EF domains without serious side effects. Improved erectile function was sustained until the mean follow-up point 62.1 months, aside the some decrease of IIEF scores in 62.1 months compared to the IIEF scores of 12 months after injection. Repeated treatment would be needed for maintaining improved erectile function (Yiou et al. 2017; Yiou et al. 2016). Al Demour et al. (2018) reported the open labeled phase I clinical trial of SCs for ED patients by DM. They delivered autologous bone marrow–derived MSCs (BM-MSCs) by intracavernosal injection to 4 refractory ED patients. Though the number of patients was small, the results showed improved sexual desire and sexual satisfaction in IIEF questionnaire. Also there were no tolerability and safety issues.

SC therapy was tried in PD on human also. Levy et al. (2015) attempted to treat PD using placental matrix–derived MSCs (PM-MSCs) in 2015. Seven out of 10 plaques of PD disappeared 3 months after SC injection. This study was the first human trial of SC therapy in PD patients. In succession, Levy et al. (2016) applied PM-MSCs to ED patients. Eight ED patients not treatable with PDE5I were enrolled in phase I clinical trial. Patients were injected PM-MSCs and followed up for 6 months with penile Duplex ultrasonography and IIEF symptom score. Patients showed improved IIEF symptom scores and increased penile arterial blood flow including peak systolic velocity. Lander et al. (2016) reported the results of clinical trial using adipose SVF combined with ESWT in 11 PD patients in 2016. The results were decrease of plaque size, improved penile curvature, and erectile function.

There are several chemical compounds enhancing the function of SCs. One of these medications is metformin, which has been used for treatment of DM approved by FDA. Administration of metformin improved endothelial progenitor cell function (Fatt et al. 2015). Phase I-II clinical trials are undergoing and some of them reported the results. Nitric oxide (NO) is a corner stone in erection. Topical gel containing glyceryl trinitrate (NO doner) was tested in 232 ED patients in phase II randomized trial. The effect of this compound is known as increasing penile blood flow, and 23.1% of patients reported improved IIEF-EF score with minimal side effects (Ralph et al. 2017). Other chemicals under clinical phase I-II trial for treatment of ED are arginine aspartate combined with adenosine monophosphate with 26 patients (Neuzillet et al. 2013) and mirabegron, the β-3 adrenoreceptor agonist dilating vessels in corpus cavernosum with 20 patients (Gur et al. 2016), and Botulinum neurotoxin-A injection therapy in 24 patients with vasculogenic ED. For the treatment of premature ejaculation, serotonin-related medications as SSRIs, compounds related to oxytocin metabolism, and folic acid are used. SCs combined with these compounds would develop the efficacy of treating ED using tissue engineering.

Combining stem cells and gene therapy would be one of the promising options in improving formation of tissue-engineered architecture. Still some investigators worry about the safety issues as oncogenic possibility that should be confirmed for clinical applications (Marks et al. 2017).

6.2 Gene Therapy for ED

Erectile dysfunction (ED) is induced by abnormal relaxation of corpus cavernosum in short. Mechanism of relaxation of corpus smooth muscle are related to various factors such as endothelial cells, neurons, nitric oxide synthase, molecules like nitric oxide, cGMP, PDE5 enzyme, calcium ions and potassium channels, RhoA/Rho-kinase pathways, etc. Aging impairs the function of tissues related to erection and similar deterioration occurs with pathologic conditions like DM, hypertension, metabolic syndrome, and surgeries around low abdomen and genital area. Sildenafil, introduced in clinical field in 1998, and other phosphodiesterase 5 inhibitors (PDE5Is) showed effect in 70–80% of ED patients but remainder of ED patients did not improve with PDE5Is. Even though guidelines recommend PDE5Is as the first-line therapy for ED, successful result of PDE5Is decrease in DM-related ED patients compared to non-DM ED patients and PDE5Is are contraindicated in patients taking medication containing NO and patients with serious cardiac disease because of the risk of hypotensive crisis. Investigations of gene therapy for ED patients includes the ideas of avoiding serious side effects of PDE5Is and alternative therapeutic modalities.

Attempts of gene therapy in ED were reported in 1990s (Christ et al. 1998; Garban et al. 1997). Since then, researchers have struggled to restore damaged functions of endothelium (growth factors as VEGF, IGF-1, cGMP, modified stem cells, DNA, RNA transfection, and gene transfer), cell to cell interaction (potassium channel, calcium channel), cavernosal fibrosis (Wnt signaling pathway, Maxi-K channel), study about gene silencing or gene augmentation using various methods conjoining vectors to deliver specific genes, growth factors or electroporation in diseased animal models (Soebadi et al. 2016).

Among viral vectors, adenovirus has been accepted as superior to other viral vectors. Recently nanoparticles have been applied in delivering genes instead of the viral vectors for effectiveness and avoiding immune reactions (Gur et al. 2018) (Fig. 4). There had been several reports about the application of labeling nanoparticles to SCs (Neri et al. 2008). The position of the penis is easily accessible for injection therapy. One of the considerations would be hemodynamic change of the cavernosal tissue. Continuous circulation results in washing the injected materials out from corporal tissue. The method of anchoring cells inside corpus cavernosum for the appropriate duration would be the task. Lin et al. reported improved erectile

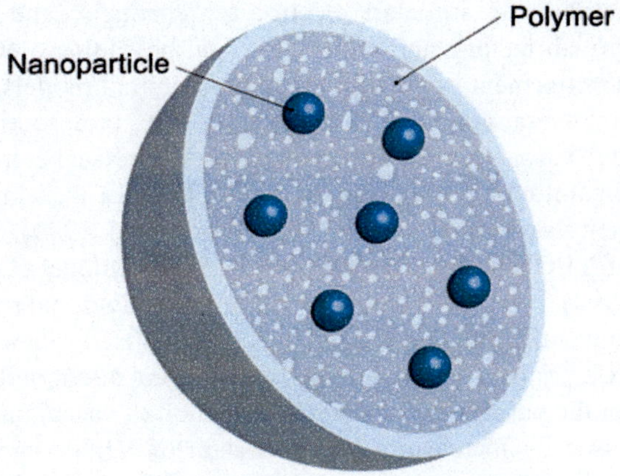

Fig. 4 Schematic figure of polymer–nanoparticle compound

function using nanoparticle with SCs in cavernosal nerve injured rat model. In order to guide the SCs to the target site and prevent the migration of injected ADSCs, they bound ADSCs with Nanoshuttle magnetic nanoparticles. Properties of ADSCs bound to nanoparticle have been maintained. These nanoparticle-ADSCs complexes were injected into corpus cavernosum of nerve injured rats, and then magnetic forces were applied for "homing" cells in one group. The nanoparticle-ADSCs injected group showed improved intracavernosal pressure, and differentiation into SMCs and ECs well. About migration of injected SCs, nanoparticle-ADSCs were staying in corpus cavernosum up to 25 days after injection (Lin et al. 2016). Zhu et al. (2017) studied another kind of nanoparticles with ADSCs. Superparamagnetic iron oxide nanoparticles (SPIONs) were labeled to ADSCs. SPIONs have been conventionally applied in MRI and investigated for application on biomedical fields (Reddy et al. 2012; Schafer et al. 2010). SPIONs labeled ADSCs maintained its characteristics in vitro. They have injected SPIONs labeled ADSCs into corpus cavernosum of diabetic rats and applied the magnetic power. Physiologic analysis showed increased intracavernosal pressure and increased SMCs and ECs compared to ADSCs without labeling SPIONs 4 weeks after injection. This study showed SPIONs did not affect viability and properties of ADSCs and external application of magnetic power was helpful in cell staying at target tissue (Zhu et al. 2017). These studies showed the possibility of application of nonorganic vehicles as nanoparticles in gene transfer for merits as immunogenic issues and cell-anchoring ability to the target tissue.

New powerful methods like electroporation, miRNA (microRNA), siRNA (small interfering RNA), and gene editing technic like CRISPR (Clustered Regularly Interspaced Short Palindromic Repeats) have been introduced in gene therapy recently. CRISPER enabled to edit target genes directly (Damian and Porteus 2013). There was a trial of reducing PDE5 enzyme by PDE5-silencer transfected siRNA. PDE5 enzyme decreased to 88.2% of control group (Lin et al. 2005). Many of these animal experiments in vivo showed favorable results as increased contents of SMCs and ECs; decreased apoptosis of cells and ECM; endothelial cell function;

normalizing ion and electrolytes channels, and improved enzyme production, growth factors, and molecules. And physiologic evaluations showed restoration or improvement of erectile function in animal models.

Several gene or molecules as VEGF, brain-derived neurotropic factor (BDNF), and vasoactive intestinal peptide (VIP) had been studied for clinical application (Shamloul and Ghanem 2013). VEGF is a well-known molecule related to angiogenesis (Burchardt et al. 1999; Dahiya et al. 1999). Several studies showed the effect of VEGF in erectile function on animal model (Park et al. 2004; Takeshita et al. 1994). Increase of smooth muscle and endothelial nitric oxide synthase (eNOS) were found after intracavernosal injection of VEGF (Park et al. 2004). To help the effect of SCs, trials of VEGF adding on SCs were attempted employing genetic engineering, for the purpose of enhancing formation of vasculatures in tissue-engineered corporal tissue by increasing the expression of VEGF in MDSC. Burchardt et al. (2005) studied about the effect of VEGF on ED therapy by DNA transfer of VEGF 165 in the rat penis. They used liposome complex of VEGF 165 expression vector instead of adenoviral vector transfected with VEGF gene. The corpus cavernosum of the rats treated with VEGF-liposome complex showed 10 folds greater VEGF concentration. MDSCs potentiated with transfection of VEGF were tested in rabbit corpus cavernosum. MDSCs were derived from gastrocnemius muscles of New Zealand white rabbits. MDSCs were transfected by the human VEGF165 lentiviral gene vector (LV-GFP-VEGF). Rabbits were implanted with non-seeded ACCM, ACCM seeded with MDSC, ACCM seeded with MDSC transfected with control vector, or ACCM seeded with MDSC expressing VEGF. The control group underwent corporal tissue excision without ACCM. MDSCs expressing higher VEGF were correlated with better distribution and growth of seeded cells on ACCM surface, higher number of nuclei, and cell adherence capacity. Also, ICP was significantly higher in this group (60% of normal), and markers for SMC and endothelium were expressed higher. Analysis including physiologic studies showed MDSC overexpressing VEGF was promoting vascular formation and cellular maturation in the engineered corpus cavernosum compared to MDSCs only (An et al. 2013). Similar work was reported by Liu et al. and the results showed improvement of erectile function accompanied by increased number of SMCs and ECs (Liu et al. 2013). Bivalacqua et al. (2007) tried to treat age-related ED using MSCs alone or combined with ex vivo modified gene with endothelial nitric oxide synthase. They modified MSCs with eNOS containing adenoviral vector and injected into corpus cavernosum of rats. The result revealed improved erectile function with increased NO pathway signals and differentiation into SMCs and ECs from MSCs.

One of the clinical trials using gene therapy for ED was reported in 2006 at first. Melman (2006) hMaxi-K gene were transferred to ED patients as a phase I clinical trial. Potassium channels act on relaxation of smooth muscle by decreasing intracellular calcium from SMCs (Christ 2002). Activation of potassium channels induces relaxation of corporal tissue resulting in erection. They have transferred naked DNA portion related to potassium channel using pVAX1 vector and injected into corpus cavernosum in 11 ED patients. Patients showed improved erectile function and improved IIEF-EF domain score up to 24 weeks after injection and there was no serious side effect. The results could not be conclusive statistically because of the small number of patients and absence of control group, but bear the potential of gene therapy for ED patients (Melman 2006).

There are some drawbacks in introducing cell therapy or gene therapy on ED patients. The uncertainty of safety is the issue of gene therapy. Vectors like virus or even transferred genes may act in uncontrolled fashion, over express the genetic function, and result in unwanted immune problems or other side effects. Nanoparticles may play an important role because it has nonorganic property.

6.3 Bioprinting

One of the emerging technics in tissue engineering is constructing tissue using 3D bioprinters. Bioprinter is automated robotic device that enables digital biofabrication of 3D functional structures. The 3D bioprinters were designed to construct biomaterial scaffold and distribute various cells in between at a time to compose ideal tissue structures (Murphy and Atala 2014). Since Professor Ralf Mulhaupt in Freiburg University introduced commercial 3D printer, bioprinters have been evolved technically and economically on the base of 2D printing technology by Hewlett-Packard company (Huang et al. 2017). Like others commercial 3D printers, 3D bioprinters reconstruct 3D structures of biologic properties by piling up biomaterials and viable cells altogether. Apart from the digital, robotic, and automatic technical hurdles, the major problem in the use of 3D printers was the high price. More than 20 companies have been developing new 3D printers to overcome this handicap. Another issue is the absence of standard system of 3D printers. There are no FDA regulations and no FDA-approved organs made by 3D printers yet.

Basic architecture of the 3D printers is composed of five parts: robotic positioning system ruling printing axis (Cartesian type robot), controlling device, nozzle dispenser like automatic syringes, collectors for 3D printed product, and sterile cabinet for 3D printed tissue or organ.

Usually 3D bioprinters have been classified into 3 types: Inkjet 3D printers, extrusion-based 3D printers, and laser-based 3D printers (Tamay et al. 2019). Inkjet type was the first 3D bioprinter as the stereotype. In tissue engineering, frequently used inkjet types are thermal and piezoelectric types. Extrusion 3D printers have been most popular among them. It could be divided into pneumatic or mechanical by dispensing system. The advantage of laser-based type would be improved cell viability by avoiding the clogging of cells. But this type is usually the most expensive among them (Mandrycky et al. 2016). Materials used for constructing scaffold in 3D bioprinters would be biocompatible, printable, and able to help cell growth, forming supportive structure of cells. The preferable material as a bioink in 3D bioprinters has been hydrogels. Sources of hydrogel are synthetic or natural (polysaccharides as alginate, chitosan, agarose and ECM composites as collagen, fibronectin, gelatin) (Mandrycky et al. 2016) (Fig. 5). Hydrogels provide environments for different cells to adhere, communicate, and differentiate to form target

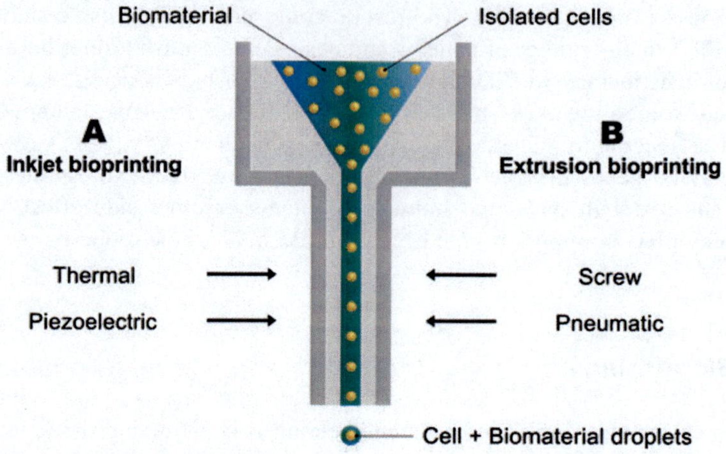

Fig. 5 Schematic figure of bioprinters

tissue as ECM. One of the drawbacks of hydrogels is low mechanical strength (Billiet et al. 2012).

Cell-biomaterial compounds would be manufactured through (A) Inkjet bioprinters with thermal or piezoelectric power, (B) Extrusion bioprinters with screw or pneumatic force.

There have been preclinical trials to reconstruct tissues using applications of 3D bioprinting. These preclinical studies included nerve, vessel, skin, bone, cartilage, and corona using various types of biomaterials and cells including SCs (Jang et al. 2017; Keriquel et al. 2017; Kim et al. 2018; Lee et al. 2017; Martínez Ávila et al. 2016; Skardal et al. 2012; Sorkio et al. 2018). But there are no reports about tissue-engineered corporal tissue using 3D bioprinting yet.

Urethral tissue was reconstructed by Zhang et al. (2017) using 3D bioprinting technology. Porous urethral scaffold was fabricated with poly (ε-caprolactone) (PCL) and Poly (lactide-co-caprolactone) (PLCL) blend using in-house designed 3D bioprinter. This 3D bioprinter, having multiple cartridges, was made in Wake Forest Institute for Regenerative Medicine (WFIRM). It is capable of printing numerous biomaterials at a time using hydrogel-based bioink. This is known as the Integrated Organ Printing (IOP) System (Kang et al. 2016). Autologous rabbit urothelial cells and SMCs were laden on hydrogel composed of fibrin, gelatin, and hyaluronic acid as bioink. Cell-laden bioinks were loaded with PCL, PLCL polymers from different nozzles. Analysis in vitro showed that this complex had acceptable mechanical properties and viable cells constructing tissue similar to the native urethra. Bioink made of fibrin hydrogel showed this material could be used in 3D bioprinting as a viable cell containing ink (Zhang et al. 2017). This 3D printer and hydrogel bioink could be utilized in producing corpus cavernosal architecture also.

Not only choosing the appropriate type of cells, but also biomaterial similar to physiologic human corporal sinusoid is necessary to create 3D bioprinted corporal tissue (Boland et al. 2006). The corporal sinusoidal trabeculae are known to be composed of elastic fiber and type I and type III collagen fibers (Costa et al. 2006). Building trabeculae structure close to natural sinusoids with these organic materials using 3D bioprinters, ECs, and SMCs could be distributed and seeded between the fibers in exact locations meticulously. Conventional methods were seeding cells at the outer surface of ready-made biomaterials like encircling biomaterials with cells. Combining cells in 3D bioprinting enables cells piled up inside the biomaterial as cells could be distributed in the biomaterial simultaneously, then the product of 3D bioprinter would be cells-biomaterial complex, seeded cells at target location. There was report about culturing SMCs and ECs on PCL scaffolds made by 3D bioprinter for applying penile reconstruction (Oh et al. 2019). Some investigators have announced the concept of 4D bioprinting, which is different from 3D bioprinting in using programmed intelligent materials to perform predesigned functions and structures as patient-specific cell-biomaterial complex (An et al. 2016; Tibbits 2014).

Approaches have been explored to combine surgical robots with 3D printing (Da Vinci Surgical System) and in situ bioprinting (Tarassoli et al. 2018). Apart from the cells and bioink, various cytokines, DNAs, and genetic materials to enhance the growth and functions could be applied on the 3D printing technology to construct better physiologic tissues.

7 Conclusions

The technics and strategies of penile reconstruction have advanced steady. Appropriate tissue in reconstructive surgery of penis was always deficit in supply. Substituting tissues bear some problems as inadequate properties or immunogenicity. Application of tissue engineering has opened the feasible supply of needed tissues. Investigations have evolved from constructing structures of tissue to the functional performances of them. Suitable biomaterials have been chosen among the numerous materials and efficient technics for dealing cells have been established. Penile reconstruction has developed from constituting primitive cavernosal tissue to properly functioning complex tissue. There are a lot of suggestions and trials to improve the quality of reconstructed penis. Stem cells, gene therapy, chemicals and cytokines, and 3D bioprinting are the representative ones. Steady development would be followed by researches applying the combinations of these technics and resources. Tissue-engineered products have advanced their positions from the bench to the clinical applications in some organs already. Engineered penis would be the next one for the clinical field.

References

Abdel Raheem A, Johnson M, Abdel-Raheem T, Capece M, Ralph D (2017) Collagenase clostridium histolyticum in the treatment of Peyronie's disease-a review of the literature and a new modified protocol. Sex Med Rev 5:529–535

Aicher A, Heeschen C, Sasaki K, Urbich C, Zeiher AM, Dimmeler S (2006) Low-energy shock wave for enhancing recruitment of endothelial progenitor cells: a new modality to increase efficacy of cell therapy in chronic hind limb ischemia. Circulation 114:2823–2830

Al Demour S, Jafar H, Adwan S, AlSharif A, Alhawari H, Alrabadi A, Zayed A, Jaradat A, Awidi A (2018) Safety and potential therapeutic effect of two intracavernous autologous bone marrow derived mesenchymal stem cells injections in diabetic patients with erectile dysfunction: an open label phase I clinical trial. Urol Int 101:358–365

Albersen M, Fandel TM, Lin G, Wang G, Banie L, Lin CS, Lue TF (2010) Injections of adipose tissue-derived stem cells and stem cell lysate improve recovery of erectile function in a rat model of cavernous nerve injury. J Sex Med 7:3331–3340

Alei G, Letizia P, Ricottilli F, Simone P, Alei L, Massoni F, Ricci S (2012) Original technique for penile girth augmentation through porcine dermal acellular grafts: results in a 69-patient series. J Sex Med 9:1945–1953

Alwaal A, Zaid UB, Lin CS, Lue TF (2015) Stem cell treatment of erectile dysfunction. Adv Drug Deliv Rev 82-83:137–144

An G, Ji C, Wei Z, Chen H, Zhang J (2013) Engineering of corpus cavernosum using vascular endothelial growth factor-expressing muscle-derived stem cells seeded on acellular corporal collagen matrices. Urology 81:424–431

An J, Chua CK, Mironov V (2016) A perspective on 4D bioprinting. Int J Bioprinting 2:3–5

Atala A (1998) Autologous cell transplantation for urologic reconstruction. J Urol 159:2–3

Atala A (2012) Regenerative medicine in urology: stem cells, tissue engineering, and cloning, 10 edn. In: Wein AJ, Kavoussi LR, Partin AW, Novick AC, Peters CA (eds) Campbell-walsh urology, 10th edn. Elsevier, Philadelpia, pp 568–588

Atala A, Guzman L, Retik AB (1999) A novel inert collagen matrix for hypospadias repair. J Urol 162:1148–1151

Atala A, Bauer SB, Soker S, Yoo JJ, Retik AB (2006) Tissue-engineered autologous bladders for patients needing cystoplasty. Lancet 367:1241–1246

Austoni E, Guarneri A, Cazzaniga A (2002) A new technique for augmentation phalloplasty: albugineal surgery with bilateral saphenous grafts–three years of experience. Eur Urol 42:245–253. discussion 252-243

Awad A, Alsaid B, Bessede T, Droupy S, Benoit G (2011) Evolution in the concept of erection anatomy. Surg Radiol Anat 33:301–312

Bahk JY, Jung JH, Han H, Min SK, Lee YS (2010) Treatment of diabetic impotence with umbilical cord blood stem cell intracavernosal transplant: preliminary report of 7 cases. Exp Clin Transplant 8:150–160

Bettocchi C, Ralph DJ, Pryor JP (2005) Pedicled pubic phalloplasty in females with gender dysphoria. BJU Int 95:120–124

Billiet T, Vandenhaute M, Schelfhout J, Van Vlierberghe S, Dubruel P (2012) A review of trends and limitations in hydrogel-rapid prototyping for tissue engineering. Biomaterials 33:6020–6041

Bivalacqua TJ, Deng W, Kendirci M, Usta MF, Robinson C, Taylor BK, Murthy SN, Champion HC, Hellstrom WJ, Kadowitz PJ (2007) Mesenchymal stem cells alone or ex vivo gene modified with endothelial nitric oxide synthase reverse age-associated erectile dysfunction. Am J Physiol Heart Circ Physiol 292:H1278–H1290

Bogoras N (1936) Über die volle plastische wiederherstellung eines zum koitus fähigen penis (Peniplastica totalis). Zentralbl Chir 22:1271–1276

Boland T, Xu T, Damon B, Cui X (2006) Application of inkjet printing to tissue engineering. Biotechnol J 1:910–917

Bora P, Majumdar AS (2017) Adipose tissue-derived stromal vascular fraction in regenerative medicine: a brief review on biology and translation. Stem Cell Res Ther 8:145

Burchardt M, Burchardt T, Chen MW, Shabsigh A, de la Taille A, Buttyan R, Shabsigh R (1999) Expression of messenger ribonucleic acid splice variants for vascular endothelial growth factor in the penis of adult rats and humans. Biol Reprod 60:398–404

Burchardt M, Burchardt T, Anastasiadis AG, Buttyan R, de la Taille A, Shabsigh A, Frank J, Shabsigh R (2005) Application of angiogenic factors for therapy of erectile dysfunction: protein and DNA transfer of VEGF 165 into the rat penis. Urology 66:665–670

Burg KJ, Holder WD Jr, Culberson CR, Beiler RJ, Greene KG, Loebsack AB, Roland WD, Eiselt P, Mooney DJ, Halberstadt CR (2000) Comparative study of seeding methods for three-dimensional polymeric scaffolds. J Biomed Mater Res 52:576

Caplan AL, Kimberly LL, Parent B, Sosin M, Rodriguez ED (2017) The ethics of penile transplantation: preliminary recommendations. Transplantation 101:1200–1205

Capogrosso P, Montorsi F, Salonia A (2018) Phase I and phase II clinical trials for the treatment of male sexual dysfunction-a systematic review of the literature. Expert Opin Investig Drugs 27:583–593

Carson MP, Saenz de Tejada I, Goldstein I, Haudenschild CC (1989) Culture of human corpus cavernosum endothelium. In Vitro Cell Dev Biol 25:248–254

Castiglione F, Hedlund P, Van der Aa F, Bivalacqua TJ, Rigatti P, Van Poppel H, Montorsi F, De Ridder D, Albersen M (2013) Intratunical injection of human adipose tissue-derived stem cells prevents fibrosis and is associated with improved erectile function in a rat model of Peyronie's disease. Eur Urol 63:551–560

Castiglione F, Hedlund P, Weyne E, Hakim L, Montorsi F, Salonia A, Bivalacqua TJ, De Ridder D, Milenkovic U, Ralph D, Garaffa G, Muneer A, Joniau S, Albersen M (2019) Intratunical injection of stromal vascular fraction prevents fibrosis in a rat model of Peyronie's disease. BJU Int 124:342–348

Chamberlain G, Fox J, Ashton B, Middleton J (2007) Concise review: mesenchymal stem cells: their phenotype, differentiation capacity, immunological features, and potential for homing. Stem Cells 25:2739–2749

Chang TS, Hwang WY (1984) Forearm flap in one-stage reconstruction of the penis. Plast Reconstr Surg 74:251–258

Chen KL, Eberli D, Yoo JJ, Atala A (2010) Bioengineered corporal tissue for structural and functional restoration of the penis. Proc Natl Acad Sci U S A 107:3346–3350

Cheng JY, Ng EM, Ko JS (2007) Depressive symptomatology and male sexual functions in late life. J Affect Disord 104:225–229

Christ GJ (2002) K channels as molecular targets for the treatment of erectile dysfunction. J Androl 23:S10–S19

Christ GJ, Rehman J, Day N, Salkoff L, Valcic M, Melman A, Geliebter J (1998) Intracorporal injection of hSlo cDNA in rats produces physiologically relevant alterations in penile function. Am J Phys 275:H600–H608

Chun JL, McGregor A, Krishnan R, Carson CC (2001) A comparison of dermal and cadaveric pericardial grafts in the modified Horton-Devine procedure for Peyronie's disease. J Urol 166:185–188

Costa WS, Carrerete FB, Horta WG, Sampaio FJ (2006) Comparative analysis of the penis corpora cavernosa in controls and patients with erectile dysfunction. BJU Int 97:567–569

Dahiya R, Chui R, Perinchery G, Nakajima K, Oh BR, Lue TF (1999) Differential gene expression of growth factors in young and old rat penile tissues is associated with erectile dysfunction. Int J Impot Res 11:201–206

Damian M, Porteus MH (2013) A crisper look at genome editing: RNA-guided genome modification. Mol Ther 21:720–722

Das ND, Song KM, Yin GN, Batbold D, Kwon MH, Kwon KD, Kim WJ, Kim YS, Ryu JK, Suh JK (2014) Xenogenic transplantation of human breast adipose-derived stromal vascular fraction enhances recovery of erectile function in diabetic mice. Biol Reprod 90:66

Davis CJ Jr (1997) The microscopic pathology of Peyronie's disease. J Urol 157:282–284

De Siati M, Selvaggio O, Di Fino G, Liuzzi G, Massenio P, Sanguedolce F, Carrieri G, Cormio L (2013) An unusual delayed complication of paraffin self-injection for penile girth augmentation. BMC Urol 13:66

de Vocht D, de Kemp V, Iljas JD, Bosch J, de Kort LMO, de Graaf P (2018) A systematic review on cell-seeded tissue engineering of penile corpora. J Tissue Eng Regen Med 12:687–694

Deng J, Zou ZM, Zhou TL, Su YP, Ai GP, Wang JP, Xu H, Dong SW (2011) Bone marrow mesenchymal stem cells can be mobilized into peripheral blood by G-CSF in vivo and integrate into traumatically injured cerebral tissue. Neurol Sci 32:641–651

Dillon BE, Chama NB, Honig SC (2008) Penile size and penile enlargement surgery: a review. Int J Impot Res 20:519–529

Drake RA, Vogl AW, Mitchell AWM (2009) Regional anatomy, 2nd edn Gray's anatomy for students, 2nd edn. Elsevier, Churchill Livingstone, pp 421–503

Dwyer ME, Salgado CJ, Lightner DJ (2011) Normal penile, scrotal, and perineal anatomy with reconstructive considerations. Semin Plast Surg 25:179–188

Eberli D, Susaeta R, Yoo JJ, Atala A (2008) A method to improve cellular content for corporal tissue engineering. Tissue Eng Part A 14:1581–1589

Egydio FM, Filho Freitas LG, Sayeg K, Laks M, Oliveira AS, Almeida FG (2015) Acellular human glans extracellular matrix as a scaffold for tissue engineering: in vitro cell support and biocompatibility. Int Braz J Urol 41:990–1001

Falke G, Yoo JJ, Kwon TG, Moreland R, Atala A (2003) Formation of corporal tissue architecture in vivo using human cavernosal muscle and endothelial cells seeded on collagen matrices. Tissue Eng 9:871–879

Fandel TM, Albersen M, Lin G, Qiu X, Ning H, Banie L, Lue TF, Lin CS (2012) Recruitment of intracavernously injected adipose-derived stem cells to the major pelvic ganglion improves erectile function in a rat model of cavernous nerve injury. Eur Urol 61:201–210

Fang JF, Huang XN, Han XY, Ouyang X, Fan L, Zhao X, Chen ZH, Wei HB (2018) Combined transplantation of mesenchymal stem cells and endothelial progenitor cells restores cavernous nerve injury-related erectile dysfunction. J Sex Med 15:284–295

Fatt M, Hsu K, He L, Wondisford F, Miller FD, Kaplan DR, Wang J (2015) Metformin acts on two different molecular pathways to enhance adult neural precursor proliferation/self-renewal and differentiation. Stem Cell Reports 5:988–995

Felici N, Felici A (2006) A new phalloplasty technique: the free anterolateral thigh flap phalloplasty. J Plast Reconstr Aesthet Surg 59:153–157

Gaertner WB, Bonsack ME, Delaney JP (2007) Experimental evaluation of four biologic prostheses for ventral hernia repair. J Gastrointest Surg 11:1275–1285

Garaffa G, Christopher NA, Ralph DJ (2010) Total phallic reconstruction in female-to-male transsexuals. Eur Urol 57:715–722

Garban H, Marquez D, Magee T, Moody J, Rajavashisth T, Rodriguez JA, Hung A, Vernet D, Rajfer J, Gonzalez-Cadavid NF (1997) Cloning of rat and human inducible penile nitric oxide synthase. Application for gene therapy of erectile dysfunction. Biol Reprod 56:954–963

Gillies H (1948) Congenital absence of the penis. Br J Plast Surg 1:8–28

Gimble JM, Katz AJ, Bunnell BA (2007) Adipose-derived stem cells for regenerative medicine. Circ Res 100:1249–1260

Gokce A, Abd Elmageed ZY, Lasker GF, Bouljihad M, Braun SE, Kim H, Kadowitz PJ, Abdel-Mageed AB, Sikka SC, Hellstrom WJ (2015) Intratunical injection of genetically modified adipose tissue-derived stem cells with human interferon alpha-2b for treatment of erectile dysfunction in a rat model of tunica albugineal fibrosis. J Sex Med 12:1533–1544

Gokce A, Peak TC, Abdel-Mageed AB, Hellstrom WJ (2016) Adipose tissue-derived stem cells for the treatment of erectile dysfunction. Curr Urol Rep 17:14

Goldstein AM, Meehan JP, Morrow JW, Buckley PA, Rogers FA (1985) The fibrous skeleton of the corpora cavernosa and its probable function in the mechanism of erection. Br J Urol 57:574–578

Goldstein I, Hartzell R, Shabsigh R (2016) The impact of Peyronie's disease on the patient: gaps in our current understanding. J Sex Marital Ther 42:178–190

Goodwin WE, Scott WW (1952) Phalloplasty. J Urol 68:903–908

Guillot-Tantay C, Phe V, Chartier-Kastler E, Mozer P, Bitker MO, Roupret M (2014) Medical and surgical treatments of congenital and acquired penile curvatures: a review. Prog Urol 24:203–211

Gur S, Peak T, Yafi FA, Kadowitz PJ, Sikka SC, Hellstrom WJ (2016) Mirabegron causes relaxation of human and rat corpus cavernosum: could it be a potential therapy for erectile dysfunction? BJU Int 118:464–474

Gur S, Abdel-Mageed AB, Sikka SC, Bartolome AR, Hellstrom WJG (2018) Destination penis? Gene therapy as a possible treatment for erectile dysfunction. Curr Gene Ther 18:225–239

Ha CH, Kim S, Chung J, An SH, Kwon K (2013) Extracorporeal shock wave stimulates expression of the angiogenic genes via mechanosensory complex in endothelial cells: mimetic effect of fluid shear stress in endothelial cells. Int J Cardiol 168:4168–4177

Haahr MK, Harken Jensen C, Toyserkani NM, Andersen DC, Damkier P, Sorensen JA, Sheikh SP, Lund L (2018) A 12-month follow-up after a single intracavernous injection of autologous adipose-derived regenerative cells in patients with erectile dysfunction following radical prostatectomy: an open-label phase I clinical trial. Urology 121:203.e206–203.e213

Hasan A, Paul A, Vrana NE, Zhao X, Memic A, Hwang YS, Dokmeci MR, Khademhosseini A (2014) Microfluidic techniques for development of 3D vascularized tissue. Biomaterials 35:7308–7325

Hinz B (2015) The extracellular matrix and transforming growth factor-beta1: tale of a strained relationship. Matrix Biol 47:54–65

Hoganson DM, Owens GE, O'Doherty EM, Bowley CM, Goldman SM, Harilal DO, Neville CM, Kronengold RT, Vacanti JP (2010) Preserved extracellular matrix components and retained biological activity in decellularized porcine mesothelium. Biomaterials 31:6934–6940

Horton CE, Dean JA (1990) Reconstruction of traumatically acquired defects of the phallus. World J Surg 14:757–762

Horton CE, Vorstman B, Teasley D, Winslow B (1987) Hidden penis release: adjunctive suprapubic lipectomy. Ann Plast Surg 19:131–134

Hotchkiss RS, Morales PA, O'Connor JJ Jr (1956) Plastic reconstructive surgery after total loss of the penis. Am J Surg 92:403–408

Hsieh CH, Liu SP, Hsu GL, Chen HS, Molodysky E, Chen YH, Yu HJ (2012) Advances in understanding of mammalian penile evolution, human penile anatomy and human erection physiology: clinical implications for physicians and surgeons. Med Sci Monit 18:Ra118–Ra125

Hu W, Lu J, Zhang L, Wu W, Nie H, Zhu Y, Deng Z, Zhao Y, Sheng W, Chao Q, Qiu X, Yang J, Bai Y (2006) A preliminary report of penile transplantation. Eur Urol 50:851–853

Huang Y, Zhang XF, Gao G, Yonezawa T, Cui X (2017) 3D bioprinting and the current applications in tissue engineering. Biotechnol J 12(8):734

Jang J, Park HJ, Kim SW, Kim H, Park JY, Na SJ, Kim HJ, Park MN, Choi SH, Park SH, Kim SW, Kwon SM, Kim PJ, Cho DW (2017) 3D printed complex tissue construct using stem cell-laden decellularized extracellular matrix bioinks for cardiac repair. Biomaterials 112:264–274

Jeon SH, Shrestha KR, Kim RY, Jung AR, Park YH, Kwon O, Kim GE, Kim SH, Kim KH, Lee JY (2016) Combination therapy using human adipose-derived stem cells on the cavernous nerve and low-energy shockwaves on the corpus cavernosum in a rat model of post-prostatectomy erectile dysfunction. Urology 88:226.e221–226.e229

Ji C, Min F, Liang W, Chen Y, Pan S, Bin L, Chen S, Wei Z, He T, Gu Y, Zhang J (2011) Construction of tissue-engineered corpus cavernosum with muscle-derived stem cells and transplantation in vivo. BJU Int 107:1638–1646

Johnston JH (1974) Lengthening of the congenital or acquired short penis. Br J Urol 46:685–688

Jones NC, Tyner KJ, Nibarger L, Stanley HM, Cornelison DD, Fedorov YV, Olwin BB (2005) The p38alpha/beta MAPK functions as a molecular switch to activate the quiescent satellite cell. J Cell Biol 169:105–116

Joo KJ, Kim BS, Han JH, Kim CJ, Kwon CH, Park HJ (2006) Porcine vesical acellular matrix graft of tunica albuginea for penile reconstruction. Asian J Androl 8:543–548

Kabalin JN, Rosen J, Perkash I (1990) Penile advancement and lengthening in spinal cord injury patients with retracted phallus who have failed penile prosthesis placement alone. J Urol 144:316–318

Kadioglu A, Sanli O, Akman T, Ersay A, Guven S, Mammadov F (2007) Graft materials in Peyronie's disease surgery: a comprehensive review. J Sex Med 4:581–595

Kajbafzadeh AM, Abbasioun R, Sabetkish S, Sabetkish N, Rahmani P, Tavakkolitabassi K, Arshadi H (2017) Future prospects for human tissue engineered urethra transplantation: decellularization and recellularization-based urethra regeneration. Ann Biomed Eng 45:1795–1806

Kang HW, Lee SJ, Ko IK, Kengla C, Yoo JJ, Atala A (2016) A 3D bioprinting system to produce human-scale tissue constructs with structural integrity. Nat Biotechnol 34:312–319

Kendirci M, Trost L, Bakondi B, Whitney MJ, Hellstrom WJ, Spees JL (2010) Transplantation of nonhematopoietic adult bone marrow stem/progenitor cells isolated by p75 nerve growth factor receptor into the penis rescues erectile function in a rat model of cavernous nerve injury. J Urol 184:1560–1566

Keriquel V, Oliveira H, Remy M, Ziane S, Delmond S, Rousseau B, Rey S, Catros S, Amedee J, Guillemot F, Fricain JC (2017) In situ printing of mesenchymal stromal cells, by laser-assisted bioprinting, for in vivo bone regeneration applications. Sci Rep 7:1778

Kershen RT, Yoo JJ, Moreland RB, Krane RJ, Atala A (2002) Reconstitution of human corpus cavernosum smooth muscle in vitro and in vivo. Tissue Eng 8:515–524

Kim JY (2013) Penile augmentation surgery using xenogenic graft and histological findings of clinical tests. J Urol 189:e7

Kim BS, Yoo JJ, Atala A (2002) Engineering of human cartilage rods: potential application for penile prostheses. J Urol 168:1794–1797

Kim JJ, Kwak TI, Jeon BG, Cheon J, Moon DG (2003) Human glans penis augmentation using injectable hyaluronic acid gel. Int J Impot Res 15:439–443

Kim BS, Kwon YW, Kong JS, Park GT, Gao G, Han W, Kim MB, Lee H, Kim JH, Cho DW (2018) 3D cell printing of in vitro stabilized skin model and in vivo pre-vascularized skin patch using tissue-specific extracellular matrix bioink: a step towards advanced skin tissue engineering. Biomaterials 168:38–53

Kwak TI, Oh M, Kim JJ, Moon du G (2011) The effects of penile girth enhancement using injectable hyaluronic acid gel, a filler. J Sex Med 8:3407–3413

Kwon TG, Yoo JJ, Atala A (2002) Autologous penile corpora cavernosa replacement using tissue engineering techniques. J Urol 168:1754–1758

Laks M, Freitas-Filho LG, Sayeg K, Leite MT, Teixeira LC (2015) Penile reconstruction using mesenchymal stem cells. Acta Cir Bras 30:529–536

Lander EB, Berman MH, See JR (2016) Stromal vascular fraction combined with shock wave for the treatment of peyronie's disease. Plast Reconstr Surg Glob Open 4:e631

Lanza R (2002) Preface. In: Atala A, Lanza R (eds) Methods of tissue engineering. Elsevier, San Diego, p xv

Laub DR, Eicher W, Laub DR, Hentz VR (1989) Penis construction in female-to-male transsexuals. In: Eicher W, Kubli F, Herms V (eds) Plastic surgery in the sexually handicapped. Springer, Berlin, Heidelberg, pp 113–128

Lee SJ, Liu J, Oh SH, Soker S, Atala A, Yoo JJ (2008) Development of a composite vascular scaffolding system that withstands physiological vascular conditions. Biomaterials 29:2891–2898

Lee SJ, Nowicki M, Harris B, Zhang LG (2017) Fabrication of a highly aligned neural scaffold via a table top stereolithography 3D printing and electrospinning. Tissue Eng Part A 23:491–502

Lever J, Frederick DA, Peplau LA (2006) Does size matter? Men's and women's views on penis size across the lifespan. Psychol Men Masculinity 7:129–143

Levine LA, Becher EF, Bella AJ, Brant WO, Kohler TS, Martinez-Salamanca JI, Trost L, Morey AF (2016) Penile prosthesis surgery: current recommendations from the international consultation on sexual medicine. J Sex Med 13:489–518

Levy JA, Marchand M, Iorio L, Zribi G, Zahalsky MP (2015) Effects of stem cell treatment in human patients with Peyronie disease. J Am Osteopath Assoc 115:e8–e13

Levy JA, Marchand M, Iorio L, Cassini W, Zahalsky MP (2016) Determining the feasibility of managing erectile dysfunction in humans with placental-derived stem cells. J Am Osteopath Assoc 116:e1–e5

Lin G, Hayashi N, Carrion R, Chang LJ, Lue TF, Lin CS (2005) Improving erectile function by silencing phosphodiesterase-5. J Urol 174:1142–1148

Lin G, Alwaal A, Zhang X, Wang J, Wang L, Li H, Wang G, Ning H, Lin CS, Xin Z, Lue TF (2015) Presence of stem/progenitor cells in the rat penis. Stem Cells Dev 24:264–270

Lin H, Dhanani N, Tseng H, Souza GR, Wang G, Cao Y, Ko TC, Jiang H, Wang R (2016) Nanoparticle improved stem cell therapy for erectile dysfunction in a rat model of cavernous nerve injury. J Urol 195:788–795

Liu G, Sun X, Bian J, Wu R, Guan X, Ouyang B, Huang Y, Xiao H, Luo D, Atala A, Zhang Y, Deng C (2013) Correction of diabetic erectile dysfunction with adipose derived stem cells modified with the vascular endothelial growth factor gene in a rodent diabetic model. PLoS One 8:e72790

Liu Y, Ma W, Liu B, Wang Y, Chu J, Xiong G, Shen L, Long C, Lin T, He D, Butnaru D, Alexey L, Zhang Y, Zhang D, Wei G (2017) Urethral reconstruction with autologous urine-derived stem cells seeded in three-dimensional porous small intestinal submucosa in a rabbit model. Stem Cell Res Ther 8:63

Lue TF (2016) Physiology of penile erection and pathophysiology of erectile dysfunction, 11th edn. In: Wein AJ, Kavoussi LR, Partin AW, Peters CA (eds) Campbell-Walsh urology, 11th edn. Elsevier, Philadelphia, pp 612–642

Lumen N, Kuehhas FE, Djakovic N, Kitrey ND, Serafetinidis E, Sharma DM, Summerton DJ (2015) Review of the current management of lower urinary tract injuries by the EAU Trauma Guidelines Panel. Eur Urol 67:925–929

Ma L, Yang Y, Sikka SC, Kadowitz PJ, Ignarro LJ, Abdel-Mageed AB, Hellstrom WJ (2012) Adipose tissue-derived stem cell-seeded small intestinal submucosa for tunica albuginea grafting and reconstruction. Proc Natl Acad Sci U S A 109:2090–2095

Machluf M, Atala A (1998) Emerging concepts for tissue and organ transplantation. Grafting 1:31–37

Mahla RS (2016) Stem cells applications in regenerative medicine and disease therapeutics. Int J Cell Biol 2016:6940283

Mandrycky C, Wang Z, Kim K, Kim DH (2016) 3D bioprinting for engineering complex tissues. Biotechnol Adv 34:422–434

Mangir N, Turkeri L (2017) Stem cell therapies in post-prostatectomy erectile dysfunction: a critical review. Can J Urol 24:8609–8619

Marks PW, Witten CM, Califf RM (2017) Clarifying stem-cell therapy's benefits and risks. N Engl J Med 376:1007–1009

Martin GR (1981) Isolation of a pluripotent cell line from early mouse embryos cultured in medium conditioned by teratocarcinoma stem cells. Proc Natl Acad Sci U S A 78:7634–7638

Martínez Ávila H, Schwarz S, Rotter N, Gatenholm P (2016) 3D bioprinting of human chondrocyte-laden nanocellulose hydrogels for patient-specific auricular cartilage regeneration. Bioprinting 1-2:22–35

Matsuda Y, Sasaki M, Kataoka-Sasaki Y, Takayanagi A, Kobayashi K, Oka S, Nakazaki M, Masumori N, Kocsis JD, Honmou O (2018) Intravenous infusion of bone marrow-derived mesenchymal stem cells reduces erectile dysfunction following cavernous nerve injury in rats. Sex Med 6:49–57

Melman A (2006) Gene transfer for the therapy of erectile dysfunction: progress in the 21st century. Int J Impot Res 18:19–25

Mittermayr R, Antonic V, Hartinger J, Kaufmann H, Redl H, Teot L, Stojadinovic A, Schaden W (2012) Extracorporeal shock wave therapy (ESWT) for wound healing: technology, mechanisms, and clinical efficacy. Wound Repair Regen 20:456–465

Miyamoto K, Inoue S, Kobayashi K, Kajiwara M, Teishima J, Matsubara A (2014) Rat cavernous nerve reconstruction with CD133+ cells derived from human bone marrow. J Sex Med 11:1148–1158

Montorsi F, Briganti A, Pozzo L, Deho F, Rigatti P, Salonia A (2003) Five years follow-up of plaque incision and vein grafting for peyronie's disease. Int J Impot Reser Suppl 15:14

Moon DG, Yoo JW, Bae JH, Han CS, Kim YK, Kim JJ (2003) Sexual function and psychological characteristics of penile paraffinoma. Asian J Androl 5:191–194
Moore KA, Lemischka IR (2006) Stem cells and their niches. Science 311:1880–1885
Muench PJ (2013) Infections versus penile implants: the war on bugs. J Urol 189:1631–1637
Munoz-Rangel CA, Fernandez-Vivar E, Banuelos-Gallo RA, Gonzalez-Ojeda A, Macias-Amezcua MD, Chavez-Tostado M, Ramirez-Campos KM, Ramirez-Arce Adel R, Cortes-Lares JA, Fuentes-Orozco C (2015) Minimally invasive therapy using intralesional onabotulinumtoxinA in Peyronie's disease. Urol J 12:2105–2110
Murphy SV, Atala A (2014) 3D bioprinting of tissues and organs. Nat Biotechnol 32:773–785
Mutaf M, Isik D, Bulut O, Buyukgural B (2006) A true one-stage nonmicrosurgical technique for total phallic reconstruction. Ann Plast Surg 57:100–106
Neri M, Maderna C, Cavazzin C, Deidda-Vigoriti V, Politi LS, Scotti G, Marzola P, Sbarbati A, Vescovi AL, Gritti A (2008) Efficient in vitro labeling of human neural precursor cells with superparamagnetic iron oxide particles: relevance for in vivo cell tracking. Stem Cells 26:505–516
Neuzillet Y, Hupertan V, Cour F, Botto H, Lebret T (2013) A randomized, double-blind, crossover, placebo-controlled comparative clinical trial of arginine aspartate plus adenosine monophosphate for the intermittent treatment of male erectile dysfunction. Andrology 1:223–228
Nishida T, Shimokawa H, Oi K, Tatewaki H, Uwatoku T, Abe K, Matsumoto Y, Kajihara N, Eto M, Matsuda T, Yasui H, Takeshita A, Sunagawa K (2004) Extracorporeal cardiac shock wave therapy markedly ameliorates ischemia-induced myocardial dysfunction in pigs in vivo. Circulation 110:3055–3061
Novosel EC, Kleinhans C, Kluger PJ (2011) Vascularization is the key challenge in tissue engineering. Adv Drug Deliv Rev 63:300–311
Nugteren HM, Balkema GT, Pascal AL, Schultz WC, Nijman JM, van Driel MF (2010) Penile enlargement: from medication to surgery. J Sex Marital Ther 36:118–123
Nukui F, Okamoto S, Nagata M, Kurokawa J, Fukui J (1997) Complications and reimplantation of penile implants. Int J Urol 4:52–54
Oh KJ, Yu HS, Park J, Lee HS, Park SA, Park K (2019) Co-culture of smooth muscle cells and endothelial cells on three-dimensional bioprinted polycaprolactone scaffolds for cavernosal tissue engineering. Aging Male 22:1–6. https://doi.org/10.1080/13685538.20191601175
Orabi H, Lin G, Ferretti L, Lin CS, Lue TF (2012) Scaffoldless tissue engineering of stem cell derived cavernous tissue for treatment of erectile function. J Sex Med 9:1522–1534
Palese MA, Burnett AL (2001) Corporoplasty using pericardium allograft (tutoplast) with complex penile prosthesis surgery. Urology 58:1049–1052
Park NC (1991) Troublesome fate of the penile augmentation with artificial oil analogues. Korean J Androl 9:49–53
Park NC (2016) General perspective: introduction. In: Park NC, Kim SW, Moon DG (eds) Penile augmentation. Springer, New York, pp 3–9
Park HJ, Yoo JJ, Kershen RT, Moreland R, Atala A (1999) Reconstitution of human corporal smooth muscle and endothelial cells in vivo. J Urol 162:1106–1109
Park K, Ahn KY, Kim MK, Lee SE, Kang TW, Ryu SB (2004) Intracavernosal injection of vascular endothelial growth factor improves erectile function in aged rats. Eur Urol 46:403–407
Perovic SV, Djordjevic ML (2000) Penile lengthening. BJU Int 86:1028–1033
Perovic SV, Djordjevic ML (2001) The penile disassembly technique in the surgical treatment of Peyronie's disease. BJU Int 88:731–738
Persson C, Diederichs W, Lue TF, Yen TS, Fishman IJ, McLin PH, Tanagho EA (1989) Correlation of altered penile ultrastructure with clinical arterial evaluation. J Urol 142:1462–1468
Pietropinto A (1986) Misconceptions about male sexuality. Med Aspects Hum Sex 20:80–85
Plaza T, Lautenschlager S (2010) Penis swelling due to foreign body reaction after injection of silicone. J Dtsch Dermatol Ges 8:689–691
Puckett CL, Montie JE (1978) Construction of male genitalia in the transsexual, using a tubed groin flap for the penis and a hydraulic inflation device. Plast Reconstr Surg 61:523–530

Qiu X, Villalta J, Ferretti L, Fandel TM, Albersen M, Lin G, Dai Y, Lue TF, Lin CS (2012) Effects of intravenous injection of adipose-derived stem cells in a rat model of radiation therapy-induced erectile dysfunction. J Sex Med 9:1834–1841

Qiu X, Lin G, Xin Z, Ferretti L, Zhang H, Lue TF, Lin CS (2013) Effects of low-energy shockwave therapy on the erectile function and tissue of a diabetic rat model. J Sex Med 10:738–746

Ralph DJ, Abdel Raheem A, Liu G (2017) Treatment of Peyronie's disease with collagenase clostridium histolyticum and vacuum therapy: a randomized, open-label pilot study. J Sex Med 14:1430–1437

Raya-Rivera A, Esquiliano DR, Yoo JJ, Lopez-Bayghen E, Soker S, Atala A (2011) Tissue-engineered autologous urethras for patients who need reconstruction: an observational study. Lancet 377:1175–1182

Reddy LH, Arias JL, Nicolas J, Couvreur P (2012) Magnetic nanoparticles: design and characterization, toxicity and biocompatibility, pharmaceutical and biomedical applications. Chem Rev 112:5818–5878

Ryu JK, Kim DH, Song KM, Yi T, Suh JK, Song SU (2014) Intracavernous delivery of clonal mesenchymal stem cells restores erectile function in a mouse model of cavernous nerve injury. J Sex Med 11:411–423

Ryu JK, Kim DH, Song KM, Ryu DS, Kim SN, Shin DH, Yi T, Suh JK, Song SU (2016) Intracavernous delivery of clonal mesenchymal stem cells rescues erectile function in the streptozotocin-induced diabetic mouse. Andrology 4:172–184

Salgado CJ, Chim H, Tang JC, Monstrey SJ, Mardini S (2011) Penile reconstruction. Semin Plast Surg 25:221–228

Santanelli F, Scuderi N (2000) Neophalloplasty in female-to-male transsexuals with the island tensor fasciae latae flap. Plast Reconstr Surg 105:1990–1996

Schafer R, Bantleon R, Kehlbach R, Siegel G, Wiskirchen J, Wolburg H, Kluba T, Eibofner F, Northoff H, Claussen CD, Schlemmer HP (2010) Functional investigations on human mesenchymal stem cells exposed to magnetic fields and labeled with clinically approved iron nanoparticles. BMC Cell Biol 11:22

Schultheiss D, Lorenz RR, Meister R, Westphal M, Gabouev AI, Mertsching H, Biancosino C, Schlote N, Wefer J, Winkler M, Stief CG, Jonas U (2004) Functional tissue engineering of autologous tunica albuginea: a possible graft for Peyronie's disease surgery. Eur Urol 45:781–786

Scott FB, Bradley WE, Timm GW (1973) Management of erectile impotence. Use of implantable inflatable prosthesis. Urology 2:80–82

Shamloul R, Ghanem H (2013) Erectile dysfunction. Lancet 381:153–165

Shan HT, Zhang HB, Chen WT, Chen FZ, Wang T, Luo JT, Yue M, Lin JH, Wei AY (2017) Combination of low-energy shock-wave therapy and bone marrow mesenchymal stem cell transplantation to improve the erectile function of diabetic rats. Asian J Androl 19:26–33

Shindel AW (2013) Sexual dysfunction: the potential of stem cell therapy for Peyronie disease. Nat Rev Urol 10:8–9

Simmons M, Montague DK (2008) Penile prosthesis implantation: past, present and future. Int J Impot Res 20:437–444

Skardal A, Mack D, Kapetanovic E, Atala A, Jackson JD, Yoo J, Soker S (2012) Bioprinted amniotic fluid-derived stem cells accelerate healing of large skin wounds. Stem Cells Transl Med 1:792–802

Snyder CC (1964) Intersex problems and hermaphroditism. In: Converse JM (ed) Reconstructive plastic surgery. Saunders, Philadelphia, pp 2078–2105

Soebadi MA, Moris L, Castiglione F, Weyne E, Albersen M (2016) Advances in stem cell research for the treatment of male sexual dysfunctions. Curr Opin Urol 26:129–139

Soebadi MA, Milenkovic U, Weyne E, Castiglione F, Albersen M (2017) Stem cells in male sexual dysfunction: are we getting somewhere? Sex Med Rev 5:222–235

Solomon MP, Komlo C, Defrain M (2013) Allograft materials in phalloplasty: a comparative analysis. Ann Plast Surg 71:297–299

Song YS, Lee HJ, Park IH, Lim IS, Ku JH, Kim SU (2008) Human neural crest stem cells transplanted in rat penile corpus cavernosum to repair erectile dysfunction. BJU Int 102:220–224. discussion 224

Song LJ, Xu YM, Li C, Fu Q, Cui L, Hu XY (2009) Construction of cavernosum smooth muscle using umbilical artery smooth muscle cells seeded on acellular corporal collagen matrices. Int J Androl 32:514–523

Song KM, Jin HR, Park JM, Choi MJ, Kwon MH, Kwon KD, Batbold D, Yin GN, Kim WJ, Koh GY, Ryu JK, Suh JK (2014) Intracavernous delivery of stromal vascular fraction restores erectile function through production of angiogenic factors in a mouse model of cavernous nerve injury. J Sex Med 11:1962–1973

Sorkio A, Koch L, Koivusalo L, Deiwick A, Miettinen S, Chichkov B, Skottman H (2018) Human stem cell based corneal tissue mimicking structures using laser-assisted 3D bioprinting and functional bioinks. Biomaterials 171:57–71

Spyropoulos E, Christoforidis C, Borousas D, Mavrikos S, Bourounis M, Athanasiadis S (2005) Augmentation phalloplasty surgery for penile dysmorphophobia in young adults: considerations regarding patient selection, outcome evaluation and techniques applied. Eur Urol 48:121–127. discussion 127-128

Sukop A, Heracek J, Mestak O, Borsky J, Bayer J, Schwarzmannova K (2013) Penis augmentation by application of silicone material: complications and surgical treatment. Acta Chir Plast 55:31–33

Sun C, Lin H, Yu W, Li X, Chen Y, Qiu X, Wang R, Dai Y (2012) Neurotrophic effect of bone marrow mesenchymal stem cells for erectile dysfunction in diabetic rats. Int J Androl 35:601–607

Takahashi K, Yamanaka S (2006) Induction of pluripotent stem cells from mouse embryonic and adult fibroblast cultures by defined factors. Cell 126:663–676

Takeshita S, Zheng LP, Brogi E, Kearney M, Pu LQ, Bunting S, Ferrara N, Symes JF, Isner JM (1994) Therapeutic angiogenesis. A single intraarterial bolus of vascular endothelial growth factor augments revascularization in a rabbit ischemic hind limb model. J Clin Invest 93:662–670

Tamay DG, Usal TD, Alagoz AS, Yucel D, Hasirci N, Hasirci V (2019) 3D and 4D printing of polymers for tissue engineering applications. Front Bioeng Biotechnol 7:164

Tan RB, Sangkum P, Mitchell GC, Hellstrom WJ (2014) Update on medical management of Peyronie's disease. Curr Urol Rep 15:415

Tarassoli SP, Jessop ZM, Al-Sabah A, Gao N, Whitaker S, Doak S, Whitaker IS (2018) Skin tissue engineering using 3D bioprinting: an evolving research field. J Plast Reconstr Aesthet Surg 71:615–623

Thompson JK, Heinberg LJ, Altabe M, Tantleff-Dunn S (1999) Exacting beauty: theory, assessment, and treatment of body image disturbance. American Psychological Association, Washington, DC

Thomson JA, Itskovitz-Eldor J, Shapiro SS, Waknitz MA, Swiergiel JJ, Marshall VS, Jones JM (1998) Embryonic stem cell lines derived from human blastocysts. Science 282:1145–1147

Tibbits S (2014) 4D printing: multi-material shape change. Archit Design 84:116–121

Vardi Y, Har-Shai Y, Gil T, Gruenwald I (2008) A critical analysis of penile enhancement procedures for patients with normal penile size: surgical techniques, success, and complications. Eur Urol 54:1042–1050

Vardi Y, Appel B, Kilchevsky A, Gruenwald I (2012) Does low intensity extracorporeal shock wave therapy have a physiological effect on erectile function? Short-term results of a randomized, double-blind, sham controlled study. J Urol 187:1769–1775

Volarevic V, Ljujic B, Stojkovic P, Lukic A, Arsenijevic N, Stojkovic M (2011) Human stem cell research and regenerative medicine–present and future. Br Med Bull 99:155–168

Wassermann RJ, Greenwald DP (1995) Debilitating silicone granuloma of the penis and scrotum. Ann Plast Surg 35:505–509. discussion 509-510

Weihs AM, Fuchs C, Teuschl AH, Hartinger J, Slezak P, Mittermayr R, Redl H, Junger WG, Sitte HH, Rünzler D (2014) Shock wave treatment enhances cell proliferation and improves wound

healing by ATP release-coupled extracellular signal-regulated kinase (ERK) activation. J Biol Chem 289:27090–27104
Wein AJ, Kavoussi LR, Partin AW, Peters CA (2016) Campbell-Walsh urology, 11th edn. Elsevier, Philadelphia
Williams JK, Dean AJ, Yoo JJ (2016) Regenerative medicine approaches to repair penile structure and function. In: Park NC, Kim SW, Moon DG (eds) Penile augmentation. Springer, Berlin, Heidelberg, pp 187–196
Woo HS, Koo SH, Ahn DS (1996) Lengthening and girth enhancement surgery for micropenis. Korean J Plast Surg 23:1432–1442
Wylie KR, Eardley I (2007) Penile size and the 'small penis syndrome'. BJU Int 99:1449–1455
Xiao YL, Riesle J, Van Blitterswijk CA (1999) Static and dynamic fibroblast seeding and cultivation in porous PEO/PBT scaffolds. J Mater Sci Mater Med 10:773–777
Xin ZC, Xu YD, Lin G, Lue TF, Guo YL (2016) Recruiting endogenous stem cells: a novel therapeutic approach for erectile dysfunction. Asian J Androl 18:10–15
Xu Y, Guan R, Lei H, Gao Z, Li H, Hui Y, Zhou F, Wang L, Lin G, Xin Z (2015) Implications for differentiation of endogenous stem cells: therapeutic effect from icariside II on a rat model of postprostatectomy erectile dysfunction. Stem Cells Dev 24:747–755
Yiou R, Hamidou L, Birebent B, Bitari D, Lecorvoisier P, Contremoulins I, Khodari M, Rodriguez AM, Augustin D, Roudot-Thoraval F, de la Taille A, Rouard H (2016) Safety of intracavernous bone marrow-mononuclear cells for postradical prostatectomy erectile dysfunction: an open dose-escalation pilot study. Eur Urol 69:988–991
Yiou R, Hamidou L, Birebent B, Bitari D, Le Corvoisier P, Contremoulins I, Rodriguez AM, Augustin D, Roudot-Thoraval F, de la Taille A, Rouard H (2017) Intracavernous injections of bone marrow mononucleated cells for postradical prostatectomy erectile dysfunction: final results of the INSTIN clinical trial. Eur Urol Focus 3:643–645
Yoo JJ, Lee I, Atala A (1998a) Cartilage rods as a potential material for penile reconstruction. J Urol 160:1164–1168. discussion 1178
Yoo JJ, Meng J, Oberpenning F, Atala A (1998b) Bladder augmentation using allogenic bladder submucosa seeded with cells. Urology 51:221–225
Yoo JJ, Park HJ, Lee I, Atala A (1999) Autologous engineered cartilage rods for penile reconstruction. J Urol 162:1119–1121
Zhang K, Fu Q, Yoo J, Chen X, Chandra P, Mo X, Song L, Atala A, Zhao W (2017) 3D bioprinting of urethra with PCL/PLCL blend and dual autologous cells in fibrin hydrogel: an in vitro evaluation of biomimetic mechanical property and cell growth environment. Acta Biomater 50:154–164
Zhou F, Hui Y, Xu Y, Lei H, Yang B, Guan R, Gao Z, Xin Z, Hou J (2016) Effects of adipose-derived stem cells plus insulin on erectile function in streptozotocin-induced diabetic rats. Int Urol Nephrol 48:657–669
Zhu LL, Zhang Z, Jiang HS, Chen H, Chen Y, Dai YT (2017) Superparamagnetic iron oxide nanoparticle targeting of adipose tissue-derived stem cells in diabetes-associated erectile dysfunction. Asian J Androl 19:425–432
Zuk PA, Zhu M, Mizuno H, Huang J, Futrell JW, Katz AJ, Benhaim P, Lorenz HP, Hedrick MH (2001) Multilineage cells from human adipose tissue: implications for cell-based therapies. Tissue Eng 7:211–228

Part IV

Musculoskeletal System

Injectable Calcium Phosphate Cements for the Reconstruction/Repair of Oral and Cranio-maxillofacial Bone Defects: Clinical Outcome and Perspectives

Hongbing Liao, Jan Willem Hoekstra, Joop Wolke, Sander Leeuwenburgh, and John Jansen

Contents

H. Liao
Department of Prosthodontics, College of Stomatology, Guangxi Medical University, Nanning, China

Department of Biomaterials, Radboud University Medical Center, Nijmegen, The Netherlands
e-mail: hongbing_liao@gxmu.edu.cn

J. W. Hoekstra · J. Wolke · S. Leeuwenburgh · J. Jansen (✉)
Department of Biomaterials, Radboud University Medical Center, Nijmegen, The Netherlands
e-mail: JanWillem.Hoekstra@radboudumc.nl; Sander.Leeuwenburgh@radboudumc.nl; John.Jansen@radboudumc.nl

D. Eberli et al. (eds.), *Organ Tissue Engineering*, Reference Series in Biomedical Engineering, https://doi.org/10.1007/978-3-030-44211-8_17

Abstract

In this chapter, the clinical performance of commercially available calcium phosphate cements (CPCs) used for reconstruction/repair of hard tissues in the oral and cranio-maxillofacial region is reviewed. Literature were collected from the electronic database of PubMed, Web of Science, and Springer Link, respectively, with various combinations of searching strategy of keywords; human preclinical/clinical studies ranging from case reports to randomized clinical trials were enrolled from the period between 1990 and 2016, highlighting the outcomes, complications, contraindications, and cautions related to the use of CPC in craniofacial reconstruction/repair. We conclude that injectable CPCs can be considered as favorable alloplasts for the reconstruction/repair of craniofacial area, when used with sufficient caution and be strict to specific clinical indications.

1 Introduction

The oral and cranio-maxillofacial region of the human being contains several organs including the brain, eyes, ears, nose, and mouth, which are united into a highly complex structure. In addition to the essential biological functions of these organs, the esthetic appearance of the face is considered increasingly important in modern society. Generally, function and form of the underlying skeleton are crucial for proper functioning of organs and attractive facial appearance. Consequently, treatment of injury and diseases or the desire for cosmetic change usually requires (re) arrangement of the underlying bony tissues.

This (re)arrangement of bony structures in the oral and cranio-maxillofacial region remains a clinical challenge for surgeons. In addition to the demanding requirements related to surgical skills and experience of surgeons, the availability of host-friendly and biologically effective synthetic bone grafts called bone substitutes is still limited despite several decades of research. To date, the best-performing bone grafts are still harvested from the patient's own body, which is however associated with drawbacks such as limited availability, increased patient morbidity, traumatic/surgical creation of donor site, and risks for infection. Although this drawback could be overcome by using allografts/xenografts as bone substitute, concerns have been expressed to allografts and xenografts related to their origin (i.e., human and bovine) and the theoretical risk for immune rejection of the graft and for transmission of infectious diseases.

Numerous synthetic bone substitutes have been developed and commercialized to date, but there is no consensus in the literature about an optimal formulation that can replace autogenous bone grafts. Generally, synthetic bone substitutes vary with respect to their chemical composition, physicochemical structure, and application form. Calcium phosphate cements (CPCs) are an emerging class of synthetic bone substitutes which exhibit chemical similarity to the mineral phase in bone and teeth and therefore are highly osteocompatible and easy to handle due to their self-setting properties. Usually CPCs consisted of powder phase (the precursor matrix) and liquid phase (the harden agent), depending on individual formulations, the powder phase containing one or more solid compounds of calcium and/or phosphate salts, and the liquid phase could be physiological saline solution or aqueous precursor solution. CPCs are prepared on site of surgery by mixing a precursor powder with a liquid phase; the mixing sets off a mild hardening reaction and becomes solid after final setting is finished. The reaction starts from the precipitation of small apatite crystal inside the mixture, finished with the entanglement of newly formed profound crystal network without radical heat release. Prior to the setting of CPC, the material is injectable/moldable which allows surgeons to manipulate with it according to specific clinical requirement/anatomical situation, such as delivering the materials in less invasive way by injection or reshaping the contour of deficit region by sculpture. Although the mechanical properties of harden CPCs are still inferior to bone tissue or highly sintered calcium phosphate ceramics to date, it is superior to hydrogel or putty forms of other inorganic salt compounds on the mechanical aspect; thus, CPCs are highly suitable for specific applications in the oral and cranio-maxillofacial complex, at the non-load-bearing region where mechanical requirements are less stringent. These unique characteristics of CPCs have therefore attracted a lot of attention and have been investigated extensively in numerous in vitro and in vivo studies; however, information on the long-term clinical performance of CPCs is still scarce or inconsistent so far. Consequently, the purpose of this book chapter is to review the clinical outcome of contemporary injectable CPCs developed for the reconstruction of hard tissues in the oral and cranio-maxillofacial region.

2 Etiology of Bone Defects in the Oral and Cranio-maxillofacial Region

Cranio-maxillofacial defects that require bone grafting can be caused by:

- Infectious diseases (osteomyelitis, periodontitis, etc.)
- Malignant or benign tumor surgery
- Ischemic conditions after radiation therapy (osteoradionecrosis)
- Traumatic injury
- Developmental/congenital diseases

Depending on the severity and complexity of the damage, the treatment approach varies from repair to reconstruction, with different requirements related for the bone graft materials to be used. In the following sections, the most common treatments are briefly addressed.

2.1 Repair of Intrabony Defects Caused by Periodontitis

Periodontitis is an infectious disease causing loss of soft and hard tissues surrounding the teeth. Depending on the progress and prognosis of the disease, the infected teeth may lose their support from the surrounding bony tissue. Therefore, repair of bony defects around infected teeth is an important aspect of the therapy in addition to treatment of the infection. To this end, various bone substitutes can be used to guide bone regeneration (GBR) into the defect area, usually assisted by the use of membranes (Reynolds et al. 2003). Ideally these criteria materials should resist the chewing force transmit by nearby tooth root and facilitate bone regenerate inside the scaffolds within 3 months for bone replacement.

2.2 Preservation of the Alveolar Volume After Tooth Extraction/Extraction Socket Management

Tooth extraction is the most commonly performed surgical intervention in the field of dentistry to date. Approximately 40–60% of the initial alveolar bone volume is lost 6 months after tooth extraction (Simion et al. 1998). When multiple teeth are extracted in the same area, bone resorption increases even more (Simion et al. 1994). Insufficient alveolar bone volume at an implant site can inhibit placement of dental implants in an optimal position that support the final prosthetic reconstruction. Immediate postextraction implant placement is a well-accepted protocol to minimize bone resorption when little pathology factor presented on the extracted socket site; however, the concept of placement of dental implants soon after removal of a tooth affected by periapical or periodontal pathology is a matter of debate. In fact, frequently, compromised teeth that are indicated for extraction are involved with infectious conditions, which conventionally contraindicate their immediate replacement with endosseous dental implants (Taschieri et al. 2010). In this situation the dilemma that clinicians face is how to manage tooth extractions to provide for the future placement of a dental implant or to maximize ridge dimensions for the fabrication of a fixed or removable prosthesis. Although all alveolar ridge preservation studies demonstrated beneficial results, no one particular grafting material has proven superior to others; the benefit of alveolar ridge grafting materials is inconclusive (Horowitz et al. 2012). Nevertheless, in both maxillary and mandibular regions, biomaterials have been used to maintain as much clinical volume as possible (Kotsakis et al. 2014; Atieh et al. 2015). This can be achieved by, e.g., GBR using particulate autografts, allografts, alloplasts, and/or xenografts with or without the additional support of resorbable or non-resorbable membranes. The bacteria/contamination resist properties of the materials could be emphasized here rather than mechanical strength since the extraction site is open to oral cavity when an advancing mucoperiosteal flap or mucogingiva soft tissue graft was not applied/available.

2.3 Reconstruction of an Atrophic Alveolar Ridge

Bone volume and the corresponding clinical contour change significantly after loss of teeth. Especially in the maxilla, the resulting atrophic alveolar ridge often does not allow for treatment with implants due to a lack of bone volume. To facilitate implant placement, the alveolar ridge often requires bone augmentation both in vertical and horizontal dimensions. At least five methods toward augmentation of the alveolar ridge can be discerned including (i) guided bone regeneration (GBR), (ii) bone splitting or spreading to expand the ridge, (iii) inlay bone grafts, (iv) onlay bone grafts, and (v) distraction osteogenesis. Several procedures such as GBR and onlay and inlay grafts can be performed using bone substitutes, membranes, or a combination thereof (Felice et al. 2009; Maestre-Ferrin et al. 2009). Distraction osteogenesis is a clinically accepted technique, but the long time needed for distraction and the discomfort of the intraoral distractor are major disadvantages (Saulacic et al. 2009; Ettl et al. 2010). In general, onlay or inlay techniques which require an additional operation at a donor site are more efficient in case of challenging conditions such as vertical augmentation of the alveolar ridge. At the area of the posterior maxilla, placement of implants is particularly complicated. At this anatomical site, bone can be augmented using a specific surgical technique called sinus augmentation (also referred to as sinus lifting or sinus floor elevation).

2.4 Reconstruction of Large Defects Caused by Trauma or Traumatic Therapy

The cranio-maxillofacial complex can be damaged severely by trauma or traumatic therapies such as surgery. In these cases, dentofacial deformity, post-traumatic bone defects, or postoperative contour irregularities are common symptoms. Depending on the complexity of damage, intensive reconstructive procedures are often required to restore oral function and esthetics. In this situation, the substitute materials should fill the defect and fix the broken bony structure rigidly and maintain the contour/volume of missed part of the complex.

2.5 Cosmetic Surgery for Developmental/Congenital Diseases at the Cranio-maxillofacial Region

Genetic deformities such as a cleft lip or palate, hemifacial macrosomia, and hemifacial atrophy are indications for repair and reconstruction of the craniofacial region. In these clinical cases, injectable bone substitutes, which can be applied less invasively, are generally preferred in view of the final prognosis and final cosmetic appearance.

3 Injectable CPCs for Reconstruction/Repair of the Oral and Cranio-maxillofacial Region

Injectable and/or moldable bone substitutes used in cranio-maxillofacial surgery include cements, pastes, putties, and hydrogels. This book chapter focuses on the use of calcium phosphate-based cements.

Calcium phosphate cements (CPCs) can be categorized according to the end product of the setting reaction, i.e., apatite or brushite (bohner 2010). Frequently, carbonate substitutes for hydroxyl groups in apatitic cements, thereby creating

Table 1 Overview of commercialized calcium phosphate cements

Product name	Chemical composition	Mechanical strength: Compression strength (MPa)	Young's modulus (MPa)	Tensile strength (MPa)	Biodegradable
BoneSource®	Tetracalcium phosphate (TTCP)/ dicalcium phosphate	6.3–34	3.6–4.7	2	Yes
Calcibon®	62.5% α-tricalcium phosphate/26.8% dicalcium phosphate anhydrous/8.9% calcium carbonate/ 1.8% hydroxyapatite	35–55	2500–3000	4.5	Yes
ChronOS Inject®	73% β-tricalcium phosphate/21% monocalcium phosphate monohydrate/5% magnesium hydrogen phosphate trihydrate	No data	No data	No date	Yes
HydroSet®	Tetracalcium phosphate/dicalcium phosphate/trisodium citrate	14–24	125–240	0.11–0.17	No date
Norian SRS®	α-Tricalcium phosphate/calcium carbonate/ monocalcium phosphate monohydrate	23–55	No data	2.1	Yes

cements with a higher degree of similarity to the mineral phase of the human bone, with which it contains a considerable amount of carbonate ions. Since their invention in the early 1980s, several apatite- and brushite-forming CPCs have been commercialized such as BoneSource®, Calcibon®, ChronOS Inject®, HydroSet®, and Norian SRS® (Van der Stok et al. 2011). The overview of commercialized calcium phosphate cements is summarized in Table 1.

4 Clinical Outcome of CPCs in Reconstruction of the Oral and Cranio-maxillofacial Region

To take a concise overview of the clinical outcome of the use of injectable CPCs for the reconstruction of the oral and cranio-maxillofacial region, literature between 1990 and 2016 were collected from the electronic database of PubMed, Web of Science, and Springer Link, respectively, with different combinations of searching strategy of keywords, i.e., the keywords of "calcium phosphate cement," "preclinical" or "clinical study," and the anatomic site of cranio-maxillofacial region such as "periodontal," "extraction," "sinus lift," "cranioplasty," "craniofacial," "skull defect," "reconstruction," "repair," etc. were combined, respectively, and an extensive electronic search was performed to identify relevant articles published up to December 2016. After the selection process, studies that met the eligibility criteria were included.

4.1 Reconstruction of Periodontal Intrabony Defects

The first human clinical trial using CPC for periodontal treatment was reported in 1998 by Brown. In this study, a hydroxyapatite cement (HAC), mainly a composition of tetracalcium phosphate and dicalcium phosphate hydrate, was tested inspired by its successful application on the reconstruction and augmentation of nonstress-bearing portion of the craniofacial skeleton. Sixteen patients with moderate to severe periodontal disease and two bilaterally similar vertical bony defects received initial therapy including scaling and root planing followed by treatment with either calcium phosphate cement, flap curettage (F/C), or debridement plus demineralized freeze-dried bone allograft (DFDBA). Standardized radiographs were taken at baseline and 12 months post-surgery for analysis; the extent of the bony defect was determined during initial and 12-month reentry surgery. The clinical attempt of HAC was not promising in comparison to the application of DFDBA: within 6 months of implant placement, 11 of 16 patients treated with calcium phosphate cement exfoliated all or most of the implant through the gingival sulcus. At all 16 test sites, the initially tight visual interface between the radiopaque calcium phosphate cement and the walls of the bony defect gave rise to a narrow radiolucent gap after 1 month post-surgery, meaning the HAC biomaterials were not integrated to the surrounding osseous tissue. This result failed to support the use of HAC for the repair and regeneration of human intrabony periodontal defect; the reason for the failure was not clear, but the authors speculated three possible mechanisms: first is the lack of sufficient flexural stress resistance of set HAC to the normal occlusal forces transmit by nearby natural tooth; the second factor was the lack of sufficient porosity in the hardened HAC to allow bone ingrowth; the innate micropores of set HAC were insufficient to allow for migration and ingrowth of cells, blood vessels, and bone; the third possible mechanism contributing to the clinical failure of HAC was bacterial contamination and colonization of the implant surface and micropores due to the loosen periodontal wound flaps closing (Brown et al. 1998).

In 2008, Shirakata et al. conducted a randomized clinical trial using Norian® cement, a novel formulation of calcium phosphate cement, mainly composed of α-tricalcium phosphate (α-$Ca_3[PO_4]_2$), monocalcium phosphate monohydrate ($Ca[H_2PO_4]_2$ H_2O), and calcium carbonate ($CaCO_3$) mixed with a solution of sodium phosphate (Constantz et al. 1995), as intrabony defect filler of periodontal bone reconstruction. Thirty patients with periodontitis and narrow intrabony defects were enrolled in the study; patients were classified randomly into the CPC graft group ($N = 15$) or the open flap debridement (OFD) alone group ($N = 15$). After routine basic periodontal treatment, mucoperiosteal full-thickness flaps were reflected, and all granulation tissues were removed completely, and the exposed root surfaces were scaled and planed with hand and ultrasonic instruments. In the graft group, the CPC was filled compactly with injectable CPC from the bottom of the defect close to the residual bone crest. After solidification of the CPC, a periosteal releasing incision was made to allow coronal repositioning of the flap, followed by suturing slightly coronal to the cementum enamel junction (CEJ). The control group was treated similarly, without placement of the CPC. Clinical measurements were performed at baseline and at 3, 6, 9, and 12 months, and radiographs were taken at baseline, 2 weeks, and 6 and 12 months after surgery. The results demonstrated that six cases in the CPC group showed exposure or loss of the graft material during the 12-month treatment, whereas the remaining nine cases (CPC-R group) showed no adverse reaction, including infection or suppuration. Overall, CPC-R and OFD treatment groups exhibited a significant reduction in probing depth and a significant gain in clinical attachment level at 3, 6, 9, and 12 months compared to baseline values. However, there were no significant differences in any of the clinical parameters between the groups. In the CPC-R group, radiographic bone level gain appeared to be greater than in the OFD group. The explanation for the lack of significant differences between CPC-R and OFD groups in any of the clinical parameters might be the fact that most of the defects were narrow three-wall, contained, intrabony defects, in which periodontal healing potentially resulted in complete clinical repair with or without periodontal regenerative therapy (Shirakata et al. 2008).

On the other hand, the setting behavior of CPCs is also crucial for its successful application, as indicated by the positive outcome of a clinical study using a newly developed washout-resistant type of CPC. The material used in the study is named "Chitra calcium phosphate cement" (Chitra-CPC); the powder phase consisted of tetracalcium phosphate (TTCP–$Ca_4(PO_4)_2O$) and dicalcium phosphate dihydrate (DCPD–$CaHPO_4 \cdot 2H_2O$) added with an optimum quantity of gelling agent in dry powder form, the water phase containing disodium hydrogen phosphate (Na_2HPO_4, in 0.2 M concentration) as the setting accelerator (Fernandez et al. 2006). The study included 60 patients and is divided into 3 groups using random number table method. There were two test groups ("CPC group" with calcium phosphate cement and "HAG group" with hydroxyapatite granules), along with a control group (with debridement only). The evaluation was focused on the periodontal soft tissue changes such as reduction in probing pocket depth, gingival recession, and gain in clinical attachment. All parameters in the control and the test groups were evaluated during the 3, 6, 9, and 12 months postoperatively. An overview of the results shows that both the test materials (calcium phosphate cement and hydroxyapatite ceramic granules) are significantly efficacious in healing the periodontal defects when compared with open flap debridement. The calcium phosphate cement formulation (Chitra-CPC) is more efficacious than the hydroxyapatite ceramic granules (Rajesh et al. 2009).

In general, the body of evidence for the use of CPC in periodontology is still limited due the lack of well-designed, randomized clinical trials.

4.2 Extraction Socket Management

Currently, no application of CPCs as sole volume filler at extraction sockets for bone volume preservation was reported. Nevertheless, in case of simultaneous placement of implants after extraction, usually a gap between implant and extraction bone wall will be created due to the mismatch between the shape/dimension of the implant fixture and the socket wall; CPCs are promising candidates to fill the space or to replace buccal bone damaged or lost during extraction, with careful selection on the indicative cases where no pathology factor exists or pathogeny of chronic infection could be eliminated. Taschieri reported a prospective study that evaluates the clinical outcome of implants immediately placed into fresh extraction sockets for the replacement of endodontically treated teeth with signs of vertical root fracture; sixteen partially edentulous patients, with one tooth scheduled for extraction and showing clinical signs and symptoms and/or radiological evidence of vertical root fracture, were included in the study. Sixteen transmucosal implants were installed immediately after extraction and careful debridement. The gap between the implant surface and the socket walls was filled using PD VitalOs Cement, an injectable synthetic bone grafting cement consisting of two calcium phosphate pastes (β-tricalcium phosphate and monocalcium phosphate monohydrate). Implant success and survival and radiographic bone loss were evaluated after 1 year of function. Overall implant success and survival rate was 100% at 1 year. All prostheses were successful and in function, 16 implants could be evaluated radiographically after 1 year of function. Peri-implant bone loss averaged 0.48–0.20 mm. Such value was not affected by implant location, lesion type, or bone quality (Taschieri et al. 2010).

4.3 Augmentation of the Maxillary Sinus

In 2000 Mazor firstly reported case series on the application of BoneSource® (hydroxyapatite cement, component of tetracalcium phosphate/dicalcium phosphate) to stabilize cylindrical hydroxyapatite-coated dental implants placed simultaneously during the augmentation of the maxillary sinus. Twenty-six HA-coated dental implants were inserted in ten grafted sinuses of ten patients with traditional lateral approach sinus floor lift surgeries. The cement was thereby placed at the superior aspect of the sinus. Implants were then fully inserted into the grafted compartment. Second-stage surgery was performed 9 months after implant placement, and core biopsies were performed in-between implant fixtures to verify the texture of graft. Prior to implant exposure, patients were evaluated radiographically. Panoramic and periapical radiographs and CT scans were used to assess the radiographic features of the graft material, the newly formed bone, and their close relation to the implants. Clinical criteria at the time of implant exposure included stability in all directions; crestal bone resorption and any reported pain or discomfort were recorded. None of the cases presented any difficulty in achieving initial stabilization; no clinical complications of the sinuses were evident. At second-stage surgery, there was no clinical evidence of crestal bone loss around the implants. All implants were clinically osseointegrated. All patients received fixed implant-supported prostheses, and the mean follow-up time was 18 months (ranging from 12 to 24 months). Histologic evaluation showed woven bone well integrated with the graft material with numerous osteocytes directly opposed to the surface. Some of the graft material was still present at the end of the core biopsy, undergoing replacement with bone by creeping substitution after 1 year (Mazor et al. 2000).

However, the clinical study of Sverzut et al. provided controversial results with the same commercialize product but different dental implant placement protocols. BoneSource® was used as filling material for maxillary sinus lifting without simultaneous dental implant insertion. Ten patients were enrolled, and traditional lateral approach sinus floor lift procedures were conducted. After a period ranging from 9 to 16 months of healing, all patients showed a gain in height radiopacity according to the radiographic evaluation and did not have postoperative complications; the secondary surgery aimed in installation of dental implant was conducted, clinical evaluation was taken, and the quantity and quality of bone formation were later verified by taking biopsy of the grafted area in the region adjacent to the axis of the implant to be inserted. However, the authors stated that the cement was friable during instrumentation and installation of the dental implant at second-stage operation. In view of these findings, the authors speculated that it was too risky to place implants due to a lack of sufficient bone formation throughout the biomaterials and decided to abort the placement of implants. The histological analysis revealed that new bone formation was minimal, mainly in direct contact with the surface of the bulk material, indicating that large amounts of cement remained in the sinus floor, possibly due to lack of vascularization inside the CPC; the authors therefore suggested that interconnected porosity should be introduced to the materials to improve the ingrowth of new bone tissue, and the success for simultaneous placement of dental implants and cement in Mazor's study are mainly give credit to the residual bone height of enrolled patient (5 mm in average), which can have provided primary stability of implanted fixtures and mechanical support to the applied CPC (Sverzut et al. 2015).

4.4 Augmentation of the Atrophic Mandible Alveolar Ridge

For this challenging situation, injectable CPC can preferably be applied with the assistance of rigid support from either implant or bone wall/chip to resist the deformed force from mastication movement. Hrefer reported cases using Norian skeletal repair system (SRS), a carbonated calcium phosphate bone cement, to augment the alveolar ridge as a single-stage procedure, with simultaneous placement of implants. Briefly a U-shaped vertical osteotomy was made in the mandible anterior region to split the atrophic alveolar ridge into two segments, and subsequently the CPC is applied as inlay filler between both segments. Thereafter, these segments are firmly fixed by either titanium miniplates or dental implants. In this way, the vertical dimension of the alveolar could be increased up to 5–7 mm (see Fig.1a, b). The prosthodontic procedure started 3 months later. In total 40 implants were inserted for 10 patients, and the follow-up period was 60 months. Overall no

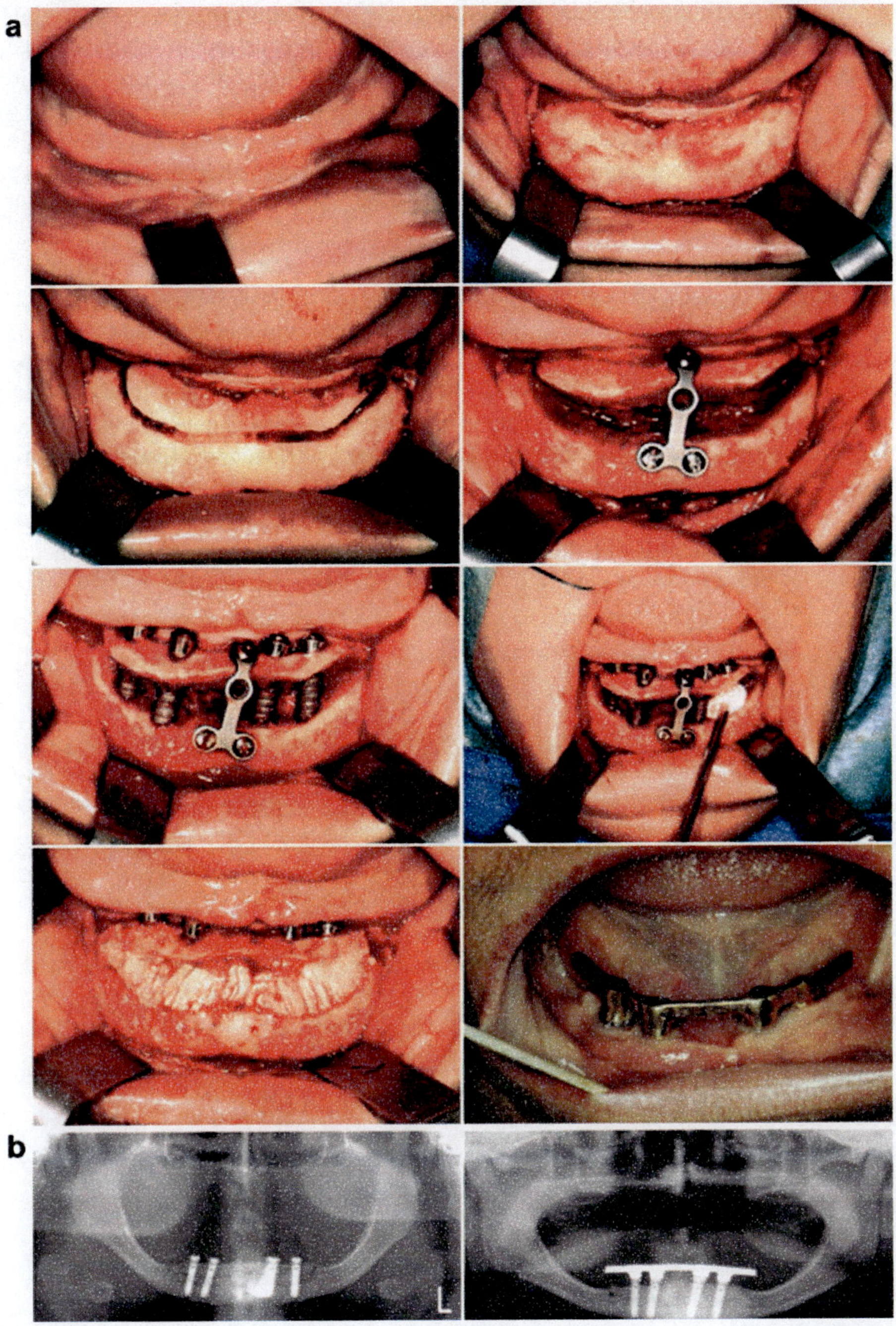

Fig. 1 (**a**) Surgical procedure related to vertical mandible reconstruction. (**b**) Postoperative panoramic examination after augmentation of the atrophic alveolar ridge. (F. Hölzle et al. 2011, reprinted with permission)

fractures or dislocations of implant developed, only one of the implants was lost, and there was one wound dehiscence, but no surgical intervention or revision was necessary. Radiographs showed good consolidation of the bony structure in all cases (Wolff et al. 2004; Hölzle et al. 2011).

In the situation where the initial fixation and subsequent protection of the CPC from this sandwich bone segment technique are not available, attempt using CPC as onlay augmentation on reconstruction of a large mandibular defect has been reported by Stanton. A 10-year-old boy suffered from odontogenic keratocysts (OKC), and a large defect presented in the mandible after giant cyst removed operation including enucleation of the cyst, extraction of all teeth involved in the lesion, and peripheral osteotomy. The remaining bony wall was eggshell thin without perforation. After smoothing and irrigating, Norian bone cement (monocalcium phosphate monohydrate, calcium carbonate, and alpha-tricalcium phosphate powder with a sodium phosphate–buffered solution) was injected to fill the void created from the ascending ramus to the midbody of the mandible, the mucosa subsequently healed, and an uneventful recovery occurred. No recurrence of OKC has been observed 3 years post-surgery. Serial panoramic radiographs have displayed progressive resorption of Norian and replacement with bone. The mandibular height and form have been successfully preserved (see Figs.2a,b and 3a,b) (Stanton et al. 2004).

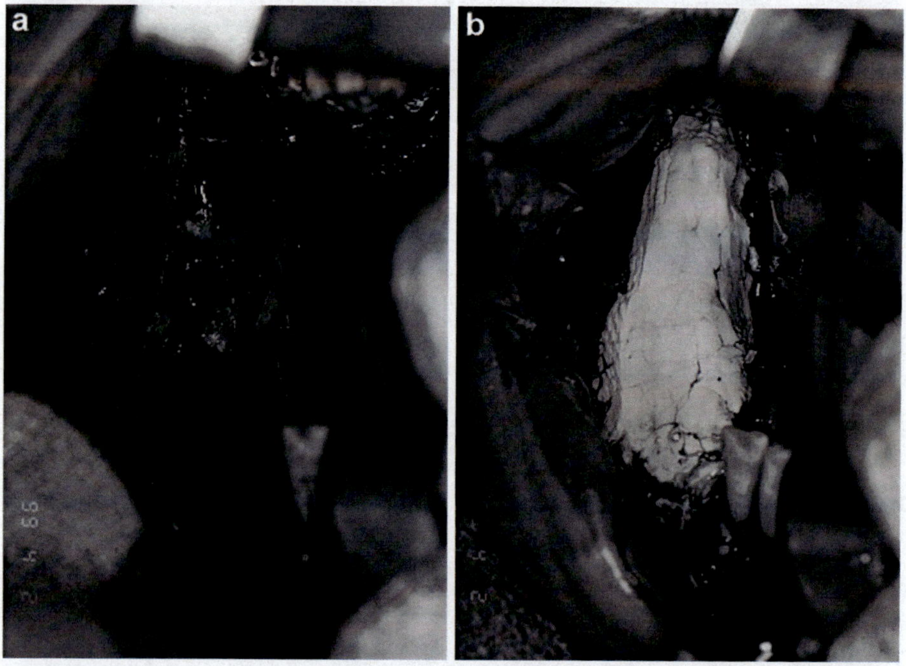

Fig. 2 (**a**) Intraoperative view of inferior border of mandible prior to application of injectable calcium phosphate cement. (**b**) Intraoperative photograph showing injectable calcium phosphate bone cement filling the void created by the lesion. (D.C. Stanton et al. 2004, reprinted with permission)

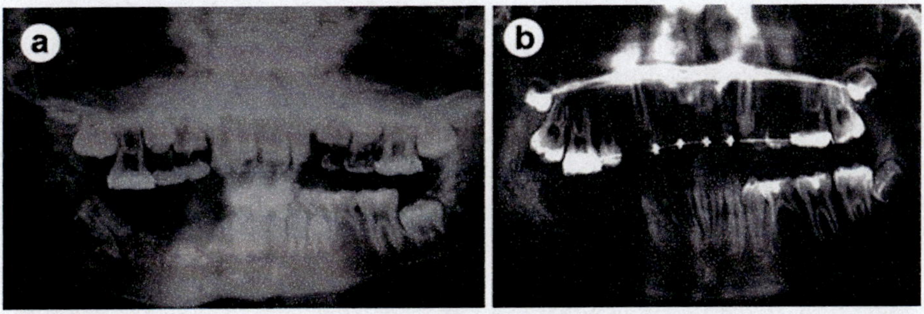

Fig. 3 (**a**) Six-month postoperative panoramic radiograph showing gradual replacement of the graft by bone. (**b**) Three-year postoperative panoramic radiograph showing complete replacement of the graft material by bone, good bone height, and contour with no recurrence of odontogenic keratocyst. (D.C. Stanton et al. 2004, reprinted with permission)

4.5 Reconstruction of the Craniofacial Region

Frequently used commercial CPCs mentioned before such as BoneSource®, Mimex®, Norian®, etc. have been clinically applied in the craniofacial complex for several decades. Successful clinical use of CPCs in this region is restricted to non-load-bearing areas on mature craniofacial skeleton, including reconstruction of full-thickness cranial defects (Mahr et al. 2000), frontal and temporal contouring (Friedman et al. 2000; Chen et al. 2004; Gosain et al. 2009), onlay grafting for augmentation and smoothing contours of skeletal irregularities (Gosain 1997; Jackson and Yavuzer 2000), depressed frontal sinus fracture (Sundaram et al. 2006; Luaces-Rey et al. 2009), nasal augmentation (Okada et al. 2004; Hatoko et al. 2005), or secondary craniofacial contouring (Magee et al. 2004; Van der Stok et al. 2011), with minimal evidence of bone replacement in relatively long-term follow-up of small sample size study design. Typically Gosain reported that cement pastes were used for onlay augmentation to the cranial vault in eight patients, hydroxyapatite (BoneSource®) used in five patients, and calcium phosphate bone cement (Norian® CRS) used in three patients. Patient mean age at implantation was 5.5 years (range, 4–8 years); the mean follow-up term was 5.7 years (range, 1–8 years). All patients had postoperative computed tomographic scans taken 1 year later that demonstrated persistent skeletal augmentation, with a computed tomographic density equivalent to that of the adjacent bone. The only complication in this group was postoperative infection in one patient necessitating partial removal of the implant. The authors concluded that when used for onlay augmentation in the patient where further skeletal growth is negligible (after age 3 in the cranial vault and after age 14 in the facial skeleton), good results could be achieved. Hydroxyapatite cement was also used in a girl with recurrent left-sided forehead recession and frontal bone irregularities. Fronto-orbital advancement at 4 years of age was performed and hydroxyapatite cement paste was applied for augmentation of irregular depressions in the frontal bone with restoration of forehead symmetry and contour. Three yeras later (age 7 years), the patient required a repeat neurosurgical procedure, at which time the hydroxyapatite onlay was found to be incorporated into the surrounding calvaria but was not been replaced with native bone. Histologic section of a biopsy specimen of the hydroxyapatite onlay demonstrated a rim of bone that had grown around the periphery of the implant, demonstrating good incorporation but no evidence of bone ingrowth within the implant (Gosain et al. 2009).

The promising safety and long-term efficacy of this bone substitute for the repair of craniofacial bone defects in the growing pediatric skull were also reported in 2005. Eight patients who underwent reconstruction of cranial defects using hydroxyapatite cement between the ages of 25 and 100 months (mean, 55 months) were followed up postoperation between 23 and 72 months (mean, 38 months). No mortalities or significant morbidities were encountered in the study population. It has been the authors' experience that hydroxyapatite cement is both biocompatible and resistant to infection when used in sites not contiguous with sinus mucosa and that it is a good alternative to autogenous bone in pediatric craniofacial reconstruction (David et al. 2005).

However, relatively longer-term follow-up and larger sample size studies revealed unfavorable outcome for the use of CPC in cranioplasty compared to the studies mentioned above. Jackson reported a retrospective study of 312 patients who had 449 procedures performed by a single surgeon to reconstruct a calvarial deformity between 1981 and 2001. The main reason for cranioplasty (32.4%) was post-tumor resection deformity, and the main surgical site was the frontal bone (53.2%). Three different materials including autogenous cranial bone, PMMA, and HA cement (BoneSource, Mimex) were used. The median postoperative follow-up was 3.3 years (range, 0.2–10 years). The following variables were assessed to evaluate the outcome: gender, age, indication for surgery, site of cranioplasty, type of material, number of surgeries performed, and complications. The eventual outcome was based on the patient's subjective evaluation of satisfaction during the follow-up period and the occurrence of complications and the need for further surgeries for either revision or improvement. The overall complication rate was 23.6%, in terms of complications and material used; autogenous bone was involved in 20.5% of the complicated cases, PMMA in 29.3%, and HA cement in 32.8%. In the HA cement group, infection and/or extrusion of the material represented 22.4% ($n = 13$ of 58) of the complications. These findings suggested that bone graft and PMMA are still the best materials in calvarial reconstruction. The application of HA cement in craniofacial reconstruction needs to be carefully considered (Moreira-Gonzalez et al. 2003). It seems the infection rate could be higher when CPCs were used for secondary surgery in growing skeleton region at Wong's retrospective chart review. Twenty patients who underwent secondary forehead cranioplasty with hydroxyapatite cement (Norian CRS) were included. Basic demographics including age, sex, and diagnosis were identified, and characteristics of the defects were recorded including size, location, and depth (full vs. partial thickness). The postoperative course was analyzed for length of follow-up and the presence of infections. The result revealed that three patients were lost to follow-up, and all patients had initially acceptable aesthetic results. Of the 17 patients, 10 (59%) ultimately had infectious complications. Infection occurred on a mean of 17.3 months after surgery (range, 4 month–4 year), and the mean amount of hydroxyapatite used was 32.5 mL (infections) versus 14.3 mL (no infections). Of the ten patients with complications, secondary forehead asymmetry was the most common presentation. Nine patients required surgical debridement and subsequent delayed reconstruction. The use of hydroxyapatite cement in secondary reconstruction has yielded unacceptably high infection rates leading to discontinuation of its use in this patient population. Calcium phosphate cements were thus considered as "off-label" used in the growing craniofacial skeleton (Wong et al. 2011).

Special caution has to be taken since complication rate is very high when CPCs are used for large craniofacial full-thickness defects. Zins summarized 16 patients who underwent correction of large, full-thickness (>25 cm^2) skull defects. The surgical technique included reconstruction of the floor of the defect with rigid fixation to the surrounding native bone, interposition of the cement to ideal contour, and closure of the defect. The mean patient age was 35 years (range, 1–69 years), and the mean defect area was 66.4 cm^2 (range, 30–150 cm^2). Cases were equally divided between BoneSource and Norian CRS. The mean amount of bone cement used was 80 g. Follow-up varied between 1 and 6 years (mean, 3 years). Major complications occurred in 8 of 16 patients, with 1 occurring as late as 6 years postoperatively (Zins et al. 2007). These implanted large amounts of CPC are hardly resorb, which could be confirmed by biopsies taken in another study's reoperations after major complication. Microfragmentation of the cement was often observed inside the recipient site, while the amount of bone replacement was only limited to the peripheral area of the materials. The amount of vascularized bone tissue inside the set CPC was generally very low (Tuncer et al. 2004).

Fragility is another problem for CPC used in large full-thickness cranial defects; to overcome this, there are also attempts to combine CPCs with reinforcements such as tough frames or meshes made by one of the following materials: polylactic acid (PLA) sheets (Cohen et al. 2004), degradable meshes (Greenberg and Schneider 2005), tantalum (Durham et al. 2003), titanium meshes and plates (Van der Stok et al. 2011), etc.

At large area where the CPC was prone to infection due to imperfect closure of soft tissue, or acute adverse events such as seroma and cellulitis formation after surgery, the use of CPC loaded with antibiotic agents, such as cephalothin, was suggested (Matic and Manson 2004; Burstein et al. 2006).

To summarize, the contraindication for the use of CPC in the reconstruction of craniofacial skeleton region, such as infected field, stress-bearing applications, inlay use for reconstruction of large, full-thickness calvarial defects, areas surrounding nonviable bone, abnormal calcium metabolism, metabolic bone disease, immunologic abnormalities, and poor wound healing (Gómez et al. 2005), has to be avoided in all clinical situation. With careful treatment plan in these challenging conditions, injectable CPC can still be considered as favorable bone substitute for the reconstruction of craniofacial area when applied with caution (Gilardino et al. 2009; Afifi et al. 2010).

5 Conclusion and Outlook

Theoretically injectable calcium phosphate cements (CPCs) are highly useful for reconstruction of bone defects in the oral and cranio-maxillofacial region. However, the evidence for successful application of CPC is not unambiguously shown for all clinical indications. Consequently, prospective long-term clinical studies are needed to obtain insight in the long-term behavior of CPC in human patients. These clinical studies have to form the translation basis of further improvements of the material properties of CPCs. In that respect, the strategic priority is firstly to develop CPC that allows for the creation of sufficient mechanical stability at the recipient site – even in non-load-bearing skeletal sites – by optimal formulation of CPC to accelerate setting time of CPC in which more strong washout resistant could be accomplished, which will certainly improve the outcome in many compromised clinical situations. Secondly, development of optimal interconnection porosity inside injectable calcium phosphate cement or optimal degradable CPC formulation which could facilitate vascularization and degradation inside set bulk CPC is also critical for the success of CPC clinical application. And thirdly, introduction of antibiotic effectively aimed at the prevention of pathologic contamination is equally important for successful clinical performance of injectable CPC.

References

Afifi AM, Gordon CR, Pryor LS, Sweeney W, Papay FA, Zins JE (2010) Calcium phosphate cements in skull reconstruction: a meta-analysis. Plast Reconstr Surg 126(4):1300–1309

Atieh MA, Alsabeeha NH, Payne AG, Duncan W, Faggion CM, Esposito M (2015) Interventions for replacing missing teeth: alveolar ridge preservation techniques for dental implant site development. Cochrane Database Syst Rev 5:Cd010176

bohner m (2010) Design of ceramic based cement and putties for bone graft substitution. Eur Cell Mater 20:1–12

Brown GD, Mealey BL, Nummikoski PV, Bifano SL, Waldrop TC (1998) Hydroxyapatite cement implant for regeneration of periodontal osseous defects in humans. J Periodontol 69(2):146–157

Burstein FD, Williams JK, Hudgins R, Boydston W, Reisner A, Stevenson K, Cohen S (2006) Hydroxyapatite cement in craniofacial reconstruction: experience in 150 patients. Plast Reconstr Surg 118(2):484–489

Chen TM, Wang HJ, Chen SL, Lin FH (2004) Reconstruction of post-traumatic frontal-bone depression using hydroxyapatite cement. Ann Plast Surg 52(3):303–308

Cohen AJ, Dickerman RD, Schneider SJ (2004) New method of pediatric cranioplasty for skull defect utilizing polylactic acid absorbable plates and carbonated apatite bone cement. J Craniofac Surg 15(3):469–472

Constantz BR, Ison IC, Fulmer MT, Poser RD, Smith ST, VanWagoner M, Ross J, Goldstein SA, Jupiter JB, Rosenthal DI (1995) Skeletal repair by in situ formation of the mineral phase of bone. Science 267(5205):1796–1799

David L, Argenta L, Fisher D (2005) Hydroxyapatite cement in pediatric craniofacial reconstruction. J Craniofac Surg 16(1):129–133

Durham SR, Gordon MJ, Levy ML (2003) Correction of large (>25 cm^2) cranial defects with "reinforced" hydroxyapatite cement: technique and complications. Neurosurgery 52(4):842–845

Ettl T, Gerlach T, Schusselbauer T, Gosau M, Reichert TE, Driemel O (2010) Bone resorption and complications in alveolar distraction osteogenesis. Clin Oral Investig 14(5):481–489

Felice P, Pistilli R, Lizio G, Pellegrino G, Nisii A, Marchetti C (2009) Inlay versus onlay iliac bone grafting in atrophic posterior mandible: a prospective controlled clinical trial for the comparison of two techniques. Clin Implant Dent Relat Res 11(Suppl 1):e69–e82

Fernandez AC, Mohanty M, Varma HK, Komath M (2006) Safety and efficacy of Chitra-CPC calcium phosphate cement as bone substitute. Curr Sci 91(12):1678–1686

Friedman CD, Costantino PD, Synderman CH, Chow LC, Takagi S (2000) Reconstruction of the frontal sinus and frontofacial skeleton with hydroxyapatite cement. Arch Facial Plast Surg 2(2):124–129

Gilardino MS, Cabiling DS, Bartlett SP (2009) Long-term follow-up experience with carbonated calcium phosphate cement (Norian) for Cranioplasty in children and adults. Plast Reconstr Surg 123(3):983–994

Gómez E, Martín M, Arias J, Carceller F (2005) Clinical applications of Norian SRS (calcium phosphate cement) in craniofacial reconstruction in children: our experience at hospital La Paz since 2001. J Oral Maxillofac Surg 63(1):8–14

Gosain AK (1997) Hydroxyapatite cement paste cranioplasty for the treatment of temporal hollowing after cranial vault remodeling in a growing child. J Craniofac Surg 8(6):506–511

Gosain AK, Chim H, Arneja JS (2009) Application-specific selection of biomaterials for pediatric craniofacial reconstruction: developing a rational approach to guide clinical use. Plast Reconstr Surg 123(1):319–330

Greenberg BM, Schneider SJ (2005) Alloplastic reconstruction of large Cranio-orbital defects. Ann Plast Surg 55(1):43–51

Hatoko M, Tada H, Tanaka A, Yurugi S, Niitsuma K, Iioka H (2005) The use of calcium phosphate cement paste for the correction of the depressed nose deformity. J Craniofac Surg 16(2):327–331; discussion 331

Hölzle F, Bauer F, Kesting MR, Mücke T, Deppe H, Wolff K-D, Swaid S (2011) Single-stage implantation in the atrophic alveolar ridge of the mandible with the Norian skeletal repair system. Br J Oral Maxillofac Surg 49(7):542–545

Horowitz R, Holtzclaw D, Rosen PS (2012) A review on alveolar ridge preservation following tooth extraction. J Evid Based Dent Pract 12(3 Suppl):149–160

Jackson IT, Yavuzer R (2000) Hydroxyapatite cement: an alternative for craniofacial skeletal contour refinements. Br J Plast Surg 53(1):24–29

Magee WP Jr, Ajkay NN, Rosenblum RS (2004) Use of fast-setting hydroxyapatite cement for secondary craniofacial contouring. Plast Reconstr Surg 114(114):289–297

Kotsakis GA, Salama M, Chrepa V, Hinrichs JE, Gaillard P (2014) A randomized, blinded, controlled clinical study of particulate anorganic bovine bone mineral and calcium phosphosilicate putty bone substitutes for socket preservation. Int J Oral Maxillofac Implants 29(1):141–151

Luaces-Rey R, Garcia-Rozado A, Crespo Escudero JL, Seijas BP, Arenaz Bua J, Lopez-Cedrun JL (2009) Use of carbonated calcium phosphate bone cement and resorbable plates for the treatment of frontal sinus fractures: two case reports. J Plast Reconstr Aesthet Surg 62(2):272–273

Maestre-Ferrin L, Boronat-Lopez A, Penarrocha-Diago M, Penarrocha-Diago M (2009) Augmentation procedures for deficient edentulous ridges, using onlay autologous grafts: an update. Med Oral Patol Oral Cir Bucal 14(8):e402–e407

Mahr MA, Bartley GB, Bite U, Clay RP, Kasperbauer JL, Holmes JM (2000) Norian craniofacial repair system bone cement for the repair of craniofacial skeletal defects. Ophthal Plast Reconstr Surg 16(5):393–398

Matic DB, Manson PN (2004) Biomechanical analysis of hydroxyapatite cement cranioplasty. J Craniofac Surg 15(3):415–422; discussion 422–413

Mazor Z, Peleg M, Garg AK, Chaushu G (2000) The use of hydroxyapatite bone cement for sinus floor augmentation with simultaneous implant placement in the atrophic maxilla. A report of 10 cases. J Periodontol 71(7):1187–1194

Moreira-Gonzalez A, Jackson IT, Miyawaki T, Barakat K, DiNick V (2003) Clinical outcome in cranioplasty: critical review in long-term follow-up. J Craniofac Surg 14(2):144–153

Okada E, Maruyama Y, Hayashi A (2004) Nasal augmentation using calcium phosphate cement. J Craniofac Surg 15(1):102–105

Rajesh JB, Nandakumar K, Varma HK, Komath M (2009) Calcium phosphate cement as a "barrier-graft" for the treatment of human periodontal intraosseous defects. Indian J Dent Res 20(4):471–479

Reynolds MA, Aichelmann-Reidy ME, Branch-Mays GL, Gunsolley JC (2003) The efficacy of bone replacement grafts in the treatment of periodontal osseous defects. A systematic review. Ann Periodontol 8(1):227–265

Saulacic N, Zix J, Iizuka T (2009) Complication rates and associated factors in alveolar distraction osteogenesis: a comprehensive review. Int J Oral Maxillofac Surg 38(3):210–217

Shirakata Y, Setoguchi T, Machigashira M, Matsuyama T, Furuichi Y, Hasegawa K, Yoshimoto T, Izumi Y (2008) Comparison of injectable calcium phosphate bone cement grafting and open flap debridement in periodontal intrabony defects: a randomized clinical trial. J Periodontol 79(1):25–32

Simion M, Trisi P, Piattelli A (1994) Vertical ridge augmentation using a membrane technique associated with osseointegrated implants. Int J Periodontics Restorative Dent 14(6):496–511

Simion M, Jovanovic SA, Trisi P, Scarano A, Piattelli A (1998) Vertical ridge augmentation around dental implants using a membrane technique and autogenous bone or allografts in humans. Int J Periodontics Restorative Dent 18(1):8–23

Stanton DC, Chou JC, Carrasco LR (2004) Injectable calcium-phosphate bone cement (Norian) for reconstruction of a large mandibular defect: a case report. J Oral Maxillofac Surg 62(2):235–240

Sundaram K, Robertson JM, Gibbons AJ (2006) Calcium phosphate cement in late reconstruction of frontal sinus fractures. J Oral Maxillofac Surg 64(3):564–565

Sverzut AT, Rodrigues DC, Lauria A, Armando RS, de Oliveira PT, Moreira RW (2015) Clinical, radiographic, and histological analyses of calcium phosphate cement as filling material in maxillary sinus lift surgery. Clin Oral Implants Res 26(6):633–638

Taschieri S, Rosano G, Weinstein T, Del Fabbro M (2010) Replacement of vertically root-fractured endodontically treated teeth with immediate implants in conjunction with a synthetic bone cement. Implant Dent 19(6):477–486

Tuncer S, Yavuzer R, Isik I, Basterzi Y, Latifoglu O (2004) The fate of hydroxyapatite cement used for cranial contouring: histological evaluation of a case. J Craniofac Surg 15(2):243–246

Van der Stok J, Van Lieshout EM, El-Massoudi Y, Van Kralingen GH, Patka P (2011) Bone substitutes in the Netherlands – a systematic literature review. Acta Biomater 7(2):739–750

Wolff KD, Swaid S, Nolte D, Bockmann RA, Holzle F, Muller-Mai C (2004) Degradable injectable bone cement in maxillofacial surgery: indications and clinical experience in 27 patients. J Craniomaxillofac Surg 32(2):71–79

Wong RK, Gandolfi BM, St-Hilaire H, Wise MW, Moses M (2011) Complications of hydroxyapatite bone cement in secondary pediatric craniofacial reconstruction. J Craniofac Surg 22(1):247–251

Zins JE, Moreira-Gonzalez A, Papay FA (2007) Use of calcium-based bone cements in the repair of large, full-thickness cranial defects: a caution. Plast Reconstr Surg 120(5):1332–1342

Tissue-Engineered Teeth

Zihan Li, Weibo Zhang, and Pamela C. Yelick

Contents

Abstract

Clinical practice in the field of dentistry has remained largely unchanged for over a century. Recently, significant advances in the fields of tissue engineering and regenerative medicine (TERM) have provided new opportunities for dental therapies to advance in ways that will provide patients with more effective therapies to regenerate living dental tissue, while at the same time preserving natural dental tissues as much as possible. Since regenerative dental therapies are based on knowledge and understanding of natural tooth development, here we first describe early tooth development, including morphogenesis of tooth crown and root structures, and review new, relevant findings in tooth development biology. With respect to regenerative approaches for dental tissue repair, we next describe the three components recognized as doctrine in tissue engineering strategies – dental stem

Z. Li · W. Zhang · P. C. Yelick (✉)
Tufts University, Boston, MA, USA
e-mail: Zihan.li@tufts.edu; Weibo.zhang@tufts.edu; Pamela.yelick@tufts.edu

D. Eberli et al. (eds.), *Organ Tissue Engineering*, Reference Series in Biomedical Engineering, https://doi.org/10.1007/978-3-030-44211-8_10

cells, scaffolds, and growth factors/signaling molecules/cytokines – and recent findings in each. We next review the use of dental stem cells for applications in tooth regeneration, including highlights on the use of innovative and promising scaffold materials and growth factors. We discuss the significant breakthrough and discovery in 2006 of induced pluripotent stem cells (iPSCs), and how this stem cell technology demonstrated the possibility of using a patient's own reprogrammed cells to regenerate new tissues and organs. We next describe promising partial tooth regeneration strategies, including regeneration of the dentin-pulp complex, the periodontium, and tooth root regeneration. Finally, we describe exciting progress in whole tooth regeneration strategies, focusing on three-dimensional (3D) tissue engineering strategies.

1 Introduction

Tooth loss, the most common organ failure in humans, can result from congenital malformations or genetic disorders, from common disorders such as dental caries and periodontal disease, and from accidental or battlefield injuries to the mouth and face (Young et al. 2002). The tooth organ itself consists of highly complex, organized, and precisely patterned mineralized tissue matrices that are functionally integrated with soft dental tissues, including dental pulp, periodontal ligament, nerves, and vasculature (Ten Cate 1998) (Fig. 1a). The major hard tissues of a tooth include enamel, dentin, and cementum, while soft dental tissues include dental pulp and periodontal ligaments. Human teeth are ectodermally derived organs that exhibit only limited regeneration potential. Mineralized enamel cannot regenerate primarily due to the fact the dental epithelium, which forms enamel in a naturally developing tooth, is lost prior to tooth eruption (Moradian-Oldak 2012), while dentin exhibits a very limited capacity for reparative dentin formation in response to injury (Song et al. 2017).

Current commonly used dental therapies consist of the following. Dental implants are the principle restorative therapy used to replace lost teeth, despite the fact that synthetic implants have none of the characteristics of natural, living teeth (Ferreira et al. 2007). Dental implants are also susceptible to a variety of insults including inflammation of the hard and soft tissues surrounding the implant, called peri-implantitis, which can potentially lead to implant failure (Smeets et al. 2014). Bone grafts, commonly used to repair and reinforce load-bearing jaw bone, have been extensively investigated although this approach does not regenerate functional dental tissues (Reynolds et al. 2003).

Another traditional surgical dental interventions used for periodontal tissue repair primarily focuses on removing diseased tissues via open flap debridement, followed by scaling and root planning (Becker et al. 1986; Brayer et al. 1989). These procedures can easily result in long-term junctional epithelium detachment, leading to a compromised gingival seal around the tooth, making it highly susceptible to bacterial infection (Ivanovski 2009). Another common dental tissue regeneration therapy used in the clinic is guided tissue regeneration (GTR), most commonly used for

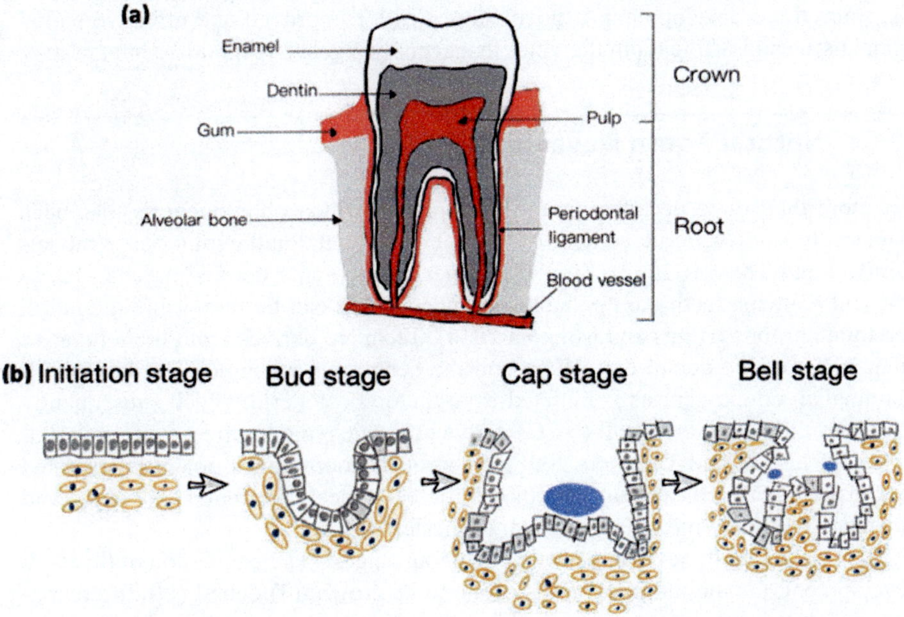

Fig. 1 Anatomy of the tooth and developmental stages of tooth morphogenesis. (**a**) Anatomy of the tooth. The enamel is the calcified tissue in the crown of the tooth. Pulp contains connective tissue, blood vessels, and nerves. (**b**) Four different developmental stages during tooth morphogenesis

periodontal tissue repair as first proposed by Melcher in the 1970s (Melcher 1976). The goal of GTR is to regenerate tight PDL tissue attachment to the tooth roots (Nyman et al. 1982a, b) via application of a tissue barrier membrane to guide the migration of cementogenic and osteogenic stem cells to the defect site (Ivanovski 2009). However, such membranes can become problematic, as early non-resorbable membranes required removal after transplantation, potentially leading to post-operational complications (Murphy 1995). A newer resorbable membrane consisting of collagen and polylactic/polyglycolic acid has also shown complications in that it can inhibit PDL tissue healing (Sculean et al. 2007).

Current approaches used to treat severe dental caries can include endodontic treatment, which involves completely removing all dental pulp tissue and subsequently replacing it with synthetic, inert cement (Parirokh et al. 2018; Torabinejad et al. 2018). During the tooth restoration step, two types of pulp capping methods can be used – direct or indirect pulp capping. Direct pulp capping involves the use of a protective material placed directly on the exposed pulp, while indirect pulp capping requires the presence of thin layer of residual dentin to avoid pulp exposure (European Society of Endodontology 2006; Hilton 2009). Unfortunately, neither method facilitates the regeneration of the dentin-pulp complex, and direct capping results in the permanent loss of the dental pulp tissue.

Together, the limited efficacy of these currently used dental tissue repair therapies has spurred the development of novel strategies for improved and effective partial dental tissue and whole tooth therapies that actually regenerate natural dental tissues.

2 Natural Tooth Development

For more than three decades, tooth development, called odontogenesis, has been extensively studied in the context of both developmental and evolutionary models (Jernvall and Thesleff 2012). Similar to other ectodermally derived organs such as hair and feathers, teeth are epithelial appendages that can be found throughout all vertebrate groups (Biggs and Mikkola 2014). Teeth are derived from the reiterative interactions of the dental epithelium and the dental mesenchyme (Thesleff 2003). Mammalian odontogenesis is initiated from the oral ectoderm, which subsequently signals to the neural crest cell (NCC) derived mesenchyme to direct tooth morphogenesis (Thesleff and Tummers 2008). Subsequent tooth development is mediated via crosstalk between two major tooth specific-cell types – the dental epithelium and the dental mesenchyme (Hurmerinta and Thesleff 1981).

Mammalian tooth development occurs in four stages: (1) the initiation of the tooth development; (2) the morphogenesis of the tooth crown; (3) dental cell differentiation; and (4) the maintenance of dental stem cells that support tooth development (Balic and Thesleff 2015). In the mouse, a simple dentition pattern consists of one central incisor and three molars in each quadrant, with continuously erupting incisors and no molar replacement teeth (Lumsden 1979; Peterkova et al. 1996; Viriot et al. 1997). Humans replace their baby, or deciduous, teeth with a second set of adult teeth that include additional tooth types including canines and premolars (Vastardis 2000). However, both mice and human dentition share similar development patterns during odontogenesis (Yu et al. 2015). The availability of a wide variety of transgenic mouse genetic knock out and reporter lines makes mice the most commonly used animal model to study tooth development (Kantarci et al. 2015). The signaling pathways regulating the complex interactions between dental epithelial and mesenchymal cells and tissues are very dynamic (Thesleff and Tummers 2008). As such, tooth organogenesis can be considered as a stepwise process where morphogenesis and cell differentiation occur through reciprocal and sequential interactions.

2.1 Morphogenesis of the Tooth Crown

The initiation stage of tooth development begins with the appearance of the *primary dental laminae*, also called odontogenic bands, which are essentially stripes of thickened epithelium that give rise to future teeth (Mina and Kollar 1987; Lumsden 1988). The thickened dental epithelium then invaginates into the underlying mesenchymal to form *placodes*. During the bud stage of tooth development, the dental papilla forms as the dental mesenchyme condenses beneath the invaginating dental epithelium around mouse embryonic day 13 (E13) (Fig. 1). The expression of dental lamina genes is restricted to the placodes, where transcription factors such as *paired like homeodomain 2* (*pitx2*) are expressed (Oosterwegel et al. 1993). The dental epithelial placodal cells express four conserved signaling molecules – *Shh*, *Wnt10*, *BMP2,* and *FGF20* – marking the sites of the future teeth (Haara et al. 2012; Jussila and Thesleff 2012).

Next, tooth bud stage dental epithelium proliferates, giving rise to cap and then bell stage tooth buds (Fig. 1). During the cap stage, a cluster of undifferentiated cells located at the inner enamel epithelium, the primary enamel knot, mark the future tooth cusp (Balic and Thesleff 2015). In multi-cusped mammalian teeth, primary and then secondary enamel knots mark the sites of multi-cusped tooth patterns. The enamel organ forms during the cap stage and is comprised of two layers of cells, the inner and outer enamel epithelium (Balic and Thesleff 2015). Demarcated by the cervical loop and the dental papilla, the basal epithelial cell layer of the cervical loop bordering the dental papilla is known as the inner enamel epithelium (iee), while the epithelial layer facing the dental follicle is known as the outer enamel epithelium (oee) (Harada et al. 1999; Tummers and Thesleff 2003; Thesleff and Tummers 2008). The inner enamel epithelium differentiates into ameloblasts that will secrete enamel matrix, and the mesenchymal cells differentiate into odontoblasts that produce dentin, all prior to tooth eruption (Jernvall and Thesleff 2012).

Tooth crown morphogenesis occurs in bell stage teeth, where the enamel knot signaling centers functions direct the height, location, and the number of the developing tooth cusps (Balic and Thesleff 2015) (Fig. 1b). In the late bell stage, dental mesenchymal cell-derived odontoblasts and dental epithelial cell-derived ameloblasts secrete matrix that will produce dentin and enamel, respectively (Balic 2018).

During tooth development, the cervical loop serves as a stem cell niche that produces daughter cells that differentiate into enamel-forming ameloblasts. Growth factors such as FGF10, expressed in the dental mesenchyme, and its receptor Fgfr2b, expressed in the dental epithelium, are both required for dental stem cell proliferation (Harada et al. 2002). Shh has been shown to be essential for dental stem cell replication and recruitment, but not for dental stem cell survival (Suomalainen and Thesleff 2010). In general, the same signaling pathway networks regulating the stem cell niche of the mouse continuously erupting incisor are also active in the stem cell niche of developing molar teeth (Tummers and Thesleff 2003), implying that largely conserved signaling pathways regulate the development of all teeth (Jernvall and Thesleff 2012).

It is worth noting that mammals have a very limited capacity to replace their teeth, as compared to other vertebrates such as reptiles and fish (Davit-Beal et al. 2009; Jernvall and Thesleff 2012). The majority of mammals, including humans, can replace their teeth only once. However, the renewal and maintenance of many dental tissues is supported by stem cells, which in turn are regulated by a variety of growth factor signaling families (Jernvall and Thesleff 2012).

2.1.1 Cells and Signaling Pathways During Tooth Development

Tooth organogenesis can be considered a stepwise process where tooth morphogenesis and dental cell differentiation occur through reciprocal and sequential dental epithelial-mesenchymal cell interactions (Thesleff and Tummers 2008). Five major signaling pathways are involved in odontogenesis including Wnt, bone morphogenetic protein (BMP), fibroblast growth factor (FGF), Sonic Hedgehog (Shh), and Ectodysplasin (Eda) – see Table 1. Together, these signaling pathways mediate the tissue interactions that lead to tooth formation.

BMP Signaling

BMPs are members of a very large family of the transforming growth factor β (TGF-β) superfamily, that consist of homodimer proteins (Valera et al. 2010). BMPs are involved in both the early and later developmental stages of tooth organogenesis (Yuan et al. 2015; Graf et al. 2016). Specifically, BMP2 is expressed in dental epithelium and BMP7 is expressed in dental mesenchyme of initiation stage tooth buds (Malik et al. 2018). Both BMP2 and BMP7 also promote early tooth mineralization (Malik et al. 2018), while BMP4 regulates the formation of the epithelial root sheath (Hosoya et al. 2008). BMP9 is notably responsible for promoting odontoblastic and osteogenic differentiation (Huang et al. 2019).

Shh Signaling

Sonic hedgehog (Shh) signaling is required for embryonic mouse tooth initiation, as well as to maintain differentiation of dental epithelial cells into ameloblasts (Gritli-Linde et al. 2002). Shh is expressed in the oral epithelium prior to invagination and in the dental epithelium during tooth development. Together with Wnt10, BMP2,

Table 1 Five major signaling pathways regulating tooth development

Signaling pathways	Key cytokines	Functions	References
BMP	BMP2	Early tooth morphogenesis and mineral secretion, found in odontoblasts	Malik et al. (2018)
	BMP7	Similar to BMP2 but also found in ameloblasts	Gao et al. (2018)
	BMP9	Promotes odontoblast differentiation and osteogenic differentiation	Huang et al. (2019)
WNT	WNT3a	Stimulates cementoblasts formation	Nemoto et al. (2016)
	WNT7b	Positions the site of tooth formation	Sarkar et al. (2000)
FGF	FGF3	Regulate proliferation of epithelial stem cell progeny	Wang et al. (2007)
	FGF8	Required for tooth initiation	Trumpp et al. (1999)
SHH	SHH	Promotes epithelial cell proliferation and are expressed throughout dental mesenchyme and epithelium	Bitgood and McMahon (1995), Cobourne et al. (2001)
EDA	EDA	Required for the development of ectodermal organ during initiation from placodes	Mustonen et al. (2004)

and FGF20, Shh expression is restricted to a cluster of placodal cells of the early signaling center (Haara et al. 2012; Jussila and Thesleff 2012). During the bud stage, the expression of Shh becomes restricted to the enamel knot, while during the cap stage, it is expressed in tissues surrounding the inner enamel epithelium (Vaahtokari et al. 1996). Loss of Shh results in arrested development of cap stage teeth and the formation of only a rudimentary tooth bud (Dassule et al. 2000). Together with the BMP signaling pathway, BMP/SHH signaling networks dictate the fate of dental epithelial stem cells in mouse molars and incisors (Li et al. 2015). Conditional Shh null mice exhibit severe craniofacial defects, including shortened height of molar placodes, indicating roles for Shh in organizing dental placodal cells. The canonical Shh pathway includes key transcription factors Gli1–3 (Hardcastle et al. 1998), and mice lacking Gli2 and Gli3 lack all molar tooth development, and form only a primitive central incisor tooth bud.

WNT Signaling

Both the canonical WNT/β-catenin pathway and the noncanonical Wnt signaling pathway play important roles in the early embryonic tooth development (Wang et al. 2014). WNT family members are expressed in the dental epithelium, and WNT7b is expressed in the oral epithelium when tooth patterns become clearly defined (Sarkar et al. 2000). Wnt/Shh signaling interactions define where the tooth is formed. In cap stage teeth, WNT 4 and WNT6 are expressed in the dental epithelium, while WNT5a, and signaling partners sFrp2 and sFrp3 are expressed in the dental mesenchyme (Sarkar and Sharpe 1999). WNT/β-catenin signaling pathways also mediate a variety of downstream signaling pathways in tooth development (Huang et al. 2020).

FGF Signaling

FGF signaling also plays important roles in odontogenesis. FGF8 is credited as the dental epithelial cell-originating factor (Trumpp et al. 1999), while FGF9 is essential for dental epithelial invagination, and initiates dental ectodermal organogenesis (Tai et al. 2012). In addition, FGF8 induces the expression of Pax9 in mouse tooth development, implying its essential role in tooth development beyond odontogenesis (Neubuser et al. 1997; Huang et al. 2020). FGF3, FGF4, FGF9, FGF15, and FGF20 are all expressed in the primary enamel knot signaling center (Porntaveetus et al. 2011).

EDA Signaling

Eda, a member of the tumor necrosis family (TNF), is a signaling molecule responsible for regulating the development of a variety of ectodermal appendages, including teeth and hair (Biggs and Mikkola 2014; Balic and Thesleff 2015). The receptor for Eda, Edar, is expressed in the dental *placode* (Haara et al. 2012; Balic and Thesleff 2015). Together with the intracellular adaptor protein Edaradd, they form a pathway leading to the downstream activation of the transcription factor NF-κB (Mikkola 2008). Eda is involved in the initial development of the tooth placode, where the Wnt/β-catenin pathway upregulates the expression of Edar, and Edar/NF-κB are required to maintain the expression of WNT10a/b (Zhang et al. 2009). In addition,

Eda has been shown to play a role in tooth bud morphogenesis, since Eda null embryos exhibit very small tooth buds (Pispa et al. 1999), although Eda is not required in later stages of tooth development (Swee et al. 2009). Eda is vital for the formation of placode during early tooth development (see Table 1), (Mustonen et al. 2004).

2.2 Tooth Root Development

2.2.1 Tooth Root Morphogenesis

Tooth root development is also regulated by crosstalk between the dental epithelium and mesenchyme (Thesleff and Sharpe 1997). Tooth root morphogenesis can be divided into root initiation and root elongation stages. After tooth crown formation, the cervical loop continues to grow and elongate after tooth crown formation, and eventually forms a double-layered epithelial structure called Hertwig's epithelial root sheath (HERS) (Ten Cate 1998). HERS is located between the dental papilla and the dental follicle and is generally thought to be the signaling center responsible for tooth root formation (Ten Cate 1998).

The cervical loop gives rise to the Hertwig's epithelial root sheath (HERS) via fusion of the outer and inner enamel epithelial cell layers, and serves as an important signaling center for tooth root formation (Huang et al. 2009a). Tooth root initiation occurs as the mesenchymal cell layer of the apical papilla comes in contact with the inner layer of the HERS, which signals mesenchymal cell differentiation into odontoblasts that form the radicular dentin that covers the tooth root (Li et al. 2017).

Interactions between the dental follicle and the newly formed dentin induce the differentiation of cementoblasts, which will produce both cementum and cementum-specific extracellular matrix such as collagen fibers (Zeichner-David 2006). HERS can give rise to cementoblasts by epithelial-to-mesenchymal transition (EMT) (Huang et al. 2009a), and also becomes the epithelial cell Rests of Malassez (ERM), which participates in cementum regeneration and repair (Xiong et al. 2013). The HERS is also responsible for the number of roots formed by a tooth, by forming protrusions downwards from the dental pulp cavity, which join horizontally to form a bridge, and then become divided to form individual tooth roots. As such, tooth root development is directed by the apical growth of HERS (Orban and Bhaskar 1980).

HERS is also heavily involved in periodontal ligament (PDL) formation, due in part to both HERS formation and degeneration (Li et al. 2017). As migrating dental follicle cells contact the HERS, PDL is formed between the tooth root and surrounding alveolar bone (Cho and Garant 2000). The PDL supports and anchors the root to the alveolar bone via collagen fibers secreted by dental follicle cells, thereby stabilizing the tooth for the forces of mastication.

2.2.2 Tooth Root Development Signaling Pathways

The discovery of a unique transcription factor, called *nuclear factor I C (Nfic),* revealed that distinct mechanisms mediate tooth crown and root formation (Steele-Perkins et al. 2003). Nfic was found to be required for tooth root formation but not crown formation, and NFIC-dependent and NFIC-independent signaling pathways have been characterized in recent years (Wang and Feng 2017). Nfic belongs to the nuclear factor I family and functions as a master regulator gene during tooth root dentin formation. The expression of Nfic is restricted to odontoblasts and pre-dontoblasts in developing molars in human and mice, where it participates in odontoblast differentiation by modulating TGFβ signaling pathway (Gao et al. 2014).

The canonical Wnt signaling pathway plays a crucial role in tooth root formation, where β-catenin-mediated Wnt signaling mediates odontoblast differentiation (Kim et al. 2013). It was determined that the integrity of Wnt signaling affects the formation of HERS, and in turn can compromise the vital epithelial-mesenchymal interactions necessary for proper tooth root development (Li et al. 2017). The Wnt signaling pathway also mediates tooth root formation through interactions with other conserved signaling pathways, including the canonical BMP signaling pathway, where BMP signaling is required to maintain the expression of Wnt signaling inhibitors (Li et al. 2011). In addition, Wnt10 was found to induce the expression of the odontoblast differentiation marker dentin sialo-phospho protein (Dspp), suggesting roles for Wnt10 in regulating tooth root dentinogenesis (Li et al. 2011, 2017). Interactions between Bmp and Wnt signaling also define the transition between tooth crown and tooth root formation (Yang et al. 2013).

Other conserved signaling pathways such as SHH and FGF are also crucial for tooth root development by their ability to mediate dental epithelial-mesenchymal cell interactions. Fgf10 regulates HERS formation during molar tooth development (Tummers and Thesleff 2003; Yokohama-Tamaki et al. 2006). Fgf2 is expressed in the apical furcation end of the tooth root where it directs odontoblast differentiation, and also in cementoblasts and fibroblasts of the PDL where it is thought to regulate tooth root development (Gao et al. 1996).

3 Cells and Scaffolds

3.1 Dental Stem Cells

Embryonic stem cells (ESCs), which are derived from the undifferentiated cells of the blastocyst, are capable of differentiating into all types of cell lineages derived from endoderm, mesoderm, and ectoderm (Thomson et al. 1998). It is not possible to use human ESCs in dental clinic applications due to serious concerns over their ethical use in nonlethal diseases (Yildirim et al. 2011). Therefore, stem cells currently available for applications in the dental clinic include dental mesenchymal cells, since the majority of dental epithelial cells disappear prior to tooth eruption (Huang et al. 2009b). Dental mesenchyme is also referred to as "ectomesenchyme," due to its derivation from the neural crest during early embryonic development (Huang et al. 2009b).

Currently available human dental stem/progenitor cells that can be harvested for applications in regenerative dentistry include: dental pulp stem cells (DPSCs); stem cells from human exfoliated deciduous teeth (SHED); periodontal ligament stem cells (PDLSCs); stem cells from apical papilla (SCAP); and dental follicle progenitor cells (DFPCs) (Huang et al. 2009b). These non-embryonic, postnatal cell populations exhibit mesenchymal-stem-like characteristics, including the ability to differentiate into multilineage cell populations (Huang et al. 2009b). Here we describe each of these five major dental stem cell types and their potential applications in regenerative dentistry.

3.1.1 Dental Pulp Stem Cells: DPSCs

DPSCs are a type of dental mesenchymal stem cell that can be isolated from dental pulp tissue (Alongi et al. 2010). The dental pulp is a highly vascularized and innervated tissue comprised of soft connective tissues, located in the central pulp cavity of the tooth crown and tooth roots (Ten Cate 1998). The dental pulp is encapsulated and protected by the highly mineralized dental tissues enamel, dentin, and cementum. DPSCs can proliferate and renew themselves, differentiate into odontoblasts, osteoblasts, and neurocytes both in vivo and in vitro under the right conditions (Gronthos et al. 2002), and were demonstrated to form both dentin and bone (Gronthos et al. 2000). One of the most prominent early studies of DPSCs by Gronthos and Batouli found that when seeded ex vivo onto hydroxyapatite/tricalcium phosphate (HA/TCP) and transplanted into immunocompromised mice, DPSCs showed formation of what closely resembled a dentine-pulp complex (Gronthos et al. 2000).

SHED, a type of stem cell isolated from human exfoliated deciduous teeth, possess similar osteogenic and odontogenic characters as DPSCs, but are more highly proliferative (Huang et al. 2009b). When expanded ex vivo and transplanted into immunocompromised mice, the formation of odontoblast-like cells associated with dentin-like structures was observed (Miura et al. 2003). In addition to higher proliferative and clonogenic ability, SHEDs can also differentiate into mesenchymal derivatives including neural cells and adipocytes (Miura et al. 2003). The neural-crest cell (NCC) origin of the dental pulp is reflected in the fact that SHED express neuronal and glial cell markers (Chai et al. 2000). Cultured SHED readily express neural cell markers such as βIII-tubulin, GAD, and NeuN, whose expression increases upon stimulation with neurogenic medium (Huang et al. 2009b). The neurogenic potential of SHED was highlighted by the fact that SHED survived transplantation into a mouse brain microenvironment for more than 10 days while expressing various neural markers (Miura et al. 2003). In addition, similar to BMSCs, SHED is capable of differentiating into neural-like cells after transplanted into mice brain (Azizi et al. 1998).

3.1.2 Periodontal Ligament Stem Cells: PDLSCs

PDLSCs, stem cells derived from human periodontal ligament, can be found in the PDL connective tissue and cementoblasts precursors, and are potential candidates to regenerate the soft tissue PDL, as well as cementoblasts and osteoblasts

(Sharpe 2016). PDLSCs can adopt adipogenic, chondrogenic, cardiomyogenic, and neurogenic cell fates (Huang et al. 2009b). When provided with lineage-specific cocktail during culture, PDLSCs have also been shown to exhibit hepatogenic cell differentiation by the expression of critical hepatic markers for glycogen storage, albumin and urea secretion, suggesting potential therapeutic roles for regenerating organs such as the liver (Vasanthan et al. 2016). A new subpopulation of hPDLSCs successfully harvested from the inner surface of alveolar bone sockets (a-PDLSCs) were reported to display enhanced multilineage differentiation potential (Wang et al. 2011). When compared with cells isolated from tooth root (r-PDLSCs), a-PDLSCs showed an increased ability to repair periodontal defects.

PSLCSs are studied clinically due to their potential roles in restoring diseased and damaged PDL tissues. For example, in a miniature swine model, surgical introduction of PDLSCs were shown to improve the restoration of periodontal lesions generated by surgical removed bone in the third molar (Liu et al. 2008). The PDLSCs used in this study were obtained from teeth extracted from miniature pigs and subsequently expanded ex vivo. In a mouse model, PDLSCs exhibited enhanced expression of the osteogenic markers ALP, BSP, OCN, and RUNX2, and transplantation of GelMA microgel encapsulated hPDLSCs into immunocompromised mice showed increased levels of vascularized, mineralized tissue formation (Chen et al. 2016a). In a recent randomized clinical trial, autologous PDLSCs in combination with bovine bone-derived mineralized tissue matrix were used to test guided tissue regeneration in periodontal intrabony defects (Chen et al. 2016b). Enrolled patients were randomly assigned to either a group treated with PDLSC-derived cell sheet, or a control group without stem cells. Although this study has yielded preliminary results, including that PDLSCs are safe to employ with no adverse reactions from patients in a 12-month follow-up, the efficacy of this therapy for repairing boney defects will require further validation.

3.1.3 Stem Cells of the Apical Papilla: SCAPs

The apical papilla, located apical to the dental epithelial diaphragm and next to the dental pulp (Fig. 1), is an abundant stem cell source that is highly proliferative and migratory (Rubio et al. 2005). SCAPs also have the potential of multilineage differentiation when expanded ex vivo, including odontogenic differentiation (Huang et al. 2008). In addition to expressing several growth factors including DSP, matrix extracellular phosphoglycoprotein (MEPE), FGFR3, the VEGF receptor 1 (Flt-1), and melanoma-associated glycoprotein (MUC18), all of which are also expressed in DPSCs. SCAPs express the unique marker CD24, which is downregulated in response to osteogenic stimulation (Huang et al. 2009b). While SCAPs exhibit similar characteristics to DPSCs in vitro, their distinction lies in the fact that the apical papilla is the natural precursor tissue of the radicular pulp, raising speculation that SCAPs may be a superior cell source for tissue regeneration (Huang et al. 2009b). The differentiation capacity of SCAP has been explored in various in vivo experiments in animal models, where the formation of a dentin-pulp-like complex was observed when SCAPs were transplanted into immunocompromised mice (Huang et al. 2006, 2008).

3.1.4 Follicle Stem Cells: DFCs

The dental follicle is a fibrous structure surrounding the developing tooth germ (Mantesso and Sharpe 2009), which gives rise to the PDL by differentiating into collagen secreting PDL fibroblasts that attach to adjacent alveolar jaw bone and cementum surrounding the tooth root (Mantesso and Sharpe 2009). DFCs can differentiate into PDL cells, osteoblasts and cementoblasts, although the critical factors regulating DFC differentiation remain to be determined (Zhai et al. 2019). DFCs express markers found in other stem cells, such as Notch-1 and Nestin (Morsczeck et al. 2005a; b). Transplanted DFCs are capable of differentiating into collagen-secreting PDL fibroblasts, and can interact with cementum and adjacent bone to generate functional cementum/PDL tissue in a nude mice model (Handa et al. 2002). Recently, in addition to roles in PDL tissue regeneration, DFCs combined with treated dentin matrix (TDM) formed tooth root-like structures, suggesting versatile and multidirectional differentiation potential in bio-root tissue engineering applications (Guo et al. 2012).

3.1.5 Induced Pluripotent Stem Cells: iPSCs

In 2007, Japanese scientist Shinya Yamanaka was awarded the Nobel Prize for the discovery of induced pluripotent stem cells (iPSCs), created by reprogramming mouse skin fibroblasts to their embryonic state in order to generate a putative patient-specific, autologous embryonic stem cell source that could be used to regenerate all tissues and organs (Takahashi et al. 2007). iPSCs are functionally superior to, and exhibit enhanced proliferation and differentiation potential, as compared to traditional somatic stem cells (Takahashi et al. 2007; Morsczeck 2012). Similar to ESCs, iPSCs can differentiate into all lineages, including endoderm, mesoderm, and ectoderm (Hu et al. 2018). iPSCs can be generated from both non-dental and dental cell types including DPSCs, SHEDs, PDLSCs, and SCAPs (Yan et al. 2010, Wada et al. 2011).

An exciting new model is that of epithelial cell sheets created from human urine cell derived-iPSCs (ifhU-iPSCs), which were shown to regenerate a whole tooth when transplanted into mouse subrental capsules (Cai et al. 2013). Furthermore, transgene-free iPSCs (TF-iPSCs), generated from the dental apical papilla, could provide an unlimited stem cell source (Zou et al. 2012). TF-iPSCs showed improved characteristics including better recovery after cryopreservation, a frequently encountered obstacle for both iPSCs and ESCs (Yan et al. 2010; Morsczeck 2012; Zou et al. 2012). Furthermore, TF-SCAP-iPSCs expressed neuro-makers without the need for added neurogenic differentiation stimuli (Zou et al. 2012), although this preferential neurogenic differentiation potential has not been further validated (Arthur et al. 2008; Vollner et al. 2009; Morsczeck 2012).

3.2 Cell Transplantation and Cell Homing

Stem-cell based tooth regeneration approaches primarily focus on cell transplantation and cell homing (Mao et al. 2010). Cell transplantation approaches rely on the regenerative potentially of stem cells such as DPSCs, with or without bioengineered scaffolds. In canonical models of cell delivery, stem cells are seeded onto bio-scaffolds as a vehicle of delivery, and the cell-seeded construct is then implanted to the host with or without added growth factors (Hu et al. 2018).

In contrast, cell homing can be broadly defined as a process in which endogenous stem/progenitor cells are actively recruited to migrate to different anatomical compartments (Mao et al. 2010). Cell homing involves local stem-cell based activation approaches, and primarily relies on the recruitment of endogenous progenitor cells to migrate to, repair and restore damaged tissue (Mao et al. 2010). The cell-homing approach offers advantages such as augmenting the host's own regenerative capacity while utilizing bio-cues such as cytokines and chemokines delivered in a single surgery (Mao et al. 2010). In contrast to cell delivery, cell homing does not require the lengthy process of cell isolation and in vitro expansion. Cell homing has been used in a variety of animal models, for example, via recruitment by a biological molecules such as vascular endothelial growth factor (VEGF), platelet-derived growth factor (PDGF), or bFGF and NGF in combination with BMP7 (Kim et al. 2010). Cell-homing approaches demonstrated the regeneration of dental-pulp tissue in an ectopic implantation model (Cordeiro et al. 2008; Huang et al. 2010). In their chemotaxis-based approach, Kim et al. reported robust re-cellularization and revascularization of teeth treated via root canal in vivo (Kim et al. 2010).

3.3 Scaffold Fabrication

Biomaterials with adequate physical strengths, porosity, and bioactivity can also be used for tooth regeneration (Sharma et al. 2014). Scaffold materials can broadly be categorized into natural versus synthetic materials, with innovative and novel 3D printing approaches employed in both types of materials. Current efforts focus on designing biomaterials that replicate native organ-specific extracellular matrix (ECM) environments, in order to facilitate the micro-environmental cell-cell and cell-scaffold interactions that are required for functional tissue regeneration (Sharma et al. 2014).

Natural biomaterials such as collagen, fibrin, and hyaluronic acid (HA) offer good cellular compatibility and biocompatibility (Lee and Kurisawa 2013). Collagen scaffolds, incorporated with growth factors such as VEGF and/or PDGF, have been shown to re-cellularize and revascularize connective tissues in the root canal; however, concerns remain with respect to batch to batch fabrication variability, potential immunogenicity, and lack of precise control over pore size (Kim et al. 2010). Fibrin-based scaffold fabrication methods allow for facile manipulation of mechanical properties, physical characteristics, and cell invasion, where increased fibrinogen content can enhance the stiffness of the scaffold while reducing porosity (Sharma et al. 2014). Another study showed that injectable fibrin hydrogels of specified 3D shapes could be used as suitable and promising materials for tooth regeneration (Linnes et al. 2007). HA also offers good biocompatibility with low immunogenicity (Delmage et al. 1986). HA hydrogels have been extensively used in regeneration studies of various organs including bone, cartilage, vocal cords, and even the brain (Sharma et al. 2014). The low mechanical strength of HA scaffolds can be overcome by the addition of stiff cross-linking chemical polymers such as alginate and RGD peptides (Lee and Kurisawa 2013).

Synthetic polymers such as polylactic acid (PLA), polyglycolic acid, (PGA) and poly(lactide-co-glycolide) (PLGA) have been broadly employed in regenerative dental applications due to their biocompatibility with a variety of dental stem cells including SHED, DPSCs, and dental pulp fibroblasts. Synthetic polymer scaffolds are highly tailorable and degrade into products that can be removed via a wide variety of metabolic pathway. However, limitations to synthetic scaffolds lie in the fact that they lack many physiological chemical components present in natural tissue ECM (Moussa and Aparicio 2019).

Both natural and synthetic scaffold materials are capable of performing basic drug delivery tasks when seeded with stem cells. However, due to the highly complex nature of human tissues and organs, we have yet to create biomaterials that are capable of enabling organ-specific, cell-cell, and cell-scaffold interactions that would be required for efficient regenerative dental therapies. Thus, there remains a strong need for novel and more advanced biomaterial fabrication methods and materials.

3.3.1 Innovative and Advanced Scaffolds

Nanotechnology has been a revolutionary force in the field of TE (Monteiro and Yelick 2017). Nanoparticles are smaller than microparticles and as such can increase the surface area per unit volume, enabling the fabrication of multiple innovative scaffolds such as composite scaffolds with nanofibrous components (Bhanja and D'Souza 2016). Nanometer-sized particles allow for increased cell migration due to increased porosity, which can also facilitate the delivery of growth factors, promote the interactions between growth factors, ECM, and cells, thereby mimicking the environments of naturally formed tissues (Bottino et al. 2017).

Recent advances have been made to more precisely control scaffold material fabrication method to facilitate applications in translational medicine and dentistry. For example, polymer nanofibers (Fig. 2) have recently emerged as new and innovative synthetic materials (Stojanov and Berlec 2020). After undergoing electrospinning and phase separation, polymer nanofibers can be modified to add growth factors, and then seeded with stem cells for tissue regeneration (Stojanov and Berlec 2020).

Advanced scaffold fabrication methods use innovative synthetic materials with exceptional abilities to mimic the native extracellular matrix (ECM), while also providing mechanical and structural support, and regulating chemical, cellular, and tissue level interactions. Novel nanofibrous spongy microspheres (NF-SMS) were used as a delivery mechanism for human DPSCs into the dental cavity (Kuang et al. 2015, 2016). These biodegradable polymer microspheres have been used to repair tissue defects based on its superior injectability. The spongy, nanoparticle structure of NF-SMS allows for self-assembly capability as well as drug delivery (Kuang et al. 2015, 2016). Dental pulp tissue formation was observed, where NF-SMS was shown

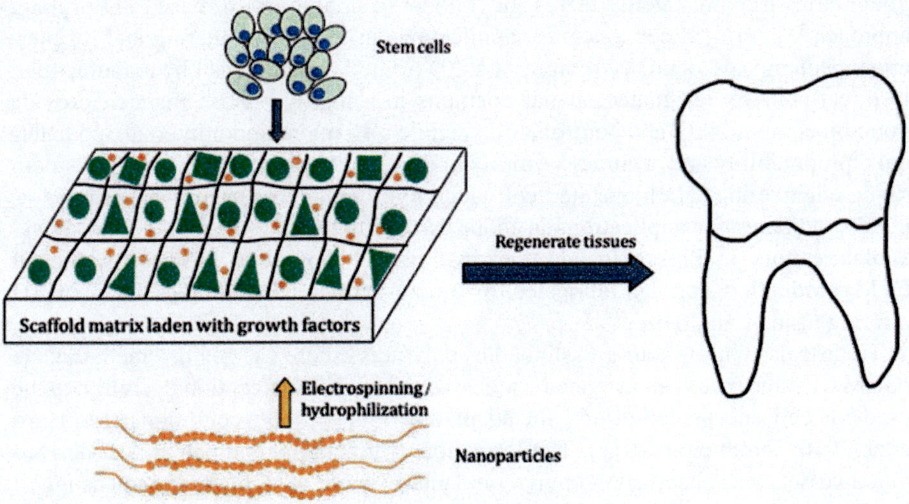

Fig. 2 Nanoparticle coated scaffold matrix stem cell delivery mechanism. Nanometer-sized particles/fibers can be incorporated into a scaffold to create a desired environment that best accommodates cell adhesion, proliferation, migration, and nutrient diffusion. Nanofibrous scaffolds provide a large surface area and high porosity, permitting efficient delivery of cells, drugs, and growth factors

to improve the attachment of and promote VEGF expression in hDPSCs cultured using 3D hypoxic conditions.

Nanostructured microspheres were also recently reported to direct growth factor release and suppress cellular inflammation while enhancing cellular differentiation (Niu et al. 2016). In addition to demonstrating growth factor-induced dentin regeneration, this group showed that microspheres could successfully deliver both hydrophobic and hydrophilic biomolecules in a controlled time-release fashion, while at the same time repressing inflammation, thereby promoting odontogenesis. More recently, nanostructured spongy microspheres demonstrated their potential as a novel delivery system to meet a variety of challenges for tooth regeneration (Bottino et al. 2017).

In addition to the whole-tooth regeneration approach, nanotechnology can also aid in the simultaneous regeneration of the complete periodontal structure including cementum, PDL, and alveolar bone (Sowmya et al. 2017). In this study, a dental follicle stem cell seeded tri-layered nanocomposite hydrogel scaffold containing tissue-specific layers to direct the formation of cementum, PDL, and alveolar bone, respectively, directed the formation of all three layers as designed, and exhibited complete defect closure and healing (Sowmya et al. 2017).

3.3.2 The Use of 3D Printing for Tooth Tissue Engineering

Three-dimensional printing is a method by which objects are fabricated by adding materials, layer by layer, in order to render a construct of predetermined 3D volumetric structure (Derby 2012). In contrast to nonbiological-based 3D printing approach, 3D printed constructs for applications in tissue engineering and regenerative medicine and dentistry requires that 3D printed construct can be manufactured to direct cells to designated spatial positions in a highly precise manner, provide good biocompatibility and controllability, while also maintaining internal and external reproducibility and accuracy (Amrollahi et al. 2016). With respect to whole tooth tissue engineering, 3D bioprinted cell-laden hydrogels have been demonstrated as good candidates for applications in tooth tissue engineering strategies (Park et al., Biofabrication, In Press). In addition, hydrogel encapsulated MSCs for efficient ECM production can be fabricated by micro-extrusion techniques used in 3D printing (Tao et al. 2019).

Hydrogels with photo-crosslinkable polymers such as gelatin methacrylate (GelMA) have been investigated extensively for their exceptional ability to be used for cell encapsulation and for adaptability for varying conditions (Monteiro et al. 2016; Smith et al. 2017). The formation of functional and patent host-derived blood vessels was observed in in vivo implanted 3D GelMA constructs consisting of human umbilical vein endothelial cells (HUVECs) and DPSCs encapsulated within GelMA (Khayat et al. 2017) (Fig. 3). Although very promising, GelMA hydrogel scaffolds present limitations with respect to polymer concentration, where higher polymer concentrations yield better mechanical properties, while encapsulated cells demonstrate increased cell proliferation in softer, lower polymer concentration scaffolds (Tao et al. 2019). To achieve better control over scaffold stiffness while also promoting cell differentiation, thermoplastic polymers such as PCL can be printed by coextrusion to act as a frame to reinforce the softer co-printed hydrogel 3D printed constructs (Morrison et al. 2018). For example, using 3D bio-printing and bioinks such as alginate, it has been found that alginate bioink composed of 3% with a mixture of low and high alginate in 1:2 ratio showed superior performance in terms of viscosity, printability, and in vitro cell responses with optimal cell proliferation and distribution (Park et al. 2017). A limitation for the use of PCL is that its

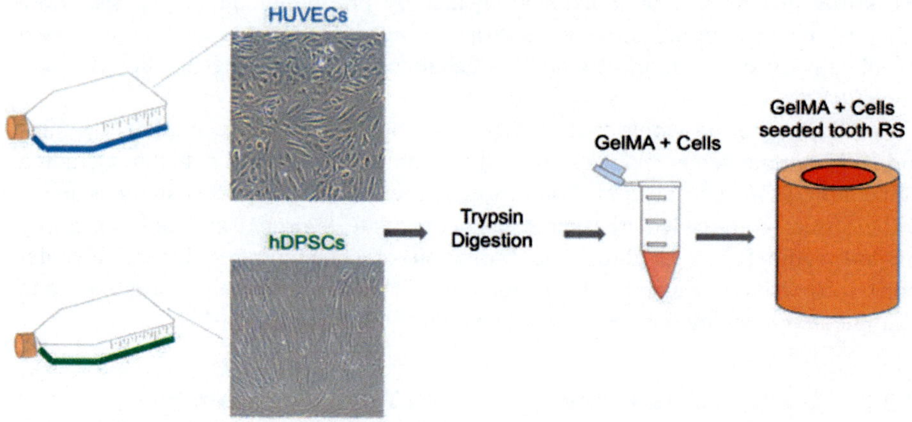

Fig. 3 Schematic of GelMA encapsulated hDPSC/HUVEC-filled tooth root segment (RS)

degradation window of 2–3 years can be a barrier to tissue formation (Tao et al. 2019).

For periodontal tissue regeneration, PCL-HA scaffolds were 3D printed using different sized microchannels to create a multiphasic scaffold (Lee et al. 2014). When seeded with hDPSCs, presumed PDL complex structures were observed. In another PDL-alveolar bone regeneration study, using an electrospun membrane on the biphasic scaffold of the periodontal compartment, scaffolds with PDL cell sheets were shown to better attach to the dentin surface as compared to those of without cell sheets (Vaquette et al. 2012).

4 Tooth Regeneration

4.1 Partial Tooth Regeneration

4.1.1 Dentin-Pulp Complex Regeneration

The dentin-pulp tissue complex is a highly organized tissue complex that offers nutrition, sensation, and defense against various pathogens (Hu et al. 2018). Irreversible pulpitis is clinically defined as vital inflamed pulp incapable of healing (Lin et al. 2020). The traditional clinical therapy for treating pulpitis is root canal therapy, which can subsequently cause tooth fragility and fractures, requiring tooth extraction (Andreasen et al. 2002). Conventional pulp chamber fillings used in endodontic treatment fail to revitalize the pulp (Ingle and Bakland 2002; Dammaschke et al. 2003). Furthermore, removal of pulp tissue results in a loss of pulpal sensation to cold/hot stimulus and the ability to detect secondary infections. (Ingle and Bakland 2002; Dammaschke et al. 2003; Caplan et al. 2005). Therefore, strategies to regenerate vital dental pulp regeneration have become the focus of many research laboratories (Kim et al. 2010). For example, using the aforementioned GelMA hydrogel seeded robust pulp-like structure has been observed in tooth root segments injected with human dental pulp stem cells (hDPSCs) and human umbilical vein endothelial cells (HUVECs) (Fig. 3).

Dentin-pulp complex regeneration strategies include scaffold-dependent stem cell therapy combined with appropriate signaling cues (Zhai et al. 2019). In addition to the aforementioned traditional scaffold materials, novel scaffold-free DPSC self-assembled spheroids have been reported (Dissanayaka et al. 2014). Using hDPSCs combined with human umbilical vein endothelial cells (HUVCEs), microtissue spheroid formation was observed in vitro (Dissanayaka et al. 2014). When microtissue spheroids were loaded within tooth root slices and implanted subcutaneously and grown in immunodeficient mice, histological analyses showed the formation of highly vascularized, dental pulp-like tissue.

Other prominent human pilot studies to regenerate dental pulp tissue include those carried out by the Nakashima group, who used mobilized hDPSCs (MDPSCs) (Nakashima et al. 2017), and the Xuan group, which used deciduous hDPSCs implanted in a randomized clinical trial (Xuan et al. 2018) (see Table 2). By using granulocyte colony-stimulating factor (G-CSF) to mobilize DPSCs harvested from

Table 2 Recent scaffold-independent tooth regeneration

Year	Author	Cell source	Regenerated tissue	Condition	Outcome
2011	(Oshima et al. 2011)	Mouse molar tooth germ cells (embryonic cells)	Whole tooth regeneration	Periodontal disease	Tooth developed periodontal tissue; alveolar bone regenerated; tooth responded to mechanical stimuli
2013	(Cai et al. 2013)	Integration-free urine-derived iPSCs (ifhU-iPSCs), mouse molar mesenchyme cells	Whole tooth regeneration/		iPSCs can functionally replace tooth germ epithelium to regenerate whole tooth-like structure
2014	(Iohara et al. 2014)	DPSCs with G-CSF (granulocyte colony-stimulating factor)	Dentin-pulp complex	Whole pulpotomy	Neurogenesis and angiogenesis
2015	(Murakami et al. 2015)	DPSCs/ BMMSCs/ ADSCs treated with G-CSF	Dentin-pulp complex	Pulpotomy	Neurogenesis and angiogenesis
2017[a]	(Nakashima et al. 2017)	DPSCs pretreated with G-CSF	Dentin-pulp complex	Pulpitis	Neurogenesis
2018[a]	(Xuan et al. 2018)	Human deciduous pulp stem cell (hDPSC)	Dentin-pulp complex	Patients with traumatized incisor tooth	Neurovascularization observed in regenerated pulp tissue with odontoblast layer and connective tissues. Incisor received implants showed functional response to stimuli

[a]Human clinical trials

discarded teeth, clinical-grade human MDPSCs were created in an isolator and expanded following good manufacturing practices (Xuan et al. 2018). All five patients involved in this clinical study had irreversible pulpits requiring pulpectomy and had undergone caries treatment including composite resin wall restoration (Xuan et al. 2018). MDPSCs were then seeded at the root position of the empty pulp chamber, followed by tooth closure and final restoration. In a 24-week follow-up, clinical and evaluation of all five patients showed no toxicity, and electric pulp test (EPT) at week 4 demonstrated a strong positive response. Magnetic resonance imaging (MRI) of the regenerated pulp tissue in the root canal was similar to the untreated control. Dentin-formation was observed in three out of five patients, as detected by cone bean computed tomography.

In addition to stem cell-based technologies, direct pulp capping with a new commercially available biomaterial, Biodentin® (Septodont, Saint-Maur-des Fossés, France), has been made available for patients needing root canal treatment (Zafar et al. 2020). Biodentin®, a synthetic material composed of calcium silicate that provides mechanical support for native dentin tissue, also supported the growth of vital DPSCs (Kaur et al. 2017). One recent case report of patients treated with Biodentin® excellent adhesion properties to the inner dentin wall where it appeared to anchor dentin microtubules, prevent microleakage and pulpal inflammation, and exhibited excellent biocompatibility with no observed cytotoxicity or DPSC cell death (Laslami et al. 2017).

4.1.2 Periodontal Bone-PDL-Cementum Complex Regeneration

Alveolar Bone Regeneration

Periodontitis is an inflammatory oral disease induced by chronic bacterial infection of the gingiva or/and the periodontium, which if left untreated can lead to tooth loss (Kinane and Marshall 2001). Periodontitis is also associated with a strong host inflammatory response that may confer elevated risk of cardiovascular disease and premature low birth weight (Holmlund et al. 2006). Clinically, the ultimate goal of periodontal treatment is to control the infection and regain the functions of the periodontal tissues (Sculean et al. 2015).

Full restoration of normal periodontal functions includes restoring structural support and functions of the alveolar bone, periodontal ligament (PDL), and the cementum, which secures the tooth via the PDL to the surrounding alveolar bone. Conventional regeneration therapies include guided tissue regeneration (GTR) to repair cementum (Goncalves et al. 2006) and PDL complex (Needleman et al. 2006), topical application of enamel matrix derivatives to treat patients with class II furcation defects (Hoffmann et al. 2006), all of which demonstrated partial regeneration of periodontal tissues. Still, challenges remain for our ability to successfully regenerate an integrated PDL apparatus, including the simultaneous formation of the entire bone-PDL-cementum complex (Sowmya et al. 2017).

Current periodontal bone regeneration therapies utilize injectable and absorbable scaffolds including calcium phosphate cement (CPCs) pastes that can subsequently harden in situ to form a solid scaffold that can also serve as a delivery mechanism (Xu et al. 2017). Recently, a tri-culture system including human-induced pluripotent stem cell-MSC (hi-PSC-MSCs), HUVECs, and pericytes delivered via CPC scaffold (Zhang et al. 2017). This novel tri-layer constructs actively promoted both angiogenesis and osteogenesis, demonstrating promise for periodontal complex regeneration.

PDL-Cementum Complex Regeneration

The PDL complex mediates mechanical forces of mastication, which in turn regulate alveolar bone remodeling (Fuks 2008). PDL tissues house immunoglobulins that form a local defense mechanism against bacteria. In one study, regenerated cementum, produced by dental follicle-derived cementoblasts, developed into a thin layer around the root neck, covering the lower part of the root up the apex (Foster et al. 2012). Cell sheet techniques have also been used for PDL-cementum complex regeneration, based on the ability to culture PDL cells to hyper-confluency which induces the formation of a cell sheet layer rich with secreted ECM that promotes extensive cell-cell interactions (Liu et al. 2019). Tri-layered PDL cell sheets grown on thin PGA scaffolds were delivered to the exposed tooth root surfaces, and any intrabony wounds were filled with microporous β-TCP. Subsequent histometric analysis showed complete periodontal regeneration with collagen fibers connecting newly formed cementum with newly generated bone (Liu et al. 2019).

In addition to stem-cell based cell delivery techniques, endogenous cell homing of resident stem cells has shown potential for recruiting endogenous stem cells to the periodontium (Yin et al. 2017). Gene therapy-based approaches have also been explored for advantages in achieving greater biocompatibility of growth factors at the defected periodontal sites (Vhora et al. 2018). Gene therapy in oral regenerative medicine rely on the delivery of genes that direct an individual's own cells to produce a therapeutic agent while minimizing patient risk (Mitsiadis and Smith 2006). For periodontal tissue repair, gene delivery vectors can be injected directly into the periodontal defect where they can be incorporated into resident stem cells (Rios et al. 2011). Gene therapy using PDGF-B has gained attention as a future promising approach for periodontal tissue engineering through modulating the microenvironment of the defect to improve periodontal regeneration (Jin et al. 2004; Cai et al. 2013; Liu et al. 2019).

4.1.3 Tooth Root Regeneration and Bio-root Engineering

In addition to PDL and dental pulp regeneration, the tooth root is probably the most important tooth structure for tooth regeneration, since it anchors and supports the tooth crown in the mouth. Ideally, bioengineered tooth roots would exhibit similar biomechanical properties such natural tooth roots, and consist of supporting periodontal ligament-like tissues, cementum, and dentin-like matrix structure, and be capable of supporting post-crown prostheses.

SCAP and PDLSCs have been extensively investigated for their utility in engineering bio-roots. Sonoyama et al. first demonstrated that a combination of SCAP and PDLSCs could generate a bio-root with periodontal ligament tissues in miniature swine model (Huang et al. 2008). In this study, HA/TCP scaffolds seeded with autologous SCAP and PDLSCs constructs were implanted to the sockets of the swine jaws, followed by placement of a porcelain crown for stability. Subsequent analyses of harvested constructs showed the formation of PDL tissues that adhered to the surrounding bone, although the tooth root showed decreased mechanical strength and integrity over time (Huang et al. 2008).

Another recent report used a sandwich model consisting of human-treated dentin matrix (hTDM) and the growth factors VEGF1, osteoclaicin (OCN), and dentin sialophosphoprotein (DSPP) (Meng et al. 2020). In this study, the authors used Matrigel, an ECM Matrigel (references) to provide a neurogenic environment for the seeded hDPSCs (Meng et al. 2020). Implanted MSCs, combined with SHED aggregates regenerated a complete dental pulp tissue with intricate and highly organized physiological patterns (Xuan et al. 2018).

4.2 Whole Tooth Regeneration

Thanks to recent advances in bioengineering technologies, the past three decades have shown great progress towards the goal to create a fully functional bioengineered tooth that meets the desired aesthetic, functional masticatory, and physiological performance of a natural tooth. Here we will review seminal studies which have brought us closer to achieving this lofty goal.

4.2.1 Scaffold-Based Methods for Whole Tooth Regeneration

Scaffold-based tooth regeneration approaches use biodegradable and biocompatible scaffolds as a delivery mechanism for cells, to facilitate the diffusion of nutrient and metabolites (Young et al. 2002; Sharma et al. 2014; Liu et al. 2020a). Ideal scaffolds are expected to facilitate not only the delivery of cells in an organized fashion but also to facilitate infiltration of a blood supply, oxygen, and nutrition (Loh and Choong 2013; Liu et al. 2020b). In the early 2000s, collagen/gelatin sponges and polyglycolic acid/poly-L-lactate-co-glycolide copolymers (PLA/PLGA) were proposed as biodegradable scaffolds that could control the size and shape of a bioengineered tooth. However, the bioengineered teeth were very small and consisted of accurately shaped tooth crown structures without tooth roots (Young et al. 2002; Honda et al. 2010).

4.2.2 Scaffold-Free Methods for Whole Tooth Regeneration

The first scaffold-free whole tooth regeneration using a tooth organ germ method was first report by Nakao group (Nakao et al. 2007). Using single-cell suspensions generated from dental epithelial and mesenchymal tissues isolated from the E14.5 mouse incisor tooth germs, the authors created a reconstitute tooth germ that generated tooth-like structures in an in vitro organ culture, and also in the oral cavity when transplanted to a mandibular incisor tooth extraction site (Nakao et al. 2007).

Building upon on Nakao group's three-dimensional organ germ method (Nakao et al. 2007), Oshima et al. coined the term "organ germ method" to describe their newly proposed 3D approach to regenerate a functional tooth (Oshima et al. 2011, 2012). In this method, embryonic dental mesenchymal (DM) cells at high density were introduced into collagen gel to form mesenchymal cell aggregates (Hirayama et al. 2013). Next, embryonic dental epithelial (DE) cells were injected to the collagen gel, and within a day, the organ culture showed a compartmentalization between the epithelial and mesenchymal layer, and DM cell compaction as seen in natural embryonic tooth development (Hirayama et al. 2013). This method yielded the subsequent creation of a full-sized bioengineered tooth and alveolar bone. Importantly in this model, bone remodeling occurred in response to mechanical street (Ikeda et al. 2009; Oshima et al. 2011).

For human applications, such a method would require appropriate and sufficient postnatal dental cell sources, scaling up to full-sized human teeth, and a complete recapitulation of the odontogenic developmental program, as well as the ability for the resulting structure to integrate with host vasculature and nervous systems (Monteiro and Yelick 2017).

5 Conclusions

The prospect of repairing and regenerating dental tissue holds great promise due to the possibilities provided by recent advances in tooth tissue engineering. However, for both partial- and whole tooth regeneration, several fundamental challenges remain. In terms of cell manipulation, regardless of the cell source, concerns regarding the nature of cell transplantation include the fact that allogenic and xenogeneic dental cells can cause immune rejection. In order to circumvent a deleterious immunogenic response, cell-homing method is currently being extensively studied as an alternative method that does not require cells to be transplanted. Other obstacles include the high cost of cell culture and handling, ethical concerns, and challenges in fabricating GMP grade scaffolds and materials (Mao and Prockop 2012; Yildirim et al. 2011).

Although iPSCs are envisioned to acquire such odontogenic potential via reprogramming approaches (Zhang and Chen 2014), the ground-breaking discovery of induced pluripotent stem cells (iPSCs) in 2006 and their use in dental therapy applications raise concerns with respect to tumorigenicity (Bhanja and D'Souza 2016). In addition, none of the currently available human dental postnatal mesenchymal cells can form teeth on their own – they need an available and suitable dental epithelial cell source as well. Conferring cells with odontogenic potential thus remains a challenge. Clearly, further research in stem cell biology, combined with state of the art tissue engineering techniques and strategies, will eventually facilitate methods to create reliable bioengineered teeth as a new and widely applicable therapy.

References

Alongi DJ, Yamaza T, Song Y, Fouad AF, Romberg EE, Shi S, Tuan RS, Huang GT (2010) Stem/progenitor cells from inflamed human dental pulp retain tissue regeneration potential. Regen Med 5(4):617–631

Amrollahi P, Shah B, Seifi A, Tayebi L (2016) Recent advancements in regenerative dentistry: a review. Mater Sci Eng C Mater Biol Appl 69:1383–1390

Andreasen JO, Farik B, Munksgaard EC (2002) Long-term calcium hydroxide as a root canal dressing may increase risk of root fracture. Dent Traumatol 18(3):134–137

Arthur A, Rychkov G, Shi S, Koblar SA, Gronthos S (2008) Adult human dental pulp stem cells differentiate toward functionally active neurons under appropriate environmental cues. Stem Cells 26(7):1787–1795

Azizi SA, Stokes D, Augelli BJ, DiGirolamo C, Prockop DJ (1998) Engraftment and migration of human bone marrow stromal cells implanted in the brains of albino rats–similarities to astrocyte grafts. Proc Natl Acad Sci U S A 95(7):3908–3913

Balic A (2018) Biology explaining tooth repair and regeneration: a mini-review. Gerontology 64(4):382–388

Balic A, Thesleff I (2015) Tissue interactions regulating tooth development and renewal. Curr Top Dev Biol 115:157–186

Becker W, Becker BE, Berg L, Samsam C (1986) Clinical and volumetric analysis of three-wall intrabony defects following open flap debridement. J Periodontol 57(5):277–285

Bhanja A, D'Souza DS (2016) Mapping the milestones in tooth regeneration: current trends and future research. Med J Armed Forces India 72(Suppl 1):S24–S30

Biggs LC, Mikkola ML (2014) Early inductive events in ectodermal appendage morphogenesis. Semin Cell Dev Biol 25-26:11–21

Bitgood MJ, McMahon AP (1995) Hedgehog and Bmp genes are coexpressed at many diverse sites of cell-cell interaction in the mouse embryo. Dev Biol 172(1):126–138

Bottino MC, Pankajakshan D, Nor JE (2017) Advanced scaffolds for dental pulp and periodontal regeneration. Dent Clin N Am 61(4):689–711

Brayer WK, Mellonig JT, Dunlap RM, Marinak KW, Carson RE (1989) Scaling and root planing effectiveness: the effect of root surface access and operator experience. J Periodontol 60(1):67–72

Cai J, Zhang Y, Liu P, Chen S, Wu X, Sun Y, Li A, Huang K, Luo R, Wang L, Liu Y, Zhou T, Wei S, Pan G, Pei D (2013) Generation of tooth-like structures from integration-free human urine induced pluripotent stem cells. Cell Regen (Lond) 2(1):6

Caplan DJ, Cai J, Yin G, White BA (2005) Root canal filled versus non-root canal filled teeth: a retrospective comparison of survival times. J Public Health Dent 65(2):90–96

Chai Y, Jiang X, Ito Y, Bringas P Jr, Han J, Rowitch DH, Soriano P, McMahon AP, Sucov HM (2000) Fate of the mammalian cranial neural crest during tooth and mandibular morphogenesis. Development 127(8):1671–1679

Chen X, Bai S, Li B, Liu H, Wu G, Liu S, Zhao Y (2016a) Fabrication of gelatin methacrylate/nanohydroxyapatite microgel arrays for periodontal tissue regeneration. Int J Nanomedicine 11:4707–4718

Chen FM, Gao LN, Tian BM, Zhang XY, Zhang YJ, Dong GY, Lu H, Chu Q, Xu J, Yu Y, Wu RX, Yin Y, Shi S, Jin Y (2016b) Treatment of periodontal intrabony defects using autologous periodontal ligament stem cells: a randomized clinical trial. Stem Cell Res Ther 7:33

Cho MI, Garant PR (2000) Development and general structure of the periodontium. Periodontol 24:9–27

Cobourne MT, Hardcastle Z, Sharpe PT (2001) Sonic hedgehog regulates epithelial proliferation and cell survival in the developing tooth germ. J Dent Res 80(11):1974–1979

Cordeiro MM, Dong Z, Kaneko T, Zhang Z, Miyazawa M, Shi S, Smith AJ, Nor JE (2008) Dental pulp tissue engineering with stem cells from exfoliated deciduous teeth. J Endod 34(8):962–969

Dammaschke T, Steven D, Kaup M, Ott KH (2003) Long-term survival of root-canal-treated teeth: a retrospective study over 10 years. J Endod 29(10):638–643

Dassule HR, Lewis P, Bei M, Maas R, McMahon AP (2000) Sonic hedgehog regulates growth and morphogenesis of the tooth. Development 127(22):4775–4785

Davit-Beal T, Tucker AS, Sire JY (2009) Loss of teeth and enamel in tetrapods: fossil record, genetic data and morphological adaptations. J Anat 214(4):477–501

Delmage JM, Powars DR, Jaynes PK, Allerton SE (1986) The selective suppression of immunogenicity by hyaluronic acid. Ann Clin Lab Sci 16(4):303–310

Derby B (2012) Printing and prototyping of tissues and scaffolds. Science 338(6109):921–926

Dissanayaka WL, Zhu L, Hargreaves KM, Jin L, Zhang C (2014) Scaffold-free prevascularized microtissue spheroids for pulp regeneration. J Dent Res 93(12):1296–1303
European Society of Endodontology (2006) Quality guidelines for endodontic treatment: consensus report of the European Society of Endodontology. Int Endod J 39(12):921–930
Ferreira CF, Magini RS, Sharpe PT (2007) Biological tooth replacement and repair. J Oral Rehabil 34(12):933–939
Foster BL, Nagatomo KJ, Nociti FH Jr, Fong H, Dunn D, Tran AB, Wang W, Narisawa S, Millan JL, Somerman MJ (2012) Central role of pyrophosphate in acellular cementum formation. PLoS One 7(6):e38393
Fuks AB (2008) Vital pulp therapy with new materials for primary teeth: new directions and treatment perspectives. Pediatr Dent 30(3):211–219
Gao J, Jordan TW, Cutress TW (1996) Immunolocalization of basic fibroblast growth factor (bFGF) in human periodontal ligament (PDL) tissue. J Periodontal Res 31(4):260–264
Gao S, Zhao YM, Ge LH (2014) Nuclear factor I-C expression pattern in developing teeth and its important role in odontogenic differentiation of human molar stem cells from the apical papilla. Eur J Oral Sci 122(6):382–390
Gao Z, Wang L, Wang F, Zhang C, Wang J, He J, Wang S (2018) Expression of BMP2/4/7 during the odontogenesis of deciduous molars in miniature pig embryos. J Mol Histol 49(5):545–553
Goncalves PF, Gurgel BC, Pimentel SP, Sallum EA, Sallum AW, Casati MZ, Nociti FH Jr (2006) Effect of two different approaches for root decontamination on new cementum formation following guided tissue regeneration: a histomorphometric study in dogs. J Periodontal Res 41(6):535–540
Graf D, Malik Z, Hayano S, Mishina Y (2016) Common mechanisms in development and disease: BMP signaling in craniofacial development. Cytokine Growth Factor Rev 27:129–139
Gritli-Linde A, Bei M, Maas R, Zhang XM, Linde A, McMahon AP (2002) Shh signaling within the dental epithelium is necessary for cell proliferation, growth and polarization. Development 129(23):5323–5337
Gronthos S, Mankani M, Brahim J, Robey PG, Shi S (2000) Postnatal human dental pulp stem cells (DPSCs) in vitro and in vivo. Proc Natl Acad Sci U S A 97(25):13625–13630
Gronthos S, Brahim J, Li W, Fisher LW, Cherman N, Boyde A, DenBesten P, Robey PG, Shi S (2002) Stem cell properties of human dental pulp stem cells. J Dent Res 81(8):531–535
Guo W, Chen L, Gong K, Ding B, Duan Y, Jin Y (2012) Heterogeneous dental follicle cells and the regeneration of complex periodontal tissues. Tissue Eng Part A 18(5–6):459–470
Haara O, Harjunmaa E, Lindfors PH, Huh SH, Fliniaux I, Aberg T, Jernvall J, Ornitz DM, Mikkola ML, Thesleff I (2012) Ectodysplasin regulates activator-inhibitor balance in murine tooth development through Fgf20 signaling. Development 139(17):3189–3199
Handa K, Saito M, Yamauchi M, Kiyono T, Sato S, Teranaka T, Sampath Narayanan A (2002) Cementum matrix formation in vivo by cultured dental follicle cells. Bone 31(5):606–611
Harada H, Kettunen P, Jung HS, Mustonen T, Wang YA, Thesleff I (1999) Localization of putative stem cells in dental epithelium and their association with Notch and FGF signaling. J Cell Biol 147(1):105–120
Harada H, Toyono T, Toyoshima K, Ohuchi H (2002) FGF10 maintains stem cell population during mouse incisor development. Connect Tissue Res 43(2–3):201–204
Hardcastle Z, Mo R, Hui CC, Sharpe PT (1998) The Shh signalling pathway in tooth development: defects in Gli2 and Gli3 mutants. Development 125(15):2803–2811
Hilton TJ (2009) Keys to clinical success with pulp capping: a review of the literature. Oper Dent 34(5):615–625
Hirayama M, Oshima M, Tsuji T (2013) Development and prospects of organ replacement regenerative therapy. Cornea 32(Suppl 1):S13–S21
Hoffmann T, Richter S, Meyle J, Gonzales JR, Heinz B, Arjomand M, Sculean A, Reich E, Jepsen K, Jepsen S, Boedeker RH (2006) A randomized clinical multicentre trial comparing enamel matrix derivative and membrane treatment of buccal class II furcation involvement in

mandibular molars. Part III: patient factors and treatment outcome. J Clin Periodontol 33(8):575–583
Holmlund A, Holm G, Lind L (2006) Severity of periodontal disease and number of remaining teeth are related to the prevalence of myocardial infarction and hypertension in a study based on 4,254 subjects. J Periodontol 77(7):1173–1178
Honda T, Tokura Y, Miyachi Y, Kabashima K (2010) Prostanoid receptors as possible targets for anti-allergic drugs: recent advances in prostanoids on allergy and immunology. Curr Drug Targets 11(12):1605–1613
Hosoya A, Kim JY, Cho SW, Jung HS (2008) BMP4 signaling regulates formation of Hertwig's epithelial root sheath during tooth root development. Cell Tissue Res 333(3):503–509
Hu L, Liu Y, Wang S (2018) Stem cell-based tooth and periodontal regeneration. Oral Dis 24(5):696–705
Huang GT, Sonoyama W, Chen J, Park SH (2006) In vitro characterization of human dental pulp cells: various isolation methods and culturing environments. Cell Tissue Res 324(2):225–236
Huang GT, Sonoyama W, Liu Y, Liu H, Wang S, Shi S (2008) The hidden treasure in apical papilla: the potential role in pulp/dentin regeneration and bioroot engineering. J Endod 34(6):645–651
Huang X, Bringas P Jr, Slavkin HC, Chai Y (2009a) Fate of HERS during tooth root development. Dev Biol 334(1):22–30
Huang GT, Gronthos S, Shi S (2009b) Mesenchymal stem cells derived from dental tissues vs. those from other sources: their biology and role in regenerative medicine. J Dent Res 88(9):792–806
Huang GT, Yamaza T, Shea LD, Djouad F, Kuhn NZ, Tuan RS, Shi S (2010) Stem/progenitor cell-mediated de novo regeneration of dental pulp with newly deposited continuous layer of dentin in an in vivo model. Tissue Eng Part A 16(2):605–615
Huang X, Wang F, Zhao C, Yang S, Cheng Q, Tang Y, Zhang F, Zhang Y, Luo W, Wang C, Zhou P, Kim S, Zuo G, Hu N, Li R, He TC, Zhang H (2019) Dentinogenesis and tooth-alveolar bone complex defects in BMP9/GDF2 knockout mice. Stem Cells Dev 28(10):683–694
Huang D, Ren J, Li R, Guan C, Feng Z, Bao B, Wang W, Zhou C (2020) Tooth regeneration: insights from tooth development and spatial-temporal control of bioactive drug release. Stem Cell Rev Rep 16(1):41–55
Hurmerinta K, Thesleff I (1981) Ultrastructure of the epithelial-mesenchymal interface in the mouse tooth germ. J Craniofac Genet Dev Biol 1(2):191–202
Ikeda E, Morita R, Nakao K, Ishida K, Nakamura T, Takano-Yamamoto T, Ogawa M, Mizuno M, Kasugai S, Tsuji T (2009) Fully functional bioengineered tooth replacement as an organ replacement therapy. Proc Natl Acad Sci U S A 106(32):13475–13480
Ingle JI, Bakland LK (2002) Endodontics. Decker, Hamilton
Iohara K, Murakami M, Nakata K, Nakashima M (2014) Age-dependent decline in dental pulp regeneration after pulpectomy in dogs. Exp Gerontol 52:39–45
Ivanovski S (2009) Periodontal regeneration. Aust Dent J 54(Suppl 1):S118–S128
Jernvall J, Thesleff I (2012) Tooth shape formation and tooth renewal: evolving with the same signals. Development 139(19):3487–3497
Jin Q, Anusaksathien O, Webb SA, Printz MA, Giannobile WV (2004) Engineering of tooth-supporting structures by delivery of PDGF gene therapy vectors. Mol Ther 9(4):519–526
Jussila M, Thesleff I (2012) Signaling networks regulating tooth organogenesis and regeneration, and the specification of dental mesenchymal and epithelial cell lineages. Cold Spring Harb Perspect Biol 4(4):a008425
Kantarci A, Hasturk H, Van Dyke TE (2015) Animal models for periodontal regeneration and peri-implant responses. Periodontol 68(1):66–82
Kaur M, Singh H, Dhillon JS, Batra M, Saini M (2017) MTA versus biodentine: review of literature with a comparative analysis. J Clin Diagn Res 11(8):ZG01–ZG05
Khayat A, Monteiro N, Smith EE, Pagni S, Zhang W, Khademhosseini A, Yelick PC (2017) GelMA-encapsulated hDPSCs and HUVECs for dental pulp regeneration. J Dent Res 96(2):192–199

Kim JY, Xin X, Moioli EK, Chung J, Lee CH, Chen M, Fu SY, Koch PD, Mao JJ (2010) Regeneration of dental-pulp-like tissue by chemotaxis-induced cell homing. Tissue Eng Part A 16(10):3023–3031

Kim TH, Bae CH, Lee JC, Ko SO, Yang X, Jiang R, Cho ES (2013) Beta-catenin is required in odontoblasts for tooth root formation. J Dent Res 92(3):215–221

Kinane DF, Marshall GJ (2001) Periodontal manifestations of systemic disease. Aust Dent J 46(1):2–12

Kuang R, Zhang Z, Jin X, Hu J, Gupte MJ, Ni L, Ma PX (2015) Nanofibrous spongy microspheres enhance odontogenic differentiation of human dental pulp stem cells. Adv Healthc Mater 4(13):1993–2000

Kuang R, Zhang Z, Jin X, Hu J, Shi S, Ni L, Ma PX (2016) Nanofibrous spongy microspheres for the delivery of hypoxia-primed human dental pulp stem cells to regenerate vascularized dental pulp. Acta Biomater 33:225–234

Laslami K, Dhoum S, Karami M, Jabri M (2017) Direct pulp capping with bioactive material: biodentine. EC Dent Sci 13:75–83

Lee F, Kurisawa M (2013) Formation and stability of interpenetrating polymer network hydrogels consisting of fibrin and hyaluronic acid for tissue engineering. Acta Biomater 9(2):5143–5152

Lee CH, Hajibandeh J, Suzuki T, Fan A, Shang P, Mao JJ (2014) Three-dimensional printed multiphase scaffolds for regeneration of periodontium complex. Tissue Eng Part A 20(7–8):1342–1351

Li J, Huang X, Xu X, Mayo J, Bringas P Jr, Jiang R, Wang S, Chai Y (2011) SMAD4-mediated WNT signaling controls the fate of cranial neural crest cells during tooth morphogenesis. Development 138(10):1977–1989

Li J, Feng J, Liu Y, Ho TV, Grimes W, Ho HA, Park S, Wang S, Chai Y (2015) BMP-SHH signaling network controls epithelial stem cell fate via regulation of its niche in the developing tooth. Dev Cell 33(2):125–135

Li J, Parada C, Chai Y (2017) Cellular and molecular mechanisms of tooth root development. Development 144(3):374–384

Lin LM, Ricucci D, Saoud TM, Sigurdsson A, Kahler B (2020) Vital pulp therapy of mature permanent teeth with irreversible pulpitis from the perspective of pulp biology. Aust Endod J 46(1):154–166

Linnes MP, Ratner BD, Giachelli CM (2007) A fibrinogen-based precision microporous scaffold for tissue engineering. Biomaterials 28(35):5298–5306

Liu Y, Zheng Y, Ding G, Fang D, Zhang C, Bartold PM, Gronthos S, Shi S, Wang S (2008) Periodontal ligament stem cell-mediated treatment for periodontitis in miniature swine. Stem Cells 26(4):1065–1073

Liu J, Ruan J, Weir MD, Ren K, Schneider A, Wang P, Oates TW, Chang X, Xu HHK (2019) Periodontal bone-ligament-cementum regeneration via scaffolds and stem cells. Cells 8(6):537

Liu H, Du Y, St-Pierre JP, Bergholt MS, Autefage H, Wang J, Cai M, Yang G, Stevens MM, Zhang S (2020a) Bioenergetic-active materials enhance tissue regeneration by modulating cellular metabolic state. Sci Adv 6(13):eaay7608

Liu W, Zhang G, Wu J, Zhang Y, Liu J, Luo H, Shao L (2020b) Insights into the angiogenic effects of nanomaterials: mechanisms involved and potential applications. J Nanobiotechnol 18(1):9

Loh QL, Choong C (2013) Three-dimensional scaffolds for tissue engineering applications: role of porosity and pore size. Tissue Eng Part B Rev 19(6):485–502

Lumsden AG (1979) Pattern formation in the molar dentition of the mouse. J Biol Buccale 7(1):77–103

Lumsden AG (1988) Spatial organization of the epithelium and the role of neural crest cells in the initiation of the mammalian tooth germ. Development 103(Suppl):155–169

Malik Z, Alexiou M, Hallgrimsson B, Economides AN, Luder HU, Graf D (2018) Bone morphogenetic protein 2 coordinates early tooth mineralization. J Dent Res 97(7):835–843

Mantesso A, Sharpe P (2009) Dental stem cells for tooth regeneration and repair. Expert Opin Biol Ther 9(9):1143–1154

Mao JJ, Prockop DJ (2012) Stem cells in the face: tooth regeneration and beyond. Cell Stem Cell 11(3):291–301

Mao JJ, Stosich MS, Moioli EK, Lee CH, Fu SY, Bastian B, Eisig SB, Zemnick C, Ascherman J, Wu J, Rohde C, Ahn J (2010) Facial reconstruction by biosurgery: cell transplantation versus cell homing. Tissue Eng Part B Rev 16(2):257–262

Melcher AH (1976) On the repair potential of periodontal tissues. J Periodontol 47(5):256–260

Meng H, Hu L, Zhou Y, Ge Z, Wang H, Wu CT, Jin J (2020) A sandwich structure of human dental pulp stem cell sheet, treated dentin matrix, and Matrigel for tooth root regeneration. Stem Cells Dev 29(8):521–532

Mikkola ML (2008) TNF superfamily in skin appendage development. Cytokine Growth Factor Rev 19(3–4):219–230

Mina M, Kollar EJ (1987) The induction of odontogenesis in non-dental mesenchyme combined with early murine mandibular arch epithelium. Arch Oral Biol 32(2):123–127

Mitsiadis TA, Smith MM (2006) How do genes make teeth to order through development? J Exp Zool B Mol Dev Evol 306(3):177–182

Miura M, Gronthos S, Zhao M, Lu B, Fisher LW, Robey PG, Shi S (2003) SHED: stem cells from human exfoliated deciduous teeth. Proc Natl Acad Sci U S A 100(10):5807–5812

Monteiro N, Yelick PC (2017) Advances and perspectives in tooth tissue engineering. J Tissue Eng Regen Med 11(9):2443–2461

Monteiro N, Smith EE, Angstadt S, Zhang W, Khademhosseini A, Yelick PC (2016) Dental cell sheet biomimetic tooth bud model. Biomaterials 106:167–179

Moradian-Oldak J (2012) Protein-mediated enamel mineralization. Front Biosci (Landmark Ed) 17:1996–2023

Morrison RJ, Nasser HB, Kashlan KN, Zopf DA, Milner DJ, Flanangan CL, Wheeler MB, Green GE, Hollister SJ (2018) Co-culture of adipose-derived stem cells and chondrocytes on three-dimensionally printed bioscaffolds for craniofacial cartilage engineering. Laryngoscope 128(7): E251–E257

Morsczeck C (2012) Transgene-free induced pluripotent dental stem cells for neurogenic differentiation. Stem Cell Res Ther 3(6):46

Morsczeck C, Gotz W, Schierholz J, Zeilhofer F, Kuhn U, Mohl C, Sippel C, Hoffmann KH (2005a) Isolation of precursor cells (PCs) from human dental follicle of wisdom teeth. Matrix Biol 24(2):155–165

Morsczeck C, Moehl C, Gotz W, Heredia A, Schaffer TE, Eckstein N, Sippel C, Hoffmann KH (2005b) In vitro differentiation of human dental follicle cells with dexamethasone and insulin. Cell Biol Int 29(7):567–575

Moussa DG, Aparicio C (2019) Present and future of tissue engineering scaffolds for dentin-pulp complex regeneration. J Tissue Eng Regen Med 13(1):58–75

Murakami M, Hayashi Y, Iohara K, Osako Y, Hirose Y, Nakashima M (2015) Trophic effects and regenerative potential of mobilized mesenchymal stem cells from bone marrow and adipose tissue as alternative cell sources for pulp/dentin regeneration. Cell Transplant 24(9):1753–1765

Murphy KG (1995) Postoperative healing complications associated with Gore-Tex periodontal material. Part I. incidence and characterization. Int J Periodontics Restorative Dent 15(4):363–375

Mustonen T, Ilmonen M, Pummila M, Kangas AT, Laurikkala J, Jaatinen R, Pispa J, Gaide O, Schneider P, Thesleff I, Mikkola ML (2004) Ectodysplasin A1 promotes placodal cell fate during early morphogenesis of ectodermal appendages. Development 131(20):4907–4919

Nakao K, Morita R, Saji Y, Ishida K, Tomita Y, Ogawa M, Saitoh M, Tomooka Y, Tsuji T (2007) The development of a bioengineered organ germ method. Nat Methods 4(3):227–230

Nakashima M, Iohara K, Murakami M, Nakamura H, Sato Y, Ariji Y, Matsushita K (2017) Pulp regeneration by transplantation of dental pulp stem cells in pulpitis: a pilot clinical study. Stem Cell Res Ther 8(1):61

Needleman IG, Worthington HV, Giedrys-Leeper E, Tucker RJ (2006) Guided tissue regeneration for periodontal infra-bony defects. Cochrane Database Syst Rev 2:CD001724

Nemoto E, Sakisaka Y, Tsuchiya M, Tamura M, Nakamura T, Kanaya S, Shimonishi M, Shimauchi H (2016) Wnt3a signaling induces murine dental follicle cells to differentiate into cementoblastic/osteoblastic cells via an osterix-dependent pathway. J Periodontal Res 51(2):164–174

Neubuser A, Peters H, Balling R, Martin GR (1997) Antagonistic interactions between FGF and BMP signaling pathways: a mechanism for positioning the sites of tooth formation. Cell 90(2):247–255

Niu X, Liu Z, Hu J, Rambhia KJ, Fan Y, Ma PX (2016) Microspheres assembled from chitosan-graft-poly(lactic acid) micelle-like core-shell nanospheres for distinctly controlled release of hydrophobic and hydrophilic biomolecules. Macromol Biosci 16(7):1039–1047

Nyman S, Gottlow J, Karring T, Lindhe J (1982a) The regenerative potential of the periodontal ligament. An experimental study in the monkey. J Clin Periodontol 9(3):257–265

Nyman S, Lindhe J, Karring T, Rylander H (1982b) New attachment following surgical treatment of human periodontal disease. J Clin Periodontol 9(4):290–296

Oosterwegel M, van de Wetering M, Timmerman J, Kruisbeek A, Destree O, Meijlink F, Clevers H (1993) Differential expression of the HMG box factors TCF-1 and LEF-1 during murine embryogenesis. Development 118(2):439–448

Orban BJ, Bhaskar SN (1980) Orban's oral histology and embryology. Mosby, St. Louis

Oshima M, Mizuno M, Imamura A, Ogawa M, Yasukawa M, Yamazaki H, Morita R, Ikeda E, Nakao K, Takano-Yamamoto T, Kasugai S, Saito M, Tsuji T (2011) Functional tooth regeneration using a bioengineered tooth unit as a mature organ replacement regenerative therapy. PLoS One 6(7):e21531

Oshima M, Ogawa M, Yasukawa M, Tsuji T (2012) Generation of a bioengineered tooth by using a three-dimensional cell manipulation method (organ germ method). Methods Mol Biol 887:149–165

Parirokh M, Torabinejad M, Dummer PMH (2018) Mineral trioxide aggregate and other bioactive endodontic cements: an updated overview – part I: vital pulp therapy. Int Endod J 51(2):177–205

Park J, Lee SJ, Chung S, Lee JH, Kim WD, Lee JY, Park SA (2017) Cell-laden 3D bioprinting hydrogel matrix depending on different compositions for soft tissue engineering: characterization and evaluation. Mater Sci Eng C Mater Biol Appl 71:678–684

Peterkova R, Lesot H, Vonesch JL, Peterka M, Ruch JV (1996) Mouse molar morphogenesis revisited by three dimensional reconstruction. I. Analysis of initial stages of the first upper molar development revealed two transient buds. Int J Dev Biol 40(5):1009–1016

Pispa J, Jung HS, Jernvall J, Kettunen P, Mustonen T, Tabata MJ, Kere J, Thesleff I (1999) Cusp patterning defect in Tabby mouse teeth and its partial rescue by FGF. Dev Biol 216(2):521–534

Porntaveetus T, Otsuka-Tanaka Y, Basson MA, Moon AM, Sharpe PT, Ohazama A (2011) Expression of fibroblast growth factors (Fgfs) in murine tooth development. J Anat 218(5):534–543

Reynolds MA, Aichelmann-Reidy ME, Branch-Mays GL, Gunsolley JC (2003) The efficacy of bone replacement grafts in the treatment of periodontal osseous defects. A systematic review. Ann Periodontol 8(1):227–265

Rios HF, Lin Z, Oh B, Park CH, Giannobile WV (2011) Cell- and gene-based therapeutic strategies for periodontal regenerative medicine. J Periodontol 82(9):1223–1237

Rubio D, Garcia-Castro J, Martin MC, de la Fuente R, Cigudosa JC, Lloyd AC, Bernad A (2005) Spontaneous human adult stem cell transformation. Cancer Res 65(8):3035–3039

Sarkar L, Sharpe PT (1999) Expression of Wnt signalling pathway genes during tooth development. Mech Dev 85(1–2):197–200

Sarkar L, Cobourne M, Naylor S, Smalley M, Dale T, Sharpe PT (2000) Wnt/Shh interactions regulate ectodermal boundary formation during mammalian tooth development. Proc Natl Acad Sci U S A 97(9):4520–4524

Sculean A, Schwarz F, Chiantella GC, Donos N, Arweiler NB, Brecx M, Becker J (2007) Five-year results of a prospective, randomized, controlled study evaluating treatment of intra-bony defects with a natural bone mineral and GTR. J Clin Periodontol 34(1):72–77

Sculean A, Chapple IL, Giannobile WV (2015) Wound models for periodontal and bone regeneration: the role of biologic research. Periodontol 68(1):7–20
Sharma S, Srivastava D, Grover S, Sharma V (2014) Biomaterials in tooth tissue engineering: a review. J Clin Diagn Res 8(1):309–315
Sharpe PT (2016) Dental mesenchymal stem cells. Development 143(13):2273–2280
Smeets R, Henningsen A, Jung O, Heiland M, Hammacher C, Stein JM (2014) Definition, etiology, prevention and treatment of peri-implantitis–a review. Head Face Med 10:34
Smith EE, Zhang W, Schiele NR, Khademhosseini A, Kuo CK, Yelick PC (2017) Developing a biomimetic tooth bud model. J Tissue Eng Regen Med 11(12):3326–3336
Song M, Yu B, Kim S, Hayashi M, Smith C, Sohn S, Kim E, Lim J, Stevenson RG, Kim RH (2017) Clinical and molecular perspectives of reparative dentin formation: lessons learned from pulp-capping materials and the emerging roles of calcium. Dent Clin North Am 61(1):93–110
Sowmya S, Mony U, Jayachandran P, Reshma S, Kumar RA, Arzate H, Nair SV, Jayakumar R (2017) Tri-layered nanocomposite hydrogel scaffold for the concurrent regeneration of cementum, periodontal ligament, and alveolar bone. Adv Healthc Mater 6(7)
Steele-Perkins G, Butz KG, Lyons GE, Zeichner-David M, Kim HJ, Cho MI, Gronostajski RM (2003) Essential role for NFI-C/CTF transcription-replication factor in tooth root development. Mol Cell Biol 23(3):1075–1084
Stojanov S, Berlec A (2020) Electrospun nanofibers as carriers of microorganisms, stem cells, proteins, and nucleic acids in therapeutic and other applications. Front Bioeng Biotechnol 8:130
Suomalainen M, Thesleff I (2010) Patterns of Wnt pathway activity in the mouse incisor indicate absence of Wnt/beta-catenin signaling in the epithelial stem cells. Dev Dyn 239(1):364–372
Swee LK, Ingold-Salamin K, Tardivel A, Willen L, Gaide O, Favre M, Demotz S, Mikkola M, Schneider P (2009) Biological activity of ectodysplasin A is conditioned by its collagen and heparan sulfate proteoglycan-binding domains. J Biol Chem 284(40):27567–27576
Tai YY, Chen RS, Lin Y, Ling TY, Chen MH (2012) FGF-9 accelerates epithelial invagination for ectodermal organogenesis in real time bioengineered organ manipulation. Cell Commun Signal 10(1):34
Takahashi K, Tanabe K, Ohnuki M, Narita M, Ichisaka T, Tomoda K, Yamanaka S (2007) Induction of pluripotent stem cells from adult human fibroblasts by defined factors. Cell 131(5):861–872
Tao O, Kort-Mascort J, Lin Y, Pham HM, Charbonneau AM, ElKashty OA, Kinsella JM, Tran SD (2019) The applications of 3D printing for craniofacial tissue engineering. Micromachines (Basel) 10(7)
Ten Cate AR (1998) Oral histology: development, structure, and function. 5th ed. Mosby, St. Louis: Mosby. xi, 497 p
Thesleff I (2003) Epithelial-mesenchymal signalling regulating tooth morphogenesis. J Cell Sci 116 (Pt 9):1647–1648
Thesleff I, Sharpe P (1997) Signalling networks regulating dental development. Mech Dev 67(2):111–123
Thesleff I, Tummers M (2008) Tooth organogenesis and regeneration. StemBook, Cambridge, MA
Thomson JA, Itskovitz-Eldor J, Shapiro SS, Waknitz MA, Swiergiel JJ, Marshall VS, Jones JM (1998) Embryonic stem cell lines derived from human blastocysts. Science 282(5391):1145–1147
Torabinejad M, Parirokh M, Dummer PMH (2018) Mineral trioxide aggregate and other bioactive endodontic cements: an updated overview – part II: other clinical applications and complications. Int Endod J 51(3):284–317
Trumpp A, Depew MJ, Rubenstein JL, Bishop JM, Martin GR (1999) Cre-mediated gene inactivation demonstrates that FGF8 is required for cell survival and patterning of the first branchial arch. Genes Dev 13(23):3136–3148
Tummers M, Thesleff I (2003) Root or crown: a developmental choice orchestrated by the differential regulation of the epithelial stem cell niche in the tooth of two rodent species. Development 130(6):1049–1057

Vaahtokari A, Aberg T, Jernvall J, Keranen S, Thesleff I (1996) The enamel knot as a signaling center in the developing mouse tooth. Mech Dev 54(1):39–43

Valera E, Isaacs MJ, Kawakami Y, Izpisua Belmonte JC, Choe S (2010) BMP-2/6 heterodimer is more effective than BMP-2 or BMP-6 homodimers as inductor of differentiation of human embryonic stem cells. PLoS One 5(6):e11167

Vaquette C, Fan W, Xiao Y, Hamlet S, Hutmacher DW, Ivanovski S (2012) A biphasic scaffold design combined with cell sheet technology for simultaneous regeneration of alveolar bone/periodontal ligament complex. Biomaterials 33(22):5560–5573

Vasanthan P, Jayaraman P, Kunasekaran W, Lawrence A, Gnanasegaran N, Govindasamy V, Musa S, Kasim NH (2016) Generation of functional hepatocyte-like cells from human deciduous periodontal ligament stem cells. Naturwissenschaften 103(7–8):62

Vastardis H (2000) The genetics of human tooth agenesis: new discoveries for understanding dental anomalies. Am J Orthod Dentofac Orthop 117(6):650–656

Vhora I, Lalani R, Bhatt P, Patil S, Patel H, Patel V, Misra A (2018) Colloidally stable small unilamellar stearyl amine lipoplexes for effective BMP-9 gene delivery to stem cells for osteogenic differentiation. AAPS Pharmscitech 19(8):3550–3560

Viriot L, Peterkova R, Vonesch JL, Peterka M, Ruch JV, Lesot H (1997) Mouse molar morphogenesis revisited by three-dimensional reconstruction. III. Spatial distribution of mitoses and apoptoses up to bell-staged first lower molar teeth. Int J Dev Biol 41(5):679–690

Vollner F, Ernst W, Driemel O, Morsczeck C (2009) A two-step strategy for neuronal differentiation in vitro of human dental follicle cells. Differentiation 77(5):433–441

Wada N, Wang B, Lin NH, Laslett AL, Gronthos S, Bartold PM (2011) Induced pluripotent stem cell lines derived from human gingival fibroblasts and periodontal ligament fibroblasts. J Periodontal Res 46(4):438–447

Wang J, Feng JQ (2017) Signaling pathways critical for tooth root formation. J Dent Res 96(11):1221–1228

Wang XP, Suomalainen M, Felszeghy S, Zelarayan LC, Alonso MT, Plikus MV, Maas RL, Chuong CM, Schimmang T, Thesleff I (2007) An integrated gene regulatory network controls stem cell proliferation in teeth. PLoS Biol 5(6):e159

Wang L, Shen H, Zheng W, Tang L, Yang Z, Gao Y, Yang Q, Wang C, Duan Y, Jin Y (2011) Characterization of stem cells from alveolar periodontal ligament. Tissue Eng Part A 17(7–8):1015–1026

Wang B, Li H, Liu Y, Lin X, Lin Y, Wang Y, Hu X, Zhang Y (2014) Expression patterns of WNT/beta-CATENIN signaling molecules during human tooth development. J Mol Histol 45(5):487–496

Xiong J, Gronthos S, Bartold PM (2013) Role of the epithelial cell rests of Malassez in the development, maintenance and regeneration of periodontal ligament tissues. Periodontol 63(1):217–233

Xu HH, Wang P, Wang L, Bao C, Chen Q, Weir MD, Chow LC, Zhao L, Zhou X, Reynolds MA (2017) Calcium phosphate cements for bone engineering and their biological properties. Bone Res 5:17056

Xuan K, Li B, Guo H, Sun W, Kou X, He X, Zhang Y, Sun J, Liu A, Liao L, Liu S, Liu W, Hu C, Shi S, Jin Y (2018) Deciduous autologous tooth stem cells regenerate dental pulp after implantation into injured teeth. Sci Transl Med 10(455):eaaf3227

Yan X, Qin H, Qu C, Tuan RS, Shi S, Huang GT (2010) iPS cells reprogrammed from human mesenchymal-like stem/progenitor cells of dental tissue origin. Stem Cells Dev 19(4):469–480

Yang Z, Hai B, Qin L, Ti X, Shangguan L, Zhao Y, Wiggins L, Liu Y, Feng JQ, Chang JY, Wang F, Liu F (2013) Cessation of epithelial Bmp signaling switches the differentiation of crown epithelia to the root lineage in a beta-catenin-dependent manner. Mol Cell Biol 33(23):4732–4744

Yildirim S, Fu SY, Kim K, Zhou H, Lee CH, Li A, Kim SG, Wang S, Mao JJ (2011) Tooth regeneration: a revolution in stomatology and evolution in regenerative medicine. Int J Oral Sci 3(3):107–116

Yin Y, Li X, He XT, Wu RX, Sun HH, Chen FM (2017) Leveraging stem cell homing for therapeutic regeneration. J Dent Res 96(6):601–609

Yokohama-Tamaki T, Ohshima H, Fujiwara N, Takada Y, Ichimori Y, Wakisaka S, Ohuchi H, Harada H (2006) Cessation of Fgf10 signaling, resulting in a defective dental epithelial stem cell compartment, leads to the transition from crown to root formation. Development 133(7):1359–1366

Young CS, Terada S, Vacanti JP, Honda M, Bartlett JD, Yelick PC (2002) Tissue engineering of complex tooth structures on biodegradable polymer scaffolds. J Dent Res 81(10):695–700

Yu T, Volponi AA, Babb R, An Z, Sharpe PT (2015) Stem cells in tooth development, growth, repair, and regeneration. Curr Top Dev Biol 115:187–212

Yuan G, Yang G, Zheng Y, Zhu X, Chen Z, Zhang Z, Chen Y (2015) The non-canonical BMP and Wnt/beta-catenin signaling pathways orchestrate early tooth development. Development 142(1):128–139

Zafar K, Jamal S, Ghafoor R (2020) Bio-active cements-Mineral Trioxide Aggregate based calcium silicate materials: a narrative review. J Pak Med Assoc 70(3):497–504

Zeichner-David M (2006) Regeneration of periodontal tissues: cementogenesis revisited. Periodontol 41:196–217

Zhai Q, Dong Z, Wang W, Li B, Jin Y (2019) Dental stem cell and dental tissue regeneration. Front Med 13(2):152–159

Zhang Y, Chen Y (2014) Bioengineering of a human whole tooth: progress and challenge. Cell Regen (Lond) 3(1):8

Zhang Y, Tomann P, Andl T, Gallant NM, Huelsken J, Jerchow B, Birchmeier W, Paus R, Piccolo S, Mikkola ML, Morrisey EE, Overbeek PA, Scheidereit C, Millar SE, Schmidt-Ullrich R (2009) Reciprocal requirements for EDA/EDAR/NF-kappaB and Wnt/beta-catenin signaling pathways in hair follicle induction. Dev Cell 17(1):49–61

Zhang C, Hu K, Liu X, Reynolds MA, Bao C, Wang P, Zhao L, Xu HHK (2017) Novel hiPSC-based tri-culture for pre-vascularization of calcium phosphate scaffold to enhance bone and vessel formation. Mater Sci Eng C Mater Biol Appl 79:296–304

Zou XY, Yang HY, Yu Z, Tan XB, Yan X, Huang GT (2012) Establishment of transgene-free induced pluripotent stem cells reprogrammed from human stem cells of apical papilla for neural differentiation. Stem Cell Res Ther 3(5):43

Generation of Ear Cartilage for Auricular Reconstruction

Yu Liu and Yilin Cao

Contents

Abstract

Auricular reconstruction is among the most challenging surgical procedures in plastic and reconstructive surgery because of the lack of an ideal auricle substitute that can guarantee a long-lasting outcome while involving minimal donor site morbidity. Tissue-engineered cartilage may provide an ideal autologous solution. After the first report on generation of human ear-shaped cartilage in a nude mouse model in 1997, cartilages with human ear shape have been engineered in vitro, in nude mice, and in immunocompetent animals using various cells and scaffolds. Recently, our group reported a pilot clinical trial of in vitro engineered human ear-shaped cartilage and its clinical translation for auricular reconstruction. This bench-to-bed process spanning over the past two decades can provide an in-

Y. Liu
National Tissue Engineering Center of China, Shanghai, People's Republic of China

Y. Cao (✉)
Shanghai Tissue Engineering Research Key Laboratory, Shanghai Ninth People's Hospital, Shanghai Jiao Tong University School of Medicine, Shanghai, People's Republic of China

D. Eberli et al. (eds.), *Organ Tissue Engineering*, Reference Series in Biomedical Engineering, https://doi.org/10.1007/978-3-030-44211-8_6

depth understanding of the development of cartilage tissue engineering for auricular reconstruction. The current chapter will illustrate the research achievements and application challenges inherent in generating translational tissue-engineered cartilage for auricular reconstruction, particularly focusing on the global trends and new research directions of each associated building block or step, namely, seed cells, scaffolds, three-dimensional printing, in vitro microenvironment simulation, preclinical evaluation, and clinical translation.

1 Introduction

Auricular defects can be caused by congenital diseases such as microtia or by acquired defects due to burns, trauma, animal bites, tumor removal, or radiotherapy performed as a treatment for cancer. Microtia is a congenital malformation of the external ear, with a varied regional prevalence rate of 0.83–17.4 per 10,000 births worldwide (Paput et al. 2012; Bly et al. 2016). Compared to microtia, ear deformities caused by acquired defects, occurring in more than 1 per 500 people, may be more common (Jessop et al. 2016). The auricle is an important identifying feature of the human face, and hence, its deformity has a profound effect on both the level of self-confidence and continued psychological development in afflicted patients. The current cosmetic procedures for total auricular reconstruction mainly include the wearing of auricular prostheses or implantation of nonabsorbable auricular frame materials or an autologous rib cartilage framework (Bly et al. 2016; Jessop et al. 2016; Wiggenhauser et al. 2017; Zhou et al. 2018). Nonabsorbable frames composed of materials such as silastic or high-density polyethylene (Medpor) can generate an excellent ear shape without donor site morbidity, but they lack bioactivity and can lead to extrusion and infections. Alloplastic implants can be easily tailored with respect to shape and mechanical properties, but often lead to infection, incompatibility, and extrusion. Autologous rib cartilage transplantation is the current gold-standard treatment approach for external ear reconstruction owing to its good long-term durability as well as the potential to grow with the patient (Rosa et al. 2014). However, harvesting rib cartilage inevitably leads to donor site injury, and patients must have a sufficient supply of donor cartilage available to be eligible candidates for this procedure (Rosa et al. 2014). Moreover, replicating the complex three-dimensional (3D) ear structure is hard to achieve using surgeons' hand skills alone, which are highly dependent on their training background and clinical experience (Zhou et al. 2018).

Tissue-engineered auricles constitute a promising alternative to current ear reconstructive options. In 1997, the generation of engineered cartilage with a human auricular shape in a nude mouse model was reported by Cao et al. (1997). The aesthetic effect coupled with press coverage of this work gave people the impression that a tissue engineering-based solution for auricular reconstruction is just at the next corner, but only recently, our group led by Cao reported the first clinical translation of human ear-shaped cartilage (Zhou et al. 2018). The major efforts during this more than 20-year-long, bench-to-bed process lay in solving a number of problems, including the lack of a proper cell source, the difficulty in generating ear-shaped cartilage with a predesigned 3D structure, the insufficient mechanical properties for shape maintenance, and the unfavorable host response to the engineered graft after its transplantation in vivo.

Developments in current cell biology, materials science, and advance manufacturing have contributed tremendously to solving the aforementioned problems and realizing the clinical translation of human ear-shaped cartilage. Although this tissue engineering-based approach still cannot shift the paradigm of conventional reconstructive methods, the lessons learned from the bench-to-bed process may provide deeper understanding and valuable experience to support further improvements in the quality of the engineered cartilage to better suit clinical situations by refining each building block of tissue engineering. The current chapter will outline the global trends and new research achievements of each building block or step in generating translational tissue-engineered cartilage for auricular reconstruction, including seed cell strategy, scaffold development, 3D printing, in vitro microenvironment simulation, and in vivo preclinical evaluations. Issues raised during clinical application that may direct future basic investigations to improve the quality of tissue-engineered cartilage are also discussed.

2 Traditional Approaches for Auricular Reconstruction

Many contemporary treatment options can be used for auricular reconstruction, including the wear of auricular prostheses, alloplastic framework implantation, and autologous cartilage transplantation. Alloplastic frameworks that are usually made from silastic or porous polyethylene (Medpor) are readily available with consistent levels of quality, are easy to work with, and require only a short operation time for placement. However, these materials lack bioactivity, and their mechanical properties are unmatched to those of the native auricular cartilage. Patients undergoing repair with Medpor implants experience complications at a rate ranging from 4% to 6.31% even when treated by the most experienced surgeons. Complications can include postoperative infections, skin perforations, framework fracture, compression ischemia, capsule fibrosis, and framework dislocations (Reighard et al. 2018). Meanwhile, the complication rate is even higher when using silastic implants (Berghaus 2007).

Autologous costal cartilage is currently the most biocompatible material available for total auricular reconstruction. The fundamental principles of current autologous cartilage transplantation strategies were first described in the English language by Gillies (1920). Tanzer revolutionized the surgical procedure by paring down the original six-stage method to encompass only three to four total surgeries, with subsequent technique modifications (Tanzer 1959; Reighard et al. 2018). Thereafter, Brent (1992, 2002), Park (Park et al. 1991; Park 2000), Nagata (1993), Firmin (1998), and Walton and Beahm (2002) contributed significantly to refining the surgical techniques used during this treatment (Olshinka et al. 2017). To date, this surgical approach has witnessed several notable advancements, including a transition to a two- or even one-stage surgery, adoption of a minimally invasive approach to harvest rib cartilage, and imaging-assisted design of a more accurate patient-specific ear shape. However, although the procedure has advantages such as a high level of biocompatibility, long-term stability, immunocompatibility, and the potential for the implant to grow with the patient (Bichara et al. 2012), its conduct still requires special surgical experience, necessitates a considerable amount of operation time during which to shape the cartilage, and involves several reconstruction steps. In fact, it is among the most challenging procedures in plastic and reconstructive surgery, and only the most talented surgeons can consistently achieve satisfactory long-term postoperative outcomes (Jessop et al. 2016). Moreover, the age range of patients eligible to receive this surgery is also strictly restricted: the amount of costal cartilage can be insufficient in those who are younger than 6 years of age, whereas in patients who are too old, the costal cartilage may undergo calcification, and bending the cartilage without breaking it could become impossible (Kawanabe and Nagata 2006). The harvesting of the costal cartilage may also involve large donor site morbidity and induce significant pain, pneumonia, pneumothorax, atelectasis, and deformities of the rib cage and unattractive scars later (Ohara et al. 1997; Uppal et al. 2008). As a result, surgeons and researchers alike continue to search for alternative approaches by which to reconstruct the auricle.

3 Tissue Engineering Approach for Auricular Reconstruction

As a means to overcome the treatment obstacles of available surgical options, the reconstruction of human ear-shaped cartilage using tissue engineering techniques has attracted tremendous attention. Theoretically, a tissue-engineered auricle combines the merits of both an alloplastic framework and sculptured costal cartilage and possesses several advantages, including minimal donor site morbidity, accurate shape control, superb biocompatibility and bioactivity, short operation time (i.e., no need to scalpel the implant), and independence from the surgeon's level of experience. The first report on tissue-engineered human ear-shaped cartilage was published by Çao et al. (1997) who used bovine chondrocytes as seed cell, polyglycolic acid (PGA)/polylactic acid (PLA) as scaffold, and the immunodeficient nude mouse as an animal model (Cao et al. 1997). Although clinical translation took significant time thereafter to manifest, the aesthetic prominence of this work attracted massive media coverage, which led people to believe that the tissue-engineered auricle lay just around the next corner. In fact, however, it took about 20 years for Cao's group to realize the first clinical translation of human ear-shaped cartilage in five patients with microtia (Zhou et al. 2018), and still, this work retains some drawbacks such as long-term in vitro procedures and high costs, hindering its further and widespread clinical application and commercialization. To improve the quality of the engineered cartilage for auricular reconstruction, technologies relating to seed cell manipulation, scaffold preparation, in vitro cartilage regeneration, and in vivo preclinical evaluations need to be consistently refined. Similarly, special surgical procedures for handling the engineered tissue also need to be developed.

3.1 Seed Cell Sources for Auricular Regeneration

The generation of a real-sized human ear-shaped cartilage requires the gathering of a significant amount of seed cells with a stable chondrogenic phenotype while simultaneously ensuring minimal donor site morbidity. Since the implantation site of the reconstructed auricle is a subcutaneous environment characterized by acute immune activity, seed cells of an autologous origin are needed to avoid immune rejection. Currently, the most intensively investigated seed cells for auricular reconstruction are auricular chondrocytes and mesenchymal stem cells (MSCs). Induced pluripotent stem cells (iPSCs) are under investigation but have yet to be used in translational studies.

3.1.1 Chondrocytes

Chondrocytes are the resident cells in the cartilage and therefore are the first choice as seed cells for cartilage tissue engineering. Chondrocytes can be divided into three types – elastic chondrocyte, hyaline chondrocyte, and fibrochondrocyte – according to the three types of cartilage where they are derived from, which are elastic cartilage, hyaline cartilage, and fibrocartilage, respectively. By adopting a developmental view, chondrocytes are derived from two embryological origins: the cranial neural crest (CNC) and the mesoderm (Taihi et al. 2019). More specifically, auricular (elastic cartilage) and nasal (hyaline cartilage) chondrocytes are derived from the CNC, whereas costal (fibrocartilage) and articular (hyaline cartilage) chondrocytes are derived from the mesoderm. Early studies on cartilage regeneration usually employed articular chondrocytes as seed cell candidates and largely found that they are nonproliferative, are prone to aging, and dedifferentiate after passage (Brittberg et al. 1994). Recently, accumulating evidence has suggested that CNC-derived chondrocytes (nasal or auricular chondrocytes) possess promising properties such as robust proliferation ability (Tay et al. 2004; Taihi et al. 2019). Moreover, these cells are more responsive to environmental cues than those derived from the mesoderm (Pelttari et al. 2014). After being treated with appropriate sources of stimulations such as basic fibroblast growth factor (bFGF) or low oxygen tension, the proliferation ability of those chondrocytes can be further enhanced, and a sufficient number of chondrocytes can be gained from a small biopsy sample collected from nasal or auricular cartilage (Zhou et al. 2018). In conditions with chondrogenic factors such as transforming growth factor-β1 (TGF-β1), those chondrocytes can regain their chondrogenic phenotype lost during extensive expansion and form stable cartilage in a nonchondrogenic subcutaneous environment (Zhou et al. 2018).

For auricular reconstruction, the application of auricular chondrocytes that can generate the elastic type of cartilage of the ear seems to be a reasonable choice. Particularly in patients with microtia, the microtic ear, which is usually discarded, can instead be reused to provide seed cells for auricular regeneration (Kamil et al. 2004). In fact, all of the current clinical studies on auricular reconstruction based on cell therapy or tissue engineering report the use of auricular chondrocytes from the microtic ear (Yanaga et al. 2009, 2013; Zhou et al. 2018), although further investigations must be conducted to systematically compare microtic chondrocytes and normal healthy auricular chondrocytes. Recently, cartilage progenitor cells have been isolated from the perichondrium of the auricular cartilage, and these cells express stem cell markers such as CD44 and CD90 and possess even higher proliferation and chondrogenic capacities than those of chondrocytes isolated from auricular cartilage (Kobayashi et al. 2011; Kagimoto et al. 2016). This may allow an even smaller biopsy sample to be taken from the auricle perichondrium (without harming the ear cartilage), which can regenerate back after the biopsy, to yield sufficient chondrogenic cells to reconstruct the auricle.

3.1.2 Mesenchymal Stem Cells

Mesenchymal stem cells (MSCs) were first reported by Friedenstein et al. (1968) and they are currently the most commonly used stem cells in human therapy and regenerative medicine. MSCs have gained considerable attention as seed cells for cartilage regeneration because of their ease of isolation (i.e., they are readily available), high proliferation potential, and great chondrogenic differentiation capacity (Huselstein et al. 2012). Moreover, MSCs have recently been found to possess immunomodulatory and anti-inflammatory properties, modulate lymphocyte cell function through the secretome, and exhibit several growth factors and cytokines, including TGF-β1, nitric oxide, interleukin-1 (IL)-1, and IL-10 (De Miguel et al. 2012). This immunomodulation is especially important when considering the acute immune reaction of the subcutaneous implantation site of the regenerated auricle.

MSCs can be derived from many sources, including bone marrow, adipose tissue, synovium, periosteum, umbilical cord blood, peripheral blood, skeletal muscle, and synovial fluid. Among the different types of MSCs, bone marrow-derived MSCs (BMSCs) have been considered to demonstrate the highest chondrogenic potential and are therefore the most frequently studied among seed cell candidates for auricular reconstruction. However, the cartilage engineered from BMSCs tends to undergo terminal ossification in the vascularized subcutaneous implantation site (Ichinose et al. 2005). In fact, ectopic ossification is a major problem restricting the application of MSCs to generate subcutaneous cartilage (De Bari et al. 2004). One possible reason for the phenotypic shift in MSC-derived cartilage is insufficient chondrogenic induction (Pelttari et al. 2006). To address this issue, a prolonged preinduction time, more robust chondrogenic growth factors, oxygen tension adjustments, and mechanical stimuli were applied to generate more sufficiently differentiated cartilage prior to implantation into the nonchondrogenic subcutaneous implantation site (Liu et al. 2008). However, it remains difficult to generate such a large piece of homogenous cartilage with a complex enough structure for auricular reconstruction through the chondrogenic induction of MSCs. Another important factor inducing ectopic ossification is the blood vessel invasion into the MSC-engineered graft (Liu et al. 2008). Evidence has revealed that BMSC-engineered cartilage shows much higher expression levels of angiogenic factors, such as vascular endothelial growth factor (VEGF), and much lower expression levels of antiangiogenic factors such as chondromodulin-I (Chm-I), relative to its counterpart engineered from chondrocytes, inducing ingrowth of blood vessels into the MSC-engineered cartilage and triggering the ossification progress (Liu et al. 2008; Zhu et al. 2015). Approaches such as wrapping of a blood vessel-blocking membrane around the MSC-engineered cartilages or adopting scaffolds that can release antiangiogenic factors to engineer the cartilage have been studied to date (Li et al. 2017). However, these methods may also block nutrition transfer or have other negative effects on the seed cells, thus hampering the formation of homogenous cartilage with the required degree of quality.

3.1.3 Co-culture of Chondrocytes with MSCs

To address the issue of terminal ossification of MSCs, the co-culture of MSCs with mature chondrocytes has been investigated and found to be effective in generating phenotypically stable cartilage in the subcutaneous implantation site for the regenerated auricle (Zhang et al. 2014). Besides phenotype maintenance, co-culturing can effectively reduce the quantity demand for chondrocytes, which are limited in supply, while exert the merit of MSCs which are more readily available (Kang et al. 2013; Zhang et al. 2014). Kang et al. found that a co-culture arrangement of 30% chondrocytes and 70% BMSCs could generate ear-shaped cartilage in vitro (Kang et al. 2013). Zhang et al. reported the generation of stable subcutaneous human ear-shaped cartilage engineered through co-culturing of a 30% human microtia chondrocytes with 70% human BMSCs in a nude mouse model (Zhang et al. 2014). Moreover, co-culturing can generate specific types of cartilage: one study has detected elastic expression, the marker of the elastic cartilage of the auricle, in cartilage engineered by the co-culturing of BMSCs and auricular chondrocytes (Kang et al. 2012). An in vivo synergistic effect of MSCs and chondrocytes has also been reported by several groups, where a mixture of MSCs and chondrocytes proved to be more beneficial than chondrocytes or MSCs alone, and the researchers speculated that MSCs' immune regulation ability played a major role in the result (Kang et al. 2012; Ahmed et al. 2014).

There are currently several co-culture models being used – including mixed co-culture, separated co-culture, and both – to investigate the mechanism of co-culture capable of supporting stable chondrogenesis (Levorson et al. 2014; Morita et al. 2015). However, overall, the mechanism is still under debate. One possible mechanism is the chondrocytes' expression of the chondrogenic growth factors and antiangiogenic factors to support the chondrogenesis of BMSCs while suppressing hypertrophy and ossification. Some studies have observed a direct differentiation of chondrocyte-like cells in the cartilage lacuna of red fluorescent protein (RFP)-labeled MSCs (Zhang et al. 2014). Chondrogenic factors such as TGF-β1, bone morphogenetic protein-2 (BMP-2), and insulin growth factor-I (IGF-I) were found in the supernatant of chondrocyte-engineered constructs (Liu et al. 2010a). Other research, however, contends that the trophic factors secreted by MSCs support chondrocyte proliferation and cartilage formation (Wu et al. 2011, 2012; Wang et al. 2013). These discrepancies in the literature may be explained by the variations in chondrocyte sources, scaffolds, and animal models used. A better understanding of these mechanisms is still needed to assess the efficacy of these co-culture systems, with particular attention paid to relative cell death in direct co-culture models. The state of the chondrocytes is also crucial to comprehend in this process (Graceffa et al. 2019). Aside from all of the intensive investigations, in clinical translation, the involvement of two types of cells together can induce complications and make standards and regulations difficult to establish. Therefore, translational studies remain more focused on identifying single-cell types for use as seed cells.

3.1.4 Pluripotent Stem Cells

Pluripotent stem cells (PSCs), including embryonic stem cells (ESCs) and iPSCs, hold immense potential for regenerative purposes because of their properties of unlimited self-renewal and pluripotency. Further, they can represent a continuous sufficient source of seed cells of a committed lineage suitable for a larger-scale production for clinical applications (Cheng et al. 2014). ESCs were first isolated from the inner cell mass of mouse blastocyst-stage embryos and cultured as cell lines by Evans and Kaufman (1981) and Martin (1981), and the first established human ESC lines were derived from human blastocyst in 1998 (Thomson et al. 1998). In 2006, Takahashi et al. achieved a significant breakthrough by generating iPSCs through cell reprogramming (Takahashi and Yamanaka 2006), which brought into being an easily accessible source of pluripotent cells for autologous application while bypassing the ethical concerns related to the harvesting of human embryos, thus opening many gateways to progress in regenerative medicine research in cartilage tissue engineering.

The differentiation of PSCs into chondrocyte-like cells has been accomplished through several methods, including the formation of embryoid bodies, co-culturing, conditioned medium, and ESC- or iPSC-derived MSCs (Cheng et al. 2014; Gibson et al. 2017). Currently, chemically defined culture conditions have been established to generate a relatively pure chondrogenic population (without off-target differentiation or any residual PSCs) suitable for large-scale production (Cheng et al. 2014). However, one major concern that arises when using PSCs to engineer cartilage for auricular reconstruction is related to subcutaneous ossification. As mentioned above, MSCs with osteochondral-lineage plasticity are prone to undergoing terminal ossification and generally cannot maintain a stable cartilage phenotype in the subcutaneous implantation site. It is unknown at this time whether cartilage engineered from PSCs would also face the same problem (Castro-Vinuelas et al. 2018). Although studies have demonstrated that PSC-derived cartilage can be more phenotypically stable than MSC-derived cartilage (Castro-Vinuelas et al. 2018), the effects of these techniques in the setting of a subcutaneous implantation site remain relatively unexplored. Moreover, the use of iPSCs as seed cells to generate cartilage for clinical application still boasts safety issues and is currently under intensive investigation. Therefore, existing efforts are more focused on assessing more clinically relevant cell sources such as MSCs and chondrocytes for auricular reconstruction.

3.2 Scaffolds for Auricular Cartilage Regeneration

Scaffold as the structural building block of an engineered tissue provides a 3D template in which seed cells can grow and produce extracellular matrix (ECM). Hence, tissue engineering can control the shape of an engineered tissue mainly by controlling the shape of its scaffold. Moreover, some hydrogel or nanoscale scaffolds can provide a 3D microenvironment to help the chondrogenic seed cells maintain or regain the spherical chondrogenic characteristics (Benya and Shaffer 1982; Kimura et al. 1984; Bonaventure et al. 1994).

Since the auricle is a sophisticated 3D structure ultimately placed in a subcutaneous environment, scaffolds for engineering an auricle should meet certain requirements, including (1) high biocompatibility to support seed cell growth and cartilage formation in the subcutaneous implantation site characterized by acute immune reaction, (2) ease of accurate manufacturing into a complex 3D structure (even better if the scaffold can be created via 3D printing), and (3) strong mechanical properties for long-term maintenance of the predesigned 3D shape, particularly under skin tension after implantation.

Many types of materials have been used as scaffold to generate cartilage for auricular reconstruction, including synthetic materials, nature-derived materials, and combinations of different types of materials.

3.2.1 Synthetic Polymers as Scaffolds for Auricular Cartilage Engineering

Synthetic polymers such as PGA, PLA, poly(ε-caprolactone) (PCL), polyurethane (Chetty et al. 2008), and their copolymers in the format of sponges, fibrous mats, hydrogels, or electrospun fibrils have been used as scaffold materials for auricular cartilage engineering (Nayyer et al. 2012). Synthetic polymers can be massively produced with batch consistency, and their material properties can be custom-tailored via chemical and physical modifications. Further, they can easily be formed into the human ear shape through molding or other techniques and have relatively strong mechanical properties by which to maintain the predesigned shape. However, the biocompatibility of these synthetic polymer scaffolds is usually unfavorable for subcutaneous cartilage formation, and they lack the surface characteristics that favor cellular attachment and growth and are prone to inducing foreign-body reactions in the subcutaneous site of immunocompetent species, hence interfering with cartilage formation (Nayyer et al. 2012). Surface modifications of these scaffolds incorporating biosignaling molecules, cell-adhesion proteins, ECM components, growth factors, and anti-inflammatory factors have all been explored to optimize the biocompatibility of the said scaffolds. However, the majority of this work has focused on articular cartilage reconstruction, and studies testing these strategies in the setting of auricular reconstruction are lacking (Nayyer et al. 2012). In vitro precultivation, which facilitates sufficient degradation of scaffold before implantation, was demonstrated to be effective in reducing the foreign-body reaction induced by biodegradable synthetic scaffolds (Liu et al. 2016), but long-term in vitro culture is time-consuming and expensive and can increase the risk of contamination. Therefore, attention has been paid to developing more biocompatible scaffolds to support subcutaneous cartilage regeneration, and nature-derived materials are generally considered to be more biocompatible than synthetic polymers.

3.2.2 Nature-Derived Materials as Scaffolds for Auricular Cartilage Engineering

The nature-derived materials that have been applied for auricular reconstruction include collagen, agarose, alginate, and decellularized cartilage ECM (Nayyer et al. 2012). These materials are usually found in the forms of sponge or hydrogel. The major advantage of using nature-derived materials is their superb biocompatibility by which to support cell attachment and low propensity to induce foreign-body reactions in the subcutaneous environment (Haisch 2010). For hydrogel scaffolds, the additional merits of the uniform encapsulation of seed cells and the ability to be formed into an accurate ear shape by injection molding exist (Faust et al. 2019). Cell-loaded hydrogels can also serve as a bio-ink for direct 3D printing of auricular-shaped constructs.

Among the nature-derived scaffold materials, calcium alginate and collagen have been successfully adopted to generate human ear-shaped cartilage in immunocompetent animals (Kamil et al. 2004). However, the mechanical properties of these scaffolds alone are too weak to maintain the sophisticated auricle shape, and stents or molds made from metal or other mechanically strong materials are usually applied to assist in shape maintenance. Although approaches such as surface modification and cross-linking can promote the mechanical properties of nature-derived scaffolds, the cross-linking media might increase the immunogenicity and cytotoxicity, and the mechanical properties of the reenforced scaffold are usually still too low for clinical application.

3.2.3 Combined Application of Materials as Scaffold for Auricular Cartilage Engineering

Since synthetic polymers generally lack biocompatibility and nature-derived materials generally lack mechanical properties, the combined application of different types of materials has been proposed to produce ear-shaped scaffolds with the desired biocompatibility and mechanical properties (Pedde et al. 2017). To maintain the reconstructed ear shape, the regenerated cartilage needs to possess a significantly high level of strength to resist tensions from the surrounding skin and scarring tissue. In this case, nondegradable inert metals such as gold or titanium have been used to assist mechanically weak materials to generate and maintain the human ear shape. Kamil et al. encapsulated mixtures of autologous chondrocyte and biodegradable polymers (i.e., calcium alginate, Pluronic F-127, and PGA) inside a human ear-shaped hollow gold mold prior to subcutaneous implantation into immunocompetent autologous hosts (swine or goat) and generated human ear-shaped cartilage with a normal anatomic definition (Kamil et al. 2004). An ear-shaped titanium stent was also used to assist mechanically inferior sheet materials (decellularized cartilage sheet or electrospun PCL sheet) to produce and maintain the ear shape (Gong et al.

2011; Xue et al. 2013). Pomerantseva et al. embedded a titanium wire into porous collagen to withstand mechanical forces and prevent shrinkage and distortion of the ear shape (Pomerantseva et al. 2016). However, these metal molds or stents cannot be regarded as intrinsic parts of the ear scaffolds given that they only play the role of physical support and do not have the ability to degrade upon the regeneration of cartilage tissue. Therefore, they either require additional surgery to be removed or present a high risk of extrusion as a metal foreign body even when they had been encased in the scaffold.

Recently, our group proposed the application of a 3D-printed PCL mesh as an inner core to replace the metal stent yet still endow the scaffold with the required mechanical strength and support for the ear shape (Zhou et al. 2018; Yin et al. 2020). PCL is a slowly degrading thermoplastic material with a low melting point (55–65 °C). Its mechanical strength can be finely tuned to approach that of a mature native cartilage by controlling the bar diameter of the 3D-printed PCL mesh. The PGA/PLA scaffold, demonstrated to support cell attachment and cartilage formation, can be easily wrapped around the PCL core through hot compression molding. It is worth noting that, owing to the low melting point of PCL, some portions of PGA fibers can fuse themselves into the PCL grids, making the PCL porous for cell infiltration after PGA/PLA degradation, which facilitates the replacement of the PCL core with the regenerated cartilage (Zhou et al. 2018). Besides PGA/PLA, nature-derived materials with better biocompatibility, such as decellularized cartilage matrix, collagen, or gelatin, can also be produced to bury the PCL core through freeze-drying and mold casting (Jia et al. 2020). These hybrid materials have shown positive results by inheriting suitable properties from both their natural precursors and PCL. Moreover, since PCL degrades slowly (the degradation time of PCL is mainly determined by its molecular weight, and the total degradation of PCL takes 3–5 years), the engineered cartilage may have sufficient time to mature and gain mechanical properties while gradually replacing the degrading inner core, which would avoid the trouble of removing the PCL stent and significantly lower the risk of stent extrusion. The combined application of different types of scaffolding materials such as PCL and hydrogel is also widely incorporated in the state-of-the-art 3D printing scenario, with PCL used to support the shape and hydrogel acting as a cell carrier.

3.3 3D Printing for Auricular Cartilage Engineering

3D printing is a process used to construct 3D objects by relying on the layer-by-layer deposition of materials onto a computer-controlled build platform (Pedde et al. 2017). In tissue engineering, 3D printing can facilitate the generation of tailor-made tissues with patient-specific geometry, through the combined use of medical imaging modalities, computer-aided design, and computer-aided manufacturing. Owing to its aesthetic impression and complex structure, human ear-shaped cartilage is frequently used as a research model to demonstrate the efficacy and superiority of 3D printing.

3D printing was originally introduced in auricular cartilage engineering to fabricate the scaffold into a patient-specific ear shape, with chondrogenic cells subsequently loaded into the scaffold for cartilage regeneration. Currently, 3D bioprinting has shown promise in the direct creation of tissue constructs, recapitulating the structural and cytoarchitectural complexities of native tissue through precise placement of cell-laden hydrogels in a layer-by-layer fashion (Kang et al. 2016). Many bioprinting techniques have been developed based on jetting, extrusion, laser-induced forward transfer (LIFT), and stereolithography (SLA) (Kang et al. 2016). The jetting method produces picoliter-scale drops with a printing resolution of 20–100 μm at high speeds (up to 10,000 droplets/s) and with a low cost, but the thickness of the printed constructs is limited because of the inadequate structural support garnered from the low-concentration, liquid-phase hydrogel being ejected (Pedde et al. 2017). Extrusion methods, which incorporate an air-pressure controller, a piston-assisted system, or a screw-assisted mechanism to continuously extrude biomaterials for layer-by-layer fabrication, can produce more stable 3D cell-laden structures using high-viscosity materials with higher cell densities (Malda et al. 2013; Wang et al. 2015), although the high printing resolution and speed achieved by involving high driving pressures and narrow nozzle diameters would create high nozzle shear forces that may reduce the cell viability (Murphy and Atala 2014; Kang et al. 2016). LIFT is a method originally developed to pattern metals and other inorganic materials onto a substrate (Pedde et al. 2017). The application of LIFT in tissue engineering has been reported to date in the fabrication of cellularized skin (Koch et al. 2012; Michael et al. 2013) and bone (Catros et al. 2011). Although LIFT is less commonly applied relative to other bioprinting systems because of its high cost, limited material versatility, low flow rates, and unwanted deposition of metallic residue, it has advantages, including high cell viabilities (exceeding 95%; Hopp et al. 2005), the precise printing of cells in relatively small constructs, and minimized clogging issues (Kang et al. 2016; Pedde et al. 2017). SLA is the highest resolution bioprinting approach currently available and relies on the irradiation of photopolymerizable macromer solutions using laser rastering or a dynamically projected light source to cross-link high-resolution patterns in the polymerization plane (Pedde et al. 2017; Melchels et al. 2010; Wang et al. 2015). Besides its high resolution, SLA is nozzle-free and can be low cost, but the widely used ultraviolet light inherent with this approach may reduce cell viability (although efforts have been made to replace the ultraviolet light with visible light), and the choices of photocurable materials with appropriate viscosities (<5 Pas) are limited (Melchels et al. 2010).

For the past 10 years, 3D printing has assisted in producing patient-specific ear-shaped scaffolds for subsequent chondrogenic cell seeding. Our group has used 3D printing-assisted approach to generate patient-specific ear-shaped scaffolds for the regeneration of human ear-shaped cartilage in vitro (Liu et al. 2010b), in nude mice (Zhang et al. 2014; Yin et al. 2020), and, recently, in a clinical setting (Zhou et al. 2018). Direct bioprinting of human ear-shaped cartilage has also been reported by a number of research groups using different scaffolds and bioprinting strategies (Otto et al. 2015; Kang et al. 2016), but their protocols have not yet been applied in preclinical animal models or clinical scenarios. The translational value of these bioprinting techniques for auricular cartilage engineering needs to be further tested.

3.4 In Vitro Generation of Human Ear-Shaped Cartilage

One primary goal of cartilage tissue engineering is to provide surgeons with a piece of high-quality cartilage that is engineered in vitro and is thus readily available. As compared with the direct implantation of freshly seeded scaffolds, in vitro culturing may offer several advantages, including the following:

1. It allows sufficient ex vivo degradation of scaffolds that are otherwise prone to induce foreign-body reaction after implantation (Liu et al. 2016).
2. It allows stem cell preinduction to commit a stable chondrogenic lineage using exogenous growth factors (Liu et al. 2008).
3. It facilitates the generation of large grafts using a low density of seed cells (Liu et al. 2008).
4. The in vitro deposited ECM may enhance attachment of cells to the scaffold and may stabilize and protect the cells, improving seed cell retention and the maturation of engineered cartilage after transplantation (Ball et al. 2004; Deponti et al. 2012; Moretti et al. 2005).
5. It may simplify surgical handling, fixation, and postoperative treatment (Moretti et al. 2005).

Fortunately, the anural, aliphatic character of native cartilage makes it an ideal target tissue for in vitro cultivation. The following section will first introduce biochemical and mechanical stimuli – two of the most important factors influencing in vitro culture conditions for auricular cartilage regeneration. Other factors such as oxygen tension and in vitro culture duration will also be covered.

3.4.1 In Vitro Biochemical Stimuli

Biochemical stimuli in culture media can affect the quality of in vitro cultured cartilage by adjusting the status of seed cells and hence can affect in vivo cartilage formation after implantation (Okubo et al. 2019). To convert seed cells from the state of self-replication to that of chondrogenic differentiation or redifferentiation, culture media are usually divided into proliferation media and chondrogenic media, respectively, by applying different types of growth factors and other ingredients such as serum, hormones, vitamins, and other chemicals. Proliferation media are mainly applied to prime cells to reduce the cell doubling time, maintain the undifferentiated phenotype of stem cells, or recover the chondrogenic phenotype postexpansion, and chondrogenic media are mainly applied to increase the synthesis of cartilage-specific ECM, accelerate cartilage formation, and maintain a stable cartilage phenotype.

Growth factors play a pivotal role in regulating cell proliferation and chondrogenesis. In auricular cartilage engineering, the most widely used growth factors include members in TGF-β superfamily (such as TGF-β1, TGF-β3, BMP-2, BMP-7), IGF-I, and bFGF (Okubo et al. 2019). TGF-β, BMPs, and IGFs are potent anabolic factors that can modulate cartilage metabolism and increase the deposition of cartilaginous ECM such as type II collagen and glycosaminoglycan (GAG). TGF-β can also maintain a chondrocyte phenotype and promote proliferation (Horton et al. 1989; Loeser and Shanker 2000; Grimaud et al. 2002). The chondroinductive actions of IGF-I were demonstrated to be equally potent to those of TGF-β1 for MSCs (Longobardi et al. 2006). IGF-I can stimulate proliferation, preserve chondrogenic potential, regulate cell apoptosis, and induce the expression of chondrocyte markers (Chiu et al. 2019; Shakibaei et al. 2006; Guenther et al. 1982; Loeser and Shanker 2000). Studies also suggest that IGF-I can improve the formation and localization of elastin (Rosa et al. 2014), which is a marker of elastic cartilage in the auricle (Chiu et al. 2019). However, IGF-I may promote the expression of collagen X, a marker of chondrocyte hypertrophy and mineralizing cartilages (Rosa et al. 2014). Meanwhile, bFGF mainly modulates cartilage metabolism (Loeser and Shanker 2000) and has been shown to elicit dose-dependent effects on chondrocyte mitotic activity but may also suppress proteoglycan synthesis (Sah et al. 1994). A synergic effect has been observed between these growth factors. TGF-β1 has been demonstrated to be able to synergistically catalyze the effect of IGF-I, and the impact of IGF-I on chondrogenesis was independent from that of TGF-β1 as indicated by the persistence of IGF-I's actions in MSCs lacking TGF-β1 signaling (Mauck et al. 2003; Longobardi et al. 2006). The combination of bFGF with insulin or IGF-I synergistically enhanced the proliferation of chondrocytes and MSCs (Munirah et al. 2010). Among these growth factors, bFGF and IGF-I are usually used in proliferation media, and TGF-β, BMPs, and IGF-I are usually used in chondrogenic media.

Besides growth factors, other medium supplements, including serum, insulin-transferrin-selenious acid premix (ITS premix), vitamin C, and dexamethasone, are usually relied upon to provide nutrition and support the growth factor function. Notably, conditions such as cell type, mechanical condition, and oxygen concentration (Jonitz et al. 2012) will influence the effects of the culture medium ingredients, and the diverse roles of ingredients may increase the level of difficulty when attempting to implement them in cartilage engineering (Trippel 1995; Faust et al. 2019).

3.4.2 Mechanical Stimuli in the In Vitro Culture Condition

Unlike articular cartilage, native auricular cartilage does not need to bear loads. However, the engineered auricular cartilage must possess a significant level of mechanical strength (more approximate to that of costal cartilage rather than that of native auricular cartilage) to effectively maintain the ear shape by resisting contractile forces stemming from the surrounding scarring soft tissue. As mentioned above, one direct way to significantly improve the immediate mechanical properties of the whole graft is by incorporating a mechanically strong scaffold. Another more bioinspired approach is the application of mechanical stimuli during in vitro culture to gradually improve the mechanical property of the regenerated cartilage tissue.

Various bioreactors, including a rotating-wall vessel, direct perfusion bioreactor, compression bioreactor, and spinner flask, have been introduced to provide static or dynamic mechanical stimuli in the forms of perfusion (laminar flow or turbulent flow), shear stress, compressive force, hydrostatic/hydrodynamic pressure, or their combinations (Concaro et al. 2009). These mechanical stimuli appear beneficial in promoting mass transfer within constructs (Moretti et al. 2005), activating growth factor signals (Yang and Barabino 2011), maintaining the spherical chondrocyte morphology, improving cell growth, regulating cell distribution, and permitting cartilage-specific ECM secretion (Graceffa et al. 2019). For the in vitro cultivation of a cartilaginous graft with a large volume and complex structure such as the auricle, these beneficial effects may translate to improved structural integrity (i.e., less void region in the central part) and enhanced mechanical properties (Vunjak-Novakovic et al. 1999; Faust et al. 2019).

However, for the current in vitro engineered cartilage, parameters such as GAG content, cell viability, and water content usually can be reached up to the level of native tissue, but the production and accumulation of some chondrocyte-specific ECM macromolecules – particularly collagen type II in cultured constructs – are insufficient when compared with native cartilage tissue, leading to the low mechanical properties of the in vitro engineered cartilage (Yan et al. 2009; Chen et al. 2014; Graceffa et al. 2019). Therefore, novel approaches need to be developed to allow more homogenous and mechanically robust cartilage regeneration in vitro.

3.4.3 Oxygen Tension

Physiologically speaking, articular cartilage is bathed in synovial fluid with low oxygen tension in joint capsule. Many studies have also reported the beneficial effect of a hypoxia environment (1–5%) on cartilage regeneration by enhancing cartilage-specific gene or protein expression, regulating apoptosis, and preventing terminal differentiation (Thoms et al. 2013; Browe et al. 2019), although the presence of hypoxia may also increase the expression of collagen type I (Lee et al. 2013). Nevertheless, the physiological subcutaneous environment for auricular cartilage is rich in nutrients and oxygen from the adjacent vascularized tissues (Moretti et al. 2005), and the effect of hypoxia is less emphasized in the context for auricular regeneration.

3.4.4 In Vitro Culture Duration

In vitro culture duration is an important factor that addresses the key question of how developed an engineered graft should be to support optimal cartilage repair or reconstruction (Moretti et al. 2005). However, no consensus has yet been reached on this matter. In a nude mouse study employing septal chondrocytes and PGA/PLA scaffold, Rotter et al. observed only minor differences in subcutaneous cartilage formation when the engineered constructs were precultured for either 1 day or 3 weeks (Rotter et al. 2002). In immunocompetent animal models (swine osteochondral model and rabbit or goat subcutaneous model) using autologous BMSCs or auricular chondrocytes as seed cells and PGA/PLA as scaffolds, our group found that the in vitro engineered cartilage displayed a time-dependent maturation process and prolonged in vitro precultivation (more than 2 weeks for the rabbit subcutaneous model, more than 4 weeks for the swine osteochondral model, and more than 8 weeks for goat subcutaneous model), which could alleviate postimplantation inflammation and support stable cartilage formation (Luo et al. 2009; Liu et al. 2016; He et al. 2017). However, other groups (using nude mouse or autologous goat subcutaneous or osteochondral models, auricular or articular chondrocytes as seed cells, and fibrin, hyaluronic acid, or collagen as scaffolds) suggested that a short-term culture period (1–3 weeks) improved in vivo chondrogenesis whereas too long of a culture time led to worse results in spite of the existence of better maturation in vitro (Deponti et al. 2012; Miot et al. 2012; Bichara et al. 2014). This disparity could be explained by variables, including the different animal models (immunodeficient nude mouse vs. immunocompetent large animal), implantation sites (osteochondral vs. subcutaneous), scaffold types (synthetic scaffold vs. nature-derived scaffold), differentiation stages at the time of cell seeding, and cell-scaffold interactions (Moretti et al. 2005). Generally speaking, in an immunodeficient in vivo environment such as the nude mouse, or when using a biocompatible scaffold that would only induce mild foreign-body reactions in vivo, long-term precultivation is not necessary. Conversely, in immune-hostile environments such as the subcutaneous environment of immunocompetent species or when using a scaffold with suboptimal biocompatibility, prolonged in vitro precultivation could promote stable in vivo cartilage formation by allowing time for sufficient degradation of the inflammation inducing scaffolds. Moreover, if the seed cells appear in a suitable in vitro condition with good cell-scaffold interaction, prolonged in vitro culture would be more likely to promote in vivo cartilage development. Otherwise, too long of an in vitro culture could be harmful. Nevertheless, from a therapeutic perspective, long-term in vitro cultivation is labor-intensive and cost-prohibitive, may increase the risk of contamination, and may complicate the application process for approvals from regulatory agencies such as the US Food and Drug Administration. Therefore, future studies are encouraged to find ways to shorten the in vitro culture duration while generating mature in vitro cartilage for clinical application.

3.5 In Vivo and Preclinical Evaluations

Although in vitro studies reveal the potential of tissue-engineered auricles based on various cell sources, scaffolds, 3D printing, and biochemical and biomechanical stimuli, more in-depth knowledge regarding their in vivo fate must be gathered. For auricular reconstruction, a large number of in vivo studies continue to use the immunodeficient nude mouse as animal model (Nayyer et al. 2012). The nude mouse model plays an important role as an initial stage in evaluating the chondrogenic capacity of constructs in vivo and may provide a crucial link between in vitro investigations and complex and costly preclinical large animal studies (Moretti et al. 2005). Moreover, the nude mouse model supports the investigation of the in vivo fate of cartilaginous grafts engineered from human cells (Moretti et al. 2005). Following the first "ear-mouse" report in 1997 (Cao et al. 1997), the successful generation of human ear-shaped cartilage in the nude mouse model using different types of seed cells and scaffolds has been reported (Nayyer et al. 2012; Yin et al. 2020).

However, the subsequent efforts in a preclinical large animal model have been less successful because of the occurrence of immune attacks against the foreign scaffolds used to provide an auricular shape in the subcutaneous implantation site (Saim et al. 2000; Kamil et al. 2004; Shieh et al. 2004). To address this issue, hydrogel or nature-derived scaffolds with superb biocompatibility by which to support large animal subcutaneous cartilage formation were tested regarding the potential to generate human ear-shaped cartilage. Saim et al. injected Pluronic F-127 hydrogel loaded with chondrocytes into a skin-fold channel positioned in the shape of the auricle helix on the ventral surface of an autologous swine and generated helix-shaped cartilage (Saim et al. 2000). Chang et al. created chondrocyte-loaded alginate implants with specific shapes through an injection-molding process that formed mature cartilage after 6 months of subcutaneous implantation into autologous sheep (Chang et al. 2003). Kamil et al. delivered autologous chondrocytes inside an ear-shaped and ear-sized hollow gold mold in combination with calcium alginate or Pluronic F-127 and created a complete, anatomically refined auricle in a large animal model (Kamil et al. 2004). Pomerantseva et al. reported the growth of a stable, human ear-shaped cartilage in a preclinical sheep model using expanded chondrocytes and a titanium wire-supported collagen scaffold (Pomerantseva et al. 2016). However, the scaffolds adopted in these studies present weaker mechanical properties and may require the involvement of a metal mold or inner wire stent to support the ear shape, which will introduce additional complexity when removing the mold or will raise the risk of metal extrusion.

Recently, our group has developed a scaffold-free approach to generate a cartilage sheet in vitro. Without the involvement of the scaffold as a potential foreign body, these cartilage sheets exhibit superb biocompatibility, and we formed homogenous cartilage blocks of a large size (10 cm in diameter) and thickness (9 mm) by stacking several layers of sheets together and subcutaneously implanting them into an autologous goat. The regenerated cartilage block was then hand-carved into the auricle shape and reimplanted into the autologous goat for auricular reconstruction (Fig. 1, unpublished data). This study may address the issue of large donor site morbidity by using autologous auricular chondrocytes to generate the cartilage sheets since these cells are highly proliferative owing to their CNC origin (Tay et al. 2004; Taihi et al. 2019). To date, approaches for chondrogenic redifferentiation of postexpanded chondrocytes have been established by several groups (Yanaga et al. 2009; Pomerantseva et al. 2016; Zhou et al. 2018). However, the generation of the ear shape still relies upon the old-fashioned hand-carving skill.

3.6 Clinical Translation of Tissue-Engineered Cartilage for Auricular Reconstruction

Challenges restricting the bench-to-bed progress of tissue-engineered cartilage for auricular reconstruction are tremendous (Haisch 2010), and thus literature reports on this subject matter are very few and far between, especially when compared with those on cartilage engineering for orthopedic repair (Liu et al. 2017). Yanaga et al. described a two-stage approach where autologous chondrocytes (expanded in vitro)

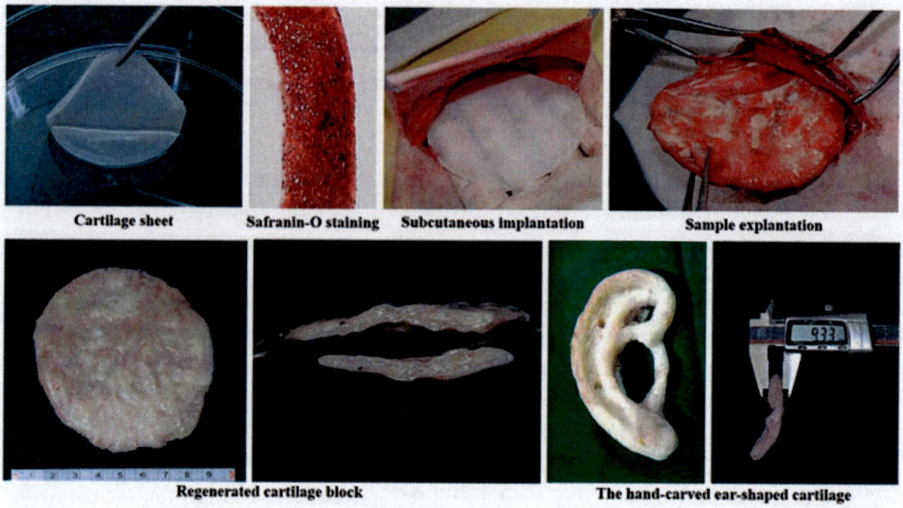

Fig. 1 Generation of human ear-shaped cartilage in an autologous goat model based on scaffold-free cartilage sheets. Cartilage sheets are generated by a scaffold-free approach. Five layers of the cartilage sheets are stacked and subcutaneously implanted into an autologous goat to generate a homogenous cartilage block of a large size (10 cm in diameter) and thickness (9 mm). The regenerated cartilage block was then hand-carved into the auricle shape

derived from the patient's microtic ear were injected into the lower abdomen of the patient to form cartilage blocks, which were then hand-carved into an ear framework to reconstruct the auricle for the patient (Yanaga et al. 2013). Recently, our group engineered patient-specific ear-shaped cartilage with proper mechanical strength in vitro based on expanded autologous microtia chondrocytes and a biodegradable PGA/PLA-PCL scaffold (Zhou et al. 2018). The regenerated cartilage was used for auricle reconstruction in five microtia patients using three different surgical approaches according to the specific condition of each patient. Postsurgical magnetic resonance imaging revealed the gradual degradation of the PCL inner core, and biopsies taken at 6 months postsurgery revealed mature elastic cartilage formation in all five patients. The achievement of this clinical translation can be attributed to the integration and innovation of several strategies, including using expanded autologous auricular chondrocytes as seed cell sources, establishing a chemically defined in vitro culture condition to alleviate the host's response toward the implanted graft, incorporating computer-aided design and manufacturing technique for patient-specific cartilage shape control, and adopting a PCL inner core for both immediate and long-term postoperative shape maintenance (Zhou et al. 2018).

Nevertheless, we still have a long way to go before tissue-engineered cartilage can be accepted as a clinically available treatment for auricular reconstruction. At present, the in vitro engineered ear graft (neocartilage) was still more delicate and fragile than the graft carved from fully developed rib cartilage, and the acute inflammatory trauma environment and the excessive handling during surgery, together with fibrogenesis of the surrounding soft tissue during wound healing, may reduce the viability of the resident chondrocytes of the engineered ear graft, thus hindering the subsequent chondrogenesis and shape maintenance efforts after implantation. Therefore, surgical procedures need to be refined to work with the fragile tissue-engineered graft at this stage. To promote the widespread application of engineered ear grafts in the future, a backward bed-to-bench process to engineer a cartilaginous graft more proximate to the native cartilage is necessary (Martin et al. 2018).

4 Conclusions

Total reconstruction of the external ear is one of the most challenging procedures performed in the realm of plastic and reconstructive surgery because of the lack of an ideal graft presenting an accurate patient-specific ear shape, biocompatibility, and appropriate mechanical properties while inducing only minimal donor site morbidity (Nayyer et al. 2012). With the ongoing fast development in cell biology, materials science, engineering, and advanced manufacturing techniques such as 3D printing, tissue-engineered cartilage may constitute an alternative scheme for traditional ear reconstruction. The 1997 inaugural report coupled with the press coverage gave people the impression that a tissue engineering-based solution for auricular reconstruction is just around the next corner (Cao et al. 1997), but the reality is different. Obstacles restricting clinical translation still exist, including the lack of autologous seed cell sources with sufficient quantity and chondrogenic function and the lack of scaffold that is biocompatible to support subcutaneous cartilage regeneration and mechanically robust enough to resist skin tension. Recently, our group reported the in vitro generation of tissue-engineered cartilage and its clinical translation (Zhou et al. 2018), yet this tissue engineering solution still cannot replace the traditional approach. Even when we see the light of clinical translation, so to speak, new issues arise, such as how to better evaluate the seed cells and associated ex vivo expansion procedures to ensure long-term safety and function, how to generate a cartilage graft ex vivo with qualities and mechanical properties comparable to those of the native tissue, and how to refine the surgical procedures to better preserve both the viability and shape of the graft during surgical handling and after being implanted in the acute inflammatory trauma environment coupled with fibrogenesis of the surrounding soft tissue during wound healing. These issues support the need for a backward bed-to-bench process with the aim of further improving the quality of the engineered cartilage to better suit the clinical situation by refining each building block of tissue engineering (Martin et al. 2018). Endeavors in bioengineering and more in-depth collaborations between cell biologists, developmental scientists, materials experts, engineers, surgeons, and officers who conduct clinical and market entrance regulations are warranted to help tissue-engineered cartilage become a commercially viable and widespread alternative to autologous reconstruction.

Acknowledgments The authors appreciate the support from the Program of Shanghai Academic/Technology Research Leader (19XD1431100).

References

Ahmed MR, Mehmood A et al (2014) Combination of ADMSCs and chondrocytes reduces hypertrophy and improves the functional properties of osteoarthritic cartilage. Osteoarthr Cartil 22(11):1894–1901

Ball ST, Goomer RS et al (2004) Preincubation of tissue engineered constructs enhances donor cell retention. Clin Orthop Relat Res 420:276–285

Benya PD, Shaffer JD (1982) Dedifferentiated chondrocytes reexpress the differentiated collagen phenotype when cultured in agarose gels. Cell 30(1):215–224

Berghaus A (2007) Implants for reconstructive surgery of the nose and ears. GMS Curr Top Otorhinolaryngol Head Neck Surg 6:Doc06

Bichara DA, O'Sullivan NA et al (2012) The tissue-engineered auricle: past, present, and future. Tissue Eng Part B Rev 18(1):51–61

Bichara DA, Pomerantseva I et al (2014) Successful creation of tissue-engineered autologous auricular cartilage in an immunocompetent large animal model. Tissue Eng Part A 20(1–2): 303–312

Bly RA, Bhrany AD et al (2016) Microtia reconstruction. Facial Plast Surg Clin North Am 24 (4):577–591

Bonaventure J, Kadhom N et al (1994) Reexpression of cartilage-specific genes by dedifferentiated human articular chondrocytes cultured in alginate beads. Exp Cell Res 212(1):97–104

Brent B (1992) Auricular repair with autogenous rib cartilage grafts: two decades of experience with 600 cases. Plast Reconstr Surg 90(3):355–374; discussion 375–376

Brent B (2002) Microtia repair with rib cartilage grafts: a review of personal experience with 1000 cases. Clin Plast Surg 29(2):257–71, vii

Brittberg M, Lindahl A et al (1994) Treatment of deep cartilage defects in the knee with autologous chondrocyte transplantation. N Engl J Med 331(14):889–895

Browe DC, Coleman CM et al (2019) Hypoxia activates the PTHrP–MEF2C pathway to attenuate hypertrophy in mesenchymal stem cell derived cartilage. Sci Rep 9(1):13274

Cao Y, Vacanti JP et al (1997) Transplantation of chondrocytes utilizing a polymer-cell construct to produce tissue-engineered cartilage in the shape of a human ear. Plast Reconstr Surg 100 (2):297–302; discussion 303–304

Castro-Vinuelas R, Sanjurjo-Rodriguez C et al (2018) Induced pluripotent stem cells for cartilage repair: current status and future perspectives. Eur Cell Mater 36:96–109

Catros S, Fricain JC et al (2011) Laser-assisted bioprinting for creating on-demand patterns of human osteoprogenitor cells and nano-hydroxyapatite. Biofabrication 3(2):025001

Chang SC, Tobias G et al (2003) Tissue engineering of autologous cartilage for craniofacial reconstruction by injection molding. Plast Reconstr Surg 112(3):793–799; discussion 800–801

Chen JL, Duan L et al (2014) Extracellular matrix production in vitro in cartilage tissue engineering. J Transl Med 12:88

Cheng A, Hardingham TE et al (2014) Generating cartilage repair from pluripotent stem cells. Tissue Eng Part B Rev 20(4):257–266

Chetty A, Steynberg T et al (2008) Hydroxyapatite-coated polyurethane for auricular cartilage replacement: an in vitro study. J Biomed Mater Res A 84(2):475–482

Chiu L, Weber JF et al (2019) Engineering of scaffold-free tri-layered auricular tissues for external ear reconstruction. Laryngoscope 129(8):E272–E283

Concaro S, Gustavson F et al (2009) Bioreactors for tissue engineering of cartilage. Adv Biochem Eng Biotechnol 112:125–143

De Bari C, Dell'Accio F et al (2004) Failure of in vitro-differentiated mesenchymal stem cells from the synovial membrane to form ectopic stable cartilage in vivo. Arthritis Rheum 50(1):142–150

De Miguel MP, Fuentes-Julian S et al (2012) Immunosuppressive properties of mesenchymal stem cells: advances and applications. Curr Mol Med 12(5):574–591

Deponti D, Di Giancamillo A et al (2012) Fibrin-based model for cartilage regeneration: tissue maturation from in vitro to in vivo. Tissue Eng Part A 18(11–12):1109–1122

Evans MJ, Kaufman MH (1981) Establishment in culture of pluripotential cells from mouse embryos. Nature 292(5819):154–156

Faust HJ, Guo Q et al (2019) Chapter 53. Cartilage tissue engineering. In: Atala A, Lanza R, Mikos AG, Nerem R (eds) Principles of regenerative medicine, 3rd edn. Academic Press, Boston, pp 937–952

Firmin F (1998) Ear reconstruction in cases of typical microtia. Personal experience based on 352 microtic ear corrections. Scand J Plast Reconstr Surg Hand Surg 32(1):35–47

Friedenstein AJ, Petrakova KV et al (1968) Heterotopic of bone marrow. Analysis of precursor cells for osteogenic and hematopoietic tissues. Transplantation 6(2):230–247

Gibson JD, O'Sullivan MB et al (2017) Regeneration of articular cartilage by human ESC-derived mesenchymal progenitors treated sequentially with BMP-2 and Wnt5a. Stem Cells Transl Med 6(1):40–50

Gillies H (1920) Plastic surgery of the face. H. Frowde, Hodder & Sougton, London

Gong YY, Xue JX et al (2011) A sandwich model for engineering cartilage with acellular cartilage sheets and chondrocytes. Biomaterials 32(9):2265–2273

Graceffa V, Vinatier C et al (2019) Chasing chimeras – the elusive stable chondrogenic phenotype. Biomaterials 192:199–225

Grimaud E, Heymann D et al (2002) Recent advances in TGF-beta effects on chondrocyte metabolism. Potential therapeutic roles of TGF-beta in cartilage disorders. Cytokine Growth Factor Rev 13(3):241–257

Guenther HL, Guenther HE et al (1982) Effect of insulin-like growth factor on collagen and glycosaminoglycan synthesis by rabbit articular chondrocytes in culture. Experientia 38 (8):979–981

Haisch A (2010) Ear reconstruction through tissue engineering. Adv Otorhinolaryngol 68:108–119

He A, Liu L et al (2017) Repair of osteochondral defects with in vitro engineered cartilage based on autologous bone marrow stromal cells in a swine model. Sci Rep 7:40489

Hopp B, Smausz T et al (2005) Survival and proliferative ability of various living cell types after laser-induced forward transfer. Tissue Eng 11(11–12):1817–1823

Horton WJ, Higginbotham JD et al (1989) Transforming growth factor-beta and fibroblast growth factor act synergistically to inhibit collagen II synthesis through a mechanism involving regulatory DNA sequences. J Cell Physiol 141(1):8–15

Huselstein C, Li Y et al (2012) Mesenchymal stem cells for cartilage engineering. Biomed Mater Eng 22(1–3):69–80

Ichinose S, Yamagata K et al (2005) Detailed examination of cartilage formation and endochondral ossification using human mesenchymal stem cells. Clin Exp Pharmacol Physiol 32(7):561–570

Jessop ZM, Javed M et al (2016) Combining regenerative medicine strategies to provide durable reconstructive options: auricular cartilage tissue engineering. Stem Cell Res Ther 7:19

Jia L, Zhang Y et al (2020) Regeneration of human-ear-shaped cartilage with acellular cartilage matrix-based biomimetic scaffolds. Appl Mater Today 20:100639

Jonitz A, Lochner K et al (2012) TGF-beta1 and IGF-1 influence the re-differentiation capacity of human chondrocytes in 3D pellet cultures in relation to different oxygen concentrations. Int J Mol Med 30(3):666–672

Kagimoto S, Takebe T et al (2016) Autotransplantation of monkey ear perichondrium-derived progenitor cells for cartilage reconstruction. Cell Transplant 25(5):951–962

Kamil SH, Vacanti MP et al (2004) Microtia chondrocytes as a donor source for tissue-engineered cartilage. Laryngoscope 114(12):2187–2190

Kang N, Liu X et al (2012) Effects of co-culturing BMSCs and auricular chondrocytes on the elastic modulus and hypertrophy of tissue engineered cartilage. Biomaterials 33(18):4535–4544

Kang N, Liu X et al (2013) Different ratios of bone marrow mesenchymal stem cells and chondrocytes used in tissue-engineered cartilage and its application for human ear-shaped substitutes in vitro. Cells Tissues Organs 198(5):357–366

Kang HW, Lee SJ et al (2016) A 3D bioprinting system to produce human-scale tissue constructs with structural integrity. Nat Biotechnol 34(3):312–319

Kawanabe Y, Nagata S (2006) A new method of costal cartilage harvest for total auricular reconstruction: part I. Avoidance and prevention of intraoperative and postoperative complications and problems. Plast Reconstr Surg 117(6):2011–2018

Kimura T, Yasui N et al (1984) Chondrocytes embedded in collagen gels maintain cartilage phenotype during long-term cultures. Clin Orthop Relat Res 186:231–239

Kobayashi S, Takebe T et al (2011) Reconstruction of human elastic cartilage by a CD44+ CD90+ stem cell in the ear perichondrium. Proc Natl Acad Sci U S A 108(35):14479–14484

Koch L, Deiwick A et al (2012) Skin tissue generation by laser cell printing. Biotechnol Bioeng 109 (7):1855–1863

Lee HH et al (2013) Hypoxia enhances chondrogenesis and prevents terminal differentiation through P13K/Akt/FoxO dependent anti-apoptotic effect. Sci Rep 3:2683

Levorson EJ, Santoro M et al (2014) Direct and indirect co-culture of chondrocytes and mesenchymal stem cells for the generation of polymer/extracellular matrix hybrid constructs. Acta Biomater 10(5):1824–1835

Li D, Zhu L et al (2017) Stable subcutaneous cartilage regeneration of bone marrow stromal cells directed by chondrocyte sheet. Acta Biomater 54:321–332

Liu K, Zhou GD et al (2008) The dependence of in vivo stable ectopic chondrogenesis by human mesenchymal stem cells on chondrogenic differentiation in vitro. Biomaterials 29(14):2183–2192

Liu X, Sun H et al (2010a) In vivo ectopic chondrogenesis of BMSCs directed by mature chondrocytes. Biomaterials 31(36):9406–9414

Liu Y, Zhang L et al (2010b) In vitro engineering of human ear-shaped cartilage assisted with CAD/CAM technology. Biomaterials 31(8):2176–2183

Liu Y, Li D et al (2016) Prolonged in vitro precultivation alleviates post-implantation inflammation and promotes stable subcutaneous cartilage formation in a goat model. Biomed Mater 12(1):015006

Liu Y, Zhou G et al (2017) Recent progress in cartilage tissue engineering – our experience and future directions. Engineering 3(1):28–35

Loeser RF, Shanker G (2000) Autocrine stimulation by insulin-like growth factor 1 and insulin-like growth factor 2 mediates chondrocyte survival in vitro. Arthritis Rheum 43(7):1552–1559

Longobardi L, O'Rear L et al (2006) Effect of IGF-I in the chondrogenesis of bone marrow mesenchymal stem cells in the presence or absence of TGF-beta signaling. J Bone Miner Res 21(4):626–636

Luo X, Zhou G et al (2009) In vitro precultivation alleviates post-implantation inflammation and enhances development of tissue-engineered tubular cartilage. Biomed Mater 4(2):025006

Malda J, Visser J et al (2013) 25th Anniversary article: engineering hydrogels for biofabrication. Adv Mater 25(36):5011–5028

Martin GR (1981) Isolation of a pluripotent cell line from early mouse embryos cultured in medium conditioned by teratocarcinoma stem cells. Proc Natl Acad Sci U S A 78(12):7634–7638

Martin I, Jakob M et al (2018) From tissue engineering to regenerative surgery. EBioMedicine 28:11–12

Mauck RL, Nicoll SB et al (2003) Synergistic action of growth factors and dynamic loading for articular cartilage tissue engineering. Tissue Eng 9(4):597–611

Melchels FP, Feijen J et al (2010) A review on stereolithography and its applications in biomedical engineering. Biomaterials 31(24):6121–6130

Michael S, Sorg H et al (2013) Tissue engineered skin substitutes created by laser-assisted bioprinting form skin-like structures in the dorsal skin fold chamber in mice. PLoS One 8(3): e57741

Miot S, Brehm W et al (2012) Influence of in vitro maturation of engineered cartilage on the outcome of osteochondral repair in a goat model. Eur Cell Mater 23:222–236

Moretti M, Wendt D et al (2005) Effects of in vitro preculture on in vivo development of human engineered cartilage in an ectopic model. Tissue Eng 11(9–10):1421–1428

Morita Y, Yamamoto S et al (2015) Development of a new co-culture system, the "separable-close co-culture system," to enhance stem-cell-to-chondrocyte differentiation. Biotechnol Lett 37 (9):1911–1918

Munirah S, Samsudin OC et al (2010) Expansion of human articular chondrocytes and formation of tissue-engineered cartilage: a step towards exploring a potential use of matrix-induced cell therapy. Tissue Cell 42(5):282–292

Murphy SV, Atala A (2014) 3D bioprinting of tissues and organs. Nat Biotechnol 32(8):773–785

Nagata S (1993) A new method of total reconstruction of the auricle for microtia. Plast Reconstr Surg 92(2):187–201

Nayyer L, Patel KH et al (2012) Tissue engineering: revolution and challenge in auricular cartilage reconstruction. Plast Reconstr Surg 129(5):1123–1137

Ohara K, Nakamura K et al (1997) Chest wall deformities and thoracic scoliosis after costal cartilage graft harvesting. Plast Reconstr Surg 99(4):1030–1036

Okubo R, Asawa Y et al (2019) Proliferation medium in three-dimensional culture of auricular chondrocytes promotes effective cartilage regeneration in vivo. Regen Ther 11:306–315

Olshinka A, Louis M et al (2017) Autologous ear reconstruction. Semin Plast Surg 31(3):146–151

Otto IA, Melchels FP et al (2015) Auricular reconstruction using biofabrication-based tissue engineering strategies. Biofabrication 7(3):032001

Paput L, Czeizel AE et al (2012) Possible multifactorial etiology of isolated microtia/anotia – a population-based study. Int J Pediatr Otorhinolaryngol 76(3):374–378

Park C (2000) Subfascial expansion and expanded two-flap method for microtia reconstruction. Plast Reconstr Surg 106(7):1473–1487

Park C, Lee TJ et al (1991) A single-stage two-flap method of total ear reconstruction. Plast Reconstr Surg 88(3):404–412

Pedde RD, Mirani B et al (2017) Emerging biofabrication strategies for engineering complex tissue constructs. Adv Mater 29(19):1606061

Pelttari K, Winter A et al (2006) Premature induction of hypertrophy during in vitro chondrogenesis of human mesenchymal stem cells correlates with calcification and vascular invasion after ectopic transplantation in SCID mice. Arthritis Rheum 54(10):3254–3266

Pelttari K, Pippenger B et al (2014) Adult human neural crest-derived cells for articular cartilage repair. Sci Transl Med 6(251):251ra119

Pomerantseva I, Bichara DA et al (2016) Ear-shaped stable auricular cartilage engineered from extensively expanded chondrocytes in an immunocompetent experimental animal model. Tissue Eng Part A 22(3–4):197–207

Reighard CL, Hollister SJ et al (2018) Auricular reconstruction from rib to 3D printing. J 3D Print Med 2(1):35–41

Rosa RG, Joazeiro PP et al (2014) Growth factor stimulation improves the structure and properties of scaffold-free engineered auricular cartilage constructs. PLoS One 9(8):e105170

Rotter N, Bonassar LJ et al (2002) Age dependence of biochemical and biomechanical properties of tissue-engineered human septal cartilage. Biomaterials 23(15):3087–3094

Sah RL, Chen AC et al (1994) Differential effects of bFGF and IGF-I on matrix metabolism in calf and adult bovine cartilage explants. Arch Biochem Biophys 308(1):137–147

Saim AB, Cao Y et al (2000) Engineering autogenous cartilage in the shape of a helix using an injectable hydrogel scaffold. Laryngoscope 110(10 Pt 1):1694–1697

Shakibaei M, Seifarth C et al (2006) Igf-I extends the chondrogenic potential of human articular chondrocytes in vitro: molecular association between Sox9 and Erk1/2. Biochem Pharmacol 72 (11):1382–1395

Shieh SJ, Terada S et al (2004) Tissue engineering auricular reconstruction: in vitro and in vivo studies. Biomaterials 25(9):1545–1557

Taihi I, Nassif A et al (2019) Head to knee: cranial neural crest-derived cells as promising candidates for human cartilage repair. Stem Cells Int 2019:9310318

Takahashi K, Yamanaka S (2006) Induction of pluripotent stem cells from mouse embryonic and adult fibroblast cultures by defined factors. Cell 126(4):663–676

Tanzer RC (1959) Total reconstruction of the external ear. Plast Reconstr Surg Transplant Bull 23 (1):1–15

Tay AG, Farhadi J et al (2004) Cell yield, proliferation, and postexpansion differentiation capacity of human ear, nasal, and rib chondrocytes. Tissue Eng 10(5–6):762–770

Thoms BL, Dudek KA et al (2013) Hypoxia promotes the production and inhibits the destruction of human articular cartilage. Arthritis Rheum 65(5):1302–1312

Thomson JA, Itskovitz-Eldor J et al (1998) Embryonic stem cell lines derived from human blastocysts. Science 282(5391):1145–1147

Trippel SB (1995) Growth factor actions on articular cartilage. J Rheumatol Suppl 43:129–132

Uppal RS, Sabbagh W et al (2008) Donor-site morbidity after autologous costal cartilage harvest in ear reconstruction and approaches to reducing donor-site contour deformity. Plast Reconstr Surg 121(6):1949–1955

Vunjak-Novakovic G, Martin I et al (1999) Bioreactor cultivation conditions modulate the composition and mechanical properties of tissue-engineered cartilage. J Orthop Res 17(1):130–138

Walton RL, Beahm EK (2002) Auricular reconstruction for microtia: part II. Surgical techniques. Plast Reconstr Surg 110(1):234–249; quiz 250–251, 387

Wang M, Rahnama R et al (2013) Trophic stimulation of articular chondrocytes by late-passage mesenchymal stem cells in coculture. J Orthop Res 31(12):1936–1942

Wang Z, Abdulla R et al (2015) A simple and high-resolution stereolithography-based 3D bioprinting system using visible light crosslinkable bioinks. Biofabrication 7(4):045009

Wiggenhauser PS, Schantz JT et al (2017) Cartilage engineering in reconstructive surgery: auricular, nasal and tracheal engineering from a surgical perspective. Regen Med 12(3):303–314

Wu L, Leijten JC et al (2011) Trophic effects of mesenchymal stem cells increase chondrocyte proliferation and matrix formation. Tissue Eng Part A 17(9–10):1425–1436

Wu L, Prins HJ et al (2012) Trophic effects of mesenchymal stem cells in chondrocyte co-cultures are independent of culture conditions and cell sources. Tissue Eng Part A 18(15–16):1542–1551

Xue J, Feng B et al (2013) Engineering ear-shaped cartilage using electrospun fibrous membranes of gelatin/polycaprolactone. Biomaterials 34(11):2624–2631

Yan D, Zhou G et al (2009) The impact of low levels of collagen IX and pyridinoline on the mechanical properties of in vitro engineered cartilage. Biomaterials 30(5):814–821

Yanaga H, Imai K et al (2009) Generating ears from cultured autologous auricular chondrocytes by using two-stage implantation in treatment of microtia. Plast Reconstr Surg 124(3):817–825

Yanaga H, Imai K et al (2013) Two-stage transplantation of cell-engineered autologous auricular chondrocytes to regenerate chondrofat composite tissue: clinical application in regenerative surgery. Plast Reconstr Surg 132(6):1467–1477

Yang YH, Barabino GA (2011) Requirement for serum in medium supplemented with insulin-transferrin-selenium for hydrodynamic cultivation of engineered cartilage. Tissue Eng Part A 17 (15–16):2025–2035

Yin Z, Li D et al (2020) Regeneration of elastic cartilage with accurate human-ear shape based on PCL strengthened biodegradable scaffold and expanded microtia chondrocytes. Appl Mater Today 20:100724

Zhang L, He A et al (2014) Regeneration of human-ear-shaped cartilage by co-culturing human microtia chondrocytes with BMSCs. Biomaterials 35(18):4878–4887

Zhou G, Jiang H et al (2018) In vitro regeneration of patient-specific ear-shaped cartilage and its first clinical application for auricular reconstruction. EBioMedicine 28:287–302

Zhu Y, Zhang Y et al (2015) The influence of Chm-I knockout on ectopic cartilage regeneration and homeostasis maintenance. Tissue Eng Part A 21(3–4):782–792

Stem Cell-Based and Tissue Engineering Approaches for Skeletal Muscle Repair

Seraina A. Domenig, Andrew S. Palmer, and Ori Bar-Nur

Contents

S. A. Domenig · A. S. Palmer · O. Bar-Nur (✉)
Laboratory of Regenerative and Movement Biology, Institute of Human Movement Sciences and Sport, Department of Health Sciences and Technology, Swiss Federal Institute of Technology (ETH) Zurich, Schwerzenbach, Switzerland
e-mail: seraina.domenig@hest.ethz.ch; ori.bar-nur@hest.ethz.ch

D. Eberli et al. (eds.), *Organ Tissue Engineering*, Reference Series in Biomedical Engineering, https://doi.org/10.1007/978-3-030-44211-8_19

Abstract

Skeletal muscle tissue exhibits significant regeneration capacity upon injury or disease. This intrinsic regeneration potential is orchestrated by stem cells termed satellite cells, which undergo activation and differentiation in response to muscle insult, giving rise to fusion-competent myogenic progenitors responsible for tissue rejuvenation. Skeletal muscle diseases such as Duchenne muscular dystrophy are characterized by progressive loss of muscle mass which precipitates reduced motility, paralysis, and in some occurrences untimely death. A manifold of muscle pathologies involve a failure to efficiently regenerate the muscle tissue, rendering stem cell-based approaches an attractive therapeutic strategy. Here we will present past and contemporary methods to treat skeletal muscle degeneration by stem cell therapy, covering prominent challenges facing this technology and potential means to overcome current hurdles. A primary focus of this chapter is directed toward illustrating innovative ways to utilize stem cells alone or in conjunction with biomaterials and tissue engineering techniques to remedy Duchenne muscular dystrophy or volumetric muscle loss.

1 Introduction

Human locomotion is carried out by skeletal muscle, a soft tissue composed of postmitotic multinucleated myofibers that contract to generate movement. In addition, skeletal muscle tissue contains mononucleated muscle resident cells in the form of fibroadipogenic progenitors, satellite, and endothelial and hematopoietic cells which play pivotal roles in skeletal muscle tissue homeostasis and repair (Fig. 1) (Giordani et al. 2019; Dell'Orso et al. 2019; Wosczyna and Rando 2018; Rubenstein et al. 2020).

Adult skeletal muscle harbors a high propensity for regeneration, which is dependent on resident stem cells within the skeletal muscle tissue known as satellite cells (Mauro 1961). These unique cells are located between the sarcolemma and basal lamina and are typically quiescent under homeostatic conditions, undergoing activation and proliferation during injury or disease to repair the tissue (Snow 1978; Schultz et al. 1978). Skeletal muscle regeneration follows a well-defined process via a myogenic differentiation program. In adult skeletal muscle, this process initiates when quiescent satellite cells (QSCs) undergo activation to form transient-amplifying activated satellite cells (ASCs) also known as myoblasts. These progenitors can further differentiate into myocytes, which are fusion-competent cells that can merge with muscle fibers to regenerate the tissue (Comai and Tajbakhsh 2014; Yin et al. 2013). This unique process is governed by key transcription factors specific to the myogenic program. QSCs express high levels of the transcription factor Paired Box Protein 7 (Pax7) as well as other unique genes such as Sprouty1 and Notch receptors (Seale et al. 2000; Shea et al. 2010; Mourikis and Tajbakhsh 2014). Upon activation, satellite cells downregulate quiescent genes while upregulating activation genes such as myoblast determination protein 1 (MyoD) and Myf5 (Comai and Tajbakhsh

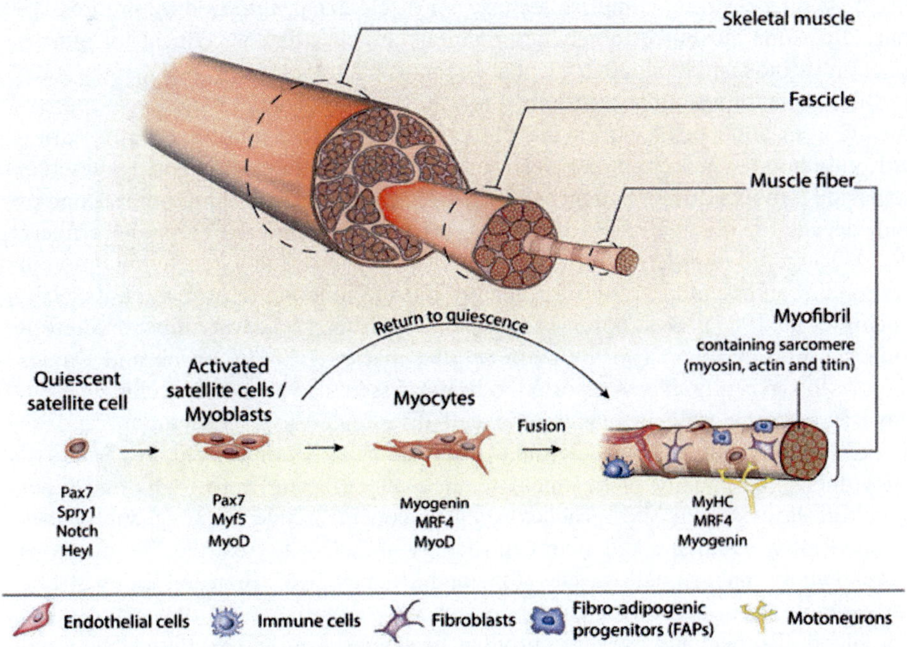

Fig. 1 Skeletal muscle tissue is composed of multinucleated aligned muscle fibers as well as other muscle resident cells. Regeneration of skeletal muscle is a step-wise process initiated by activation of quiescent satellite cells, which give rise to highly proliferative myoblasts and fusion-competent myocytes that merge with each other or with damaged myofibers to regenerate the tissue. In addition to muscle cells, a milieu of resident cells in the form of fibroadipogenic progenitors, endothelial, white blood cells, and other cell types reside in skeletal muscle tissue, playing key roles in its homeostasis and regeneration

2014; Yin et al. 2013). Myocytes further upregulate myogenin and MRF4, whereas differentiated multinucleated myofibers express various myosin heavy chain (MyHC) isoforms and a plethora of other differentiation-associated genes (Fig. 1) (Yin et al. 2013; Comai and Tajbakhsh 2014). The capacity to regenerate skeletal muscle is dependent on satellite cells, as their ablation completely abrogates skeletal muscle regeneration following injury (Murphy et al. 2011; Lepper et al. 2011; Sambasivan et al. 2011). In addition, mice lacking Pax7 exhibit growth retardation and small muscle mass size and typically die several weeks to months post birth (Seale et al. 2000; Oustanina et al. 2004). Recent research utilizing single cell sequencing methods have demonstrated that satellite cells display a large heterogeneity during homeostasis or regeneration, which is mirrored in their transcriptional expression profile (Yartseva et al. 2020; De Micheli et al. 2020).

Muscle regeneration is impaired in a wide array of pathological conditions such as traumatic muscle injury in the form of volumetric muscle loss (VML)(Corona et al. 2016), cancer-associated cachexia (Baracos et al. 2018), age-associated sarcopenia (Dennison et al. 2017), or muscular dystrophies (Motohashi et al.

2019). The latter group comprise a cluster of diseases characterized by progressive muscle wasting attributed to genetic mutations in genes that are critical for muscle function (Motohashi et al. 2019). Duchenne muscular dystrophy (DMD) is the most prevalent form of muscular dystrophy. It is an X-linked disorder that affects approximately 1 in 5000 boys worldwide, inflicting progressive muscle atrophy during early childhood, which eventually leads to complete muscle wasting and dysfunction (Yiu and Kornberg 2015). Patients with DMD experience death during the second or third decade of life as a result of cardiac or respiratory failure (Yiu and Kornberg 2015). The rapid muscle loss in DMD patients is attributed to a genetic mutation in dystrophin, an essential protein for healthy muscle function (Hoffman et al. 1987; Koenig et al. 1987). Dystrophin is expressed in differentiated myofibers, where it connects the sarcolemma to the extra cellular matrix (ECM) (Nowak and Davies 2004). More recently, it was reported to be expressed also in satellite cells, where it plays an essential role in asymmetric cell division during regeneration, and its absence significantly impairs satellite cell self-renewal (Dumont et al. 2015). As of today, there is a paucity of treatment for patients suffering from DMD and other muscle diseases such as VML, cachexia, and sarcopenia, rendering novel approaches in regenerative medicine and stem cell therapy attractive to treat these ailments as was shown for other somatic tissues (Mao and Mooney 2015; Blau and Daley 2019). For example, stem cell therapy via bone marrow engraftment is routinely used to treat blood disorders and more recently has been successfully used to treat skin and ocular diseases, raising hope for future stem cell therapies for skeletal muscle diseases (Blau and Daley 2019).

In this book chapter, we will present various stem cell-based and tissue engineering approaches to treat muscle diseases, with emphasis on treating patients who suffer from DMD or VML. We will further present past and contemporary challenges in utilizing stem cells or tissue engineering to treat muscle diseases and discuss innovative ways to resolve them. Lastly, we will highlight future research directions spearheaded by these technologies for skeletal muscle reparative therapy.

2 Cell Replacement Therapy for the Treatment of Muscle Disorders

Cell replacement therapy is an attractive approach to treat muscle-associated diseases characterized by cell death such as muscular dystrophies. The ultimate goal of this approach is to repair muscle damage via engraftment of healthy donor-derived cells into a disease-afflicted muscle tissue (Blau and Daley 2019). Candidate cell types proposed for this therapeutic approach include muscle stem or progenitor cells in the form of myoblasts (Partridge et al. 1989) or satellite cells (Montarras et al. 2005; Collins et al. 2005). Alternatively, other cell types can also be utilized via systemic delivery (Cossu et al. 2015; Sampaolesi et al. 2003; Torrente et al. 2004). The candidate cells can be healthy donor-derived allogeneic cells or autologous cells which are derived from the patient and therefore may require an ex vivo gene correction procedure (Fig. 2). To date, a majority of completed clinical trials have

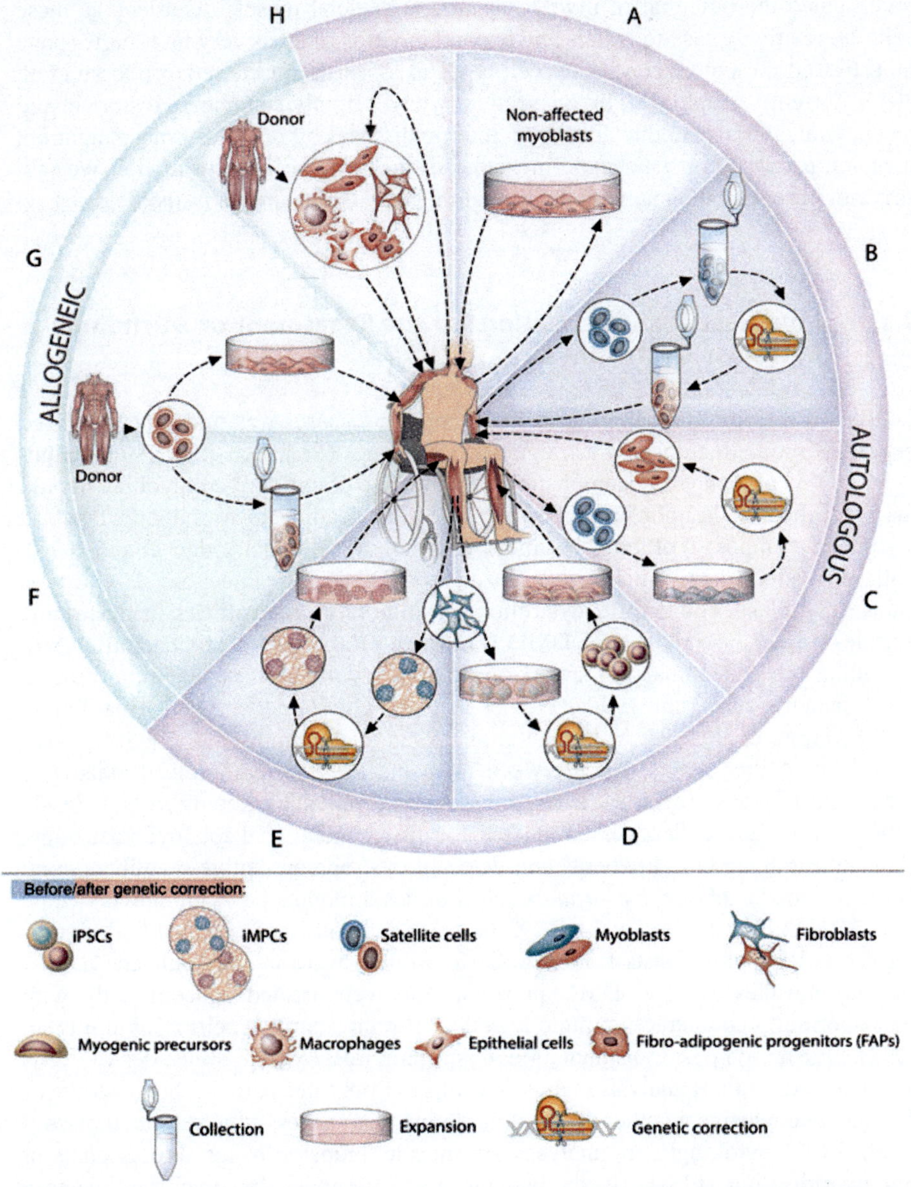

Fig. 2 Allogeneic and autologous cell replacement therapy approaches to treat muscular dystrophies. (**a**) Autologous myoblasts derived from healthy muscles are engrafted into dystrophic muscles. (**b**) Autologous satellite cells are genetically corrected during a transient ex vivo phase followed by engraftment. (**c**) Autologous myoblasts are transplanted back into muscles after expansion and ex vivo genetic correction phase. (**d**) Patient-derived induced pluripotent stem cells (iPSCs) are genetically corrected and differentiated into disease-free muscle precursors, which are engrafted back into the patient. (**e**) Patient-derived induced myogenic progenitor cells (iMPCs) are genetically corrected and engrafted back into the patient. (**f**) Engraftment of allogeneic

investigated the potential of myoblasts to treat skeletal muscle disorders, as these cells are relatively easy to isolate and expand in vitro, whereas very few studies have investigated the potential of other cell types to assist in muscle repair (Cossu et al. 2015; Torrente et al. 2007). In this section, we will briefly summarize historical and recent clinical trials conducted to treat muscle diseases by myoblast transplantation, focusing primarily on Duchenne muscular dystrophy (DMD) as a model. We will strive to introduce the main challenges and recent advancements in the field.

2.1 Myoblast Transplantation for the Treatment of Duchenne Muscular Dystrophy

Cell-based therapy as an approach to treat muscle diseases was first conceptualized in a 1978 seminal study by Partridge and colleagues (Partridge et al. 1978). Highlights of this pivotal study demonstrated that mouse donor-derived myoblasts could fuse with host cells upon intramuscular injection, leading to formation of hybrid myofibers composed of both recipient and donor cells (Partridge et al. 1978). Follow-up studies utilizing this approach established that injection of wild-type mouse myoblasts could form dystrophin-positive hybrid myofibers in dystrophic muscles of a mouse model for DMD (Partridge et al. 1989; Karpati et al. 1989), providing the conceptual framework to attempt myoblast transplantation trials in DMD patients during the 1990s (Law et al. 1991; Gussoni et al. 1992; Huard et al. 1992; Karpati et al. 1993; Tremblay et al. 1993a, b; Morandi et al. 1995; Mendell et al. 1995; Miller et al. 1997). Between 1990 and 1997, several clinical trials were conducted to assess myoblast transplantation as a potential therapy to treat DMD patients. Law and colleagues conducted the first clinical trial for myoblast transplantation in three DMD patients and detected dystrophin-positive myofibers in all of the patients' *extensor digitorum brevis* muscles 3 months post cell delivery (Law et al. 1990, 1991). In the notable 1992 Gussoni and colleagues clinical trial, donor-derived cultured myoblasts from a father or sibling were injected into the *tibialis anterior* muscles of eight DMD patients, who were treated concomitantly with cyclosporine to counteract immune rejection of donor-derived cells (Gussoni et al. 1992). One month post transplantation, dystrophin mRNA from donor-derived cells was detected via PCR analysis in muscle samples from three patients; however it was deemed inconclusive whether dystrophin protein was successfully restored (Gussoni et al. 1992). Strikingly, re-analyses of muscle biopsies using fluorescence in situ hybridization (FISH) of six patients after intramuscular myoblast injection revealed fused and unfused donor nuclei in all examined muscles of the patients

Fig. 2 (continued) satellite cells derived from a healthy donor into dystrophic muscles. (**g**) Engraftment of allogeneic myoblasts derived from a healthy donor into dystrophic muscles. (**h**) Co-engraftment of muscle cells with other cell types involved in muscle repair could enhance muscle regeneration and remedy dystrophic symptoms

(Gussoni et al. 1997). In parallel and following the Gussoni study, additional clinical trials utilizing similar experimental designs were completed, however differed in the use of immunosuppressive drugs, injection strategies, and interpretation of outcome (Karpati et al. 1993; Huard et al. 1992; Tremblay et al. 1993a, b; Morandi et al. 1995; Mendell et al. 1995; Miller et al. 1997). An extensive comparison of these studies has been previously reported (Skuk 2004). Collectively, these studies confirmed that myoblast transplantation in humans is feasible and safe; however the clinical outcome was deemed deficient (Karpati et al. 1993; Gussoni et al. 1992; Huard et al. 1992; Tremblay et al. 1993a, b; Morandi et al. 1995; Mendell et al. 1995; Miller et al. 1997). Of note, a few studies reported slight beneficial effects on muscle strength following myoblast transplantation (Huard et al. 1992; Karpati et al. 1993; Tremblay et al. 1993a; Law et al. 1991), however the majority of studies did not asses or report on functional improvement and failed to detect dystrophin-positive myofibers in patients (Morandi et al. 1995; Tremblay et al. 1993b; Mendell et al. 1995; Miller et al. 1997). Possible explanations for the unfavorable clinical outcome have been extensively discussed (Negroni et al. 2016; Skuk and Tremblay 2014), with attributing factors deemed to be (i) poor myoblast survival in host muscles; (ii) immune rejection against donor-derived myoblasts; (iii) failure of myoblasts to efficiently migrate from the injection site; and (iv) inability of cultured myoblasts to restore the muscle stem cell reservoir and elicit robust regeneration (Karpati et al. 1993; Wilschut et al. 2012; Maffioletti et al. 2014; Fan et al. 1996; Guerette et al. 1995; Qu et al. 1998). The rather disappointing human clinical trials conducted during the 1990s impeded subsequent clinical trials and their ramifications linger to some extent to this day.

2.2 Addressing Translational Roadblocks for Myoblast Transplantation

Despite the results of the unsuccessful clinical trials conducted during the 1990s, more recent studies performed on non-human primates have addressed a portion of the aforementioned challenges with the objective of improving myoblast transplantation (Table 1) (Skuk and Tremblay 2014).

One such approach is to modulate the type of immunosuppressive drugs administered to counteract immunological responses in patients. Administration of immunosuppressive drugs for successful allogeneic myoblast transplantation has been rigorously demonstrated in both rodent and non-human primate studies (Kinoshita et al. 1994; Pavlath et al. 1994; Kinoshita et al. 1996). Historically, cyclosporine was the preferred immunosuppressant used in the majority of early human trials (Gussoni et al. 1992; Morandi et al. 1995; Mendell et al. 1995; Miller et al. 1997). However, more recent studies have shown that cyclosporine impedes myoblast fusion and differentiation (Hong et al. 2002). Alternative treatment of non-human primates with the immunosuppressant tacrolimus have led to long-term engraftment of allogeneic myoblasts (Kinoshita et al. 1996), resulting in tacrolimus replacing cyclosporine as the immunosuppressant of choice.

Table 1 Select studies that have investigated myoblast transplantation in non-human primates

Study/topic	Primary experiments	Results/findings
Kinoshita et al. (1996) **Immune system**	Intramuscular injection of allogeneic myoblasts (*LacZ*) into NHPs treated with tacrolimus or without immunosuppressant	Lymphocyte infiltration and increase in cytokines 7 days post MT in the absence of tacrolimus. ß-Gal-positive fibers were not detected 4 weeks post MT Detection of ß-Gal-positive fibers 1, 4, and 12 weeks post MT with significant reduced lymphocyte infiltration and cytokines with tacrolimus compared to no tacrolimus administration
Skuk et al. (1999) **Notexin**	Allogeneic myoblasts (*LacZ*) were injected with or without notexin and varying cell number (4×10^6, 8×10^6, 24×10^6) with 35–40 injections (separated by 1 mm)	8×10^6 led to most ß-Gal-positive fibers 4 weeks post MT Injection of 8×10^6 myoblasts suspended in notexin led to a 50% increase of ß-Gal fibers 4 weeks post MT
Skuk et al. (2002) **Injection, immune system**	Injection of allogeneic myoblasts (*LacZ*) with varying inter-injection distances (1 mm, 1–1.5 mm, 2 mm) and different concentrations of tacrolimus alone or combined with mycophenolate mofetil	Detection of 25–67% ß-Gal-positive fibers 1 month post MT (injection distance 1 mm) Combination of tacrolimus with mycophenolate mofetil enabled a decreased dosage of tacrolimus
Lafreniere et al. (2009) **Growth factors**	Intramuscular injection of allogeneic myoblasts (*LacZ*) combined either with bFGF or IGF-1	Growth factors increased myoblast migration 60 h post MT No significant differences detected in ß-Gal-positive fiber formation 4 weeks post MT
Skuk et al. (2011) **Migration**	Allogeneic myoblasts (*LacZ*) were subcutaneously injected above a needle-damaged muscle	Detection of migrated myoblasts that formed ß-Gal-positive myofibers within a damaged muscle along the needle trajectories 1 cm apart from injected site, 1 month post subcutaneous MT
Skuk et al. (2013) **Electroporation**	A 1 mm^2 muscle area was electroporated with 3 pulses of 400 V/cm before allogeneic myoblasts (*LacZ*) were injected either in a matrix consisting of 25 injections or radial injection	Myofiber necrosis 1 day and maximal tissue distortion 7 days post electroporation Complete muscle regeneration 2–3 months post electroporation Increased ß-Gal-positive fibers after electroporation compared to injection alone (1.6–6.3-fold for matrices and 2.5–9 for radial injections)
Skuk et al. (2014) **Injection**	Intramuscular injection of allogeneic myoblasts (*LacZ*) with varying needle sizes (27 G, 22 G, 18 G), amount of cells (10^2, 10^3, 10^4, 10^5, 10^6/cm injection trajectory, and injected volume (1 ul, 5 ul, 25 ul/cm injection trajectory)	Muscle sections revealed that an injection of at least 100,000 cells/cm in 1 μl/cm trajectory with a 27 G generated highest amounts of ß-Gal-positive fibers 1 or 2 months post MT

(continued)

Table 1 (continued)

Study/topic	Primary experiments	Results/findings
Skuk and Tremblay (2017a) **Cell death**	Injection of radiolabeled autologous or allogeneic myoblasts (with or without *LacZ*) into NHPs (with or without immunosuppressant)	Majority of cells (80–92%) died within the first 7 days regardless of whether 1), they were autologous or allogeneic or 2) if the recipient NHP received an immunosuppressant drug No further cell death 3 weeks post injection
Skuk and Tremblay (2017b) **Engraftment**	Intramuscular injection of myoblasts (*LacZ*) and subsequent assessment of cell engraftment (after 1 h, 1 day, 3 days, 7 days and 3 weeks)	Detection of majority of myoblasts in between the muscle bundle and the perimysium 1 h and 1 day post MT Detection of ß-Gal-positive myofibers along the injection trajectories 3 weeks post MT

NHPs non-human primates, *MT* myoblast transplantation, *ß-Gal* ß-galactosidase

Inflicting localized muscle damage by irradiation, mechanical pressure, freezing, or chemicals is a commonly used method to enhance muscle stem and progenitor cell-engraftment in rodent transplantation models (Mueller and Bloch 2019; Hardy et al. 2016). Moreover, studies performed in non-human primates reported partial increase in engrafted myofibers when myoblasts were transplanted into injured muscle tissue by notexin or electroporation (Skuk et al. 1999, 2000, 2013). However, inflicting muscle damage prior to transplantation is both ethically and technically challenging to translate to human patients due to potential adverse side effects (Skuk et al. 2002). Therefore, myoblast engraftment has been investigated without inducing muscle damage with the exception of localized insult caused by the injection needle (Skuk et al. 2002). Of note, the non-human primate studies demonstrated migration of myoblasts toward damaged myofibers and engraftment at the injection site (Skuk et al. 2002, 2011). Based on these observations, it was suggested that in the absence of extensive muscle damage, multiple close proximity injections will be required to achieve substantial engraftment of transplanted myoblasts (Skuk et al. 2002, 2006b, 2011). This hypothesis has led to two clinical trials involving human patients that utilized several novel strategies (Table 2) (Skuk et al. 2004, 2006a, 2007, 2010; Skuk and Tremblay 2016). In the first study, nine DMD patients were injected with high-density allogeneic myoblasts in the *tibialis anterior* muscle while being treated in parallel with the immunosuppressant tacrolimus (Skuk et al. 2006a). In contrast to previous trials, between 3.5% and 26% dystrophin-positive myofibers were detected in sampled muscles, while patients did not develop a major immune reaction (Skuk et al. 2006a). In a separate study, an entire *thenar eminence* muscle was injected with myoblasts to assess the safety of high-density injections (Skuk et al. 2007). The procedure was well tolerated by the patient without severe adverse effects or complications. The patient further showed an increase in force generation in comparison to pre-transplantation (Skuk et al. 2007). Remarkably, donor-derived muscle cells

Table 2 Myoblast transplantation trials in DMD and oculopharyngeal muscular dystrophy (OPMD) patients between 2004 and 2014

Study	**Patient cohort** (Number/age demographic)	**Donor**	**Study protocol:** (a) Quantity of injected myoblasts/muscle type (b) Injection method (c) Immunosuppressive drug (d) Blinded vs. non-blinded	**Results:** (a) Molecular findings (b) Functional findings
Skuk et al. (2004, 2006a, 2010), Skuk and Tremblay (2016)	9 patients/age: 8–17 years	Father or mother	(a) 18.5–30 × 10^6 into TA (b) 0.25 cm^3 with 25 injections 1–2 mm apart with 25- or 22-gauge needle; or 1 cm^3 with 25 or 100 injections 1–2 mm apart with 27-, 26- or 25-gauge needle (c) Tacrolimus (d) Non-blinded	(a) Dystrophin-positive myofibers detected in TA of 8 patients (3.5–26%), specific antibody staining showed that dystrophin-positive myofibers were donor-derived in 7 patients. Detection of two groups of dystrophin-positive myofibers: Large fibers derived from fusion with existing fibers and small fibers derived from fusion of donor cells alone, donor mRNA was detected in 9 patients, detection of mononuclear donor-derived myoblasts in the satellite cell niche 4 weeks post MT, detection of antibodies against donor cells in two patients (already detected pre-transplantation) (b) Not assessed
Skuk et al. (2007)	1 patient/age: 26 years	Father	(a) 275 × 10^6 into TE (whole muscle), 538 × 10^6 into BB (whole muscle), 135 × 10^6 into G (b) High-density injections (separated by 1 mm) with 27- or 26-gauge needle (c) Tacrolimus (d) Non-blinded	(a) Dystrophin-positive myofibers detected in G (27.5% 1 month, 34.5% 18 months post MT), massive fibrosis and adipose tissue in BB and only 2 dystrophin-positive myofibers detected, detection of donor-derived mRNA in BB and G (b) Increased strength of first metacarpal flexion post MT (70% 1 month, 100% 2 months, 50% 15 months post MT compared to pre-transplantation)
Perie et al. (2014)	12 patients/age: 18–75 years	Cells derived from patient's unaffected Q or SCM	(a) 178 × 10^6 (median) into pharyngeal muscles (b) Myotomy before injection, 12 injections, 10 cm^2 (c) Not administered (d) Non-blinded	(a) Myoblast engraftment was not assessed (b) Upper esophageal sphincter function improved in 8 patients 2 years post MT, increased quality of life score in 12 patients, no degradation in swallowing function in 10 patients

MT myoblast transplantation, *TA* tibialis anterior, *BB* biceps brachii, *TE* thenar eminence, *G* gastrocnemius, *Q* quadriceps, *SCM* sternocleidomastoid remained at the engraftment site 18 months post transplantation, suggesting a long-term engraftment potential (Skuk et al. 2007). Currently, safety and functional outcome of myoblast transplantation throughout extensor *carpi radialis* muscles of DMD patients are being tested in a clinical phase I/II trial (NCT02196467).

Inflicting confined muscle damage to provide a niche for cell engraftment carries potential significant health risks for human DMD patients. However in the context of blood diseases, bone marrow ablation by irradiation prior to blood stem cell infusion is a well-established treatment procedure (Blau and Daley 2019). Therefore, it may be of interest to assess if localized ablation of muscle cells by irradiation followed by high-proximity injection of myoblasts into dystrophic muscles could elicit migration beyond the injection site, achieving robust regeneration. Given evident experimental success in animal models, there is potential to adapt this approach in human patients during future trials.

2.3 Clinical Procedures Involving Autologous Myoblast Transplantation

To date, only allogeneic myoblasts with or without immunosuppressive drug administration have been used to treat human DMD patients (Gussoni et al. 1992; Huard et al. 1992; Karpati et al. 1993; Tremblay et al. 1993a, b; Morandi et al. 1995; Mendell et al. 1995; Miller et al. 1997; Skuk et al. 2006a, 2007; Law et al. 1991). In contrast, autologous myoblast transplantation has been successfully employed in animal models and several clinical trials for urinary incontinence (Mitterberger et al. 2008; Sebe et al. 2011; Blaganje and Lukanovic 2012; Stangel-Wojcikiewicz et al. 2014; Jankowski et al. 2018; Eberli et al. 2012), fecal incontinence (Frudinger et al. 2015; Romaniszyn et al. 2015; Boyer et al. 2018; Bisson et al. 2015), as well as oculopharyngeal muscular dystrophy (OPMD)(Perie et al. 2014). For further reading on these procedures, we refer the readers to recently published reviews (Hillary et al. 2020; Simillis et al. 2019; Trebol et al. 2018). During these trials, immunosuppressive drugs were not administered, thereby simplifying the procedure and reducing the risk of adverse health complications. Notably, a recent success in utilizing autologous myoblasts to treat OPMD has been reported (Perie et al. 2014) (Fig. 2a). This type of muscular dystrophy is characterized by progressive degradation and weakening of the ocular and pharyngeal muscles, leading to ptosis and dysphagia (Brais et al. 1998). OPMD is caused by an autosomal gene mutation that results in amplification of GCG nucleotides in the PABPN1 gene (Brais et al. 1998). Currently, symptoms are treated by a cricopharyngeal myotomy; however, the long-term benefits of this procedure remain controversial (Duranceau et al. 1980; Perie et al. 2014). In rare disparity to other muscular dystrophies, not all muscles of OPMD patients are affected (Perie et al. 2006). Comparisons of myoblasts derived from affected and non-affected muscles of OPMD patients have shown that the latter exhibited higher proliferation and myogenic regeneration capacity (Perie et al. 2006). This observation was followed by a non-placebo controlled clinical trial that tested myoblast transplantation in OPMD patients (Perie et al. 2014).

Autologous myoblasts derived from unaffected *quadriceps* or *sternocleidomastoid* muscles were injected into pharyngeal muscles of 12 OPMD patients after myotomy (Perie et al. 2014). Remarkably, this procedure successfully ceased further degradation of pharyngeal muscles in patients 2 years post-surgery in addition to improving quality of life criteria factors in all patients (Perie et al. 2014). To date, this study is one of a handful to demonstrate a beneficial therapeutic outcome of myoblast transplantation. In the future, it will be central to corroborate these promising results with additional experiments that include placebo-controlled groups and patients subjected to myoblast engraftment with or without prior myotomy. Autologous myoblast transplantation from non-afflicted muscles may be applicable to other muscular dystrophies such as facioscapulohumeral muscular dystrophy and could provide a beneficial method to treat muscle diseases (Fig. 2a) (Vilquin et al. 2005).

Autologous myoblasts expressing a shorter but functional micro-dystrophin have been successfully transplanted into non-human primates (Quenneville et al. 2007). However, autologous myoblast transplantation in human cell engraftment patients has not been reported, although rapid advances in genome engineering techniques may pave way for investigating such an approach in the future (Doudna and Charpentier 2014; Briggs and Morgan 2013; Min et al. 2019) (Fig. 2c). Harnessing the CRISPR-Cas9 genome editing system to correct the dystrophin mutation in the Dmd^{mdx} mouse model was recently reported by several seminal studies, demonstrating functional restoration of the dystrophin protein in myofibers (Long et al. 2016; Tabebordbar et al. 2016; Nelson et al. 2016) and potentially muscle stem cells (Tabebordbar et al. 2016; Nance et al. 2019). Dmd^{mdx} myoblasts were also genetically corrected in vitro using CRISPR-Cas9 and reported to give rise to dystrophin-positive fibers upon transplantation (Zhu et al. 2017; Matre et al. 2019; Ousterout et al. 2015). With respect to correction of human dystrophin mutations, Young and colleagues recently developed a strategy to fix a mutational hot spot by a CRISPR-Cas9 exon skipping-based approach, which encompasses approximately 60% of all DMD patient mutations (Young et al. 2016). In this elegant study, DMD patient-derived fibroblasts were first reprogrammed into induced pluripotent stem cells (iPSCs) (Young et al. 2016). Correction of the dystrophin mutation was performed at the pluripotent state, followed by directed differentiation of the iPSCs into dystrophin-positive muscle cells (Young et al. 2016). By taking a similar experimental approach, myoblasts could be expanded from patients, corrected by genome engineering, and then transplanted back into patients (Fig. 2c). Similarly, it might be feasible to correct satellite cells during a short-lived ex vivo phase followed by engraftment (Fig. 2b). However, it is important to note that restoration of dystrophin expression by transplantation of autologous myoblasts carries the risk of eliciting an immune reaction against a "foreign" protein, necessitating potential use of immunosuppressants (Selvaraj et al. 2019).

2.4 Further Considerations for Improvement of Myoblast Transplantation

The overarching goal of cell replacement therapy for muscle diseases is to establish the long-term regeneration potential of transplanted cells, which requires contribution to the muscle stem cell reservoir. However, it remains unclear whether myoblasts carry an efficient propensity to induce such long-term regeneration capacity, albeit some studies indicate feasibility (Yao and Kurachi 1993; Heslop et al. 2001; Skuk et al. 2010). Notably, a clinical trial in DMD human patients detected a presence of donor-derived mononuclear cells in muscles, suggestive of contribution to the satellite cell niche (Skuk et al. 2006a, 2010). In support of this observation, dystrophin-positive myofibers have been detected in DMD patients 18 months post myoblast transplantatio n (Skuk et al. 2007). Similar observations have been documented in myoblast transplantation trials for the treatment of fecal or urinary incontinences (Jankowski et al. 2018; Frudinger et al. 2015; Romaniszyn et al. 2015), although cell engraftment was not thoroughly assessed (Jankowski et al. 2018; Frudinger et al. 2015).

Whereas the contribution of transplanted myoblasts to long-term muscle regeneration has yet to be fully elucidated, satellite cells exhibit a high regeneration capacity and can efficiently contribute to the muscle stem cell reservoir (Montarras et al. 2005; Collins et al. 2005; Cerletti et al. 2008). In the next sections, we will discuss advances in direct isolation and transplantation of satellite cells or engraftment after an ex vivo propagation phase. We will further discuss new techniques that could allow partial retainment of satellite-cell characteristics in vitro. These novel approaches could pave the way for utilizing satellite cells instead of myoblasts as the preferred cells for engraftment in clinical settings.

3 Engraftment of Freshly Isolated Satellite Cells for Muscle Regeneration

Despite the immense therapeutic potential of utilizing satellite cells to treat muscle-associated diseases, unresolved challenges have limited their use in cell-based therapies. Shortly after QSCs are mechanically and enzymatically dissociated from muscle fibers they undergo activation, a process characterized by rapid upregulation of myoblast-associated genes and commitment to a myogenic differentiation program (Machado et al. 2017; Pietrosemoli et al. 2017; van Velthoven et al. 2017). This renders the therapeutic use of satellite cells for muscle transplantation challenging, as their engraftment potential decreases once expanded in vitro (Montarras et al. 2005; Collins et al. 2005; Quarta et al. 2016; Xu et al. 2015b; Sacco et al. 2008). To overcome this challenge, satellite cells can be isolated by a fluorescence-activated cell sorting (FACS) machine and injected directly into muscles. A manifold of reports have shown that by utilizing this approach, injected mouse or human satellite cells could significantly contribute to muscle regeneration (Montarras et al. 2005; Quarta et al. 2016; Xu et al. 2015b; Sherwood et al. 2004; Kuang et al. 2007). Moreover, engrafted satellite cells can enter their respective niche under the basal lamina (Xu et al. 2015b; Kuang et al. 2007), sustain multiple rounds of injury-induced regeneration (Collins et al. 2005; Rocheteau et al. 2012), and in addition be subjected to serial transplantation for up to seven rounds in mice (Rocheteau et al. 2012). Furthermore, engrafted myofiber fragments containing adjoined satellite cells can promote robust muscle regeneration and repair (Collins et al. 2005; Hall et al.

2010; Marg et al. 2014). Strikingly, a seminal study reported that a single satellite cell can regenerate a significant segment of skeletal muscle tissue in mice (Sacco et al. 2008).

Satellite cells comprise a subset of the mononucleated cell fraction in skeletal muscle tissue (Dell'Orso et al. 2019; Giordani et al. 2019; Rubenstein et al. 2020). This necessitates usage of genetic reporters or identification of surface markers for prospective isolation and transplantation of homogenous cell populations. Genetic reporters for Pax3 and Pax7 have been successfully employed to isolate murine satellite cells (Fig. 3) (Bosnakovski et al. 2008; Sambasivan et al. 2009; Montarras et al. 2005); however isolation of humans satellite cells using this approach is not feasible. Alternatively, surface markers can be applied to isolate satellite cells by a FACS machine. To this end, several murine-positive surface markers have been identified including CD34, Cxcr4/ITGB1, Vcam1, and alpha-7 integrin (Sherwood et al. 2004; Liu et al. 2013; Beauchamp et al. 2000; Fukada et al. 2004; Cerletti et al. 2008; Kuang et al. 2007). Similarly, several positive surface markers have been successfully identified that allow prospective isolation of human muscle stem cells including CD56/CD29 (Xu et al. 2015b), CD56/CD29/CXCR4 (Garcia et al. 2018), CD56/ITGA7 (Castiglioni et al. 2014), CD82/CD318 (Uezumi et al. 2016), and CD82 (Alexander et al. 2016). These efforts to identify, isolate, and characterize muscle stem cells have provided a plethora of new knowledge in respect to transplantation methods and assessment of satellite cell regenerative potential. For

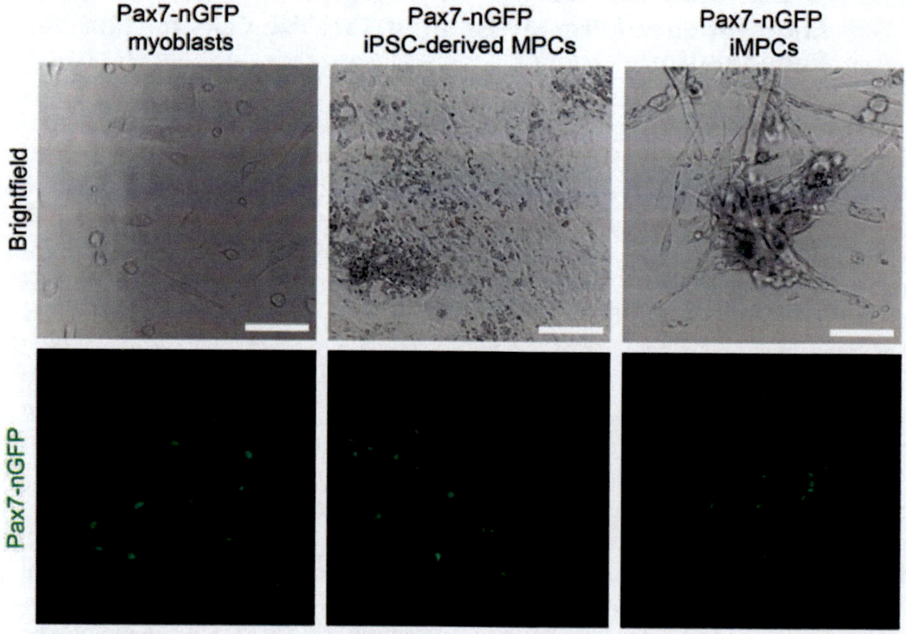

Fig. 3 Pax7 nuclear GFP reporter expression in cultured myoblasts, iPSC-derived myogenic precursors and fibroblast-derived induced myogenic progenitors (iMPCs). Scale bar, 100 μm

example, studies have reported that freshly isolated human satellite cells efficiently fuse with myofibers of immunodeficient or dystrophic mice (Alexander et al. 2016; Uezumi et al. 2016; Xu et al. 2015b). Furthermore, 5 weeks post muscle transplantation, human spectrin, and dystrophin-positive myofibers were detected in mouse muscles, and putative human satellite cells were identified under the basal lamina, suggesting contribution to the satellite cell niche (Xu et al. 2015b). Remarkably, engrafted human satellite cells were also shown to sustain two rounds of serial transplantation in mice (Garcia et al. 2018).

Akin to myoblasts, satellite cells cannot be delivered systemically and do not efficiently migrate across the muscle beyond the injection area. Due to this limitation, Garcia and colleagues opted to transplant satellite cells across multiple injection sites within the same muscle, demonstrating that as few as 2000 human satellite cells have led to the formation of approximately 155 human spectrin-positive myofibers (Garcia et al. 2018). Additionally, they detected increased quantities of human donor-derived $Pax7^{+}$ cells around the injection sites (Garcia et al. 2018). The authors suggest that satellite cell-delivery utilizing multiple injection sites is essential and more important than the number of injected cells (Garcia et al. 2018). This observation is reminiscent of the enhanced myoblast engraftment potential seen in human patients transplanted using close-proximity injections (Skuk et al. 2006a).

An additional limiting factor of using satellite cells for regenerative medicine purposes is low extraction yield of cells from allogeneic muscle donors. An alternative source for obtaining high quantities of satellite cells could be cadavers, as quiescent satellite cells have been shown to maintain their regenerative capacity in muscles for up to 17 days post mortem (Latil et al. 2012) or following hypothermic storage (Marg et al. 2014). These intriguing observations suggest that muscles of deceased individuals could potentially be used for satellite cell extraction, similar to donor-derived human organs purposed for transplantation.

Although pure myogenic stem cells can be isolated from muscles via a suite of surface markers, no clinical trials to date have investigated the potential of freshly isolated human satellite cells to remedy muscle-associated diseases in humans. Taking into consideration evidence from transplantation experiments of human myoblasts and satellite cells, one can cautiously postulate that engraftment of satellite cells into dystrophic muscles of DMD patients will be superior to that of myoblasts. To test this hypothesis, it will be of interest to assess the capacity of freshly isolated human satellite cells and myoblasts to engraft into muscles of non-human primates, similar to reports that have assessed myoblast engraftment in non-human primates (Skuk et al. 2011; Skuk and Tremblay 2017a; Skuk et al. 2002). Skeletal muscle of primates may provide a more receptive host niche for engraftment of human muscle stem cells due to species similarity and size. Successes in generating human-primate hybrid muscles will provide salient insights into the propensity of freshly isolated human satellite cells to engraft, migrate, and regenerate muscles, in addition to bearing implications for adapting this technique in humans.

Notable challenges of satellite cell utilization in cell therapy include low cell extraction yield, poor cell migration, and an inability to deliver the cells systemically. In consideration of these limitations, it might be beneficial to exploit satellite cells in clinical settings for muscles that are of small size and in conjunction with close-proximity injections. For example, localized satellite cell transplantation into finger muscles of DMD patients might curtail muscle degeneration and thus provide substantial improvement in life quality measures, despite not treating the underlying cause of mortality (Fig. 2f).

Besides allogeneic satellite cells, autologous satellite cells could be isolated from a patient's biopsy followed by correction of the dystrophin gene and subsequent engraftment into dystrophic muscles (Fig. 2b). While it was reported that quiescent satellite cells cannot be transduced by adeno-associated viral (AAV) vectors (Arnett et al. 2014), recent studies suggest that muscle stem cells can be targeted and gene corrected in vivo using AAV serotype 9 (Tabebordbar et al. 2016; Goldstein et al. 2019; Nance et al. 2019). However, in vitro correction of satellite cells remains challenging, mainly due to the conversion of satellite cells into myoblasts shortly after isolation and culture.

In summary, obtaining sufficient quantities of satellite cells for muscle injections still remains a formidable challenge. This hurdle highlights the need for generating expandable myogenic stem cells apt for regenerative medicine purposes. In the next section, we will introduce alternative methods for generating copious amounts of expandable muscle precursor cells in vitro.

4 Generation of Myogenic Precursors from Pluripotent Stem Cells

An alternative cellular source for production of myogenic precursors are pluripotent stem cells (PSCs) in the form of blastocyst-derived embryonic stem cells (ESCs) or somatic cell-derived induced pluripotent stem cells (iPSCs) (Selvaraj et al. 2019; Magli and Perlingeiro 2017). An advantage for using PSCs is their ability to expand indefinitely in vitro, thus enabling the generation of high quantities of muscle precursors (Selvaraj et al. 2019). The following section will briefly recount recent works which reported on successful differentiation of PSCs into myogenic precursors. For a comprehensive overview of this approach, we refer readers to recently published literature (Chal and Pourquie 2017; Kodaka et al. 2017; Selvaraj et al. 2019; Magli and Perlingeiro 2017).

ESCs are pluripotent cells that form the inner cell mass of blastocyst-stage embryos. These undifferentiated cells can expand indefinitely in vitro while maintaining the potential to differentiate into multiple cell types from the three embryonic germ layers (Martin 1981; Evans and Kaufman 1981; Thomson et al. 1998; Itskovitz-Eldor et al. 2000). Unlike ESCs, iPSCs are generated from adult somatic cells by overexpression of defined transcription factors (Takahashi and Yamanaka 2006; Takahashi et al. 2007). Since iPSCs can originate from an autologous cellular source, they could bypass immune rejection issues and ethical roadblocks associated with use of fetus-derived allogeneic ESCs (Takahashi and Yamanaka 2006).

Early studies revealed that both mouse and human PSCs can successfully differentiate into cells with myogenic potential (Barberi et al. 2007; Ozasa et al. 2007; Zheng et al. 2006; Mizuno et al. 2010; Dekel et al. 1992). Two major approaches have since been successfully employed to differentiate PSCs into myogenic cells: (i) transgene-dependent overexpression of canonical myogenic transcription factors and (ii) treatment with small molecules involved in differentiation-related signaling pathways (Kodaka et al. 2017; Chal and Pourquie 2017). Typical differentiation protocols occur in a two-dimensional monolayer or three-dimensional embryoid bodies. Given the highly heterogenous nature of these differentiated cultures, cell purification by surface markers indicative of muscle stem or progenitor cells is often employed (Kodaka et al. 2017; Chal et al. 2015).

Darabi and colleagues pioneering study first demonstrated that ectopic over-expression of the early developmental myogenic transcription factor Pax3 can form myogenic cells with capacity to engraft muscle in vivo (Darabi et al. 2008). A subsequent notable work further demonstrated that conditional Pax7 expression induces differentiation of human ESCs and iPSCs into expandable myogenic precursors (Darabi et al. 2012). Engrafted cells restored dystrophin expression in dystrophic murine muscles and donor-derived Pax7 cells contributed to the satellite cell pool, highlighting the immense potential of PSC-derived myogenic precursors to remedy dystrophic symptoms (Darabi et al. 2011, 2012). A follow-up study further demonstrated that ESC-derived myogenic precursors engraft and can fuse to form dystrophin-positive myofibers in a model of severe muscular dystrophy (Filareto et al. 2012). Continued research followed a similar trajectory demonstrating efficient differentiation of ESCs and iPSCs into myogenic precursors or skeletal myocytes by overexpression of MyoD (Goudenege et al. 2012; Tanaka et al. 2013; Pawlowski et al. 2017; Maffioletti et al. 2015), Pax3, or Pax7 (Rao et al. 2018; Filareto et al. 2013). Whereas this approach was shown to be successful with respect to production of engraftable myogenic cells, the use of ectopic transgenes renders it less favorable for clinical application.

An alternative and safer method to facilitate conversion of PSCs into myogenic precursors is via small molecule treatment. Several recent studies have reported on PSC-differentiation into myogenic progenitors solely by small molecules which include bFGF, Forskolin and GSK3-β inhibitor (GSK3-βi) (Xu et al. 2013), GSK3-βi and FGF2 (Shelton et al. 2014; van der Wal et al. 2018), WNT3A (Hwang et al. 2014), GSK3-βi and DAPT (Choi et al. 2016), BMP4, TGF-β and GSK3-β inhibitors (Xi et al. 2017; Wu et al. 2018). A notable study by Chal and colleagues uncovered genetic pathways upregulated during paraxial mesoderm differentiation during mouse embryonic development, assisting in development of a step-wise differentiation protocol for derivation of myogenic precursors from PSCs by BMP4 and GSK3-β inhibitor treatment (Chal et al. 2015). A follow-up study further refined this protocol by allowing for long-term capture of myogenic precursors in vitro when serum was added as a media supplement (Fig. 3) (Chal et al. 2018). The myogenic precursors both express Pax7 and promote muscle regeneration in dystrophic murine muscles, further contributing Pax7 positive cells to the satellite cell pool (Chal et al. 2015; Chal et al. 2018). Due to the heterogeneity of PSC-derived differentiated cultures reported using this protocol, myogenic precursors can be further purified by cell surface markers (Kim et al. 2017). Of note, such FACS-sorting techniques have been successfully employed by multiple studies to efficiently purify muscle precursors from PSC-differentiated cultures (Magli et al. 2017; Borchin et al. 2013; Wu et al. 2018; Kim et al. 2017; Hicks et al. 2018).

Patient-specific iPSCs from Duchenne patients carry the same mutation as their somatic parental cells (Park et al. 2008). Recent developments of the CRISPR-Cas9 genome editing system enable correction of the dystrophin mutation in iPSCs, followed by directed differentiation into dystrophin-positive muscle cells (Young et al. 2016; Li et al. 2015). Upon transplantation, corrected cells fuse to form dystrophin-positive myofibers within host murine muscles (Young et al. 2016). These findings illuminate the powerful therapeutic potential of combining genome engineering and stem cell differentiation methods to streamline new therapeutic treatments (Fig. 2d). However, the generation of gene-corrected iPSC-derived muscle stem cells in vitro remains a challenge (Ortiz-Vitali and Darabi 2019). For a comprehensive overview detailing PSC treatment methods of muscular dystrophies, we refer readers to recent reviews (Selvaraj et al. 2019; Ortiz-Vitali and Darabi 2019).

An additional approach for generating PSC-derived myogenic precursors is via in vivo developmental processes that can help further mature engrafted myogenic cells (Incitti et al. 2019). PSCs form tumors commonly known as teratomas upon in vivo engraftment, which contain differentiated cells indicative of all three embryonic germ layers (Ben-David and Benvenisty 2011). In a recent captivating study, Chan and colleagues capitalized on the PSC haphazard differentiation process during teratoma formation to detect and purify myogenic progenitors with an astounding regenerative capacity (Chan et al. 2018). FACS-sorted teratoma-derived α7-Integrin^{+}/VCAM-1^{+} myogenic progenitors promoted robust regeneration upon engraftment into irradiated and cardiotoxin-injured dystrophic leg muscles in mice, restoring dystrophin expression in 80% of the *tibialis anterior* muscle and contributing up to 50% of the total DNA content (Chan et al. 2018). The teratoma-derived myogenic stem cells could further sustain regeneration in repeated injury and serial transplantation assays, demonstrating bona fide stem cell attributes. These cells exhibited a superior engraftment potential in comparison to previously published PSC-derived myogenic precursor cells (Chan et al. 2018). This remarkable contribution to muscle regeneration unleashes the vast potential of engrafted myogenic precursors to regenerate large muscle areas, which to date has only been achieved with partial success.

From a clinical perspective, a risk for teratoma formation by residual PSCs precludes facile adoption of this approach for therapy, and as such it will be of high interest to recapitulate this developmental process by a directed differentiation assay. To this end, single-cell sequencing could decipher the molecular trajectory PSC undergo during their differentiation into myogenic progenitors in teratomas. Such analysis may uncover candidate genes and signaling pathways that could be tested to produce teratoma-derived myogenic precursors in the culture dish, as was recently demonstrated for in vitro paraxial mesoderm and somite formation (Chal et al. 2018; Chal et al. 2015; Xi et al. 2017).

Prominent advances have been made in recent years in respect to generating PSC-derived myogenic precursors with therapeutic applicability. However additional research is vital to translate bench lab findings to bedside treatments. Ethical limitations involving use of embryo-derived cells, risks for teratoma formation, and spurious differentiation of PSCs inhibit efforts to utilize PSC-derived myogenic cells to treat muscle-associated diseases. In respect to myogenic precursors, further investigation is certainly required to characterize the cells both molecularly and functionally, as well as devise safe and efficient protocols for their induction, purification, and in vitro maintenance. Future and ongoing human trials involving PSC-derived cells to treat diseases such as macular degeneration and Parkinson's disease hold paramount implications for assessing the therapeutic potential of PSC derived cells in clinical settings (Blau and Daley 2019). Successes are expected to expand the appetite for utilizing PSC-derived cells to treat other diseases, strengthen the nexus between scientists and clinicians, and may eventually destine PSC-based therapy to become a leading technology in regenerative medicine.

5 Direct Reprogramming of Somatic Cells into Myogenic Progenitors

Limitations of utilizing PSC-derived myogenic progenitors include both ethical and safety concerns and challenges associated with inducing PSCs to differentiate into adult mature cells. The aforementioned hurdles significantly hamper their utility in clinical settings, necessitating the search for additional approaches to produce expandable myogenic cells. An alternative method that can circumvent these challenges is direct lineage reprogramming, also known as "transdifferentiation." This approach denotes the conversion of one cell type into another by either ectopic overexpression of cell-type-specific transcription factors or small molecule treatment (Xu et al. 2015a). Historical pioneering studies spearheading this technique have shown that forced overexpression of the myogenic transcription factor MyoD in fibroblasts can convert them into skeletal myocytes via a short-lived myoblast stage (Davis et al. 1987; Weintraub et al. 1989; Tapscott et al. 1988). Since the landmark work by Davis and colleagues (Davis et al. 1987), a plethora of studies have reported on successful generation of various cell types by direct lineage reprogramming methods as reviewed (Xu et al. 2015a), including direct conversion into tissue-specific progenitors such as cardiac and neural precursors (Lalit et al. 2016; Lujan et al. 2012; Ring et al. 2012). Despite the formation of skeletal myocytes by MyoD overexpression being the first representation of direct lineage reprogramming, surprisingly only in recent years successful conversion into myogenic progenitors has been achieved (Ito et al. 2017; Bar-Nur et al. 2018; Lee et al. 2018; Sato et al. 2019; Bansal et al. 2019). We will next briefly introduce these new works, further highlighting unresolved issues and desired research directions spanning this technology.

Ito and colleagues were first to report that overexpressing various combinations of transcription factors such as Pax3, Pax7, Pitx1, Mef2b, and MyoD in mouse fibroblasts could give rise to myogenic progenitors (Ito et al. 2017). Reprogrammed myogenic progenitors upregulate satellite cell markers such as endogenous Pax3 and

Myf5, although intriguingly did not upregulate endogenous Pax7 expression (Ito et al. 2017). Additionally, the myogenic progenitors generated dystrophin-positive myofibers when injected into dystrophic murine muscles; however it is yet to be determined if these cells can contribute to the satellite cell pool in vivo (Ito et al. 2017). Another recent work reported on the conversion of fibroblasts into induced myogenic progenitor cells (iMPCs) by transient overexpression of MyoD in concert with three small molecules treatment (Bar-Nur et al. 2018). In stark contrast to overexpression of MyoD alone in fibroblasts, which generates myocytes (Davis et al. 1987), the combined treatment with Forskolin, a cyclic-AMP agonist, RepSox, a TGF-β inhibitor, and CHIR-99021, a GSK3-β inhibitor, surprisingly formed heterogenous cultures containing both undifferentiated progenitors and differentiated contractile myofibers (Bar-Nur et al. 2018). These progenitors can expand extensively in vitro, express high levels of endogenous Pax7 and Myf5, and engraft into both wild-type and dystrophic murine muscles (Fig. 3) (Bar-Nur et al. 2018). Remarkably, iMPCs efficiently contribute to the satellite cell pool and sustain multiple rounds of regeneration by repeated injury assay (Bar-Nur et al. 2018). Two additional reports have recently further demonstrated reprogramming of fibroblasts into induced myogenic precursor cells (Lee et al. 2018; Sato et al. 2019). Lee and colleagues have shown that overexpression of Six1, Eya1, Esrrb, and Pax3 can generate myogenic precursors that express endogenous Pax7 and fuse to form dystrophin-positive myofibers in dystrophic mice (Lee et al. 2018). In addition, Sato and colleagues established that overexpression of Pax3, Heyl, and KLF4 in concert with transient MyoD expression can generate myogenic precursors from both mouse and human fibroblasts (Sato et al. 2019), these precursors can expand long-term, upregulate satellite cell markers such as Pax7, Spry1, and Sdc4, and efficiently engraft into dystrophic muscles including the satellite cell niche (Sato et al. 2019). It will be of interest to assess if the induced precursors are true myogenic stem cells that can both sustain regeneration in a repeated injury assay as well as maintain engraftment by serial transplantation. Lastly, a new report demonstrated that small molecule treatment alone in fibroblasts can elicit skeletal myocyte formation via a short-lived progenitor state (Bansal et al. 2019). Of note, several of the small molecules (i.e., Forskolin, RepSox and CHIR-99021) are identical to the ones used in conjunction with transient MyoD expression to form iMPCs (Bar-Nur et al. 2018) or maintain quiescence of satellite cells in vitro (Quarta et al. 2016). However, myogenic cells produced by small molecule treatment alone did not permanently capture a myogenic progenitor state in vitro nor was the capacity of the induced cells to contribute to muscle regeneration in vivo investigated.

Generation of induced myogenic precursor/progenitor cells by direct lineage reprogramming is an attractive approach to produce expandable cells for therapy; however notable challenges and unresolved questions remain. An interesting question that still requires further interrogation is how similar to one another are the induced cells generated by the aforementioned methods. Likewise, it will be of interest to assess how akin induced myogenic precursor/progenitor cells are molecularly and functionally to both myoblasts and satellite cells, and whether they are true myogenic stem cells that can sustain muscle regeneration by a serial transplantation assay. For translational purposes, future works will need to assess if these cells can be generated without integration of viral vectors, thus mitigating risks for insertional mutagenesis. To this end, synthetic mRNA transfection or protein transduction could be adapted to generate induced myogenic precursor/progenitor cells as was shown for iPSCs (Malik and Rao 2013). Similarly, production of these cells by small molecule treatment alone may provide a safer and more scalable approach to generate cells for translational purposes. Additional aspects of interest include disease modeling of muscular dystrophies using these cells and investigating their adoption for drug screens. In summary, this promising and nascent technique is in its infancy, and as such a litany of remaining questions will need to be thoroughly addressed prior to considering its potential use in clinical applications.

6 Maintaining Satellite Cell Potency In Vitro

Isolation of QSCs from skeletal muscle tissue is commonly performed by enzymatic digestion and mechanical breakage vital to dislodge the cells from their association with myofibers and the extracellular matrix (ECM). This harsh treatment mimics to an extent dissociation of satellite cells from their natural environment during injury or disease and likewise elicits rapid activation of myoblast-associated genes (Machado et al. 2017; Pietrosemoli et al. 2017; van Velthoven et al. 2017). Typically, freshly isolated satellite cells are separated from other mononucleated cell types by a pre-plating procedure or by FACS-sorting followed by seeding onto culture dishes coated with basement membrane substrates and in medium containing high serum levels and basic FGF (Rando and Blau 1994). Under these conditions, it was shown that nascent myoblasts rapidly proliferate and can fuse to form multinucleated myotubes in low serum conditions (Rando and Blau 1994). Upon further in vitro expansion, activated satellite cells promptly lose regenerative capacity as they form a population of proliferative myoblasts (Montarras et al. 2005). In the following section, we will outline various techniques that have been shown to ebb precocious differentiation of isolated satellite cells cultured in vitro and further augment their engraftment potential (Fig. 4a). We will focus on unique basement membrane substrates, treatment with small molecules, and induction of genes that promote an undifferentiated satellite cell state.

6.1 Basement Membrane Proteins that Support an Undifferentiated Satellite Cell State

During homeostasis native satellite cells are located between the sarcolemma and a structure of the basement membrane known as the basal lamina, providing an anatomical environment known as the "satellite cell niche." An intricate signaling crosstalk between the cells and their surrounding governs the satellite cell quiescent state (Rayagiri et al. 2018; Baghdadi et al. 2018). External stimuli such as the ones experienced during muscle injury and disease induce a signaling cascade that

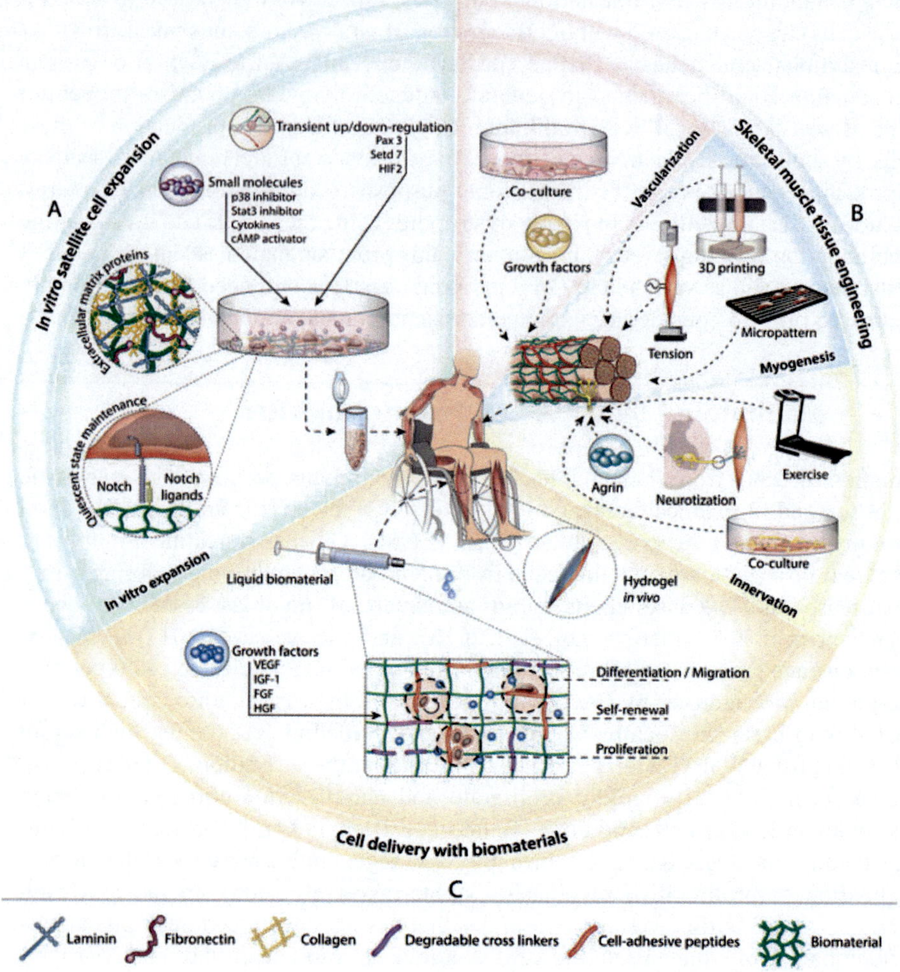

Fig. 4 Biomaterial and tissue engineering strategies to enhance satellite cell expansion, engraftment and reconstruction of skeletal muscle tissue. (**a**) Select approaches to expand satellite cells in vitro while maintaining in vivo engraftment potential. Different methods can be utilized including expansion on unique basement membrane proteins, treatment with small molecules and induction or repression of genetic pathways. (**b**) Tissue engineering approaches to treat severe muscle dystrophy or VML using in vitro generated muscle bioconstructs which consist of satellite cells, MRCs, and aligned, multinucleated myofibers. Bioconstruct can be further innervated or vascularized to facilitate engraftment and retention of grafts in muscles. (**c**) Strategies to enhance satellite cell engraftment using injectable biomaterials which extend satellite cell viability and retention at the injury site

promote satellite cell activation and proliferation, which are imperative for muscle repair (Dumont and Rudnicki 2017). In their endogenous microenvironment, satellite cells are embedded in ECM, which encapsulate myofibers and is associated with different proteins during quiescence and activation, including laminins

(Yao et al. 1996), fibronectin (Bentzinger et al. 2013), and collagen (Urciuolo et al. 2013). Conventional basement membrane proteins such as laminin-111, collagen I, Matrigel, and fibronectin are commonly used to expand satellite cells as myoblasts in vitro (Urciuolo et al. 2013; Maley et al. 1995; Grefte et al. 2012; Wilschut et al. 2010; Boonen et al. 2009; Duffy et al. 2016; Kuhl et al. 1986; Foster et al. 1987; Ocalan et al. 1988). These matrices are essential for attachment of freshly isolated satellite cells to the culture dish; however basement membrane proteins such as Matrigel are secreted by murine tumors and are composed of thousands of peptides, rendering them unsuitable for therapeutic applications in humans (Hughes et al. 2010). In addition, other basement membrane proteins confer on myoblasts reduced differentiation potential and loss of engraftment capacity during extended passaging. We will next discuss recent studies that set to uncover basement membrane proteins that could support satellite cell expansion while maintaining robust differentiation in vitro and high engraftment capacity in vivo.

Laminins (LMs) are active protein components of the ECM and belong to a glycoprotein family consisting of different chains including α, β, and γ. (Colognato and Yurchenco 2000). These proteins can support in vitro growth of myoblasts, albeit which laminin components form the satellite cell niche or are upregulated during injury was only recently thoroughly assessed (Ishii et al. 2018; Rayagiri et al. 2018). In one recent study, extensive analysis of the different laminin isoforms in skeletal muscle revealed that the satellite cell niche is composed of LM-α2, LM-α3, LM-α4, and LM-α5, which could then be mimicked in vitro by using Laminin E8 (LM-E8) fragments, which are truncated proteins composed of the α, β, and γ chains C-terminus domain (Ishii et al. 2018). Freshly isolated satellite cells cultured on LM-E8 derivative substrates remained undifferentiated in comparison to satellite cells cultured on Matrigel as indicated by higher quantities of proliferative $Pax7^{+}/MyoD^{-}$ cells (Ishii et al. 2018). Furthermore, mouse and human satellite cells expanded on LM-E8 could robustly contribute to muscle regeneration in vivo and restore dystrophin expression in mouse dystrophic muscles (Ishii et al. 2018). Another recent study examined conventional substrates in the form of Matrigel and LM-111 to biologically relevant laminins present in muscles such as adult LM-211, embryonic LM-521, and others (Penton et al. 2016). This meticulous analysis demonstrated that freshly isolated mouse and human satellite cells cultured as myoblasts on LM-521 demonstrate a robust differentiation potential as assessed by myotube size and fusion index in comparison to myoblasts cultured on conventional basement membrane proteins (Penton et al. 2016). However, the number of Pax7 and MyoD positive cells was similar among all tested conditions, and in vivo engraftment utilizing LM-521 cultured myoblasts was not assessed (Penton et al. 2016).

In concordance with these findings, a recent study reported that satellite cells regulate their quiescence via secretion of LM-α1, and genetic ablation of this protein impairs satellite cell self-renewal (Rayagiri et al. 2018). Similarly, another study demonstrated that collagen V (COLV) is produced by satellite cells and is key for their regeneration capacity, as blocking its production severely precludes satellite cell proliferation and leads to depletion of the satellite cell pool (Baghdadi et al. 2018). It will be of interest to assess if overexpression of LM-α1, LM-521, or COLV could maintain satellite cell characteristics in vitro and augment muscle engraftment of cultured cells. Determining which basement membrane protein can best support satellite cell proliferation, and differentiation is expected to yield widespread implications with respect to the potential of human satellite cells cultured on these unique matrices to regenerate muscles in vivo.

6.2 Inhibition of Satellite Cell Differentiation by Small Molecules, Ligands, and Cytokines

Disruption of satellite cell quiescence by injury or disease is accompanied by activation of cellular signaling cascades that promote proliferation and regeneration. However, these pathways are altered in high-passage myoblast cultures, which is one attribute to their loss of engraftment potential. Small molecules that affect signaling pathways could mitigate this effect and promote maintenance of satellite cell characteristics in vitro. We will briefly describe recent efforts to harness small molecules to culture and propagate satellite cells.

The p38 mitogen-activated protein kinase (MAPK) pathway is upregulated during activation of mouse satellite cells (Jones et al. 2005; Simone et al. 2004). Furthermore, previous studies documented that p38 inhibitor treatment of satellite cells attenuated muscle aging while increasing functionality of muscles in aged mice (Cosgrove et al. 2014; Bernet et al. 2014). Charville and colleague recently corroborated this observation by demonstrating that human satellite cells similarly upregulate p38 pathway in vitro, and its inhibition blunted the differentiation of satellite cells into committed myoblasts (Charville et al. 2015). Treatment of human ASCs with p38 MAPK inhibitor SB03580 reportedly increased proliferation rate while mitigating precocious differentiation into myoblasts and myotubes (Charville et al. 2015). Remarkably, SB03580-treated satellite cells robustly engrafted into healthy and dystrophic muscles in comparison to non-treated cells or freshly isolated satellite cell controls and could contribute cells to the muscle stem cell reservoir (Charville et al. 2015). Another study corroborated this finding by demonstrating that p38 inhibition maintains satellite cell-like phenotype and proliferation of freshly isolated bovine satellite cells (Ding et al. 2018).

Another well-studied pathway associated with satellite cell quiescence, activation, and differentiation is the Notch pathway(Mourikis and Tajbakhsh 2014), first reported to be upregulated during activation of satellite cells (Conboy and Rando 2002). Subsequent studies have documented that Notch and its downstream targets Hey1, HeyL, and Hes1 are highly expressed in QSCs and downregulated upon differentiation (Mourikis et al. 2012; Bjornson et al. 2012; Brohl et al. 2012). Abrogation of Notch in satellite cells mitigates their proliferation rate and elicits precocious differentiation into mature muscle cells (Mourikis et al. 2012; Bjornson et al. 2012; Brohl et al. 2012). In line with these observations, overexpression of Notch-1 abolishes expression of muscle-associated differentiation genes (Conboy and Rando 2002), and overexpression of constitutively active intracellular domain of Notch1 (NICD1) can rescue precocious differentiation in satellite cells lacking Pax7 (Pasut et al. 2016). Different Notch ligands have been implicated in regulating satellite cell quiescence and self-renewal (Baghdadi et al. 2018; Low et al. 2018; Verma et al. 2018). Of note, a recent study highlighted the role of endothelial cells in maintenance of satellite cell quiescence through the Notch ligand Dll4 (Verma et al. 2018).

These collective findings suggest that manipulating the Notch pathway could impact proliferation and differentiation potential of cultured satellite cells and myoblasts. Indeed, Parker and colleagues were first to report that culturing canine freshly isolated satellite cells in conditions that promote Notch activation blunted differentiation and facilitated muscle engraftment, strikingly in equal levels to those of freshly isolated canine satellite cells (Parker et al. 2012). Similarly, mouse and human satellite cells expanded on Notch ligands reportedly proliferated better, elevated expression level of Notch downstream targets, and reduced expression of differentiation-associated genes (Sakai et al. 2017). However, contribution of cells to engraftment in host muscles was similar between treated and control cells (Sakai et al. 2017). In another recent work, Gerli and colleagues cultured freshly isolated satellite cells and myoblasts on Dll4 ligand in concert with PDGF-BB supplementation (Gerli et al. 2019). This treatment attenuated proliferation and differentiation of cultured satellite cells, increased the proportion of $Pax7^+/MyoD^-$ cells and elevated expression of Notch and its downstream targets (Gerli et al. 2019). The authors hypothesize that the combined treatment augments perivascular and satellite cell attributes in cultured myoblasts (Gerli et al. 2019). However, the in vivo capacity of treated myoblasts to engraft into muscles in vivo is yet to be determined (Gerli et al. 2019). Collectively, conflicting reports exist with respect to the capacity of Notch ligands to enhance engraftment of cultured myoblasts in vivo; however this pathway undisputedly helps in sustaining satellite cell attributes in vitro. More work is required to resolve if alteration of this pathway can enhance engraftment of myogenic cells in vivo.

Forskolin is a small molecule that has been recently implicated in enhancement of muscle stem or progenitor cell proliferation and engraftment potential (Xu et al. 2013; Bar-Nur et al. 2018; Quarta et al. 2016). Forskolin increases cAMP intracellular levels that further activate the protein kinase A (PKA) pathway, which subsequently phosphorylates the transcription factor cAMP response element-binding protein (CREB). Activation of this pathway was previously shown to be essential for WNT-related induction of embryonic myogenesis (Chen et al. 2005). Recent works implicated Forskolin in enhancing expansion of mouse satellite cells in vitro (Xu et al. 2013). Freshly isolated satellite cells cultured in the presence of Forskolin retained satellite cell characteristics in vitro, and treated cultures could engraft robustly into dystrophic mouse muscles (Xu et al. 2013). Forskolin is also a component of a small molecule cocktail which reportedly maintains muscle stem cell quiescence in vitro in conjunction with engineered muscles fibers (Quarta et al. 2016). This artificial construct can regenerate muscle in vivo upon engraftment (Quarta et al. 2016). Lastly, forskolin is one of three small molecules that have been reported as critical for induction and proliferation of induced myogenic progenitor cells from fibroblasts in concert with MyoD (Bar-Nur et al. 2018).

Muscle damage initiates infiltration of white blood cells to the injury site, where these cells play important role in muscle repair via secretion of cytokines (Wosczyna and Rando 2018). Growing amount of evidence suggest that cytokines can directly affect satellite cells and myoblasts. Notable examples include leukemia inhibitory factor (LIF), which reportedly increases the proliferation of cultured myoblasts (Austin and Burgess 1991; Austin et al. 1992; Alter et al. 2008). Accordingly, LIF was recently shown to effectively enhance the transplantation potential of freshly isolated myoblasts in a DMD mouse model, presumably by attenuating apoptosis (Ito et al. 2016). LIF is one of several members of the IL-6 cytokine family, which includes IL-6, IL-27, IL-11, and Oncostatin (OSM) (Stefan Rose-John 2018). It functions by binding to the LIfr/gp130 receptor which activates JAK1 signaling cascade leading to phosphorylation and activation of the transcription factor Stat3 (Nicola and Babon 2015). Similarly, IL-6 is a proinflammatory cytokine which is upregulated during the initial phase of muscle regeneration and mediates its effect through Stat3 downstream activation (Munoz-Canoves et al. 2013). Stat3 pathway is crucial for myogenic differentiation, and absence of Stat3 signaling perturbs efficient muscle regeneration (Tierney et al. 2014; Wang et al. 2008; Sun et al. 2007). Accordingly, IL-6 supplementation upregulates MyoD expression through Stat3 signaling, promoting muscle regeneration (Tierney et al. 2014). Attenuating Stat3 expression enhances myoblasts proliferation; however engraftment potential of these cells was not assessed (Tierney et al. 2014). In accordance with this observation, another cytokine of the IL-6 pathway that activate the Stat3 pathway is OSM, which reportedly blocks differentiation of myoblasts and keeps them in a progenitor-like state (Xiao et al. 2011). A recent study revealed that OSM is secreted by muscle fibers, exerting a quiescent phenotype on muscle stem cells (Sampath et al. 2018). OSM treatment of cultured myoblasts maintains their stemness in vitro as judged by Pax7-GFP reporter expression, and this treatment dramatically enhances myoblast engraftment potential and further enables serial transplantation of cells (Sampath et al. 2018). Wnt4 is another recently reported niche-specific factor secreted from myofibers that regulates satellite cell quiescence (Eliazer et al. 2019). It will be of interest to assess if akin to OSM treatment of freshly isolated satellite cells, administration of Wnt4 can help support in vitro muscle stem cell quiescence and enhance in vivo engraftment potential. Building upon the notable effects cytokines have on muscle stem cells, Fu and colleagues screened multiple proinflammatory cytokines to find combinations of cytokines that promote satellite cell expansion (Fu et al. 2015). This effort uncovered culture conditions that consist of the T-cell secreted cytokines IL-1α, IL-13, TNF-α, and IFN-γ to support extensive expansion of satellite-like cells (Fu et al. 2015). Expanded cells treated in these conditions could very efficiently engraft into muscles and further replenish the satellite cell reservoir (Fu et al. 2015).

6.3 Genetic Alteration that Support Satellite Cell Self-Renewal and Regeneration

Aside from basement membrane proteins and small molecule treatment, alteration of genes and genetic pathways is an additional method to augment satellite cell expansion potential. An elegant manifestation of this approach was recently reported by Filareto and colleagues who demonstrated that transient overexpression of the transcription factor Pax3 in freshly isolated satellite cells abates spontaneous differentiation into myoblasts, and further enhances cell proliferation without compromising satellite cell capacity to give rise to myotubes in response to differentiation cues (Filareto et al. 2015). Moreover, treated satellite cells could robustly restore dystrophin expression in DMD mice and contribute cells to the satellite cell pool (Filareto et al. 2015). To further investigate the translation potential of this approach, freshly isolated satellite cells extracted from DMD mice were subjected to transient Pax3 overexpression and infected with a construct encoding human micro-dystrophin (Filareto et al. 2015). Following this procedure, cells were engrafted into dystrophic mouse muscles and could restore dystrophin expression and remedy symptoms of the affliction, demonstrating an elegant autologous satellite cell-based approach to treat DMD in a mouse model (Filareto et al. 2015), which can be adapted to humans (Fig. 2b).

A few additional studies have recently reported means to sustain satellite cell expansion in vitro by altering genetic pathways. One study demonstrated that deletion of the lysine methyltransferase Setd7 enables in vitro expansion of undifferentiated freshly isolated human and mouse satellite cells and could further enhance their engraftment potential in vivo (Judson et al. 2018). This phenotype is mediated via interaction of Setd7 with β-catenin and modification of its methylation state, which subsequently induces a differentiation myogenic gene commitment program that is abated in Setd7 deleted cells (Judson et al. 2018). Inhibition of Setd7 can be mimicked by pharmacological treatment, providing a facile approach to utilize this method for therapeutic interventions (Judson et al. 2018). Another study reported that inhibition of HIF2A support satellite cell proliferation (Xie et al. 2018). Quiescent satellite cells express the hypoxia-inducible factor 2A (HIF2A) and are in a hypoxic state during homeostasis (Xie et al. 2018). This state is rapidly altered in normoxic state during in vitro growth (Xie et al. 2018). Capitalizing on this unique observation, it was shown that short-lived inhibition of HIF2A by pharmacological modulation accelerates muscle stem cell proliferation, while deletion of HIF2A abrogates muscle regeneration due to depletion of the satellite cell reservoir (Xie et al. 2018). It will be of interest to assess whether satellite cells subjected to this short-lived treatment can better engraft into muscles in vivo.

Culturing satellite cells using the aforementioned treatments has been demonstrated to mitigate precocious differentiation and enhance engraftment potential of cells. Further research is required to investigate the mode of action by which these treatments work. Additionally, a manifold of these studies have been reported in mouse cells, and it is of interest to assess which method can efficiently adapt to expand human satellite cells. Lastly, it will be of interest to investigate the effects of combinatorial administration of small molecules on human satellite cells to assess if combined treatment can facilitate the expansion and engraftment of cells in comparison to individual treatment.

7 Enhancing Muscle Stem Cell Engraftment Using Biomaterials

Delivery methods of myogenic cells into recipient muscles could impact translational outcomes by increasing engraftment efficacy and decreasing the requirement for high quantities of donor-derived satellite cells (Fig. 4c) (Boldrin et al. 2007).

In early myoblast transplantation trials, engrafted cells were typically resuspended in HBSS or PBS prior to injections (Huard et al. 1992; Tremblay et al. 1993a; Mendell et al. 1995). However, aftereffects of these trials were disappointing, and low cell survival may have accounted for the unfavorable outcome (Beauchamp et al. 1999). This observation raises an interesting question whether delivery of muscle stem or progenitors cells in conjunction with unique substances such as biomaterials could manifest better therapeutic outcomes.

Biomaterials are defined as a synthetic or natural substance, or combination of substances, that can remedy or replace tissues and organs in the body (National Institutes of Health Consensus Development 1983). To date, numerous natural and synthetic biomaterials have been widely investigated for their potential to boost skeletal muscle regeneration (Cezar and Mooney 2015; Qazi et al. 2015). Biomaterials can encapsulate cells, for instance, in the form of soft hydrogels that are created by the crosslinking of hydrophilic polymers or 3D scaffolds, which are generated by various techniques including chemical crosslinking, freeze-drying, or electrospinning (Qazi et al. 2015; Wolf et al. 2015; Hu et al. 2019). An injectable or implantable biomaterial can protect the engrafted cells from apoptosis and increase cell proliferation while negating immunological responses (Han et al. 2017). Moreover, degradable biomaterials can be replaced over time by a regenerated tissue, which is important once a therapeutic objective is achieved (Han et al. 2017). Commonly used biomaterials which have been reported to sustain cell delivery into skeletal muscle tissue include natural alginate (Wang et al. 2014; Hill et al. 2006), hyaluronic acid (HA) (Rossi et al. 2011; Davoudi et al. 2018), ECM (Rao et al. 2017), or fibrin (Matthias et al. 2018; Page et al. 2011; Gerard et al. 2012), as well as synthetic scaffolds in the form of poly(ethylene glycol) (PEG) (Han et al. 2018, 2019), poly(lactic-co-glycolic acid) (PLGA) (Boldrin et al. 2007, 2008), or a combined PEG-fibrinogen constructs (Fuoco et al. 2012, 2014, 2015).

Several studies have reported on the beneficial effects of utilizing synthetic or natural biomaterials during myoblast engraftment into murine muscles (Boldrin et al. 2007, 2008; Hill et al. 2006; Wang et al. 2014; Rao et al. 2017). For example, high quantities of donor-derived myofibers were detected when myoblasts were encapsulated inside micropatterned PGA scaffolds and engrafted into pre-injured mouse muscles (Boldrin et al. 2007, 2008). A subsequent study demonstrated enhanced muscle regenerative capacity upon delivery of myoblasts embedded in alginate scaffolds compared to injection of cells alone (Hill et al. 2006). In these studies, scaffolds containing myoblasts in their native size were implanted into muscles (Boldrin et al. 2007, 2008; Hill et al. 2006). However, an injectable biomaterial could additionally enable a less invasive cell delivery method (Qazi et al. 2015). This can be achieved by using shape-memory scaffolds (Wang et al. 2014) or injectable biomaterials that can be polymerized into gels in situ due to a temperature shift (Davoudi et al. 2018), light (Rossi et al. 2011; Fuoco et al. 2015), or ions (Sleep et al. 2017). Shape-memory scaffolds can be compressed during transplantation and then expand back to their original size in situ (Lendlein and Langer 2002). An alginate-based shape-memory scaffold has been suggested as a vehicle for myoblast delivery, as it enables a minimally invasive transplantation procedure (Wang et al. 2014).

Upon engraftment into pre-injured muscles, increased myoblast viability, muscle regeneration, and reduced fibrosis were reported (Wang et al. 2014). Further, robust myoblast proliferation and engraftment efficacy were demonstrated upon assembly of myogenic cells inside an ECM hydrogel and injection into ischemic muscles in conjunction with fibroblasts (Rao et al. 2017).

Prolonged in vitro expansion of myoblasts reduces engraftment potential as compared to freshly isolated satellite cells (Montarras et al. 2005). As such, biomaterial-based delivery of satellite cells has been recently investigated with hope of increasing engraftment efficacies (Sleep et al. 2017; Davoudi et al. 2018; Han et al. 2018, 2019; Rossi et al. 2011). In a notable study, Rossi and colleagues delivered myoblasts or satellite cells embedded in a photocrosslinkable hyaluronan hydrogel into pre-injured mouse muscles, reporting both increased detection of donor-derived myofibers and improved contraction forces (Rossi et al. 2011). Recently, Sleep and colleagues reported on a method to embed satellite cells in a liquid crystalline scaffold prior to intramuscular injections (Sleep et al. 2017). The scaffold stiffness was adjusted to physiological values, in agreement with previously published works highlighting the effects of mechanical properties on myogenesis (Engler et al. 2004; Gilbert et al. 2010; Sleep et al. 2017). Utilizing a unique apparatus that forms aligned scaffolds in vivo, freshly isolated satellite cells were encapsulated in a biomimetic scaffold and engrafted into muscles, demonstrating robust regeneration in comparison to control non-encapsulated cells (Sleep et al. 2017). Of interest, the scaffold degradation pace was akin to the typical time period necessary for muscle tissue regeneration, leading the authors to suggest that this temporal concordance is key for vascularization and innervation of the regenerating tissue (Sleep et al. 2017).

In a recent study, a hydrogel-based cell delivery system promoting cell viability, proliferation, and differentiation has been reported (Han et al. 2018). The synthetic hydrogel consisted of both four-arm PEG and maleimide groups (PEG-4MAL) and was further functionalized with arginine, glycine, and aspartic acid (RGD) peptides (Han et al. 2018). RGD motifs are found in fibronectin and other extracellular matrix proteins that are essential for both satellite cell expansion and myoblast differentiation (Ruoslahti and Pierschbacher 1987; Garcia et al. 1999; Lukjanenko et al. 2016) and are frequently used to support cell adhesion (Wang et al. 2014; Hill et al. 2006; Han et al. 2018, 2019; Borselli et al. 2011). Utilizing this delivery system, RGD functionalized hydrogels increased freshly isolated satellite cell viability, proliferation, and migration, and the RGD adhesive sites were key for cell fusion in vitro (Han et al. 2018). Of interest, utilizing a protease-degradable crosslinker enabled degradation of the hydrogel and further enhanced cell viability and proliferation in vitro (Han et al. 2018). Lastly, satellite cell survival and proliferation were also improved via hydrogel delivery during "supra-muscular" injections onto the muscle surface of aged or dystrophic mice (Han et al. 2018). In another notable study, Davoudi and colleagues demonstrated that satellite cell delivery in a hyaluronan and methylcellulose (HAMC) hydrogel resulted in a 45% increase of donor-derived myofiber engraftment compared to saline control (Davoudi et al. 2018). The authors noted improved dispersion of regenerated myofibers, suggesting that cell migration was enhanced by use of the hydrogel (Davoudi et al. 2018). They further conducted several experiments to elucidate the mechanisms governing the enhanced engraftment observed with HAMC usage (Davoudi et al. 2018). For instance, hydrogels have been shown to protect various cells from mechanical forces during the injection process (Aguado et al. 2012). To note, HAMC did not affect the viability of extruded satellite cells however led to a 6% increase in cell number (Davoudi et al. 2018). Collectively, the authors suggest that the enhanced engraftment could be attributed to elevated muscle stem cell proliferation, protection from active clearance by the immune system and detaining precocious differentiation (Davoudi et al. 2018).

An additional advantage of cell encapsulation inside biomaterials entails the potential of cell delivery with incorporation of various growth factors. Whereas growth factors mixed in a saline solution tend to diffuse rapidly inside a recipient tissue, a scaffold can support gradual release of growth factors into the tissue (Qazi et al. 2015; Han et al. 2017; Sleep et al. 2017). To this end, Borselli and colleagues encapsulated myoblasts in conjunction with the growth factors VEGF and IGF-1 inside an alginate scaffold and delivered the bioconstruct into pre-injured murine muscles (Borselli et al. 2011). This technique reportedly improved cell engraftment and further stimulated vascularization and myogenesis in the host tissue (Borselli et al. 2011). In another study, Hill and colleagues embedded myoblasts in an alginate scaffold in the presence of HGF and FGF2 prior to delivery (Hill et al. 2006). They reported increased cell viability, prevention of terminal differentiation, and improved regeneration in muscles engrafted with this construct (Hill et al. 2006). Additional studies further demonstrated that sustained release of bFGF from engineered constructs facilitated myoblast or satellite cell engraftment (Hagiwara et al. 2016; Sleep et al. 2017). Lastly, another recent report demonstrated the assembly of a PEG-4MAL hydrogel with muscle stem cells and the protein Wnt7a (Han et al. 2019), a pro-myogenic factor involved in muscle repair and stem cell migration (von Maltzahn et al. 2011; Bentzinger et al. 2014). This effort has led to an increase in muscle mass and regeneration following construct delivery into pre-injured muscles and strikingly also elevated stem cell migration (Han et al. 2019). Collectively, the aforementioned reports suggest that the delivery of biomaterials together with unique small molecules and growth factors could augment myogenic engraftment efficacy.

The various studies discussed in this section indicate that synthetic or natural materials can enhance muscle stem cell engraftment. However, the molecular mechanisms that mediate this process are not completely understood and additional research is warranted. Furthermore, it will be of interest to assess whether combining different cell types inside a biomaterial could enhance regeneration capacity. Ideal candidates include cell types that are associated with skeletal muscle repair (Wosczyna and Rando 2018), such as resident endothelial cells, fibro-adipogenic progenitors, or macrophages which reportedly promote muscle regeneration (Lukjanenko et al. 2019; Wosczyna et al. 2019; Latroche et al. 2017; Joe et al. 2010). Collectively, amassing inside a biomaterial multiple cell types native to muscle tissue and essential for its repair could boost the therapeutic competence of this tissue engineering approach.

8 Stem Cell-Based Approaches to Engineer Skeletal Muscle Tissue

Volumetric muscle loss (VML) denotes the irrecoverable structural and functional deficits of skeletal muscle tissue induced by acute injury or surgical removal (Greising et al. 2019). Due to extensive muscle loss, satellite and other muscle resident cells fail to repair the tissue damage which is often times replaced by fibrotic tissue (Greising et al. 2019; Corona et al. 2019). Harvest of healthy muscle fragments known as "muscle flaps" followed by engraftment into VML-afflicted muscle sites can partially treat this muscle loss; however this complicated surgery is often associated with donor site morbidity (Lin et al. 2007; Klinkenberg et al. 2013; Gilbert-Honick and Grayson 2019). Several late-stage muscular dystrophies such as DMD share resemblances to VML, as the muscle tissue is both severely damaged and typically replaced by adipogenic and fibrotic tissues (Klingler et al. 2012; Hooijmans et al. 2017; Skuk et al. 2007). If an entire muscle segment is acutely damaged, muscle stem cells cannot efficiently regenerate skeletal muscle, and myogenic cell engraftment is expected to fail in the absence of a supportive tissue. Notably, this has previously been shown for DMD patients following allogeneic myoblast transplantation, which was reportedly effective when cells were engrafted into an intact muscle tissue, however failed to engraft following injections into highly dystrophic muscles devoid of parenchyma (Skuk et al. 2007). An alternative therapeutic approach to cell transplantation involves reconstructing a functional skeletal muscle tissue in vitro that could incorporate into VML-afflicted muscle sites or acutely dystrophic muscles (Fig. 4b) (Juhas and Bursac 2013). In the following section, we will highlight attempts to engineer tissue-like muscle constructs in vitro utilizing myogenic stem or progenitor cells seeded on designated scaffolds prior to engraftment. For further reading into other approaches that involve tissue engineering of skeletal muscle, we refer the readers to recently published literature (Kwee and Mooney 2017).

8.1 Engineering Skeletal Muscle Tissue In Vitro Using Myogenic Precursors

Historically, early attempts to engineer bioartificial skeletal muscle tissue utilized avian myotubes or immortalized mouse myoblast cell lines such as C2C12 (Vandenburgh and Kaufman 1979; Vandenburgh et al. 1988, 1998, 1996). The latter have been oftentimes used as "biological vehicles" to deliver growth factors into muscle tissue (Vandenburgh et al. 1996, 1998). Vandenburg and colleagues reported on a method to produce a bioartificial muscle (BAM) consisting of parallel myotubes that upon muscle engraftment secreted recombinant human growth factors and attenuated muscle atrophy in mice (Vandenburgh et al. 1998). The BAM was formed by coalescing collagen and Matrigel with C2C12 myoblasts, which were genetically engineered to express a specific growth factor (Vandenburgh et al. 1988). This pre-gel solution was allowed to solidify in long casts, thus forming aligned myotubes which were successfully subcutaneously transplanted (Vandenburgh et al. 1996, 1998). Multiple groups have since utilized C2C12 to produce muscle constructs in vitro by various tissue engineering techniques (Engler et al. 2004; Costantini et al. 2017; Zhang and Guo 2017; Kang et al. 2016). Furthermore, C2C12 has also been incorporated into engineered muscle constructs to replace muscle tissue in vivo (Levenberg et al. 2005; Koffler et al. 2011; Kaufman et al. 2019). While these studies provided the experimental framework to engineer skeletal muscle cells, C2C12 is an immortalized cell line which is physiologically distinct from normal myogenic cells, necessitating usage of physiologically relevant and therapeutically acceptable primary cells (Yaffe and Saxel 1977; Morgan et al. 1992; Wernig et al. 1991). To this end, skeletal muscle constructs have also been generated from primary myogenic cells derived from rodent (Juhas et al. 2014; Corona et al. 2014; Machingal et al. 2011) or human (Gholobova et al. 2015; Quarta et al. 2017; Madden et al. 2015; Powell et al. 1999) sources. To induce muscle construct formation, parallel alignment of differentiated myotubes is typically generated by patterned scaffolds and substrates (Neal et al. 2014; Kang et al. 2016; Takahashi et al. 2013; Bian et al. 2009; Nakayama et al. 2019), or platforms that induce unidirectional tension in muscle bundles (Powell et al. 1999; Gholobova et al. 2015; Vandenburgh et al. 1998). A uniform muscle architecture has been elegantly demonstrated using parallel multinucleated myofibers and sarcomeric striations (Juhas et al. 2014; Madden et al. 2015). Moreover, several studies reported on functional muscle contractions involving active calcium signaling in vitro and contribution to force generation in vivo (Madden et al. 2015; Quarta et al. 2017; Vandenburgh et al. 2008). Interrogating the regeneration potential of an engineered muscle construct in vitro is key for predicting therapeutic outcomes in vivo. This can be achieved by inflicting muscle damage and recording the regeneration capacity of the engineered tissue in vitro (Juhas et al. 2014). In a notable study, Juhas and colleagues generated 3D engineered muscle bundles from neonatal rat myogenic cells and intriguingly observed Pax7 satellite-like cells under the basal lamina in these constructs (Juhas et al. 2014). In concordance with this observation, the engineered construct could regenerate muscle fibers in vitro following localized cardiotoxin muscle injury (Juhas et al. 2014).

An alternative myogenic cell source for tissue engineering of skeletal muscle are iPSCs, as they can potentially differentiate into an unlimited number of myogenic precursors that can be used for muscle reconstruction (Pantelic and Larkin 2018; Del Carmen Ortuno-Costela et al. 2019). Rao and colleagues recently reported on the first generation of functional skeletal muscle tissue entirely from human iPSCs (Rao et al. 2018). Human iPSCs were exposed to the GSK3-β inhibitor CHIR-99201 in conjunction with transient Pax7 overexpression, resulting in differentiation into myogenic precursors that express endogenous Pax7 (Rao et al. 2018). These cells were further cultured in 2D dishes or in 3D fibrin hydrogels (Rao et al. 2018). The authors demonstrated that using a 2D culture system, skeletal muscle was partially formed, while usage of the 3D culture system gave rise to mature muscle bundles that expressed adult myosin heavy chain isoforms (Rao et al. 2018). Strikingly, the muscle bundles were functional in vitro and in vivo as demonstrated by induction of contractions and calcium transits using electrical and chemical stimuli (Rao et al. 2018).

Collectively, the findings described thus far illustrate eminent advancements in skeletal muscle cell organization in vitro; however engineering functional muscle tissue is expected to require vascularization and innervation of the muscle constructs. In the next sections we will briefly describe methods to vascularize or innervate in vitro engineered muscle cells.

8.2 Vascularization of Skeletal Muscle Tissue Constructs

Skeletal muscle is a highly vascularized tissue, as blood vessels are essential for in vitro formation of large-sized tissue since they enable nutrient and oxygen supply across large tissue segments (Kaully et al. 2009). Several studies have therefore investigated the ability to vascularize tissue-engineered muscle constructs by in vivo endothelial cell recruitment or in vitro pre-vascularization (Gilbert-Honick and Grayson 2019). With respect to the first approach, some studies reported on host vessel infiltration into small muscle constructs (Juhas et al. 2014; Kang et al. 2016); albeit for larger muscle constructs this is deemed a limiting factor. To address this limitation, several studies documented recruitment of host blood vessels by engineering cells or scaffolds to secrete growth factors such as VEGF to facilitate vessel migration (Zhou et al. 2015; Koffler et al. 2011; Osaki et al. 2018a; Borselli et al. 2010). In one notable study, muscle constructs consisting of myoblasts that ectopically express VEGF have been reported to induce robust vascularization in vivo (Zhou et al. 2015). Furthermore, interactions between C2C12-derived muscle bundles that ectopically express angiopoietin-1 and endothelial cells have been studied (Osaki et al. 2018a). Angiopoietin-1 secretion evoked endothelial cell sprouting toward the muscle fiber bundles, and this interaction increased myofiber contraction force strength (Osaki et al. 2018a).

Whereas enhancement of host cell infiltration is a promising research direction, replacing a muscle tissue in vivo may be more advantageous for therapy if the tissue is composed of multiple cell types with different physiological roles. Co-culturing myoblasts together with endothelial cells is an established method to improve vascularization of muscle grafts (Levenberg et al. 2005; Koffler et al. 2011; Gholobova et al. 2019; Nakayama et al. 2019). For example, co-culturing murine myoblasts with endothelial cells and fibroblasts have been reported to form muscle constructs consisting of vascular networks in vitro (Levenberg et al. 2005), and these findings were further recapitulated using human cells (Gholobova et al. 2015, 2019; Maffioletti et al. 2018). In a recent study, Kaufman and colleagues investigated vascularization of muscle grafts consisting of C2C12, endothelial cells, and fibroblasts in mice afflicted with injured abdominal walls (Kaufman et al. 2019). Utilizing this approach, robust graft integration into host muscles has been documented, consisting of parallel donor-derived myotubes and absence of fibrotic structures (Kaufman et al. 2019). Similarly, using human umbilical vein endothelial cells (HUVECs) with human myoblasts reportedly created endothelial cell networks in

BAM constructs, highlighting the importance of vascularization of myotubes for production of larger bio-artificial muscle constructs (Gholobova et al. 2019). Lastly, patterned scaffolds seeded with myoblasts and vascular endothelial cells have been recently demonstrated to positively guide parallel myotube alignment and organize microvasculature of engineered skeletal muscle constructs (Nakayama et al. 2019). In comparison to randomly aligned myotubes, these parallel aligned myofibers gave rise to coordinated contractions, upregulation of myogenic genes, and exceedingly improved integration into VML lesions in a mouse model (Nakayama et al. 2019). This work highlights the potential of combining endothelial cells in concert with unique scaffolds to induce vascularization in addition to muscle regeneration upon engraftment.

8.3 Innervation of Skeletal Muscle Tissue Constructs

Motor neurons induce muscle contractions via secretion of neurotransmitters in neuromuscular junctions (Slater 2017). This intricate molecular crosstalk allows muscle innervation by motor neurons, a process which plays a pivotal role in skeletal muscle function, and its derailment rapidly elicits muscle atrophy (Afshar Bakooshli et al. 2019; Osaki et al. 2018b; Kaufman et al. 2019). Generation of tissue-engineered muscle constructs therefore may require extensive innervation to enable optimal graft function inside host muscles. To this end, multiple factors have been assessed with respect to their capacity to innervate muscle constructs including neurotization (Dhawan et al. 2007; Kaufman et al. 2019), chemicals (Wang et al. 2013; Ko et al. 2013), co-culture with neural cell types (Larkin et al. 2006; Morimoto et al. 2013; Afshar Bakooshli et al. 2019), and exercise (Quarta et al. 2017).

Neurotization is a well-established method that can elicit tissue innervation via transfer of host nerves into a de-innervated tissue. For example, attaching an engineered muscle construct to femoral nerves reportedly increases innervation of flaps (Kaufman et al. 2019). Innervation of muscle tissue is key for achieving therapeutic success as lack of innervation in muscle flaps culminates in adipogenic or fibrotic tissue replacement (Chang et al. 2018). Furthermore, proper innervation of muscle constructs has also been associated with enhanced muscle maturation and increased force production (Dhawan et al. 2007).

Aside from neurotization, "preconditioning" innervation in vitro could also provide means to accelerate innervation in vivo, particularly for large muscle constructs (Ko et al. 2013). This can be achieved by inducing expression of acetylcholine receptors which are pivotal for neuromuscular junction function (Ko et al. 2013). To this end, multiple studies have reported on beneficial effects for utilizing the motor neuron secreted factor agrin on skeletal muscle cells (Bian and Bursac 2012; Bentzinger et al. 2005; Bezakova and Ruegg 2003). Agrin is an acetylcholine receptor cluster-inducing factor and accordingly muscle constructs that were pre-treated with agrin prior to transplantation demonstrated an increase in acetylcholine receptors as well as large quantities of neuromuscular junctions both in vitro and in vivo (Ko et al. 2013). In addition to chemical stimulation, co-culturing muscle and neural cells was also studied as a method to enhance innervation of muscle constructs (Morimoto et al. 2013; Cvetkovic et al. 2017; Happe et al. 2017; Martin et al. 2015; Larkin et al. 2006). In one such study, Morimoto and colleagues reported on a technique to co-culture mouse muscle bundles with neural stem cells that were further differentiated into motor neurons (Morimoto et al. 2013). This co-culture system enabled formation of neuromuscular junctions and muscle contraction following chemical stimulation (Morimoto et al. 2013).

Human iPSC-derived neural and muscle cells have similarly been co-cultured in 3D, reportedly enabling maturation of neuromuscular junctions by such technique (Osaki et al. 2018b; Afshar Bakooshli et al. 2019; Maffioletti et al. 2018). In a recent study, innervated skeletal muscle constructs have been generated by co-culturing human muscle progenitors cells and pluripotent stem cell-derived motor neurons inside a fibrin/Geltrex hydrogel (Afshar Bakooshli et al. 2019). Following 2 weeks of culture, motor neuron clusters were detected at the periphery of the newly formed muscle bundles (Afshar Bakooshli et al. 2019). Importantly, an epsilon acetylcholine receptor subunit, indicative of adult synapse formation, was detected in the newly formed neuromuscular junctions in 3D but not in a 2D co-culture system (Afshar Bakooshli et al. 2019). In line with these results, the authors detected calcium transits in response to motor neuron stimulation in the fibers, which suggests functional innervation (Afshar Bakooshli et al. 2019). Collectively, these results highlight the importance of 3D co-culture systems for production of functional neuromuscular junctions in vitro.

8.4 Integrative Tissue Engineering Approaches to Treat VML

The extent by which current therapeutic interventions assist in treating VML in humans is still not completely understood. However, a recently published large-scale meta-analysis comparing different VML regenerative medicine-based treatment options in animal models concluded that interventions improve functional recovery in comparison to untreated injury (Greising et al. 2019). Importantly, of all examined interventions, delivery of acellular biomaterials seeded with cells was deemed the most beneficial therapeutic approach (Greising et al. 2019). However, previous works have revealed the potential of ECM based-scaffolds to treat VML. Acellular scaffolds alone have been implanted in rodents and humans to counteract VML; however, these constructs solely depend on infiltration of host cells, which presents a challenge (Greising et al. 2019; Sicari et al. 2014; Dziki et al. 2016; Corona and Greising 2016). An alternative approach is engineering skeletal muscle constructs in vitro for the treatment of VML or late stage DMD. Notably, success of this approach depends on integration of different techniques and approaches, such as combining different cell types with various biomaterials. Such an approach involves selection of an appropriate biomaterial, the type of muscle precursor cell to utilize, as well as methods to innervate and vascularize the muscle constructs. In the last section, we will introduce several studies that successfully assimilated different techniques to produce engineered skeletal muscle tissue constructs in vitro that further demonstrated functional therapeutic outcome in vivo.

A powerful approach to engineer muscle tissue involves combination of acellular scaffolds with live cells prior to implantation, thus circumventing the dependency on host cell-infiltration. Therapeutic applicability of this approach has been successfully investigated in rodent VML models (Mintz et al. 2019; Quarta et al. 2017; Nakayama et al. 2019). In one recent study, engineered muscle bundles have been created from adult rat myogenic cells seeded with bone marrow-derived macrophages, an essential cellular component for muscle regeneration (Novak et al. 2014; Bencze et al. 2012; Lesault et al. 2012; Juhas et al. 2018). The incorporation of macrophages into engineered muscle constructs was deemed a critical step following cardiotoxin injury in vitro, enabling extensive structural and functional tissue repair (Juhas et al. 2018).

Directed differentiation of iPSCs is an elegant approach to produce muscle cells or other cell types that comprise skeletal muscle tissue. Indeed, a few recent studies reported on new disease models for muscular dystrophies using iPSC-derived muscle cells (Long et al. 2018; Maffioletti et al. 2018). In one notable study, Maffioletti and colleagues cultured human iPSC-derived muscle progenitors together with endothelial cells and pericytes inside a fibrin hydrogel, demonstrating formation of vessel-like structures inside muscle bundles (Maffioletti et al. 2018). Furthermore, iPSC-derived neural precursors were incorporated into these engineered muscle bundles and gave rise to motor neurons (Maffioletti et al. 2018). This artificial skeletal muscle tissue-like construct consist of multiple human iPSC-derived cell types that conjunctionally mimic physiological muscle tissue, offering an elegant platform for disease modeling or a tissue source to treat VML lesions (Maffioletti et al. 2018).

A growing number of new studies have unearthed the roles of muscle resident cells (MRCs) in mediating skeletal muscle tissue regeneration and repair (Wosczyna and Rando 2018). In a recent elegant study, Quarta and colleagues assessed the MRCs capacity to assist engineered muscle constructs in repairing VML lesions in a mouse model(Quarta et al. 2017). As first step, freshly isolated satellite cells were suspended in collagen I or ECM hydrogel solutions and seeded onto decellularized muscle scaffolds (Quarta et al. 2017). To test the effects of MRCs on muscle regeneration, FACS-purified MRCs including hematopoietic, endothelial, fibro-adipogenic, and fibroblast-like cells were coalesced with muscle stem cells and engineered muscle constructs and further delivered into VML lesions (Quarta et al. 2017). Remarkably, combination of muscle stem cells and MRCs dramatically increased the regeneration potential of the engrafted constructs, culminating in increased muscle mass and force production of treated lesions compared to control mice treated with constructs consisting of muscle stem cells alone (Quarta et al. 2017). Moreover, this bioconstructs restored biomechanical properties in a murine model of VML and reduced fibrosis (Quarta et al. 2018). Lastly the potential of running exercise to assist in the regeneration process was further assessed, demonstrating improved innervation of the muscle constructs in lesions, which facilitated functional recovery of treated mice (Quarta et al. 2017). It will be of interest to further delineate the molecular mechanisms that enable exercise to potentiate muscle regeneration in this model and whether it can be harnessed to assist other engraftment paradigms such as stem cell delivery into dystrophic muscles.

Reconstructing skeletal muscle tissue in vitro is a formidable feat that involves a plethora of cellular and acellular components as well as engineering techniques. The breadth of recent research in the tissue engineering field is spearheading efforts to achieve this goal. Novel approaches to grow cells on scaffolds or hydrogels, improved methods to vascularize or innervate engineered tissues as well as harnessing muscle resident cells for repair streamline novel assistive means to engineer muscle tissue. This melting pot of different techniques and approaches is expected to yield new therapeutic applications to treat VML and other muscle-associated diseases.

9 Conclusions

Healthy skeletal muscle tissue harbors a prodigious regenerative capacity to restore muscle loss during injury or disease states. However, this capability is often derailed in patients with severe pathological conditions such as muscular dystrophies, sarcopenia, cachexia, and VML. Intramuscular transplantation of muscle stem or progenitor cells provides an attractive approach to treat some of the aforementioned conditions, albeit significant challenges still inhibit use of this approach in clinical settings. Notably, for therapeutic interventions to succeed, it will be imperative to formulate methods to match a treatment to a specific disease condition, while taking into consideration the severity grade of the affliction. For example, muscle stem cells are expected to engraft better when delivered into early-stage dystrophic muscles which contain sufficient amount of myofibers that could enable donor-derived cells to engraft. Contrastingly, late-stage DMD and VML is characterized by extensive muscle lesions and replacement with fibrotic or adipogenic tissues, rendering muscle stem cell transplantation futile in the absence of sufficient muscle volume. In such occurrences, incorporation of 3D tissue-engineered skeletal muscle constructs could provide a more valuable therapeutic method to restore muscle mass.

An additional important criterion to consider for cell replacement therapy is the choice of engrafted cells. To date, a majority of research studies have focused primarily on relatively undefined myoblast cultures for human transplantation. Rapid advances in isolation and ex vivo culture of human satellite cells could provide a superior cell source for engraftment, thus allowing a long-term therapeutic effect. It is therefore of interest to assess whether direct injection of freshly isolated satellite cells, or in vitro expanded satellite-like cells, could promote dystrophin restoration and potential recovery in DMD patient's muscles. Building upon decades of myoblast transplantation trials in humans can lend useful insights when embarking on such new translational endeavors and assist in overcoming current translational roadblocks. Moreover, novel studies have also recently illustrated ways to generate unlimited number of therapeutically applicable myogenic precursors, either from PSCs or by direct reprogramming of somatic cells. It is of interest to further investigate these new conversion paradigms and compare the generated myogenic precursors molecularly and functionally to satellite cells and myoblasts. Such comparison could assist in finding the most suitable cell type for therapeutic applications.

Lastly, the role of muscle resident cells in mediating muscle tissue repair has been increasingly elucidated in recent years. Single cell sequencing technologies have helped map the various cell types that comprise muscle tissue, shedding new light on the cellular and molecular pathways involved in muscle regeneration. From a translational perspective, it will be of interest to investigate in detail whether injections of MRCs with muscle stem cells could better alleviate dystrophic symptoms and increase therapeutic outcomes in animal models. It is foreseeable that regenerating a patient's dystrophic muscles may require injections of three to five different cell types, each carrying a unique role in mediating tissue repair. Further research is also warranted in respect to tissue engineering of bioconstructs by incorporating MRCs and muscle stem cells. The majority of research to date has incorporated myoblasts into muscle constructs, whereas recent work nicely implemented muscles stem cells and MRCs to achieve muscle regeneration utilizing tissue-like constructs. Transplantation of an in vitro engineered muscle-like tissue that harbors satellite cell niches and consists of multiple cell types could provide a promising treatment strategy for patients suffering from VML or late-stage muscular dystrophies.

In conclusion, muscle-associated diseases inflict tremendous burden and pain on patients, their families, and society. New research in stem cell biology, biomaterials and tissue engineering immensely elevates our ability to remedy degenerative muscle tissue, bringing new hope for novel therapeutic treatments. It is anticipated that a synergistic approach combining different disciplines and assimilating various techniques will provide the most suitable way for achieving this goal.

Acknowledgments We wish to thank Dr. Ines Soro Arnáiz and Dr. Gommaar D'Hulst as well as all members of the Regenerative and Movement Biology laboratory for their constructive comments and feedback. We are further grateful to Dr. Shahragim Tajbakhsh for providing the Pax7-nGFP mouse strain and Xhem Qabrati, Inseon Kim and Ajda Lenardic for the Pax7-nGFP myoblasts, iMPCs and iPSC-derived MPCs.

References

Afshar Bakooshli M, Lippmann ES, Mulcahy B, Iyer N, Nguyen CT, Tung K, Stewart BA, van den Dorpel H, Fuehrmann T, Shoichet M, Bigot A, Pegoraro E, Ahn H, Ginsberg H, Zhen M, Ashton RS, Gilbert PM (2019) A 3D culture model of innervated human skeletal muscle enables studies of the adult neuromuscular junction. Elife 8. https://doi.org/10.7554/eLife.44530

Aguado BA, Mulyasasmita W, Su J, Lampe KJ, Heilshorn SC (2012) Improving viability of stem cells during syringe needle flow through the design of hydrogel cell carriers. Tissue Eng Part A 18(7–8):806–815. https://doi.org/10.1089/ten.TEA.2011.0391

Alexander MS, Rozkalne A, Colletta A, Spinazzola JM, Johnson S, Rahimov F, Meng H, Lawlor MW, Estrella E, Kunkel LM, Gussoni E (2016) CD82 is a marker for prospective isolation of human muscle satellite cells and is linked to muscular dystrophies. Cell Stem Cell 19(6):800–807. https://doi.org/10.1016/j.stem.2016.08.006

Alter J, Rozentzweig D, Bengal E (2008) Inhibition of myoblast differentiation by tumor necrosis factor alpha is mediated by c-Jun N-terminal kinase 1 and leukemia inhibitory factor. J Biol Chem 283(34):23224–23234. https://doi.org/10.1074/jbc.M801379200

Arnett AL, Konieczny P, Ramos JN, Hall J, Odom G, Yablonka-Reuveni Z, Chamberlain JR, Chamberlain JS (2014) Adeno-associated viral (AAV) vectors do not efficiently target muscle satellite cells. Mol Ther Methods Clin Dev 1. https://doi.org/10.1038/mtm.2014.38

Austin L, Burgess AW (1991) Stimulation of myoblast proliferation in culture by leukaemia inhibitory factor and other cytokines. J Neurol Sci 101(2):193–197. https://doi.org/10.1016/0022-510x(91)90045-9

Austin L, Bower J, Kurek J, Vakakis N (1992) Effects of leukaemia inhibitory factor and other cytokines on murine and human myoblast proliferation. J Neurol Sci 112(1–2):185–191. https://doi.org/10.1016/0022-510x(92)90149-f

Baghdadi MB, Castel D, Machado L, Fukada SI, Birk DE, Relaix F, Tajbakhsh S, Mourikis P (2018) Reciprocal signalling by Notch-Collagen V-CALCR retains muscle stem cells in their niche. Nature 557(7707):714–718. https://doi.org/10.1038/s41586-018-0144-9

Bansal V, De D, An J, Kang TM, Jeong HJ, Kang JS, Kim KK (2019) Chemical induced conversion of mouse fibroblasts and human adipose-derived stem cells into skeletal muscle-like cells. Biomaterials 193:30–46. https://doi.org/10.1016/j.biomaterials.2018.11.037

Baracos VE, Martin L, Korc M, Guttridge DC, Fearon KCH (2018) Cancer-associated cachexia. Nat Rev Dis Primers 4:17105. https://doi.org/10.1038/nrdp.2017.105

Barberi T, Bradbury M, Dincer Z, Panagiotakos G, Socci ND, Studer L (2007) Derivation of engraftable skeletal myoblasts from human embryonic stem cells. Nat Med 13(5):642–648. https://doi.org/10.1038/nm1533

Bar-Nur O, Gerli MFM, Di Stefano B, Almada AE, Galvin A, Coffey A, Huebner AJ, Feige P, Verheul C, Cheung P, Payzin-Dogru D, Paisant S, Anselmo A, Sadreyev RI, Ott HC, Tajbakhsh S, Rudnicki MA, Wagers AJ, Hochedlinger K (2018) Direct reprogramming of mouse fibroblasts into functional skeletal muscle progenitors. Stem Cell Reports 10(5):1505–1521. https://doi.org/10.1016/j.stemcr.2018.04.009

Beauchamp JR, Morgan JE, Pagel CN, Partridge TA (1999) Dynamics of myoblast transplantation reveal a discrete minority of precursors with stem cell-like properties as the myogenic source. J Cell Biol 144(6):1113–1122. https://doi.org/10.1083/jcb.144.6.1113

Beauchamp JR, Heslop L, Yu DS, Tajbakhsh S, Kelly RG, Wernig A, Buckingham ME, Partridge TA, Zammit PS (2000) Expression of CD34 and Myf5 defines the majority of quiescent adult skeletal muscle satellite cells. J Cell Biol 151(6):1221–1234. https://doi.org/10.1083/jcb.151.6.1221

Bencze M, Negroni E, Vallese D, Yacoub-Youssef H, Chaouch S, Wolff A, Aamiri A, Di Santo JP, Chazaud B, Butler-Browne G, Savino W, Mouly V, Riederer I (2012) Proinflammatory macrophages enhance the regenerative capacity of human myoblasts by modifying their kinetics of proliferation and differentiation. Mol Ther 20(11):2168–2179. https://doi.org/10.1038/mt.2012.189

Ben-David U, Benvenisty N (2011) The tumorigenicity of human embryonic and induced pluripotent stem cells. Nat Rev Cancer 11(4):268–277. https://doi.org/10.1038/nrc3034

Bentzinger CF, Barzaghi P, Lin S, Ruegg MA (2005) Overexpression of mini-agrin in skeletal muscle increases muscle integrity and regenerative capacity in laminin-alpha2-deficient mice. FASEB J 19(8):934–942. https://doi.org/10.1096/fj.04-3376com

Bentzinger CF, Wang YX, von Maltzahn J, Soleimani VD, Yin H, Rudnicki MA (2013) Fibronectin regulates Wnt7a signaling and satellite cell expansion. Cell Stem Cell 12(1):75–87. https://doi.org/10.1016/j.stem.2012.09.015

Bentzinger CF, von Maltzahn J, Dumont NA, Stark DA, Wang YX, Nhan K, Frenette J, Cornelison DD, Rudnicki MA (2014) Wnt7a stimulates myogenic stem cell motility and engraftment resulting in improved muscle strength. J Cell Biol 205(1):97–111. https://doi.org/10.1083/jcb.201310035

Bernet JD, Doles JD, Hall JK, Kelly Tanaka K, Carter TA, Olwin BB (2014) p38 MAPK signaling underlies a cell-autonomous loss of stem cell self-renewal in skeletal muscle of aged mice. Nat Med 20(3):265–271. https://doi.org/10.1038/nm.3465

Bezakova G, Ruegg MA (2003) New insights into the roles of agrin. Nat Rev Mol Cell Biol 4(4):295–308. https://doi.org/10.1038/nrm1074

Bian W, Bursac N (2012) Soluble miniagrin enhances contractile function of engineered skeletal muscle. FASEB J 26(2):955–965. https://doi.org/10.1096/fj.11-187575

Bian W, Liau B, Badie N, Bursac N (2009) Mesoscopic hydrogel molding to control the 3D geometry of bioartificial muscle tissues. Nat Protoc 4(10):1522–1534. https://doi.org/10.1038/nprot.2009.155

Bisson A, Freret M, Drouot L, Jean L, Le Corre S, Gourcerol G, Doucet C, Michot F, Boyer O, Lamacz M (2015) Restoration of anal sphincter function after myoblast cell therapy in incontinent rats. Cell Transplant 24(2):277–286. https://doi.org/10.3727/096368913X674053

Bjornson CR, Cheung TH, Liu L, Tripathi PV, Steeper KM, Rando TA (2012) Notch signaling is necessary to maintain quiescence in adult muscle stem cells. Stem Cells 30(2):232–242. https://doi.org/10.1002/stem.773

Blaganje M, Lukanovic A (2012) Intrasphincteric autologous myoblast injections with electrical stimulation for stress urinary incontinence. Int J Gynaecol Obstet 117(2):164–167. https://doi.org/10.1016/j.ijgo.2011.11.029

Blau HM, Daley GQ (2019) Stem cells in the treatment of disease. N Engl J Med 380(18):1748–1760. https://doi.org/10.1056/NEJMra1716145

Boldrin L, Elvassore N, Malerba A, Flaibani M, Cimetta E, Piccoli M, Baroni MD, Gazzola MV, Messina C, Gamba P, Vitiello L, De Coppi P (2007) Satellite cells delivered by micro-patterned scaffolds: a new strategy for cell transplantation in muscle diseases. Tissue Eng 13(2):253–262. https://doi.org/10.1089/ten.2006.0093

Boldrin L, Malerba A, Vitiello L, Cimetta E, Piccoli M, Messina C, Gamba PG, Elvassore N, De Coppi P (2008) Efficient delivery of human single fiber-derived muscle precursor cells via biocompatible scaffold. Cell Transplant 17(5):577–584. https://doi.org/10.3727/096368908785095980

Boonen KJ, Rosaria-Chak KY, Baaijens FP, van der Schaft DW, Post MJ (2009) Essential environmental cues from the satellite cell niche: optimizing proliferation and differentiation. Am J Physiol Cell Physiol 296(6):C1338–C1345. https://doi.org/10.1152/ajpcell.00015.2009

Borchin B, Chen J, Barberi T (2013) Derivation and FACS-mediated purification of PAX3+/PAX7+ skeletal muscle precursors from human pluripotent stem cells. Stem Cell Reports 1(6):620–631. https://doi.org/10.1016/j.stemcr.2013.10.007

Borselli C, Storrie H, Benesch-Lee F, Shvartsman D, Cezar C, Lichtman JW, Vandenburgh HH, Mooney DJ (2010) Functional muscle regeneration with combined delivery of angiogenesis and myogenesis factors. Proc Natl Acad Sci U S A 107(8):3287–3292. https://doi.org/10.1073/pnas.0903875106

Borselli C, Cezar CA, Shvartsman D, Vandenburgh HH, Mooney DJ (2011) The role of multifunctional delivery scaffold in the ability of cultured myoblasts to promote muscle regeneration. Biomaterials 32(34):8905–8914. https://doi.org/10.1016/j.biomaterials.2011.08.019

Bosnakovski D, Xu Z, Li W, Thet S, Cleaver O, Perlingeiro RC, Kyba M (2008) Prospective isolation of skeletal muscle stem cells with a Pax7 reporter. Stem Cells 26(12):3194–3204. https://doi.org/10.1634/stemcells.2007-1017

Boyer O, Bridoux V, Giverne C, Bisson A, Koning E, Leroi AM, Chambon P, Dehayes J, Le Corre S, Jacquot S, Bastit D, Martinet J, Houivet E, Tuech JJ, Benichou J, Michot F, The Study Group of Myoblast Therapy for Faecal I (2018) Autologous myoblasts for the treatment of fecal incontinence: results of a phase 2 randomized placebo-controlled study (MIAS). Ann Surg 267(3):443–450. https://doi.org/10.1097/SLA.0000000000002268

Brais B, Bouchard JP, Xie YG, Rochefort DL, Chretien N, Tome FM, Lafreniere RG, Rommens JM, Uyama E, Nohira O, Blumen S, Korczyn AD, Heutink P, Mathieu J, Duranceau A, Codere F, Fardeau M, Rouleau GA (1998) Short GCG expansions in the PABP2 gene cause oculopharyngeal muscular dystrophy. Nat Genet 18(2):164–167. https://doi.org/10.1038/ng0298-164

Briggs D, Morgan JE (2013) Recent progress in satellite cell/myoblast engraftment - relevance for therapy. FEBS J 280(17):4281–4293. https://doi.org/10.1111/febs.12273

Brohl D, Vasyutina E, Czajkowski MT, Griger J, Rassek C, Rahn HP, Purfurst B, Wende H, Birchmeier C (2012) Colonization of the satellite cell niche by skeletal muscle progenitor cells depends on Notch signals. Dev Cell 23(3):469–481. https://doi.org/10.1016/j.devcel.2012.07.014

Castiglioni A, Hettmer S, Lynes MD, Rao TN, Tchessalova D, Sinha I, Lee BT, Tseng YH, Wagers AJ (2014) Isolation of progenitors that exhibit myogenic/osteogenic bipotency in vitro by fluorescence-activated cell sorting from human fetal muscle. Stem Cell Reports 2(1):92–106. https://doi.org/10.1016/j.stemcr.2013.12.006

Cerletti M, Jurga S, Witczak CA, Hirshman MF, Shadrach JL, Goodyear LJ, Wagers AJ (2008) Highly efficient, functional engraftment of skeletal muscle stem cells in dystrophic muscles. Cell 134(1):37–47. https://doi.org/10.1016/j.cell.2008.05.049

Cezar CA, Mooney DJ (2015) Biomaterial-based delivery for skeletal muscle repair. Adv Drug Deliv Rev 84:188–197. https://doi.org/10.1016/j.addr.2014.09.008

Chal J, Pourquie O (2017) Making muscle: skeletal myogenesis in vivo and in vitro. Development 144(12):2104–2122. https://doi.org/10.1242/dev.151035

Chal J, Oginuma M, Al Tanoury Z, Gobert B, Sumara O, Hick A, Bousson F, Zidouni Y, Mursch C, Moncuquet P, Tassy O, Vincent S, Miyanari A, Bera A, Garnier JM, Guevara G, Hestin M, Kennedy L, Hayashi S, Drayton B, Cherrier T, Gayraud-Morel B, Gussoni E, Relaix F, Tajbakhsh S, Pourquie O (2015) Differentiation of pluripotent stem cells to muscle fiber to model Duchenne muscular dystrophy. Nat Biotechnol 33(9):962–969. https://doi.org/10.1038/nbt.3297

Chal J, Al Tanoury Z, Oginuma M, Moncuquet P, Gobert B, Miyanari A, Tassy O, Guevara G, Hubaud A, Bera A, Sumara O, Garnier JM, Kennedy L, Knockaert M, Gayraud-Morel B, Tajbakhsh S, Pourquie O (2018) Recapitulating early development of mouse musculoskeletal precursors of the paraxial mesoderm in vitro. Development 145(6). https://doi.org/10.1242/dev.157339

Chan SS, Arpke RW, Filareto A, Xie N, Pappas MP, Penaloza JS, Perlingeiro RCR, Kyba M (2018) Skeletal muscle stem cells from PSC-derived teratomas have functional regenerative capacity. Cell Stem Cell 23(1):74–85 e76. https://doi.org/10.1016/j.stem.2018.06.010

Chang H, Hwang S, Lim S, Eo S, Minn KW, Hong KY (2018) Long-term fate of denervated skeletal muscle after microvascular flap transfer. Ann Plast Surg 80(6):644–647. https://doi.org/10.1097/SAP.0000000000001397

Charville GW, Cheung TH, Yoo B, Santos PJ, Lee GK, Shrager JB, Rando TA (2015) Ex vivo expansion and in vivo self-renewal of human muscle stem cells. Stem Cell Reports 5(4):621–632. https://doi.org/10.1016/j.stemcr.2015.08.004

Chen AE, Ginty DD, Fan CM (2005) Protein kinase A signalling via CREB controls myogenesis induced by Wnt proteins. Nature 433(7023):317–322. https://doi.org/10.1038/nature03126

Choi IY, Lim H, Estrellas K, Mula J, Cohen TV, Zhang Y, Donnelly CJ, Richard JP, Kim YJ, Kim H, Kazuki Y, Oshimura M, Li HL, Hotta A, Rothstein J, Maragakis N, Wagner KR, Lee G (2016) Concordant but varied phenotypes among duchenne muscular dystrophy patient-specific myoblasts derived using a human iPSC-based model. Cell Rep 15(10):2301–2312. https://doi.org/10.1016/j.celrep.2016.05.016

Collins CA, Olsen I, Zammit PS, Heslop L, Petrie A, Partridge TA, Morgan JE (2005) Stem cell function, self-renewal, and behavioral heterogeneity of cells from the adult muscle satellite cell niche. Cell 122(2):289–301. https://doi.org/10.1016/j.cell.2005.05.010

Colognato H, Yurchenco PD (2000) Form and function: the laminin family of heterotrimers. Dev Dyn 218(2):213–234. https://doi.org/10.1002/(SICI)1097-0177(200006)218:2<213::AID-DVDY1>3.0.CO;2-R

Comai G, Tajbakhsh S (2014) Molecular and cellular regulation of skeletal myogenesis. Curr Top Dev Biol 110:1–73. https://doi.org/10.1016/B978-0-12-405943-6.00001-4

Conboy IM, Rando TA (2002) The regulation of Notch signaling controls satellite cell activation and cell fate determination in postnatal myogenesis. Dev Cell 3(3):397–409. https://doi.org/10.1016/s1534-5807(02)00254-x

Corona BT, Greising SM (2016) Challenges to acellular biological scaffold mediated skeletal muscle tissue regeneration. Biomaterials 104:238–246. https://doi.org/10.1016/j.biomaterials.2016.07.020

Corona BT, Ward CL, Baker HB, Walters TJ, Christ GJ (2014) Implantation of in vitro tissue engineered muscle repair constructs and bladder acellular matrices partially restore in vivo skeletal muscle function in a rat model of volumetric muscle loss injury. Tissue Eng Part A 20 (3–4):705–715. https://doi.org/10.1089/ten.TEA.2012.0761

Corona BT, Wenke JC, Ward CL (2016) Pathophysiology of volumetric muscle loss injury. Cells Tissues Organs 202(3–4):180–188. https://doi.org/10.1159/000443925

Corona BT, Rivera JC, Dalske KA, Wenke JC, Greising SM (2019) Pharmacological mitigation of fibrosis in a porcine model of volumetric muscle loss injury. Tissue Eng Part A. https://doi.org/10.1089/ten.TEA.2019.0272

Cosgrove BD, Gilbert PM, Porpiglia E, Mourkioti F, Lee SP, Corbel SY, Llewellyn ME, Delp SL, Blau HM (2014) Rejuvenation of the muscle stem cell population restores strength to injured aged muscles. Nat Med 20(3):255–264. https://doi.org/10.1038/nm.3464

Cossu G, Previtali SC, Napolitano S, Cicalese MP, Tedesco FS, Nicastro F, Noviello M, Roostalu U, Natali Sora MG, Scarlato M, De Pellegrin M, Godi C, Giuliani S, Ciotti F, Tonlorenzi R, Lorenzetti I, Rivellini C, Benedetti S, Gatti R, Marktel S, Mazzi B, Tettamanti A, Ragazzi M, Imro MA, Marano G, Ambrosi A, Fiori R, Sormani MP, Bonini C, Venturini M, Politi LS, Torrente Y, Ciceri F (2015) Intra-arterial transplantation of HLA-matched donor mesoangioblasts in Duchenne muscular dystrophy. EMBO Mol Med 7(12):1513–1528. https://doi.org/10.15252/emmm.201505636

Costantini M, Testa S, Mozetic P, Barbetta A, Fuoco C, Fornetti E, Tamiro F, Bernardini S, Jaroszewicz J, Swieszkowski W, Trombetta M, Castagnoli L, Seliktar D, Garstecki P, Cesareni G, Cannata S, Rainer A, Gargioli C (2017) Microfluidic-enhanced 3D bioprinting of aligned myoblast-laden hydrogels leads to functionally organized myofibers in vitro and in vivo. Biomaterials 131:98–110. https://doi.org/10.1016/j.biomaterials.2017.03.026

Cvetkovic C, Rich MH, Raman R, Kong H, Bashir R (2017) A 3D-printed platform for modular neuromuscular motor units. Microsyst Nanoeng 3:17015. https://doi.org/10.1038/micronano.2017.15

da Cruz L, Fynes K, Georgiadis O, Kerby J, Luo YH, Ahmado A, Vernon A, Daniels JT, Nommiste B, Hasan SM, Gooljar SB, Carr AF, Vugler A, Ramsden CM, Bictash M, Fenster M, Steer J, Harbinson T, Wilbrey A, Tufail A, Feng G, Whitlock M, Robson AG, Holder GE, Sagoo MS, Loudon PT, Whiting P, Coffey PJ (2018) Phase 1 clinical study of an embryonic stem cell-derived retinal pigment epithelium patch in age-related macular degeneration. Nat Biotechnol 36(4):328–337. https://doi.org/10.1038/nbt.4114

Darabi R, Gehlbach K, Bachoo RM, Kamath S, Osawa M, Kamm KE, Kyba M, Perlingeiro RC (2008) Functional skeletal muscle regeneration from differentiating embryonic stem cells. Nat Med 14(2):134–143. https://doi.org/10.1038/nm1705

Darabi R, Santos FN, Filareto A, Pan W, Koene R, Rudnicki MA, Kyba M, Perlingeiro RC (2011) Assessment of the myogenic stem cell compartment following transplantation of Pax3/Pax7-induced embryonic stem cell-derived progenitors. Stem Cells 29(5):777–790. https://doi.org/10.1002/stem.625

Darabi R, Arpke RW, Irion S, Dimos JT, Grskovic M, Kyba M, Perlingeiro RC (2012) Human ES- and iPS-derived myogenic progenitors restore DYSTROPHIN and improve contractility upon transplantation in dystrophic mice. Cell Stem Cell 10(5):610–619. https://doi.org/10.1016/j.stem.2012.02.015

Davis RL, Weintraub H, Lassar AB (1987) Expression of a single transfected cDNA converts fibroblasts to myoblasts. Cell 51(6):987–1000

Davoudi S, Chin CY, Cooke MJ, Tam RY, Shoichet MS, Gilbert PM (2018) Muscle stem cell intramuscular delivery within hyaluronan methylcellulose improves engraftment efficiency and dispersion. Biomaterials 173:34–46. https://doi.org/10.1016/j.biomaterials.2018.04.048

De Micheli AJ, Laurilliard EJ, Heinke CL, Ravichandran H, Fraczek P, Soueid-Baumgarten S, De Vlaminck I, Elemento O, Cosgrove BD (2020) Single-cell analysis of the muscle stem cell hierarchy identifies heterotypic communication signals involved in skeletal muscle regeneration. Cell Rep 30(10):3583–3595 e3585. https://doi.org/10.1016/j.celrep.2020.02.067

Dekel I, Magal Y, Pearson-White S, Emerson CP, Shani M (1992) Conditional conversion of ES cells to skeletal muscle by an exogenous MyoD1 gene. New Biol 4(3):217–224

Del Carmen Ortuno-Costela M, Garcia-Lopez M, Cerrada V, Gallardo ME (2019) iPSCs: a powerful tool for skeletal muscle tissue engineering. J Cell Mol Med 23(6):3784–3794. https://doi.org/10.1111/jcmm.14292

Dell'Orso S, Juan AH, Ko KD, Naz F, Perovanovic J, Gutierrez-Cruz G, Feng X, Sartorelli V (2019) Single cell analysis of adult mouse skeletal muscle stem cells in homeostatic and regenerative conditions. Development 146(12). https://doi.org/10.1242/dev.174177

Dennison EM, Sayer AA, Cooper C (2017) Epidemiology of sarcopenia and insight into possible therapeutic targets. Nat Rev Rheumatol 13(6):340–347. https://doi.org/10.1038/nrrheum.2017.60

Dhawan V, Lytle IF, Dow DE, Huang YC, Brown DL (2007) Neurotization improves contractile forces of tissue-engineered skeletal muscle. Tissue Eng 13(11):2813–2821. https://doi.org/10.1089/ten.2007.0003

Ding S, Swennen GNM, Messmer T, Gagliardi M, Molin DGM, Li C, Zhou G, Post MJ (2018) Maintaining bovine satellite cells stemness through p38 pathway. Sci Rep 8(1):10808. https://doi.org/10.1038/s41598-018-28746-7

Doudna JA, Charpentier E (2014) Genome editing. The new frontier of genome engineering with CRISPR-Cas9. Science 346(6213):1258096. https://doi.org/10.1126/science.1258096

Duffy RM, Sun Y, Feinberg AW (2016) Understanding the role of ECM protein composition and geometric micropatterning for engineering human skeletal muscle. Ann Biomed Eng 44 (6):2076–2089. https://doi.org/10.1007/s10439-016-1592-8

Dumont NA, Rudnicki MA (2017) Characterizing satellite cells and myogenic progenitors during skeletal muscle regeneration. Methods Mol Biol 1560:179–188. https://doi.org/10.1007/978-1-4939-6788-9_12

Dumont NA, Wang YX, von Maltzahn J, Pasut A, Bentzinger CF, Brun CE, Rudnicki MA (2015) Dystrophin expression in muscle stem cells regulates their polarity and asymmetric division. Nat Med 21(12):1455–1463. https://doi.org/10.1038/nm.3990

Duranceau A, Forand MD, Fauteux JP (1980) Surgery in oculopharyngeal muscular dystrophy. Am J Surg 139(1):33–39. https://doi.org/10.1016/0002-9610(80)90226-3

Dziki J, Badylak S, Yabroudi M, Sicari B, Ambrosio F, Stearns K, Turner N, Wyse A, Boninger ML, Brown EHP, Rubin JP (2016) An acellular biologic scaffold treatment for volumetric muscle loss: results of a 13-patient cohort study. NPJ Regen Med 1:16008. https://doi.org/10.1038/npjregenmed.2016.8

Eberli D, Aboushwareb T, Soker S, Yoo JJ, Atala A (2012) Muscle precursor cells for the restoration of irreversibly damaged sphincter function. Cell Transplant 21(9):2089–2098. https://doi.org/10.3727/096368911X623835

Eliazer S, Muncie JM, Christensen J, Sun X, D'Urso RS, Weaver VM, Brack AS (2019) Wnt4 from the niche controls the mechano-properties and quiescent state of muscle stem cells. Cell Stem Cell 25(5):654–665 e654. https://doi.org/10.1016/j.stem.2019.08.007

Engler AJ, Griffin MA, Sen S, Bonnemann CG, Sweeney HL, Discher DE (2004) Myotubes differentiate optimally on substrates with tissue-like stiffness: pathological implications for soft or stiff microenvironments. J Cell Biol 166(6):877–887. https://doi.org/10.1083/jcb.200405004

Evans MJ, Kaufman MH (1981) Establishment in culture of pluripotential cells from mouse embryos. Nature 292(5819):154–156

Fan Y, Maley M, Beilharz M, Grounds M (1996) Rapid death of injected myoblasts in myoblast transfer therapy. Muscle Nerve 19(7):853–860. https://doi.org/10.1002/(SICI)1097-4598(199607)19:7<853::AID-MUS7>3.0.CO;2-8

Filareto A, Darabi R, Perlingeiro RC (2012) Engraftment of ES-derived myogenic progenitors in a severe mouse model of muscular dystrophy. J Stem Cell Res Ther 10(1). https://doi.org/10.4172/2157-7633.S10-001

Filareto A, Parker S, Darabi R, Borges L, Iacovino M, Schaaf T, Mayerhofer T, Chamberlain JS, Ervasti JM, McIvor RS, Kyba M, Perlingeiro RC (2013) An ex vivo gene therapy approach to treat muscular dystrophy using inducible pluripotent stem cells. Nat Commun 4:1549. https://doi.org/10.1038/ncomms2550

Filareto A, Rinaldi F, Arpke RW, Darabi R, Belanto JJ, Toso EA, Miller AZ, Ervasti JM, McIvor RS, Kyba M, Perlingeiro RC (2015) Pax3-induced expansion enables the genetic correction of dystrophic satellite cells. Skelet Muscle 5:36. https://doi.org/10.1186/s13395-015-0061-7

Foster RF, Thompson JM, Kaufman SJ (1987) A laminin substrate promotes myogenesis in rat skeletal muscle cultures: analysis of replication and development using antidesmin and anti-BrdUrd monoclonal antibodies. Dev Biol 122(1):11–20. https://doi.org/10.1016/0012-1606(87)90327-7

Frudinger A, Pfeifer J, Paede J, Kolovetsiou-Kreiner V, Marksteiner R, Halligan S (2015) Autologous skeletal-muscle-derived cell injection for anal incontinence due to obstetric trauma: a 5-year follow-up of an initial study of 10 patients. Colorectal Dis 17(9):794–801. https://doi.org/10.1111/codi.12947

Fu X, Xiao J, Wei Y, Li S, Liu Y, Yin J, Sun K, Sun H, Wang H, Zhang Z, Zhang BT, Sheng C, Wang H, Hu P (2015) Combination of inflammation-related cytokines promotes long-term muscle stem cell expansion. Cell Res 25(6):655–673. https://doi.org/10.1038/cr.2015.58

Fukada S, Higuchi S, Segawa M, Koda K, Yamamoto Y, Tsujikawa K, Kohama Y, Uezumi A, Imamura M, Miyagoe-Suzuki Y, Takeda S, Yamamoto H (2004) Purification and cell-surface marker characterization of quiescent satellite cells from murine skeletal muscle by a novel monoclonal antibody. Exp Cell Res 296(2):245–255. https://doi.org/10.1016/j.yexcr.2004.02.018

Fuoco C, Salvatori ML, Biondo A, Shapira-Schweitzer K, Santoleri S, Antonini S, Bernardini S, Tedesco FS, Cannata S, Seliktar D, Cossu G, Gargioli C (2012) Injectable polyethylene glycol-fibrinogen hydrogel adjuvant improves survival and differentiation of transplanted meso-angioblasts in acute and chronic skeletal-muscle degeneration. Skelet Muscle 2(1):24. https://doi.org/10.1186/2044-5040-2-24

Fuoco C, Sangalli E, Vono R, Testa S, Sacchetti B, Latronico MV, Bernardini S, Madeddu P, Cesareni G, Seliktar D, Rizzi R, Bearzi C, Cannata SM, Spinetti G, Gargioli C (2014) 3D hydrogel environment rejuvenates aged pericytes for skeletal muscle tissue engineering. Front Physiol 5:203. https://doi.org/10.3389/fphys.2014.00203

Fuoco C, Rizzi R, Biondo A, Longa E, Mascaro A, Shapira-Schweitzer K, Kossovar O, Benedetti S, Salvatori ML, Santoleri S, Testa S, Bernardini S, Bottinelli R, Bearzi C, Cannata SM, Seliktar D, Cossu G, Gargioli C (2015) In vivo generation of a mature and functional artificial skeletal muscle. EMBO Mol Med 7(4):411–422. https://doi.org/10.15252/emmm.201404062

Garcia AJ, Vega MD, Boettiger D (1999) Modulation of cell proliferation and differentiation through substrate-dependent changes in fibronectin conformation. Mol Biol Cell 10(3):785–798. https://doi.org/10.1091/mbc.10.3.785

Garcia SM, Tamaki S, Lee S, Wong A, Jose A, Dreux J, Kouklis G, Sbitany H, Seth R, Knott PD, Heaton C, Ryan WR, Kim EA, Hansen SL, Hoffman WY, Pomerantz JH (2018) High-yield purification, preservation, and serial transplantation of human satellite cells. Stem Cell Reports 10(3):1160–1174. https://doi.org/10.1016/j.stemcr.2018.01.022

Gardner RJ, Bobrow M, Roberts RG (1995) The identification of point mutations in Duchenne muscular dystrophy patients by using reverse-transcription PCR and the protein truncation test. Am J Hum Genet 57(2):311–320

Gerard C, Forest MA, Beauregard G, Skuk D, Tremblay JP (2012) Fibrin gel improves the survival of transplanted myoblasts. Cell Transplant 21(1):127–137. https://doi.org/10.3727/096368911X576018

Gerli MFM, Moyle LA, Benedetti S, Ferrari G, Ucuncu E, Ragazzi M, Constantinou C, Louca I, Sakai H, Ala P, De Coppi P, Tajbakhsh S, Cossu G, Tedesco FS (2019) Combined Notch and PDGF signaling enhances migration and expression of stem cell markers while inducing perivascular cell features in muscle satellite cells. Stem Cell Reports 12(3):461–473. https://doi.org/10.1016/j.stemcr.2019.01.007

Gholobova D, Decroix L, Van Muylder V, Desender L, Gerard M, Carpentier G, Vandenburgh H, Thorrez L (2015) Endothelial network formation within human tissue-engineered skeletal muscle. Tissue Eng Part A 21(19–20):2548–2558. https://doi.org/10.1089/ten.TEA.2015.0093

Gholobova D, Gerard M, Terrie L, Desender L, Shansky J, Vandenburgh H, Thorrez L (2019) Coculture method to obtain endothelial networks within human tissue-engineered skeletal muscle. Methods Mol Biol 1889:169–183. https://doi.org/10.1007/978-1-4939-8897-6_10

Gilbert PM, Havenstrite KL, Magnusson KE, Sacco A, Leonardi NA, Kraft P, Nguyen NK, Thrun S, Lutolf MP, Blau HM (2010) Substrate elasticity regulates skeletal muscle stem cell self-renewal in culture. Science 329(5995):1078–1081. https://doi.org/10.1126/science.1191035

Gilbert-Honick J, Grayson W (2019) Vascularized and innervated skeletal muscle tissue engineering. Adv Healthc Mater:e1900626. https://doi.org/10.1002/adhm.201900626

Giordani L, He GJ, Negroni E, Sakai H, Law JYC, Siu MM, Wan R, Corneau A, Tajbakhsh S, Cheung TH, Le Grand F (2019) High-dimensional single-cell cartography reveals novel skeletal muscle-resident cell populations. Mol Cell 74(3):609–621 e606. https://doi.org/10.1016/j.molcel.2019.02.026

Goldstein JM, Tabebordbar M, Zhu K, Wang LD, Messemer KA, Peacker B, Ashrafi Kakhki S, Gonzalez-Celeiro M, Shwartz Y, Cheng JKW, Xiao R, Barungi T, Albright C, Hsu YC, Vandenberghe LH, Wagers AJ (2019) In situ modification of tissue stem and progenitor cell genomes. Cell Rep 27(4):1254–1264 e1257. https://doi.org/10.1016/j.celrep.2019.03.105

Goudenege S, Lebel C, Huot NB, Dufour C, Fujii I, Gekas J, Rousseau J, Tremblay JP (2012) Myoblasts derived from normal hESCs and dystrophic hiPSCs efficiently fuse with existing muscle fibers following transplantation. Mol Ther 20(11):2153–2167. https://doi.org/10.1038/mt.2012.188

Grefte S, Vullinghs S, Kuijpers-Jagtman AM, Torensma R, Von den Hoff JW (2012) Matrigel, but not collagen I, maintains the differentiation capacity of muscle derived cells in vitro. Biomed Mater 7(5):055004. https://doi.org/10.1088/1748-6041/7/5/055004

Greising SM, Corona BT, McGann C, Frankum JK, Warren GL (2019) Therapeutic approaches for volumetric muscle loss injury: a systematic review and meta-analysis. Tissue Eng Part B Rev. https://doi.org/10.1089/ten.TEB.2019.0207

Guerette B, Asselin I, Vilquin JT, Roy R, Tremblay JP (1995) Lymphocyte infiltration following allo- and xenomyoblast transplantation in mdx mice. Muscle Nerve 18(1):39–51. https://doi.org/10.1002/mus.880180107

Gussoni E, Pavlath GK, Lanctot AM, Sharma KR, Miller RG, Steinman L, Blau HM (1992) Normal dystrophin transcripts detected in Duchenne muscular dystrophy patients after myoblast transplantation. Nature 356(6368):435–438. https://doi.org/10.1038/356435a0

Gussoni E, Blau HM, Kunkel LM (1997) The fate of individual myoblasts after transplantation into muscles of DMD patients. Nat Med 3(9):970–977. https://doi.org/10.1038/nm0997-970

Hagiwara K, Chen G, Kawazoe N, Tabata Y, Komuro H (2016) Promotion of muscle regeneration by myoblast transplantation combined with the controlled and sustained release of bFGFcpr. J Tissue Eng Regen Med 10(4):325–333. https://doi.org/10.1002/term.1732

Hall JK, Banks GB, Chamberlain JS, Olwin BB (2010) Prevention of muscle aging by myofiber-associated satellite cell transplantation. Sci Transl Med 2(57):57ra83. https://doi.org/10.1126/scitranslmed.3001081

Han WM, Jang YC, Garcia AJ (2017) Engineered matrices for skeletal muscle satellite cell engraftment and function. Matrix Biol 60–61:96–109. https://doi.org/10.1016/j.matbio.2016.06.001

Han WM, Anderson SE, Mohiuddin M, Barros D, Nakhai SA, Shin E, Amaral IF, Pego AP, Garcia AJ, Jang YC (2018) Synthetic matrix enhances transplanted satellite cell engraftment in dystrophic and aged skeletal muscle with comorbid trauma. Sci Adv 4(8):eaar4008. https://doi.org/10.1126/sciadv.aar4008

Han WM, Mohiuddin M, Anderson SE, Garcia AJ, Jang YC (2019) Co-delivery of Wnt7a and muscle stem cells using synthetic bioadhesive hydrogel enhances murine muscle regeneration and cell migration during engraftment. Acta Biomater 94:243–252. https://doi.org/10.1016/j.actbio.2019.06.025

Happe CL, Tenerelli KP, Gromova AK, Kolb F, Engler AJ (2017) Mechanically patterned neuromuscular junctions-in-a-dish have improved functional maturation. Mol Biol Cell 28(14):1950–1958. https://doi.org/10.1091/mbc.E17-01-0046

Hardy D, Besnard A, Latil M, Jouvion G, Briand D, Thepenier C, Pascal Q, Guguin A, Gayraud-Morel B, Cavaillon JM, Tajbakhsh S, Rocheteau P, Chretien F (2016) Comparative study of injury models for studying muscle regeneration in mice. PLoS One 11(1):e0147198. https://doi.org/10.1371/journal.pone.0147198

Heslop L, Beauchamp JR, Tajbakhsh S, Buckingham ME, Partridge TA, Zammit PS (2001) Transplanted primary neonatal myoblasts can give rise to functional satellite cells as identified using the Myf5nlacZl+ mouse. Gene Ther 8(10):778–783. https://doi.org/10.1038/sj.gt.3301463

Hicks MR, Hiserodt J, Paras K, Fujiwara W, Eskin A, Jan M, Xi H, Young CS, Evseenko D, Nelson SF, Spencer MJ, Handel BV, Pyle AD (2018) ERBB3 and NGFR mark a distinct skeletal muscle progenitor cell in human development and hPSCs. Nat Cell Biol 20(1):46–57. https://doi.org/10.1038/s41556-017-0010-2

Hill E, Boontheekul T, Mooney DJ (2006) Regulating activation of transplanted cells controls tissue regeneration. Proc Natl Acad Sci U S A 103(8):2494–2499. https://doi.org/10.1073/pnas.0506004103

Hillary CJ, Roman S, MacNeil S, Aicher WK, Stenzl A, Chapple CR (2020) Regenerative medicine and injection therapies in stress urinary incontinence. Nat Rev Urol. https://doi.org/10.1038/s41585-019-0273-4

Hirsch T, Rothoeft T, Teig N, Bauer JW, Pellegrini G, De Rosa L, Scaglione D, Reichelt J, Klausegger A, Kneisz D, Romano O, Secone Seconetti A, Contin R, Enzo E, Jurman I, Carulli S, Jacobsen F, Luecke T, Lehnhardt M, Fischer M, Kueckelhaus M, Quaglino D, Morgante M, Bicciato S, Bondanza S, De Luca M (2017) Regeneration of the entire human epidermis using transgenic stem cells. Nature 551(7680):327–332. https://doi.org/10.1038/nature24487

Hoffman EP, Brown RH Jr, Kunkel LM (1987) Dystrophin: the protein product of the Duchenne muscular dystrophy locus. Cell 51(6):919–928. https://doi.org/10.1016/0092-8674(87)90579-4

Hong F, Lee J, Song JW, Lee SJ, Ahn H, Cho JJ, Ha J, Kim SS (2002) Cyclosporin A blocks muscle differentiation by inducing oxidative stress and inhibiting the peptidyl-prolyl-cis-trans isomerase activity of cyclophilin A: cyclophilin A protects myoblasts from cyclosporin A-induced cytotoxicity. FASEB J 16(12):1633–1635. https://doi.org/10.1096/fj.02-0060fje

Hooijmans MT, Niks EH, Burakiewicz J, Anastasopoulos C, van den Berg SI, van Zwet E, Webb AG, Verschuuren J, Kan HE (2017) Non-uniform muscle fat replacement along the proximodistal axis in Duchenne muscular dystrophy. Neuromuscul Disord 27(5):458–464. https://doi.org/10.1016/j.nmd.2017.02.009

Hu W, Wang Z, Xiao Y, Zhang S, Wang J (2019) Advances in crosslinking strategies of biomedical hydrogels. Biomater Sci 7(3):843–855. https://doi.org/10.1039/c8bm01246f

Huard J, Bouchard JP, Roy R, Malouin F, Dansereau G, Labrecque C, Albert N, Richards CL, Lemieux B, Tremblay JP (1992) Human myoblast transplantation: preliminary results of 4 cases. Muscle Nerve 15(5):550–560. https://doi.org/10.1002/mus.880150504

Hughes CS, Postovit LM, Lajoie GA (2010) Matrigel: a complex protein mixture required for optimal growth of cell culture. Proteomics 10(9):1886–1890. https://doi.org/10.1002/pmic.200900758

Hwang Y, Suk S, Shih YR, Seo T, Du B, Xie Y, Li Z, Varghese S (2014) WNT3A promotes myogenesis of human embryonic stem cells and enhances in vivo engraftment. Sci Rep 4:5916. https://doi.org/10.1038/srep05916

Incitti T, Magli A, Darabi R, Yuan C, Lin K, Arpke RW, Azzag K, Yamamoto A, Stewart R, Thomson JA, Kyba M, Perlingeiro RCR (2019) Pluripotent stem cell-derived myogenic progenitors remodel their molecular signature upon in vivo engraftment. Proc Natl Acad Sci U S A 116(10):4346–4351. https://doi.org/10.1073/pnas.1808303116

Ishii K, Sakurai H, Suzuki N, Mabuchi Y, Sekiya I, Sekiguchi K, Akazawa C (2018) Recapitulation of extracellular LAMININ environment maintains stemness of satellite cells in vitro. Stem Cell Reports 10(2):568–582. https://doi.org/10.1016/j.stemcr.2017.12.013

Ito N, Shimizu N, Tanaka H, Takeda S (2016) Enhancement of satellite cell transplantation efficiency by leukemia inhibitory factor. J Neuromuscul Dis 3(2):201–207. https://doi.org/10.3233/JND-160156

Ito N, Kii I, Shimizu N, Tanaka H, Takeda S (2017) Direct reprogramming of fibroblasts into skeletal muscle progenitor cells by transcription factors enriched in undifferentiated subpopulation of satellite cells. Sci Rep 7(1):8097. https://doi.org/10.1038/s41598-017-08232-2

Itskovitz-Eldor J, Schuldiner M, Karsenti D, Eden A, Yanuka O, Amit M, Soreq H, Benvenisty N (2000) Differentiation of human embryonic stem cells into embryoid bodies compromising the three embryonic germ layers. Mol Med 6(2):88–95

Jankowski RJ, Tu LM, Carlson C, Robert M, Carlson K, Quinlan D, Eisenhardt A, Chen M, Snyder S, Pruchnic R, Chancellor M, Dmochowski R, Kaufman MR, Carr L (2018) A double-blind, randomized, placebo-controlled clinical trial evaluating the safety and efficacy of autologous muscle derived cells in female subjects with stress urinary incontinence. Int Urol Nephrol 50 (12):2153–2165. https://doi.org/10.1007/s11255-018-2005-8

Joe AW, Yi L, Natarajan A, Le Grand F, So L, Wang J, Rudnicki MA, Rossi FM (2010) Muscle injury activates resident fibro/adipogenic progenitors that facilitate myogenesis. Nat Cell Biol 12(2):153–163. https://doi.org/10.1038/ncb2015

Jones NC, Tyner KJ, Nibarger L, Stanley HM, Cornelison DD, Fedorov YV, Olwin BB (2005) The p38alpha/beta MAPK functions as a molecular switch to activate the quiescent satellite cell. J Cell Biol 169(1):105–116. https://doi.org/10.1083/jcb.200408066

Judson RN, Quarta M, Oudhoff MJ, Soliman H, Yi L, Chang CK, Loi G, Vander Werff R, Cait A, Hamer M, Blonigan J, Paine P, Doan LTN, Groppa E, He W, Su L, Zhang RH, Xu P, Eisner C, Low M, Barta I, Lewis CB, Zaph C, Karimi MM, Rando TA, Rossi FM (2018) Inhibition of methyltransferase Setd7 allows the in vitro expansion of myogenic stem cells with improved therapeutic potential. Cell Stem Cell 22(2):177–190 e177. https://doi.org/10.1016/j.stem.2017.12.010

Juhas M, Bursac N (2013) Engineering skeletal muscle repair. Curr Opin Biotechnol 24(5):880–886. https://doi.org/10.1016/j.copbio.2013.04.013

Juhas M, Engelmayr GC Jr, Fontanella AN, Palmer GM, Bursac N (2014) Biomimetic engineered muscle with capacity for vascular integration and functional maturation in vivo. Proc Natl Acad Sci U S A 111(15):5508–5513. https://doi.org/10.1073/pnas.1402723111

Juhas M, Abutaleb N, Wang JT, Ye J, Shaikh Z, Sriworarat C, Qian Y, Bursac N (2018) Incorporation of macrophages into engineered skeletal muscle enables enhanced muscle regeneration. Nat Biomed Eng 2(12):942–954. https://doi.org/10.1038/s41551-018-0290-2

Kang HW, Lee SJ, Ko IK, Kengla C, Yoo JJ, Atala A (2016) A 3D bioprinting system to produce human-scale tissue constructs with structural integrity. Nat Biotechnol 34(3):312–319. https://doi.org/10.1038/nbt.3413

Karpati G, Pouliot Y, Zubrzycka-Gaarn E, Carpenter S, Ray PN, Worton RG, Holland P (1989) Dystrophin is expressed in mdx skeletal muscle fibers after normal myoblast implantation. Am J Pathol 135(1):27–32

Karpati G, Ajdukovic D, Arnold D, Gledhill RB, Guttmann R, Holland P, Koch PA, Shoubridge E, Spence D, Vanasse M et al (1993) Myoblast transfer in Duchenne muscular dystrophy. Ann Neurol 34(1):8–17. https://doi.org/10.1002/ana.410340105

Kaufman T, Kaplan B, Perry L, Shandalov Y, Landau S, Srugo I, Ad-El D, Levenberg S (2019) Innervation of an engineered muscle graft for reconstruction of muscle defects. Am J Transplant 19(1):37–47. https://doi.org/10.1111/ajt.14957

Kaully T, Kaufman-Francis K, Lesman A, Levenberg S (2009) Vascularization-the conduit to viable engineered tissues. Tissue Eng Part B Rev 15(2):159–169. https://doi.org/10.1089/ten.teb.2008.0193

Kim J, Magli A, Chan SSK, Oliveira VKP, Wu J, Darabi R, Kyba M, Perlingeiro RCR (2017) Expansion and purification are critical for the therapeutic application of pluripotent stem cell-derived myogenic progenitors. Stem Cell Reports 9(1):12–22. https://doi.org/10.1016/j.stemcr.2017.04.022

Kinoshita I, Vilquin JT, Guerette B, Asselin I, Roy R, Tremblay JP (1994) Very efficient myoblast allotransplantation in mice under FK506 immunosuppression. Muscle Nerve 17(12):1407–1415. https://doi.org/10.1002/mus.880171210

Kinoshita I, Roy R, Dugre FJ, Gravel C, Roy B, Goulet M, Asselin I, Tremblay JP (1996) Myoblast transplantation in monkeys: control of immune response by FK506. J Neuropathol Exp Neurol 55(6):687–697. https://doi.org/10.1097/00005072-199606000-00002

Klingler W, Jurkat-Rott K, Lehmann-Horn F, Schleip R (2012) The role of fibrosis in Duchenne muscular dystrophy. Acta Myol 31(3):184–195

Klinkenberg M, Fischer S, Kremer T, Hernekamp F, Lehnhardt M, Daigeler A (2013) Comparison of anterolateral thigh, lateral arm, and parascapular free flaps with regard to donor-site morbidity and aesthetic and functional outcomes. Plast Reconstr Surg 131(2):293–302. https://doi.org/10.1097/PRS.0b013e31827786bc

Ko IK, Lee BK, Lee SJ, Andersson KE, Atala A, Yoo JJ (2013) The effect of in vitro formation of acetylcholine receptor (AChR) clusters in engineered muscle fibers on subsequent innervation of constructs in vivo. Biomaterials 34(13):3246–3255. https://doi.org/10.1016/j.biomaterials.2013.01.029

Kodaka Y, Rabu G, Asakura A (2017) Skeletal muscle cell induction from pluripotent stem cells. Stem Cells Int 2017:1376151. https://doi.org/10.1155/2017/1376151

Koenig M, Hoffman EP, Bertelson CJ, Monaco AP, Feener C, Kunkel LM (1987) Complete cloning of the Duchenne muscular dystrophy (DMD) cDNA and preliminary genomic organization of the DMD gene in normal and affected individuals. Cell 50(3):509–517. https://doi.org/10.1016/0092-8674(87)90504-6

Koffler J, Kaufman-Francis K, Shandalov Y, Egozi D, Pavlov DA, Landesberg A, Levenberg S (2011) Improved vascular organization enhances functional integration of engineered skeletal muscle grafts. Proc Natl Acad Sci U S A 108(36):14789–14794. https://doi.org/10.1073/pnas.1017825108

Kuang S, Kuroda K, Le Grand F, Rudnicki MA (2007) Asymmetric self-renewal and commitment of satellite stem cells in muscle. Cell 129(5):999–1010. https://doi.org/10.1016/j.cell.2007.03.044

Kuhl U, Ocalan M, Timpl R, von der Mark K (1986) Role of laminin and fibronectin in selecting myogenic versus fibrogenic cells from skeletal muscle cells in vitro. Dev Biol 117(2):628–635. https://doi.org/10.1016/0012-1606(86)90331-3

Kwee BJ, Mooney DJ (2017) Biomaterials for skeletal muscle tissue engineering. Curr Opin Biotechnol 47:16–22. https://doi.org/10.1016/j.copbio.2017.05.003

Lafreniere JF, Caron MC, Skuk D, Goulet M, Cheikh AR, Tremblay JP (2009) Growth factor coinjection improves the migration potential of monkey myogenic precursors without affecting cell transplantation success. Cell Transplant 18(7):719–730. https://doi.org/10.3727/096368909X470900

Lalit PA, Salick MR, Nelson DO, Squirrell JM, Shafer CM, Patel NG, Saeed I, Schmuck EG, Markandeya YS, Wong R, Lea MR, Eliceiri KW, Hacker TA, Crone WC, Kyba M, Garry DJ, Stewart R, Thomson JA, Downs KM, Lyons GE, Kamp TJ (2016) Lineage reprogramming of fibroblasts into proliferative induced cardiac progenitor cells by defined factors. Cell Stem Cell 18(3):354–367. https://doi.org/10.1016/j.stem.2015.12.001

Larkin LM, Van der Meulen JH, Dennis RG, Kennedy JB (2006) Functional evaluation of nerve-skeletal muscle constructs engineered in vitro. In Vitro Cell Dev Biol Anim 42(3–4):75–82. https://doi.org/10.1290/0509064.1

Latil M, Rocheteau P, Chatre L, Sanulli S, Memet S, Ricchetti M, Tajbakhsh S, Chretien F (2012) Skeletal muscle stem cells adopt a dormant cell state post mortem and retain regenerative capacity. Nat Commun 3:903. https://doi.org/10.1038/ncomms1890

Latroche C, Weiss-Gayet M, Muller L, Gitiaux C, Leblanc P, Liot S, Ben-Larbi S, Abou-Khalil R, Verger N, Bardot P, Magnan M, Chretien F, Mounier R, Germain S, Chazaud B (2017) Coupling between myogenesis and angiogenesis during skeletal muscle regeneration is stimulated by restorative macrophages. Stem Cell Reports 9(6):2018–2033. https://doi.org/10.1016/j.stemcr.2017.10.027

Law PK, Bertorini TE, Goodwin TG, Chen M, Fang QW, Li HJ, Kirby DS, Florendo JA, Herrod HG, Golden GS (1990) Dystrophin production induced by myoblast transfer therapy in Duchenne muscular dystrophy. Lancet 336(8707):114–115. https://doi.org/10.1016/0140-6736(90)91628-n

Law PK, Goodwin TG, Fang QW, Chen M, Li HJ, Florendo JA, Kirby DS (1991) Myoblast transfer therapy for Duchenne muscular dystrophy. Acta Paediatr Jpn 33(2):206–215. https://doi.org/10.1111/j.1442-200x.1991.tb01545.x

Lee EJ, Kim M, Kim YD, Chung MJ, Elfadl A, Ulah HMA, Park D, Lee S, Park HS, Kim TH, Hwang D, Jeong KS (2018) Establishment of stably expandable induced myogenic stem cells by four transcription factors. Cell Death Dis 9(11):1092. https://doi.org/10.1038/s41419-018-1114-8

Lendlein A, Langer R (2002) Biodegradable, elastic shape-memory polymers for potential biomedical applications. Science 296(5573):1673–1676. https://doi.org/10.1126/science.1066102

Lepper C, Partridge TA, Fan CM (2011) An absolute requirement for Pax7-positive satellite cells in acute injury-induced skeletal muscle regeneration. Development 138(17):3639–3646. https://doi.org/10.1242/dev.067595

Lesault PF, Theret M, Magnan M, Cuvellier S, Niu Y, Gherardi RK, Tremblay JP, Hittinger L, Chazaud B (2012) Macrophages improve survival, proliferation and migration of engrafted myogenic precursor cells into MDX skeletal muscle. PLoS One 7(10):e46698. https://doi.org/10.1371/journal.pone.0046698

Levenberg S, Rouwkema J, Macdonald M, Garfein ES, Kohane DS, Darland DC, Marini R, van Blitterswijk CA, Mulligan RC, D'Amore PA, Langer R (2005) Engineering vascularized skeletal muscle tissue. Nat Biotechnol 23(7):879–884. https://doi.org/10.1038/nbt1109

Li HL, Fujimoto N, Sasakawa N, Shirai S, Ohkame T, Sakuma T, Tanaka M, Amano N, Watanabe A, Sakurai H, Yamamoto T, Yamanaka S, Hotta A (2015) Precise correction of the dystrophin gene in duchenne muscular dystrophy patient induced pluripotent stem cells by TALEN and CRISPR-Cas9. Stem Cell Reports 4(1):143–154. https://doi.org/10.1016/j.stemcr.2014.10.013

Lin CH, Lin YT, Yeh JT, Chen CT (2007) Free functioning muscle transfer for lower extremity posttraumatic composite structure and functional defect. Plast Reconstr Surg 119(7):2118–2126. https://doi.org/10.1097/01.prs.0000260595.85557.41

Liu L, Cheung TH, Charville GW, Hurgo BM, Leavitt T, Shih J, Brunet A, Rando TA (2013) Chromatin modifications as determinants of muscle stem cell quiescence and chronological aging. Cell Rep 4(1):189–204. https://doi.org/10.1016/j.celrep.2013.05.043

Long C, Amoasii L, Mireault AA, McAnally JR, Li H, Sanchez-Ortiz E, Bhattacharyya S, Shelton JM, Bassel-Duby R, Olson EN (2016) Postnatal genome editing partially restores dystrophin expression in a mouse model of muscular dystrophy. Science 351(6271):400–403. https://doi.org/10.1126/science.aad5725

Long C, Li H, Tiburcy M, Rodriguez-Caycedo C, Kyrychenko V, Zhou H, Zhang Y, Min YL, Shelton JM, Mammen PPA, Liaw NY, Zimmermann WH, Bassel-Duby R, Schneider JW, Olson EN (2018) Correction of diverse muscular dystrophy mutations in human engineered heart muscle by single-site genome editing. Sci Adv 4(1):eaap9004. https://doi.org/10.1126/sciadv.aap9004

Low S, Barnes JL, Zammit PS, Beauchamp JR (2018) Delta-like 4 activates Notch 3 to regulate self-renewal in skeletal muscle stem cells. Stem Cells 36 (3):458–466. https://doi.org/10.1002/stem.2757

Lujan E, Chanda S, Ahlenius H, Sudhof TC, Wernig M (2012) Direct conversion of mouse fibroblasts to self-renewing, tripotent neural precursor cells. Proc Natl Acad Sci U S A 109 (7):2527–2532. https://doi.org/10.1073/pnas.1121003109

Lukjanenko L, Jung MJ, Hegde N, Perruisseau-Carrier C, Migliavacca E, Rozo M, Karaz S, Jacot G, Schmidt M, Li L, Metairon S, Raymond F, Lee U, Sizzano F, Wilson DH, Dumont NA, Palini A, Fassler R, Steiner P, Descombes P, Rudnicki MA, Fan CM, von Maltzahn J, Feige JN, Bentzinger CF (2016) Loss of fibronectin from the aged stem cell niche affects the regenerative capacity of skeletal muscle in mice. Nat Med 22(8):897–905. https://doi.org/10.1038/nm.4126

Lukjanenko L, Karaz S, Stuelsatz P, Gurriaran-Rodriguez U, Michaud J, Dammone G, Sizzano F, Mashinchian O, Ancel S, Migliavacca E, Liot S, Jacot G, Metairon S, Raymond F, Descombes

P, Palini A, Chazaud B, Rudnicki MA, Bentzinger CF, Feige JN (2019) Aging disrupts muscle stem cell function by impairing matricellular WISP1 secretion from fibro-adipogenic progenitors. Cell Stem Cell 24(3):433–446 e437. https://doi.org/10.1016/j.stem.2018.12.014

Machado L, Esteves de Lima J, Fabre O, Proux C, Legendre R, Szegedi A, Varet H, Ingerslev LR, Barres R, Relaix F, Mourikis P (2017) In situ fixation redefines quiescence and early activation of skeletal muscle stem cells. Cell Rep 21(7):1982–1993. https://doi.org/10.1016/j.celrep.2017.10.080

Machingal MA, Corona BT, Walters TJ, Kesireddy V, Koval CN, Dannahower A, Zhao W, Yoo JJ, Christ GJ (2011) A tissue-engineered muscle repair construct for functional restoration of an irrecoverable muscle injury in a murine model. Tissue Eng Part A 17(17–18):2291–2303. https://doi.org/10.1089/ten.TEA.2010.0682

Madden L, Juhas M, Kraus WE, Truskey GA, Bursac N (2015) Bioengineered human myobundles mimic clinical responses of skeletal muscle to drugs. Elife 4:e04885. https://doi.org/10.7554/eLife.04885

Maffioletti SM, Noviello M, English K, Tedesco FS (2014) Stem cell transplantation for muscular dystrophy: the challenge of immune response. Biomed Res Int 2014:964010. https://doi.org/10.1155/2014/964010

Maffioletti SM, Gerli MF, Ragazzi M, Dastidar S, Benedetti S, Loperfido M, VandenDriessche T, Chuah MK, Tedesco FS (2015) Efficient derivation and inducible differentiation of expandable skeletal myogenic cells from human ES and patient-specific iPS cells. Nat Protoc 10(7):941–958. https://doi.org/10.1038/nprot.2015.057

Maffioletti SM, Sarcar S, Henderson ABH, Mannhardt I, Pinton L, Moyle LA, Steele-Stallard H, Cappellari O, Wells KE, Ferrari G, Mitchell JS, Tyzack GE, Kotiadis VN, Khedr M, Ragazzi M, Wang W, Duchen MR, Patani R, Zammit PS, Wells DJ, Eschenhagen T, Tedesco FS (2018) Three-dimensional human iPSC-derived artificial skeletal muscles model muscular dystrophies and enable multilineage tissue engineering. Cell Rep 23(3):899–908. https://doi.org/10.1016/j.celrep.2018.03.091

Magli A, Perlingeiro RRC (2017) Myogenic progenitor specification from pluripotent stem cells. Semin Cell Dev Biol 72:87–98. https://doi.org/10.1016/j.semcdb.2017.10.031

Magli A, Incitti T, Kiley J, Swanson SA, Darabi R, Rinaldi F, Selvaraj S, Yamamoto A, Tolar J, Yuan C, Stewart R, Thomson JA, Perlingeiro RCR (2017) PAX7 targets, CD54, integrin alpha9beta1, and SDC2, allow isolation of human ESC/iPSC-derived myogenic progenitors. Cell Rep 19(13):2867–2877. https://doi.org/10.1016/j.celrep.2017.06.005

Maley MA, Davies MJ, Grounds MD (1995) Extracellular matrix, growth factors, genetics: their influence on cell proliferation and myotube formation in primary cultures of adult mouse skeletal muscle. Exp Cell Res 219(1):169–179. https://doi.org/10.1006/excr.1995.1217

Malik N, Rao MS (2013) A review of the methods for human iPSC derivation. Methods Mol Biol 997:23–33. https://doi.org/10.1007/978-1-62703-348-0_3

Mao AS, Mooney DJ (2015) Regenerative medicine: current therapies and future directions. Proc Natl Acad Sci U S A 112(47):14452–14459. https://doi.org/10.1073/pnas.1508520112

Marg A, Escobar H, Gloy S, Kufeld M, Zacher J, Spuler A, Birchmeier C, Izsvak Z, Spuler S (2014) Human satellite cells have regenerative capacity and are genetically manipulable. J Clin Invest 124(10):4257–4265. https://doi.org/10.1172/JCI63992

Martin GR (1981) Isolation of a pluripotent cell line from early mouse embryos cultured in medium conditioned by teratocarcinoma stem cells. Proc Natl Acad Sci U S A 78(12):7634–7638. https://doi.org/10.1073/pnas.78.12.7634

Martin NR, Passey SL, Player DJ, Mudera V, Baar K, Greensmith L, Lewis MP (2015) Neuromuscular junction formation in tissue-engineered skeletal muscle augments contractile function and improves cytoskeletal organization. Tissue Eng Part A 21(19–20):2595–2604. https://doi.org/10.1089/ten.TEA.2015.0146

Matre PR, Mu X, Wu J, Danila D, Hall MA, Kolonin MG, Darabi R, Huard J (2019) CRISPR/Cas9-based dystrophin restoration reveals a novel role for dystrophin in bioenergetics and stress resistance of muscle progenitors. Stem Cells 37(12):1615–1628. https://doi.org/10.1002/stem.3094

Matthias N, Hunt SD, Wu J, Lo J, Smith Callahan LA, Li Y, Huard J, Darabi R (2018) Volumetric muscle loss injury repair using in situ fibrin gel cast seeded with muscle-derived stem cells (MDSCs). Stem Cell Res 27:65–73. https://doi.org/10.1016/j.scr.2018.01.008

Mauro A (1961) Satellite cell of skeletal muscle fibers. J Biophys Biochem Cytol 9:493–495

Mendell JR, Kissel JT, Amato AA, King W, Signore L, Prior TW, Sahenk Z, Benson S, McAndrew PE, Rice R et al (1995) Myoblast transfer in the treatment of Duchenne's muscular dystrophy. N Engl J Med 333(13):832–838. https://doi.org/10.1056/NEJM199509283331303

Miller RG, Sharma KR, Pavlath GK, Gussoni E, Mynhier M, Lanctot AM, Greco CM, Steinman L, Blau HM (1997) Myoblast implantation in Duchenne muscular dystrophy: the San Francisco study. Muscle Nerve 20(4):469–478. https://doi.org/10.1002/(sici)1097-4598(199704)20:4<469::aid-mus10>3.0.co;2-u

Min YL, Bassel-Duby R, Olson EN (2019) CRISPR correction of duchenne muscular dystrophy. Annu Rev Med 70:239–255. https://doi.org/10.1146/annurev-med-081117-010451

Mintz EL, Passipieri JA, Franklin IR, Toscano VM, Afferton EC, Sharma PR, Christ GJ (2019) Long-term evaluation of functional outcomes following rat volumetric muscle loss injury and repair. Tissue Eng Part A. https://doi.org/10.1089/ten.TEA.2019.0126

Mitterberger M, Pinggera GM, Marksteiner R, Margreiter E, Fussenegger M, Frauscher F, Ulmer H, Hering S, Bartsch G, Strasser H (2008) Adult stem cell therapy of female stress urinary incontinence. Eur Urol 53(1):169–175. https://doi.org/10.1016/j.eururo.2007.07.026

Mizuno Y, Chang H, Umeda K, Niwa A, Iwasa T, Awaya T, Fukada S, Yamamoto H, Yamanaka S, Nakahata T, Heike T (2010) Generation of skeletal muscle stem/progenitor cells from murine induced pluripotent stem cells. FASEB J 24(7):2245–2253. https://doi.org/10.1096/fj.09-137174

Montarras D, Morgan J, Collins C, Relaix F, Zaffran S, Cumano A, Partridge T, Buckingham M (2005) Direct isolation of satellite cells for skeletal muscle regeneration. Science 309 (5743):2064–2067. https://doi.org/10.1126/science.1114758

Morandi L, Bernasconi P, Gebbia M, Mora M, Crosti F, Mantegazza R, Cornelio F (1995) Lack of mRNA and dystrophin expression in DMD patients three months after myoblast transfer. Neuromuscul Disord 5(4):291–295

Morgan JE, Moore SE, Walsh FS, Partridge TA (1992) Formation of skeletal muscle in vivo from the mouse C2 cell line. J Cell Sci 102(Pt 4):779–787

Morimoto Y, Kato-Negishi M, Onoe H, Takeuchi S (2013) Three-dimensional neuron-muscle constructs with neuromuscular junctions. Biomaterials 34(37):9413–9419. https://doi.org/10.1016/j.biomaterials.2013.08.062

Motohashi N, Shimizu-Motohashi Y, Roberts TC, Aoki Y (2019) Potential therapies using myogenic stem cells combined with bio-engineering approaches for treatment of muscular dystrophies. Cells 8(9). https://doi.org/10.3390/cells8091066

Mourikis P, Tajbakhsh S (2014) Distinct contextual roles for Notch signalling in skeletal muscle stem cells. BMC Dev Biol 14:2. https://doi.org/10.1186/1471-213X-14-2

Mourikis P, Sambasivan R, Castel D, Rocheteau P, Bizzarro V, Tajbakhsh S (2012) A critical requirement for notch signaling in maintenance of the quiescent skeletal muscle stem cell state. Stem Cells 30(2):243–252. https://doi.org/10.1002/stem.775

Mueller AL, Bloch RJ (2019) Skeletal muscle cell transplantation: models and methods. J Muscle Res Cell Motil. https://doi.org/10.1007/s10974-019-09550-w

Munoz-Canoves P, Scheele C, Pedersen BK, Serrano AL (2013) Interleukin-6 myokine signaling in skeletal muscle: a double-edged sword? FEBS J 280(17):4131–4148. https://doi.org/10.1111/febs.12338

Murphy MM, Lawson JA, Mathew SJ, Hutcheson DA, Kardon G (2011) Satellite cells, connective tissue fibroblasts and their interactions are crucial for muscle regeneration. Development 138 (17):3625–3637. https://doi.org/10.1242/dev.064162

Nakayama KH, Quarta M, Paine P, Alcazar C, Karakikes I, Garcia V, Abilez OJ, Calvo NS, Simmons CS, Rando TA, Huang NF (2019) Treatment of volumetric muscle loss in mice using nanofibrillar scaffolds enhances vascular organization and integration. Commun Biol 2:170. https://doi.org/10.1038/s42003-019-0416-4

Nance ME, Shi R, Hakim CH, Wasala NB, Yue Y, Pan X, Zhang T, Robinson CA, Duan SX, Yao G, Yang NN, Chen SJ, Wagner KR, Gersbach CA, Duan D (2019) AAV9 edits muscle stem cells in normal and dystrophic adult mice. Mol Ther 27(9):1568–1585. https://doi.org/10.1016/j.ymthe.2019.06.012

National Institutes of Health Consensus Development (1983) Clinical applications of biomaterials. Natl Inst Health Consens Dev Conf Summ 4(5).:10 p

Neal D, Sakar MS, Ong LL, Harry Asada H (2014) Formation of elongated fascicle-inspired 3D tissues consisting of high-density, aligned cells using sacrificial outer molding. Lab Chip 14 (11):1907–1916. https://doi.org/10.1039/c4lc00023d

Negroni E, Bigot A, Butler-Browne GS, Trollet C, Mouly V (2016) Cellular therapies for muscular dystrophies: frustrations and clinical successes. Hum Gene Ther 27(2):117–126. https://doi.org/10.1089/hum.2015.139

Nelson CE, Hakim CH, Ousterout DG, Thakore PI, Moreb EA, Castellanos Rivera RM, Madhavan S, Pan X, Ran FA, Yan WX, Asokan A, Zhang F, Duan D, Gersbach CA (2016) In vivo genome editing improves muscle function in a mouse model of Duchenne muscular dystrophy. Science 351(6271):403–407. https://doi.org/10.1126/science.aad5143

Nicola NA, Babon JJ (2015) Leukemia inhibitory factor (LIF). Cytokine Growth Factor Rev 26 (5):533–544. https://doi.org/10.1016/j.cytogfr.2015.07.001

Novak ML, Weinheimer-Haus EM, Koh TJ (2014) Macrophage activation and skeletal muscle healing following traumatic injury. J Pathol 232(3):344–355. https://doi.org/10.1002/path.4301

Nowak KJ, Davies KE (2004) Duchenne muscular dystrophy and dystrophin: pathogenesis and opportunities for treatment. EMBO Rep 5(9):872–876. https://doi.org/10.1038/sj.embor.7400221

Ocalan M, Goodman SL, Kuhl U, Hauschka SD, von der Mark K (1988) Laminin alters cell shape and stimulates motility and proliferation of murine skeletal myoblasts. Dev Biol 125(1):158–167. https://doi.org/10.1016/0012-1606(88)90068-1

Ortiz-Vitali JL, Darabi R (2019) iPSCs as a platform for disease modeling, drug screening, and personalized therapy in muscular dystrophies. Cells 8(1). https://doi.org/10.3390/cells8010020

Osaki T, Sivathanu V, Kamm RD (2018a) Crosstalk between developing vasculature and optogenetically engineered skeletal muscle improves muscle contraction and angiogenesis. Biomaterials 156:65–76. https://doi.org/10.1016/j.biomaterials.2017.11.041

Osaki T, Uzel SGM, Kamm RD (2018b) Microphysiological 3D model of amyotrophic lateral sclerosis (ALS) from human iPS-derived muscle cells and optogenetic motor neurons. Sci Adv 4 (10):eaat5847. https://doi.org/10.1126/sciadv.aat5847

Oustanina S, Hause G, Braun T (2004) Pax7 directs postnatal renewal and propagation of myogenic satellite cells but not their specification. EMBO J 23(16):3430–3439. https://doi.org/10.1038/sj.emboj.7600346

Ousterout DG, Kabadi AM, Thakore PI, Majoros WH, Reddy TE, Gersbach CA (2015) Multiplex CRISPR/Cas9-based genome editing for correction of dystrophin mutations that cause Duchenne muscular dystrophy. Nat Commun 6:6244. https://doi.org/10.1038/ncomms7244

Ozasa S, Kimura S, Ito K, Ueno H, Ikezawa M, Matsukura M, Yoshioka K, Araki K, Yamamura KI, Abe K, Niwa H, Miike T (2007) Efficient conversion of ES cells into myogenic lineage using the gene-inducible system. Biochem Biophys Res Commun 357(4):957–963. https://doi.org/10.1016/j.bbrc.2007.04.032

Page RL, Malcuit C, Vilner L, Vojtic I, Shaw S, Hedblom E, Hu J, Pins GD, Rolle MW, Dominko T (2011) Restoration of skeletal muscle defects with adult human cells delivered on fibrin microthreads. Tissue Eng Part A 17(21–22):2629–2640. https://doi.org/10.1089/ten.TEA.2011.0024

Pantelic MN, Larkin LM (2018) Stem cells for skeletal muscle tissue engineering. Tissue Eng Part B Rev 24(5):373–391. https://doi.org/10.1089/ten.TEB.2017.0451

Park IH, Arora N, Huo H, Maherali N, Ahfeldt T, Shimamura A, Lensch MW, Cowan C, Hochedlinger K, Daley GQ (2008) Disease-specific induced pluripotent stem cells. Cell 134 (5):877–886. https://doi.org/10.1016/j.cell.2008.07.041

Parker MH, Loretz C, Tyler AE, Duddy WJ, Hall JK, Olwin BB, Bernstein ID, Storb R, Tapscott SJ (2012) Activation of Notch signaling during ex vivo expansion maintains donor muscle cell engraftment. Stem Cells 30(10):2212–2220. https://doi.org/10.1002/stem.1181

Partridge TA, Grounds M, Sloper JC (1978) Evidence of fusion between host and donor myoblasts in skeletal muscle grafts. Nature 273(5660):306–308. https://doi.org/10.1038/273306a0

Partridge TA, Morgan JE, Coulton GR, Hoffman EP, Kunkel LM (1989) Conversion of mdx myofibres from dystrophin-negative to -positive by injection of normal myoblasts. Nature 337 (6203):176–179. https://doi.org/10.1038/337176a0

Pasut A, Chang NC, Gurriaran-Rodriguez U, Faulkes S, Yin H, Lacaria M, Ming H, Rudnicki MA (2016) Notch signaling rescues loss of satellite cells lacking Pax7 and promotes brown adipogenic differentiation. Cell Rep 16(2):333–343. https://doi.org/10.1016/j.celrep.2016.06.001

Pavlath GK, Rando TA, Blau HM (1994) Transient immunosuppressive treatment leads to long-term retention of allogeneic myoblasts in hybrid myofibers. J Cell Biol 127(6 Pt 2):1923–1932. https://doi.org/10.1083/jcb.127.6.1923

Pawlowski M, Ortmann D, Bertero A, Tavares JM, Pedersen RA, Vallier L, Kotter MRN (2017) Inducible and deterministic forward programming of human pluripotent stem cells into neurons, skeletal myocytes, and oligodendrocytes. Stem Cell Reports 8(4):803–812. https://doi.org/10.1016/j.stemcr.2017.02.016

Penton CM, Badarinarayana V, Prisco J, Powers E, Pincus M, Allen RE, August PR (2016) Laminin 521 maintains differentiation potential of mouse and human satellite cell-derived myoblasts during long-term culture expansion. Skelet Muscle 6(1):44. https://doi.org/10.1186/s13395-016-0116-4

Perie S, Mamchaoui K, Mouly V, Blot S, Bouazza B, Thornell LE, St Guily JL, Butler-Browne G (2006) Premature proliferative arrest of cricopharyngeal myoblasts in oculo-pharyngeal muscular dystrophy: therapeutic perspectives of autologous myoblast transplantation. Neuromuscul Disord 16(11):770–781. https://doi.org/10.1016/j.nmd.2006.07.022

Perie S, Trollet C, Mouly V, Vanneaux V, Mamchaoui K, Bouazza B, Marolleau JP, Laforet P, Chapon F, Eymard B, Butler-Browne G, Larghero J, St Guily JL (2014) Autologous myoblast transplantation for oculopharyngeal muscular dystrophy: a phase I/IIa clinical study. Mol Ther 22(1):219–225. https://doi.org/10.1038/mt.2013.155

Pietrosemoli N, Mella S, Yennek S, Baghdadi MB, Sakai H, Sambasivan R, Pala F, Di Girolamo D, Tajbakhsh S (2017) Comparison of multiple transcriptomes exposes unified and divergent features of quiescent and activated skeletal muscle stem cells. Skelet Muscle 7(1):28. https://doi.org/10.1186/s13395-017-0144-8

Powell C, Shansky J, Del Tatto M, Forman DE, Hennessey J, Sullivan K, Zielinski BA, Vandenburgh HH (1999) Tissue-engineered human bioartificial muscles expressing a foreign recombinant protein for gene therapy. Hum Gene Ther 10(4):565–577. https://doi.org/10.1089/10430349950018643

Qazi TH, Mooney DJ, Pumberger M, Geissler S, Duda GN (2015) Biomaterials based strategies for skeletal muscle tissue engineering: existing technologies and future trends. Biomaterials 53:502–521. https://doi.org/10.1016/j.biomaterials.2015.02.110

Qu Z, Balkir L, van Deutekom JC, Robbins PD, Pruchnic R, Huard J (1998) Development of approaches to improve cell survival in myoblast transfer therapy. J Cell Biol 142(5):1257–1267. https://doi.org/10.1083/jcb.142.5.1257

Quarta M, Brett JO, DiMarco R, De Morree A, Boutet SC, Chacon R, Gibbons MC, Garcia VA, Su J, Shrager JB, Heilshorn S, Rando TA (2016) An artificial niche preserves the quiescence of muscle stem cells and enhances their therapeutic efficacy. Nat Biotechnol 34(7):752–759. https://doi.org/10.1038/nbt.3576

Quarta M, Cromie M, Chacon R, Blonigan J, Garcia V, Akimenko I, Hamer M, Paine P, Stok M, Shrager JB, Rando TA (2017) Bioengineered constructs combined with exercise enhance stem cell-mediated treatment of volumetric muscle loss. Nat Commun 8:15613. https://doi.org/10.1038/ncomms15613

Quarta M, Cromie Lear MJ, Blonigan J, Paine P, Chacon R, Rando TA (2018) Biomechanics show stem cell necessity for effective treatment of volumetric muscle loss using bioengineered constructs. NPJ Regen Med 3:18. https://doi.org/10.1038/s41536-018-0057-0

Quenneville SP, Chapdelaine P, Skuk D, Paradis M, Goulet M, Rousseau J, Xiao X, Garcia L, Tremblay JP (2007) Autologous transplantation of muscle precursor cells modified with a lentivirus for muscular dystrophy: human cells and primate models. Mol Ther 15(2):431–438. https://doi.org/10.1038/sj.mt.6300047

Rando TA, Blau HM (1994) Primary mouse myoblast purification, characterization, and transplantation for cell-mediated gene therapy. J Cell Biol 125(6):1275–1287. https://doi.org/10.1083/jcb.125.6.1275

Rao N, Agmon G, Tierney MT, Ungerleider JL, Braden RL, Sacco A, Christman KL (2017) Engineering an injectable muscle-specific microenvironment for improved cell delivery using a nanofibrous extracellular matrix hydrogel. ACS Nano 11(4):3851–3859. https://doi.org/10.1021/acsnano.7b00093

Rao L, Qian Y, Khodabukus A, Ribar T, Bursac N (2018) Engineering human pluripotent stem cells into a functional skeletal muscle tissue. Nat Commun 9(1):126. https://doi.org/10.1038/s41467-017-02636-4

Rayagiri SS, Ranaldi D, Raven A, Mohamad Azhar NIF, Lefebvre O, Zammit PS, Borycki AG (2018) Basal lamina remodeling at the skeletal muscle stem cell niche mediates stem cell self-renewal. Nat Commun 9(1):1075. https://doi.org/10.1038/s41467-018-03425-3

Ring KL, Tong LM, Balestra ME, Javier R, Andrews-Zwilling Y, Li G, Walker D, Zhang WR, Kreitzer AC, Huang Y (2012) Direct reprogramming of mouse and human fibroblasts into multipotent neural stem cells with a single factor. Cell Stem Cell 11(1):100–109. https://doi.org/10.1016/j.stem.2012.05.018

Rocheteau P, Gayraud-Morel B, Siegl-Cachedenier I, Blasco MA, Tajbakhsh S (2012) A subpopulation of adult skeletal muscle stem cells retains all template DNA strands after cell division. Cell 148(1–2):112–125. https://doi.org/10.1016/j.cell.2011.11.049

Romaniszyn M, Rozwadowska N, Malcher A, Kolanowski T, Walega P, Kurpisz M (2015) Implantation of autologous muscle-derived stem cells in treatment of fecal incontinence: results of an experimental pilot study. Tech Coloproctol 19(11):685–696. https://doi.org/10.1007/s10151-015-1351-0

Rossi CA, Flaibani M, Blaauw B, Pozzobon M, Figallo E, Reggiani C, Vitiello L, Elvassore N, De Coppi P (2011) In vivo tissue engineering of functional skeletal muscle by freshly isolated satellite cells embedded in a photopolymerizable hydrogel. FASEB J 25(7):2296–2304. https://doi.org/10.1096/fj.10-174755

Rubenstein AB, Smith GR, Raue U, Begue G, Minchev K, Ruf-Zamojski F, Nair VD, Wang X, Zhou L, Zaslavsky E, Trappe TA, Trappe S, Sealfon SC (2020) Single-cell transcriptional profiles in human skeletal muscle. Sci Rep 10(1):229. https://doi.org/10.1038/s41598-019-57110-6

Ruoslahti E, Pierschbacher MD (1987) New perspectives in cell adhesion: RGD and integrins. Science 238(4826):491–497. https://doi.org/10.1126/science.2821619

Sacco A, Doyonnas R, Kraft P, Vitorovic S, Blau HM (2008) Self-renewal and expansion of single transplanted muscle stem cells. Nature 456(7221):502–506. https://doi.org/10.1038/nature07384

Sakai H, Fukuda S, Nakamura M, Uezumi A, Noguchi YT, Sato T, Morita M, Yamada H, Tsuchida K, Tajbakhsh S, Fukada SI (2017) Notch ligands regulate the muscle stem-like state ex vivo but are not sufficient for retaining regenerative capacity. PLoS One 12(5):e0177516. https://doi.org/10.1371/journal.pone.0177516

Sambasivan R, Gayraud-Morel B, Dumas G, Cimper C, Paisant S, Kelly RG, Tajbakhsh S (2009) Distinct regulatory cascades govern extraocular and pharyngeal arch muscle progenitor cell fates. Dev Cell 16(6):810–821. https://doi.org/10.1016/j.devcel.2009.05.008

Sambasivan R, Yao R, Kissenpfennig A, Van Wittenberghe L, Paldi A, Gayraud-Morel B, Guenou H, Malissen B, Tajbakhsh S, Galy A (2011) Pax7-expressing satellite cells are indispensable for adult skeletal muscle regeneration. Development 138(17):3647–3656. https://doi.org/10.1242/dev.067587

Sampaolesi M, Torrente Y, Innocenzi A, Tonlorenzi R, D'Antona G, Pellegrino MA, Barresi R, Bresolin N, De Angelis MG, Campbell KP, Bottinelli R, Cossu G (2003) Cell therapy of alpha-sarcoglycan null dystrophic mice through intra-arterial delivery of mesoangioblasts. Science 301(5632):487–492. https://doi.org/10.1126/science.1082254

Sampath SC, Sampath SC, Ho ATV, Corbel SY, Millstone JD, Lamb J, Walker J, Kinzel B, Schmedt C, Blau HM (2018) Induction of muscle stem cell quiescence by the secreted niche factor Oncostatin M. Nat Commun 9(1):1531. https://doi.org/10.1038/s41467-018-03876-8

Sato T, Higashioka K, Sakurai H, Yamamoto T, Goshima N, Ueno M, Sotozono C (2019) Core transcription factors promote induction of PAX3-positive skeletal muscle stem cells. Stem Cell Reports 13(2):352–365. https://doi.org/10.1016/j.stemcr.2019.06.006

Schultz E, Gibson MC, Champion T (1978) Satellite cells are mitotically quiescent in mature mouse muscle: an EM and radioautographic study. J Exp Zool 206(3):451–456. https://doi.org/10.1002/jez.1402060314

Schwartz SD, Hubschman JP, Heilwell G, Franco-Cardenas V, Pan CK, Ostrick RM, Mickunas E, Gay R, Klimanskaya I, Lanza R (2012) Embryonic stem cell trials for macular degeneration: a preliminary report. Lancet 379(9817):713–720. https://doi.org/10.1016/S0140-6736(12)60028-2

Seale P, Sabourin LA, Girgis-Gabardo A, Mansouri A, Gruss P, Rudnicki MA (2000) Pax7 is required for the specification of myogenic satellite cells. Cell 102(6):777–786

Sebe P, Doucet C, Cornu JN, Ciofu C, Costa P, de Medina SG, Pinset C, Haab F (2011) Intrasphincteric injections of autologous muscular cells in women with refractory stress urinary incontinence: a prospective study. Int Urogynecol J 22(2):183–189. https://doi.org/10.1007/s00192-010-1255-5

Selvaraj S, Kyba M, Perlingeiro RCR (2019) Pluripotent stem cell-based therapeutics for muscular dystrophies. Trends Mol Med 25(9):803–816. https://doi.org/10.1016/j.molmed.2019.07.004

Shea KL, Xiang W, LaPorta VS, Licht JD, Keller C, Basson MA, Brack AS (2010) Sprouty1 regulates reversible quiescence of a self-renewing adult muscle stem cell pool during regeneration. Cell Stem Cell 6(2):117–129. https://doi.org/10.1016/j.stem.2009.12.015

Shelton M, Metz J, Liu J, Carpenedo RL, Demers SP, Stanford WL, Skerjanc IS (2014) Derivation and expansion of PAX7-positive muscle progenitors from human and mouse embryonic stem cells. Stem Cell Reports 3(3):516–529. https://doi.org/10.1016/j.stemcr.2014.07.001

Sherwood RI, Christensen JL, Conboy IM, Conboy MJ, Rando TA, Weissman IL, Wagers AJ (2004) Isolation of adult mouse myogenic progenitors: functional heterogeneity of cells within and engrafting skeletal muscle. Cell 119(4):543–554. https://doi.org/10.1016/j.cell.2004.10.021

Sicari BM, Rubin JP, Dearth CL, Wolf MT, Ambrosio F, Boninger M, Turner NJ, Weber DJ, Simpson TW, Wyse A, Brown EH, Dziki JL, Fisher LE, Brown S, Badylak SF (2014) An acellular biologic scaffold promotes skeletal muscle formation in mice and humans with volumetric muscle loss. Sci Transl Med 6(234):234ra258. https://doi.org/10.1126/scitranslmed.3008085

Simillis C, Lal N, Pellino G, Baird D, Nikolaou S, Kontovounisios C, Smith JJ, Tekkis PP (2019) A systematic review and network meta-analysis comparing treatments for faecal incontinence. Int J Surg 66:37–47. https://doi.org/10.1016/j.ijsu.2019.04.007

Simone C, Forcales SV, Hill DA, Imbalzano AN, Latella L, Puri PL (2004) p38 pathway targets SWI-SNF chromatin-remodeling complex to muscle-specific loci. Nat Genet 36(7):738–743. https://doi.org/10.1038/ng1378

Skuk D (2004) Myoblast transplantation for inherited myopathies: a clinical approach. Expert Opin Biol Ther 4(12):1871–1885. https://doi.org/10.1517/14712598.4.12.1871

Skuk D, Tremblay JP (2014) Clarifying misconceptions about myoblast transplantation in myology. Mol Ther 22(5):897–898. https://doi.org/10.1038/mt.2014.57

Skuk D, Tremblay JP (2016) Confirmation of donor-derived dystrophin in a duchenne muscular dystrophy patient allotransplanted with normal myoblasts. Muscle Nerve 54(5):979–981. https://doi.org/10.1002/mus.25129

Skuk D, Tremblay JP (2017a) Cell therapy in myology: dynamics of muscle precursor cell death after intramuscular administration in non-human primates. Mol Ther Methods Clin Dev 5:232–240. https://doi.org/10.1016/j.omtm.2017.05.002

Skuk D, Tremblay JP (2017b) The process of engraftment of myogenic cells in skeletal muscles of primates: understanding clinical observations and setting directions in cell transplantation research. Cell Transplant 26(11):1763–1779. https://doi.org/10.1177/0963689717724798

Skuk D, Roy B, Goulet M, Tremblay JP (1999) Successful myoblast transplantation in primates depends on appropriate cell delivery and induction of regeneration in the host muscle. Exp Neurol 155(1):22–30. https://doi.org/10.1006/exnr.1998.6973

Skuk D, Goulet M, Roy B, Tremblay JP (2000) Myoblast transplantation in whole muscle of nonhuman primates. J Neuropathol Exp Neurol 59(3):197–206. https://doi.org/10.1093/jnen/59.3.197

Skuk D, Goulet M, Roy B, Tremblay JP (2002) Efficacy of myoblast transplantation in nonhuman primates following simple intramuscular cell injections: toward defining strategies applicable to humans. Exp Neurol 175(1):112–126. https://doi.org/10.1006/exnr.2002.7899

Skuk D, Roy B, Goulet M, Chapdelaine P, Bouchard JP, Roy R, Dugre FJ, Lachance JG, Deschenes L, Helene S, Sylvain M, Tremblay JP (2004) Dystrophin expression in myofibers of Duchenne muscular dystrophy patients following intramuscular injections of normal myogenic cells. Mol Ther 9(3):475–482. https://doi.org/10.1016/j.ymthe.2003.11.023

Skuk D, Goulet M, Roy B, Chapdelaine P, Bouchard JP, Roy R, Dugre FJ, Sylvain M, Lachance JG, Deschenes L, Senay H, Tremblay JP (2006a) Dystrophin expression in muscles of duchenne muscular dystrophy patients after high-density injections of normal myogenic cells. J Neuropathol Exp Neurol 65(4):371–386. https://doi.org/10.1097/01.jnen.0000218443.45782.81

Skuk D, Goulet M, Tremblay JP (2006b) Use of repeating dispensers to increase the efficiency of the intramuscular myogenic cell injection procedure. Cell Transplant 15(7):659–663. https://doi.org/10.3727/000000006783981648

Skuk D, Goulet M, Roy B, Piette V, Cote CH, Chapdelaine P, Hogrel JY, Paradis M, Bouchard JP, Sylvain M, Lachance JG, Tremblay JP (2007) First test of a "high-density injection" protocol for myogenic cell transplantation throughout large volumes of muscles in a Duchenne muscular dystrophy patient: eighteen months follow-up. Neuromuscul Disord 17(1):38–46. https://doi.org/10.1016/j.nmd.2006.10.003

Skuk D, Paradis M, Goulet M, Chapdelaine P, Rothstein DM, Tremblay JP (2010) Intramuscular transplantation of human postnatal myoblasts generates functional donor-derived satellite cells. Mol Ther 18(9):1689–1697. https://doi.org/10.1038/mt.2010.128

Skuk D, Goulet M, Tremblay JP (2011) Transplanted myoblasts can migrate several millimeters to fuse with damaged myofibers in nonhuman primate skeletal muscle. J Neuropathol Exp Neurol 70(9):770–778. https://doi.org/10.1097/NEN.0b013e31822a6baa

Skuk D, Goulet M, Tremblay JP (2013) Electroporation as a method to induce myofiber regeneration and increase the engraftment of myogenic cells in skeletal muscles of primates. J Neuropathol Exp Neurol 72(8):723–734. https://doi.org/10.1097/NEN.0b013e31829bac22

Skuk D, Goulet M, Tremblay JP (2014) Intramuscular transplantation of myogenic cells in primates: importance of needle size, cell number, and injection volume. Cell Transplant 23 (1):13–25. https://doi.org/10.3727/096368912X661337

Slater CR (2017) The structure of human neuromuscular junctions: some unanswered molecular questions. Int J Mol Sci 18(10). https://doi.org/10.3390/ijms18102183

Sleep E, Cosgrove BD, McClendon MT, Preslar AT, Chen CH, Sangji MH, Perez CMR, Haynes RD, Meade TJ, Blau HM, Stupp SI (2017) Injectable biomimetic liquid crystalline scaffolds enhance muscle stem cell transplantation. Proc Natl Acad Sci U S A 114(38):E7919–E7928. https://doi.org/10.1073/pnas.1708142114

Snow MH (1978) An autoradiographic study of satellite cell differentiation into regenerating myotubes following transplantation of muscles in young rats. Cell Tissue Res 186(3):535–540. https://doi.org/10.1007/bf00224941

Stangel-Wojcikiewicz K, Jarocha D, Piwowar M, Jach R, Uhl T, Basta A, Majka M (2014) Autologous muscle-derived cells for the treatment of female stress urinary incontinence: a 2-year follow-up of a Polish investigation. Neurourol Urodyn 33(3):324–330. https://doi.org/10.1002/nau.22404

Sun L, Ma K, Wang H, Xiao F, Gao Y, Zhang W, Wang K, Gao X, Ip N, Wu Z (2007) JAK1-STAT1-STAT3, a key pathway promoting proliferation and preventing premature differentiation of myoblasts. J Cell Biol 179(1):129–138. https://doi.org/10.1083/jcb.200703184

Stefan Rose-John (2018) Interleukin-6 Family Cytokines. Cold Spring Harbor Perspectives in Biology 10(2):a028415

Tabebordbar M, Zhu K, Cheng JKW, Chew WL, Widrick JJ, Yan WX, Maesner C, Wu EY, Xiao R, Ran FA, Cong L, Zhang F, Vandenberghe LH, Church GM, Wagers AJ (2016) In vivo gene editing in dystrophic mouse muscle and muscle stem cells. Science 351(6271):407–411. https://doi.org/10.1126/science.aad5177

Takahashi K, Yamanaka S (2006) Induction of pluripotent stem cells from mouse embryonic and adult fibroblast cultures by defined factors. Cell 126(4):663–676. https://doi.org/10.1016/j.cell.2006.07.024

Takahashi K, Tanabe K, Ohnuki M, Narita M, Ichisaka T, Tomoda K, Yamanaka S (2007) Induction of pluripotent stem cells from adult human fibroblasts by defined factors. Cell 131(5):861–872. https://doi.org/10.1016/j.cell.2007.11.019

Takahashi H, Shimizu T, Nakayama M, Yamato M, Okano T (2013) The use of anisotropic cell sheets to control orientation during the self-organization of 3D muscle tissue. Biomaterials 34 (30):7372–7380. https://doi.org/10.1016/j.biomaterials.2013.06.033

Tanaka A, Woltjen K, Miyake K, Hotta A, Ikeya M, Yamamoto T, Nishino T, Shoji E, Sehara-Fujisawa A, Manabe Y, Fujii N, Hanaoka K, Era T, Yamashita S, Isobe K, Kimura E, Sakurai H (2013) Efficient and reproducible myogenic differentiation from human iPS cells: prospects for modeling Miyoshi myopathy in vitro. PLoS One 8(4):e61540. https://doi.org/10.1371/journal.pone.0061540

Tapscott SJ, Davis RL, Thayer MJ, Cheng PF, Weintraub H, Lassar AB (1988) MyoD1: a nuclear phosphoprotein requiring a Myc homology region to convert fibroblasts to myoblasts. Science 242(4877):405–411. https://doi.org/10.1126/science.3175662

Thomson JA, Itskovitz-Eldor J, Shapiro SS, Waknitz MA, Swiergiel JJ, Marshall VS, Jones JM (1998) Embryonic stem cell lines derived from human blastocysts. Science 282(5391):1145–1147. https://doi.org/10.1126/science.282.5391.1145

Tierney MT, Aydogdu T, Sala D, Malecova B, Gatto S, Puri PL, Latella L, Sacco A (2014) STAT3 signaling controls satellite cell expansion and skeletal muscle repair. Nat Med 20(10):1182–1186. https://doi.org/10.1038/nm.3656

Torrente Y, Belicchi M, Sampaolesi M, Pisati F, Meregalli M, D'Antona G, Tonlorenzi R, Porretti L, Gavina M, Mamchaoui K, Pellegrino MA, Furling D, Mouly V, Butler-Browne GS, Bottinelli R, Cossu G, Bresolin N (2004) Human circulating AC133(+) stem cells restore dystrophin expression and ameliorate function in dystrophic skeletal muscle. J Clin Invest 114(2):182–195. https://doi.org/10.1172/JCI20325

Torrente Y, Belicchi M, Marchesi C, D'Antona G, Cogiamanian F, Pisati F, Gavina M, Giordano R, Tonlorenzi R, Fagiolari G, Lamperti C, Porretti L, Lopa R, Sampaolesi M, Vicentini L, Grimoldi N, Tiberio F, Songa V, Baratta P, Prelle A, Forzenigo L, Guglieri M, Pansarasa O, Rinaldi C, Mouly V, Butler-Browne GS, Comi GP, Biondetti P, Moggio M, Gaini SM, Stocchetti N, Priori A, D'Angelo MG, Turconi A, Bottinelli R, Cossu G, Rebulla P, Bresolin N (2007) Autologous transplantation of muscle-derived CD133+ stem cells in Duchenne muscle patients. Cell Transplant 16(6):563–577

Trebol J, Carabias-Orgaz A, Garcia-Arranz M, Garcia-Olmo D (2018) Stem cell therapy for faecal incontinence: current state and future perspectives. World J Stem Cells 10(7):82–105. https://doi.org/10.4252/wjsc.v10.i7.82

Tremblay JP, Bouchard JP, Malouin F, Theau D, Cottrell F, Collin H, Rouche A, Gilgenkrantz S, Abbadi N, Tremblay M et al (1993a) Myoblast transplantation between monozygotic twin girl carriers of Duchenne muscular dystrophy. Neuromuscul Disord 3(5–6):583–592

Tremblay JP, Malouin F, Roy R, Huard J, Bouchard JP, Satoh A, Richards CL (1993b) Results of a triple blind clinical study of myoblast transplantations without immunosuppressive treatment in young boys with Duchenne muscular dystrophy. Cell Transplant 2(2):99–112. https://doi.org/10.1177/096368979300200203

Tremblay JP, Skuk D, Palmieri B, Rothstein DM (2009) A case for immunosuppression for myoblast transplantation in duchenne muscular dystrophy. Mol Ther 17(7):1122–1124. https://doi.org/10.1038/mt.2009.125

Uezumi A, Nakatani M, Ikemoto-Uezumi M, Yamamoto N, Morita M, Yamaguchi A, Yamada H, Kasai T, Masuda S, Narita A, Miyagoe-Suzuki Y, Takeda S, Fukada S, Nishino I, Tsuchida K (2016) Cell-surface protein profiling identifies distinctive markers of progenitor cells in human skeletal muscle. Stem Cell Reports 7(2):263–278. https://doi.org/10.1016/j.stemcr.2016.07.004

Urciuolo A, Quarta M, Morbidoni V, Gattazzo F, Molon S, Grumati P, Montemurro F, Tedesco FS, Blaauw B, Cossu G, Vozzi G, Rando TA, Bonaldo P (2013) Collagen VI regulates satellite cell self-renewal and muscle regeneration. Nat Commun 4:1964. https://doi.org/10.1038/ncomms2964

van der Wal E, Herrero-Hernandez P, Wan R, Broeders M, In 't Groen SLM, van Gestel TJM, van IWFJ, Cheung TH, van der Ploeg AT, Schaaf GJ, Pijnappel W (2018) Large-scale expansion of human iPSC-derived skeletal muscle cells for disease modeling and cell-based therapeutic strategies. Stem Cell Reports 10 (6):1975–1990. https://doi.org/10.1016/j.stemcr.2018.04.002

van Velthoven CTJ, de Morree A, Egner IM, Brett JO, Rando TA (2017) Transcriptional profiling of quiescent muscle stem cells in vivo. Cell Rep 21(7):1994–2004. https://doi.org/10.1016/j.celrep.2017.10.037

Vandenburgh H, Kaufman S (1979) In vitro model for stretch-induced hypertrophy of skeletal muscle. Science 203(4377):265–268. https://doi.org/10.1126/science.569901

Vandenburgh HH, Karlisch P, Farr L (1988) Maintenance of highly contractile tissue-cultured avian skeletal myotubes in collagen gel. In Vitro Cell Dev Biol Anim 24(3):166–174. https://doi.org/10.1007/bf02623542

Vandenburgh H, Del Tatto M, Shansky J, Lemaire J, Chang A, Payumo F, Lee P, Goodyear A, Raven L (1996) Tissue-engineered skeletal muscle organoids for reversible gene therapy. Hum Gene Ther 7(17):2195–2200. https://doi.org/10.1089/hum.1996.7.17-2195

Vandenburgh H, Del Tatto M, Shansky J, Goldstein L, Russell K, Genes N, Chromiak J, Yamada S (1998) Attenuation of skeletal muscle wasting with recombinant human growth hormone secreted from a tissue-engineered bioartificial muscle. Hum Gene Ther 9(17):2555–2564. https://doi.org/10.1089/hum.1998.9.17-2555

Vandenburgh H, Shansky J, Benesch-Lee F, Barbata V, Reid J, Thorrez L, Valentini R, Crawford G (2008) Drug-screening platform based on the contractility of tissue-engineered muscle. Muscle Nerve 37(4):438–447. https://doi.org/10.1002/mus.20931

Verma M, Asakura Y, Murakonda BSR, Pengo T, Latroche C, Chazaud B, McLoon LK, Asakura A (2018) Muscle satellite cell cross-talk with a vascular niche maintains quiescence via VEGF and Notch signaling. Cell Stem Cell 23(4):530–543 e539. https://doi.org/10.1016/j.stem.2018.09.007

Vilquin JT, Marolleau JP, Sacconi S, Garcin I, Lacassagne MN, Robert I, Ternaux B, Bouazza B, Larghero J, Desnuelle C (2005) Normal growth and regenerating ability of myoblasts from unaffected muscles of facioscapulohumeral muscular dystrophy patients. Gene Ther 12 (22):1651–1662. https://doi.org/10.1038/sj.gt.3302565

von Maltzahn J, Bentzinger CF, Rudnicki MA (2011) Wnt7a-Fzd7 signalling directly activates the Akt/mTOR anabolic growth pathway in skeletal muscle. Nat Cell Biol 14(2):186–191. https://doi.org/10.1038/ncb2404

von Maltzahn J, Jones AE, Parks RJ, Rudnicki MA (2013) Pax7 is critical for the normal function of satellite cells in adult skeletal muscle. Proc Natl Acad Sci U S A 110(41):16474–16479. https://doi.org/10.1073/pnas.1307680110

Wang K, Wang C, Xiao F, Wang H, Wu Z (2008) JAK2/STAT2/STAT3 are required for myogenic differentiation. J Biol Chem 283(49):34029–34036. https://doi.org/10.1074/jbc.M803012200

Wang L, Shansky J, Vandenburgh H (2013) Induced formation and maturation of acetylcholine receptor clusters in a defined 3D bio-artificial muscle. Mol Neurobiol 48(3):397–403. https://doi.org/10.1007/s12035-013-8412-z

Wang L, Cao L, Shansky J, Wang Z, Mooney D, Vandenburgh H (2014) Minimally invasive approach to the repair of injured skeletal muscle with a shape-memory scaffold. Mol Ther 22 (8):1441–1449. https://doi.org/10.1038/mt.2014.78

Weintraub H, Tapscott SJ, Davis RL, Thayer MJ, Adam MA, Lassar AB, Miller AD (1989) Activation of muscle-specific genes in pigment, nerve, fat, liver, and fibroblast cell lines by forced expression of MyoD. Proc Natl Acad Sci U S A 86(14):5434–5438. https://doi.org/10.1073/pnas.86.14.5434

Wernig A, Irintchev A, Hartling A, Stephan G, Zimmermann K, Starzinski-Powitz A (1991) Formation of new muscle fibres and tumours after injection of cultured myogenic cells. J Neurocytol 20(12):982–997. https://doi.org/10.1007/bf01187916

Wilschut KJ, Haagsman HP, Roelen BA (2010) Extracellular matrix components direct porcine muscle stem cell behavior. Exp Cell Res 316(3):341–352. https://doi.org/10.1016/j.yexcr.2009.10.014

Wilschut KJ, Ling VB, Bernstein HS (2012) Concise review: stem cell therapy for muscular dystrophies. Stem Cells Transl Med 1(11):833–842. https://doi.org/10.5966/sctm.2012-0071

Wolf MT, Dearth CL, Sonnenberg SB, Loboa EG, Badylak SF (2015) Naturally derived and synthetic scaffolds for skeletal muscle reconstruction. Adv Drug Deliv Rev 84:208–221. https://doi.org/10.1016/j.addr.2014.08.011

Wosczyna MN, Rando TA (2018) A muscle stem cell support group: coordinated cellular responses in muscle regeneration. Dev Cell 46(2):135–143. https://doi.org/10.1016/j.devcel.2018.06.018

Wosczyna MN, Konishi CT, Perez Carbajal EE, Wang TT, Walsh RA, Gan Q, Wagner MW, Rando TA (2019) Mesenchymal stromal cells are required for regeneration and homeostatic maintenance of skeletal muscle. Cell Rep 27(7):2029–2035 e2025. https://doi.org/10.1016/j.celrep.2019.04.074

Wu J, Matthias N, Lo J, Ortiz-Vitali JL, Shieh AW, Wang SH, Darabi R (2018) A myogenic double-reporter human pluripotent stem cell line allows prospective isolation of skeletal muscle progenitors. Cell Rep 25(7):1966–1981 e1964. https://doi.org/10.1016/j.celrep.2018.10.067

Xi H, Fujiwara W, Gonzalez K, Jan M, Liebscher S, Van Handel B, Schenke-Layland K, Pyle AD (2017) In vivo human somitogenesis guides somite development from hPSCs. Cell Rep 18 (6):1573–1585. https://doi.org/10.1016/j.celrep.2017.01.040

Xiao F, Wang H, Fu X, Li Y, Ma K, Sun L, Gao X, Wu Z (2011) Oncostatin M inhibits myoblast differentiation and regulates muscle regeneration. Cell Res 21(2):350–364. https://doi.org/10.1038/cr.2010.144

Xie L, Yin A, Nichenko AS, Beedle AM, Call JA, Yin H (2018) Transient HIF2A inhibition promotes satellite cell proliferation and muscle regeneration. J Clin Invest 128(6):2339–2355. https://doi.org/10.1172/JCI96208

Xu C, Tabebordbar M, Iovino S, Ciarlo C, Liu J, Castiglioni A, Price E, Liu M, Barton ER, Kahn CR, Wagers AJ, Zon LI (2013) A zebrafish embryo culture system defines factors that promote vertebrate myogenesis across species. Cell 155(4):909–921. https://doi.org/10.1016/j.cell.2013.10.023

Xu J, Du Y, Deng H (2015a) Direct lineage reprogramming: strategies, mechanisms, and applications. Cell Stem Cell 16(2):119–134. https://doi.org/10.1016/j.stem.2015.01.013

Xu X, Wilschut KJ, Kouklis G, Tian H, Hesse R, Garland C, Sbitany H, Hansen S, Seth R, Knott PD, Hoffman WY, Pomerantz JH (2015b) Human satellite cell transplantation and regeneration from diverse skeletal muscles. Stem Cell Reports 5(3):419–434. https://doi.org/10.1016/j.stemcr.2015.07.016

Yablonka-Reuveni Z, Rivera AJ (1994) Temporal expression of regulatory and structural muscle proteins during myogenesis of satellite cells on isolated adult rat fibers. Dev Biol 164(2):588–603. https://doi.org/10.1006/dbio.1994.1226

Yaffe D, Saxel O (1977) Serial passaging and differentiation of myogenic cells isolated from dystrophic mouse muscle. Nature 270(5639):725–727. https://doi.org/10.1038/270725a0

Yao SN, Kurachi K (1993) Implanted myoblasts not only fuse with myofibers but also survive as muscle precursor cells. J Cell Sci 105(Pt 4):957–963

Yao CC, Ziober BL, Sutherland AE, Mendrick DL, Kramer RH (1996) Laminins promote the locomotion of skeletal myoblasts via the alpha 7 integrin receptor. J Cell Sci 109(Pt 13):3139–3150

Yartseva V, Goldstein LD, Rodman J, Kates L, Chen MZ, Chen YJ, Foreman O, Siebel CW, Modrusan Z, Peterson AS, Jovicic A (2020) Heterogeneity of satellite cells implicates DELTA1/NOTCH2 signaling in self-renewal. Cell Rep 30(5):1491–1503 e1496. https://doi.org/10.1016/j.celrep.2019.12.100

Yin H, Price F, Rudnicki MA (2013) Satellite cells and the muscle stem cell niche. Physiol Rev 93(1):23–67. https://doi.org/10.1152/physrev.00043.2011

Yiu EM, Kornberg AJ (2015) Duchenne muscular dystrophy. J Paediatr Child Health 51(8):759–764. https://doi.org/10.1111/jpc.12868

Young CS, Hicks MR, Ermolova NV, Nakano H, Jan M, Younesi S, Karumbayaram S, Kumagai-Cresse C, Wang D, Zack JA, Kohn DB, Nakano A, Nelson SF, Miceli MC, Spencer MJ, Pyle AD (2016) A single CRISPR-Cas9 deletion strategy that targets the majority of DMD patients restores dystrophin function in hiPSC-derived muscle cells. Cell Stem Cell 18(4):533–540. https://doi.org/10.1016/j.stem.2016.01.021

Zammit PS, Golding JP, Nagata Y, Hudon V, Partridge TA, Beauchamp JR (2004) Muscle satellite cells adopt divergent fates: a mechanism for self-renewal? J Cell Biol 166(3):347–357. https://doi.org/10.1083/jcb.200312007

Zhang M, Guo B (2017) Electroactive 3D scaffolds based on silk fibroin and water-borne polyaniline for skeletal muscle tissue engineering. Macromol Biosci 17(9). https://doi.org/10.1002/mabi.201700147

Zheng JK, Wang Y, Karandikar A, Wang Q, Gai H, Liu AL, Peng C, Sheng HZ (2006) Skeletal myogenesis by human embryonic stem cells. Cell Res 16(8):713–722. https://doi.org/10.1038/sj.cr.7310080

Zhou W, He DQ, Liu JY, Feng Y, Zhang XY, Hua CG, Tang XF (2015) Angiogenic gene-modified myoblasts promote vascularization during repair of skeletal muscle defects. J Tissue Eng Regen Med 9(12):1404–1416. https://doi.org/10.1002/term.1692

Zhu P, Wu F, Mosenson J, Zhang H, He TC, Wu WS (2017) CRISPR/Cas9-mediated genome editing corrects dystrophin mutation in skeletal muscle stem cells in a mouse model of muscle dystrophy. Mol Ther Nucleic Acids 7:31–41. https://doi.org/10.1016/j.omtn.2017.02.007

Ligament Tissue Engineering: The Anterior Cruciate Ligament

Thomas Nau and Andreas Teuschl

Contents

Abstract

Due to ongoing problems with reconstructive surgery and recent advancements in the field of musculoskeletal tissue engineering, regeneration of ligaments in general and of the anterior cruciate ligament (ACL) in particular has gained an increasing research interest. It is the aim of this chapter to review the current research efforts and highlight promising ligament engineering strategies.

The four main components of tissue engineering also apply to several ligament regeneration research efforts. Scaffolds are made of biologic materials, biodegradable polymers, and composite materials. The main cell sources are

T. Nau (✉)
Ludwig Boltzmann Institute for Experimental and Clinical Traumatology, AUVA Research Center, Vienna, Austria

The Austrian Cluster for Tissue Regeneration, Vienna, Austria

Mohammed Bin Rashid University and Health Sciences – MBRU, Dubai, United Arab Emirates

A. Teuschl
The Austrian Cluster for Tissue Regeneration, Vienna, Austria

Department Life Science Engineering, University of Applied Sciences Technikum Wien, Vienna, Austria
e-mail: andreas.teuschl@technikum-wien.at

D. Eberli et al. (eds.), *Organ Tissue Engineering*, Reference Series in Biomedical Engineering, https://doi.org/10.1007/978-3-030-44211-8_7

mesenchymal stem cells and ACL fibroblasts. In addition, growth factors and mechanical stimuli are applied. So far, the regenerated ACL constructs have been tested in few in vivo studies, and the results are encouraging although the real breakthrough has not been achieved thus far.

The different strategies reaching from in vitro ACL regeneration in bioreactor systems to bio-enhanced repair and true regeneration are under constant development. We expect considerable progress in the near future that will result in a realistic option for ligament reconstructive surgery soon.

1 Introduction

Tendons and ligaments are important connective tissues which enable the motion of the musculoskeletal system. Tendons connect muscle to bone, whereas ligaments connect bone to bone to provide static stability as well as support the transmission of loads between the bones. Considering the composition, ligaments are densely packed bundles of collagenous fibers. Collagen makes up approximately 70–80% of the tissue's dry weight. Type 1 collagen accounts for 95%; type 3 and type 5 collagens are found as well. The extracellular matrix (ECM) also contains elastin, glycoproteins, proteoglycans, and water, which accounts for 65–75% of the wet weight (Rumian et al. 2007). This ECM is organized in a highly hierarchical and specified structure in order to provide its mechanical strength (Alberts et al. 2014). Collagen molecules have a triple-helical alpha domain, consisting of three alpha chains (two α1-chains and one α2-chain) with glycine at every third residue. Apart from glycine, collagens are also abundant in the amino acids proline and hydroxyproline. Collagen molecules are cross-linked and form fibrils with a wavy pattern. Fibrils are assembled to form a fiber with an average diameter of 10 μm. Fibers are organized into fiber bundles or fascicles, and many fascicles are in turn forming the ligament. Collagen fibers of ligaments are not as parallel as those of tendons, because the more random orientation allows multiaxial loading patterns (Rumian et al. 2007).

The cell content of ligaments is relatively low. Fibroblasts are the predominant type and have a role in normal tissue turnover, repair, and regeneration. They are producing collagen and other ECM components, but also cytokines in case of injury and healing. Stem cells are found in ligament tissue as well. The existence of mesenchymal stem cells (MSCs) in ACL in low numbers was just recently reported. In comparison with other MSCs, ACL-derived MSCs show a higher proliferative capacity and stay undifferentiated when treated with basic fibroblast growth factor (bFGF) (Fu et al. 2016).

The ACL is definitely the most prominent and intensively debated ligament in the medical literature. Due to its unique hierarchical structure, it has ligament-specific properties, such as high mechanical strength and a stress-strain as well as a stress-relaxation behavior. These properties are crucial for the ACL to withstand high loads under physiological conditions. At low strain rates, the ligament is more easily deformed and can absorb more mechanical energy. At higher strain rates, it becomes stiffer and more resistant to deformation (Wang et al. 2012). A classic stress-strain curve of the ACL, usually done in a uniaxial pattern, demonstrates three phases, the toe region, the linear region, and the yield region (Fig. 1). Immediately after applying strain, the crimped collagen fibers are stretched out in the direction of the force applied. As strain increases, elastic deformation occurs during the linear phase. Microscopic failure of the fibrils usually starts at stretching beyond 4% and continues to macroscopically visible failure at around 10%.

Whereas the classic stress-strain behavior is tested in a uniaxial pattern, the native human ACL is a much more complex structure. It is approximately 27–32 mm long and has a cross-sectional area of 44.4–57.5 mm^2 (Freeman et al. 2007). Functionally and macroscopically, the ACL has two distinguished bundles, the anteromedial (AM) and the posterolateral (PL) bundles. Their names derive from the location at the tibial insertion area. Throughout the course of the ligament, from distal-anterior-medial to proximal-posterior-lateral, the ACL undergoes a 180° twist between its bony attachments (Duthon et al. 2006). The two bundles behave biomechanically different, depending on the movement of the knee joint, and are therefore crucial for the functional integrity of the joint motion. The tensile strength of the native human ACL has been reported to be 1800–2196 N (Noyes et al. 1984). The properties of the femur-ACL-tibia complex showed a tensile strength of 2160 N and a mean stiffness of 242 N/mm for specimen in the age of 22–35 years (Woo et al. 1991).

Knee injuries frequently result in ruptured ligaments, with high-pivoting sporting activities such as skiing, football, and basketball as the predominant causes. In 2005, around 400,000 physician office visits in the United States were related to knee injuries (Goodwin 2005). The worldwide estimation of young sports players that require surgery following a knee injury lies between 17% and 61% (Louw et al. 2008). The ACL as a main stabilizing structure of the knee is one of the most common injured ligaments. In the United States alone, around 350,000

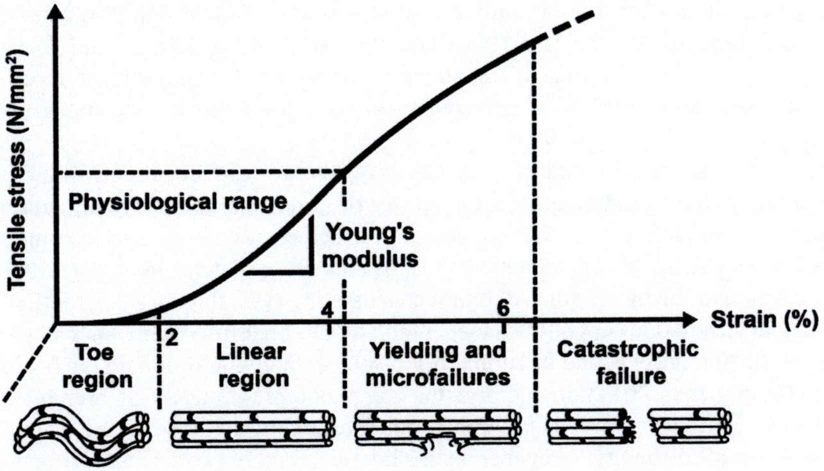

Fig. 1 Stress-strain curve for tensile testing of ligament tissues. Collagen fibril straightening and failures in relation to different regions (toe region, linear region, yielding and microfailures and catastrophic failure) of the stress-strain curve are schematically depicted

reconstructive surgeries of the ACL are performed annually. According to the National Center for Health Statistics, the annual costs for the acute care of these injuries are around $6 billion. Besides the immediate functional deficit, ACL injuries carry the considerable long-term risk for knee osteoarthritis.

Historically, different approaches have been used to treat ACL injuries, reaching from nonoperative care to several surgical procedures (Seitz et al. 2013). Simple primary suturing of the torn ligament in the 1970s had been abandoned due to bad functional results. Although augmented ACL repair using natural as well as synthetic grafts showed improved results compared to the simple suturing, the failure rate still remained unacceptably high. From the early 1990s onwards, ACL reconstruction with autograft or allograft material has gained popularity and become the widely accepted method of choice for most surgeons. However, despite the good results of autografts, problems mostly associated with donor site morbidity remain, such as anterior knee pain, infrapatellar contracture, tendinitis, patellar fracture, muscle weakness, and limited graft availability (Ma et al. 2012). Allograft material on the other hand still carries the theoretical risk for disease transmission; also the delayed biological intracorporation is known as a major disadvantage (Jackson et al. 1977). In addition, high failure rates of ACL reconstruction, especially in young and active patients, have been reported for allografts (Kaeding et al. 2011). The incidence of posttraumatic arthritis after injury and reconstruction of the ACL still lies around 50% within 7–14 years and thus is considered to be the main drawback of this surgical strategy, resulting in enormous ongoing research interest on that topic (Roos 2005; Lohmander et al. 2004).

Regenerating musculoskeletal tissues has become a very popular topic in orthopedic research. Structures that are injured or lost due to trauma and disease represent the ideal candidates to be engineered. Tissue engineering, as a multidisciplinary effort, includes strategies of engineering, material science, and biology with the aim to regenerate tissues that are not only recreating the morphology but also restoring the normal function. In the late 1980s and the early 1990s, Langer and Vacanti emphasized the classic four basic components that are needed in tissue engineering (Fig. 2): a structural scaffold, a cell source, biologic modulators, and mechanical stimuli (Langer and Vacanti 1993).

The ACL has only limited healing capacity; therefore, reconstructive surgery remains the method of choice for the majority of cases. Consequently, it certainly represents an attractive but also challenging structure for tissue engineering. In contrast to extraarticular ligaments, such as the medial collateral ligament (MCL), the intra-articular location of the ACL prevents its primary healing. The disruption of the synovial sheath does not allow a local blood clot to be formed which is known to be crucial for the onset of the inflammatory response that would initiate the healing process (Leong et al. 2013). In addition, the complex three-dimensional architecture of the ACL, with different tensioning patterns throughout the knee path of motion, contributes to the difficulty to regenerate this ligament in terms of form and function.

In the following, the current strategies in ACL regeneration are presented, to provide an overview of the approaches and their limitations and to present future directions of this evolving research technology.

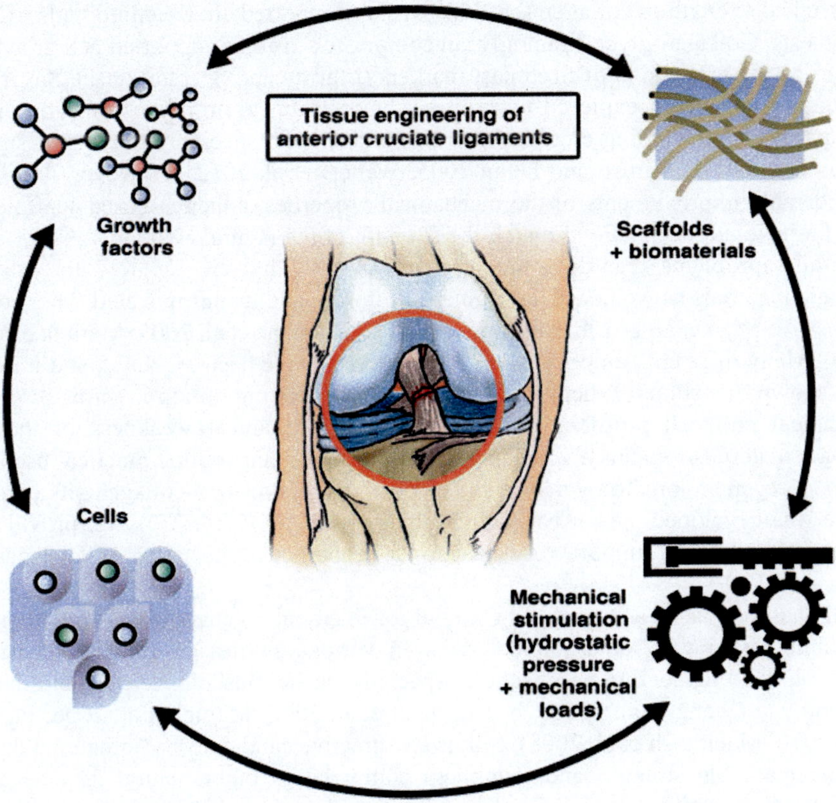

Fig. 2 Generally, tissue engineering strategies of the anterior cruciate ligament comprise at least one of the four basic components: (1) structural scaffolds made of synthetic or native biomaterials; (2) cells, mainly mesenchymal stem cells or primary ligament/tendon cells; (3) growth factors as biological modulators; or (4) mechanical stimulation via bioreactors or physical therapies; or multiple combinations thereof are used

2 Scaffolds for ACL Regeneration

A variety of biomaterials have been introduced as potential scaffold for ACL tissue engineering. The ideal scaffold is biocompatible, and its mechanical properties should mimic the natural ACL as close as possible. It also needs to be biodegradable to enable tissue ingrowth, which is crucial for the new ligament to form. Biologic materials, biodegradable polymers, and composite materials have all been or still are under evaluation for ACL regeneration (Dunn et al. 1995).

The first biological scaffolds for ACL regeneration were made of collagen fibrils. It was shown that ACL fibroblasts adhered to these scaffolds and remained viable in vitro as well as in vivo. Unfortunately, after 6 weeks, the constructs were completely resorbed (Dunn et al. 1995; Bellincampi et al. 1998). The decreased mechanical strength of collagen scaffolds was also reported after seeding with ACL fibroblasts. Collagen-glycosaminoglycan composite scaffolds supported cell growth as well as the expression of fibroblast markers (Murray and Spector 2001). Several techniques have been explored to improve the mechanical properties of collagen-based scaffolds, including cross-linking the collagen or a special braid-twist design (Koob et al. 2001; Caruso and Dunn 2005; Walters et al. 2012). However, despite considerable improvements of the mechanical properties, collagen-based scaffolds thus far have not been able to match the strength of the natural ACL.

Similar problems regarding the mechanical strength were reported for other biologic materials as well, such as alginate, chitosan, and hyaluronic acid (Hansson et al. 2012; Shao et al. 2010b; Masuko et al. 2005; Yamane et al. 2005; Majima et al. 2005). Many different composites of these materials have been explored, and it has been shown that some of them may represent an interesting option in terms of cell attachment and cell proliferation. However, the mechanical weakness of these biologic materials remains a considerable problem for their routine practical use in ligament regeneration. To overcome the mechanical weakness, a collagen-silk composite was developed, and it was shown that a scaffold with >25% silk provides sufficient mechanical support, very close to the properties of the native ACL (Panas-Perez et al. 2013).

Silk has been used as a scaffold material for ligament regeneration even without the combination with other biomaterials. In various studies its functionality in diverse tissue engineering approaches, especially in the musculoskeletal field, has been proven (Wang et al. 2006a, b; Hofmann et al. 2006; Meinel et al. 2006; Park et al. 2010; Macintosh et al. 2008). Silk is an attractive candidate as biomaterial due to its remarkable strength and toughness compared to other natural as well as synthetic biomaterials (Jin and Kaplan 2003; Rockwood et al. 2011). Most studies dealing with silk as raw material for scaffold production use fibers from cocoons of the mulberry silkworm *Bombyx mori*. Due to biocompatibility requirements, silk requires the removal of the surface protein layer sericin, which can cause adverse immune responses (Vepari and Kaplan 2007; Teuschl et al. 2013). Once sericin is removed, the remaining silk fibroin fibers are non-immunogenic, biocompatible, and capable of promoting cell adhesion, growth, and differentiation in the case of progenitor cells such as MSCs. The classical way to remove this protein layer is to boil raw silk fibers in alkaline solutions such as sodium carbonate. Recently, our group developed a procedure to remove sericin from a compact and highly ordered raw *Bombyx mori* silk fiber scaffold using borate buffer-based solutions (Teuschl et al. 2013). The removal of sericin after the textile engineering process eases the production of complex 3D structures in tissue engineering applications because the gliding properties of the silk fiber due to the gumlike sericin assist during textile engineering steps such as braiding and weaving. The pioneers in using silk fibers as raw material for ACL scaffolds are Altman and Kaplan, who could demonstrate that the mechanical properties of their twisted fiber scaffolds match that of the native human ACL (Altman et al. 2002b). Furthermore, the same group demonstrated the processability of silk fibers with a large number of different textile engineering techniques enabling the generation of complex hierarchical structures with defined properties (Horan et al. 2009). An additional characteristic that makes silk an attractive candidate for ACL regeneration is its slow rate of degradation (proteolytic degradation). Thus, silk fiber-based ACL scaffolds can provide the primary stability over an extended period of time, allowing ingrowing cells to regenerate tissue without exposing the knee joint to periods of instability. Moreover, the gradual transfer of stabilizing properties from the silk scaffold to the new forming tissue should allow a neotissue formation similar to the initial native tissue regarding collagen alignment, vascularization, etc.

A number of silk-based ligament grafts have been tested in animal models (Chen et al. 2008; Fan et al. 2008; Liu et al. 2008; Altman et al. 2008). Historically, former ACL studies with synthetic materials have shown that the eventual translation of findings from animal data to humans needs large animal studies, using goat, sheep, or pig models. In a pig model, Fan et al. were the first to show that their woven silk ligament scaffold in conjunction with seeded MSCs supported ligament regeneration after 24 weeks post implantation period. In conclusion, these very promising in vivo studies suggest that ACL scaffolds fabricated from silk fibroin have a great potential for the translation into clinic applications. Our group developed and introduced a novel silk fiber-based scaffold for ACL regeneration more recently (Teuschl et al. 2016, 2019). The scaffold was tested in an in vivo study in sheep with an observation period of 12 months. Ligament regeneration was initiated and in parallel, silk degradation was observed. In one experimental group, additional cell seeding with the stromal vascular fraction including adipose derived stem cells showed increased regeneration for 6 months. Interestingly, after 12 months, this effect was not present anymore. The hard tissue histology revealed excellent osteointegration of the construct. A bony capsule was formed, and a fibrous interzone between the scaffold and this capsule was seen. Sharpey-like fibers spread from the interzone into the bone to anchor the construct. About 50% of the silk was degraded, and the joints did not show signs of foreign body reaction, inflammation, or advanced degeneration. Based on these encouraging results, we initiated another study using the same sheep model, which includes a longer observation period of 24 months as well as a thorough biomechanical workup, which is currently under way and will be presented shortly.

Apart from biologic materials, synthetic biodegradable polymers have been introduced in ligament tissue engineering. The advantages of synthetic polymers, such as proper selection and different manufacturing techniques, allow for an exact adaptation of the mechanical properties, cellular response, and degradation rate (Petrigliano et al. 2006). In 1999, a scaffold composed of polyglycolic acid coated with polycaprolactone was introduced (Lin et al. 1999). A braided polydioxanone scaffold in an in vivo animal study reported on an early loss of mechanical properties (Buma et al. 2004). In a comparison of different synthetic braided materials, poly-L-lactic acid (PLLA) scaffolds had the best results in terms of mechanical properties as well as in terms of fibroblast proliferation (Lu et al. 2005). Another PLLA scaffold, braided in a three-dimensional fashion, with distinct regions for the bony portions and the intra-articular portion of the construct was introduced (Laurencin and Freeman 2005). The same group subsequently compared different PLLA scaffolds with different manufacturing techniques and demonstrated that a braid-twist scaffold had the most favorable viscoelastic properties (Freeman et al. 2007, 2009). In another study, a polyethylene glycol hydrogel was added to the PLLA scaffold, which resulted in even better viscoelastic performance of the construct, but on the other hand also in a decreased pore size of the scaffold which may negatively influence cell proliferation (Freeman et al. 2010).

More recently, electrospinning has been used for the development of ligament scaffolds (Cardwell et al. 2012). With this technique, very thin fibers in the nanometer to micron range can be produced, to allow for a more exact adaptation of the mechanical properties. Some of the studies using electrospinning reported on better cell proliferation and extracellular matrix production (Cardwell et al. 2012; James et al. 2011). However, these techniques are under constant investigation, and while early in vitro studies show interesting results, the overall biological and mechanical performance has still to be examined further to draw any conclusions for a later clinical use of these materials.

3 Cell Sources for ACL Regeneration

In general, two different cell types are considered to be the primary choice for ACL regeneration: MSCs and ACL fibroblasts (Chen et al. 2003).

MSCs are present in almost all tissue types of the body (Sinclair et al. 2013; Al-Nbaheen et al. 2013). However, for cell therapeutic purposes, bone marrow and adipose tissue are known to be the main feasible sources to isolate MSCs (Pendleton et al. 2013; Ong and Sugii 2013). MSCs have the potential to differentiate into various mesenchymal lineages, including fibroblastic, osteogenic, chondrogenic, and myogenic. Various applications of MSCs to enhance repair in different musculoskeletal tissues, in particular in the bone and ligament, have been introduced (Fan et al. 2008; Keibl et al. 2011; Peterbauer-Scherb et al. 2010; Butler et al. 2010).

Harvesting ACL fibroblasts certainly involves the risk of local infection in the knee. Considering that the seeded cells should rebuild the ligament tissue by ECM deposition, ACL fibroblasts would be the most appropriate cell type, as they represent the native cell type in the intact ligament. Therefore, they are used as control cells for cell behavior such as protein expression, especially in in vitro studies. Different studies could also show that the ACL tissue contains populations of cells sharing MSC characteristics such as cluster of differentiation markers or multipotency (Cheng et al. 2009; Steinert et al. 2011). Stem cells are present in the ACL tissue, but their regenerative capacity is too low to be capable of healing ruptured ligaments. As ACL tissue can only be harvested reasonably in diagnostic arthroscopic procedures following ACL rupture, other ligament fibroblast sources have been suggested, such as the medial collateral ligament (Nagineni et al. 1992). However, the majority of studies involving cell therapy strategies in ACL tissue engineering uses MSCs as cell source, since they can be obtained easily in higher numbers as well as show higher proliferation and collagen production rates compared to ligament fibroblasts (Huang et al. 2008; Ge et al. 2005).

The knee joint comprises different sources of cells (ligament tissue, synovium, etc.) that have been shown to participate during the ligament regeneration process such as the above-described ACL fibroblasts or MSCs that are natively recruited after ligament ruptures or tears (Morito et al. 2008). The activation and recruitment of regenerating cells can be augmented mechanically, for instance by the surgical procedure (e.g., drilling of bone holes for the graft which gives access to the vasculature of bone tissue) or biochemically via growth factors or gene therapeutic approaches.

4 Growth Factors and Gene Therapy

Growth factors can be directly applied either via inserted cells, which produce these biochemical signal molecules in situ, via local delivery of growth factors or via gene therapeutic approaches, where vehicles are encoding the chosen growth factor.

The most frequently used factors belong to proteins that directly stimulate the deposition of extracellular matrix proteins such as the bone morphogenic proteins (BMPs) or the degradation of ECM components assisting in remodeling impaired tissue. BMPs are part of the TGFβ superfamily. They are known to induce the differentiation of MSCs into the osteogenic or chondrogenic lineage. One special class of BMPs, however, the growth and differentiation factors (GDFs) 5/6 and 7, has been shown to ectopically induce neotendon/ligament formation in vivo (Wolfman et al. 1997). Aspenberg and Forslund (1999) could demonstrate the enhanced regenerative effect of GDFs 5 and 6 in an Achilles tendon rat model. Interestingly, the effects of GDFs depend on the mechanical loading of the injection site. The injection of GDF 6 in unloaded Achilles tendon defects induced bone formation, which in contrast was not observed in control groups of loaded tendons (Forslund and Aspenberg 2002). This clearly indicates the interaction of growth factors and mechanical stimulation.

Other factors have also been shown to enhance the repair of tendon/ligament structures but are not directly associated with ECM turnover, such as insulin-like growth factor 1 (IGF1) (Letson and Dahners 1994), vascular endothelial growth factor (VEGF) (Wei et al. 2011), epidermal growth factor (EGF) (Yasuda et al. 2004), or platelet-derived growth factor (PDGF) (Li et al. 2007; Nakamura et al. 1998; Thomopoulos et al. 2009; Hildebrand et al. 1998). VEGF, for instance, is well known to stimulate angiogenesis, or IGF1 is a good example for having an anti-inflammatory effect (Kurtz et al. 1999), since functional analysis revealed a decreased recovery time but no biomechanical improvement in an Achilles tendon injury model.

A clinically widely established approach to augment tendon and ligament healing with growth factors is the use of platelet-rich plasma (PRP). PRP is obtained by plasma separation and consists of platelets, blood proteins such as fibrinogen, and a mixture of diverse growth factors (PDGF, VEGF, TGFβ, IGF, etc.), which are parts of general healing processes (de Mos et al. 2008). The major advantage of PRP is its autologous nature as well as its combination of growth factors in native proportions

(Marx 2004). This feature of PRP is important, as various studies have proven the synergistic effects of different growth factor combinations. Although beneficial effects of PRP were demonstrated in cell culture (Chen et al. 2012) as well as in in vivo models (Mastrangelo et al. 2011; Fernández-Sarmiento et al. 2013) on tendon/ligament regeneration, the effectiveness of PRP in clinical uses is still debated due to varying outcomes (Figueroa et al. 2013; Yuan et al. 2013). These variances are mainly attributed to non-optimized treatment protocols and the still relatively low number of randomized clinical trials.

Another strategy to improve the regenerative capacity is to deliver therapeutic genes, either in vivo with vehicles or ex vivo in cells which are subsequently implanted. In an ACL rabbit model, it was shown that autologous graft remodeling can be enhanced by local administration of TGFβ-1/VEGF165 gene-transduced bone MSCs leading to superior mechanical properties compared to solely TGFβ-1 gene-transduced cells (Wei et al. 2011). MSCs have been genetically modified to co-express Smad8 and BMP2 (Hoffmann et al. 2006) and consequently enhanced the regeneration of the Achilles tendon in mice. Single-gene therapeutic approaches seem to be less efficient compared to the co-expression of growth factors.

5 Mechanical Stimulation in ACL Regeneration

Mechanical stimuli and dynamic loading are necessary for ligaments to maintain their strengths. Numerous studies have shown that immobilization leads to weakened mechanical properties of ligaments (Woo et al. 1982, 1987, 1999). On a cellular level, it is known that cells react to mechanical stimuli via integrin-mediated focal adhesions and cytoskeleton deformation (Tetsunaga et al. 2009; Henshaw et al. 2006; Berry et al. 2003b). Mechanical stimuli are able to influence stem cell differentiation and ECM production (Altman et al. 2002b). MSCs differentiated into fibroblast-like cells, as seen by the upregulation of ligament ECM markers including tenascin C and collagen types I and III and the formation of collagen fibers, were observed (Vunjak-Novakovic et al. 2004). Petrigliano et al. (2007) showed that uniaxial cyclic strain of a three-dimensional polymer scaffold seeded with MSCs resulted in upregulated tenascin C and collagen types I and III. Berry et al. (2003b) demonstrated the proliferative effect of uniaxial strain on young fibroblasts. Another study showed that 8% cyclic strain in ligament fibroblasts resulted in higher cell proliferation and collagen production compared to 4% strain and unloaded controls (Park et al. 2006). Despite the well-accepted fact that mechanical stimuli play an important role in ligament tissue engineering, the timing, direction, and magnitude of the stimuli remain largely unclear (Leong et al. 2013). Another study demonstrated that immediate mechanical stimulation of MSCs after seeding results in inhibited expression of collagens I and II. In contrast, the opposite effect was observed when the mechanical loading was applied at the peak of MSC proliferation (Moreau et al. 2008). Leong et al. mentioned that in case of ACL tissue engineering, the mechanotransduction pathways that are necessary for tissue formation and maintenance still need to be explored in more detail. They also stated that up to this date, it is not known if any mechanical stimulation is required prior to implantation of tissue-engineered ACL constructs.

6 Future Directions in ACL Regeneration

In a questionnaire study, 300 orthopedic surgeons were asked if they would consider using a tissue-engineered ACL if it were an available option (Rathbone et al. 2012). In 86% the answer was positive, if the construct demonstrates biological and mechanical success. For 63%, improved patient satisfaction was important, and 76% of the participants mentioned that a tissue-engineered ACL would be superior to any of the currently used autograft materials. A fully load-bearing construct for early weight bearing is needed at the time of implantation, and several tissue engineering strategies should address this need for mechanical integrity. The currently used ACL reconstruction techniques with autograft or allograft material provide an immediate load-bearing environment. Therefore, any tissue-engineered solution will have to meet these standards as well. Any type of regenerated ACL that would require a prolonged period of immobilization or non-weight bearing would most probably not represent a feasible option for the orthopedic surgical community (Nau and Teuschl 2015).

For a true translation into clinical practice, different strategies seem to be promising. One approach focusses on the ex vivo engineering process with subsequent implantation, wherein (?) the optimal timing remains an open question. Our group presented that mechanical stimulation of silk grafts with a custom-made bioreactor system could increase the maturity of cell-loaded grafts prior to implantation (Hohlrieder et al. 2013). In accordance to a study of Altman et al., MSCs can be stimulated to produce layers of ECM on silk-based grafts. Our hypothesis is that the applied mechanical stimulation triggers MSCs into ligamentous cells, which – in conjunction with the secreted ECM – leads to improved functionality of the cell/scaffold construct and therefore will fulfill its tasks, once it is implanted. Future studies, using a procedure starting with an in vitro ligament engineering step using a bioreactor followed by subsequent in vivo implantation, are certainly needed to gain a deeper insight into this complex topic.

On the other hand, engineering scaffolds with adequate mechanical properties that are implantable at any time also seem to be a good option. Future research efforts may also demonstrate which cell type seeded on those scaffolds is the ideal candidate for direct in vivo implantation.

Furthermore, there is also some interest to explore the regenerative potential of solely implanting scaffolds that would recruit cells in vivo when provided with the appropriate mechanical and physiological environment. Our own strategy has shifted somehow away from using additional cell seeding of a newly developed silk fiber-based scaffold. The results of our 12-month study in sheep indicate that cell seeding was beneficial only for 6 months post implantation, whereas after 12 months this advantage was not seen anymore. Apparently, in the case of implantation of the scaffold alone, cells are recruited and ligament regeneration is initiated. This seem to be an even more important finding, as any eventual translation into clinical practice will pass through the official regulatory bodies and receive approval for clinical use in humans in a much easier way compared to any strategy that requires the harvesting and manipulation of human cells.

Murray et al. (2010) introduced the strategy of repair and regeneration. Here, tissue engineering efforts are undertaken to overcome the obstacles to native ACL healing. This group proposed a bio-enhanced ACL repair (BEAR) technique that uses a collagen scaffold saturated with PRP. In a number of large animal studies, they could demonstrate improved mechanical and biological healing of the ACL (Vavken et al. 2012; Murray et al. 2010; Joshi et al. 2009; Murray and Fleming 2013b). In a randomized trial in pigs, BEAR achieved equal results compared with ACL reconstruction. It was also shown that the knees treated with enhanced ACL repair had significantly less signs of osteoarthritis in contrast to those treated with ACL reconstruction after 1 year (Murray and Flemming 2013b). This BEAR technique has already advanced to a first-in-human study with 2-year results just recently published (Murray et al. 2019). BEAR demonstrated similar subjective and objective clinical results compared to conventional ACL reconstruction. Despite these encouraging results, it has to be mentioned that BEAR seems to be suitable only for certain types of ACL ruptures, in which the tear occurs at the proximal end of the ligament with a long stump (>50%) of good tissue quality remaining. However, the results are definitely encouraging, and future research will show if a "repair and regenerate" strategy for all types of ACL tears can be developed.

7 Conclusions

There is a growing research interest in the tissue engineering of the ACL, and the clinical need seems obvious. Different strategies reaching from in vitro engineering of ACL grafts, with or without additional cell seeding, to bio-enhanced repair and regeneration are followed. For the surgical community, any type of engineered ACL may represent a future option, provided that it is easy to implant, does allow for at least the same aggressive rehabilitation protocol as currently used, and will lead to better patient satisfaction and long-term outcome.

Acknowledgments The financial support by the City of Vienna (MA 27 – Project 12-06), by the Austrian's working compensation board (AUVA), and by the Austrian Research Agency FFG (Bridge-Project #815471) is gratefully acknowledged. Furthermore, we want to thank Johannes Zipperle for his outstanding artwork.

References

Alberts B, Bray D, Hopkin K, Johnson AD, Roberts K, Lewis J (2014) Essential cell biology, 4th edn. Garland Science, New York

Al-Nbaheen M, Vishnubalaji R, Ali D, Bouslimi A, Al-Jassir F, Megges M, Prigione A, Adjaye J, Kassem M, Aldahmash A (2013) Human stromal (mesenchymal) stem cells from bone marrow, adipose tissue and skin exhibit differences in molecular phenotype and differentiation potential. Stem Cell Rev 9:32–43

Altman GH, Horan RL, Lu HH, Moreau J, Martin I, Richmond JC, Kaplan DL (2002a) Silk matrix for tissue engineered anterior cruciate ligaments. Biomaterials 23:4131–4141

Altman GH, Horan RL, Martin I, Farhadi J, Stark PRH, Volloch V, Richmond JC, Vunjak-Novakovic G, Kaplan DL (2002b) Cell differentiation by mechanical stress. FASEB J 16:270–272

Altman GH, Lu HH, Horan RL, Calabro T, Ryder D, Kaplan DL, Stark P, Martin I, Richmond JC, Vunjak-Novakovic G (2002c) Advanced bioreactor with controlled application of multi-dimensional strain for tissue engineering. J Biomech Eng 124:742–749

Altman GH, Horan RL, Weitzel P, Richmond JC (2008) The use of long-term bioresorbable scaffolds for anterior cruciate ligament repair. J Am Acad Orthop Surg 16:177–187

Aspenberg P, Forslund C (1999) Enhanced tendon healing with GDF 5 and 6. Acta Orthop Scand 70:51–54

Bellincampi LD, Closkey RF, Prasad R, Zawadsky JP, Dunn MG (1998) Viability of fibroblast-seeded ligament analogs after autogenous implantation. J Orthop Res 16:414–420

Berry CC, Cacou C, Lee DA, Bader DL, Shelton JC (2003a) Dermal fibroblasts respond to mechanical conditioning in a strain profile dependent manner. Biorheology 40:337–345

Berry CC, Shelton JC, Bader DL, Lee DA (2003b) Influence of external uniaxial cyclic strain on oriented fibroblast-seeded collagen gels. Tissue Eng 9:613–624

Buma P, Kok HJ, Blankevoort L, Kuijpers W, Huiskes R, Van Kampen A (2004) Augmentation in anterior cruciate ligament reconstruction – a histological and biomechanical study on goats. Int Orthop 28:91–96

Butler DL, Gooch C, Kinneberg KRC, Boivin GP, Galloway MT, Nirmalanandhan VS, Shearn JT, Dyment NA, Juncosa-Melvin N (2010) The use of mesenchymal stem cells in collagen-based scaffolds for tissue-engineered repair of tendons. Nat Protoc 5:849–863

Cardwell RD, Dahlgren LA, Goldstein AS (2012) Electrospun fibre diameter, not alignment, affects mesenchymal stem cell differentiation into the tendon/ligament lineage. J Tissue Eng Regen Med. https://doi.org/10.1002/term.1589

Caruso AB, Dunn MG (2005) Changes in mechanical properties and cellularity during long-term culture of collagen fiber ACL reconstruction scaffolds. J Biomed Mater Res A 73:388–397

Chen J, Altman GH, Karageorgiou V, Horan R, Collette A, Volloch V, Colabro T, Kaplan DL (2003) Human bone marrow stromal cell and ligament fibroblast responses on RGD-modified silk fibers. J Biomed Mater Res A 67:559–570

Chen X, Qi Y-Y, Wang L-L, Yin Z, Yin G-L, Zou X-H, Ouyang H-W (2008) Ligament regeneration using a knitted silk scaffold combined with collagen matrix. Biomaterials 29:3683–3692

Chen L, Dong S-W, Tao X, Liu J-P, Tang K-L, Xu J-Z (2012) Autologous platelet-rich clot releasate stimulates proliferation and inhibits differentiation of adult rat tendon stem cells towards nontenocyte lineages. J Int Med Res 40:1399–1409

Cheng M-T, Yang H-W, Chen T-H, OK-S L (2009) Isolation and characterization of multipotent stem cells from human cruciate ligaments. Cell Prolif 42:448–460. https://doi.org/10.1111/j.1365-2184.2009.00611.x. PMID: 19489981

De Mos M, van der Windt AE, Jahr H, van Schie HTM, Weinans H, Verhaar JAN, van Osch GJVM (2008) Can platelet-rich plasma enhance tendon repair? A cell culture study. Am J Sports Med 36:1171–1178

Dunn MG, Liesch JB, Tiku ML, Zawadsky JP (1995) Development of fibroblast-seeded ligament analogs for ACL reconstruction. J Biomed Mater Res 29:1363–1371

Duthon VB, Barea C, Abrassart S, Fasel JH, Fritschy D, Menetrey J (2006) Anatomy of the anterior cruciate ligament. Knee Surg Sports Traumatol Arthrosc 14:204–213

Fan H, Liu H, Wong EJW, Toh SL, Goh JCH (2008) In vivo study of anterior cruciate ligament regeneration using mesenchymal stem cells and silk scaffold. Biomaterials 29:3324–3337

Fernández-Sarmiento JA, Domínguez JM, Granados MM, Morgaz J, Navarrete R, Carrillo JM, Gómez-Villamandos RJ, Muñoz-Rascón P, Martín de Las Mulas J, Millán Y, García-

Balletbó M, Cugat R (2013) Histological study of the influence of plasma rich in growth factors (PRGF) on the healing of divided Achilles tendons in sheep. J Bone Joint Surg Am 95:246–255
Figueroa PD, Figueroa BF, Ximena AP, Calvo RR, Vaisman BA (2013) Use of platelet rich plasma in knee ligament surgery. Rev Med Chil 141:1315–1320
Forslund C, Aspenberg P (2002) CDMP-2 induces bone or tendon-like tissue depending on mechanical stimulation. J Orthop Res 20:1170–1174
Freeman JW, Woods MD, Laurencin CT (2007) Tissue engineering of the anterior cruciate ligament using a braid-twist scaffold design. J Biomech 40:2029–2036
Freeman JW, Woods MD, Cromer DA, Wright LD, Laurencin CT (2009) Tissue engineering of the anterior cruciate ligament: the viscoelastic behavior and cell viability of a novel braid-twist scaffold. J Biomater Sci Polym Ed 20:1709–1728
Freeman JW, Woods MD, Cromer DA, Ekwueme EC, Andric T, Atiemo EA, Bijoux CH, Laurencin CT (2010) Evaluation of a hydrogel-fiber composite for ACL tissue engineering. J Biomech 44:694–699
Fu W, Li Q, Tang X, Chen G, Zhang C, Li J (2016) Mesenchymal stem cells reside in interior cruciate ligament remnants in situ. Int Orthop 40:1523–1530
Ge Z, Goh JCH, Lee EH (2005) Selection of cell source for ligament tissue engineering. Cell Transplant 14:573–583
Goodwin E (2005) The American Orthopaedic Society for sports medicine conference on allografts in orthopaedic sports medicine, Keystone
Hansson A, Hashom N, Falson F, Rousselle P, Jordan O, Borchard G (2012) In vitro evaluation of an RGD-functionalized chitosan derivative for enhanced cell adhesion. Carbohydr Polym 90:1494–1500
Henshaw DR, Attia E, Bhargava M, Hannafin JA (2006) Canine ACL fibroblast integrin expression and cell alignment in response to cyclic tensile strain in three-dimensional collagen gels. J Orthop Res 24:481–490
Hildebrand KA, Woo SL, Smith DW, Allen CR, Deie M, Taylor BJ, Schmidt CC (1998) The effects of platelet-derived growth factor-BB on healing of the rabbit medial collateral ligament. An in vivo study. Am J Sports Med 26:549–554
Hoffmann A, Pelled G, Turgeman G, Eberle P, Zilberman Y, Shinar H, Keinan-Adamsky K, Winkel A, Shahab S, Navon G, Gross G, Gazit D (2006) Neotendon formation induced by manipulation of the Smad8 signalling pathway in mesenchymal stem cells. J Clin Invest 116:940–952
Hofmann S, Knecht S, Langer R, Kaplan DL, Vunjak-Novakovic G, Merkle HP, Meinel L (2006) Cartilage-like tissue engineering using silk scaffolds and mesenchymal stem cells. Tissue Eng 12:2729–2738
Hohlrieder M, Teuschl AH, Cicha K, van Griensven M, Redl H, Stampfl J (2013) Bioreactor and scaffold design for the mechanical stimulation of anterior cruciate ligament grafts. Biomed Mater Eng 23:225–237
Horan RL, Toponarski I, Boepple HE, Weitzel PP, Richmond JC, Altman GH (2009) Design and characterization of a scaffold for anterior cruciate ligament engineering. J Knee Surg 22:82–92
Huang T-F, Chen Y-T, Yang T-H, Chen L-L, Chiou S-H, Tsai T-H, Tsai C-C, Chen M-H, Ma H-L, Hung S-C (2008) Isolation and characterization of mesenchymal stromal cells from human anterior cruciate ligament. Cytotherapy 10:806–814. https://doi.org/10.1080/14653240802474323. PMID: 19023768
Jackson DW, Grood ES, Goldstein JD, Rosen MA, Kurzweil PR, Cummings JF, Simon TM (1977) A comparison of patellar tendon autograft and allograft used for anterior cruciate ligament reconstruction in the goat model. Am J Sports Med 21:176–185
James R, Toti US, Laurencin CT, Kumbar SG (2011) Electrospun nanofibrous scaffolds for engineering soft connective tissues. Methods Mol Biol 726:243–258
Jin HJ, Kaplan DL (2003) Mechanism of silk processing in insects and spiders. Nature 424:1057–1061
Joshi SM, Mastrangelo AN, Magarian EM, Fleming BC, Murray MM (2009) Collagen–platelet composite enhances biomechanical and histologic healing of the porcine anterior cruciate ligament. Am J Sports Med 37:2401–2410

Kaeding CC, Aros B, Pedroza A, Pifel E, Amendola A, Andrish JT, Dunn WR, Marx RG, McCarty EC, Parker RD, Wright RW, Spindler KP (2011) Allograft versus autograft anterior cruciate ligament reconstruction: predictors of failure from a MOON prospective longitudinal cohort. Sports Health 3:73–81

Keibl C, Fügl A, Zanoni G, Tangl S, Wolbank S, Redl H, van Griensven M (2011) Human adipose derived stem cells reduce callus volume upon BMP-2 administration in bone regeneration. Injury 42:814–820

Koob TJ, Willis TA, Qiu YS, Hernandez DJ (2001) Biocompatibility of NDGA-polymerized collagen fibers. II. Attachment, proliferation, and migration of tendon fibroblasts in vitro. J Biomed Mater Res 56:40–48

Kurtz CA, Loebig TG, Anderson DD, DeMeo PJ, Campbell PG (1999) Insulin-like growth factor I accelerates functional recovery from Achilles tendon injury in a rat model. Am J Sports Med 27:363–369

Langer R, Vacanti JP (1993) Tissue engineering. Science 260:920–926

Laurencin CT, Freeman JW (2005) Ligament tissue engineering: an evolutionary materials science approach. Biomaterials 26:7530–7536

Leong NL, Petrigliano FA, McAllister DR (2013) Current tissue engineering strategies in anterior cruciate ligament reconstruction. J Biomed Mater Res A. https://doi.org/10.1002/jbm.a.34820

Letson AK, Dahners LE (1994) The effect of combinations of growth factors on ligament healing. Clin Orthop Relat Res 308:207–212

Li F, Jia H, Yu C (2007) ACL reconstruction in a rabbit model using irradiated Achilles allograft seeded with mesenchymal stem cells or PDGF-B gene-transfected mesenchymal stem cells. Knee Surg Sports Traumatol Arthrosc 15:1219–1227

Lin VS, Lee MC, O'Neal S, McKean J, Sung KL (1999) Ligament tissue engineering using synthetic biodegradable fiber scaffolds. Tissue Eng 5:443–452

Liu H, Fan H, Toh SL, Goh JCH (2008) A comparison of rabbit mesenchymal stem cells and anterior cruciate ligament fibroblasts responses on combined silk scaffolds. Biomaterials 29:1443–1453

Lohmander LS, Ostenberg A, Englund M, Roos H (2004) High prevalence of knee osteoarthritis, pain, and functional limitations in female soccer players twelve years after anterior cruciate ligament injury. Arthritis Rheum 50:3145–3152

Louw QA, Manilall J, Grimmer KA (2008) Epidemiology of knee injuries among adolescents: a systematic review. Br J Sports Med 42:2–10

Lu HH, Cooper JA, Manuel S, Freeman JW, Attawia MA, Ko FK, Laurencin CT (2005) Anterior cruciate ligament regeneration using braided biodegradable scaffolds: in vitro optimization studies. Biomaterials 26:4805–4816. https://doi.org/10.1016/j.biomaterials.2004.11.050. PMID: 15763260

Ma J, Smietana MJ, Kostrominova TY, Wojtys EM, Larkin LM, Arruda EM (2012) Three-dimensional engineered bone–ligament–bone constructs for anterior cruciate ligament replacement. Tissue Eng Part A 18:103–116

Macintosh AC, Kearns VR, Crawford A, Hatton PV (2008) Skeletal tissue engineering using silk biomaterials. Tissue Eng 2:71–80

Majima T, Funakosi T, Iwasaki N, Yamane S-T, Harada K, Nonaka S, Minami A, Nishimura S (2005) Alginate and chitosan polyion complex hybrid fibers for scaffolds in ligament and tendon tissue engineering. J Orthop Sci 10:302–307

Marx RE (2004) Platelet-rich plasma: evidence to support its use. J Oral Maxillofac Surg 62:489–496

Mastrangelo AN, Vavken P, Fleming BC, Harrison SL, Murray MM (2011) Reduced platelet concentration does not harm PRP effectiveness for ACL repair in a porcine in vivo model. J Orthop Res 29:1002–1007

Masuko T, Iwasaki N, Yamane S, Funakoshi T, Majima T, Minami A, Ohsuga N, Ohta T, Nishimura S-I (2005) Chitosan-RGDSGGC conjugate as a scaffold material for musculoskeletal tissue engineering. Biomaterials 26:5339–5347

Meinel L, Betz O, Fajardo R, Hofmann S, Nazarian A, Cory E, Hilbe M, McCool J, Langer R, Vunjak-Novakovic G, Merkle HP, Rechenberg B, Kaplan DL, Kirker-Head C (2006) Silk based biomaterials to heal critical sized femur defects. Bone 39:922–931

Moreau JE, Bramono DS, Horan RL, Kaplan DL, Altman GH (2008) Sequential biochemical and mechanical stimulation in the development of tissue-engineered ligaments. Tissue Eng Part A 14:1161–1172

Morito T, Muneta T, Hara K, Ju Y-J, Mochizuki T, Makino H, Umezawa A, Sekiya I (2008) Synovial fluid-derived mesenchymal stem cells increase after intra-articular ligament injury in humans. Rheumatology 47:1137–1143

Murray MM, Fleming BC (2013a) Biology of anterior cruciate ligament injury and repair: Kappa Delta Ann Doner Vaughn award paper. J Orthop Res 31:1501–1506

Murray MM, Fleming BC (2013b) Use of a bioactive scaffold to stimulate anterior cruciate ligament healing also minimizes posttraumatic osteoarthritis after surgery. Am J Sports Med 41:1762–1770

Murray MM, Spector M (2001) The migration of cells from the ruptured human anterior cruciate ligament into collagen–glycosaminoglycan regeneration templates in vitro. Biomaterials 22:2393–2402

Murray MM, Magarian E, Zurakowski D, Fleming BC (2010) Bone-to-bone fixation enhances functional healing of the porcine anterior cruciate ligament using a collagen–platelet composite. Arthroscopy 26:S49–S57

Murray MM, Kalish LA, Fleming BC et al (2019) Bridged-enhanced anterior cruciate ligament repair – two-year results of a first-in-human study. Orthop J Sports Med 22:2325967118824356

Nagineni CN, Amiel D, Green MH, Berchuck M, Akeson WH (1992) Characterization of the intrinsic properties of the anterior cruciate and medial collateral ligament cells: an in vitro cell culture study. J Orthop Res 10:465–475

Nakamura N, Shino K, Natsuume T, Horibe S, Matsumoto N, Kaneda Y, Ochi T (1998) Early biological effect of in vivo gene transfer of platelet-derived growth factor (PDGF)-B into healing patellar ligament. Gene Ther 5:1165–1170

Nau T, Teuschl A (2015) Regeneration of the anterior cruciate ligament: current strategies in tissue engineering. World J Orthop 6:127–136

Noyes FR, Butler DL, Grood ES, Zernicke RF, Hefzy MS (1984) Biomechanical analysis of human ligament grafts used in knee-ligament repairs and reconstructions. J Bone Joint Surg Am 66:344–352

Ong WK, Sugii S (2013) Adipose-derived stem cells: fatty potentials for therapy. Int J Biochem Cell Biol 45:1083–1086

Panas-Perez E, Gatt CJ, Dunn MG (2013) Development of a silk and collagen fiber scaffold for anterior cruciate ligament reconstruction. J Mater Sci Mater Med 24:257–265

Park SA, Kim IA, Lee YJ, Shin JW, Kim C-R, Kim JK, Yang Y-I, Shin J-W (2006) Biological responses of ligament fibroblasts and gene expression profiling on micropatterned silicone substrates subjected to mechanical stimuli. J Biosci Bioeng 102:402–412

Park SY, Ki CS, Park YH, Jung HM, Woo KM, Kim HJ (2010) Electrospun silk fibroin scaffolds with macropores for bone regeneration: an in vitro and in vivo study. Tissue Eng Part A 16:1271–1279

Pendleton C, Li Q, Chesler DA, Yuan K, Guerrero-Cazares H, Quinones-Hinojosa A (2013) Mesenchymal stem cells derived from adipose tissue vs bone marrow: in vitro comparison of their tropism towards gliomas. PLoS One 8:e58198

Peterbauer-Scherb A, van Griensven M, Meinl A, Gabriel C, Redl H, Wolbank S (2010) Isolation of pig bone marrow mesenchymal stem cells suitable for one-step procedures in chondrogenic regeneration. J Tissue Eng Regen Med 4:485–490

Petrigliano FA, McAllister DR, Wu BM (2006) Tissue engineering for anterior cruciate ligament reconstruction: a review of current strategies. Arthroscopy 22:441–451

Petrigliano FA, English CS, Barba D, Esmende S, Wu BM, McAllister DR (2007) The effects of local bFGF release and uniaxial strain on cellular adaptation and gene expression in a 3D environment: implications for ligament tissue engineering. Tissue Eng 13:2721–2731

Rathbone S, Maffulli N, Cartmell SH (2012) Most British surgeons would consider using a tissue-engineered anterior cruciate ligament: a questionnaire study. Stem Cells Int 2012:303724

Rockwood DN, Preda RC, Yücel T, Wang X, Lovett ML, Kaplan DL (2011) Materials fabrication from *Bombyx mori* silk fibroin. Nat Protoc 6:1612–1631

Roos EM (2005) Joint injury causes knee osteoarthritis in young adults. Curr Opin Rheumatol 17:195–200

Rumian AP, Wallace AL, Birch HL (2007) Tendons and ligaments are anatomically distinct but overlap in molecular and morphological features – a comparative study in an ovine model. J Orthop Res 25:458–464

Seitz H, Pichl W, Matzi V, Nau T (2013) Biomechanical evaluation of augmented and non-augmented primary repair of the anterior cruciate ligament: an in vivo animal study. Int Orthop 37:2305–2311

Shao H-J, Chen CS, Lee Y-T, Wang J-H, Young T-H (2010a) The phenotypic responses of human anterior cruciate ligament cells cultured on poly(epsilon-caprolactone) and chitosan. J Biomed Mater Res A 93:1297–1305

Shao H-J, Lee Y-T, Chen C-S, Wang J-H, Young T-H (2010b) Modulation of gene expression and collagen production of anterior cruciate ligament cells through cell shape changes on polycaprolactone/chitosan blends. Biomaterials 31:4695–4705

Sinclair K, Yerkovich ST, Chambers DC (2013) Mesenchymal stem cells and the lung. Respirology 18:397–411

Steinert AF, Kunz M, Prager P, Barthel T, Jakob F, Nöth U, Murray MM, Evans CH, Porter RM (2011) Mesenchymal stem cell characteristics of human anterior cruciate ligament outgrowth cells. Tissue Eng Part A 17:1375–1388

Tetsunaga T, Furumatsu T, Abe N, Nishida K, Naruse K, Ozaki T (2009) Mechanical stretch stimulates integrin alphaVbeta3-mediated collagen expression in human anterior cruciate ligament cells. J Biomech 42:2097–2103

Teuschl AH, van Griensven M, Redl H (2013) Sericin removal from raw *Bombyx mori* silk scaffolds of high hierarchical order. Tissue Eng Part C Methods. https://doi.org/10.1089/ten.TEC.2013.0278

Teuschl A, Heimel P, Nuernberger S, van Griensven M, Redl H, Nau T (2016) A novel silk fiber-based scaffold for regeneration of the anterior cruciate ligament: histological results from a study in sheep. Am J Sports Med 44:1547–1557

Teuschl A, Tangl S, Heimel P, Schwarze UY, Monforte X, Redl H, Nau T (2019) Osteointegration of a novel silk fiber-based ACL scaffold by formation of a ligament–bone interface. Am J Sports Med 47:620–627

Thomopoulos S, Das R, Silva MJ, Sakiyama-Elbert S, Harwood FL, Zampiakis E, Kim HM, Amiel D, Gelberman RH (2009) Enhanced flexor tendon healing through controlled delivery of PDGF-BB. J Orthop Res 27:1209–1215

Vavken P, Fleming BC, Mastrangelo AN, Machan JT, Murray MM (2012) Biomechanical outcomes after bioenhanced anterior cruciate ligament repair and anterior cruciate ligament reconstruction are equal in a porcine model. Arthroscopy 28:672–680

Vepari C, Kaplan DL (2007) Silk as a biomaterial. Prog Polym Sci 32:991–1007

Vunjak-Novakovic G, Altman G, Horan R, Kaplan DL (2004) Tissue engineering of ligaments. Annu Rev Biomed Eng 6:131–156

Walters VI, Kwansa AL, Freeman JW (2012) Design and analysis of braid-twist collagen scaffolds. Connect Tissue Res 53:255–266

Wang Y, Blasioli DJ, Kim H-J, Kim H, Kaplan DL (2006a) Cartilage tissue engineering with silk scaffolds and human articular chondrocytes. Biomaterials 27:4434–4442. https://doi.org/10.1016/j.biomaterials.2006.03.050. PMID: 16677707

Wang Y, Kim H-JJ, Vunjak-Novakovic G, Kaplan DL (2006b) Stem cell-based tissue engineering with silk biomaterials. Biomaterials 27:6064–6082

Wang J, Guo Q, Li B (2012) Tendon biomechanics and mechanobiology – a mini review of basic concepts and recent advancements. J Hand Ther 25:133–141

Wei X, Mao Z, Hou Y, Lin L, Xue T, Chen L, Wang H, Yu C (2011) Local administration of TGFβ-1/VEGF165 gene-transduced bone mesenchymal stem cells for Achilles allograft replacement of the anterior cruciate ligament in rabbits. Biochem Biophys Res Commun 406:204–210

Wolfman NM, Hattersley G, Cox K, Celeste a J, Nelson R, Yamaji N, Dube JL, DiBlasio-Smith E, Nove J, Song JJ, Wozney JM, Rosen V (1997) Ectopic induction of tendon and ligament in rats by growth and differentiation factors 5, 6, and 7, members of the TGF-beta gene family. J Clin Invest 100:321–330

Woo SL, Gomez MA, Woo YK, Akeson WH (1982) Mechanical properties of tendons and ligaments. II. The relationships of immobilization and exercise on tissue remodeling. Biorheology 19:397–408

Woo SL, Gomez MA, Sites TJ, Newton PO, Orlando CA, Akeson WH (1987) The biomechanical and morphological changes in the medial collateral ligament of the rabbit after immobilization and remobilization. J Bone Joint Surg Am 69:1200–1211

Woo SL, Hollis JM, Adams DJ, Lyon RM, Takai S (1991) Tensile properties of the human femur–anterior cruciate ligament–tibia complex. The effects of specimen age and orientation. Am J Sports Med 19:217–225

Woo SL, Debski RE, Withrow JD, Janaushek MA (1999) Biomechanics of knee ligaments. Am J Sports Med 27:533–543

Yamane S, Iwasaki N, Majima T, Funakoshi T, Masuko T, Harada K, Minami A, Monde K, Nishimura S-I (2005) Feasibility of chitosan-based hyaluronic acid hybrid biomaterial for a novel scaffold in cartilage tissue engineering. Biomaterials 26:611–619

Yasuda K, Tomita F, Yamazaki S, Minami A, Tohyama H (2004) The effect of growth factors on biomechanical properties of the bone–patellar tendon–bone graft after anterior cruciate ligament reconstruction: a canine model study. Am J Sports Med 32:870–880

Yuan T, Zhang C, Wang JH (2013) Augmenting tendon and ligament repair with platelet-rich plasma (PRP). Muscles Ligaments Tendons J 3:139–149

Multiscale Multifactorial Approaches for Engineering Tendon Substitutes

Ana I. Gonçalves, Márcia T. Rodrigues, Ana M. Matos, Helena Almeida, Manuel Gómez-Florit, Rui M. A. Domingues, and Manuela E. Gomes

Contents

Abstract

The physiology of tendons and the continuous strains experienced daily make tendons very prone to injury. Excessive and prolonged loading forces and aging also contribute to the onset and progression of tendon injuries, and conventional treatments have limited efficacy in restoring tendon biomechanics. Tissue engineering and regenerative medicine (TERM) approaches hold the promise to provide therapeutic solutions for injured or damaged tendons despite the challenging cues of tendon niche and the lack of tendon-specific factors to guide cellular responses and tackle regeneration. The roots of engineering tendon substitutes lay in multifactorial approaches from adequate stem cells sources and environmental stimuli to the construction of multiscale 3D scaffolding systems.

A. I. Gonçalves · M. T. Rodrigues · A. M. Matos · H. Almeida · M. Gómez-Florit · R. M. A. Domingues · M. E. Gomes (✉)
3B's Research Group, I3Bs – Research Institute on Biomaterials, Biodegradables and Biomimetics, University of Minho, Headquarters of the European Institute of Excellence on Tissue Engineering and Regenerative Medicine, Barco/Guimarães, Portugal

ICVS/3B's–PT Government Associate Laboratory, Braga/Guimarães, Portugal
e-mail: ana.goncalves@i3bs.uminho.pt; mrodrigues@i3bs.uminho.pt; amfmatos@gmail.com; helena-mca@hotmail.com; mgflorit@i3bs.uminho.pt; rui.domingues@i3bs.uminho.pt; megomes@i3bs.uminho.pt

D. Eberli et al. (eds.), *Organ Tissue Engineering*, Reference Series in Biomedical Engineering, https://doi.org/10.1007/978-3-030-44211-8_8

To achieve such advanced tendon substitutes, incremental strategies have been pursued to more closely recreate the native tendon requirements providing structural as well as physical and chemical cues combined with biochemical and mechanical stimuli to instruct cell behavior in 3D architectures, pursuing mechanically competent constructs with adequate maturation before implantation.

1 Introduction

Tendon is composed mainly of water (70% of dry weight) and a solid matrix (30% of dry weight) (Sharma and Maffulli 2005) mostly of collagen (70–80%), which is the force-transmitting unit of tendon (Magnusson et al. 2003; de Aro et al. 2012). Collagen type I is the main type of collagen within tendons (95% of total collagen and 60–85% of matrix dry weight (Screen et al. 2015)) followed by collagen type III and collagen type V. The alignment of collagen type I is recognized as the principal structural feature of healthy tendons (Lipman et al. 2018). Together with collagen type I, type V, and type XI, collagen type III is a fibril-forming collagen and widely distributed in collagen I containing tissues. Upon injury, collagen type III increases. This has been associated to its rapid cross-linking contributing to stabilize the repair site and to the remodeling process of the matrix highlighting the importance of collagen type III in tendon healing processes. However, collagen type III forms smaller and less organized fibrils than the ones of collagen type I. Abundant collagen type III results in a tendon with inferior mechanical properties, which has been related to a degenerative process (Millar et al. 2015). Collagen type V is important for regulation and stabilization of collagen type I structures during self-assembly (Wenstrup et al. 2004), while collagen type VI regulates collagen I fibrillogenesis to form a functional and mature tendon (Izu et al. 2011).

Tendons are mainly composed of tenoblasts and tenocytes approximately 90–95% of tendon cells (Schneider et al. 2017) whose major function relies in the synthesis and maintenance of the tendon extracellular matrix. Tenoblasts are roundly shaped cells with a large, ovoid nucleus that mature to spindle-shaped tenocytes with elongated nuclei in response to mechanical regulators and to growth factors during tendon development. These cells are discriminated based on morphology; however, precise identification based on specific markers is still lacking. Nevertheless, there are tendon-associated markers that may assist the establishment of tenogenic protocols and offer new perspectives for studying tendon biology.

The transcription factor scleraxis (Scx) is expressed from immature to fully differentiated stages of tendon cells. Other factors, namely, Mowahk (Mwk), Egr1, and Egr2 are detected at the stage of tenoblasts. Both tendon-progenitor cells and tenoblasts also express collagen I, Tenascin (TNC), Thrombospondin (Thbs4), and Tenomodulin (Tnmd). Mature tenocytes typically express Scx and Mwk as well as Egr1 and Smad3. Developmental and in vitro studies have been pointing Scx and Mwk as tendon-specific factors despite the fact they can be detected in other tissues. The ECM matrix produced by tenocytes is rich in collagen I, III, V, and XIV; Decorin; Fibromodulin; Lumican; and Tnmd.

The tenocytes are organized in longitudinal rows between collagen fibers and respond to mechanical forces. These cells establish communication with adjacent cells usually through connexins 26, 32, and 43 in gap junctions. Tendon cell activity declines with age leading to unbalanced processes thus contributing to tissue degeneration and to incremental severity of injuries.

More recently, stem progenitor cells (TSPCs) were identified in tendons shown to possess regenerative capabilities (Salingcarnboriboon et al. 2003; Bi et al. 2007; de Mos et al. 2007; Zhang and Wang 2010). TSPCs exhibit the typical surface antigens of adult mesenchymal stem cells (MSCs), self-renewal, clonogenicity, and tri-lineage differentiation, fitting the classical MSCs criteria, and highly express tendon-specific factors, namely, Scx and Tnmd (Bi et al. 2007).

Paratenon, epitenon, and endotenon are the non-tenogenic components of tendons. The paratenon allows tendon to move freely and continues to the fascia, while epitenon is defined as a dense connective tissue sheath covering tendon. Both are vascularized and enervated. Endotenon is formed by collagen fibers surrounding the tertiary fascicles of the tendon. Not all tendons are sheathed by paratenon although there is evidence of its involvement in tendon healing (Dyment et al. 2013; Muller et al. 2018).

During the last decade, tremendous effort has been done pursuing the key features of tendon niches to recapitulate tendon biology enabling the identification of specific tissue requirements guiding the mechanisms underlying healing and regeneration phenomena. Undoubtedly, the mechanical forces applied to tendons are critical for the maintenance of the homeostasis in healthy tendons, and that misuse or overuse can deregulate the physiological balance inflicting pathological alterations at a biomolecular and cell levels, thus contributing to impaired healing and abnormal tissue functionality. Understanding how mechanical forces regulate (stem) cell behavior and matrix features may provide key insights on tendon biology and assist the development of regenerative therapies (Vining and Mooney 2017).

This chapter overviews tendon engineering approaches to mimic tendon niches envisioning sophisticated alternatives to stimulate tendon healing and regenerative processes, anticipating functional contributions for improved tendon-oriented therapies and treatments.

2 Tendon Injuries and Repair Mechanisms

Along with the socioeconomic burden, tendon pathologies are a significant cause for disability for both working and ageing populations. The incidence of tendon injuries is rising and becoming a paramount problem without effective therapeutics. Despite tendons are designed to withstand mechanical forces, they are prone to lesions that aggravate with the time, compromising articular performance and the quality of movements and physical activities. Damage in the joint may eventually lead to morbidity, pain, and osteoarthritis (Wang et al. 2012). Although genetic predisposition factors make individuals more or less susceptible to injury (Riley 2004; Rees et al. 2006; Docheva et al. 2015), tendons become more prone to degeneration and to injuries with advanced stages of life. Moreover, aging and age-related pathogenesis are also associated to impaired healing, presenting a huge opportunity to address the unmet needs for new therapies and customized treatments for tendon diseases.

Excessive mechanical loading resulting from small repetitive strains may lead to accumulation of tendon microinjuries (Soslowsky et al. 2000; Willett et al. 2007) contributing for the progression of tendinopathic conditions. Tendinopathy typically defines a non-rupture injury in the tendon or paratenon, which is exacerbated by mechanical loading and typically accompanied by pain/swelling of injured tendon. Pain and functional limitation are frequent symptoms conventionally treated with nonsteroidal anti-inflammatory drugs (NSAIDs), corticosteroids, and physical rehabilitation. However, the prolonged systemic action of these drugs has been reported to cause significant long-term side effects as changes in blood pressure, myocardial infarction and strokes, increased bleeding, osteoporosis, or peptic ulcers. The more severe cases typically require surgical intervention, which rely on tissue replacement with auto- or allografts (Rodrigues et al. 2013). However, these are often accompanied with donor site morbidity, pain, inferior functionalities, and eventually graft failure. The limited long-term success of surgical procedures compromises tendon functionality impacting the quality of life of patients.

The healing capacity of tendons varies according to severity of injury and anatomical location, and it is classically divided in three main stages. The first stage is a short inflammatory phase that initiates after injury. The vascular permeability increases as well as the influx of inflammatory cells (platelets, macrophages, monocytes, and neutrophils) to the site of injury. These cells release chemotactic agents responsible for the recruitment of blood vessels, fibroblasts, and intrinsic tenocytes that together with inflammatory cells form a hematoma (Lin et al. 2004). As part of the inflammatory response, macrophage phagocyte injured tissue fragments and has an important role on tenocytes proliferation and angiogenesis (Woo et al. 1999). The second stage is characterized by an abundant proliferation of tenocytes at the site of injury with concomitant production of collagen. This phase lasts from a few days to a couple of weeks. The remodeling phase is the third and final stage and can last several months (Sharma and Maffulli 2005). The extracellular matrix is reorganized with a gradual change in the direction of collagen fibers from fibrous to scar-like tissue accompanied by a cellular decrease and reduction of vascular vessels (Sharma and Maffulli 2006).

Tendon healing was traditionally described to be intrinsic or extrinsic in nature. The intrinsic process was mediated by tenocytes from epitenon and endotenon, and the repaired tissue evidenced higher mechanical properties and fewer complications. Extrinsic healing was dependent on cell migration from adjacent tissues resulting in scar tissue leading to adhesions formation (Sharma and Maffulli 2005; Zhao et al. 2015). Although the repair process depends on the anatomy and physiology of each tendon, tendon healing is now described to occur from the interplay of local tendon cells and external cells, including immune, vascular, synovial, and mesenchymal stem cells.

Understanding and modulating the mechanisms underlying healing will allow more effective treatments and rehabilitation. Moreover, as a key process in tendon tissues, insights tendon healing and the research of tendon mechanobiology will likely contribute to advanced clinical solutions resourcing to tissue engineering and regenerative medicine strategies.

3 Mechanoregulation Mechanisms

As mechanosensitive and mechanoresponsive tissue, the main function of tendon is transmitting tensile loads that varies with the anatomical location and can reach to several tens of megapascal (MPa). This incredible feature of tendons is extremely dependent on its structure and cellular organization, and it is critical for the proper function of joints (Killian et al. 2012; Lavagnino et al. 2015). The unique structure and composition confer tendon a characteristic mechanical behavior which is classically described in a stress-strain curve (Fig. 1). The initial "toe" region represents the flattening of crimp pattern when tendon is strained up to 2%. In the linear region, tendon is stretched up to 4%, and collagen fibers lose the crimp pattern. Altogether, these regions represent the physiological response of tendon. If tendon is stretched more than 4%, a microscopic failure of collagen fibers occurs. However, when the

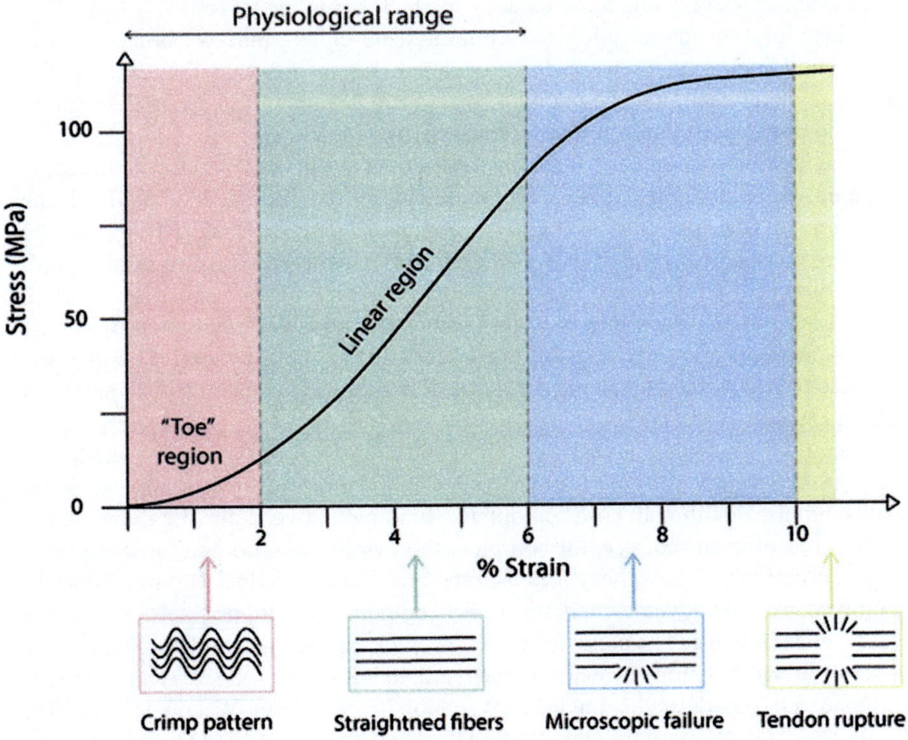

Fig. 1 Typical stress-strain curve of tendon tissues exposed to increasing mechanical loads

forces applied are below this value, tendon remains in the elastic zone and can return to its original shape when a load is removed. Macroscopic failure of collagen fibers is observed in loads between 8 and 10%, while rupture of collagen fibers is verified with mechanical loads higher than 10%.

Thus, the sensing of mechanical forces by the cells and further translation of the cellular response to microenvironmental cues is of utmost importance in deciphering precise mechanisms of action toward tenogenic recipes and opens avenues of research seeking to develop efficient therapies.

Mechanotransduction is the ability of cells to respond to mechanical stimuli through biochemical signals (Santos et al. 2015) being important to maintain musculoskeletal tissue development, homeostasis, repair, and regeneration (Wang 2006). Cells are able to sense distinct mechanical stimulus within the matrix and to respond accordingly and rapidly by adjusting physical ligands or rearranging its cytoskeleton. These adjustments affect the perception of loading in the mechanosensory elements of cells, present in the cellular membrane and nucleus.

Signaling cascades are the main routes of communication between the membrane and intracellular regulatory targets. The two main signaling pathways identified as being involved in tendon development are TGFβ-SMAD2/3 and FGF-ERK/MAPK pathways (Havis et al. 2014a). FGF signaling from the myotome was firstly associated to induction of *Scx*-expressing tendon progenitors in adjacent somatic subcompartment of developing axial tendons in chick (Brent et al. 2003).

Interestingly, in pharmacologically immobilized chick embryos, both FGF and TGF-β signaling cascades were downregulated, suggesting that FGF and TGF-β ligands regulate tendon differentiation acting downstream to mechanical forces present in developing embryo (Havis et al. 2016).

In mammalians, there are three isoforms of TGF-β, and all are involved in tenogenic differentiation. Studies reported that in the presence of TGF-β1 the expression of Scx and Mkx is highly upregulated (Farhat et al. 2012), and the decline of mechanical properties was significantly hindered (Katsura et al. 2006). The isoform TGF-β2 increased the expression of *Col1a1* and *Scx* genes (Havis et al. 2014b; Liu et al. 2015), while TGF-β3 promoted tendon differentiation of equine embryo-derived stem cells (Barsby and Guest 2013). Furthermore, TGF-β3 was suggested to be an essential element of a recently proposed tenogenic differentiation cocktail (Stanco et al. 2019).

Generally, in a signaling cascade, the ligand requires two types of serine/threonine kinases receptors, a type I and a type II that form a receptor complex. Some ligands require additional co-receptors for binding between the ligand and the complex receptor. In the receptor complex, the cytoplasmic domain of the type II receptor is active and phosphorylates the type I receptor on serines and threonines in a highly conserved glycine- and serine-rich domain, neighboring the region that crosses the membrane. The activated type I receptor supplies a ligand binding site for the downstream substrates, the receptor-regulated SMADs (R-SMADs) that will transduce the signal to the nucleus (Wrighton et al. 2009; Wakefield and Hill 2013). The TGF-βs, activins, and NODAL signal through SMAD2 and SMAD3 (Wakefield and Hill 2013; Zhang et al. 2018a). In humans, there are seven type I receptors known as activin receptor like-kinases (ALK) each being responsible to regulate the phosphorylation of different R-SMADs. ALK1, ALK2, ALK3, and ALK6 phosphorylate SMAD1, SMAD5, and SMAD8, while ALK4, ALK5, and ALK7 phosphorylate SMAD2 and SMAD3 (Schmierer and Hill 2007).

Furthermore, the synergistic effect of mechanical stimulation and the activation of TGF-β/SMAD2/3 cascade has an important role on the regulation of mechanical and biochemical signaling pathways that regulate *Scx* expression (Maeda et al. 2011; Goncalves et al. 2018b).

4 Tendon Tissue Engineering

4.1 Challenges of Cell-Based Approaches

Tendon tissue engineering has been challenged by a significant lack of understanding in the identification and characterization of tendon niches. As functional living entities, cells are of critical importance on the interplay of biological responses leading to tissue repair and regeneration. Despite the growing knowledge, the precise culture conditions and environment cues to stimulate tendon resident cells and guide cell fate require further developments.

In 2007, Bi et al. reported a novel stem/progenitor cells (TSPCs) population within the tendon fascicle (Bi et al. 2007). Despite the efforts, the origin of resident tendon-progenitor/stem cells and the factors that influence their differentiation is still poorly understood, compromising the design and development of novel repair and regenerative strategies.

TSPC subsets have been identified in peritenon (Cadby et al. 2014; Mienaltowski et al. 2014) holding unique signatures yet both capable of tenogenic differentiation and forming collagen-rich structures (Cadby et al. 2014, Mienaltowski et al. 2014). Mienaltowski MJ et al. reported that stem cells from tendon proper are more suited for regeneration of tendon structure, while stem cells from peritendon secreted tendon-promoting factors that bolster expression of tendon markers in tendon proper stem cell and tenocytes (Mienaltowski et al. 2014). Cadby JA et al. observed that cells from the peritenon migrated faster replicate more quickly holding higher expression of progenitor cell markers (Cadby et al. 2014).

Within paratenon tissues, fibroblastic, vascular, synovial, neural, and fat cells may be detected. TSPC subsets expressing vascular (Mienaltowski et al. 2014) markers have also been described which emphases the intercellular role in homeostasis and in the healing processes. In particular, the vicinity of vascular cell populations of pericytes or perivascular cells has attracted a lot of attention in tendon biology not only because perivascular niche may be source for epitenon-derived stem cells, known to express vascular markers but the fact that extrinsic healing in tendons rely on the migration of cells from surrounding tissues, including blood vessels. These mechanisms are central to our understanding of the onset and development of tissue pathologies and to guide cellular responses with precise regenerative action.

The origin of stem cells populations relates to distinctive signature profiles yet with complementary roles in tendon biology. Moreover, cell plasticity of tendon core and peritendinous sources may provide particular contributions for tendon repair mechanisms and for improved therapeutic solutions for tendon pathologies.

Despite the therapeutic promise of TSPCs, the limited number of local TSPC populations, especially in adult and elder patients, and the invasiveness of the harvesting procedures from a relatively hypocellular tissue can compromise their availability and use in the clinical practice. Moreover, local cell sources require the patient to wait for cell expansion procedures until a suffice number of cells is achieved for implantation. Thus, alternative cell sources have been investigated for tissue engineering and regenerative medicine strategies. Mesenchymal stem cells from bone marrow and adipose tissue are heterogeneous sources described not to incite immunologically adverse responses or to raise ethical constrains as pluripotent stem cells (ESCs and iPSCs) and have been pursued for tenogenic phenotype with evidence of improving tendon healing in in vivo models of tendon injuries (Gonçalves et al. 2013; Shen et al. 2018). Unlike TSPCs, bone marrow and adipose tissue-derived stem cell sources are considerably available and have recognized potential to be used for the regeneration of different types of tissues including tendons (Yin et al. 2016; Perucca Orfei et al. 2019). Tendon-oriented strategies employing non-tenogenic MSCs cultures typically rely on inductive factors to stimulate tenogenic cues such as TGF-β (Gonçalves et al. 2013), GDF-5, and connective tissue growth factor (CTGF), despite the fact that cocktails have not been fully established (Yin et al. 2016, Perucca Orfei et al. 2019). Moreover, previous studies by our research group (Rada et al. 2011a, b, 2012; Mihaila et al. 2013, 2014, 2015; Goncalves et al. 2018a) and others (Miranville et al. 2004; Sengenes et al. 2007) show that human adipose stem cells (hASCs) are composed of subpopulations with distinct differentiation potential, which highlights the differential role of subpopulations to lineage commitment and toward tissue-oriented applications. A Tenomodulin positive subpopulations demonstrated increased tenogenic differentiation potential (Goncalves et al. 2018a) when compared to the crude population and also when compared to SSEA-4+ hASCs subpopulation (Gonçalves et al. 2019).

The control over stem cell fate, either from tendinous or non-tendinous tissues for cell-based therapies or as part of a multiscale approach offers new possibilities into improved healing and treatment intervention of tendon pathologies. However, finding tendon-specific markers enabling standardization of protocols and proper characterization of the cellular roles within tendon tissues are still unmet features to be addressed in the following years.

4.2 Biomaterial Approaches for Tendon Tissue Engineering

Tendon TE strategies where the tenogenic niche is being mimicked are prone to induce a more adequate response regarding cell phenotype and fate. Different studies have emerged in this field in which several aspects of tendons' ECM structure and composition were recapitulated into mechanically competent constructs and cell behavior assessed, aiming to induce endogenous tissue repair.

4.2.1 Fibrous Scaffolds for Tendon Tissue Engineering

Two- and three-dimensional (2D and 3D) structures designed to resemble native tendons have been developed over the years (Chainani et al. 2013). Attending to their architecture, some authors developed systems mainly based on fibrous scaffolds where the fiber diameter, alignment, and overall stiffness were considered as topographic beacons to guide cell fate. Others valued the mechanical demand these tissues are subjected to and considered mechanical stimulation.

Strategies Involving Topographic Cues

Concerning 2D systems, these are mostly developed to evaluate how different stimuli influence cell behavior, once they usually lack appropriate mechanical properties for potential clinical application (Spanoudes et al. 2014, Santos et al. 2017). Fiber diameter and alignment have been proven to modulate cell's behavior conjointly. Systems with topographic cues in the nanoscale seem to be more effective in allowing cell growth and synthesis of glycosaminoglycans (GAGs) and collagen overtime, in either random 4 or aligned (Erisken et al. 2013a; Lee et al. 2017) conformations. Micro-cues, on the other hand, have been shown to suppress cell growth and collagen synthesis (Gilchrist et al. 2014; Lee et al. 2017).

Regarding cell morphology, both tendon cells (Erisken et al. 2013b; English et al. 2015) and MSCs (Bashur et al. 2009) present higher cell aspect ratio when exposed to substrates with microfeatures compared to nanocues (Yu et al. 2013). Additionally, anisotropic constructs are more prone to induce the tenocyte-like spindle-shape morphology and alignment along nanofibers' axis in the seeded cells, as opposed to the arbitrarily oriented cell morphologies acquired in random substrates (Moffat et al. 2009; Xie et al. 2010). Nonetheless, grooves' depths as small as 40 nm did not induce tenocyte alignment despite their parallel organization, suggesting that there is a minimum size cells can perceive (English et al. 2015). Accordingly, Domingues et al. used a solution of poly-ϵ-caprolactone (PCL) and chitosan (CHT) reinforced with rod-shape nanofillers, cellulose nanocrystals (CNCs), to produce random and aligned meshes through electrospinning with increased mechanical properties. Tenocytes seeded onto the anisotropic constructs maintained their elongated phenotype with aligned arrangement, as opposed to the cells on the random meshes (Domingues et al. 2016).

Furthermore, the expression of tendon-related markers as Scx, Tnmd, ColI, and Mkx by stem cells is enhanced in aligned surfaces, indicating that tenogenesis is encouraged, while in randomly oriented membranes, osteogenic genes are upregulated (Zhang et al. 2015). Schoenenberger et al. also correlated substrate alignment with ECM turnover, demonstrating that tendon cells on PCL random nanomats are polygonal and show decreased Mkx, ColI, and ColIII expression getting further from a tenogenic profile, as well as decreased MMP1 suggesting less matrix remodeling, as opposed to aligned constructs (Schoenenberger et al. 2018).

Submicron 2D systems typically display limited dimensions and pore size, do not respond properly to the mechanical demand necessary to support tendon healing, and are generally too weak to be surgically implanted (Barber et al. 2013; Hakimi et al. 2015; Zheng et al. 2017). Thus, 3D systems that mechanically sustain tissue regeneration and mimic tendon architecture have been considered (Santos et al. 2017). By stacking multiple electrospun PCL aligned membranes, clinically relevant sized constructs can be generated, allowing hASCs infiltration, alignment, and higher *Tnmd* and *ColIII* expression than random multilayered scaffolds. In addition, the tensile mechanical properties are improved upon cell seeding for anisotropic constructs (Orr et al. 2015).

Mimicking the native tendon highly anisotropic and hierarchic architecture has been a research priority (Brennan et al. 2018). Textile techniques, as braiding, twisting, and weaving, enable the construction of 3D tendon scaffolds of incremental organization and mechanical performance, from threads or yarns (Freeman et al. 2007, Czaplewski et al. 2014, Santos et al. 2017). Electrochemically aligned collagen (ELAC) threads have shown to induce tenogenic differentiation of human MSCs, unlike random collagen threads, evidenced by the superior expression of Scx, Tnmd, TNC, and ColIII, accompanied by enhanced ECM deposition and alignment (Kishore et al. 2012). When assembled into yarns, these natural-based woven textiles exhibited a significant increase in their mechanical properties, reaching an average tensile Young's modulus between 500 and 600 MPa and ultimate tensile strength (UTS) of 60–70 MPa, within the range of the native tendon (Younesi et al. 2014). Additionally, the yarns maintained their tenoinductive ability, with MSCs showing an increased expression of Tnmd and ColI overtime, parallel to the inhibition of osteogenic markers (Kishore et al. 2012).

Following the same principle, Laranjeira et al. recreated the nano-to-macro hierarchical and anisotropic structure of the native tendon by assembling woven scaffolds of continuous and aligned electrospun nanofibrous threads, made of PCL and CHT mechanically reinforced with CNCs (Domingues et al. 2016). The weaved structures exhibited the native tendon nonlinear mechanical behavior with evidenced toe region. The woven scaffolds presented a Young's modulus smaller than 200 MPa and UTS of approximately 40 MPa, whereas the yarn 12 of 300–400 MPa and 60 MPa, respectively (Laranjeira et al. 2017). As outcome, these structures not only avoided the tenogenic phenotypic drift of human tenocytes but also triggered the tenogenic differentiation of hASCs without biochemical supplementation, increasing the expression of tendon-related markers TNC and SCX.

Strategies Involving Mechanical Stimulation

Besides topographic cues, cells are sensitive to applied mechanical forces within the physiological range, capable of modulating their lineage commitment and ECM biosynthesis and degradation (Wang et al. 2013; Gonçalves et al. 2018). Tendons high physical demand and mechanosensitive behavior led to the development of bioreactors as exogenous sources of mechanical stimulation, to recapitulate the in vivo microenvironment native cells are exposed to (Wang et al. 2013; Spanoudes et al. 2014; Gonçalves et al. 2018).

Subramony et al. assessed the role of nanofiber alignment and mechanical stimulation on MSCs differentiation, in the absence of biochemical supplementation, using electrospun unaligned and aligned poly(lactide-co-glycolic acid) (PLGA) nanofibrous membranes, placed on a bioreactor that applied uniaxial tensile strain. The authors stated that tenogenic differentiation was induced on aligned substrates with load application, whereas loading nonaligned scaffolds or exposing cells just to aligned substrates did not induce proper differentiation, reinforcing the importance of mechanical stimulation (Subramony et al. 2013).

Nonetheless, the use of bioreactors faces some shortcomings, as the inability to use upon implantation and the cell impairment that may result from their physical presence in culture environments (Riehl et al. 2012). Therefore, scaffolds incorporating magnetic responsiveness, which can be remotely actuated by the application of an external magnetic field, allowing cell guidance and stimulation following implantation have been recently developed (Goncalves et al. 2016). Considering this, Tomás et al. used the setup proposed in a previous study (Laranjeira et al. 2017) to produce yarns of continuous and aligned electrospun threads of PCL and CNCs coated with iron oxide magnetic nanoparticles (MNPs) (Fig. 2). Cell studies revealed that hASCs express tendon-related genes both in static and dynamic conditions after 11 days of culture (Tomás et al. 2019), in agreement with previous studies (Laranjeira et al. 2017; Almeida et al. 2019). Furthermore, mechanical stimulation lead to marked upregulation of Scx and Tnmd at the gene and protein levels (Fig. 2) and downregulation of Runx2, suggesting a synergistic effect of nanotopography and mechanical actuation on the tenogenic commitment.

4.2.2 3D Printing for Tendon Tissue Engineering

Over the past decade, 3D printing has evolved from layer-by-layer deposition of materials to fabricate 3D scaffolds to bioprinting, which allows patterning and assembly of cells and materials with a defined 3D organization to produce bioengineered structures serving in regenerative medicine, pharmacokinetics, and basic cell biology studies. Although there are many promising examples of the application of this technology in various tissue engineering and regenerative medicine strategies (Bracaglia et al. 2017; Lim et al. 2017; Derakhshanfar et al. 2018; Moroni et al. 2018a, b), its use for the fabrication of tendon TE scaffolds is a field under development, and very few studies have been published. One of the first works in the field was developed by Merceron et al. to fabricate a muscle-tendon unit construct (Merceron et al. 2015). The combination of polymeric materials to print the structural component with cell-laden bioink to print the cellular component resulted in customizable hybrid/multimaterial constructs with anisotropic patterns, mimicking tendon and muscle characteristics. Thermoplastic polyurethane (PU) and C2C12 myoblasts were used for the muscle side and PCL and NIH/3 T3 fibroblasts for the tendon side. These constructs showed over 80% cell viability 1 week after printing, and the cell organization was consistent with the native tissue interface. In a more recent study, 3D musculoskeletal-tendon-like tissues were printed alternating layers of gelatin methacryloyl and cells (tenocytes and myoblasts) around and between

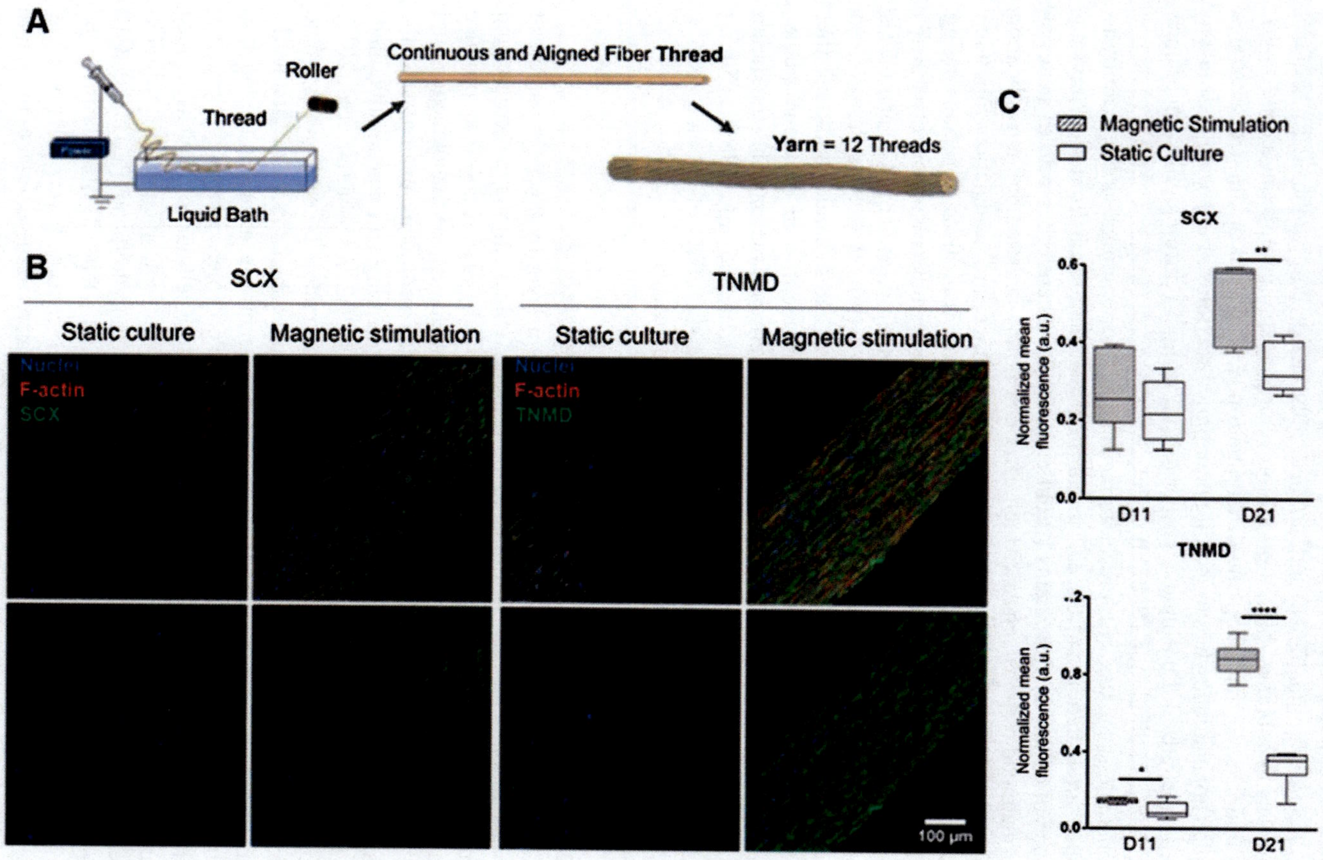

Fig. 2 (**a**) Schematic representation of the fabrication process of PCL/(CNCs with MNPs) continuous and aligned fiber threads, further assembled into yarns of 12; (**b**) confocal microscopy images of tendon-related markers (SCX and TNMD, green) expression by hASCs after 21 days of culture, with stained nuclei

posts created in a culture well, envisioning a screening platform. The cells showed high viability in culture and tissue differentiation markers (Laternser et al. 2018).

The development of a magnetic scaffold by 3D printing for tendon tissue applications was described by us (Goncalves et al. 2016), based on a polymeric blend of starch and PCL (SPCL) incorporating magnetic nanoparticles. The purpose was to combine structural features of the 3D scaffold with mechanomagnetic actuation to study the differentiation of hASCs to the tenogenic phenotype and assist tendon regeneration. The anisotropic scaffolds promoted the tenogenic differentiation of hASCs under magneto-stimulation with evidence of good biocompatibility and integration in an ectopic rat model (Goncalves et al. 2016). In a different strategy, anticipating patient-specific therapies, 3D printed scaffold sleeves made of PCL-PLGA-β-TCP were developed considering the size and shape of the tendon and bone tunnel. Then, these scaffolds were seeded with MSCs and tested in vivo for up to 12 weeks in an anterior cruciate ligament (ACL) reconstruction model in rabbits (Park et al. 2018). The construct enhanced osteointegration between the tendon and tunnel bone in the ACL reconstruction (Park et al. 2018).

4.3 Role of Biological Cues in Tendon Tissue Engineering Strategies

Substrates providing instructive cues through cell-surface contact guidance have been associated with biochemical signals to add functionality, creating microenvironments for cell differentiation that closely resemble native tissues (Santos et al. 2017). Several bioinductive elements have been investigated as intermediaries of cell destiny, as media supplements or surface modifications (Spanoudes et al. 2014; Lin et al. 2018).

4.3.1 Strategies Involving Medium Supplementation with Growth Factors

Strategies based on growth factors and other small cell-signaling molecules have emerged in the context of tendon tissue regeneration. These have been identified as potent modulators of cell fate, acting on chemotaxis, proliferation, matrix synthesis, and/or differentiation (Baldwin et al. 2018; Dalby et al. 2018).

TGF-β3 is a modulator of tenocyte function and tendon development, since its signaling is known to induce the expression of Scx, a potent tenogenic marker (Leung et al. 2013). When used as a medium supplement in combination with aligned PCL/CHT meshes, TGF-β3 has a synergistic effect on the expression of tenogenic markers by BM-MSCs, suggesting that its individual action cannot trigger

Fig. 2 (continued) (blue) and actin filaments (red). Scale bar: 100 μm; and (**c**) normalized mean fluorescence intensity quantification of SCX and TNMD. Adapted from Tomás et al. (2019) with permission from the Royal Society of Chemistry the lineage commitment as structural cues do, but helps in tenocyte maturation (Leung et al. 2013). Therefore, this growth factor is often used as a component of tenogenic differentiation medium (Yang et al. 2016; Rothrauff et al. 2017; Wu et al. 2017). Accordingly, Yang et al. studied the combined effect of native tendon-derived ECM obtained from decellularized tissues with TGF-β3 on the tenogenic commitment of hASCs. They observed that on tissue culture plastic, the dual action of TGF-β3 and decellularized tendon ECM markedly enhanced the expression of Scx and Tnc when compared to their individual effect. Furthermore, this same trend was observed on PCL aligned scaffolds (Yang et al. 2017). Moreover, hASCs seeded in decellularized tendon ECM-supplemented collagen gels showed upregulation of tendon-related genes and downregulation of bone-related genes, in comparison to pure collagen scaffolds (Yang et al. 2013).

A different approach was considered by Engebretson et al. that used tendon lysates as a form of supplementation together with cyclic mechanical stimulation of human umbilical vein scaffolds with seeded MSCs. The tendon lysates used in a static system increased the expression of tendon-related genes more than in non-supplemented conditions, while when associated with cyclic mechanical stimulation, it resulted in tissue enhanced tensile strength and fibril alignment, as in the native tissue (Engebretson et al. 2017).

Several studies have demonstrated the positive impact of bone morphogenetic protein-12 (BMP-12) on the tenogenic differentiation of MSCs and tissue healing (Shen et al. 2013; Dale et al. 2018). Aligned gelatin methacryloyl/alginate yarns loaded with BM-MSCs were exposed to static mechanical stretching and BMP-12 supplementation both independently and simultaneously (Rinoldi et al. 2019). This combination enhanced stem cells tenogenic commitment although it inhibited collagen gene expression (Rinoldi et al. 2019).

4.3.2 Biofunctionalization of Scaffolds with Growth Factors

Growth factors may present a very quick inactivation and short half-life when used as medium supplements (Sahoo et al. 2010; Font Tellado et al. 2018). Therefore, systems in which these therapeutic agents are incorporated for in situ action and gradual delivery while ensuring their biological activity and stability hold numerous advantages.

Knitted silk scaffolds coated with (FGF-2)-releasing ultrafine PLGA promoted superior MSCs viability and proliferation although its role on tenogenesis was not fully assessed (Sahoo et al. 2010). Likewise, tenocytes cultured on braided PCL/collagen-FGF-2 scaffolds exhibited enhanced proliferation and expression of COLI, COLIII, and TNC, and scaffolds subjected to dynamic stimulation and further implanted in vivo enabled the deposition of aligned collagen after 12 weeks (Jayasree et al. 2019).

The conjugation of connective tissue growth factor (CTGF) in hierarchically electrospun scaffolds encouraged the proliferation of MSCs and the deposition of COLI and III in vitro and in vivo; however stem cell differentiation was not completely evaluated (Pauly et al. 2017). Even though CTGF has been positively correlated with tendon regeneration, further studies should be conducted to better understand its influence on tenogenic commitment (Lee et al. 2015; Shen et al. 2018).

A PDGF delivery system was developed using PLGA-monomethoxy-poly(ethylene glycol) (PLGA-m-PEG) nanoparticles. These were then associated with collagen to form aligned fibers through an electrochemical process. The alignment of the collagen fibrils resulted in increased Tnmd gene expression by hASCs, with the controlled release of PDGF boosting this effect and leading to higher Scx expression, in comparison with randomly oriented fibers (Cheng et al. 2014). Accordingly, hASCs seeded on porous membranes with reverse gradients of PDGF-BB and BMP-2 exhibited a tenogenesis-like behavior in the sections with higher PDGF-BB and lower BMP-2 content, correspondent to the immunohistochemical observation of higher TNMD and lower bone sialoprotein expression, while osteogenesis was promoted on the inverse situation (Min et al. 2014). More recently, PDGF was immobilized on the surface of electrospun aligned structures of poly(l-lactic acid) (PLLA) through a polydopamine (PDA) coating, forming a gradient of concentrations. hASCs adhered and spread similarly along the gradient, proliferating better than in the random scaffolds. Regarding the tenogenic differentiation, these cells had the highest protein expression of SCX and TNMD in the sections with highest amount of surface-bound PGDF, gradually decreasing with lower PDGF content (Madhurakkat Perikamana et al. 2018). Both strategies are particularly interesting for tendon to bone repair.

Silk fibroin scaffolds with isotropic (bone-like) and anisotropic (tendon-like) sections were functionalized with heparin to immobilize TGF-β2 and growth differentiation factor-5 (GDF-5). In the aligned segment, the presence of TGF-β2 allowed the highest deposition of COLI and slightly increased ColI, Mkx, and Tnc expressions, while in the isotropic region, the expression of the chondrogenic markers Sox9, ColII, and aggrecan was upregulated in the simultaneous presence of TGF-β2 and GDF-5, therefore unveiling new possibilities for enthesis-targeting strategies (Font Tellado et al. 2018).

Considering the reduced expression of histone deacetylases (HDACs) in native tendon stem/progenitor cells, Zhang et al. tried to enhance their tenogenesis on PLLA/poly(ethylene oxide) (PEO) aligned scaffolds using an HDAC inhibitor, trichostatin A. This small molecule incorporated on the aligned PLLA fibers led to increased mRNA expression of Scx and Mkx by TSPCs and superior SCX, COLI, COLV, and TNMD protein expression. Random fibers, on the other hand, proved to be the less efficient on TSPC tenoinduction (Zhang et al. 2018c).

4.3.3 Biofunctionalization of Scaffolds with ECM Components

Tendons are characterized by a scarce cell population within a very dense ECM composed of several macromolecules, each playing a distinct role on the mechanical function and biochemical signaling of the tissue (Yang et al. 2013). Hence, scaffolds' biofunctionality in terms of cell adhesion and stem cell tenogenic commitment might be enhanced with surface coatings of ECM components, to resemble the native musculoskeletal tissue (Lin et al. 2018; Jenkins and Little 2019).

Collagen, the main component of tendon's ECM, has been shown to improve the adhesion and growth of tenocytes, when used either as a bulk material (Theisen et al. 2010) or as a coating (Qin et al. 2005; Czaplewski et al. 2014). Tong et al. replicated native tendon's microenvironment, namely, surface topography and elasticity, using bio-imprinted substrates of polydimethylsiloxane (PDMS). When the surface chemistry was modified using a collagen type I coating, MSCs showed increased TNMD expression without using exogenous growth factors (Tong et al. 2012). Additional tendon-related markers needed to be evaluated to better understand the impact on cell differentiation.

Fibronectin, another well-defined cell adhesion enhancer, has been used along with collagen in a strategy targeting to the bone-to-tendon transition (Sharma and Snedeker 2010; Lin et al. 2018). Sharma et al. evaluated the influence of both biochemical and biomechanical gradients on bone marrow stromal cells differentiation along osteogenic and tenogenic lineages. For that, hydrogels of varying stiffness at a cell length-scale (within the kPa range) were functionalized with collagen and fibronectin. Osteoblast differentiation was observed on the stiffer (80 kPa) and fibronectin-coated substrates. Tenogenic markers, on the other hand, were upregulated on collagen-coated substrates with a stiffness of 40 kPa (Sharma and Snedeker 2010). In general, softer substrates are more prone to induce tenogenic differentiation, while rigid surfaces induce the expression of chondrogenesis and osteogenic genes (Engler et al. 2006; Zhang et al. 2018b).

Elastin is another tendon ECM fibrous protein, responsible for tissue reversible recoil (Miranda-Nieves and Chaikof 2017; Jenkins and Little 2019). In a recent

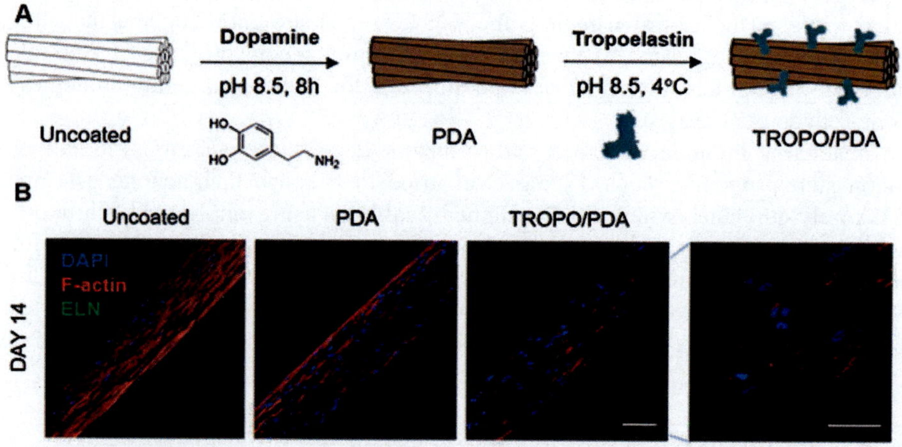

Fig. 3 (**a**) Schematic representation of the general strategy to immobilize TROPO on the surfaces of yarns mimicking tendon collagen fascicles via PDA linking; and (**b**) confocal images of immunolabeled samples against elastin (ELN – green) expressed by hASCs after 21 days of culture, with stained nuclei (blue) and cytoskeletons (red). Scale bar for low and high magnifications: 100 and 50 μm, respectively. Reprinted (adapted) with permission from Almeida et al. (2019). Copyright (2019) American Chemical Society

work, the surface of PCL/CHT/CNCs yarns (Laranjeira et al. 2017) was functionalized with tropoelastin (TROPO), the soluble precursor of elastin, through PDA linking (Fig. 3), thereby tuning its elasticity and composition to mimic the tendon native ECM (Almeida et al. 2019). Upon the decrease of surface stiffness with the incorporation of PDA and TROPO, hASCs acquired faster the tenocyte-like spindle-shape morphology and exhibited a sustained expression of *Scx* and *Tnmd*, up to 21 days of culture. Even though all conditions led to the production of a tendon-like ECM, immunocytochemistry data shows that only in the presence of TROPO, cells synthesized and deposited elastin creating a more mimetic matrix in comparison with the other conditions (Fig. 3) (Almeida et al. 2019).

5 Conclusions

The cellular interactions and mechanisms underlying tendon tissues remain to be fully elucidated, challenging strategies that would greatly benefit from the knowledge of tenogenic benchmarks as specific biomarkers for tendon cell maturation, appropriate stimuli conditions, and/or the factors leading to impaired tendon healing. As a mechanoresponsive tissue, mechanical cues are key features for tendon homeostasis and function. Several works have been emphasizing scaffolds' topographic cues and mechanical stimulation as powerful modulators of cell fate. All the proposed strategies provide insightful information about the interactions between cells and their substrates. However, these systems are often considered overly simplistic as do not consider the biological signaling present in the in vivo niche. Therefore, associating bioactive prompts to scaffolds is an increasingly trend in tendon tissue engineering.

In the years to come, cellular interactions with the different stimuli should be deeply investigated, and the *spatiotemporal* dynamics of *cellular differentiation* and tissue maturation assessed in the presence of artificial tenogenic stimuli to determine preferential rankings among conditioning factors. Another aspect to be considered in future strategies is the anatomical location of tendon tissues. Distinct locations within the body relate to dissimilar microenvironments, which likely influence different responses to stimuli and to repair mechanisms.

In summary, multiple aspects of tendon physiology in combinatorial strategies have been explored in the context of tendon TERM. The combination of both biomechanical and biochemical cues has been shown to produce biomimetic tendon scaffolds from nano- to macro-scales, although the optimal strategy to achieve tendon regeneration continues an unmet research and clinical need. The recreation of tendon niche complexity demands for the assessment of suitable cellular and environmental conditions and for the development of more sophisticated tendon substitutes with improved tenogenic functions for therapeutic application in tendon pathologies.

Acknowledgments Authors acknowledge the project "Accelerating tissue engineering and personalized medicine discoveries by the integration of key enabling nanotechnologies, marine-derived biomaterials and stem cells," supported by Norte Portugal Regional Operational

Programme (NORTE 2020), under the Portugal 2020 Partnership Agreement, through the European Regional Development Fund (ERDF).

Authors acknowledge the H2020 Achilles Twinning Project No. 810850, and also the European Research Council CoG MagTendon No. 772817, and the FCT Project MagTT PTDC/CTM-CTM/29930/2017 (POCI-01-0145-FEDER-29930).

References

Almeida H, Domingues RMA, Mithieux SM, Pires RA, Gonçalves AI, Gómez-Florit M, Reis RL, Weiss AS, Gomes ME (2019) Tropoelastin-coated tendon biomimetic scaffolds promote stem cell Tenogenic commitment and deposition of elastin-rich matrix. ACS Appl Mater Interfaces 11(22):19830–19840

Baldwin M, Snelling S, Dakin S, Carr A (2018) Augmenting endogenous repair of soft tissues with nanofibre scaffolds. J R Soc Interface 15(141)

Barber JG, Handorf AM, Allee TJ, Li WJ (2013) Braided nanofibrous scaffold for tendon and ligament tissue engineering. Tissue Eng Part A 19(11–12):1265–1274

Barsby T, Guest D (2013) Transforming growth factor Beta3 promotes tendon differentiation of equine embryo-derived stem cells. Tissue Eng A 19(19–20):2156–2165

Bashur CA, Shaffer RD, Dahlgren LA, Guelcher SA, Goldstein AS (2009) Effect of fiber diameter and alignment of electrospun polyurethane meshes on mesenchymal progenitor cells. Tissue Eng Part A 15(9):2435–2445

Bi Y, Ehirchiou D, Kilts TM, Inkson CA, Embree MC, Sonoyama W, Li L, Leet AI, Seo BM, Zhang L, Shi S, Young MF (2007) Identification of tendon stem/progenitor cells and the role of the extracellular matrix in their niche. Nat Med 13(10):1219–1227

Bracaglia LG, Smith BT, Watson E, Arumugasaamy N, Mikos AG, Fisher JP (2017) 3D printing for the design and fabrication of polymer-based gradient scaffolds. Acta Biomater 56:3–13

Brennan DA, Conte AA, Kanski G, Turkula S, Hu X, Kleiner MT, Beachley V (2018) Mechanical considerations for electrospun nanofibers in tendon and ligament repair. Adv Healthc Mater 7(12):e1701277

Brent AE, Schweitzer R, Tabin CJ (2003) A somitic compartment of tendon progenitors. Cell 113(2):235–248

Cadby JA, Buehler E, Godbout C, van Weeren PR, Snedeker JG (2014) Differences between the cell populations from the peritenon and the tendon core with regard to their potential implication in tendon repair. PLoS One 9(3):e92474

Chainani A, Hippensteel KJ, Kishan A, Garrigues NW, Ruch DS, Guilak F, Little D (2013) Multilayered electrospun scaffolds for tendon tissue engineering. Tissue Eng Part A 19(23–24):2594–2604

Cheng X, Tsao C, Sylvia VL, Cornet D, Nicolella DP, Bredbenner TL, Christy RJ (2014) Platelet-derived growth-factor-releasing aligned collagen-nanoparticle fibers promote the proliferation and tenogenic differentiation of adipose-derived stem cells. Acta Bio 10(3):1360–1369

Czaplewski SK, Tsai TL, Duenwald-Kuehl SE, Vanderby R Jr, Li WJ (2014) Tenogenic differentiation of human induced pluripotent stem cell-derived mesenchymal stem cells dictated by properties of braided submicron fibrous scaffolds. Biomaterials 35(25):6907–6917

Dalby MJ, García AJ, Salmeron-Sanchez M (2018) Receptor control in mesenchymal stem cell engineering. Nat Rev Mater 3:17091

Dale TP, Mazher S, Webb WR, Zhou J, Maffulli N, Chen GQ, El Haj AJ, Forsyth NR (2018) Tenogenic differentiation of human embryonic stem cells. Tissue Eng Part A 24(5–6):361–368

de Aro AA, Vidal BD, Pimentel ER (2012) Biochemical and anisotropical properties of tendons. Micron 43(2–3):205–214

de Mos M, Koevoet WJ, Jahr H, Verstegen MM, Heijboer MP, Kops N, van Leeuwen JP, Weinans H, Verhaar JA, van Osch GJ (2007) Intrinsic differentiation potential of adolescent human tendon tissue: an in-vitro cell differentiation study. BMC Musculoskelet Disord 8:16

Derakhshanfar S, Mbeleck R, Xu K, Zhang X, Zhong W, Xing M (2018) 3D bioprinting for biomedical devices and tissue engineering: a review of recent trends and advances. Bioact Mater 3(2):144–156
Docheva D, Muller SA, Majewski M, Evans CH (2015) Biologics for tendon repair. Adv Drug Deliv Rev 84:222–239
Domingues RM, Chiera S, Gershovich P, Motta A, Reis RL, Gomes ME (2016) Enhancing the biomechanical performance of anisotropic Nanofibrous scaffolds in tendon tissue engineering: reinforcement with cellulose nanocrystals. Adv Healthc Mater 5(11):1364–1375
Dyment NA, Liu C-F, Kazemi N, Aschbacher-Smith LE, Kenter K, Breidenbach AP, Shearn JT, Wylie C, Rowe DW, Butler DL (2013) The paratenon contributes to scleraxis-expressing cells during patellar tendon healing. PLoS One 8(3):e59944
Engebretson B, Mussett ZR, Sikavitsas VI (2017) Tenocytic extract and mechanical stimulation in a tissue-engineered tendon construct increases cellular proliferation and ECM deposition. Biotechnol J 12(3):1–26
Engler AJ, Sen S, Sweeney HL, Discher DE (2006) Matrix elasticity directs stem cell lineage specification. Cell 126(4):677–689
English A, Azeem A, Spanoudes K, Jones E, Tripathi B, Basu N, McNamara K, Tofail SAM, Rooney N, Riley G, O'Riordan A, Cross G, Hutmacher D, Biggs M, Pandit A, Zeugolis DI (2015) Substrate topography: a valuable in vitro tool, but a clinical red herring for in vivo tenogenesis. Acta Biomater 27:3–12
Erisken C, Zhang X, Moffat KL, Levine WN, Lu HH (2013a) Scaffold fiber diameter regulates human tendon fibroblast growth and differentiation. Tissue Eng Part A 19(3–4):519–528
Erisken C, Zhang X, Moffat KL, Levine WN, Lu HH (2013b) Scaffold Fiber diameter regulates human tendon fibroblast growth and differentiation. Tissue Eng A 19(3–4):519–528
Farhat YM, Al-Maliki AA, Chen T, Juneja SC, Schwarz EM, O'Keefe RJ, Awad HA (2012) Gene expression analysis of the pleiotropic effects of TGF-beta 1 in an in vitro model of flexor tendon healing. PLoS One 7(12):12
Font Tellado S, Chiera S, Bonani W, Poh PSP, Migliaresi C, Motta A, Balmayor ER, van Griensven M (2018) Heparin functionalization increases retention of TGF-β2 and GDF5 on biphasic silk fibroin scaffolds for tendon/ligament-to-bone tissue engineering. Acta Biomater 72:150–166
Freeman JW, Woods MD, Laurencin CT (2007) Tissue engineering of the anterior cruciate ligament using a braid – twist scaffold design. 40:2029–2036
Gilchrist CL, Ruch DS, Little D, Guilak F (2014) Micro-scale and meso-scale architectural cues cooperate and compete to direct aligned tissue formation. Biomaterials 35(38):10015–10024
Gonçalves AI, Rodrigues MT, Lee SJ, Atala A, Yoo JJ, Reis RL, Gomes ME (2013) Understanding the role of growth factors in modulating stem cell tenogenesis. PLoS One 8(12):e83734
Goncalves AI, Rodrigues MT, Carvalho PP, Banobre-Lopez M, Paz E, Freitas P, Gomes ME (2016) Exploring the potential of starch/Polycaprolactone aligned magnetic responsive scaffolds for tendon regeneration. Adv Healthc Mater 5(2):213–222
Gonçalves AI, Berdecka D, Rodrigues MT, Reis RL, Gomes ME (2018) Bioreactors for tendon tissue engineering: challenging mechanical demands towards tendon regeneration. Bioreactors for stem cell expansion and differentiation. CRC Press, Boca Raton, pp 269–300
Goncalves AI, Gershovich PM, Rodrigues MT, Reis RL, Gomes ME (2018a) Human adipose tissue-derived tenomodulin positive subpopulation of stem cells: a promising source of tendon progenitor cells. J Tissue Eng Regen Med 12(3):762–774
Goncalves AI, Rotherham M, Markides H, Rodrigues MT, Reis RL, Gomes ME, El Haj AJ (2018b) Triggering the activation of Activin a type II receptor in human adipose stem cells towards tenogenic commitment using mechanomagnetic stimulation. Nanomed Nanotechnol Biol Med 14(4):1149–1159
Goncalves AI, Berdecka D, Rodrigues MT, Eren AD, de Boer J, Reis RL, Gomes ME (2019) Evaluation of tenogenic differentiation potential of selected subpopulations of human adipose-derived stem cells. J Tissue Eng Regen Med 13(12):2204–2217
Hakimi O, Mouthuy PA, Zargar N, Lostis E, Morrey M, Carr A (2015) A layered electrospun and woven surgical scaffold to enhance endogenous tendon repair. Acta Biomater 26:124–135

Havis E, Bonnin M-A, Olivera-Martinez I, Nazaret N, Ruggiu M, Weibel J, Durand C, Guerquin M-J, Bonod-Bidaud C, Ruggiero F (2014a) Transcriptomic analysis of mouse limb tendon cells during development. Development 141(19):3683–3696

Havis E, Bonnin MA, Olivera-Martinez I, Nazaret N, Ruggiu M, Weibel J, Durand C, Guerquin MJ, Bonod-Bidaud C, Ruggiero F, Schweitzer R, Duprez D (2014b) Transcriptomic analysis of mouse limb tendon cells during development. Development 141(19):3683–3696

Havis E, Bonnin MA, de Lima JE, Charvet B, Milet C, Duprez D (2016) TGFbeta and FGF promote tendon progenitor fate and act downstream of muscle contraction to regulate tendon differentiation during chick limb development. Development 143(20):3839–3851

Izu Y, Ansorge HL, Zhang GY, Soslowsky LJ, Bonaldo P, Chu ML, Birk DE (2011) Dysfunctional tendon collagen fibrillogenesis in collagen VI null mice. Matrix Biol 30(1):53–61

Jayasree A, Kottappally Thankappan S, Ramachandran R, Sundaram MN, Chen C-H, Mony U, Chen J-P, Jayakumar R (2019) Bioengineered braided micro–Nano (multiscale) fibrous scaffolds for tendon reconstruction. ACS Biomater Sci Eng 5(3):1476–1486

Jenkins TL, Little D (2019) Synthetic scaffolds for musculoskeletal tissue engineering: cellular responses to fiber parameters. npj Regen Med 4(1):1–14

Katsura T, Tohyama H, Kondo E, Kitamura N, Yasuda K (2006) Effects of administration of transforming growth factor (TGF)-beta1 and anti-TGF-beta1 antibody on the mechanical properties of the stress-shielded patellar tendon. J Biomech 39(14):2566–2572

Killian ML, Cavinatto L, Galatz LM, Thomopoulos S (2012) The role of mechanobiology in tendon healing. J Shoulder Elb Surg 21(2):228–237

Kishore V, Bullock W, Sun X, Van Dyke WS, Akkus O (2012) Tenogenic differentiation of human MSCs induced by the topography of electrochemically aligned collagen threads. Biomaterials 33(7):2137–2144

Laranjeira M, Domingues RMA, Costa-Almeida R, Reis RL, Gomes ME (2017) 3D mimicry of native-tissue-Fiber architecture guides tendon-derived cells and adipose stem cells into artificial tendon constructs. Small 13(31)

Laternser S, Keller H, Leupin O, Rausch M, Graf-Hausner U, Rimann M (2018) A novel microplate 3D bioprinting platform for the engineering of muscle and tendon tissues. SLAS Technol 23(6):599–613

Lavagnino M, Wall ME, Little D, Banes AJ, Guilak F, Arnoczky SP (2015) Tendon Mechanobiology: current knowledge and future research opportunities. J Orthop Res 33(6):813–822

Lee CH, Yang G, Mao JJ, Lee CH, Lee FY, Tarafder S, Kao K, Jun Y, Yang G, Mao JJ (2015) Harnessing endogenous stem/progenitor cells for tendon regeneration find the latest version: harnessing endogenous stem/progenitor cells for tendon regeneration. 125(7):2690–2701

Lee NM, Erisken C, Iskratsch T, Sheetz M, Levine WN, Lu HH (2017) Polymer fiber-based models of connective tissue repair and healing. Biomaterials 112:303–312

Leung M, Jana S, Tsao C-T, Zhang M (2013) Tenogenic differentiation of human bone marrow stem cells via a combinatory effect of aligned chitosan-poly-caprolactone nanofibers and TGF-β3. J Mater Chem B 1:6516–6524

Lim PN, Wang Z, Wang D, Konishi T, San Thian E (2017) Development in additive printing for tissue-engineered bone and tendon regeneration. Curr Opin Biomed Eng 2:99–104

Lin TW, Cardenas L, Soslowsky LJ (2004) Biomechanics of tendon injury and repair. J Biomech 37(6):865–877

Lin J, Zhou W, Han S, Bunpetch V, Zhao K, Liu C, Yin Z, Ouyang H (2018) Cell-material interactions in tendon tissue engineering. Acta Biomater 70:1–11

Lipman K, Wang CC, Ting K, Soo C, Zheng Z (2018) Tendinopathy: injury, repair, and current exploration. Drug Des Devel Ther 12:591–603

Liu HH, Zhang C, Zhu SA, Lu P, Zhu T, Gong XN, Zhang ZW, Hu JJ, Yin Z, Heng BC, Chen X, Ouyang HW (2015) Mohawk promotes the Tenogenesis of mesenchymal stem cells through activation of the TGF beta signaling pathway. Stem Cells 33(2):443–455

Madhurakkat Perikamana SK, Lee J, Ahmad T, Kim EM, Byun H, Lee S, Shin H (2018) Harnessing biochemical and structural cues for tenogenic differentiation of adipose derived stem cells

(ADSCs) and development of an in vitro tissue interface mimicking tendon-bone insertion graft. Biomaterials 165:79–93

Maeda T, Sakabe T, Sunaga A, Sakai K, Rivera AL, Keene DR, Sasaki T, Stavnezer E, Iannotti J, Schweitzer R, Ilic D, Baskaran H, Sakai T (2011) Conversion of mechanical force into TGF-beta-mediated biochemical signals. Curr Biol 21(11):933–941

Magnusson SP, Hansen P, Kjaer M (2003) Tendon properties in relation to muscular activity and physical training. Scand J Med Sci Sports 13(4):211–223

Merceron TK, Burt M, Seol YJ, Kang HW, Lee SJ, Yoo JJ, Atala A (2015) A 3D bioprinted complex structure for engineering the muscle-tendon unit. Biofabrication 7(3):035003

Mienaltowski MJ, Adams SM, Birk DE (2014) Tendon proper- and peritenon-derived progenitor cells have unique tenogenic properties. Stem Cell Res Ther 5(4):86

Mihaila SM, Frias AM, Pirraco RP, Rada T, Reis RL, Gomes ME, Marques AP (2013) Human adipose tissue-derived SSEA-4 subpopulation multi-differentiation potential towards the endothelial and osteogenic lineages. Tissue Eng Part A 19(1–2):235–246

Mihaila SM, Gaharwar AK, Reis RL, Khademhosseini A, Marques AP, Gomes ME (2014) The osteogenic differentiation of SSEA-4 sub-population of human adipose derived stem cells using silicate nanoplatelets. Biomaterials 35(33):9087–9099

Mihaila SM, Resende MF, Reis RL, Gomes ME, Marques AP (2015) Interactive endothelial phenotype maintenance and osteogenic differentiation of adipose tissue stromal vascular fraction SSEA-4+-derived cells. J Tissue Eng Regen Med

Millar NL, Gilchrist DS, Akbar M, Reilly JH, Kerr SC, Campbell AL, Murrell GAC, Liew FY, Kurowska-Stolarska M, McInnes IB (2015) MicroRNA29a regulates IL-33-mediated tissue remodelling in tendon disease. Nat Commun 6:13

Min HK, Oh SH, Lee JM, Im GI, Lee JH (2014) Porous membrane with reverse gradients of PDGF-BB and BMP-2 for tendon-to-bone repair: in vitro evaluation on adipose-derived stem cell differentiation. Acta Biomater 10(3):1272–1279

Miranda-Nieves D, Chaikof EL (2017) Collagen and elastin biomaterials for the fabrication of engineered living tissues. ACS Biomater Sci Eng 3(5):694–711

Miranville A, Heeschen C, Sengenes C, Curat CA, Busse R, Bouloumie A (2004) Improvement of postnatal neovascularization by human adipose tissue-derived stem cells. Circulation 110(3):349–355

Moffat KL, Kwei ASP, Spalazzi JP, Doty SB, Levine WN, Lu HH (2009) Novel nanofiber-based scaffold for rotator cuff repair and augmentation. Tissue Eng A 15(1):115–126

Moroni L, Boland T, Burdick JA, De Maria C, Derby B, Forgacs G, Groll J, Li Q, Malda J, Mironov VA, Mota C, Nakamura M, Shu W, Takeuchi S, Woodfield TBF, Xu T, Yoo JJ, Vozzi G (2018a) Biofabrication: a guide to technology and terminology. Trends Biotechnol 36(4):384–402

Moroni L, Burdick JA, Highley C, Lee SJ, Morimoto Y, Takeuchi S, Yoo JJ (2018b) Biofabrication strategies for 3D in vitro models and regenerative medicine. Nat Rev Mater 3:5

Muller SA, Evans CH, Heisterbach PE, Majewski M (2018) The role of the Paratenon in Achilles tendon healing: a study in rats. Am J Sports Med 46(5):1214–1219

Orr SB, Chainani A, Hippensteel KJ, Kishan A, Gilchrist C, Garrigues NW, Ruch DS, Guilak F, Little D (2015) Aligned multilayered electrospun scaffolds for rotator cuff tendon tissue engineering. Acta Biomater 24:117–126

Park SH, Choi YJ, Moon SW, Lee BH, Shim JH, Cho DW, Wang JH (2018) Three-dimensional bioprinted scaffold sleeves with mesenchymal stem cells for enhancement of tendon-to-bone healing in anterior cruciate ligament reconstruction using soft-tissue tendon graft. Arthroscopy 34(1):166–179

Pauly HM, Sathy BN, Olvera D, McCarthy HO, Kelly DJ, Popat KC, Dunne NJ, Haut Donahue TL (2017) Hierarchically structured electrospun scaffolds with chemically conjugated growth factor for ligament tissue engineering. Tissue Eng Part A 23(15–16):823–836

Perucca Orfei C, Vigano M, Pearson JR, Colombini A, De Luca P, Ragni E, Santos-Ruiz L, de Girolamo L (2019) In vitro induction of tendon-specific markers in tendon cells, adipose- and bone marrow-derived stem cells is dependent on TGFbeta3, BMP-12 and ascorbic acid stimulation. Int J Mol Sci 20(1)

Qin TW, Yang ZM, Wu ZZ, Xie HQ, Qin J, Cai SX (2005) Adhesion strength of human tenocytes to extracellular matrix component-modified poly(DL-lactide-co-glycolide) substrates. Biomaterials 26(33):6635–6642
Rada T, Gomes ME, Reis RL (2011a) A novel method for the isolation of subpopulations of rat adipose stem cells with different proliferation and osteogenic differentiation potentials. J Tissue Eng Regen Med 5(8):655–664
Rada T, Reis RL, Gomes ME (2011b) Distinct stem cells subpopulations isolated from human adipose tissue exhibit different chondrogenic and osteogenic differentiation potential. Stem Cell Rev 7(1):64–76
Rada T, Santos TC, Marques AP, Correlo VM, Frias AM, Castro AG, Neves NM, Gomes ME, Reis RL (2012) Osteogenic differentiation of two distinct subpopulations of human adipose-derived stem cells: an in vitro and in vivo study. J Tissue Eng Regen Med 6(1):1–11
Rees JD, Wilson AM, Wolman RL (2006) Current concepts in the management of tendon disorders. Rheumatology 45(5):508–521
Riehl BD, Park JH, Kwon IK, Lim JY (2012) Mechanical stretching for tissue engineering: two-dimensional and three-dimensional constructs. Tissue Eng Part B Rev 18(4):288–300
Riley G (2004) The pathogenesis of tendinopathy. A molecular perspective. Rheumatology 43(2):131–142
Rinoldi C, Costantini M, Kijeńska-Gawrońska E, Testa S, Fornetti E, Heljak M, Ćwiklińska M, Buda R, Baldi J, Cannata S, Guzowski J, Gargioli C, Khademhosseini A, Swieszkowski W (2019) Tendon tissue engineering: effects of mechanical and biochemical stimulation on stem cell alignment on cell-laden hydrogel yarns. Adv Healthc Mater 8(7):1–10
Rodrigues MT, Reis RL, Gomes ME (2013) Engineering tendon and ligament tissues: present developments towards successful clinical products. J Tissue Eng Regen Med 7(9):673–686
Rothrauff BB, Lauro BB, Yang G, Debski RE, Musahl V, Tuan RS (2017) Braided and stacked electrospun Nanofibrous scaffolds for tendon and ligament tissue engineering. Tissue Eng Part A 23(9–10):378–389
Sahoo S, Toh SL, Goh JCH (2010) A bFGF-releasing silk/PLGA-based biohybrid scaffold for ligament/tendon tissue engineering using mesenchymal progenitor cells. Biomaterials 31(11):2990–2998
Salingcarnboriboon R, Yoshitake H, Tsuji K, Obinata M, Amagasa T, Nifuji A, Noda M (2003) Establishment of tendon-derived cell lines exhibiting pluripotent mesenchymal stem cell-like property. Exp Cell Res 287(2):289–300
Santos LJ, Reis RL, Gomes ME (2015) Harnessing magnetic-mechano actuation in regenerative medicine and tissue engineering. Trends Biotechnol 33(8):471–479
Santos ML, Rodrigues MT, Domingues RM, Reis RL, Gomes ME (2017) Biomaterials as tendon and ligament substitutes: current developments. In: Regenerative strategies for the treatment of knee joint disabilities, vol 21. Springer, pp 349–371
Schmierer B, Hill CS (2007) TGF beta-SMAD signal transduction: molecular specificity and functional flexibility. Nat Rev Mol Cell Biol 8(12):970–982
Schneider M, Angele P, Jarvinen TAH, Docheva D (2017) Rescue plan for Achilles: therapeutics steering the fate and functions of stem cells in tendon wound healing. Adv Drug Deliv Rev
Schoenenberger AD, Foolen J, Moor P, Silvan U, Snedeker JG (2018) Substrate fiber alignment mediates tendon cell response to inflammatory signaling. Acta Biomater 71:306–317
Screen HRC, Berk DE, Kadler KE, Ramirez F, Young MF (2015) Tendon Functional Extracellular Matrix. J Orthop Res 33(6):793–799
Sengenes C, Miranville A, Maumus M, de Barros S, Busse R, Bouloumie A (2007) Chemotaxis and differentiation of human adipose tissue CD34+/CD31- progenitor cells: role of stromal derived factor-1 released by adipose tissue capillary endothelial cells. Stem Cells 25(9):2269–2276
Sharma P, Maffulli N (2005) Current concepts review tendon injury and tendinopathy: healing and repair. J Bone Joint Surg Am 87A(1):187–202
Sharma P, Maffulli N (2006) Biology of tendon injury: healing, modeling and remodeling. J Musculoskelet Nueronal Interact 6(2):181

Sharma RI, Snedeker JG (2010) Biochemical and biomechanical gradients for directed bone marrow stromal cell differentiation toward tendon and bone. Biomaterials 31(30):7695–7704

Shen H, Gelberman RH, Silva MJ, Sakiyama-Elbert SE, Thomopoulos S (2013) BMP12 induces tenogenic differentiation of adipose-derived stromal cells. PLoS One 8(10):1–14

Shen H, Jayaram R, Yoneda S, Linderman SW, Sakiyama-Elbert SE, Xia Y, Gelberman RH, Thomopoulos S (2018) The effect of adipose-derived stem cell sheets and CTGF on early flexor tendon healing in a canine model. Sci Rep 8(1):11078

Soslowsky LJ, Thomopoulos S, Tun S, Flanagan CL, Keefer CC, Mastaw J, Carpenter JE (2000) Neer award 1999 – overuse activity injures the supraspinatus tendon in an animal model: a histologic and biomechanical study. J Shoulder Elb Surg 9(2):79–84

Spanoudes K, Gaspar D, Pandit A, Zeugolis DI (2014) The biophysical, biochemical, and biological toolbox for tenogenic phenotype maintenance in vitro. Trends Biotechnol 32 (9):474–482

Stanco D, Caprara C, Ciardelli G, Mariotta L, Gola M, Minonzio G, Soldati G (2019) Tenogenic differentiation protocol in xenogenic-free media enhances tendon-related marker expression in ASCs. PLoS One 14(2)

Subramony SD, Dargis BR, Castillo M, Azeloglu EU, Tracey MS, Su A, Lu HH (2013) The guidance of stem cell differentiation by substrate alignment and mechanical stimulation. Biomaterials 34(8):1942–1953

Theisen C, Fuchs-Winkelmann S, Knappstein K, Efe T, Schmitt J, Paletta JRJ, Schofer MD (2010) Influence of nanofibers on growth and gene expression of human tendon derived fibroblast. Biomed Eng Online 9:1–12

Tomás AR, Gonçalves AI, Paz E, Freitas PP, Domingues RMA, Gomes ME (2019) Magneto-mechanical actuation of magnetic responsive fibrous scaffolds boosts Tenogenesis of human adipose stem cells. Nanoscale

Tong WY, Shen W, Yeung CW, Zhao Y, Cheng SH, Chu PK, Chan D, Chan GC, Cheung KM, Yeung KW, Lam YW (2012) Functional replication of the tendon tissue microenvironment by a bioimprinted substrate and the support of tenocytic differentiation of mesenchymal stem cells. Biomaterials 33(31):7686–7698

Vining KH, Mooney DJ (2017) Mechanical forces direct stem cell behaviour in development and regeneration. Nat Rev Mol Cell Biol 18(12):728–742

Wakefield LM, Hill CS (2013) Beyond TGF beta: roles of other TGF beta superfamily members in cancer. Nat Rev Cancer 13(5):328–341

Wang JH-C (2006) Mechanobiology of tendon. J Biomech 39(9):1563–1582

Wang JHC, Guo QP, Li B (2012) Tendon biomechanics and Mechanobiology-A Minireview of basic concepts and recent advancements. J Hand Ther 25(2):133–140

Wang T, Gardiner BS, Lin Z, Rubenson J, Kirk TB, Wang A, Xu J, Smith DW, Lloyd DG, Zheng MH (2013) Bioreactor Design for Tendon/ligament engineering. Tissue Eng Part B Rev 19(2):133–146

Wenstrup RJ, Florer JB, Brunskill EW, Bell SM, Chervoneva I, Birk DE (2004) Type V collagen controls the initiation of collagen fibril assembly. J Biol Chem 279(51):53331–53337

Willett TL, Labow RS, Avery NC, Lee JM (2007) Increased proteolysis of collagen in an in vitro tensile overload tendon model. Ann Biomed Eng 35(11):1961–1972

Woo SLY, Hildebrand K, Watanabe N, Fenwick JA, Papageorgiou CD, Wang JHC (1999) Tissue engineering of ligament and tendon healing. Clin Orthop Relat Res 367:S312–S323

Wrighton KH, Lin X, Feng XH (2009) Phospho-control of TGF-beta superfamily signaling. Cell Res 19(1):8–20

Wu S, Wang Y, Streubel PN, Duan B (2017) Living nanofiber yarn-based woven biotextiles for tendon tissue engineering using cell tri-culture and mechanical stimulation. Acta Biomater 62:102–115

Xie J, Li X, Lipner J, Manning CN, Schwartz AG, Thomopoulos S, Xia Y (2010) "Aligned-to-random" nanofiber scaffolds for mimicking the structure of the tendon-to-bone insertion site. Nanoscale 2(6):923–923

Yang G, Rothrauff BB, Lin H, Gottardi R, Alexander PG, Tuan RS (2013) Enhancement of tenogenic differentiation of human adipose stem cells by tendon-derived extracellular matrix. Biomaterials 34:9295–9306

Yang G, Lin H, Rothrauff BB, Yu S, Tuan RS (2016) Multilayered polycaprolactone/gelatin fiber-hydrogel composite for tendon tissue engineering. Acta Biomater 35:68–76

Yang G, Rothrauff BB, Lin H, Yu S, Tuan RS (2017) Tendon-derived extracellular matrix enhances transforming growth factor-β3-induced Tenogenic differentiation of human adipose-derived stem cells. Tissue Eng A 23(3–4):166–176

Yin Z, Guo J, Wu TY, Chen X, Xu LL, Lin SE, Sun YX, Chan KM, Ouyang H, Li G (2016) Stepwise differentiation of mesenchymal stem cells augments tendon-like tissue formation and defect repair in vivo. Stem Cells Transl Med 5(8):1106–1116

Younesi M, Islam A, Kishore V, Anderson JM, Akkus O (2014) Tenogenic induction of human MSCs by Anisotropically aligned collagen biotextiles. Adv Funct Mater 24(36):5762–5770

Yu H, Lim KP, Xiong S, Tan LP, Shim W (2013) Functional morphometric analysis in cellular behaviors: shape and size matter. Adv Healthc Mater 2(9):1188–1197

Zhang J, Wang JH-C (2010) Characterization of differential properties of rabbit tendon stem cells and tenocytes. BMC Musculoskelet Disord 11:10

Zhang C, Yuan H, Liu H, Chen X, Lu P, Zhu T, Yang L, Yin Z, Heng BC, Zhang Y, Ouyang H (2015) Well-aligned chitosan-based ultrafine fibers committed Teno-lineage differentiation of human induced pluripotent stem cells for Achilles tendon regeneration. Biomaterials 53:716–730

Zhang YJ, Chen X, Li G, Chan KM, Heng BC, Yin Z, Ouyang HW (2018a) Concise review: stem cell fate guided by bioactive molecules for tendon regeneration. Stem Cells Transl Med 7(5):404–414

Zhang H, Liu M-F, Liu R-C, Shen W-L, Yin Z, Chen X (2018b) Physical microenvironment-based inducible scaffold for stem cell differentiation and tendon regeneration. Tissue Eng Part B Rev 24(6):443–453

Zhang C, Wang X, Zhang E, Yang L, Yuan H, Tu W, Zhang H, Yin Z, Shen W, Chen X, Zhang Y, Ouyang H (2018c) An epigenetic bioactive composite scaffold with well-aligned nanofibers for functional tendon tissue engineering. Acta Biomater 66:141–156

Zhao X, Jiang SC, Liu S, Chen S, Lin ZY, Pan GQ, He F, Li FF, Fan CY, Cui WG (2015) Optimization of intrinsic and extrinsic tendon healing through controllable water-soluble mitomycin-C release from electrospun fibers by mediating adhesion-related gene expression. Biomaterials 61:61–74

Zheng Z, Ran J, Chen W, Hu Y, Zhu T, Chen X, Yin Z, Heng BC, Feng G, Le H, Tang C, Huang J, Chen Y, Zhou Y, Dominique P, Shen W, Ouyang HW (2017) Alignment of collagen fiber in knitted silk scaffold for functional massive rotator cuff repair. Acta Biomater 51:317–329

Meniscus Regeneration Strategies

Johannes Zellner and Peter Angele

Contents

J. Zellner
Department of Trauma Surgery, Caritas Hospital St. Josef, Regensburg, Germany

Department of Trauma Surgery, University Hospital of Regensburg, Regensburg, Germany
e-mail: jzellner@caritasstjosef.de

P. Angele (✉)
Department of Trauma Surgery, University Hospital of Regensburg, Regensburg, Germany

Sporthopaedicum Regensburg, Regensburg, Germany
e-mail: peter.angele@ukr.de

D. Eberli et al. (eds.), *Organ Tissue Engineering*, Reference Series in Biomedical Engineering, https://doi.org/10.1007/978-3-030-44211-8_16

Abstract

Due to the important role of the meniscus concerning the integrity of the knee joint, meniscus tissue preserving techniques for the therapy of meniscal injuries seem to be reasonable. Due to a diminished self-healing capacity of meniscus particularly in the avascular zone, meniscus restoration is not always possible. Meniscus deficiency leads to functional restrictions and osteoarthritic changes in mid- and long term. Regenerative treatment augmentation for meniscal therapy and Tissue Engineering are promising approaches to preserve and restore as much meniscus tissue as possible. This chapter gives insights on current preclinical and clinical treatment strategies of meniscus and emphasizes on future perspectives like biological treatment augmentation for meniscal regeneration and Tissue Engineering of meniscus.

1 Introduction

Meniscal lesions represent one of the most common intra-articular knee injuries, and are therefore the most frequent cause of surgical procedures in orthopedic surgery. The mean annual incidence of meniscal lesions has been reported to be 66/100,000 inhabitants, 61 of which result in partial or subtotal meniscectomy (Makris et al. 2011). The changes in "pivoting" sports activities in the past few decades have resulted in increased injury rates of the meniscus (1.5 million injuries in Germany/year) (Stein et al. 2010). Especially in combination with anterior cruciate ligament injuries a high incidence of acute meniscal lesions (40–80%) can be detected.

Additionally, an increasing number of degenerative meniscus lesions can be detected over the last decades. Although it is still under debate whether these meniscus lesions are better treated conservatively or operatively, there is no discussion about the fact that such degenerative meniscal changes promote and accelerate the development of osteoarthritis of the knee.

Meniscus integrity is the key for joint health of the knee. Untreated meniscus tears cause intermittent pain, joint swelling, recurrent mechanical symptoms (clicking, catching, giving way), and, therefore, significant reduction in quality of life in predominately young and active patients (McDermott 2011).

In the long-term, meniscus tears can result in the onset of joint degeneration and, finally, knee osteoarthritis with all its consequences including pain, immobility, and knee arthroplasty (Lohmander et al. 2007; Stein et al. 2010; Borchers et al. 2011; Jeong et al. 2012; Badlani et al. 2013). In a recent published case-control study (Level of evidence 3), specific meniscus tear morphologies (meniscus extrusion, complex tears, tears with large radial involvement) have shown to be significantly more common in patients with progressive development of osteoarthritic changes in a 2-year follow-up indicating that these meniscus tears represent a negative prognostic risk factor for later development of osteoarthritis (Badlani et al. 2013).

Removal of meniscus tears lead to short term relief of clinical symptoms, but also to knee osteoarthritis in long-term (Salata et al. 2010; Papalia et al. 2011; Paxton et al. 2011; Jeong et al. 2012). Especially the amount meniscus removed, lateral meniscectomy, concomitant injuries like ACL ruptures, axis deviation, high BMI, and longer duration of clinical symptoms preoperatively have been identified as negative prognostic risk factors for the onset of osteoarthritis in systematic reviews (Papalia et al. 2011; Jeong et al. 2012). Elevated expression levels of arthritis-related markers in meniscus tears in patients under 40 years old, compared to patients over 40 years, and in patients with meniscus and anterior cruciate ligament tears, compared to patients with isolated meniscus tears, indicate an increased catabolic response suggesting a higher risk for progression of osteoarthritis following partial meniscectomy (Brophy and Matava 2012).

Knowing the risk for the onset of osteoarthritis after meniscectomy, the majority of meniscus tears are still treated with partial meniscectomy as shown in a huge cohort of more than 1000 young patients undergoing anterior cruciate ligament reconstruction (Fetzer et al. 2009).

Therefore, the main goal of every meniscus treatment should be the maintenance of as much meniscus tissue as possible (Fetzer et al. 2009; Starke et al. 2009; Stein et al. 2010; Abrams et al. 2013). This includes repair of meniscus tears and regeneration of meniscus defects after meniscectomy with regenerative treatment approaches like biological augmentation or Tissue Engineering.

In recent years, there has been a growing interest in using mesenchymal stem cells or other cell types to enhance meniscal healing or to regenerate damaged or lost meniscus tissue. Especially pluripotent cells are able to fulfill a dual role for musculoskeletal repair, because they have the potential to differentiate into the repair cells themselves and to produce special trophic factors like growth factors for its repair (Caplan and Dennis 2006). This chapter focuses on the present and future treatment options especially for meniscus repair and meniscus regeneration given by Tissue Engineering.

2 Clinical Aspects

The meniscus plays a decisive role for the integrity of the knee joint. This includes shock absorption and transmission, but also joint stabilization, proprioception, lubrication, and nutrition of the articular cartilage (Makris et al. 2011). Biomechanical studies have shown that a loss of meniscus integrity leads to remarkable changes in kinematics and load distribution in the knee joint. The pressure on the surrounding native articular cartilage subsequently increases. Even a resection of only 15–34% of meniscus tissue enhances the load on the surrounding hyaline cartilage up to 350% (Radin et al. 1984).

In accordance to that, osteoarthritis of the knee, as a resulting effect of meniscectomy, has already been described a long time ago (Fairbank 1948). According to the current literature partial meniscectomy is also well known to predispose the knee for the development and an early onset of osteoarthritis (McDermott and Amis 2006; Petty and Lubowitz 2011). Especially the following criteria are defined as risk factors for the development of degenerative changes in context to meniscus injuries (according to Mordecai et al. 2014):

- Partial meniscectomy of the lateral meniscus
- Resection of larger portions of meniscus tissue
- Radial tears reducing or cancelling the meniscus ring tension (functional meniscectomy)
- Preexisting cartilage lesions
- Persisting ligamentous joint instability
- Axis deviation (varus-medial, valgus-lateral)
- Obesity
- Age >40 years
- Low activity level

According to the increasing knowledge concerning the biology and function of the meniscus there is a consensus to preserve as much meniscus tissue as possible in the treatment of meniscus injuries. Thus, different techniques for the therapy of meniscus tears have been developed over time. Today, the meniscus suture matters as gold standard for the regenerative treatment of meniscus lesions particularly in vascularized portion. Whereas initially this procedure was performed as an open procedure, up to now it is almost exclusively performed arthroscopically. Different techniques for meniscus suturing have to be distinguished: all-inside, outside-in, inside-out.

The vascularization and nutritional situation of the injured meniscus area as well as the type of meniscus tear are decisive of the success of a meniscus reconstruction.

While the inner 2/3 of the meniscus ("white-white") is nourished by diffusion from the synovial fluid, the periphery in the so-called red-red zone has a vascular supply. Between the white-white zone and the vascularized portion a red-white transition zone is located. Especially the outer third and, to a lesser extent, the red-white transition zone show a regenerative potential with good conditions for a successful meniscus suturing (Arnoczky 1999).

However, meniscus still remains a challenging structure for repair and restoration. The question that arises is whether the diminished self-healing capacity mainly in the inner thirds of the meniscus can be overcome by innovative treatment options or modern treatment strategies like for example cell-based treatment augmentation. Additionally, over the last decades different Tissue Engineering approaches came in the focus of research to enhance the healing potential in order to save or to rebuild as much meniscus tissue as possible to improve long-term outcome after meniscus treatment and to prevent the onset of osteoarthritis. This chapter summarizes different aspects of Tissue Engineering of Meniscus and gives an overview on the variety of treatment strategies and their specific aims.

2.1 Endogenous Repair Cells in Case of a Meniscus Injury

Currently, a number of potential repair cells for meniscus regeneration are available. Repair cells of meniscus injury can either be located in the meniscus tissue itself or entering the meniscus predominately via circulation. In this section, we will focus on the availability of repair cells and repair potential of the meniscus in clinical reality in case of a meniscal injury.

Endogenous repair of meniscus injury seems to be dependent of the different vascularization of the outer and the inner zone of the meniscus (Scotti et al. 2013). Repair in the vascularized outer zone can be achieved, but fail to encourage healing in the avascular inner zone of the meniscus. However, in several studies also regeneration could be seen in the inner zone of the meniscus indicating regenerative potential independently from the vascularization (Pabbruwe et al. 2010). Hennerbichler et al. have shown in an experimental setup that punch defects, which were directly filled with the removed punches, showed no significant difference in healing potential between the vascularized and avascularized meniscus zone (Hennerbichler et al. 2007). Croutze and coworkers could demonstrate equivalent differentiation potential toward chondrogenic phenotype and extracellular matrix production of isolated human meniscus cells from the inner and the outer zone (Croutze et al. 2013). Different studies stated that milieu factors like oxygen tension and growth factors may play a key role in modulating redifferentiation of meniscal fibrochondrocytes at least in vitro.

From a clinical standpoint, meniscus cells from meniscectomized tissue would be the ideal cell source for repair. However, is this approach applicable? Nakata et al. have shown feasibility of meniscus regeneration by expansion of human meniscus cells from meniscectomized tissue (Nakata et al. 2001). With the same cell source Baker et al. could achieve repair tissue with equivalent mechanical properties than native meniscus tissue (Baker et al. 2009). However, the low proliferation rate and matrix production limit the use of this cell type. In addition, potential candidates for meniscus transplantation have already undergone partial meniscectomy with the necessity to find alternative cell sources.

Stem cells are characterized by self-renewal capacity and multilineage differentiation potential to a variety of cell types of mesenchymal tissue like bone, cartilage, or fat (Caplan and Dennis 2006). Recently, the stem cell perspective has changed by identification of pericytes around almost every blood vessel in the body, which contain stem cell characteristics (Crisan et al. 2011). The existing traditional view, which focuses on the multipotent differentiation capacity of these cells, has been expanded to include their equally interesting role as cellular modulators that bring them into a broader therapeutic scenario.

With this background, it is not surprising that Osawa et al. have identified in the vascularized region of the meniscus more blood-vessel derived stem cells (CD34- and CD146-positive cells) than in the avascular region. These meniscal-derived stem cells were multipotent and contributed to meniscal regeneration, which they proof via transplantation in knee joints of athymic rats (Osawa et al. 2013).

Does also the avascular inner zone of the meniscus contain stem cells? Stem cells have been identified in the surface zone of other avascular tissue, mainly in the articular cartilage (Williams et al. 2010), but little is known about meniscus stem cells. Mauck et al. describes regional multilineage differentiation potential of meniscus cells in both zones available, however showed differences in pluripotency between the zones (Mauck et al. 2007). Especially, pluripotent cells from the avascular tissue seem to lack osteogenic differentiation potential, which could be of clinical interest for meniscus regenerative approaches.

Besides existence of mesenchymal stem cells in the meniscus, the synovium and the synovial fluid contain stem cells for meniscus regeneration (Matsukura et al. 2014). Matsukura and coworkers found elevated levels of mesenchymal stem cells in the synovium fluid after meniscus injury compared to normal knee joints suggesting that mesenchymal stem cells in the fluid may play a role in regeneration of meniscus. Horie and coworkers showed promotion of meniscal regeneration in rat massive meniscal defects by intra-articular injection of synovial stem cells. The injected cells adhere to the lesion site, directly differentiate in meniscus cells, and promote meniscal regeneration without mobilization to distant organs (Horie et al. 2009).

In summary, local or systemic stem cells seem to play a fundamental and essential role in the regeneration of meniscus injury, either as direct repair cells or as a source for secretion of bioactive modulators or immunomodulation.

2.2 Meniscus Reconstruction Improves the Knee Function in Long-Term

The first description of a meniscus suture technique was published by Annandale in 1885 (Di Matteo et al. 2013). Since then, the treatment options for the reconstructive therapy of meniscus lesions have been significantly advanced, especially by the development of arthroscopic techniques. Regarding studies and meta-analysis describing the long-term outcome after meniscal reconstructive therapy the technical development of the treatment options (open versus arthroscopic procedures) have to be considered.

Tengrootenhuysen et al. retrospectively compared the clinical outcome after successful and failed meniscus suture in 119 patients after a mean follow-up of more than 5 years (Tengrootenhuysen et al. 2011). The successful reconstruction of the meniscus was associated with a significant improvement of the knee function according to the IKDC and Lysholm score.

Xu et al. evaluated the long-term outcome of meniscus reconstruction in comparison to the long-term outcome after partial meniscectomy (Xu and Zhao 2015). According to the inclusion criteria, 367 patients of 7 studies were included in this meta-analysis. After a mean follow-up of 84 months, they detected a significant improvement of the IKDC and Lysholm score in the group of patients receiving a meniscus suture in contrast to the group of patients who have had partial meniscectomy. Regarding the Tegner score, the results of both groups were reduced in comparison to the activity level before the meniscal injury. Nevertheless, further differentiation of both groups showed a reduced loss of function over the time in the group of patients who had received a meniscus suture. So, they summarized that the preservation of meniscus tissue is associated with an improved clinical and functional outcome over a mid- and long-term period.

In addition to that, Stein et al. showed that 96.2% of the patients who had a meniscus reconstruction were able to restore their pre-injury activity level within a mean follow-up of almost 9 years in comparison to 50% of the patients who had a partial meniscectomy. These results go along with the observations of Majewski et al. who analyzed isolated and longitudinal meniscus tears in stable knee joints after a mean follow-up of 10 years (Stein et al. 2010).

Overall, the current literature shows a significant positive effect of a meniscus tissue preserving therapy on the knee joint function in long-term. However, the question remains to what extend meniscus preserving techniques, such as meniscus suture, are able to positively influence the development of degenerative changes within the knee joint.

2.3 Prevention of Osteoarthritis by Meniscus Suturing in Long-Term

The integrity of the meniscus is of impact for the prevention of osteoarthritis, such as shown by (partial) meniscectomy. It usually goes along with a loss of symptoms and functional improvement in short-term (Mezhov et al. 2014). However, the long-term outcome after (partial) meniscectomy shows a trend to degenerative effects. Englund and Lohmander described an association between the degenerative effect and the amount of lost meniscus tissue (Englund and Lohmander 2004). Even if the partial meniscectomy does not show that extended destructive effect, osteoarthritic changes are also documented after a follow up of 16 years after partial meniscectomy (Englund et al. 2003). So, Papalia et al. defined the amount of resected meniscus tissue as a predictive factor for the development of osteoarthritis (Papalia et al. 2011).

In a systematic review concerning the outcome after arthroscopically performed partial meniscectomy with a minimum follow-up of 8 years and a mean age of 36 years. Satisfying results concerning the functional outcome (i.e., Lysholm scoring, Tegner scoring,or IKDC scoring) were found by Petty and Lubowitz (2011). Nevertheless, all included studies evaluating radiologically based signs of osteoarthritis in the index and contralateral site detected significantly enhanced signs of osteoarthritis in the partially meniscectomized knee. Comparing the medial and lateral meniscus, especially partial meniscectomy of the lateral meniscus shows a negative influence on the development of degenerative changes (Salata et al. 2010). In this context, Lee et al. examining 49 patients after subtotal resection of the lateral meniscus and having lateral meniscus replacement throughout after a mean of 4.5 years. The authors observed a significant development of signs of osteoarthritis according to the Kellgren-Lawrence classification and a progressive loss of the joint line. Though, the process of progressive joint degeneration could have positively been influenced by meniscus replacement (Lee et al. 2016).

In contrast to the (partial) meniscectomy meniscus preserving techniques such as meniscus suturing show a cartilage protective effect in long-term. Noyes et al. evaluated the meniscal status of 33 patients having meniscus suture after a mean follow-up of 16.8 years by MRI scan. No degenerative changes in the operated compartment or differences concerning the status of degeneration in comparison to the healthy, contralateral site were found in patients after having successful meniscus reconstruction (Noyes et al. 2011). Further studies described a progress of radiological signs of osteoarthritis after meniscus suture in long-term; however, these results showed just a mild progress of degenerative changes (Nepple et al. 2012). Johnson et al. compared the injured and contralateral knee joint 10 years after meniscus suture on a radiological base (Johnson et al. 1999). Only 8% of these patients developed osteoarthritic signs on the operated site while degenerative changes were also found in even 3% of the contralateral, intact knee joints. Furthermore, Tengrootenhuysen et al. analyzed differences between patients after a successful meniscus suture and patients in whom the meniscus suture failed (Tengrootenhuysen et al. 2011). In 14% of the patients having a successful reconstruction of the meniscus signs of osteoarthritis were documented in X-ray. In contrast to that, in more than 80% of the patients with a failed meniscus preserving therapy signs of an osteoarthritis were seen.

Regarding the development of osteoarthritis of the knee, techniques preserving a functional intact meniscus tissue show advantages in comparison to partial meniscectomy. Stein et al. showed no progress of radiological signs of osteoarthritis in 81% of the evaluated patients after almost 9 years after meniscus suturing, whereas a reduction of a degenerative progress was only seen in 40% of the patients after partial meniscectomy (Stein et al. 2010). Similar results were found by Paxton et al. (2011). While 78% of the patients had no progress of the osteoarthritic status according to the X-ray after having reconstruction of the meniscus, just in 64% of the patients, who had partial meniscectomy, no further development of osteoarthritis was detected. Especially in younger patients further studies showed also clear advantages of the meniscus preserving techniques in contrast to the partial meniscectomy regarding osteoarthrosis preventing qualities (Sommerlath 1991).

Despite promising results for successful meniscus reconstruction regarding functional outcome and prevention of osteoarthritis, there is still the need to improve the healing rate after meniscus suture.

2.4 Stimulation of the Regenerative Potential of the Meniscus Tissue

The aim of any meniscus therapy should be to preserve as much meniscus tissue as possible, such as performed by reconstructive techniques like meniscus suture. The regenerative potential of the meniscus can be further supported by various measures.

Refreshing of the margins of the meniscus tears is an obligate procedure before each meniscus suture (Fig. 1). Different further techniques, such as the trephination of the meniscus margins by awls or K-wires as well as roughening of the defect sites by special meniscus tissue rasps, are available. In a comparative study, Zhang et al. analyzed the effect of such a refreshment of the meniscus defect site by trephination before meniscus suturing (Zhang and Arnold 1996). They found a significantly lower failure rate of the meniscus suture when, in addition to the suture, a trephination was performed before.

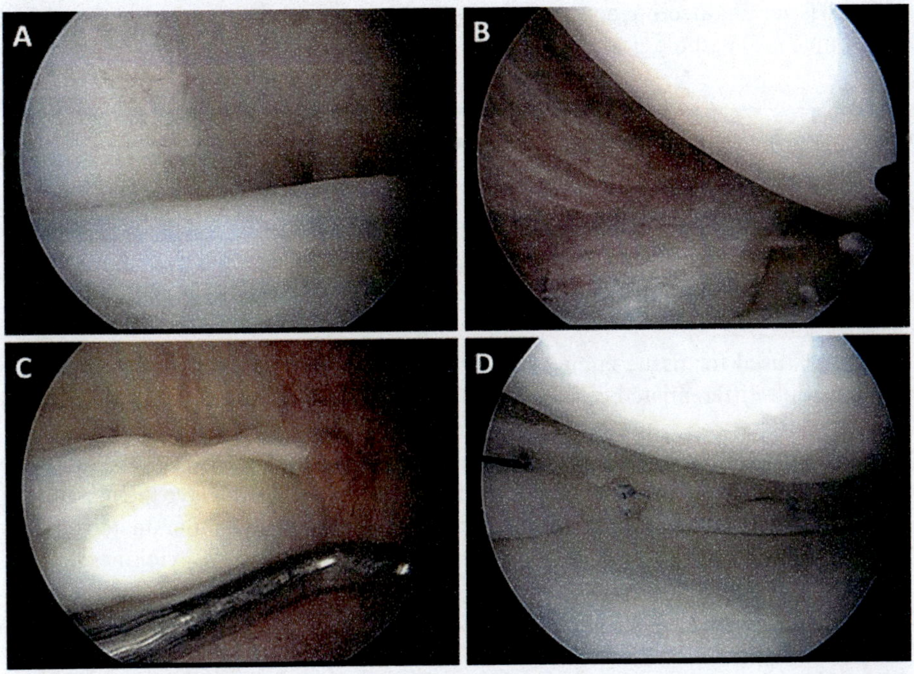

Fig. 1 Meniscus reconstruction: Arthroscopic images of an 18-year-old man with a bucket handle tear of his left medial meniscus. (**a**) The "bucket handle part" of the meniscus tear in the notch lateral to the medial femoral condyle. (**b**) The tear is located in the red-red and the red-white zone of the meniscus from the posterior horn until the far anterior part of the pars intermedia. (**c**) Arthroscopical reposition of the meniscal tear with a probe. (**d**) Reconstruction of the medial meniscus with outside-in sutures (purple) and all-inside sutures (blue)

In addition to that, a beneficial joint milieu can positively influence the meniscus regeneration. Cannon et al. detected an increased healing rate of 93% in patients after meniscus suture and simultaneous anterior cruciate ligament (ACL) reconstruction in comparison to a healing rate of 50% in patients, who had an isolated meniscus suture without simultaneous ACL replacement (Cannon and Vittori 1992). This fact has led to a marked increase of the number of meniscus sutures in combination with an ACL replacement in recent years. However, the positive effect presumably can be ascribed to on the opening of the bone marrow space by drilling the femoral and tibial tunnels for the ACL reconstruction. Via these medullary tunnels mesenchymal stem cells as well as bioactive substances, which support the meniscus regeneration, may arrive to the meniscus defect site and influence the joint milieu. To imitate this effect, some authors also recommend a trephination of the notch before meniscus suture to support meniscus healing (Mordecai et al. 2014).

However, although improvements regarding successful regeneration after meniscus reconstruction could be seen in recent years, there is still a lack of treatment options for meniscal injuries particularly in the avascular zone and for critical size defects. Tissue Engineering approaches might be possible future perspective for this kind of meniscal pathologies.

3 Meniscus Tissue Engineering

3.1 Cell Sources

The classical strategy for Tissue Engineering is to seed cells on a scaffold and after a period of preconditioning in culture the cell matrix construct is placed in the meniscal defect.

The cells used for tissue engineering of meniscus can be classified in precursor cells and adult differentiated cells (Haddad et al. 2013; Bilgen et al. 2018).

3.1.1 Stem Cells

In addition to the capacity for chondrogenic differentiation, MSCs have the potential to improve local microenvironment, immunoregulation, and anti-inflammatory biological activities through the secretion of exosomes, growth factors, cytokines, anti-inflammatory factors, and other bioactive substances. Besides direct differentiation to the repair cells studies suggest that MSCs contribute to joint regeneration by regulation of local inflammation, apoptosis, and proliferation of cells through paracrine mechanisms. Additionally, stem cells are also self-renewing and highly proliferative. Therefore, several sources of stem and progenitor cells have demonstrated promise for use in meniscus repair.

3.1.2 Mesenchymal Stem Cells

Bone Marrow Derived Stem Cells

Bone marrow derived stem cells are the most commonly explored source of MSCs used for meniscal repair. These cells can be extracted autologously through minimally invasive bone marrow aspiration procedures. They have been analyzed in many preclinical studies for meniscus tissue engineering with promising results. However also disadvantages like osteogenic differentiation or the propensity for hypertrophy have been described. To address these problems, coculture with meniscal cells seem to be beneficial. However, a large number of meniscal cells are needed to reduce the tendency of hypertrophy of MSCs during the culture period. Therefore, one of the main focus of research on bone marrow derived MSCs for meniscal regeneration should be to address the limitation of hypertrophy to overcome the burden to translate MSC-based meniscal repair techniques in daily clinical practice.

Due to their potential for differentiation, trophic modulation, and good availability, MSCs appear to be the best cell source to support meniscal healing (Zellner et al. 2017). In different animal models, MSCs showed promising results regarding the development of differentiated meniscus-like repair tissue in small or large meniscus defects or meniscus tears, even in the avascular zone (Angele et al. 2008; Zellner et al. 2010, 2013; Koch et al. 2018). Vangsness et al. injected allogenic MSCs for meniscal treatment following partial meniscectomy in a clinical setting and detected meniscus regeneration and improvement in knee pain (Vangsness Jr. et al. 2014).

Synoviocytes

Lineage tracing studies in mice have demonstrated that articular cartilage and synovium have a common developmental origin. Therefore, in vitro studies have discussed the superiority of MSCs from the synovium for cartilage formation and meniscal repair. These cells have a similar gene expression profile to meniscus cells. Preclinical meniscal repair studies have investigated the use of synovium-derived MSCs with moderate success (Niu et al. 2016).

Adipose Tissue-Derived Stem Cells

Another source for harvesting MSCs is adipose tissue obtained from liposuction or subpatellar fat pad, centrifugated and digested by collagenase I to prepare concentrated adipose-derived MSCs. These adipose-derived MSCs have demonstrated positive effects on pain reduction and knee function without cell-depended adverse events after injection in osteoarthritic knees. Jo et al. showed a significant dose dependency of the outcome after adipose-derived MSCs with the need of further studies to confirm clinical advantages of high-dose injection. However, in a comparative study, adipose-derived MSCs reveal a lower chondrogenic potential, lower cartilage specificity of matrix protein production, and lower expression rate of collagen I gene than bone marrow-derived MSCs. Additionally, comparable to synovium-derived cells isolation of ASCs requires initial surgery for tissue extraction. However, ASCs were reported to improve healing of lesions in the avascular meniscal region in a rabbit model dependent on the defect size and the time between surgical creation of the lesion and treatment.

Meniscus-Derived Stem Cells

Meniscus-derived stem cells are a target for enhancement of meniscus healing or restoration. After a meniscal injury, the level of MSCs in the synovial fluid is elevated and a migration of MSCs in the meniscus towards the meniscal injury can be detected. Meniscal cells may be isolated from the tissue and then reinserted within a carrier back into the patient. These cells have been evaluated in vitro, while also demonstrating multilineage differentiation potential (Hennerbichler et al. 2007; Mauck et al. 2007; Zellner et al. 2015). However, there is a lack of in vivo data concerning the use of meniscal cells for tissue repair in previous studies. Regional differences regarding their chondrogenic potential exist, as meniscal cells derived from the outer vascularized zone show a higher chondrogenic capacity than meniscal cells from the inner nonvascularized part (Zellner et al. 2015). Thus, meniscal cells alone are not wholly responsible for the reduced intrinsic self-healing capacity. Hennerbichler et al. (2007) showed that reinserted meniscal plugs in the outer and inner zone of the meniscus reintegrated into the surrounding meniscal tissue in vitro with stable connecting fibers between the meniscal cells. Explants from the avascular inner zone and vascular outer zone of the meniscus exhibit similar healing potential and repair strength in vitro. In an own study we have shown the repair capacity of the meniscus cells in an in vivo situation for meniscal punch defects in the avascular zone (Zellner et al. 2017). The reason for the enhanced repair in these previous investigations may be the existence of progenitor populations within the meniscus particularly from the outer meniscus (Mauck et al. 2007; Makris et al. 2011). The problem with meniscal cells or meniscal-derived stem cells are their availability. However, the use of allogenic meniscus derived stem cells might be a practical approach that has shown to promote of meniscal healing.

Cartilage-Derived Stem Cells

MSCs can also be found in the articular cartilage called cartilage-derived chondrogenic progenitor cells (CPCs). These cells can be isolated from mature cartilage, cultured and differentiated in different mesenchymal linages (Williams et al. 2010). Like MMSCs, CPC also migrate to cartilage injuries or trauma. They are fibrochondrocyte-like in appearance and can therefore be considered as a potential repair cell for meniscus regeneration. In comparison to BM-MSC CPCs from healthy cartilage are more resistant to terminal differentiation and hypertrophy. However, for a clinical reason, again the bioavailability might be critical as they make up less than 1% of all cells in articular cartilage. According to the clinically established method of autologous chondrocyte transplantation, it is conceivable to harvest cartilage from a non-weight bearing area, isolate and differentiate CPCs, and treat meniscal injuries with these cartilage-derived MSCs.

3.2 Mature Cells

3.2.1 Meniscus Fibrochondrocytes

Meniscus cells are often used for seeding and testing of scaffolds designed for meniscus replacement. Meniscal fibrochondrocytes from the inner and from the outer zone show a high capacity for differentiation and regeneration which can be improved by the application of dynamic hydrostatic pressure in an in vitro meniscus cell pellet model (Zellner et al. 2015). Similar to meniscus-derived MSCs also adult meniscus fibrochondrocytes have the disadvantage of limited bioavailability. Comparison of different cell types for Tissue Engineering of meniscus showed favorable results for MSCs compared to adult meniscal cells. Zellner et al. revealed regenerative potential of the meniscus by an autologous cell-based Tissue Engineering approach even in a challenging setting of early osteoarthritis (Zellner et al. 2017). Autologous MSCs and meniscal cells were found to have improved meniscal healing in an animal model, thus demonstrating its feasibility in a clinical setting. However, donor site morbidity, less availability, and reduced chondrogenic differentiation of human meniscal cells from debris of meniscal tears favors autologous MSCs for clinical use for cell-based meniscus regeneration. However, the use of allogenic meniscal fibrochondrocytes for meniscus repair or coculture of MSCs seems feasible.

3.2.2 Articular Chondrocytes

Autologous chondrocyte transplantation has become the standard of care for cartilage defects of the knee. The technique of cell harvest, culture, and matrix-based transplantation is routinely used in daily clinical practice. A limited amount of studies has been performed to evaluate the use of adult chondrocytes also for meniscal repair in conjunction with scaffolds, showing positive results. Further studies are needed for better understanding of the positive effect of allogenic or autologous mature chondrocytes for cell-based meniscal repair (Bilgen et al. 2018).

4 Biomaterials for Meniscus Tissue Engineering

Biomaterials for meniscus Tissue Engineering can be synthesized from natural or synthetic components. The aims of a scaffolds are to provide a 3D structure, biomechanical support, and to foster differentiation into meniscus-like repair tissue. Cell-free as well as cell-loaded scaffolds are described for Tissue Engineering which can be applied in different forms like gels or solid stiff structures. Solid matrices can provide extended mechanical force as an important function of meniscus tissue and can protect repair cells at the defect site (Rongen et al. 2014).

4.1 Acellular Biomaterials

Cell-free matrices often provide 3D shape support and could be bioactive or bioinert. Many different biomaterials and components have been tested or are still under investigation in search of the optimal scaffold for meniscus Tissue Engineering. Nondegradable materials may cause friction and damage the surrounding articular cartilage. However, degradable matrices may be too weak and shrink over time if it is not enforced by developed repair tissue.

Silk scaffolds are natural biomaterials that were developed for partial meniscal regeneration currently undergoing clinical trials. In animal models, the biomaterial showed chondroprotective effects.

Fibrous scaffolds like polycaprolactone polymers have been electrospun to create cell-free biodegradable biomaterials for Tissue Engineering of meniscus. They were shown to provide early mechanical function and slow degradation during the time of meniscus tissue regeneration.

Non-biodegradable scaffolds for meniscus tissue engineering were, e.g., fabricated of poly-vinyl alcohol hydrogels imitating the mechanical meniscal properties. However, animal models showed more cartilage degeneration over time than meniscal allograft transplantation.

An alternative approach for biomimetic cell free bioscaffolds is the decellularization of meniscus tissue having the advantage of preserving the special zonal organization of meniscus ultrastructure and extracellular matrix. A major issue for repair tissue development is to facilitate cell infiltration into these scaffolds due to its dense extracellular matrix structure. As frozen allograft often has dead cell debris with potential adverse events on cell ingrowth and differentiation, the decellularization process may make up this disadvantage by cleaning up the cell debris.

Other examples for biomaterials which were tested for meniscal regeneration without cell loading are synthetic scaffolds like carbon fibers, PLLA, polyurethane, polyethylene, and PCL-polyurethane. Natural cell-free scaffolds are hyaluronan-collagen composite matrices, fibrin glues, and small intestinal submucosa.

4.2 Clinical Use of Cell-Free Scaffolds

Due to regulatory burdens, implementation of cell-free scaffolds in clinical practice can be easily achieved than cell-based approaches. Striving for optimal restoration of meniscal tissue biocompatible scaffolds came in the focus for treatment of large meniscal defects in the last decade. The rationale for using such cell-free biomaterials after extensive loss of meniscal tissue is based on repopulation of the scaffold by host cells recruited from the synovium or the meniscal remnants (Scotti et al. 2013).

At least only one biomaterial is currently in clinical use. The collagen meniscus implant (CMI) was the first scaffold for meniscus replacement. It is porous, derived from bovine collagen I, molded in shape of a medial or lateral meniscus, and available off the shelf (Rodkey et al. 2008). Some authors report of satisfactory clinical outcome while MRI and histological results are controversial. Some cell infiltration can be seen over time; however, Steadman et al. saw a tendency of shrinking over time and no histological remnants of the collagen scaffold 5–6 years after implantation (Steadman and Rodkey 2005). Predominantly, scar tissue formation instead of fibrocartilage was described after implantation. Another clinically used biomaterial was a polyurethane-based scaffold called Actifit with clinical results similar to CMI. The first 12-month report showed tissue ingrowth into the scaffold at this time and consistent MRI and histology data (Verdonk et al. 2011). In an animal model, this biomaterial showed an accelerated in growth in the surrounding native cartilage when seeded with mesenchymal stem cells (Koch et al. 2018). However, at the moment this biomaterial is not available for clinical use. Despite promising short-term results of meniscal implants, none of them has currently demonstrated regeneration of a functional, long-lasting meniscal tissue.

In order to restore the biomechanical environment for the knee joint, a cell free anisotropic synthetic biomaterial called NuSurface was developed. The idea of this implant is not the ingrowth or the induction of meniscal repair tissue but more a replacement of lost meniscus tissue to provide shock absorption for the medial knee compartment of the knee. The implant is designed for free floating in the medial compartment and requires an intact meniscal peripheral rim for keeping the implant in place (Shemesh et al. 2014).

As biomaterials for meniscal substitution need a stable and intact rim of the native meniscus, meniscal transplantation is actually the only biologic treatment option available for totally meniscectomized knee joints (Lubowitz et al. 2007). A recently published meta-analysis concluded that this procedure is an effective technique for these very selected patients (Elattar et al. 2011). However, fixation of the allograft, the limited availability, and the frequent mismatch of graft and host tissue are still problems of this procedure. Additionally, there is no consensus in the optimal storage protocols for meniscal allografts which leads to variations in cell viability in meniscus allografts potentially affecting the outcome. Therefore, development of biomaterials that can be used for total meniscus replacement are areas of improvement.

4.3 Scaffolds for Cell-Based Meniscus Therapy – Preclinical

Cell-based Tissue Engineering of meniscus has the advantage to deliver potential repair cells and bioactive agents directly at the defect site. Scaffolds with different natural or synthetic components have been used as a supportive cell carrier. Injectable biomaterials like hydrogels are often applied with cells because of the ease with which they can be seeded with cells and applied in the knee or at the defect. In contrast solid scaffolds provide a biomechanically more robust 3D structure that protects the cells at the defect during the regeneration process.

Examples for natural matrix components are: collagen type I or II, decellularized or devitalized meniscus, hyaluronan, chitosan.

Examples for synthetic matrix components for meniscus Tissue Engineering are: polylactic acid, PGA, agarose, and alginate (Hasan et al. 2014).

Most biomaterials used for Tissue Engineering of meniscus try to match the biochemical and mechanical properties of the meniscus. An alternative approach is the self-assembling of different allogenic or autologous cells to create a zonally differentiated tissue-engineered construct. Alternatively, cocultured cells can be used for such a self-assembling scaffold-free approach to mimic the zonal organization of meniscal matrices. The need for high amounts of cells and subsequent use of matrix degrading enzymes are reasons that might inhibit clinical application in the future (Niu et al. 2016).

5 Potential Ways for Healing Enhancement by Suture Augmentation

5.1 Augmentation of Meniscus Suture with Mesenchymal Stem Cells

Preclinical trials by ourselves and others have shown enhanced healing of meniscal lesions with the application of mesenchymal-based cells (Izuta et al. 2005; Angele et al. 2008; Zellner et al. 2010, 2013; Makris et al. 2011; Hasan et al. 2013). Locally applied expanded mesenchymal stem cells from the bone marrow have achieved regeneration of longitudinal meniscus tears in the avascular zone in the lateral meniscus of New Zealand White Rabbits. Control groups with untreated tears, treatment with meniscus suture alone, or meniscus suture in combination with implanted cell-free biomaterials revealed no recognizable healing indicating the development of a critical size meniscus tear model. In contrast, mesenchymal stem cells from the bone marrow in combination with hyaluronan collagen carriers resulted in meniscal repair with differentiated meniscus-like tissue detected by histology, immunohistochemistry, and biomechanical analysis.

It is not clear whether this is a direct action of the mesenchymal-based cells or is rather mediated by secretion of certain stimulating factors (Starke et al. 2009). Despite the fact that meniscus regeneration seems to be feasible by growth factors and mononucleated cells, not many of the cell-based strategies has entered clinical practice to date (Scotti et al. 2013). The implementation of cell-based strategies is mainly limited by the necessity to expand cells prior to transplantation resulting in high treatment costs.

Whitehouse et al. conducted a first in human safety study of five patients with a critical avascular meniscal tear. Autologous MSCs were taken from the iliac crest, expanded, cultured, and seeded on a collagen scaffold. These MSC-scaffold constructs were implanted in the meniscal tears and secured in the defect with sutures. At 2-years post-op, three patients were asymptomatic with functional improvement and no signs of a re-tear in the MRI. Two patients required subsequent meniscectomy due to nonhealing after approximately 15 months (Whitehouse et al. 2017). Further clinical studies are needed to show the benefit of cell-based treatment for meniscus injuries. As the meniscus plays an essential role for joint integrity of the knee, its restoration is a key factor in a whole joint approach for cell-based regenerative treatment of the knee.

5.2 Regeneration of Large-Size Meniscus Defects

In experimental trials, different settings have been tested for mesenchymal stem cell-based Tissue Engineering approaches for treatment of meniscus defects. Ischimura et al. showed a faster and improved healing of avascular meniscal defects in a rabbit model by using bone marrow fibrin clot constructs compared to fibrin clot alone (Ishimura et al. 1991). They postulated that the benefit of this treatment is due to the pluripotential stem cells in the bone marrow. In a scaffold-free engineered meniscal tissue, AufderHeide et al. found a high-tensile modulus and improved mechanical properties by a high density coculture of fibrochondrocytes and mesenchymal stem cells in ring-shaped molds (AufderHeide and Athanasiou 2004). However, for in vivo application of mesenchymal stem cells scaffolds seem to be a useful tool to ensure a lasting effect of the cells directly at the defect site. Yamasaki et al. using cell-seeded rat decellularized meniscus scaffolds concluded that constructs are more effective than scaffolds alone (Yamasaki et al. 2008). Multiple natural and synthetic scaffolds have been used to deliver cells to the meniscal injury (Buma et al. 2004). A hyaluronan collagen composite matrix was found to be suitable for chondrogenic differentiation of mesenchymal stem cells (Angele et al. 2009). Treatment of meniscal full-size defects with this scaffold seeded with autologous mesenchymal stem cells after resection of the pars intermedia of the medial meniscus in a rabbit model resulted in a complete defect filling after 3 months in vivo. Only treatment with mesenchymal stem cells was able to repair this critical size meniscal defects with stable differentiated meniscus like tissue compared to untreated defects or the treatment with a cell-free hyaluronan collagen scaffold (Angele et al. 2008). Similar results were detected for treatment of isolated avascular meniscal punch defects in the pars intermedia of the lateral meniscus in a rabbit model (Zellner et al. 2010). After 3 months, in vivo meniscal defects were filled with differentiated repair tissue after treatment with a hyaluronan collagen composite matrix seeded with mesenchymal stem cells. Interestingly, precultured stem cell matrix constructs showed a reduced integration in the native meniscus while non precultured stem cell matrix constructs resulted in a completely integrated repair tissue indicating that also the grade of differentiation of the stem cell matrix construct seems to be an important factor for successful treatment of meniscus with stem cell-based Tissue Engineering. In the same model, growth factors applied by autologous PRP failed to repair the avascular meniscal punch defect. So, mesenchymal stem cells seem to play an essential role for treatment of meniscal defects.

5.3 Growth Factors

In preclinical trials and in vitro studies various growth factors have been identified to have therapeutic positive effect and the potential to enhance meniscal repair. PDGF, FGF-2, IGF-I, and TGF-ß have shown positive effect on activation of cell proliferation and survival. TGF-ß and SDF-I also revealed influence and cell migration. In different studies, growth factors like PDGF, TGF-ß, BMP7 HGF, FGF-2, and IGF-I stimulated anabolic pathways, while IL-I receptor antagonist, TNF antibody, inhibitors of MMPs, and TGF-ß inhibit inflammatory and catabolic pathways. The activation of biomechanical signaling pathways are also pro-anabolic or anti-catabolic.

In daily clinical practice, a single stage regenerative treatment would be preferable for meniscus injuries. Especially, clinically applicable bioactive substances or isolated growth factors like Platelet Rich Plasma (PRP) as a cocktail of bioactive substances or Bone Morphogenic Protein 7 (BMP7) are in the focus of interest. However, in literature results of the use of, e.g., PRP or isolated growth factors are ambiguous in preclinical and first clinical studies.

In an own study, the effects of PRP and BMP7 on the regeneration of avascular meniscal defects were evaluated. In vitro analysis showed that PRP secretes multiple growth factors over a period of 8 days. BMP7 enhances the collagen II deposition in an aggregate culture model of MSCs. However, applied to different-shaped meniscal defects in vivo PRP or BMP7 in combination with a composite matrix failed to improve meniscus healing in the avascular zone in a rabbit model. In a similar model, Koch et al. saw no effect by additional application of PRP to the suture for repair of a vascular meniscal tear. However, the augmentation of a meniscal suture with autologous bone marrow concentrate showed improved healing of tears in the avascular zone of the meniscus in a rabbit model (Koch et al. 2019) (Fig. 2).

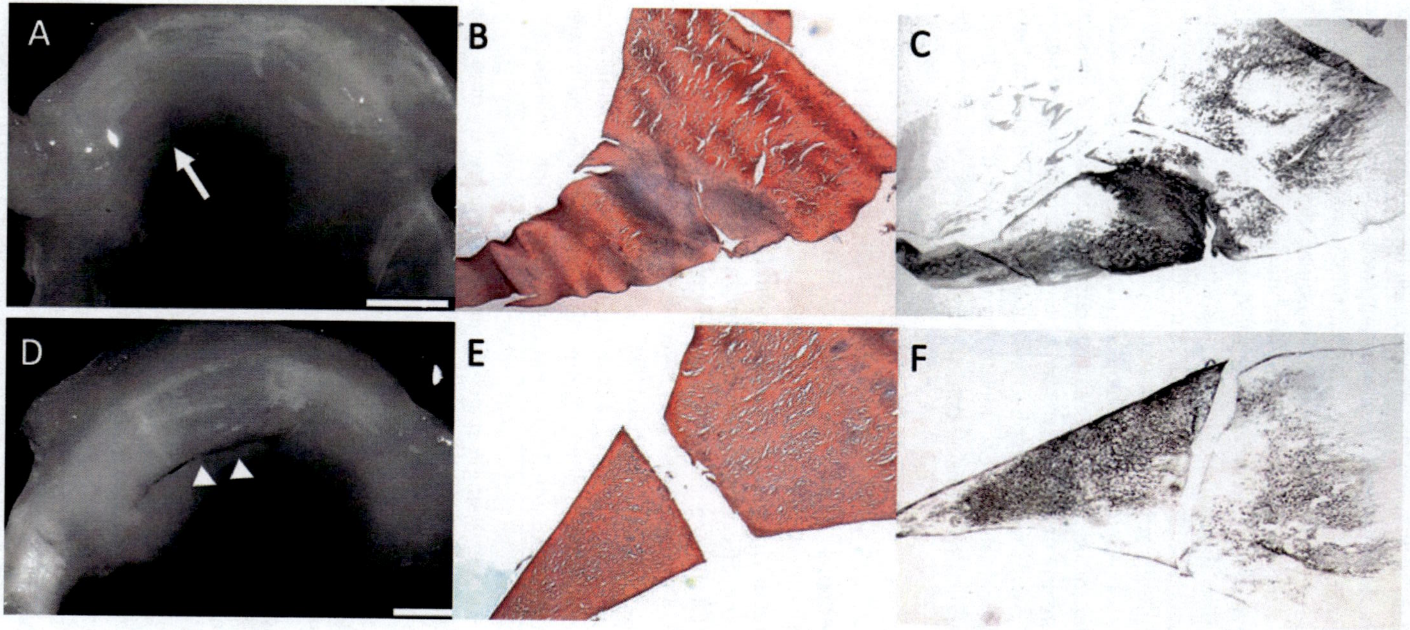

Fig. 2 Rabbit model for analyzing the repair potential of different biological augmentations to support meniscal sutures: macroscopic (**a**) and histological (**b**: HE staining, **c**: collagen II immunostaining) images of lateral meniscus 12 weeks after treatment of an avascular meniscal tear with suture and autologous bone marrow concentrate showing stable meniscal healing in the avascular zone. (**d**–**f**) Control: suture alone without biological augmentation with no meniscal healing

Theoretically, a highly angiogenic growth factor like VEGF might have a positive effect on the regeneration of an avascular tissue like the inner zone of the meniscus. However, there are reports that VEGF-coated PDLLA sutures failed and showed even worse results than uncoated sutures when meniscal tears in the avascular zone of meniscus were reconstructed in a rabbit model (Petersen et al. 2007).

Further information of the repair mechanism at the defect site is needed to develop special release systems or carriers for the appropriate application of growth factors to support biological augmentation of meniscus regeneration.

5.4 Gene Transfer

As direct application of recombinant factors to meniscal lesions is potentially hindered by their short half-lives, gene delivery procedures may help to improve the therapeutic effect.

Different gene vehicles are currently available to genetically modify relevant target cells and tissues like nonviral vectors, adenoviral vectors, retro-/lentiviral vectors, Herpes simplex virus vectors, and recombinant adeno-associated virus vectors. All vectors show different transduction efficiency for meniscal cells or progenitor cells. Strategies for application are direct injection, administrating modified cells, implanting biocompatible materials that deliver recombinant factors, or a gene transfer vector or implantation of a cell-scaffold construct with genetically modified cells.

Therapeutic gene transfer in vivo has been performed by transplantation of meniscal cells modified by an HGF adenoviral vector using a PGA scaffold in an athymic nude mouse model or by progenitor cells (MSCs) modified with an IGF-I nonviral vector using alginate in goat meniscal lesions, leading to an enhanced repair of the treated lesions for up to 16 weeks. In general, more information will be needed on the possible deleterious effects of the different approaches especially when viral vectors are being manipulated in vivo for a safe future application in a clinical setting (Cucchiarini et al. 2016).

5.5 Animal Models for Tissue Engineering of Meniscus

As already mentioned above, many animal models were tested for the evaluation of Tissue Engineering products for meniscus regeneration in vivo. Described models vary in the biomaterials and the cells used. Models for many defect situations exist like complete resection, critical size defects like resection of parts of the meniscus (e.g., pars intermedia) with or without an intact rim at the periphery or meniscal lesions in the vascularized or avascular zone of the meniscus like tears or punch defects. The defect model used depends on the intention of a potential later indication in clinical use or the question that needs to be answered for the specific research issue.

The development of new treatment options for meniscus injuries is directly related to the appropriateness of animal models for their investigation. Data suggest that the main structural features that influence meniscal repair like cellularity, vascularity, and collagen structure differ between animals and human. High similarity was seen between sheep and human meniscus (Chevrier et al. 2009). Other large animal models for investigation of meniscus issues are horse, calf, swine, and goat (Deponti et al. 2015). For specific questions also small animal models are described in literature. Rats, mice, rabbits, and dogs were also used for testing Tissue Engineering approaches in vivo (Pereira et al. 2011). Therefore, the need and the appropriateness of an animal model should be taken into intensive consideration before planning an in vivo meniscus research project.

5.6 3D Printing

Three-dimensional printing is a promising method to fabricate scaffolds for meniscus replacement to address special requirements of shape, size-matching, and ultrastructure of the meniscus. A 3D printed scaffold made of fibrous polycaprolactone was combined with polymer microspheres providing growth factor release for successful cell differentiation and induction of cellular homeostasis. In vitro studies with this scaffold also showed that seeded MSCs differentiated into fibrochondrocytes. An animal model revealed zone-specific deposition of collagen type I and II. Therefore, 3D printing is a promising approach to address the meniscus-specific challenge of special shape and zonal variations (Shimomura et al. 2018).

6 Conclusions

Meniscus integrity is the key for joint health. Therefore, the main goal of every meniscus treatment should be the maintenance of as much meniscus tissue as possible. Meniscus preserving techniques to obtain a functional intact meniscus after meniscus injury in long-term are of great importance for the prevention of the development of osteoarthritis in the knee joint. In recent years, huge steps have been taken regarding the awareness of the importance of meniscus tissue for knee joint integrity. Healing rates for meniscus reconstruction have been improved. However, strategies for a successful meniscus regeneration should be developed for every meniscal zone and all meniscal defect situations. Tissue Engineering of meniscus is a promising approach and important future perspective for meniscus regeneration. Due to growing knowledge in recent years and improved techniques for application, we have the chance to implement meniscus Tissue Engineering in daily clinical practice. Efforts in all research fields should be taken to translate these approaches in clinical practice as the standard of care for meniscus regeneration where needed.

References

Abrams GD, Frank RM, Gupta AK, Harris JD, McCormick FM, Cole BJ (2013) Trends in meniscus repair and meniscectomy in the United States, 2005–2011. Am J Sports Med 41:2333–2339

Angele P, Johnstone B, Kujat R, Zellner J, Nerlich M, Goldberg V et al (2008) Stem cell based tissue engineering for meniscus repair. J Biomed Mater Res A 85:445–455

Angele P, Muller R, Schumann D, Englert C, Zellner J, Johnstone B et al (2009) Characterization of esterified hyaluronan-gelatin polymer composites suitable for chondrogenic differentiation of mesenchymal stem cells. J Biomed Mater Res A 91:416–427

Arnoczky SP (1999) Building a meniscus. Biologic considerations. Clin Orthop Relat Res 367: S244–S253

AufderHeide AC, Athanasiou KA (2004) Mechanical stimulation toward tissue engineering of the knee meniscus. Ann Biomed Eng 32:1161–1174

Badlani JT, Borrero C, Golla S, Harner CD, Irrgang JJ (2013) The effects of meniscus injury on the development of knee osteoarthritis: data from the osteoarthritis initiative. Am J Sports Med 41:1238–1244

Baker BM, Nathan AS, Huffman GR, Mauck RL (2009) Tissue engineering with meniscus cells derived from surgical debris. Osteoarthr Cartil 17:336–345

Bilgen B, Jayasuriya CT, Owens BD (2018) Current concepts in meniscus tissue engineering and repair. Adv Healthc Mater 7:e1701407

Borchers JR, Kaeding CC, Pedroza AD, Huston LJ, Spindler KP, Wright RW (2011) Intra-articular findings in primary and revision anterior cruciate ligament reconstruction surgery: a comparison of the MOON and MARS study groups. Am J Sports Med 39:1889–1893

Brophy RH, Matava MJ (2012) Surgical options for meniscal replacement. J Am Acad Orthop Surg 20:265–272

Buma P, Ramrattan NN, van Tienen TG, Veth RP (2004) Tissue engineering of the meniscus. Biomaterials 25:1523–1532

Cannon WD Jr, Vittori JM (1992) The incidence of healing in arthroscopic meniscal repairs in anterior cruciate ligament-reconstructed knees versus stable knees. Am J Sports Med 20:176–181

Caplan AI, Dennis JE (2006) Mesenchymal stem cells as trophic mediators. J Cell Biochem 98:1076–1084

Chevrier A, Nelea M, Hurtig MB, Hoemann CD, Buschmann MD (2009) Meniscus structure in human, sheep, and rabbit for animal models of meniscus repair. J Orthop Res 27:1197–1203

Crisan M, Corselli M, Chen CW, Peault B (2011) Multilineage stem cells in the adult: a perivascular legacy? Organogenesis 7:101–104

Croutze R, Jomha N, Uludag H, Adesida A (2013) Matrix forming characteristics of inner and outer human meniscus cells on 3D collagen scaffolds under normal and low oxygen tensions. BMC Musculoskelet Disord 14:353

Cucchiarini M, McNulty AL, Mauck RL, Setton LA, Guilak F, Madry H (2016) Advances in combining gene therapy with cell and tissue engineering-based approaches to enhance healing of the meniscus. Osteoarthr Cartil 24:1330–1339

Deponti D, Di Giancamillo A, Scotti C, Peretti GM, Martin I (2015) Animal models for meniscus repair and regeneration. J Tissue Eng Regen Med 9:512–527

Di Matteo B, Tarabella V, Filardo G, Vigano A, Tomba P, Marcacci M (2013) Thomas Annandale: the first meniscus repair. Knee Surg Sports Traumatol Arthrosc 21:1963–1966

Elattar M, Dhollander A, Verdonk R, Almqvist KF, Verdonk P (2011) Twenty-six years of meniscal allograft transplantation: is it still experimental? A meta-analysis of 44 trials. Knee Surg Sports Traumatol Arthrosc 19:147–157

Englund M, Lohmander LS (2004) Risk factors for symptomatic knee osteoarthritis fifteen to twenty-two years after meniscectomy. Arthritis Rheum 50:2811–2819

Englund M, Roos EM, Lohmander LS (2003) Impact of type of meniscal tear on radiographic and symptomatic knee osteoarthritis: a sixteen-year followup of meniscectomy with matched controls. Arthritis Rheum 48:2178–2187

Fairbank TJ (1948) Knee joint changes after meniscectomy. J Bone Joint Surg Br 30B:664–670

Fetzer GB, Spindler KP, Amendola A, Andrish JT, Bergfeld JA, Dunn WR et al (2009) Potential market for new meniscus repair strategies: evaluation of the MOON cohort. J Knee Surg 22:180–186

Haddad B, Pakravan AH, Konan S, Adesida A, Khan W (2013) A systematic review of tissue engineered meniscus: cell-based preclinical models. Curr Stem Cell Res Ther 8:222–231

Hasan J, Fisher J, Ingham E (2013) Current strategies in meniscal regeneration. J Biomed Mater Res B Appl Biomater. https://doi.org/10.1002/jbm.b.33030

Hasan J, Fisher J, Ingham E (2014) Current strategies in meniscal regeneration. J Biomed Mater Res B Appl Biomater 102:619–634
Hennerbichler A, Moutos FT, Hennerbichler D, Weinberg JB, Guilak F (2007) Repair response of the inner and outer regions of the porcine meniscus in vitro. Am J Sports Med 35:754–762
Horie M, Sekiya I, Muneta T, Ichinose S, Matsumoto K, Saito H et al (2009) Intra-articular injected synovial stem cells differentiate into meniscal cells directly and promote meniscal regeneration without mobilization to distant organs in rat massive meniscal defect. Stem Cells 27:878–887
Ishimura M, Tamai S, Fujisawa Y (1991) Arthroscopic meniscal repair with fibrin glue. Arthroscopy 7:177–181
Izuta Y, Ochi M, Adachi N, Deie M, Yamasaki T, Shinomiya R (2005) Meniscal repair using bone marrow-derived mesenchymal stem cells: experimental study using green fluorescent protein transgenic rats. Knee 12:217–223
Jeong HJ, Lee SH, Ko CS (2012) Meniscectomy. Knee Surg Relat Res 24:129–136
Johnson MJ, Lucas GL, Dusek JK, Henning CE (1999) Isolated arthroscopic meniscal repair: a long-term outcome study (more than 10 years). Am J Sports Med 27:44–49
Koch M, Achatz FP, Lang S, Pfeifer CG, Pattappa G, Kujat R et al (2018) Tissue engineering of large full-size meniscus defects by a polyurethane scaffold: accelerated regeneration by mesenchymal stromal cells. Stem Cells Int 2018:8207071
Koch M, Hammer S, Fuellerer J, Lang S, Pfeifer CG, Pattappa G et al (2019) Bone marrow aspirate concentrate for the treatment of avascular meniscus tears in a one-step procedure – evaluation of an in vivo model. Int J Mol Sci 20:1120
Lee BS, Bin SI, Kim JM (2016) Articular cartilage degenerates after subtotal/total lateral meniscectomy but radiographic arthrosis progression is reduced after meniscal transplantation. Am J Sports Med 44:159–165
Lohmander LS, Englund PM, Dahl LL, Roos EM (2007) The long-term consequence of anterior cruciate ligament and meniscus injuries: osteoarthritis. Am J Sports Med 35:1756–1769
Lubowitz JH, Verdonk PC, Reid JB 3rd, Verdonk R (2007) Meniscus allograft transplantation: a current concepts review. Knee Surg Sports Traumatol Arthrosc 15:476–492
Makris EA, Hadidi P, Athanasiou KA (2011) The knee meniscus: structure–function, pathophysiology, current repair techniques, and prospects for regeneration. Biomaterials 32:7411–7431
Matsukura Y, Muneta T, Tsuji K, Koga H, Sekiya I (2014) Mesenchymal stem cells in synovial fluid increase after meniscus injury. Clin Orthop Relat Res 472:1357–1364
Mauck RL, Martinez-Diaz GJ, Yuan X, Tuan RS (2007) Regional multilineage differentiation potential of meniscal fibrochondrocytes: implications for meniscus repair. Anat Rec 290:48–58
McDermott I (2011) Meniscal tears, repairs and replacement: their relevance to osteoarthritis of the knee. Br J Sports Med 45:292–297
McDermott ID, Amis AA (2006) The consequences of meniscectomy. J Bone Joint Surg Br 88:1549–1556
Mezhov V, Teichtahl AJ, Strasser R, Wluka AE, Cicuttini FM (2014) Meniscal pathology – the evidence for treatment. Arthritis Res Ther 16:206
Mordecai SC, Al-Hadithy N, Ware HE, Gupte CM (2014) Treatment of meniscal tears: an evidence based approach. World J Orthop 5:233–241
Nakata K, Shino K, Hamada M, Mae T, Miyama T, Shinjo H et al (2001) Human meniscus cell: characterization of the primary culture and use for tissue engineering. Clin Orthop Relat Res 391:S208–S218
Nepple JJ, Dunn WR, Wright RW (2012) Meniscal repair outcomes at greater than five years: a systematic literature review and meta-analysis. J Bone Joint Surg Am 94:2222–2227
Niu W, Guo W, Han S, Zhu Y, Liu S, Guo Q (2016) Cell-based strategies for meniscus tissue engineering. Stem Cells Int 2016:4717184
Noyes FR, Chen RC, Barber-Westin SD, Potter HG (2011) Greater than 10-year results of red-white longitudinal meniscal repairs in patients 20 years of age or younger. Am J Sports Med 39:1008–1017

Osawa A, Harner CD, Gharaibeh B, Matsumoto T, Mifune Y, Kopf S et al (2013) The use of blood vessel-derived stem cells for meniscal regeneration and repair. Med Sci Sports Exerc 45:813–823
Pabbruwe MB, Kafienah W, Tarlton JF, Mistry S, Fox DJ, Hollander AP (2010) Repair of meniscal cartilage white zone tears using a stem cell/collagen-scaffold implant. Biomaterials 31:2583–2591
Papalia R, Del Buono A, Osti L, Denaro V, Maffulli N (2011) Meniscectomy as a risk factor for knee osteoarthritis: a systematic review. Br Med Bull 99:89–106
Paxton ES, Stock MV, Brophy RH (2011) Meniscal repair versus partial meniscectomy: a systematic review comparing reoperation rates and clinical outcomes. Arthroscopy 27:1275–1288
Pereira H, Frias AM, Oliveira JM, Espregueira-Mendes J, Reis RL (2011) Tissue engineering and regenerative medicine strategies in meniscus lesions. Arthroscopy 27:1706–1719
Petersen W, Pufe T, Starke C, Fuchs T, Kopf S, Neumann W et al (2007) The effect of locally applied vascular endothelial growth factor on meniscus healing: gross and histological findings. Arch Orthop Trauma Surg 127:235–240
Petty CA, Lubowitz JH (2011) Does arthroscopic partial meniscectomy result in knee osteoarthritis? A systematic review with a minimum of 8 years' follow-up. Arthroscopy 27:419–424
Radin EL, de Lamotte F, Maquet P (1984) Role of the menisci in the distribution of stress in the knee. Clin Orthop Relat Res 185:290–294
Rodkey WG, DeHaven KE, Montgomery WH III, Baker CL Jr, Beck CL Jr, Hormel SE et al (2008) Comparison of the collagen meniscus implant with partial meniscectomy. A prospective randomized trial. J Bone Joint Surg Am 90:1413–1426
Rongen JJ, van Tienen TG, van Bochove B, Grijpma DW, Buma P (2014) Biomaterials in search of a meniscus substitute. Biomaterials 35:3527–3540
Salata MJ, Gibbs AE, Sekiya JK (2010) A systematic review of clinical outcomes in patients undergoing meniscectomy. Am J Sports Med 38:1907–1916
Scotti C, Hirschmann MT, Antinolfi P, Martin I, Peretti GM (2013) Meniscus repair and regeneration: review on current methods and research potential. Eur Cell Mater 26:150–170
Shemesh M, Asher R, Zylberberg E, Guilak F, Linder-Ganz E, Elsner JJ (2014) Viscoelastic properties of a synthetic meniscus implant. J Mech Behav Biomed Mater 29:42–55
Shimomura K, Hamamoto S, Hart DA, Yoshikawa H, Nakamura N (2018) Meniscal repair and regeneration: current strategies and future perspectives. J Clin Orthop Trauma 9:247–253
Sommerlath KG (1991) Results of meniscal repair and partial meniscectomy in stable knees. Int Orthop 15:347–350
Starke C, Kopf S, Petersen W, Becker R (2009) Meniscal repair. Arthroscopy 25:1033–1044
Steadman JR, Rodkey WG (2005) Tissue-engineered collagen meniscus implants: 5- to 6-year feasibility study results. Arthroscopy 21:515–525
Stein T, Mehling AP, Welsch F, von Eisenhart-Rothe R, Jager A (2010) Long-term outcome after arthroscopic meniscal repair versus arthroscopic partial meniscectomy for traumatic meniscal tears. Am J Sports Med 38:1542–1548
Tengrootenhuysen M, Meermans G, Pittoors K, van Riet R, Victor J (2011) Long-term outcome after meniscal repair. Knee Surg Sports Traumatol Arthrosc 19:236–241
Vangsness CT Jr, Farr J II, Boyd J, Dellaero DT, Mills CR, LeRoux-Williams M (2014) Adult human mesenchymal stem cells delivered via intra-articular injection to the knee following partial medial meniscectomy: a randomized, double-blind, controlled study. J Bone Joint Surg Am 96:90–98
Verdonk R, Verdonk P, Huysse W, Forsyth R, Heinrichs EL (2011) Tissue ingrowth after implantation of a novel, biodegradable polyurethane scaffold for treatment of partial meniscal lesions. Am J Sports Med 39:774–782
Whitehouse MR, Howells NR, Parry MC, Austin E, Kafienah W, Brady K et al (2017) Repair of torn avascular meniscal cartilage using undifferentiated autologous mesenchymal stem cells: from in vitro optimization to a first-in-human study. Stem Cells Transl Med 6:1237–1248
Williams R, Khan IM, Richardson K, Nelson L, McCarthy HE, Analbelsi T et al (2010) Identification and clonal characterisation of a progenitor cell sub-population in normal human articular cartilage. PLoS One 5:e13246

Xu C, Zhao J (2015) A meta-analysis comparing meniscal repair with meniscectomy in the treatment of meniscal tears: the more meniscus, the better outcome? Knee Surg Sports Traumatol Arthrosc 23:164–170

Yamasaki T, Deie M, Shinomiya R, Yasunaga Y, Yanada S, Ochi M (2008) Transplantation of meniscus regenerated by tissue engineering with a scaffold derived from a rat meniscus and mesenchymal stromal cells derived from rat bone marrow. Artif Organs 32:519–524

Zellner J, Mueller M, Berner A, Dienstknecht T, Kujat R, Nerlich M et al (2010) Role of mesenchymal stem cells in tissue engineering of meniscus. J Biomed Mater Res A 94:1150–1161

Zellner J, Hierl K, Mueller M, Pfeifer C, Berner A, Dienstknecht T et al (2013) Stem cell-based tissue-engineering for treatment of meniscal tears in the avascular zone. J Biomed Mater Res B Appl Biomater 101:1133–1142

Zellner J, Mueller M, Xin Y, Krutsch W, Brandl A, Kujat R et al (2015) Dynamic hydrostatic pressure enhances differentially the chondrogenesis of meniscal cells from the inner and outer zone. J Biomech. https://doi.org/10.1016/j.jbiomech.2015.02.003

Zellner J, Pattappa G, Koch M, Lang S, Weber J, Pfeifer CG et al (2017) Autologous mesenchymal stem cells or meniscal cells: what is the best cell source for regenerative meniscus treatment in an early osteoarthritis situation? Stem Cell Res Ther 8:225

Zhang Z, Arnold JA (1996) Trephination and suturing of avascular meniscal tears: a clinical study of the trephination procedure. Arthroscopy 12:726–731

Part V

Ocular System

Bioengineered Corneas Entering the Clinical Realm

Victor H. Hu, Pushpinder Kanda, Kamal Malhotra, Emilio I. Alarcon, Miguel Gonzalez-Andrades, Matthew Burton, and May Griffith

Contents

Victor H. Hu and Pushpinder Kanda contributed equally.

V. H. Hu · M. Burton
International Centre for Eye Health, London School of Hygiene and Tropical Medicine, London, UK
e-mail: Victor.Hu@lshtm.ac.uk; Matthew.Burton@lshtm.ac.uk

P. Kanda
Department of Ophthalmology, University of Ottawa Eye Institute, The Ottawa Hospital, Ottawa, ON, Canada

K. Malhotra
Département d'ophtalmologie, Université de Montréal and Centre de Recherche de l'Hopital Maisonneuve-Rosemont, Montréal, QC, Canada

E. I. Alarcon
BioEngineering and Therapeutic Solutions (BEaTS), Division of Cardiac Surgery, University of Ottawa Heart Institute, Ottawa, ON, Canada

Department of Biochemistry, Microbiology, and Immunology, Faculty of Medicine, University of Ottawa, Ottawa, ON, Canada
e-mail: ealarcon@ottawaheart.ca

M. Gonzalez-Andrades
Maimonides Biomedical Research Institute of Cordoba (IMIBIC), Department of Ophthalmology, Reina Sofia University Hospital and University of Cordoba, Cordoba, Spain
e-mail: miguel.gonzalez@imibic.org

D. Eberli et al. (eds.), *Organ Tissue Engineering*, Reference Series in Biomedical Engineering, https://doi.org/10.1007/978-3-030-44211-8_9

Abstract

Rapid progress is being made in developing bioengineered corneas to restore vision in patients with corneal opacity or irregularities that result in blindness. Human donor corneas have served well in the past for corneal transplant surgery and surgical innovation, especially lamellar endothelial replacement which has helped to improve outcomes. However, the massive shortage of suitable donor tissue combined with the expense and technical challenges of eye-banking leaves a substantial proportion of the world's population without access to treatment. This is mainly the case in lower-income nations, where corneal blindness is most prevalent. This chapter reviews the many different advances in the development of bioengineered substitutes for improving or replacing the human cornea for vision restoration. Cell-based therapies are part of mainstream clinical practice for replacing the corneal epithelium. Recent trials show promise for endothelial replacement, and 3D printing of the corneal stroma has begun. Traditional keratoprostheses that are well-established but lack bio-integration with host tissue are now being designed to encourage seamless integration. Cell-free pro-regeneration implants have been used successfully in human trials and allow incorporation of bioactive substances. Finally, the regulatory pathway should be a consideration in bioengineering corneal substitutes or implants.

1 Introduction

1.1 The Human Cornea and Corneal Blindness

The cornea is the clear window at the front of the eye and constitutes the major refractive component of the eye, helping to focus light onto the retina. Around 11–12 mm in diameter and 550 μm thick centrally, its unique structure and

M. Griffith (✉)
Département d'ophtalmologie, Université de Montréal and Centre de Recherche de l'Hopital Maisonneuve-Rosemont, Montréal, QC, Canada

Institut de génie biomédical, Université de Montréal and Centre de recherche du Centre hospitalier de l'Université de Montréal, Montréal, QC, Canada
e-mail: may.griffith@umontreal.ca

configuration of mainly collagenous fibers makes the cornea highly transparent with a high tensile strength. It also functions as an effective barrier to microorganisms, is richly innervated, and has the ability to rapidly regenerate the epithelium layer in response to injury.

Over 90% of the thickness of the cornea consists of mainly extracellular matrix, of which type I collagen (70% of the dry weight of the cornea) is predominant, with types III, V, and VI also present. Collagen in the cornea is organized into fibrils with a nearly uniform diameter of 30 nm and an interfibrillary distance of 60 nm (Brunette et al. 2017). This highly ordered arrangement, with distances much less than half the wavelength of visible light (which is 400–700 nm), means that any scattering of light by the collagen fibrils is largely cancelled out by interference. Parallel collagen fibrils form 200 to 300 lamellae, each around 2 μm thick. Cellular components constitute around only 3–5% of the corneal stromal volume. These comprise stromal cells or keratocytes with a stellate morphology, connected by extended cellular processes (Lakshman et al. 2010). Normally quiescent in the stroma, keratocytes transform into a myofibroblast-like phenotype following corneal injury and migrate toward the injured area. Transformed stromal cells produce significant amounts of disordered collagen that contributes to hazing and opacity during ulceration or scarring when there is pathological inflammation leading to vision loss.

The stroma is sandwiched by two cellular layers. The epithelium or outer layer of the cornea comprises 5–6 layers of non-keratinized, stratified epithelial cells. Basal epithelial cells at the limbus (far periphery of the cornea) have stem cell characteristics with a high proliferative capacity. Rapid proliferation and migration of cells from the limbus characterizes repair of the epithelial surface. By contrast, the inner surface of the stroma is lined by a single layer of endothelial cells, which under normal physiological conditions have minimal, if any, proliferative potential. A decline in endothelial cell density occurs with age while the integrity of this layer is maintained by spreading and migration of neighboring cells. The endothelium plays a critical role in maintaining corneal transparency by actively keeping the stroma relatively dehydrated. Marked loss of endothelial cells, somewhere below 1000 cells/mm^2, results in corneal edema and clouding.

A recent systematic review and meta-analysis, using Bayesian hierarchical modeling, estimated that globally there are 36 million people blind, and another 217 million with moderate to severe vision impairment (MSVI) (Bourne et al. 2017; Flaxman et al. 2017; World Health Organization 2018a). The majority of those blind patients are found in developing countries and around 80% of blindness is estimated to be avoidable. Corneal opacity (apart from trachoma) was estimated to account for blindness in 1.3 million and MSVI in 2.9 million, with corneal opacity from trachoma estimated to affect an additional 0.9 and 1.6 million, respectively. Corneal visual impairment has been named by the WHO as a priority eye disease (World Health Organization 2018b). By definition, blindness requires loss of vision in both eyes; however, there is huge underreporting of vision loss in only one eye. While robust data is lacking, it has been estimated that there are well over 1.5 million new cases of unilateral blindness each year worldwide, mainly in the developing world (Whitcher and Srinivasan 1997). Major causes of corneal opacity have traditionally been attributed to trachoma, xerophthalmia, measles, neonatal ophthalmia, and leprosy. However, ocular trauma with subsequent corneal ulceration and infection is being increasingly recognized as a "silent epidemic" (Whitcher et al. 2001). Vision loss from corneal disease in developed countries tends to be more from degenerative or inherited disorders such as keratoconus and Fuchs' Endothelial Dystrophy, as well as corneal infections, with contact lens use playing an important role in this context.

1.2 Corneal Transplantation and Challenges

Corneas are the most commonly transplanted tissue globally, with 185,000 transplant surgeries estimated to have been performed in 2012 (Gain et al. 2016). Donor corneal transplantation can replace selective layers of the cornea (lamellar keratoplasty), or involve a full-thickness replacement (penetrating keratoplasty, PK). It was first performed at the beginning of the twentieth century, but only gained popularity a few decades later with improvements in surgical technique and the use of steroid eye drops (Tan et al. 2012). Eye retrieval is performed within 24 h post-mortem and transported to an eye bank where the cornea can be stored for up to 4 weeks in organ culture media (Joint United Kingdom (UK) Blood Transfusion and Tissue Transplantation Services Professional Advisory Committee 2016). Until recently, PK was the only real option for corneal transplant surgery. Over the last couple of decades, however, lamellar corneal transplant surgery has become the preferred option. Endothelial keratoplasty (EK) is now commonly used where endothelial dysfunction is present causing corneal edema, but the host cornea is otherwise healthy, such as with Fuchs' Endothelial Dystrophy. Only the corneal endothelium, with or without a thin layer of stroma, is transplanted. This is greatly advantageous in avoiding many of the complications associated with PK, such as astigmatism, and markedly reducing the postoperative recovery time. If, on the other hand, the host endothelium is healthy and there is disease affecting only the anterior cornea, as is commonly the case in keratoconus or with scarring after infectious keratitis, then an anterior lamellar keratoplasty (ALK) can be performed. This has the great benefit of avoiding endothelial allograft rejection, a major cause of transplant failure after PK (George and Larkin 2004).

The most common indications for PK between 1980 and 2014 were keratoconus (in Europe, Australia, the Middle East, Africa, and South America), pseudophakic bullous keratopathy/aphakic bullous keratopathy (in North America), and keratitis (in Asia) (Matthaei et al. 2017). Corneal grafts have a short-term survival rate of 86% at 1 year which does not withstand the readily believed notion that the cornea is an immune privileged organ (Williams et al. 2006, 2008). The graft survival rate steadily declines to 73%, 62%, and 55% at 5, 10, and 15 years post-operation respectively, reaching rates similar to other transplanted organs (Williams et al. 2006). The success rate for corneal transplantation is not equal among patients, with excellent outcomes of >90% survival at 10 years seen with "low-risk" patients that have an avascular and non-inflammatory corneal disease like keratoconus (Di Zazzo et al. 2017; Inoue et al. 2000; Williams et al. 2008). Patients with disease resulting in inflammation and vascularization of the cornea are considered "high-risk" with rejection rates greater than 55% ; rates that are worse than heart and kidney transplantation (Bartels et al. 2003; Coster and Williams 2003; Yu et al. 2016). Although the definition of a high-risk patient is a relative term (often compared to grafts indicated for uncomplicated keratoconus), a consensus (but non-comprehensive list) of characteristics defining high-risk transplantation is summarized below (Coster and Williams 2003; Di Zazzo et al. 2017; Williams et al. 2008; Yu et al. 2016).

1.2.1 Risk Factors Defining High-Risk Corneal Transplantation

- Indications for graft: Bullous keratopathy, herpetic eye disease, scar, and opacifications
- Previous graft rejection (40% risk with first rejection and risk increases ~1.2-fold with every subsequent rejection.)
- Previous and current keratitis
- Past or current ocular inflammation
- Neovascularization of the host cornea or a previous graft (increasing risk with a greater number of quadrants involved. Risk doubles with all four quadrants involved.)
- Aphakia
- Increased intraocular pressure after graft
- Larger graft diameter (>8 mm)
- Limbal stem cell deficiency

Although PK provides a means to improved visual outcomes, ocular pain relief, or both, graft failure in high-risk patients requires re-transplantation. The rejection risk doubles with every subsequent transplantation (survival rates <20% after the third rejection) (Claesson and Armitage 2013; Williams et al. 2008).

Healthy corneas and those in low-risk patients are immune-privileged for a number of reasons (Di Zazzo et al. 2017; Hori et al. 2019; Qazi and Hamrah 2013):

1. Low expression of antigen-presenting major histocompatibility complex (MHC) I and lack of MHC II.
2. Lack of lymphatics (afferent arm of the immune system which carries antigen presenting cells (APC) to local lymph node to activate and promote clonal expansion of T cells).
3. Lack of vessels (efferent arm of immune system which transports immune cells to the graft).
4. Lack of activated APCs in allograft corneas in turn minimizes activated T cells via direct allorecognition pathway.
5. Anterior chamber-associated immune deviation (ACAID), a phenomenon in which the shedding of alloantigen into the anterior chamber suppresses the delayed-type hypersensitivity reaction resulting in immune tolerance (Vendomele et al. 2017).

6. Expression of proapoptotic *Fas*-ligand by corneal endothelium and epithelium mediate immune cell destruction.
7. Corneal cells express programmed death ligand-1 that suppresses T-cell proliferation and production of inflammatory mediators (e.g., interferon gamma (IFNγ)) (Liechtenstein et al. 2012; Shen et al. 2007).

These protective mechanisms are abrogated in high-risk patients due to ocular inflammation promoting neovascularization, lymphangiogenesis, and recruitment of APCs to the cornea (Di Zazzo et al. 2017).

The success rate for corneal grafts in high-risk patients remains disappointing despite the use of immunosuppressants, pharmacological agents to inhibit neovascularization, optimizing surgical techniques to minimize inflammation, and performing partial-thickness corneal transplantation (lamellar keratoplasty) to decrease the exposure to alloantigens (Di Zazzo et al. 2017).

1.3 Alloimmune Corneal Graft Rejection

Corneal graft rejection is primarily mediated by a delayed-type hypersensitivity immune response via the indirect allorecognition pathway (Fig. 1). Allogenic sensitization begins with corneal grafting, which triggers inflammation and the release of various inflammatory chemokines/cytokines including interleukin (IL)-1, IL-6, chemokine ligand 2, IFNγ, and tumor necrosis factor alpha (TNF-α) (Coster et al. 2005; Coster et al. 2009). Inflammation mediates the recruitment and activation of neutrophils, macrophages, lymphocytes, and APC to the cornea (Coster et al. 2005; Di Zazzo et al. 2017). Host APCs are exposed to shed donor antigen as they mobilize to the corneal limbus or migrate into the graft. As APCs process the alloantigens, they drain via lymphatic system to local lymph nodes where they activate naïve T cells (Coster et al. 2009). Conversely, in the direct allorecognition pathway, donor APCs housed within the graft directly activate naïve host T cells, although this pathway plays minor role given the lack of APCs in the cornea (Qazi and Hamrah 2013; Williams and Coster 2007). Inflammation also promotes angiogenesis and lymphangiogenesis which enhance the sensitization to alloantigens (Niederkorn 2010).

Corneal graft rejection is primary mediated by CD4+ Th1 cells (Chauhan et al. 2015). The exact mechanism by which CD4+ Th1 cells mediate graft rejection is unclear. However, evidence suggests that CD4+ Th1 cells mediate corneal endothelial apoptosis via direct cell-cell interaction independent of the Fas ligand pathway (Chauhan et al. 2015; Hegde et al. 2005). Corneal endothelial apoptosis is also mediated by prolonged exposure to cytotoxic cytokines (e.g., IL-1, TNF-α, IFNγ) produced by CD4+ Th1 cells. In addition to CD4+ Th1 cells, multiple effector cells may play a role in corneal graft rejection including T-cells with a Th2 phenotype, and Th17 cells which also promote lymphangiogenesis (Chauhan et al. 2011; Chen et al. 2009; Qazi and Hamrah 2013). Inevitably, the irreversible destruction of the corneal endothelium results in corneal edema, opacification, and vision loss. As such,

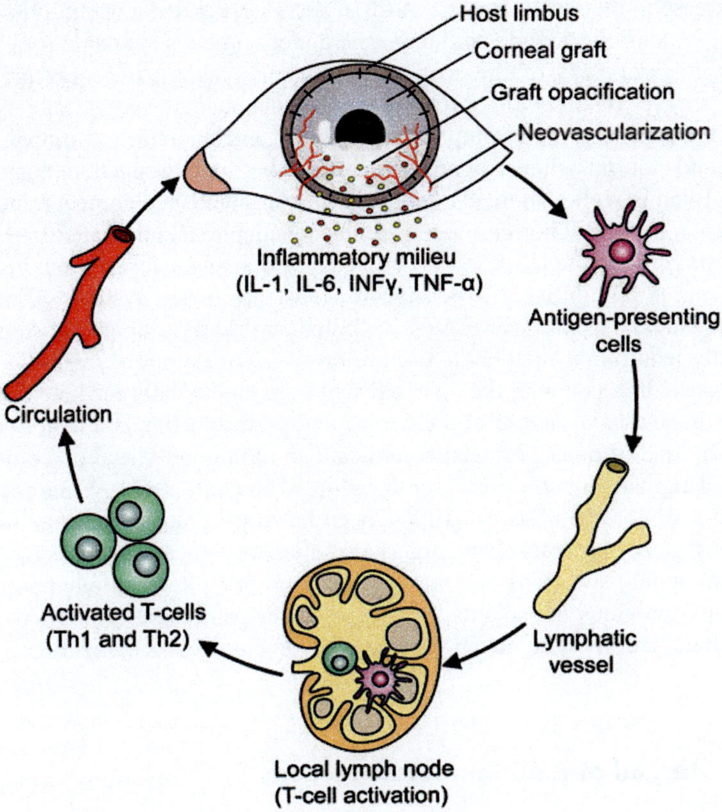

Fig. 1 Diagram showing the mechanism of corneal allograft rejection

various strategies have been adopted clinically to minimize the sensitization to alloantigens including the use of topical or systemic immunosuppressants (e.g., corticosteroids, cyclosporine A, tacrolimus), anti-vascular endothelial growth factor (VEGF) drugs, human leukocyte antigen matching, and intraoperative optimization (e.g., suturing techniques to minimize inflammation); this topic is covered in detail by another review (Di Zazzo et al. 2017).

Despite these perioperative optimizations, improving the survival of corneal graft surgery in high-risk patients remains a challenge and hence the need for bioengineered corneas and/or therapies designed to avoid or suppress adverse immune responses.

1.4 Limitations to Human Corneal Transplantation

The overwhelming shortage of donor corneas, particularly in the developing world, is a significant drawback to relying on donor corneas for reducing blindness from corneal disease in the global context. A 2016 survey estimated a waiting list of 12.7 million individuals worldwide, and only one donor cornea is available for every 70 needed, with over 53% of the world's population having no access to corneal transplant surgery (Gain et al. 2016). Eye banks face increasingly stringent regulation and are expensive and complex to establish and maintain (Armitage 2011). Religious and cultural beliefs can present a serious barrier to tissue donation in some countries. Even in well-resourced countries with an established corneal transplantation service, a lack of donor corneas presents a challenge (Gain et al. 2016). A few countries, principally the USA, export corneas for transplant surgery, but procuring these corneas is prohibitively expensive for most healthcare systems (Gain et al. 2016). The surgery itself requires highly skilled staff and prolonged postoperative care, usually requiring the patient to use eye drops for one or more years. The use of allogenic tissue brings it with the possibility of immunologically mediated rejection even with the avascular nature of the cornea and graft rejection is a major cause of failure. A myriad of other potential complications limit good visual outcomes such as corneal astigmatism and primary graft failure (Tan et al. 2012). While the risk is considered low, corneal transplants have been known to transmit a range of infections (Armitage 2011). Transplant surgery using donor corneal tissue has a high risk of failure in certain situations and may not be appropriate to even attempt in cases with multiple previous transplants, or severe ocular pathologies complicated with marked inflammation of the ocular surface and corneal neovascularization.

1.5 Potential of Bioengineered Corneas

Bioengineering refers to the use of engineering principles to address issues in biology and medicine. As such, bioengineering covers a very wide area that includes genetic engineering to medical imaging and other biological disease diagnostic devices, to tissue engineered organs. Here, we define "Bioengineered Corneas" as substitutes for human donor corneal tissue that are designed to replace part or the full thickness of damaged or diseased corneas. These therefore range from prosthetic devices that are designed to exclusively replace the cornea's function to transmit light into the eye, to tissue-engineered hydrogels and fully reconstructed tissues that allow regeneration of the host tissue. Furthermore, there are also implantable lenticules that may be implanted into the cornea to improve vision by altering the refractive properties of the eye as alternatives to wearing spectacles or undergoing refractive surgery.

In recent years, there have been very significant developments in many areas of bioengineered corneas from the addition of regenerative capacities to prosthetic devices to the use of stem cells to repopulate pathologic corneas. In keeping with the theme of this book on Organ Tissue Engineering, however, we will focus on the design requirements for bioengineering substitutes to improve or replace portions or the entire human cornea. We will also focus on technologies designed to restore vision rather than correction of refractive errors, and in particular, bioengineered corneal implants developed to restore eyesight to patients who are at high-risk of rejecting conventional donor cornea transplantation.

2 Cell-Based Implants

An excellent and comprehensive review of stem cell technologies for eye tissue restoration, including derivation of therapeutic cells, can found in Stern et al. (2018), so we will only provide a synopsis relating to corneal regeneration here. Briefly, the combination of bioengineering with therapeutic cells (Stevens and Murry 2018) has allowed for the reconstruction of different corneal layers. Expansion of corneal-derived stem cells, particularly from autologous sources has been used to offset the shortage of donor corneas as well as to circumvent immune challenges, as we show below. However, despite successes, it would be pertinent to note the regulatory and scientific challenges of cell-based therapies and the specialized cleanrooms and staffing needed, as well as adverse immune reactions against donor cells from allogeneic sources (Burdick et al. 2013).

2.1 Bioengineered Corneal Epithelium

The corneal epithelium benefits from having excellent regenerative capacity. So long as the limbal stem cells are intact, epithelial replacement is unnecessary. However, in situations where the stem cells are damaged or dysfunctional, such as after a severe chemical injury or Stevens-Johnson syndrome, a challenging situation is faced. Adjacent conjunctival epithelium, which is translucent and vascularized, is able to invade the corneal surface resulting in severe visual loss. While not normally considered under the umbrella of corneal transplant surgery, ocular surface reconstruction can be attempted in cases of limbal stem cell failure with techniques such as conjunctival limbal autograft (CLAU) from the contralateral eye if it is healthy, or from a living relative or donor eye. Cultivated limbal epithelial transplantation (CLET), using ex vivo expansion of limbal stem cells, either autologous or allogeneic, is successful and gaining popularity (Buznyk et al. 2015; Pellegrini et al. 1997; Rama et al. 2010).

Various biomaterials have been tested as substrates for stem cell transplantation including human amniotic membranes, fibrin, silk, and various electrospun meshes that allow the cultivation of entire epithelial sheets. Figure 2 shows a CLET implant performed in Canada using a fibrin substrate (Le-Bel et al. 2019). More recently, a simple limbal epithelial transplantation (SLET) has been developed which bypasses the need for expansion of stem cells in expensive cleanrooms (Sangwan et al. 2012).

When the limbal stem cells of both eyes are damaged or nonfunctional, then an alternative solution is needed. Allogeneic transplantations have not been as successful as autografts. Oral mucosa from the patient as an autologous cell source has been tested and is used in clinical practice, termed cultivated oral mucosa epithelial transplantation (COMET) (Utheim 2015). More recently, allogeneic bone marrow

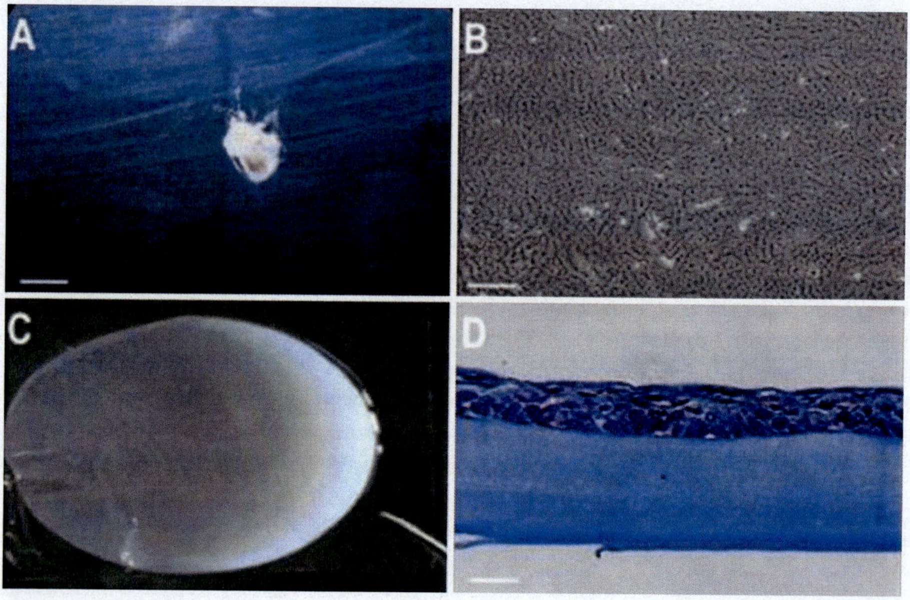

Fig. 2 Preparation of a corneal epithelial limbal graft, showing the steps. (**a**) Limbal biopsy from the patient's contralateral eye; bar: 1 mm. (**b**) expansion by co-culturing with irradiated human dermal fibroblasts; bar: 200 μm. (**c**) Cells on fibrin. (**d**) Histology (Masson's Trichrome staining) of the tissue-engineered epithelium, showing a differentiated epithelium on a fibrin substrate; bar: 10 μm. Reproduced from Le-Bel et al. (2019) with permission from Elsevier.

mesenchymal stroma/stem cells grown on human amniotic membrane have been tested clinically for treating epithelial stem cell deficiency (Calonge et al. 2019).

2.2 Bioengineered Corneal Endothelium

The cells in the monolayered corneal endothelium are post-mitotic and damage to these cells, for example, after surgery or in Fuchs Endothelial Dystrophy, leads to corneal edema and vision loss. However, a dormant population of replication-competent endothelial progenitor cells has been identified (Amano et al. 2006). A number of different cell sources and substrates have been used in an attempt to activate the dormant endothelial progenitor cells for expansion, including amniotic membrane, decellularized cornea, and different collagen types (Levis et al. 2015; Navaratnam et al. 2015).

Approaches to activate the dormant endothelial progenitor cells also include the use of Rho-associated kinase (ROCK) inhibitor Y-27632 (Koizumi et al. 2014). ROCK signaling is involved in a wide range of cellular events including adhesion, motility, differentiation, proliferation, and apoptosis, and also in a variety of different diseases. ROCK inhibitors have been found to enhance corneal endothelial cell proliferation (Okumura et al. 2017). Preliminary studies in humans have shown promising effects in clearing the cornea in cases of endothelial dysfunction, potentially combined with injection of corneal endothelial cells into the anterior chamber of the eye (Okumura et al. 2017). Injection of corneal endothelial cells in combination with a ROCK inhibitor has now been demonstrated in human clinical trials (Kinoshita et al. 2018).

2.3 Bioengineered Corneal Stroma

Stem cell replacement within the stroma by direct infusion of mesenchymal stromal cells is now entering preclinical and early clinical trials with promising results (Basu et al. 2014; Shukla et al. 2019). Most recently, extracellular vesicles from the mesenchymal stromal cells have been shown to restore clarity to hazy corneas by reducing the fibrosis in mouse corneas (Shojaati et al. 2019). Alió del Barrio and Alió (2018), Brunette et al. (2017). Matthyssen et al. (2018) provide a thorough and comprehensive review of cornea stromal bioengineering.

Recent advances in stromal bioengineering also include the 3D bioprinting of stromal constructs. The first report by Connon and colleagues (Isaacson et al. 2018) was of a 3D bioprinted stroma that used human corneal stromal keratocytes, sodium alginate, and methacrylated collagen type 1 as an "ink" for printing corneal stromal implants. The team printed their implants into formed molds (Fig. 3). Subsequently, Duarte Campos et al. (2019) reported the printing of a stroma comprising human corneal stromal keratocytes incorporated into an "ink" comprising 0.5% agarose and 0.2% type 1 rat tail collagen. A drop-on-demand technique was used to print the corneal stromas on a custom-made 3D bioprinter.

The Connon group has also produced corneal stroma using a tissue-templating approach based on poly-tetrafluoroethylene nanogrooves to direct the orientation of keratocytes and their deposited extracellular matrix, together with a smart peptide amphiphile coating that allows the self-assembled stroma to release into a curved tissue with aligned collagen fibrils (Gouveia et al. 2017a, 2017b).

2.4 Bioengineered Multilayered Corneal Replacements

There are several cell-based tissue-engineered corneal replacements. The most widely known ones are those that are self-assembled from sheets of extracellular matrix elaborated by cells under the influence of ascorbic acid. A review that includes these engineered constructs can be found in Simpson et al. (2019).

Gonzalez-Andrades and colleagues (Gonzalez-Andrades et al. 2017, 2018) reported on the clinical trial of patients with ulcers that were nonresponsive to conventional treatment who received combination corneal constructs comprising epithelial and stromal cells in a fibrin-0.1% agarose scaffold. The five initial patients performed well enough in the initial safety evaluation to merit continuation of the trial for a total of 24 months in another 15 patients. Results are expected in 2020.

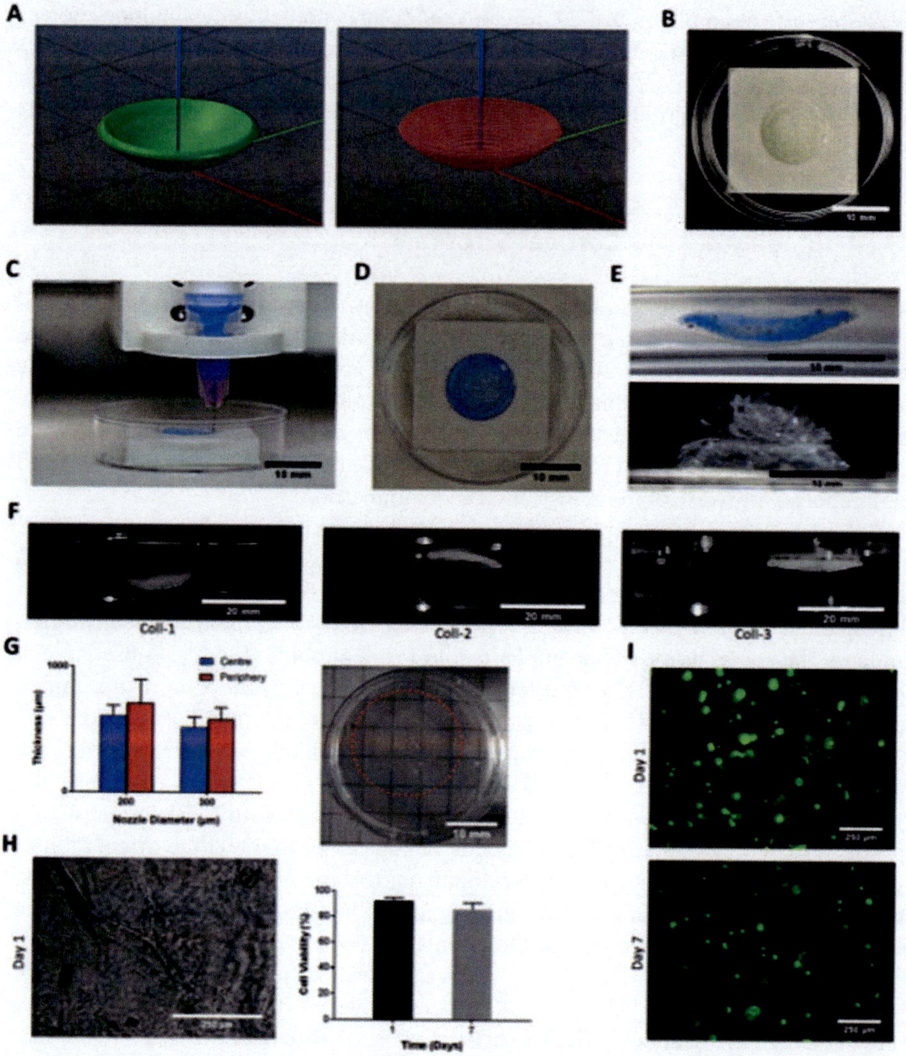

Fig. 3 Using support structure to facilitate the printing of a corneal structure with 3% alginate (nozzle diameter = 200 μm) and optimization of bio-inks for corneal 3D bioprinting. (**a**) Digital cornea is imported to the computer driving the 3D printer software slic3r and a preview of the concentric directionality of print is displayed. (**b**) The support structure is coated with FRESH to facilitate the 3D bioprinting of corneal structures. (**c**) View of the 3D bioprinting process. Corneal structures were printed with 3% alginate bio-ink stained with trypan blue to increase visibility. (**d**) Image of 3D bioprinted corneal structure captured prior to incubation. (**e**) FRESH is aspirated after 8 min of incubation and corneal structure is carefully removed from support, but begins to unravel 1-day post-printing once keratocytes were combined with the alginate bio-ink. (**f**) Images of corneal structures 3D bioprinted from composite bio-inks. (**g**) Relationship between nozzle diameter and printed thickness of corneal structures (left) and depiction of transparency of corneal structure 3D bioprinted from Coll-1 bio-ink. (**h**) Brightfield image of 3D bioprinted corneal structure containing cells at day 1 (left) and cell viability measurements over 7 days (right). (**i**) Representative live/dead stain images using fluorescence microscopy at days 1 and 7 after 3D bioprinting in Coll-1. Reproduced from Isaacson et al. (2018), licensed under CC BY 4.0

3 Keratoprostheses

The earliest bioengineered corneal replacements were prostheses known as keratoprostheses (KPros), designed to replace the minimal function of the human cornea. Although a plethora of KPros have been designed and tested in vitro and in vivo, only a handful are used in clinical practice. Traditional KPros were designed as core-and-skirt models, with an optically transparent core that allowed light transmission into the eye for vision and a skirt that allowed for integration with the hosts tissues. The Boston Keratoprosthesis is the most widely used KPro today, with over 12,000 units transplanted worldwide since 2015 (Salvador-Culla et al. 2016). The device combines PMMA and titanium. Critically, however, it still requires a donor corneal tissue skirt for attachment to the host cornea, thereby failing to overcome the shortage of donor tissue. The osteo-odonto-keratoprosthesis (OOKP) uses a piece of tooth to house an optical portion of PMMA. It requires multiple, complex operations and is performed in only a handful of locations worldwide. Other keratoprostheses have been also been developed. All of the different designs, however, tend to come with a high risk of severe complications including glaucoma, retroprosthetic membrane formation, retinal detachment and extrusion, leading to a high risk of failure (Matthyssen et al. 2018). Several recent reviews of the performance and advances in keratoprostheses can be found (Klufas et al. 2015; Salvador-Culla and Kolovou 2016), including a policy article from the NHS in the UK on when KPros would be used (Specialised Commissioning Team 2018).

3.1 Recent Advances in Keratoprosthes is Design for Improved Bio-integration

While traditional keratoprostheses restore minimal function, bio-integration between the carrier cornea and the optic stem is critical for maintaining the longevity of the KPros. Although PMMA is an inert, nontoxic material suitable for implantation, corneal cells adhere poorly to the surface owing to its hydrophobicity, neutral charge, and lack of bioactive ligands (e.g., collagen) needed for cell attachment (Riau et al. 2017). Lack of adhesion causes the formation of gaps between the cornea–optic stem interface resulting in: 1) stromal downgrowth; 2) leakage of aqueous humour; 3) migration of microorganisms into the eye causing corneal necrosis/melt and/or endophthalmitis; 4) loosing of the optic stem or device extrusion. Therefore, it is essential to allow epithelization to restore cellular barriers of the eye.

Corneal epithelization can be promoted by using regenerative KPros that support the adhesion of cells and allow the penetration of nutrients and primary glucose to maintain the healthy environment whereas cellular barriers prevent the entry to contaminants to prevent infections (Griffith et al. 2012; Myung et al. 2008). Additionally, different sterilization methods such as gamma irradiation could facilitate corneal epithelization of PMMA compared to ethylene oxide (the most commonly used method for sterilizing KPros) (Gonzalez-Andrades et al. 2018).

Various PMMA surface modifications have been investigated to enhance bio-integration. For example, Patel and colleagues (Patel et al. 2006) developed a two-prong approach to modify the PMMA surface to either promote or abrogate cell adhesion at a desired location. PMMA surface modified with di-amino-PEG conjugated to a peptide motif enhanced corneal cell adhesion while the non-conjugated di-amino-PEG prevented adhesion. This strategy promoted bio-integration between the cornea and the KPros while inhibiting RPM growth over the optics. Recently, hydroxyapatite (HAp), a major component in bone composed of complexed calcium and phosphate molecules, has been widely used to enhance bio-integration of PMMA. HAp coated surfaces promote cell adhesion or binding to collagen (major component of corneal stroma) via electrostatic interaction (Riau et al. 2017; Wang et al. 2011). Wang and colleagues (Wang et al. 2011) deposited a HAp coat using a dopamine-activated PMMA surface. Dopamine acts as a versatile adhesive macromolecule with excellent affinity to a wide variety of inert materials. They also incorporated an 11-mercaptoundecanoic acid (11-MUA) coat to provide additional carboxyl functional groups to enhance HAp deposition. The HAp-coated PMMA was superior at promoting corneal fibroblast growth and improving cell viability compared to uncoated PMMA. In addition, HAp-coated PMMA cylinders adhered strongly to explanted porcine cornea and required a force 14.7-fold greater to detach it from the cornea compared to the uncoated PMMA. Implantation of the HAp-coated PMMA into rabbit cornea lamellae was well tolerated and generated a milder inflammatory response compared to uncoated material. Similarly, Riau and colleagues (Riau et al. 2015) demonstrated cell growth and improved long-term collagen hydrogel (major component of corneal stroma) adhesion to HAp-coated PMMA after surface activation with dopamine and 11-MUA. The same group produced an alternative method of dip coating HAp onto PMMA without having to activate surface chemistry. Instead, they used chloroform to introduce cavities on PMMA surface which trapped titanium or HAp nanoparticles. This resulted in a HAp layer that was strongly resistant to delamination and capable of cell growth and adhesion (Riau et al. 2016).

Alongside bio-integration, some researchers have attempted to develop KPros with antimicrobial properties in order to prevent infection of the implanted material. Baino and colleagues (Baino et al. 2016) used a radio-frequency sputtering technique to deposit a coat of silver/silica nanoparticles onto PMMA. The coated surface released silver ions with bactericidal effects on *Staphylococcus aureus*. Unfortunately, antimicrobial properties are normally short lived, dissipating once all the silver ions have been released. For a more permanent solution, surfaces can be coated with titanium oxide which improves cell adhesion (Salvador-Culla et al. 2016) and also exhibit antimicrobial properties by limiting bacterial adhesion (Tan et al. 2011). Similarly, Behlau et al. (2011) demonstrated that N, N-hexyl,methyl-polyethylenimine (HMPEI)-coated PMMA or titanium inhibited biofilm formation by *S. aureus*. Compared to plain PMMA or titanium, the HMPEI-coated materials did not exert additional cytotoxic effects on cultured corneal cells or produce adverse inflammatory reaction in vivo after implantation into rabbit corneas.

4 Cell-Free Pro-Regeneration Corneal Implants

Despite the successes of cell-based therapies, it faces is various regulatory and scientific challenges such as the need for specialized cleanrooms requiring highly trained personnel and, adverse immune reactions against donor cells from allogeneic sources (Burdick et al. 2013; Pashuck and Stevens 2012). Acellular biomaterials that can serve as scaffolds or templates to stimulate endogenous repair mechanisms and regeneration of the target organ would potentially circumvent adverse immune reactions directed against allogeneic cells. They are also advantageous regarding the strict regulatory control of advanced therapy medicinal products as cell-free implants are considered medical devices. To date, most biomaterials that have pro-regeneration properties are based on or inspired by the macromolecules of the extracellular matrix or synthetic derivatives and analogs.

4.1 Decellularized Corneas

A number of methods can be used to remove all cellular components from a donor cornea, leaving only extracellular matrix. The decellularized corneas have the potential advantage of reducing immunological rejection.

There have been several clinical studies using glycerol-preserved human corneal tissue compared to fresh donor corneas. In one study of 68 patients in a randomized trial, high-risk patients (with viral, bacterial or fungal keratitis or ocular burns) were grafted with either acellular glycerol cryopreserved (stored at −78 °C) corneas or fresh donor corneas (in Optisol-GS eye banking medium) with deep anterior lamellar keratoplasty (DALK) (Li et al. 2011). The glycerol preserved corneas were stored for an average of 9.6 ± 2.8 months prior to use. At 2 years post-operation, 100% of acellular corneal grafts were still stable while only 78.8% of fresh donor corneas remained. Better survival in glycerol-preserved corneas was attributed to the lack of antigen-presenting cells in the preserved tissues (Gupta and Upadhyay 2017). However, the shortage of donor corneal tissue still needs to be faced, even if poor quality tissue unsuitable for standard transplant surgery is considered.

Xenogeneic alternatives have also been proposed to address the donor tissue shortage, although caution is needed to avoid adverse immune reactions. Aamodt and Grainger (2016) provide a comprehensive review of the host response to extracellular matrix-based biomaterials, with a focus on decellularization and methods for ensuring proper removal of unwanted contaminants. Decellularization of porcine corneas has been achieved by using ionic detergent, such as SDS, which produce biocompatible acellular matrices with biomechanical and optical properties similar to that of native cornea (Gonzalez-Andrades et al. 2015, 2011; Islam et al. 2019; Yoeruek et al. 2012). The use of nonionic detergents such as Triton X-100 was encouraged as it maintains the biocompatibility, tensile strength, and optical characteristics of the matrix (Xu et al. 2014). To minimize the toxic effects of chemicals, high-hydrostatic pressurization (HHP) has also been used to decellularize porcine corneas, which in turn were successfully grafted into the corneas of rabbits without any adverse immune reaction and achieving optical clarity (Hashimoto et al. 2010).

The decellularizing agent used and the sterilization method are both critical as they will determine the transparency of the implant and the stromal architecture of the matrix (Islam et al. 2019). Recent zoonotic viral transmission leading to a pandemic indicates that caution with xenogeneic corneas is paramount (Coronaviridae Study Group of the International Committee on Taxonomy of Viruses 2020).

Decellularized porcine corneal stromas have been used for anterior lamellar keratoplasty (ALK) in 47 patients with fungal keratitis failing to respond to medical therapy (Zhang et al. 2015). Epithelialization successfully occurred in 43 patients, with gradual improvements in transparency seen over time, and regression of preoperative corneal neovascularization. No rejection episodes were seen, although follow-up was only a minimum 6 months. Improvement in more than two lines of best corrected visual acuity was noted in 34 eyes. Further studies will help clarify the potential role for acellular porcine corneas.

Meanwhile, more controversial methods for derivation of potential donor organs have been explored, such as the recent report of the development of monkey-human hybrids by the Belmonte group in China (MIT Technology Review 2019).

4.2 Recombinant Human Collagen Implants

In 2009, the initial report was published of 10 patients in Sweden who had undergone corneal transplantation with cell-free, bioresponsive recombinant human collagen type III (RHCIII) implants, with the 24-month and 4-year follow-up data subsequently released (Fagerholm et al. 2009, 2010, 2014). Eight of the patients were male, and the age range was 18–75 years. Nine of the patients had advanced keratoconus and were contact-lens intolerant, and the other patient had a central corneal scar from previous bacterial keratitis. Inclusion criteria included being on the waiting list for a first corneal transplant surgery, having a clear posterior corneal stroma and a normal endothelium, and with a clear, non-vascularized peripheral cornea. Implants were produced following Good Manufacturing Practice guidelines using RHCIII produced in the yeast *Pichia pastoris*. Cross-linking was performed with EDC-NHS, with these reactants not incorporated into the final product. Implants were thoroughly characterized showing excellent light transmission of $95.1 \pm 0.05\%$, compared to >87% in human cornea, but a much lower tensile strength of 0.286 ± 0.062 MPa, compared to 3.81 ± 0.40 MPa in human cornea. The mean central corneal thickness of the implants was 493 ± 27 μm.

ALK was performed with a trephination diameter of 6.0–6.5 mm at a depth of 370–400 μm, using a manual excision to leave residual posterior stroma and an intact endothelium. Implants of 6.25–6.75 mm diameter were secured in place with three or four 10–0 nylon overlying mattress sutures and a bandage contact lens (implants were not sufficiently robust for sutures at the host-graft interface). Postoperatively, steroid and antibiotic drops were used three times daily for 3 weeks, at which point the bandage contact lens and sutures were typically removed and drops were continued for a further 4 weeks only. An additional nine patients who underwent penetrating keratoplasty with human donor cornea with a peripheral continuous suture were used for comparison (5 with keratoconus, 2 with endothelial decompensation, 1 with a deep central scar, and 1 with pseudophakic bullous keratopathy). Twenty volunteers with healthy corneas were also included for further comparison. Postoperative assessments were conducted at 1, 3, 6, and 9 months and 1, 2, and 4 years and included slit-lamp examination, corneal sensitivity with Cochet-Bonnet esthesiometry, anterior segment ocular coherence tomography (OCT), and in vivo confocal microscopy (IVCM).

Full epithelialization of the biosynthetic implants took a mean of 2.5 months. Initial epithelial cell migration was impeded by the mattress sutures and removal of these facilitated epithelial coverage (Fagerholm et al. 2010). One patient received an amniotic membrane patch graft at 3 months to aid epithelialization which was complete at 5 months. The shape and thickness of the biosynthetic implants remained largely stable over 4 years with no signs of wound dehiscence or implant extrusion and good optical clarity (Fig. 4).

The mean central corneal thickness after 4 years in patients with a biosynthetic implant was reduced, at 358 μm ± 101 compared to 534 μm ± 30 in controls, and 576 μm ± 50 in patients with donor corneal transplants. Biosynthetic substitute corneas showed flattening of the preoperative keratoconic cone and mild improvement in central astigmatism. However, there was still significant surface irregularity with a tendency towards hexagonal, paracentral bulging corresponding with tension from the overlying mattress sutures in these early implants. Preoperative best spectacle corrected visual acuity was counting fingers in two patients and 0.74 logMAR in the rest (range 0.5–1.0). After 4 years the mean BSCVA for all 10 patients with biosynthetic implants was 0.88 logMAR (range 0.4–1.0). Most patients became tolerant of rigid contact lenses after surgery, and the best contact-lens corrected visual acuity was 0.44 logMAR (0.1–1.0).

The major achievement, however, was the immune compatibility due to the induced in situ tissue regeneration without addition of exogenous cells. No episodes of immune rejection were observed in the biosynthetic implant patient cohort, despite having steroid drops for less than 2 months after surgery. One patient with in the control donor cornea group developed a rejection episode after a year that resolved with topical dexamethasone treatment. Biosynthetic implants had similar numbers of dendritic cells to control corneas, while donor corneas had significantly greater numbers of mature and immature dendritic cells. Biosynthetic implanted corneas showed partial population with keratocytes and reinnervation to the central cornea. Touch sensitivity was better than in donor corneas, but reduced compared to unoperated eyes. One patient with a biosynthetic implant was unable to tolerate contact lenses and underwent further transplantation surgery. Histological sections through the operated cornea shows the implant to have undergone marked remodeling with only a thin remnant left and surrounding deposition of what looks like healthy, lamellar collagen. A healthy, stratified epithelium had been formed and the host endothelium is present.

This pioneering piece of research was the first to use a cell-free biosynthetic analog of human collagen for stimulating endogenous progenitor cells from within the patients' own eyes to regrow corneal tissue. The implants remained largely clear with improvements in visual acuity over the 4-year follow-up period. Steroid drops were only used for the first few weeks, in marked contrast to donor corneal surgery, with no episodes of rejection. A major limitation of the implants was the weak tensile

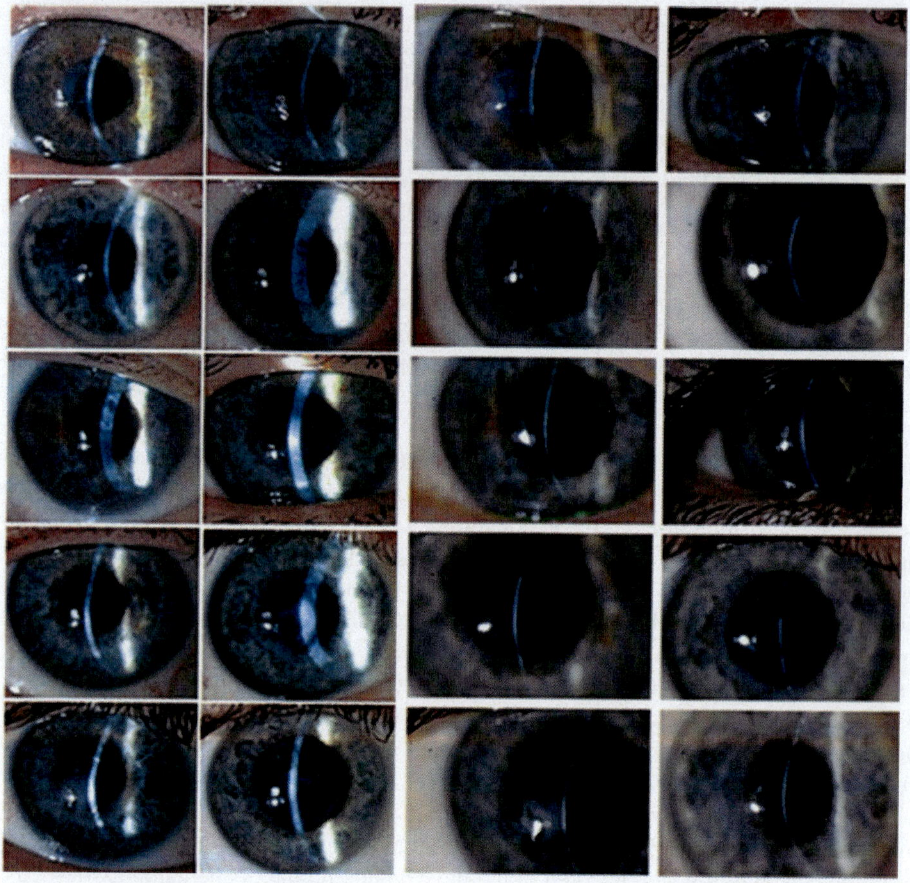

Fig. 4 Gross appearance of the regenerated cornea of all 10 patients who received bioengineered corneal implants comprising recombinant human collagen. Left two columns: patients at 2 years post-operation. Right two columns: The same patients at 4 years post-surgery. Compiled figures adapted from Fagerholm et al. (2010, 2014), with permission from AAAS and Elsevier

strength compared to donor cornea. This required overlying mattress sutures with resulting surface irregularity and impedance of epithelial cell migration. Overall, however, this was a very encouraging piece of work, with the initial implant design providing a template for further improvement and development as seen below.

4.3 Recombinant Human Collagen Implants for High-Risk Patients

A synthetic phospholipid, 2-methacryloyloxyethyl phosphorylcholine (MPC), has been shown to have inflammation suppressing properties. It has previously been shown to reduce neovascularization in a rabbit model of severe pathology with corneal alkali burn when incorporated into the RHCIII implants (Hackett et al.

2011). Two case series of patients considered high risk for rejection and/or failure have received these RHCIII-MPC implants (Buznyk et al. 2015; Islam et al. 2018). The first three pilot patients were from Ukraine, two with chemical burns and one with a rejected PK, who all had painful recurrent/persistent corneal ulceration and vision ranging from Snellen 6/600 to light perception (Buznyk et al. 2015). Surgery was performed to remove the ulcerated areas. The patients were then given tectonic patches. Surgery were performed in a similar manner to the initial Swedish cohort with a smaller 4–5 mm trephine of host tissue and a lamellar dissection at 250–300 μm. Postoperative steroid drops were discontinued after 6 weeks. Epithelialization occurred after a mean of 6 weeks with implants remaining free of neovascularization or erosions in the 9–12 months follow-up period, and all patients experiencing improvement in pain. Two of the patients had improvement in vision to 6/38 and 6/75 with the third patient developing conjunctivalization of the corneal surface and no change in vision.

A clinical trial of seven patients from Ukraine and India with a mixture of previous microbial keratitis, chemical/thermal burn, graft rejection, and neurotrophic keratitis received the RHCIII-MPC implants by ALK (Islam et al. 2018). All patients had severe vision loss in the operated eye and persistent/recurrent episodes of pain were present in five of the patients. Surgical technique in India included the use of a femtosecond laser to perform the lamellar dissection at 350 μm with an 8 mm trephine; fibrin glue was used to help secure the implants as well as overlying sutures. All patients received bandage contact lenses in the initial postoperative period with sutures removed between 3 and 12 weeks and steroid drops used only for the first month. One patient developed a fungal infection after implantation that was unrelated to the biosynthetic implant which was not affected by the infection. The patient was treated by PK and excluded from the study. Follow-up ranged from 14 to 35 months, with a mean of 24 months. Figure 5 shows the pre- and post-

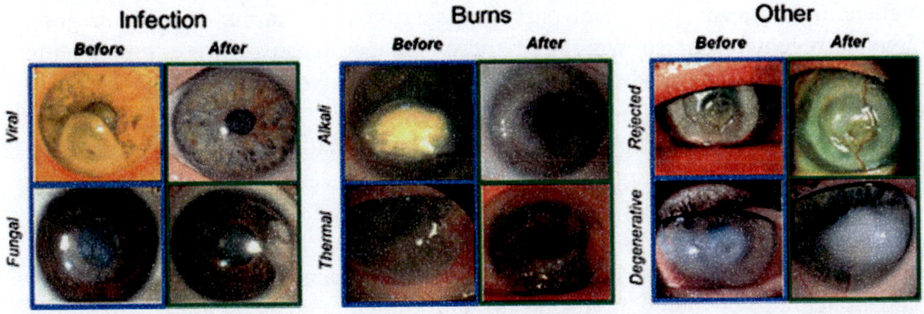

Fig. 5 Clinical trial of RHC-MPC implants in six high-risk patients, before and after grafting at last follow-up. Patients were divided into three groups based on their preoperative diagnoses: infection (herpes simplex viral and fungal keratitis), burns (alkali and thermal), and others (failed graft and post stroke neurotrophic keratitis). Patients with scarred or ulcerated corneas from severe infection showed better vision improvement, followed by corneas with burns. Grafting promoted nerve regeneration in all patients. These two case series show impressive results, both for improving vision, especially in relatively uncomplicated cases with simple corneal scarring, and for improving pain in complex cases with recurrent corneal ulceration. Figure reproduced from Islam et al. (2018), licensed under CC BY 4.0

operation images of the treated corneas. Full epithelial cell migration took place between 4 and 50 weeks, being slower in patients with stem cell deficiency. However, all patients were reported to be free from pain and irritation at 1–2 weeks postoperatively. Some notable improvements in visual acuities were seen, especially in those patients who only had scarring from previous microbial keratitis without stem cell deficiency or neovascularization (cataract and end stage glaucoma also limited visual improvement in some patients). The more complex patients with stem cell deficiency and previously rejected donor corneal transplant developed some neovascularization of the biosynthetic implant, but they remained pain-free as above. Surprisingly good improvements in corneal sensation were noted.

4.4 Collagen-Like Peptides (CLPs) as Collagen Analogs

Human collagens are large proteins that are difficult to manipulate and modify, leading to interest in short peptides that retain the properties of the full-length macromolecules. Short collagen mimetic peptides (CMPs) or collagen-like peptides (CLPs) have been developed by various groups. Like collagen, they can self-assemble and form triple helical structures. They are also readily prepared and customized, for example, conjugated to polymers for increased mechanical strength, making them potentially attractive alternatives to collagen (Rubert Perez et al. 2015). For example, CLPs can be conjugated to polymers such as polyethylene glycol (PEG) (Rubert Perez et al. 2014; Stahl et al. 2010). Most importantly, they have been shown to be successful in promoting regeneration in a number of different tissue and organ systems (Rubert Perez et al. 2015). Implants made from a CLP designed by the Hartgerink group (O'Leary et al. 2011) conjugated to a polyethylene glycol (PEG) backbone through a short amino acid linker have been used for ALK in rabbit and mini-pig animal models in a similar technique to the RHCIII-MPC human studies above (Islam et al. 2016; Jangamreddy et al. 2018). The CLP-PEG implants showed excellent light transmission with stable integration into animal eyes with good epithelial regeneration, showing that these collagen analogs were functionally equivalent to implants made from RHCIII-MPC that were successfully tested in patients in clinical testing (Fig. 6).

Interestingly, Jangamreddy et al. (2018) found that the regenerated corneal epithelium had secreted a large number of extracellular vesicles that were positively stained for exosome markers into the stroma. Among the contents of these vesicles was collagen type V that was enriched within the cornea. The tensile strength of the CLP-PEG was even lower than RHCIII-MPC implants, but this was partially compensated for by being more elastic and showing greater elongation during surgical handling. Overall however, the CLP-PEG and RHCIII-MPC implants performed in a very similar manner. The main advantage of using short CLPs over full-length recombinant human collagen is the ease in manufacturing. Short sequences can be synthesized on solid support by any contract research laboratory that produces clinical grade materials, and do not require the complex manufacturing processes of having to recombinantly co-produce enzymes like prolyl-4-hydroxylase and pepsin needed to obtain the final product, to lengthy purification processes

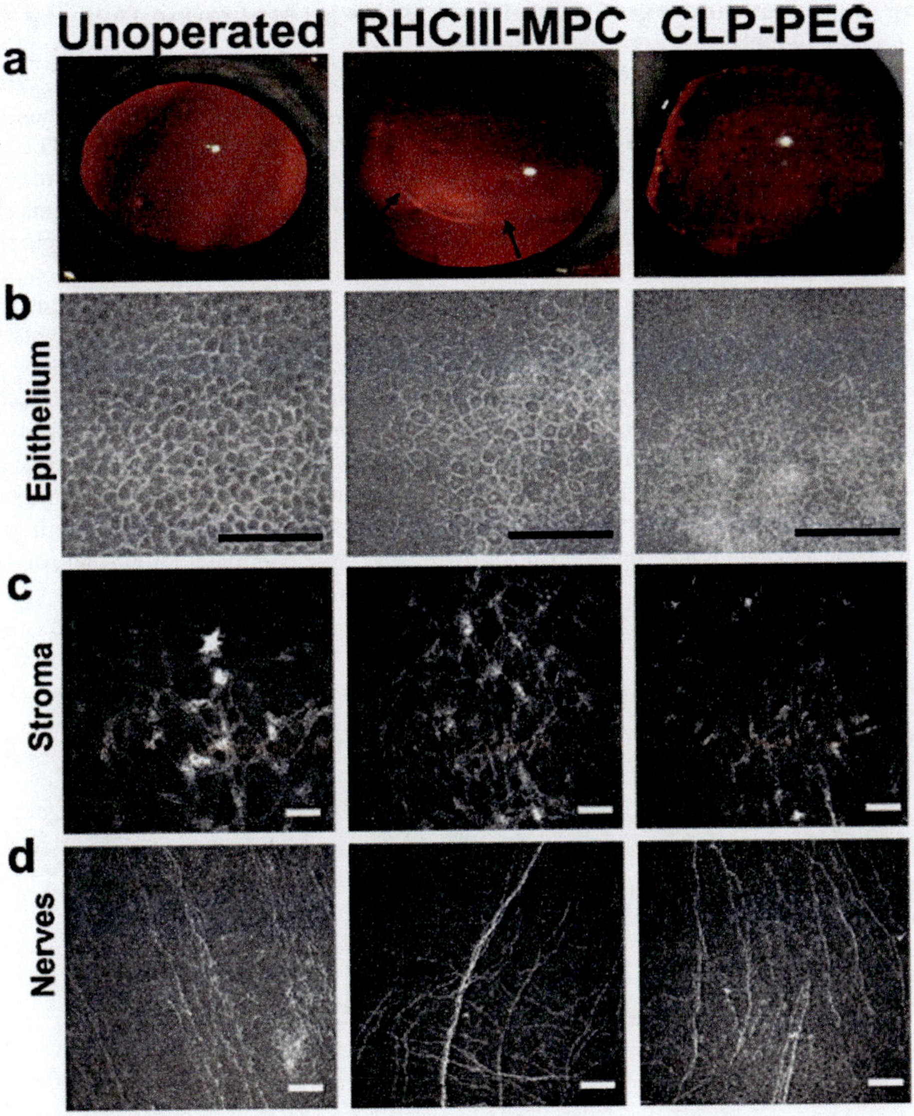

Fig. 6 RHCIII-MPC and CLP-PEG corneal implants and their performance in the corneas of Göttingen mini-pigs, showing optical clarity at 12 months post-implantation, like the unoperated healthy cornea (**a**). The boundaries of the implants are indicated by arrows. In vivo confocal microscopy shows that both RHCIII-MPC and CLP-PEG implanted corneas had regenerated their epithelium (**b**), stroma (**c**), and subepithelial nerve plexus (**d**) to resemble their counterparts with normal, healthy corneas. Bars, 150 mm. Adapted from Jangamreddy et al. (2018), licensed under CC BY 4.0

(Yang et al. 2004), and hence are less costly. They are fully customizable and hence can incorporate bioactive groups as desired. They can also be blended or hybridized to polymeric groups for additive manufacturing. Overall, the use of CLP allows for versatility.

5 Bioengineered Corneal Constructs Incorporating Delivery Systems

Bioengineered corneal implants are now being developed that incorporate nanoparticles or other carriers of drugs or other bioactive molecules, allowing them to be tailored to specific corneal indications. To help bioengineered implants remain free from infection during the early healing process in particular, anti-infective agents have been incorporated, for example. One potential advantage of this is to help overcome poor post-surgical compliance to multiple drops.

Riau et al. (2015) reported the inclusion of the antibiotic, vancomycin, within a collagen-based hydrogel. When implanted into the corneal stromas of rabbits, the implants were found to have prophylactic activity against infection by *S. aureus* bacteria. Because of increased bacterial resistance to antibiotics, silver nanoparticles have been proposed as alternatives to antibiotics. In Alarcon et al. (2016), silver nanoparticles (AgNPs) with antibacterial properties were incorporated into corneal constructs made from collagen (Fig. 7). Two out of three different constructs with silver nanoparticles showed in vitro efficacy in blocking *Pseudomonas aeruginosa* growth that was comparable to that of gentamicin.

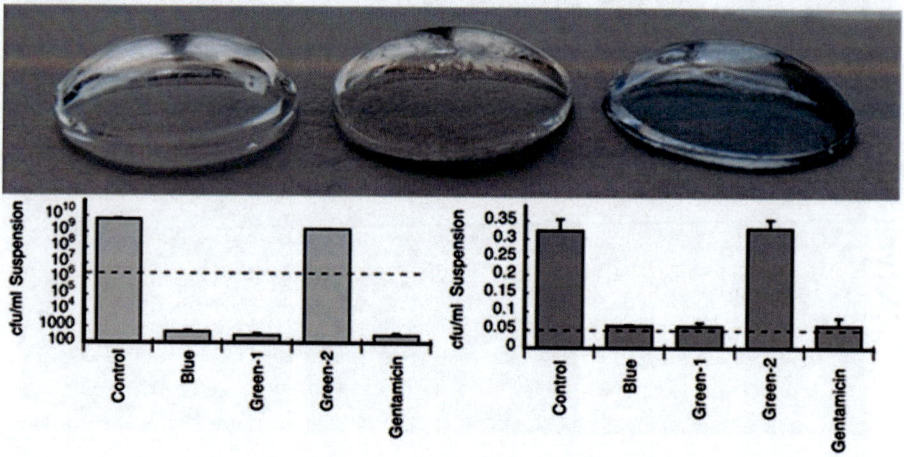

Fig. 7 (Top) From left to right: collagen-based corneal implant; green and blue implants comprising different shaped nanoparticles within a collagen matrix. The differentially shaped nanoparticles confer different absorbance properties, and hence, different colors. (Bottom) Left: Absorption at 600 nm for bacterial suspension of *Pseudomonas aeruginosa* (strain PAO1) in the presence of hydrogels without and with the different types of AgNPs assessed in this work measured after 24 h. The dashed line in the plot indicates the absorbance of the samples measured at time 0. (Right) Survival colonies cultured after 24 h incubation of incubation of *P. aeruginosa* cultures in the presence of hydrogels containing the different types of AgNPs employed in this work. Dashed line shows the initial bacteria density. Adapted from Alarcon et al. (2016), with permission from the Royal Society of Chemistry

6 From Bench to Bedside

The regulatory pathway of different bioengineered corneas or corneal implants differ according to whether they contain cells or are cell-free, and whether they contain any drugs or bioactives that confer pharmacologic properties. They may be classified as advanced therapy medicinal products (ATMPs), biologics, or medical devices. There are various national regulatory agencies and a central European Medicines Agency, but most are harmonized following various standards; those from the International Standards Organization (ISO) being the most widely used from guiding safety and efficacy testing to cleanroom requirements for manufacturing to conduct of clinical trials. Pellegrini et al. (2016) and Rico-Sanchez et al. (2019) provide thorough reviews of the process from research to regulatory approval for clinical use of corneal products containing stem cells. Brunette et al. (2017) discuss the regulatory requirements for translation of corneal stromal replacements, and in particular, cell-free implants that are regulated as medical devices.

If clinical translation is the goal, researchers need to carefully consider regulatory requirements in order to design biomaterials compatible for clinical applications. For a medical device such as a cell-free, pro-regeneration corneal implant for example, there are a number of steps from the bench to the bedside.

Development of novel biomaterials, whether biomimetics, designer polymers chemically synthesized from monomers to mimic ECM molecules, designer recombinant proteins, etc., the choice of raw materials used must take into consideration patient safety and desired time to clinical application. Use of approved biomaterials as a starting point enables faster regulatory approval as the biocompatibility of the material in patients has been documented. Novel materials can result in high impact research papers, but if the materials are difficult to purify or to produce in large quantities for widespread use, there will be issues getting these into regular clinical application. Testing according ISO 10993 standards is needed to determine cytotoxicity, genotoxicity, acute and systemic toxicity, and biocompatibility within the target site and/or bodily fluids. The likelihood of allergic reactions against the materials and presence of any known contaminants that are hard to remove, ease of procurement or production (scalability) and costs are also considerations.

Next, the manufacturing methods should be considered. For example, whether molding or additive manufacturing would be more appropriate to obtain a corneal implant would need to take into account the optics of the implant and the ability of that manufacturing method to produce an implant that allows for maximal light transmission with minimal back scatter. Selection of the manufacturing method should also consider the ease of incorporation of stem cells and micro- and nanoparticles for delivery of drugs or bioactive factors.

Characterization of the various physical, mechanical, and chemical properties will help determine the appropriateness of the manufacturing methods in order to control the implantation, stability, degradation, and ultimate biointegration of the construct. In parallel, in vitro testing of biomaterials for bio- and immune-compatibility toxicology is performed. This can be done on cell lines as 2D cultures or cell-based organotypic models. Many organ-on-a-chip models are being developed for in vitro testing. Bennet et al. (2018), for example, have developed a "corneal epithelium" on a chip, albeit for eye drop evaluation. The steps from biomaterials design through in vitro testing are generally iterative. To ensure that the 3Rs of Humane Animal Experimentation – Replacement, Reduction, and Refinement – are being followed, most of the testing to identify the most optimal implants should be performed in vitro. The best two or three models would then proceed to testing in small animals, for example, rodents. Finally, the best or most optimal implant is tested for toxicology and biocompatibility under Good Laboratory Practice (GLP) at a certified Contract Research Organization (i.e., third-party unbiased laboratory) and also for efficacy in a large animal model such as a mini-pig, also under GLP. GLP testing is recommended for results that are relevant for submission to the ethics boards and regulatory authorities for authorization for clinical trials.

We recommend early pre-submission meetings, wherever possible, with the relevant national regulatory authority to ensure that the necessary testing is conducted to ensure safety of the products patients, and that all clinical studies planned take into consideration the appropriate inclusion-exclusion criteria, primary and secondary endpoint, trial design, endpoints, and numbers of patients enrolled for statistical relevance.

7 Conclusions

A watershed point in corneal transplant surgery is rapidly being approached. Donor cornea has served well in the past to help restore vision in those with corneal opacity, and innovations in lamellar surgery have helped to improve outcomes, especially for endothelial keratoplasty. However, the massive shortage of donor tissue, combined with the expense and technical challenges of eye-banking, means operations with donor corneal tissue are prohibitive for much of the world's population, particularly in lower income areas.

Bioengineering corneas has now become a sizeable field in tissue engineering, regenerative medicine, and ophthalmology. Over recent years, there have been a plethora of approaches to develop biosynthetic alternatives to donor corneas, both cell-based and scaffold-based. It was not possible to discuss each innovation, so we have selected the ones that have now reached clinical evaluation, clinical application, or will soon enter this arena for discussion in this review. Other references have been included as appropriate.

Cornea bioengineering remains a fast-growing and exciting field. Technologies developed for the cornea are now being applied to other target organs in regenerative medicine. For the patients in countries without eye banking or with a deficit of available, good quality corneas, or those high-risk patients who are not amenable to conventional donor cornea transplantation, bioengineered corneas represent a hope for eyesight restoration in corneal blind patients.

Acknowledgments We thank Dr. Elle Edin, Ms. Yasmina El-Khoury and Mr. Marc Groleau, Centre de Recherche de l'Hopital Maisonneuve-Rosemont, Montréal, for help with researching the topic. MG holds a Canada Research Chair Tier 1 in Biomaterials and Stem Cells in Ophthalmology and the Caroline Durand Foundation Chair for Stem Cell Therapy in the Eye. EIA thanks the grant support from NSERC, CIHR, NFRF, and of the Ministry of Economic Development, Job Creation and Trade for an Early Researcher Award.

References

Aamodt JM, Grainger DW (2016) Extracellular matrix-based biomaterial scaffolds and the host response. Biomaterials 86:68–82. https://doi.org/10.1016/j.biomaterials.2016.02.003

Alarcon EI, Vulesevic B, Argawal A, Ross A, Bejjani P, Podrebarac J, ... Griffith M (2016) Coloured cornea replacements with anti-infective properties: expanding the safe use of silver nanoparticles in regenerative medicine. Nanoscale 8(12):6484–6489. https://doi.org/10.1039/c6nr01339b

Alió del Barrio JL, Alió JL (2018) Cellular therapy of the corneal stroma: a new type of corneal surgery for keratoconus and corneal dystrophies. Eye Vis 5(1):28. https://doi.org/10.1186/s40662-018-0122-1

Amano S, Yamagami S, Mimura T, Uchida S, Yokoo S (2006) Corneal stromal and endothelial cell precursors. Cornea 25(10 Suppl 1):S73–S77. https://doi.org/10.1097/01.ico.0000247218.10672.7e

Armitage WJ (2011) Preservation of human cornea. Transfus Med Hemother 38(2):143–147. https://doi.org/10.1159/000326632

Baino F, Ferraris S, Miola M, Perero S, Verne E, Coggiola A, ... Ferraris M (2016) Novel antibacterial ocular prostheses: proof of concept and physico-chemical characterization. Mater Sci Eng C Mater Biol Appl 60:467–474. https://doi.org/10.1016/j.msec.2015.11.075

Bartels MC, Doxiadis II, Colen TP, Beekhuis WH (2003) Long-term outcome in high-risk corneal transplantation and the influence of HLA-A and HLA-B matching. Cornea 22(6):552–556

Basu S, Hertsenberg AJ, Funderburgh ML, Burrow MK, Mann MM, Du Y, ... Funderburgh JL (2014) Human limbal biopsy-derived stromal stem cells prevent corneal scarring. Sci Transl Med 6(266):266ra172. https://doi.org/10.1126/scitranslmed.3009644

Behlau I, Mukherjee K, Todani A, Tisdale AS, Cade F, Wang L, ... Klibanov AM (2011) Biocompatibility and biofilm inhibition of N,N-hexyl,methyl-polyethylenimine bonded to Boston Keratoprosthesis materials. Biomaterials 32(34):8783–8796. https://doi.org/10.1016/j.biomaterials.2011.08.010

Bennet D, Estlack Z, Reid T, Kim J (2018) A microengineered human corneal epithelium-on-a-chip for eye drops mass transport evaluation. Lab Chip 18(11):1539–1551. https://doi.org/10.1039/c8lc00158h

Bourne RRA, Flaxman SR, Braithwaite T, Cicinelli MV, Das A, Jonas JB, ... Vision Loss Expert Group (2017) Magnitude, temporal trends, and projections of the global prevalence of blindness and distance and near vision impairment: a systematic review and meta-analysis. Lancet Glob Health 5(9):e888–e897. https://doi.org/10.1016/S2214-109X(17)30293-0

Brunette I, Roberts CJ, Vidal F, Harissi-Dagher M, Lachaine J, Sheardown H, ... Griffith M (2017) Alternatives to eye bank native tissue for corneal stromal replacement. Prog Retin Eye Res 59:97–130. https://doi.org/10.1016/j.preteyeres.2017.04.002

Burdick JA, Mauck RL, Gorman JH 3rd, Gorman RC (2013) Acellular biomaterials: an evolving alternative to cell-based therapies. Sci Transl Med 5(176):176ps174. https://doi.org/10.1126/scitranslmed.3003997

Buznyk O, Pasyechnikova N, Islam MM, Iakymenko S, Fagerholm P, Griffith M (2015) Bioengineered corneas grafted as alternatives to human donor corneas in three high-risk patients. Clin Transl Sci 8(5):558–562. https://doi.org/10.1111/cts.12293

Calonge M, Perez I, Galindo S, Nieto-Miguel T, Lopez-Paniagua M, Fernandez I, ... Herreras JM (2019) A proof-of-concept clinical trial using mesenchymal stem cells for the treatment of

corneal epithelial stem cell deficiency. Transl Res 206:18–40. https://doi.org/10.1016/j.trsl.2018.11.003
Chauhan SK, Jin Y, Goyal S, Lee HS, Fuchsluger TA, Lee HK, Dana R (2011) A novel pro-lymphangiogenic function for Th17/IL-17. Blood 118(17):4630–4634. https://doi.org/10.1182/blood-2011-01-332049
Chauhan SK, Jurkunas U, Funaki T, Dastjerdi M, Dana R (2015) Quantification of allospecific and nonspecific corneal endothelial cell damage after corneal transplantation. Eye (Lond) 29(1):136–144. https://doi.org/10.1038/eye.2014.248
Chen H, Wang W, Xie H, Xu X, Wu J, Jiang Z, . . . Zheng S (2009) A pathogenic role of IL- 17 at the early stage of corneal allograft rejection. Transpl Immunol 21(3):155–161. https://doi.org/10.1016/j.trim.2009.03.006
Claesson M, Armitage WJ (2013) Clinical outcome of repeat penetrating keratoplasty. Cornea 32(7):1026–1030. https://doi.org/10.1097/ICO.0b013e31828a2810
Coronaviridae Study Group of the International Committee on Taxonomy of Viruses (2020) The species severe acute respiratory syndrome-related coronavirus: classifying 2019-nCoV and naming it SARS-CoV-2. Nat Microbiol 5(4):536–544. https://doi.org/10.1038/s41564-020-0695-z
Coster DJ, Williams KA (2003) Management of high-risk corneal grafts. Eye (Lond) 17(8):996–1002. https://doi.org/10.1038/sj.eye.6700634
Coster DJ, Jessup CF, Biotech B, Williams KA (2005) Mechanisms of corneal allograft rejection. Ocul Surf 3(4 Suppl):S165–S168
Coster DJ, Jessup CF, Williams KA (2009) Mechanisms of corneal allograft rejection and regional immunosuppression. Eye (Lond) 23(10):1894–1897. https://doi.org/10.1038/eye.2009.17
Di Zazzo A, Kheirkhah A, Abud TB, Goyal S, Dana R (2017) Management of high-risk corneal transplantation. Surv Ophthalmol 62(6):816–827. https://doi.org/10.1016/j.survophthal.2016.12.010
Duarte Campos DF, Rohde M, Ross M, Anvari P, Blaeser A, Vogt M, . . . Fuest M (2019) Corneal bioprinting utilizing collagen-based bioinks and primary human keratocytes. J Biomed Mater Res A 107(9):1945–1953. https://doi.org/10.1002/jbm.a.36702
Fagerholm P, Lagali NS, Carlsson DJ, Merrett K, Griffith M (2009) Corneal regeneration following implantation of a biomimetic tissue-engineered substitute. Clin Transl Sci 2(2):162–164. https://doi.org/10.1111/j.1752-8062.2008.00083.x
Fagerholm P, Lagali NS, Merrett K, Jackson WB, Munger R, Liu Y, . . . Griffith M (2010) A biosynthetic alternative to human donor tissue for inducing corneal regeneration: 24-month follow-up of a phase 1 clinical study. Sci Transl Med 2(46):46ra61. https://doi.org/10.1126/scitranslmed.3001022
Fagerholm P, Lagali NS, Ong JA, Merrett K, Jackson WB, Polarek JW, . . . Griffith M (2014) Stable corneal regeneration four years after implantation of a cell-free recombinant human collagen scaffold. Biomaterials 35(8):2420–2427. https://doi.org/10.1016/j.biomaterials.2013.11.079
Flaxman SR, Bourne RRA, Resnikoff S, Ackland P, Braithwaite T, Cicinelli MV, . . . Vision Loss Expert Group of the Global Burden of Disease Study (2017) Global causes of blindness and distance vision impairment 1990–2020: a systematic review and meta-analysis. Lancet Glob Health 5(12):e1221–e1234. https://doi.org/10.1016/S2214-109X(17)30393-5
Gain P, Jullienne R, He Z, Aldossary M, Acquart S, Cognasse F, Thuret G (2016) Global survey of corneal transplantation and eye banking. JAMA Ophthalmol 134(2):167–173. https://doi.org/10.1001/jamaophthalmol.2015.4776
George AJT, Larkin DFP (2004) Corneal transplantation: the forgotten graft. Am J Transplant 4(5):678–685. https://doi.org/10.1111/j.1600-6143.2004.00417.x
Gonzalez-Andrades M, de la Cruz Cardona J, Ionescu AM, Campos A, Del Mar Perez M, Alaminos M (2011) Generation of bioengineered corneas with decellularized xenografts and human keratocytes. Invest Ophthalmol Vis Sci 52(1):215–222. https://doi.org/10.1167/iovs.09-4773
Gonzalez-Andrades M, Carriel V, Rivera-Izquierdo M, Garzon I, Gonzalez-Andrades E, Medialdea S, . . . Campos A (2015) Effects of detergent-based protocols on decellularization of corneas

with Sclerocorneal limbus. Evaluation of regional differences. Transl Vis Sci Technol 4(2):13. https://doi.org/10.1167/tvst.4.2.13

Gonzalez-Andrades M, Mata R, Gonzalez-Gallardo MDC, Medialdea S, Arias-Santiago S, Martinez-Atienza J . . . Cuende N (2017) A study protocol for a multicentre randomised clinical trial evaluating the safety and feasibility of a bioengineered human allogeneic nanostructured anterior cornea in patients with advanced corneal trophic ulcers refractory to conventional treatment. BMJ Open 7(9):e016487. https://doi.org/10.1136/bmjopen-2017-016487

Gonzalez-Andrades M, Sharifi R, Islam MM, Divoux T, Haist M, Paschalis EI, . . . Dohlman CH (2018) Improving the practicality and safety of artificial corneas: pre-assembly and gamma-rays sterilization of the Boston Keratoprosthesis. Ocul Surf 16(3):322–330. https://doi.org/10.1016/j.jtos.2018.04.002

Gouveia RM, Gonzalez-Andrades E, Cardona JC, Gonzalez-Gallardo C, Ionescu AM, Garzon I, . . . Connon CJ (2017a) Controlling the 3D architecture of self-lifting auto-generated tissue equivalents (SLATEs) for optimized corneal graft composition and stability. Biomaterials 121:205–219. https://doi.org/10.1016/j.biomaterials.2016.12.023

Gouveia RM, Koudouna E, Jester J, Figueiredo F, Connon CJ (2017b) Template curvature influences cell alignment to create improved human corneal tissue equivalents. Adv Biosyst 1(12):1700135. https://doi.org/10.1002/adbi.201700135

Griffith M, Polisetti N, Kuffova L, Gallar J, Forrester J, Vemuganti GK, Fuchsluger TA (2012) Regenerative approaches as alternatives to donor allografting for restoration of corneal function. Ocul Surf 10(3):170–183. https://doi.org/10.1016/j.jtos.2012.04.004

Gupta N, Upadhyay P (2017) Use of glycerol-preserved corneas for corneal transplants. Indian J Ophthalmol 65(7):569–573. https://doi.org/10.4103/ijo.IJO_56_17

Hackett JM, Lagali N, Merrett K, Edelhauser H, Sun Y, Gan L, . . . Fagerholm P (2011) Biosynthetic corneal implants for replacement of pathologic corneal tissue: performance in a controlled rabbit alkali burn model. Invest Ophthalmol Vis Sci 52(2):651–657. https://doi.org/10.1167/iovs.10-5224

Hashimoto Y, Funamoto S, Sasaki S, Honda T, Hattori S, Nam K, . . . Kishida A (2010) Preparation and characterization of decellularized cornea using high-hydrostatic pressurization for corneal tissue engineering. Biomaterials 31(14):3941–3948. https://doi.org/10.1016/j.biomaterials.2010.01.122

Hegde S, Beauregard C, Mayhew E, Niederkorn JY (2005) CD4(+) T-cell-mediated mechanisms of corneal allograft rejection: role of Fas-induced apoptosis. Transplantation 79(1):23–31. https://doi.org/10.1097/01.tp.0000147196.79546.69

Hori J, Yamaguchi T, Keino H, Hamrah P, Maruyama K (2019) Immune privilege in corneal transplantation. Prog Retin Eye Res. https://doi.org/10.1016/j.preteyeres.2019.04.002

Inoue K, Amano S, Oshika T, Sawa M, Tsuru T (2000) A 10-year review of penetrating keratoplasty. Jpn J Ophthalmol 44(2):139–145

Isaacson A, Swioklo S, Connon CJ (2018) 3D bioprinting of a corneal stroma equivalent. Exp Eye Res 173:188–193. https://doi.org/10.1016/j.exer.2018.05.010

Islam MM, Ravichandran R, Olsen D, Ljunggren MK, Fagerholm P, Lee CJ, . . . Phopase J (2016) Self-assembled collagen-like-peptide implants as alternatives to human donor corneal transplantation. RSC Adv 6(61):55745–55749. https://doi.org/10.1039/C6RA08895C

Islam MM, Buznyk O, Reddy JC, Pasyechnikova N, Alarcon EI, Hayes S, . . . Griffith M (2018) Biomaterials-enabled cornea regeneration in patients at high risk for rejection of donor tissue transplantation. NPJ Regen Med 3:2. https://doi.org/10.1038/s41536-017-0038-8

Islam MM, Sharifi R, Mamodaly S, Islam R, Nahra D, Abusamra DB, . . . Gonzalez-Andrades M (2019) Effects of gamma radiation sterilization on the structural and biological properties of decellularized corneal xenografts. Acta Biomater 96:330–344. https://doi.org/10.1016/j.actbio.2019.07.002

Jangamreddy JR, Haagdorens MKC, Mirazul Islam M, Lewis P, Samanta A, Fagerholm P, . . . Griffith M (2018) Short peptide analogs as alternatives to collagen in pro-regenerative corneal implants. Acta Biomater 69:120–130. https://doi.org/10.1016/j.actbio.2018.01.011

Joint United Kingdom (UK) Blood Transfusion and Tissue Transplantation Services Professional Advisory Committee (2016) Ocular tissue retrieval, processing and storage. Retrieved from https://www.transfusionguidelines.org/red-book/chapter-21-tissue-banking-tissue-retrieval-and-processing/21-12-ocular-tissue-retrieval-and-storage

Kinoshita S, Koizumi N, Ueno M, Okumura N, Imai K, Tanaka H, . . . Hamuro J (2018) Injection of cultured cells with a ROCK inhibitor for bullous keratopathy. N Engl J Med 378(11):995–1003. https://doi.org/10.1056/NEJMoa1712770

Klufas MA, Yannuzzi NA, D'Amico DJ, Kiss S (2015) Vitreoretinal aspects of permanent keratoprosthesis. Surv Ophthalmol 60(3):216–228. https://doi.org/10.1016/j.survophthal.2014.11.001

Koizumi N, Okumura N, Ueno M, Kinoshita S (2014) New therapeutic modality for corneal endothelial disease using rho-associated kinase inhibitor eye drops. Cornea 33(Suppl 11): S25–S31. https://doi.org/10.1097/ICO.0000000000000240

Lakshman N, Kim A, Petroll WM (2010) Characterization of corneal keratocyte morphology and mechanical activity within 3-D collagen matrices. Exp Eye Res 90(2):350–359. https://doi.org/10.1016/j.exer.2009.11.016

Le-Bel G, Guerin LP, Carrier P, Mouriaux F, Germain L, Guerin SL, Bazin R (2019) Grafting of an autologous tissue-engineered human corneal epithelium to a patient with limbal stem cell deficiency (LSCD). Am J Ophthalmol Case Rep 15:100532. https://doi.org/10.1016/j.ajoc.2019.100532

Levis HJ, Kureshi AK, Massie I, Morgan L, Vernon AJ, Daniels JT (2015) Tissue engineering the cornea: the evolution of RAFT. J Funct Biomater 6(1):50–65. https://doi.org/10.3390/jfb6010050

Li J, Yu L, Deng Z, Wang L, Sun L, Ma H, Chen W (2011) Deep anterior lamellar keratoplasty using acellular corneal tissue for prevention of allograft rejection in high-risk corneas. Am J Ophthalmol 152(5):762–770 e763. https://doi.org/10.1016/j.ajo.2011.05.002

Liechtenstein T, Dufait I, Bricogne C, Lanna A, Pen J, Breckpot K, Escors D (2012) PD-L1/PD-1 co-stimulation, a brake for T cell activation and a T cell differentiation signal. J Clin Cell Immunol:S12. https://doi.org/10.4172/2155-9899.S12-006

Matthaei M, Sandhaeger H, Hermel M, Adler W, Jun AS, Cursiefen C, Heindl LM (2017) Changing indications in penetrating keratoplasty: a systematic review of 34 years of global reporting. Transplantation 101(6):1387–1399. https://doi.org/10.1097/TP.0000000000001281

Matthyssen S, Van den Bogerd B, Dhubhghaill SN, Koppen C, Zakaria N (2018) Corneal regeneration: a review of stromal replacements. Acta Biomater 69:31–41. https://doi.org/10.1016/j.actbio.2018.01.023

MIT Technology Review (2019) Scientists are making human-monkey hybrids in China. Retrieved from https://www.technologyreview.com/s/614052/scientists-are-making-human-monkey-hybrids-in-china/

Myung D, Duhamel PE, Cochran JR, Noolandi J, Ta CN, Frank CW (2008) Development of hydrogel-based keratoprostheses: a materials perspective. Biotechnol Prog 24(3):735–741. https://doi.org/10.1021/bp070476n

Navaratnam J, Utheim TP, Rajasekhar VK, Shahdadfar A (2015) Substrates for expansion of corneal endothelial cells towards bioengineering of human corneal endothelium. J Funct Biomater 6(3):917–945. https://doi.org/10.3390/jfb6030917

Niederkorn JY (2010) High-risk corneal allografts and why they lose their immune privilege. Curr Opin Allergy Clin Immunol 10(5):493–497. https://doi.org/10.1097/ACI.0b013e32833dfa11

O'Leary LE, Fallas JA, Bakota EL, Kang MK, Hartgerink JD (2011) Multi-hierarchical self-assembly of a collagen mimetic peptide from triple helix to nanofibre and hydrogel. Nat Chem 3(10):821–828. https://doi.org/10.1038/nchem.1123

Okumura N, Kinoshita S, Koizumi N (2017) The role of Rho kinase inhibitors in corneal endothelial dysfunction. Curr Pharm Des 23(4):660–666. https://doi.org/10.2174/1381612822666161205110027

Pashuck ET, Stevens MM (2012) Designing regenerative biomaterial therapies for the clinic. Sci Transl Med 4(160):160sr164. https://doi.org/10.1126/scitranslmed.3002717

Patel S, Thakar RG, Wong J, McLeod SD, Li S (2006) Control of cell adhesion on poly(methyl methacrylate). Biomaterials 27(14):2890–2897. https://doi.org/10.1016/j.biomaterials.2005.12.009

Pellegrini G, Traverso CE, Franzi AT, Zingirian M, Cancedda R, De Luca M (1997) Long-term restoration of damaged corneal surfaces with autologous cultivated corneal epithelium. Lancet 349(9057):990–993. https://doi.org/10.1016/S0140-6736(96)11188-0

Pellegrini G, Lambiase A, Macaluso C, Pocobelli A, Deng S, Cavallini GM, … Rama P (2016) From discovery to approval of an advanced therapy medicinal product-containing stem cells, in the EU. Regen Med 11(4):407–420. https://doi.org/10.2217/rme-2015-0051

Qazi Y, Hamrah P (2013) Corneal allograft rejection: Immunopathogenesis to therapeutics. J Clin Cell Immunol 2013(Suppl 9). https://doi.org/10.4172/2155-9899.S9-006

Rama P, Matuska S, Paganoni G, Spinelli A, De Luca M, Pellegrini G (2010) Limbal stem-cell therapy and long-term corneal regeneration. N Engl J Med 363(2):147–155. https://doi.org/10.1056/NEJMoa0905955

Riau AK, Mondal D, Yam GH, Setiawan M, Liedberg B, Venkatraman SS, Mehta JS (2015) Surface modification of PMMA to improve adhesion to corneal substitutes in a synthetic core-skirt Keratoprosthesis. ACS Appl Mater Interfaces 7(39):21690–21702. https://doi.org/10.1021/acsami.5b07621

Riau AK, Mondal D, Setiawan M, Palaniappan A, Yam GH, Liedberg B, … Mehta JS (2016) Functionalization of the polymeric surface with bioceramic nanoparticles via a novel, nonthermal dip coating method. ACS Appl Mater Interfaces 8(51):35565–35577. https://doi.org/10.1021/acsami.6b12371

Riau AK, Venkatraman SS, Dohlman CH, Mehta JS (2017) Surface modifications of the PMMA optic of a Keratoprosthesis to improve biointegration. Cornea 36(Suppl 1):S15–S25. https://doi.org/10.1097/ICO.0000000000001352

Rico-Sanchez L, Garzon I, Gonzalez-Andrades M, Ruiz-Garcia A, Punzano M, Lizana-Moreno A, … Alaminos M (2019) Successful development and clinical translation of a novel anterior lamellar artificial cornea. J Tissue Eng Regen Med. https://doi.org/10.1002/term.2951

Rubert Perez CM, Rank LA, Chmielewski J (2014) Tuning the thermosensitive properties of hybrid collagen peptide-polymer hydrogels. Chem Commun (Camb) 50(60):8174–8176. https://doi.org/10.1039/c4cc03171g

Rubert Perez CM, Stephanopoulos N, Sur S, Lee SS, Newcomb C, Stupp SI (2015) The powerful functions of peptide-based bioactive matrices for regenerative medicine. Ann Biomed Eng 43(3):501–514. https://doi.org/10.1007/s10439-014-1166-6

Salvador-Culla B, Kolovou PE (2016) Keratoprosthesis: a review of recent advances in the field. J Funct Biomater 7(2). https://doi.org/10.3390/jfb7020013

Salvador-Culla B, Jeong KJ, Kolovou PE, Chiang HH, Chodosh J, Dohlman CH, Kohane DS (2016) Titanium coating of the Boston Keratoprosthesis. Transl Vis Sci Technol 5(2):17. https://doi.org/10.1167/tvst.5.2.17

Sangwan VS, Basu S, MacNeil S, Balasubramanian D (2012) Simple limbal epithelial transplantation (SLET): a novel surgical technique for the treatment of unilateral limbal stem cell deficiency. Br J Ophthalmol 96(7):931–934. https://doi.org/10.1136/bjophthalmol-2011-301164

Shen L, Jin Y, Freeman GJ, Sharpe AH, Dana MR (2007) The function of donor versus recipient programmed death-ligand 1 in corneal allograft survival. J Immunol 179(6):3672–3679. https://doi.org/10.4049/jimmunol.179.6.3672

Shojaati G, Khandaker I, Funderburgh ML, Mann MM, Basu R, Stolz DB, … Funderburgh JL (2019) Mesenchymal stem cells reduce corneal fibrosis and inflammation via extracellular vesicle-mediated delivery of miRNA. Stem Cells Transl Med. https://doi.org/10.1002/sctm.18-0297

Shukla S, Shanbhag SS, Tavakkoli F, Varma S, Singh V, Basu S (2019) Limbal epithelial and mesenchymal stem cell therapy for corneal regeneration. Curr Eye Res:1–13. https://doi.org/10.1080/02713683.2019.1639765

Simpson F, Alarcon EI, Hilborn J, Brunette I, Griffith M (2019) Chapter 63 – regenerative medicine in the cornea. In: Atala A, Lanza R, Mikos AG, Nerem R (eds) Principles of regenerative medicine, 3rd edn. Academic Press, Boston, pp 1115–1129

Specialised Commissioning Team C c p (2018) Keratoprosthesis for corneal blindness. Retrieved from https://www.england.nhs.uk/wp-content/uploads/2018/07/1618-keratoprosthesis.pdf

Stahl PJ, Romano NH, Wirtz D, Yu SM (2010) PEG-based hydrogels with collagen mimetic peptide-mediated and tunable physical cross-links. Biomacromolecules 11(9):2336–2344. https://doi.org/10.1021/bm100465q

Stern JH, Tian Y, Funderburgh J, Pellegrini G, Zhang K, Goldberg JL, ... Temple S (2018) Regenerating eye tissues to preserve and restore vision. Cell Stem Cell 22(6):834–849. https://doi.org/10.1016/j.stem.2018.05.013

Stevens KR, Murry CE (2018) Human pluripotent stem cell-derived engineered tissues: clinical considerations. Cell Stem Cell 22(3):294–297. https://doi.org/10.1016/j.stem.2018.01.015

Tan XW, Perera AP, Tan A, Tan D, Khor KA, Beuerman RW, Mehta JS (2011) Comparison of candidate materials for a synthetic osteo-odonto keratoprosthesis device. Invest Ophthalmol Vis Sci 52(1):21–29. https://doi.org/10.1167/iovs.10-6186

Tan DT, Dart JK, Holland EJ, Kinoshita S (2012) Corneal transplantation. Lancet 379(9827):1749–1761. https://doi.org/10.1016/S0140-6736(12)60437-1

Utheim TP (2015) Concise review: transplantation of cultured oral mucosal epithelial cells for treating limbal stem cell deficiency-current status and future perspectives. Stem Cells 33(6):1685–1695. https://doi.org/10.1002/stem.1999

Vendomele J, Khebizi Q, Fisson S (2017) Cellular and molecular mechanisms of anterior chamber-associated immune deviation (ACAID): what we have learned from knockout mice. Front Immunol 8:1686. https://doi.org/10.3389/fimmu.2017.01686

Wang L, Jeong KJ, Chiang HH, Zurakowski D, Behlau I, Chodosh J, ... Kohane DS (2011) Hydroxyapatite for keratoprosthesis biointegration. Invest Ophthalmol Vis Sci 52(10):7392–7399. https://doi.org/10.1167/iovs.11-7601

Whitcher JP, Srinivasan M (1997) Corneal ulceration in the developing world – a silent epidemic. Br J Ophthalmol 81(8):622–623. https://doi.org/10.1136/bjo.81.8.622

Whitcher JP, Srinivasan M, Upadhyay MP (2001) Corneal blindness: a global perspective. Bull World Health Organ 79(3):214–221

Williams KA, Coster DJ (2007) The immunobiology of corneal transplantation. Transplantation 84 (7):806–813. https://doi.org/10.1097/01.tp.0000285489.91595.13

Williams KA, Esterman AJ, Bartlett C, Holland H, Hornsby NB, Coster DJ (2006) How effective is penetrating corneal transplantation? Factors influencing long-term outcome in multivariate analysis. Transplantation 81(6):896–901. https://doi.org/10.1097/01.tp.0000185197.37824.35

Williams KA, Lowe M, Bartlett C, Kelly TL, Coster DJ, All C (2008) Risk factors for human corneal graft failure within the Australian corneal graft registry. Transplantation 86(12):1720–1724. https://doi.org/10.1097/TP.0b013e3181903b0a

World Health Organization (2018a) Blindness and vision impairment. Retrieved from https://www.who.int/news-room/fact-sheets/detail/blindness-and-visual-impairment

World Health Organization (2018b) Priority eye diseases: corneal opacities. Retrieved from https://www.who.int/blindness/causes/priority/en/index8.html

Xu H, Xu B, Yang Q, Li X, Ma X, Xia Q, ... Zhang Y (2014) Comparison of decellularization protocols for preparing a decellularized porcine annulus fibrosus scaffold. PLoS One 9(1): e86723. https://doi.org/10.1371/journal.pone.0086723

Yang C, Hillas P, Tang J, Balan J, Notbohm H, Polarek J (2004) Development of a recombinant human collagen-type III based hemostat. J Biomed Mater Res B Appl Biomater 69B:18–24. https://doi.org/10.1002/jbm.b.20030

Yoeruek E, Bayyoud T, Maurus C, Hofmann J, Spitzer MS, Bartz-Schmidt KU, Szurman P (2012) Decellularization of porcine corneas and repopulation with human corneal cells for tissue-engineered xenografts. Acta Ophthalmol 90(2):e125–e131. https://doi.org/10.1111/j.1755-3768.2011.02261.x

Yu T, Rajendran V, Griffith M, Forrester JV, Kuffova L (2016) High-risk corneal allografts: a therapeutic challenge. World J Transplant 6(1):10–27. https://doi.org/10.5500/wjt.v6.i1.10

Zhang MC, Liu X, Jin Y, Jiang DL, Wei XS, Xie HT (2015) Lamellar keratoplasty treatment of fungal corneal ulcers with acellular porcine corneal stroma. Am J Transplant 15(4):1068–1075. https://doi.org/10.1111/ajt.13096

Index

D. Eberli et al. (eds.), *Organ Tissue Engineering*, Reference Series in Biomedical Engineering, https://doi.org/10.1007/978-3-030-44211-8

P

U

V